W9-AOB-490

Microbial Functional Genomics

Microbial Functional Genomics

Jizhong Zhou
Dorothea K. Thompson
Ying Xu
Oak Ridge National Laboratory
Oak Ridge, Tennessee

James M. Tiedje
Michigan State University
East Lansing, Michigan

A John Wiley & Sons, Inc., Publication

Copyright © 2004 by John Wiley & Sons, Inc. All rights reserved.

Published by John Wiley & Sons, Inc., Hoboken, New Jersey.
Published simultaneously in Canada.

No part of this publication may be reproduced, stored in a retrieval system, or transmitted in any form or by any means, electronic, mechanical, photocopying, recording, scanning, or otherwise, except as permitted under Section 107 or 108 of the 1976 United States Copyright Act, without either the prior written permission of the Publisher, or authorization through payment of the appropriate per-copy fee to the Copyright Clearance Center, Inc., 222 Rosewood Drive, Danvers, MA 01923, 978-750-8400, fax 978-646-8600, or on the web at www.copyright.com. Requests to the Publisher for permission should be addressed to the Permissions Department, John Wiley & Sons, Inc., 111 River Street, Hoboken, NJ 07030, (201) 748-6011, fax (201) 748-6008.

Limit of Liability/Disclaimer of Warranty: While the publisher and author have used their best efforts in preparing this book, they make no representations or warranties with respect to the accuracy or completeness of the contents of this book and specifically disclaim any implied warranties of merchantability or fitness for a particular purpose. No warranty may be created or extended by sales representatives or written sales materials. The advice and strategies contained herein may not be suitable for your situation. You should consult with a professional where appropriate. Neither the publisher nor author shall be liable for any loss of profit or any other commercial damages, including but not limited to special, incidental, consequential, or other damages.

For general information on our other products and services please contact our Customer Care Department within the U.S. at 877-762-2974, outside the U.S at 317-572-3993 or fax 317-572-4002.

Wiley also publishes its books in a variety of electronic formats. Some content that appears in print, however, may not be available in electronic format.

Library of Congress Cataloging-in-Publication Data:

Microbial functional genomics / Jizhong Zhou ... [et al.].
    p. ; cm.
Includes bibliographical references and index.
   ISBN 0-471-07190-0 (cloth : alk. paper)
  1. Microbial genomics.
   [DNLM: 1. Genetics, Microbial. 2. Genomics. 3. Proteomics. QW 51 M6263 2004] I. Zhou, Jizhong, 1959–
   QH447.M53 2004
   579′.135–dc22

                                                        2003017519

Printed in the United States of America

10 9 8 7 6 5 4 3 2 1

Contents

8. Mutagenesis as a Genomic Tool for Studying Gene Function   207

Alexander S. Beliaev

9. Mass Spectrometry   241

Nathan VerBerkmoes, Joshua Sharp, and Robert Hettich

10. Identification of Protein–Ligand Interactions   285

Timothy Palzkill

14. Application of Microarray-based Genomic Technology to Mutation Analysis and Microbial Detection

Jizhong Zhou and Dorothea K. Thompson

Preface

The overarching aim of genomics is to provide a comprehensive, genome-level understanding of the molecular basis of the structure, functions, and evolution of biological systems using whole-genome sequence information and high-throughput genomic technologies. The goal of functional genomics is to obtain a system-level understanding of the functional aspects of biological systems, that is, gene functions and regulatory networks. This is a formidable task given that knowledge of the nucleotide blueprint of an organism is only the initial step toward understanding the dynamic nature of gene function and gene product interactions that enable a cell to grow, replicate, and adapt to changes in its environment. To address the avalanche of data generated from numerous genome sequencing projects, a variety of high-throughput, parallel genomic technologies and bioinformatic, data-mining tools have emerged that allow a comprehensive analysis of biological systems at different levels. These technologies, combined with traditional molecular approaches, are beginning to provide a physiological, evolutionary, and ecological context for genomic sequence information.

The main purpose of this book is to establish a systematic description and review of the principles, methodology, applications, and challenges in the emerging field of microbial functional genomics from the viewpoints of genetics, biochemistry, cell biology, physiology, and ecology. As the title suggests, this book emphasizes studies that focus on microorganisms; however, some studies related to higher organisms are included for comparative purposes. The book's purpose is to identify research gaps, challenges, and experimental directions in microbial functional genomics, as well as to point out the advantages and limitations of recently developed high-throughput genomic technologies, most notably DNA microarrays and proteomic tools. The impact of functional genomics on such established research areas as bacterial pathogenesis, antimicrobial drug discovery, and toxicology is also described. The intent of this book is not to present an exhaustive survey of the published literature on microbial functional genomics, but to provide the reader with representative examples illustrative of basic principles, approaches, and applications in the field, with an emphasis on genomic-scale studies using integrative multidisciplinary approaches. Clearly, there are many studies employing functional genomics that we have not been able to include or cover in sufficient detail because of space constraints. This book is intended as a guide and reference for senior undergraduate and graduate-level students, as well as for professionals interested in microbial genomics and functional genomics in general.

This book begins with a chapter that attempts to codify and distinguish various aspects of functional genomics by defining the purpose and scope of genomics and functional genomics. Chapter 1 also provides a brief overview of the history of genomics and discusses the general approaches employed in functional genomics studies. Chapter 2 discusses the microbial diversity present in natural environments from biochemical and genetic perspectives and addresses the challenges associated with describing prokaryotic diversity. This chapter also reviews recent insights into microbial diversity that have emerged from whole-genome sequencing projects. Chapter 3 explores the strategies, approaches, and tools for annotating microbial genomic sequences by focusing on comparative analyses and different computational models. The potential problems associated with assigning gene functions based on sequence homology are also addressed in this chapter. Chapter 4 provides the latest insights into molecular evolution that have recently resulted from whole-genome sequence comparisons.

Chapters 5 through 10 address the various genomic technologies that are currently being used to elucidate gene function, regulatory networks, and protein interaction maps on a global scale. Chapter 5, for example, presents bioinformatic approaches to predicting gene function and discusses *in silico* methods for constructing metabolic pathways based on genome sequence information. Chapter 6 provides a detailed description of DNA microarray technology related to fabrication, hybridization, and detection, microarray experimental design, and general approaches to using microarrays for gene expression profiling. Chapter 7 is an overview of the current methods being used for microarray data mining and analysis. Chapter 8 discusses the principles behind and applications of three major strategies for genome-wide mutagenesis to the study of gene functions: gene disruption by allelic exchange, transposon mutagenesis, and expression inhibition using antisense RNA molecules. The advantages and disadvantages of each method are also presented. Chapter 9 describes proteomic tools for analyzing gene expression, with an emphasis on mass spectrometry for protein profile analysis on a genomic scale. Chapter 10 ends this section by discussing the tools for studying DNA–protein and protein–protein interactions. Two-hybrid systems and phage display for gene functional analysis and protein arrays are reviewed.

Chapters 11 and 12 focus on studies related to the functional analysis of genomic sequences of individual microorganisms. Chapter 11, for example, assesses some of the important contributions being made by functional genomics studies to further our knowledge of the well-studied model organisms *Escherichia coli*, *Bacillus subtilis*, and *Saccharomyces cerevisiae*. Chapter 12 discusses the functional genomic analysis of selected bacterial pathogens and several environmentally important microorganisms. More specifically, this chapter describes the contribution of genome sequence and *in silico* bioinformatic analyses to virulence gene identification, the impact of comparative genomics on revealing genetic diversity and evolutionary trends among pathogenic bacteria, and genomic and proteomic approaches to elucidating bacterial gene function and host–pathogen interactions.

Chapters 13 and 14 are concerned with the more applied aspects of microbial functional genomics or associated genomic technologies. Chapter 13 discusses the impact of genomics on antimicrobial drug discovery and toxicology. This chapter presents a brief historical overview of antimicrobial drug discovery, followed by a discussion of the challenges of new drug discovery, the impact of microbial genome sequencing on target identification, and the application of genomic-scale experimental technologies to target validation and drug candidate screening. The chapter concludes by discussing the newly emerged subdiscipline of toxicogenomics. Chapter 14 reviews the basic principles and most recent advances in applying microarray technology to the analysis of genetic mutations and detection of microorganisms in natural environments. It explores various methods for the analysis of mutations using microarrays and the various types of microarrays specifically developed for analyzing microbial community structure within the context of environmental samples.

As the last chapter of the book, Chapter 15 discusses future directions in microbial genomic research and the likelihood that current understanding will guide future initiatives in productive directions. The two major areas are suggested to be further reductionistic work in the "omics" to provide a more comprehensive understanding of the cell's coordinated function, and secondly, the use of genomics information to provide a more holistic view of the microbial community and its function. Eventually genomics-based microbial science may merge with ecology so that the functioning of ecosystems can be understood and perhaps even managed from the information and molecular catalyst perspective.

Foreword

Parallel with the genomic revolution, and a part of it, was the extraordinary expansion in information technology, yielding vast amounts of data from many sources that could be rendered interoperable and rapidly processed. Systematic understanding of the full complexity of the microbial environment then became possible. Development of analytic models of microbial ecosystems arising out of the information technology explosion moved microbial ecology out of a "black box" in the early ecological models to today's "wiring diagrams" that have multiple pathways and incorporate entire sets of environmental parameters. The earlier, relatively primitive, hypotheses for water, air, and soil ecosystems gave way to the more complete analysis of the complex interactions comprising an ecosystem. From all these advances, the science of biocomplexity gained momentum and evolved into the interdisciplinary teamwork of scientists and engineers from many disciplines, mathematicians, chemists, physicists, atmospheric scientists, space scientists, geoscientists, clinicians, and social scientists, working together to provide a more profound knowledge of global ecological systems.

Into this mix has now come the capacity to determine the full genomic sequence of microorganisms and more than 100 such sequences have been published to date, with hundreds more in progress, all promising to deliver powerful information about the structure, function, and evolution of biological systems at a global level. Genomics may prove to be the most powerful tool to date. It opens up microbial diversity to inspection at details heretofore unimagined. Microbial genomes analyzed through whole genome sequencing reveal the enormity of diversity in the microbial world—a diversity that, with whole genome sequencing of total microbial populations, has been proven. A single gram of soil, a milliliter of seawater, or a cubic foot of air are now known to be the source of extraordinary diversity, suspected, but now revealed.

This book, then, provides valuable, highly technical information on the methodology of microbial genomics and offers great insight into specific aspects of microbial genomic ecology, using computational genome annotation, viewing microbial evolution from a genomic perspective, and providing details of DNA microarrays. It touches on the extraordinary capability to observe gene expression in real time and gives an entirely new perspective for microbial ecology. Functional genomics and metabolomics will certainly expand applications, as well as our understanding of microbial ecology in coming years. But, with this book, we have been given a passport to the new world of microbial ecology. . .exciting, fascinating, and rich with promise.

Rita R. Colwell
University of Maryland

Acknowledgments

This book is dedicated to our families for their encouragement and constant support.

We would like to thank Kostas Konstantinidis (Chapter 2), Alexander Beliaev (Chapter 8), Robert Hettich, Nathan VerBerkmoes, and Joshua Sharp (Chapter 9), and Timothy Palzkill (Chapter 10) for contributing chapters in their areas of expertise. We are deeply grateful to Luna Han for her editorial guidance and patience throughout the writing of this book, and the staff at John Wiley & Sons for making this project possible. We extend our sincere thanks to Kim Smith for excellent secretarial assistance, Lynn Kszos for editorial assistance, and LeJean Hardin at Creative Media Solutions (Oak Ridge National Laboratory) for her excellent work in fine-tuning the book's illustrations. In addition, we wish to thank all those who granted us permission to adopt figures and data that had been previously published. We would also like to thank Drs. Daniel Drell and Anna Palmisano of the Department of Energy's Office of Biological and Environmental Research, Office of Science, for their continuous support of this project. Finally, we acknowledge research funding from the DOE Office of Biological and Environmental Research and Oak Ridge National Laboratory, which is operated and managed by the University of Tennessee-Battelle LLC for the Department of Energy under contract DE-AC05-00OR22725.

With Contributions From

Robert Hettich, Oak Ridge National Laboratory, Oak Ridge, Tennessee

Konstantinos Konstantinidis, Michigan State University, East Lansing, Michigan

Timothy Palzkill, Department of Molecular Virology and Microbiology, Baylor College of Medicine, Houston, Texas

Joshua Sharp, Genome Science and Technology Graduate School, Oak Ridge National Laboratory, Oak Ridge, Tennessee, and The University of Tennessee, Knoxville, Tennessee

Nathan VerBerkmoes, Genome Science and Technology Graduate School, Oak Ridge National Laboratory, Oak Ridge, Tennessee, and The University of Tennessee, Knoxville, Tennessee

Alexander S. Beliaev, Pacific Northwest National Laboratory, Richland, Washington

1

Genomics: Toward A Genome-Level Understanding of the Structure, Functions, and Evolution of Biological Systems

Jizhong Zhou, Dorothea K. Thompson, and James M. Tiedje

1.1 INTRODUCTION

Biology is undergoing a revolution, driven by the availability of whole-genome sequence information for many organisms and the development of associated high-throughput genomic technologies. Genomic sequencing and analysis are currently in a period of "exponential growth." More than 100 eukaryotic and prokaryotic genomes have been completely sequenced, with more than 200 genome sequences presently underway (see http://www.jgi.doe.gov; http://www.tigr.org). The nearly complete sequence of the human genome is the cornerstone of genome-based biology and provides the richest intellectual resource in the history of biology. This is the first time that we have had access to the entire genomic content of living organisms; thus, we have been presented with an unprecedented scientific opportunity to visualize the complete and dynamic pictures of living systems.

Determination of an entire genome sequence represents only the first step toward understanding the physiology and behavior of an organism. The next critical step is to elucidate the functions of these sequences, thus providing biochemical, physiological,

Microbial Functional Genomics, Edited by Jizhong Zhou, Dorothea K. Thompson, Ying Xu, and James M. Tiedje. ISBN 0-471-07190-0 © 2004 John Wiley & Sons, Inc.

evolutionary, and ecological meaning to the nucleotide sequence data. By utilizing complete inventories of genes, a variety of high-throughput genomic technologies have been developed to allow the comprehensive analysis of biological systems at different levels. The use of such genomic technologies will shed light on a wide range of important research questions, from how cells grow, differentiate, and evolve to the medical challenges of pathogenesis, antibiotic resistance, and cancer, from agricultural issues of plant and animal breeding and pesticide resistance to the biotechnological challenges of drug discovery and the remediation of environmental contamination.

Knowledge of entire genetic sequences opens a whole new range of possibilities for more efficient research. Many laboratories are researching important questions in functional genomics by integrating genomic, proteomic, metabolomic, genetic, biochemical, and computational approaches. To keep pace with the massive data generated from genome sequencing projects, genomics and associated genomic technologies are developing very rapidly, and the former is emerging as one of the most exciting biological disciplines. Here we will systematically present the principles, methodology, applications, and challenges in this new, rapidly growing biological field. In this introduction, we will define the terminology of genomics, followed by a brief discussion of the history of genomics. We will discuss the challenges encountered in genomic studies and provide an overview of the general scope and approaches of genomic studies. Finally, we will briefly describe the importance of microbial functional genomics within the context of understanding higher eukaryotic organisms.

1.2 DEFINITIONS AND CLASSIFICATIONS

Genomics is a term that has become widely used in the scientific community and has sometimes been contrived as a slogan to attract attention (Lederberg and McCray, 2001). The term was first proposed by Thomas H. Roderick in 1986 to describe a new scientific discipline concerned with mapping, sequencing, and analyzing genomes. Genomics was derived from the word "genome," which refers to the complete set of genes and chromosomes of an organism. The word itself was first used by H. Winkler in 1920 and created by eliding the words "GENes" and "chromosOMEs" (McKusick, 1997). In 1987, it became the name of a new journal. Since 1995, genome analysis has expanded from its original focus on mapping and sequencing to include gene functions (Hieter and Boguski, 1997). To reflect this fundamental change, genomics has been proposed as a more general term signifying both structural and functional studies of genomes (McKusick, 1997).

Although the term genomics has been universally accepted for the past two decades, its exact meaning has never been clearly defined. In most cases, it still refers to the mapping, sequencing, and analysis of genomes. To reflect the focus of related studies on the functional analysis of genomic sequences and proteins, various terms continue to circulate in the literature that specify genomics as functional genomics, proteomics, and structural genomics (Fields et al., 1999; Martin and Nelson, 2001). In addition, the application of genome sequence information and genomic technologies to such diverse fields as toxicology, pharmacology, medicine, physiology, and ecology has resulted in the emergence of many new biological subdisciplines. For example, various terms associated with genomics have been proposed, such as toxicogenomics, pharmacogenomics, and medical, physiological, and ecological genomics. To accommodate such changes, genomics is used here as a more general term to signify the field in biology that aims at a genome-level understanding of the molecular basis of the structure, functions, and evolution of biological systems using whole-genome sequence information and high-throughput genomic technologies. One of the key features of this definition is that genomics focuses on understanding the molecular basis of the structure and functions of biological systems. In contrast to research employing traditional molecular biology approaches, genomics-based studies investigate biological systems at a whole-genome scale using genome sequence information and high-throughput genomic technologies. Here, biological systems may refer to cellular and subcellular systems, such as individual pathways and networks, cells, tissues, organs, or whole organisms. Biological systems can also be used to describe organizational levels above that of the cellular level such as populations, communities, and ecosystems.

Because of the broad meaning assigned to the field of genomics, genomic studies can be classified into various subdisciplines (Fig. 1.1). In the following sections, we will briefly discuss the classifications of genomic studies and definitions of various terms related to genomics.

1.2.1 Classification Based on System Attributes

Biological systems are highly self-organized, complex homeostatic systems (Kitano, 2002). Based on the general system theory, structure and function are the two primary characteristics of any system. Therefore, from the perspective of general system theory, genomics can be divided into structural genomics and functional genomics (Fig. 1.1).

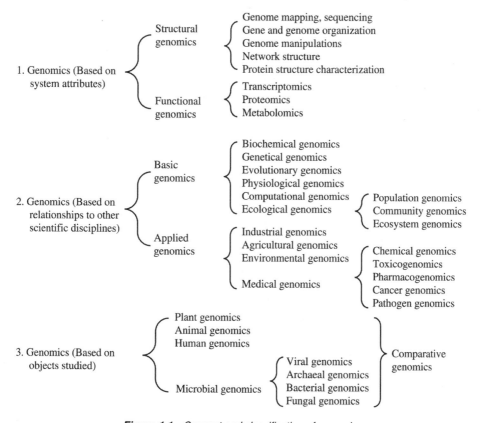

Figure 1.1 *Concept and classification of genomics.*

Structural genomics is defined here as the genome-wide structural study of genes, proteins, and other biomolecules, including genome mapping, sequencing, and organization as well as protein structure characterization. It should be noted that the term structural genomics has been used in different cases to represent the initial studies of genome analysis, that is, genome mapping, sequencing, and organization (Hieter and Boguski, 1997), or genome-wide protein structural characterization and prediction (Kim, 1998; Gaasterland, 1998). Here we apply a broader meaning to the term than the one typically presented in the literature.

Functional genomics aims at a system-level understanding of the functional aspects of biological systems, that is, gene functions and regulatory networks, using genome-wide approaches (Fields et al., 1999; Hieter and Boguski, 1997). Genes can be defined from the perspective of various levels of biological function: biochemical function (e.g., phosphorylation by protein kinases), cellular function (e.g., a role in cell division, DNA replication), developmental function (e.g., a role in cell-type differentiation), or adaptive function (the contribution of gene products to the fitness of an organism) (Bouchez and Hofte, 1998). Functional genomics is characterized by large-scale experimental approaches combined with statistical analysis, mathematical modeling, and computational analysis of the experimental results (Hieter and Boguski, 1997). Its aim is to link genome sequences to biological function, and consequently, it is expected that the field will provide new insights into the behavior and dynamics of biological systems. Like the term genomics, functional genomics has also been widely used in the literature, which has contributed to some general confusion over its exact meaning. In many cases, it is used in reference only to mRNA measurement studies incorporating microarray-based gene expression profiling.

Knowing the genetic blueprint of a cell only provides information about what might happen but not what will happen, because the functions of most genes are realized by their encoded proteins and associated metabolites. To exert their functions, most genes encoded by DNA need to be transcribed into messenger RNAs, which are then translated into proteins. The expressed proteins produce a variety of metabolites that participate in various cellular processes. RNAs, proteins, and metabolites are all context-dependent and vary not only in the types of cells, tissues, organs, or organisms, but also in their developmental status and physiological and environmental conditions. It is this context dependency and tight regulation that make them so valuable in understanding gene function (Oliver, 2002). Thus, the functions of genes cannot be understood merely by examining DNA and/or mRNA alone. Full understanding and characterization of a gene's function must include functional analyses at the levels of RNA, proteins, and metabolites. From this perspective, functional studies of the cellular proteome (proteomics) and metabolome (metabolomics) should be considered as integral parts of functional genomics. To reflect these three levels of complementary functional investigations of genes, functional genomics can be further divided into transcriptomics, proteomics, and metabolomics (Fig. 1.1).

Knowledge of when and where a gene is expressed in a cell often provides a strong clue to its biological function, and in turn, the gene expression patterns in a cell can provide detailed information about its physiological state. Although regulation of protein abundance in a cell is by no means accomplished solely by the regulation of mRNA abundance, virtually all differences in cell type or state are correlated with the changes in the mRNA levels of many genes (DeRisi et al., 1997), so that mRNA expression patterns are very useful in elucidating the functions of unknown genes, regulatory pathways, and protein networks. High-throughput studies on the expression dynamics of the transcriptome, the entire assembly of mRNAs in a cell, are referred to as transcriptomics. Because mRNAs are not the functional entities of the cell, however, transcriptome analyses can only indirectly approach gene functionality. Although the regulation of protein abundance in a cell is often accomplished by controlling the expression of mRNA, translational regulation and posttranslational modification also play important roles in determining protein levels in a cell. There is no strict linear relationship between genes and proteins (Pandey and Mann, 2000). Thus, studies focusing on protein expression dynamics and protein interactions are crucial to elucidating gene function and are complementary to transcriptomics. Proteomics is the large-scale study of expression dynamics and protein interactions of the proteome, the entire set of proteins encoded in a genome, using high-throughput approaches that directly measure or identify proteins (Pandey and Mann, 2000).

Metabolites are the end products of various cellular processes. Along with proteins, they are the functional entities within the cell. They change with time and space, both within an individual cell and among cells of multicellular organisms. Their levels can be regarded as the ultimate response of biological systems to genetic, physiological, and/or environmental changes (Fiehn, 2002). Thus, metabolite analysis provides important information about the functionality of genes. In parallel to transcriptomics and proteomics, metabolomics refers to the large-scale investigation of the dynamics and interactions of the metabolome, the complete set of metabolites (Oliver, 2002; Fiehn, 2002).

1.2.2 Classification Based on Relationships to Other Scientific Disciplines

The accumulation of vast genomic sequence information and the rapid advancement of genomic technologies allow researchers to address both basic and applied biological questions at the entire genome level. Genomic studies can therefore be classified into basic and applied genomics. Whereas the former uses whole-genome sequence data and genomics technologies to understand basic cellular processes from a genome-wide perspective, the latter applies genomic sequence information and associated high-throughput technologies to solving practical problems in various fields.

Basic genomics can be subdivided into biochemical, genetical, evolutionary, physiological, ecological, and computational genomics. The goal of biochemical genomics focuses on the large-scale identification of genes encoding biochemical activities, such as enzyme catalysis. This approach was used by Martzen et al. (1999) to identify novel biochemical activities of yeast genes. While genetical genomics uses high-throughput genomic tools to investigate genetic inheritance and variations (polymorphisms) between related individuals in segregating populations (Jansen and Nap, 2001), evolutionary genomics applies similar genomic tools to deciphering evolutionary relationships and processes among extant organisms (see Chapter 4). Physiological genomics aims at defining the genetic pathways and protein interactions that mediate the physiological responses of an organism (Dzau and Glueck, 2001; Bassingthwaighte, 2000). Computational genomics uses integral computational analysis to model *in silico* whole-genome information for the prediction of metabolic pathways, regulatory networks, phylogenetic relationships, and protein structure and interactions (Tsoka and Ouzounis, 2000). Computational genomics provides computational technologies to support experimental genomics.

The term ecological genomics has several different meanings. Nevo (2001) refers to it as the extension of ecological genetics, which focuses on understanding the genetic basis of ecological phenomena, whereas Cary and Chisholm (http://www.ocean.udel.edu/

genomics/final_draft.pdf) used it to imply the application of genomic sciences to understanding the structure and functions of ecosystems. Here we expand the meaning of ecological genomics to include studies that aim at genome-wide understanding of the molecular basis and mechanisms for determining the composition, structure, functions, and dynamics of biological systems within the context of ecology. Based on the levels of ecological systems investigated, ecological genomics can be further subdivided into population genomics, community genomics, and ecosystem genomics (Fig. 1.1). The goal of population genomics is to understand genome-wide genetic variations within a population (Jorde et al., 2001; Black et al., 2001). While community genomics seeks a genome-level understanding of the diversity, structure, functions, and evolution of a biological community, ecosystem genomics employs genomic sequence information and genomic technologies to understand the molecular mechanisms controlling the fluxes of materials and energy among various trophic levels within an ecosystem, and the genetic basis and factors controlling community stability, adaptation, and responses to environmental changes. Some examples of the questions that community or ecosystem genomics can address are as follows: (1) How do different organisms interact at the molecular level to produce the observed patterns of biological diversity of an ecosystem? (2) What is the genetic basis for functional stability and adaptation to environmental stressors in biological communities? (3) How does metabolic capacity relate to biodiversity, disturbance response, and an ecosystem's functionality? (4) Can the functional stability and future status of a biological community be predicted based on the conservation of metabolic functions and the differentiation of individual populations? (5) Can the metabolic capabilities of a biological community be manipulated to achieve a desired stable function in environments under different levels of control?

Genome sequence information and genomic technology have also been widely used to address problems related to medicine, industry, agriculture, and the environment. The emergence of genome-based studies in various fields has given birth to the subdisciplines of industrial genomics, agricultural genomics (Timberlake, 1998), medical genomics (van Ommen, 2001), and environmental genomics. Some other terminologies related to different subfields of medical sciences have also been proposed, such as cancer genomics, toxicogenomics, and pharmacogenomics. These terminologies are considerably different, but their extensions and scope of study are not mutually exclusive. Whereas cancer genomics signifies large-scale cancer studies using high-throughput genomic technologies, toxicogenomics is concerned with identifying the molecular mechanisms of action that underlie the potential toxicity of chemicals, environmental pollutants, and drug candidates through the use of genomics resources (Nuwaysir et al., 1999; see Chapter 14). Toxicogenomics fundamentally changes the way that toxicologists study the toxicity of chemicals on living organisms (Tennant, 2002). Pharmacogenomics aims to identify genetic markers that enable the prediction of drug responses in clinical diagnosis. Pharmacogenomic markers include single-nucleotide polymorphisms between individuals and quantitative differences in gene expression between normal and diseased tissues (Cockett et al., 2000). The term chemical genomics refers to studies of drug screening and the mechanisms of drug action, processes integral to drug discovery and development in the pharmaceutical industry (Cockett et al., 2000). In this context, therefore, it may be more appropriate to use the term drug genomics since the word "chemical" has a broader meaning.

The definition of environmental genomics is particularly unclear. In some cases, it refers to studies related to environmental health on how genetic variations affect responses to environmental exposures (Sharp and Barrett, 2000). It may also signify genome-based studies of organisms that impact the environment. Here we would like to refer to environmental genomics as the genomic studies important to both environmental health and environmental protection.

1.2.3 Classification Based on Types of Organisms Studied

Based on the types of organisms studied, genomics can also be classified into microbial genomics, plant genomics, animal genomics, and human genomics (Fig. 1.1). Based on the microorganisms studied, microbial genomics can be further divided into viral, archaeal, bacterial, and fungal genomics. Another term that is often used in the literature is comparative genomics. It generally refers to the comparison of genome information (e.g., genomic sequences, mRNA, and protein expression profiles) from a variety of organisms for the purpose of obtaining a genome-wide understanding of biological processes and phenomena using both computational and experimental high-throughput approaches.

1.3 HISTORICAL PERSPECTIVE OF GENOMICS

Genomics is a relatively new field of biological inquiry. It began with the proposal to map and sequence the human genome in 1985 and 1986 (McKusick, 1997). The Human Genome Project (HGP) was intensively discussed, debated, and planned between 1986 and 1990 (see Table 1.1). In February 1986, the U.S. Department of Energy (DOE) first announced that it would fund an initiative to pursue a detailed understanding of the human genome (Patrinos, 1996). In 1988, the human genome committee commissioned by the National Academy of Sciences and National Research Council (NRC) released a report (Alberts et al., 1988) that endorsed this project and provided the basis for the first joint plan by the National Institutes of Health (NIH) and the DOE. The committee concluded that the human genome project could be accomplished in 15 years at a cost of $200 million per year and recommended that the human genome first be mapped and then sequenced. The committee also recommended that genomic sequencing projects for model organisms such as yeast and the mouse be pursued in parallel with the human genome project. In 1990, the HGP was formally initiated in the United States with the aim of accomplishing its goals in 15 years.

In parallel to the human genome project, DOE formulated the Microbial Genome Initiative in 1994 to sequence microbial genomes important to energy production, environmental remediation, and biotechnology. In July 1995, the whole genome sequence (~ 1.8 Mb) for *Haemophilus influenzae* Rd, a free-living microorganism, was published by The Institute for Genomic Research (TIGR). This represented the first genome sequenced successfully using a shotgun sequencing approach, which is a fast and effective method for obtaining genomic sequences. Completion of the genome sequence took about one year and marks the beginning of the era of genomics (Venter et al., 1999). Since then more than 100 microbial genomes have been fully sequenced, and 200 microbial genome sequencing projects are underway.

Saccharomyces cerevisiae is an important experimental surrogate in the functional analysis of other organisms' genomes. Unlike humans and other eukaryotes, it is a unicellular organism with a considerably smaller genome size (~ 12 Mb). It can be grown on defined media, and hence the chemical and physical conditions for growth can be

TABLE 1.1 Important Milestones and Events in Genomics

Year	Milestones and Events	Reference
1985–1988	Discussion, debate, and planning of Human Genome Project (HGP) by National Academy of Sciences and NRC.	(Alberts et al., 1988)
1990	Formal initiation of HGP in the United States.	(Burris et al., 1998)
1995	The genome of the first free-living organism, the bacterium *Haemophilus influenzae*, was sequenced using a shotgun sequencing approach.	(Fleischmann et al., 1995)
1996	The budding yeast *Saccharomyces cerevisiae* was fully sequenced by an international team and represented the first eukaryotic organism to be completely sequenced.	(Goffeau et al., 1996)
1998	The first multicellular organism, the worm *Caenorhabditis elegans*, was completely sequenced by the *C. elegans* Sequencing Consortium.	(The *C. elegans* Sequencing Consortium, 1998)
1998	Microarrays and genomics were listed as two of the ten top breakthroughs in science.	(The News and Editorial Staffs, 1998)
2000	The fruit fly *Drosophila melanogaster* was sequenced using a whole-genome shotgun approach. It was the second and the largest animal genome to be sequenced.	(Adams et al., 2000)
2000	Former U.S. President Clinton announced the nearly complete human genome sequence (3,000 Mb) as "the most wondrous map ever produced by humankind."	(Marshall, 2000)
2000	The genome of the flowering plant *Arabidopsis thaliana* was published by an international team and represented the first plant genome to be completely sequenced.	(The *Arabidopsis* Genome Initiative, 2000)
2001	Two versions of the draft for the human genome were published in *Science* and *Nature*. This is the cornerstone of genome-based biology and provides the richest intellectual resource in the history of biology.	(Venter et al., 2001; Lander et al., 2001)
2002	The draft genome sequences of the two major subspecies of rice (*Oryza sativa*) were published. This is a milestone in agricultural research and the first economically important cereal crop whose draft sequences are available.	(Yu et al., 2002; Goff et al., 2002)

completely controlled. Also, *S. cerevisiae* has a life cycle that is ideally suited to classical genetic analysis. Efficient genetic tools for yeast have been developed that allow the replacement of any gene with a mutant allele or its complete deletion from the genome. For these reasons, the genome of *S. cerevisiae* was completely sequenced in 1996 by an international team (Goffeau et al., 1996). This was the first eukaryotic organism to be completely sequenced.

Caenorhabditis elegans, a rhabditid nematode, is a genetically amenable model organism widely used to study genetics, development, and other biological processes. In 1998, the genome (97 Mb) of this organism was published (The *C. elegans* Sequencing Consortium, 1998). This was the first multicellular organism to be completely sequenced.

The fruit fly *Drosophila melanogaster* is one of the most extensively studied organisms in biology and serves as a model system for investigating many developmental and cellular processes common to higher eukaryotes, including humans. In the year 2000, the genome (180 Mb) of this organism was sequenced using a whole-genome shotgun approach (Adams et al., 2000). This constituted a landmark achievement, marking the end of the twentieth century and heralding a new era of exploration and analysis in biology. It is the second and the largest animal genome sequenced so far (Kornberg and Krasnow, 2000). It is also the latest milestone in nine decades of research on this organism (Rubin and Lewis, 2000).

Flowering plants have organizational and physiological features that are very different from those of other multicellular organisms, such as *C. elegans* and *Drosophila*. Whole-genome sequence information from plants will be useful in understanding the genetic basis of differences between plants and animals, and for characterizing the functions and regulation of plant genes. *Arabidopsis thaliana* is an important model system for plant genome analysis due to its short generation time, large number of offspring, and relatively small genome ($\sim$125 Mb); it offers a window into the genetic makeup of all plants, including key agricultural crops. In the year 2000, the *Arabdopsis* genome sequence was determined by an international consortium (The *Arabidopsis* Genome Initiative, 2000), representing the first plant genome to be completely sequenced.

On June 26, 2000, former U.S. President Bill Clinton described the nearly complete human genome sequence (3,000 Mb) as "the most wondrous map ever produced by humankind," and British Prime Minister Tony Blair predicted that genome-based studies would lead to "a revolution in medical science whose implications will far surpass even the discovery of antibiotics." In February 2001, one version of the draft of the human genome sequence was published in the journal *Science* by a group of authors led by Craig Venter of Celera Genomics (Venter et al., 2001), whereas the other version was published in the journal *Nature* by the International Human Genome Sequencing Consortium from the publicly funded consortium of laboratories led by Francis Collins (Lander et al., 2001). The nearly complete human genome sequence is the cornerstone of genome-based biology and provides the richest intellectual resource in the history of biology. Similar to the landing of the first human on the moon and the detonation of the first atomic bomb, the determination of the human genome sequence has enormous symbolic significance because it essentially changes how we think about ourselves (Paabo, 2001). It is the first time that we have been able to view the internal genetic scaffold on which every human life is molded. The availability of the human genome sequence is a historic achievement toward understanding human biology and evolution and also marks a new age in medicine (Paabo, 2001; Jeffords and Daschle, 2001).

Rice (*Oryza sativa*) is the most important cereal crop in the world, constituting more than 60% of the total worldwide agricultural production. Most rice is consumed directly by humans, and about one-third of the human population relies on rice for more than 50% of its caloric intake. Rice is also a critical model species for studying the expression and regulation of plant genes. Rice has the smallest genome ($\sim$430 Mb) compared to other important cereal crops such as sorghum, maize, barley, and wheat. In the year 2002, drafts of the genome sequences for the two major subspecies of rice were published (Yu et al., 2002b; Goff et al., 2002). This was a milestone in agricultural research (Normile and Pennisi, 2002) and represents the first genomic sequence information available for an economically important cereal.

The availability of whole-genome sequences has greatly facilitated the development and application of genomic technologies, such as microarrays, for the functional analysis of genome sequences. In 1995, microarrays prepared by high-speed robotic printing of complementary DNAs on glass slides were first proposed for the quantitative expression measurements of corresponding genes (Schena et al., 1995). After this successful application, both DNA and oligonucleotide microarrays were used for monitoring gene expression and detecting differentially expressed genes in yeast (DeRisi et al., 1997; Lashkari et al., 1997; Wodicka et al., 1997; Spellman et al., 1998), and different human cells, tissues, and those that are diseased (Lockhart et al., 1996; Heller et al., 1997). In 1998, the news and editorial staffs of the journal *Science* listed microarrays as one of ten breakthrough technologies along with genomics (The News and Editorial Staffs, 1998). Similar to the situation in which microprocessors have accelerated computational processes, microarray-based genomic technologies have revolutionized the genetic analysis of biological systems. Microarray technology represents a powerful new tool that allows researchers to view the living cell under various physiological states from a comprehensive and dynamic molecular perspective.

Successful sequencing of the genomes of humans and other organisms has also greatly facilitated the development and application of proteomic technologies such as mass spectrometry (MS). During the 1990s, advances in MS instrumentation and techniques revolutionized protein chemistry and fundamentally changed the method of protein analysis (Aebersold and Goodlett, 2001). Although two important technical breakthroughs (electrospray ionization [ES] and matrix-assisted laser desorption/ionization [MALDI]) for proteomics were made in the late 1980s, biological MS did not emerge as a powerful analytical method for protein studies until whole genome sequences became available. With the accessibility of complete genomic sequences, proteins isolated from a given cell type no longer needed to be identified by de novo sequencing, but they could be identified by correlating the molecular masses of short peptides and/or short amino acid sequences with those predicted from sequence databases. The rapid development of biological mass spectrometry, coupled with the availability of the entire genome sequences from many different organisms, also marks the beginning of a new era of biology (Pandey and Mann, 2000).

1.4 CHALLENGES OF STUDYING FUNCTIONAL GENOMICS

The ultimate goal of functional genomics is to use genome sequence information and related genomic technologies to link sequences with functions and phenotypes and to understand how biological systems at different levels function in nature. This goal faces the following challenges.

1.4.1 Defining Gene Function

Although computational analysis based on sequence homology comparisons is useful in defining the functions of genes, it only provides clues about gene function because homology is not function and substantial portions of predicted open reading frames (ORFs) are functionally unknown. In addition, the functional assignment of genes by similarity comparison is sometimes misleading or incorrect due to the complicated evolutionary and structure-function relationships among different genes. In other words, the relationships of genes to enzymes are one to one and also many to one and one to many. Such complicated gene-enzyme relationships present difficulties for sequence-based functional annotation. Therefore, experimental analysis must be performed to understand the biological role of a gene product. However, defining the role(s) of each gene within the context of a complex cellular machine and regulatory networks is a formidable experimental task (DeRisi et al., 1997; Strauss and Falkow, 1997; Hieter and Boguski, 1997).

1.4.2 Identifying and Characterizing the Molecular Machines of Life

Cells are the basic working units of all living systems and are the biological "factories" that carry out and integrate thousands of discrete, highly specialized processes using molecular "machines" consisting of different proteins and other molecules. Proteins rarely work alone and often assemble into larger multiprotein complexes. Some of these complexes behave like complicated machines and execute fundamental cellular functions and metabolic processes (e.g., DNA replication, transcription, translation, protein degradation), mediate information flow within and among cells and with their environment (e.g., signal transduction, energy conversion, cell movement), or build cellular structures. Many protein machines appear to be highly conserved in overall composition and function; thus, the knowledge learned from one organism may be applicable to other organisms. Although the number of different types of protein machines is not known, whole-genome sequence analyses suggest that the number of protein machines could be finite, perhaps in the range of several thousand per cell.

Recent progress in whole-genome sequencing and protein-structure determination has provided information on the composition of many proteins in model organisms, but the grand challenge that remains is to fully understand and characterize the repertoire of molecular machines present in living systems and to understand how proteins confer on cells their capabilities, structure, and higher-order properties.

1.4.3 Delineating Gene Regulatory Networks

All living systems are capable of making rapid changes in space and time in response to both intracellular and environmental stimuli. Such capabilities are achieved through complex coordinated interactions of many individual proteins and protein complexes under various regulatory controls. Individual proteins and other macromolecules are "wired" together like electronic circuits to form complex genetic regulatory networks, which are the "brains" of the cells. The networks receive information from outside and inside cells by signal transduction pathways, process the information to make "decisions" on which genes are expressed and how much product is synthesized, and trigger the appropriate cellular and physiological responses. The coordinated behavior of genetic regulatory networks determines the physiological properties of a living cell and is crucial to cell survival, growth, and reproduction.

Recent genome sequence comparison indicates that a human has only two to three times as many genes as a simple worm, and only five to ten times as many as a single-cell microorganism. However, the human has many more different types of cells, tissues, and organs with distinct functions. Thus, the simple differences in gene number alone cannot explain the huge phenotypic differences between humans and a worm or microbes. The phenotypic differences could be largely attributed to the architecture and complexity of genetic regulatory networks. Multicellular organisms may have evolved from simple organisms by developing more complicated and distinct regulatory networks capable of exquisitely controlling complex combinatorial gene expression patterns, while the repertoire of genes itself has rather modestly expanded. If this hypothesis is true, changes in the connections and nodes in such regulatory networks could dramatically change the physiological properties and behaviors of living systems. However, understanding how the genome functions as a whole gives life in the complex natural history of living organisms presents an even greater challenge (DeRisi et al., 1997).

1.4.4 System-level Understanding of Biological Systems beyond Individual Cells

Organisms interact with each other and with their environments to produce more complex living systems such as populations, communities, ecosystems, and the biosphere. The challenge of bringing genomes to life is to use genomic sequence information to understand the consequences of genetic capacity and interactions at spatial scales ranging from microns to continents, and to link the genetic level of mechanistic studies with system levels of functional performance within the context of biological systems (e.g., individuals, communities, ecosystems, and the biosphere).

Microorganisms appear to be the most diverse group of life presently known, inhabiting almost every imaginable environment on Earth. The total bacterial cell count for Earth has been estimated to be 4 to 6×10^{30} cells, and the cellular carbon content to be 3.5 to 5.5×10^{14} kg (Whitman et al., 1998). Bacteria also constitute the largest living pool of nitrogen $(1.3 \times 10^{14}$ kg$)$ and phosphorous $(1.4 \times 10^{13}$ kg$)$ (Whitman et al., 1998). In contrast to plants and animals, however, the extent of microbial diversity is largely unknown. Due to the uncultivated status of the majority of microorganisms found in nature, little is known about their genetic properties, metabolic characteristics, and functions. Understanding and characterizing uncultured microorganisms present great challenges.

Functional stability and adaptation are two important properties of biological systems. The relationship between the diversity and stability of biological communities is a long-standing question in macro-community ecology. One of the most active and controversial research areas in macroorganismal ecology is the relationship between biodiversity and the functional properties of ecosystems. The fundamental question is how many species are necessary for "adequate" functioning or stability of any particular ecosystem process. The controversy revolves around conflicting interpretations of data, some of which suggest that many species function at a higher rate relative to others (i.e., biomass production), while some studies demonstrate that communities with low species diversity can be just as productive as those with high diversity. A prominent series of exchanges have presented both sides of the issue, but the controversy remains unresolved, primarily because of the inability to measure the functional contributions of different species in experiments (Huston, 1997; Huston et al., 2000; Loreau et al., 2001).

One critical issue is how the functional properties of different organisms interact to yield the total functionality and stability of a community consisting of many different types of organisms. It is generally hypothesized that the functional stability and adaptation of biological systems are determined by the genetic and metabolic diversity of individual populations, which includes their ability to respond to environmental changes. Prior to the genomic era, it was not possible to test this hypothesis because of the difficulty in accessing genetic information. With the recent revolution and rapid development in genomics and genomic technologies, we now have a unique opportunity to link subcellular metabolic processes with the functional performance of biological communities in complex natural environments.

1.4.5   Computational Challenges

Metabolism, the biochemical engine that drives cellular processes, and its components can be defined from annotated genomic sequence data. The translation of genomic information into metabolic function, physiology, growth, and behavioral potential is under direct genetic control up to the level of the whole organism. One great challenge is to use what essentially amounts to a "parts catalogue" and synthesize those parts using computer models that simulate biological cellular functions.

Above the level of a single organism, the interactions of multiple individuals (species) drive the ecological processes that influence evolution and ecosystem dynamics. The individual organism is the fundamental unit of both evolutionary and ecological processes. A second grand challenge is to synthesize the knowledge of cellular-level processes through computational modeling and simulations to understand population-, community-, and ecosystem-level dynamics in order to gain new knowledge of evolutionary and ecological processes that cannot be explained through understanding individual organisms or populations.

1.4.6   Multidisciplinary Collaborations

A full mechanistic understanding is required at all levels in order to predict the complete range of functional and behavioral dynamics of biological systems under realistic environmental conditions. This is a Herculean task that will take at least several decades (perhaps much longer). No single laboratory, institution, or government agency can encompass the breadth and depth of biological and technical expertise that will be required for a comprehensive functional characterization of genomic sequences and their relationships to the functional performance of biological systems on various scales (Oliver, 1996). Collaborations among investigators with different research capabilities from multiple institutions are needed to meet these challenges. However, forming such a large and efficient team is in itself a challenge.

1.5   SCOPE AND GENERAL APPROACHES

1.5.1   Structural Genomics

Although genome sequencing projects reveal enormous amounts of information about a particular organism, these projects only scratch the surface of biological diversity in general (Fraser and Dujon, 2000). Clearly, insight into genome diversity will require sequencing the genomes of many different species from various environments. Thus, whole-genome sequencing and functional annotation are still a critical part of structural genomics. The genome sequence differences among different organisms hold the key to understanding how nature produces and selects such diversity of organisms with a variety of forms and functions. The information on sequence conservation among distantly related organisms will be extremely useful in discerning the real functional constraints on genes and gene products (Lander, 1996). In addition, sequence comparisons among closely related organisms will directly provide insights into regulatory motifs and mechanisms (Lander, 1996). Sequence information will also provide important information on genome organization and structure.

With the significant reduction in sequencing cost and the availability of automated high-throughput capillary sequencing robots, it is now feasible to rapidly decipher whole-genome sequences from many phylogenetically representative organisms. Associated with whole-genome sequencing, more advanced computational tools for sequence alignment and analysis, and functional annotation are critical to managing and analyzing the enormous quantity and complexity of genome sequence data.

An important aspect of structural genomics is the disruption and manipulation of genome structure, which is one of the most powerful ways to understand gene function. The biological functions of genes can be examined by turning off the function of a gene using gene knockout strategies or by changing its normal pattern of expression through replacing the wild-type gene with a mutated counterpart. Generally, two common approaches can be used to disrupt or inactivate a gene (see Chapter 8). One is transposon-based random insertional mutagenesis whereby an antibiotic resistant drug cassette is randomly inserted into a genome; another approach is in-frame deletion by homologous recombination in which the target gene is completely or partially deleted in frame. Homologous recombination is normally achieved using either nonreplicating, or suicide, plasmids or plasmids with a conditionally active replicon as a delivery system.

Protein structure determination and prediction are also critical to structural genomics and require knowledge in several scientific fields such as genome sequencing, protein expression, crystallization, X-ray crystallography, nuclear magnetic resonance, computational analysis, modeling and prediction of protein structure (Gaasterland, 1998). Although protein sequences are highly diverse, the number of unique structural folds is fairly small in nature. Even in the absence of protein sequence similarities, proteins of unknown function could have three-dimensional (or 3D) structures similar to those of other proteins of known functions (Kim, 1998). Because the function of a protein is tightly correlated with its three-dimensional structure, understanding its folded structure could provide important insight into its biochemical functions or role in a pathway.

Both experimental and computational approaches have been used complementarily to determine and predict protein structure. The high-throughput experimental determination of protein structure involves protein expression, purification and crystallization, X-ray crystallography using high-flux synchotron bean lines, and computational analysis of the resulting data. However, the entire process is generally slow and costly (Gaasterland, 1998). The high-throughput experimental approach provides a rich data set for computational biology. Based on experimentally determined structures, appropriate computational models can be constructed and optimized. This will allow rapid, meaningful structure predictions of whole proteomes across many different organisms.

1.5.2 Transcriptomics

The comprehensive analysis of transcriptomes is of great value for studying gene function and regulation for the following reasons (Brown and Botstein, 1999; Brent 1999):

(i) *Correlation of Gene Expression and Function.* Natural selection dictates that genes expressed in specific cells under specific conditions must contribute to the fitness of the organism. Similar to the phenomena that natural selection has precisely fine-tuned the biochemical properties of gene products, it has also fine-tuned the regulatory networks that control when and where the gene product is made and in what quantity. Thus, it is believed that there is a strong connection between gene expression and the function and regulation of the encoded product.

(ii) *Connection between Gene Expression and Physiological States.* The set of genes expressed in a cell will determine what the cell is made of, how the cell is built, what biochemical and regulatory systems are operative, and what the cell can and cannot do. Gene expression patterns can provide information about the dynamic changes in physiological states and functional activities of a cell under different environmental conditions.

(iii) *Guilt by Association.* It is generally believed that genes expressed together may function together. Sets of genes that share similar functions can be grouped on the basis of similarities in their expression patterns. Also, common regulatory mechanisms can be inferred for genes with similar expression profiles, and conserved regulatory elements can be identified. Thus, based on the expression patterns of known genes, the function of hypothetical proteins can be predicted if such genes are coexpressed with genes of known function.

(iv) *Importance of Transcriptional Regulation.* As mentioned previously, transcriptional regulation is an important mechanism for controlling the cellular levels of proteins and metabolites. Thus, mRNA levels can be used as a measure of gene expression and activity.

(v) *Technological Advances.* In contrast to proteins and metabolites, mRNA molecules can be analyzed in a fully comprehensive manner using massive parallel hybridization-based techniques. High-density microarrays containing whole-genome sequence information are powerful and indispensable tools for the analysis of gene expression and regulation. Once we learn the biological consequences of gene expression patterns as a result of our growing knowledge of the functions of individual genes, we should be able to use microarrays as a "microscope" for visualizing the complex and dynamic nature of living cells (Brown and Botstein, 1999). However, mRNAs are not the functional entities within the cell but just the transmitters for synthesizing proteins. There are several potential problems associated with using mRNA coexpression profiles for understanding gene function and regulatory networks. First, although there is a strong connection between coexpression and gene function, some functionally unrelated genes might have similar expression patterns. Thus, any measurement based on mRNA expression levels may result in misleading interpretations. Second, not all functionally related genes are expressed together, and hence, approaches based on mRNA expression profiles may miss important, functionally related genes. Finally, translational regulation and posttranslational modification are also important in determining gene function and regulation (Pradet-Balade et al. 2001). Because of these potential problems, key coexpressed genes should be validated using conventional molecular methods, such as Northern blot hybridization and real-time PCR. Also, proteomic and metabolic analyses should be used for comprehensively understanding gene function and regulatory networks.

1.5.3 Proteomics

Comprehensive analysis of the proteome is central in functional genomics for several reasons (Pandey and Mann, 2000). First, unlike mRNAs, proteins are the active reagents in cells. Proteome analysis is both qualitatively and quantitatively distinct and complementary to transcriptome analysis. Second, the existence of an ORF predicted from sequence similarity does not necessarily mean the existence of a functional protein. Despite advances in computational genomics, it is still difficult to accurately predict gene function based on sequence homology (Eisenberg et al., 2000; Pandey and Mann, 2000). This is particularly true when the predicted genes are small or the hypothetical genes have little or no homology to other known genes. For instance, at least 8% of the annotations for 340 genes from the smallest sequenced genome, that of *Mycoplasma genitalium*, were incorrect (Brenner, 1999), thus emphasizing the need to validate gene products using proteomic methods. Third, important posttranscriptional mechanisms for regulating gene function cannot be measured using gene expression profiling. Fourth, protein decays far more slowly than mRNA, and the mechanisms for regulating protein abundance and therefore function by proteolysis, recycling, and sequestration in cell compartments only affect gene products rather than genes. Thus, protein levels may or may not correlate with mRNA levels. Fifth, proteins can be modified after translation, and such modifications can be determined only by proteomic analysis. Sixth, it is generally difficult to predict the cellular locations of gene products based on sequence data, and protein-based experimental approaches must be used to determine their locations. Finally, a protein-level approach is the only way to understand protein–protein interactions and the molecular composition of cellular structure. Although proteomics is important in understanding gene functions and genetic regulatory networks, proteomic analysis is fraught with technical difficulties. Most currently used proteomic techniques are neither comprehensive in their scope nor massively parallel in their execution (Delneri et al., 2001).

Proteomic studies can generally be classified into three main categories. The first area is protein identification and analysis. The use of MS is currently one of the most important tools being used in proteomic studies. It is the method of choice for protein identification, because it can deal with complex protein mixtures and offers much higher sensitivity and throughput (Pandey and Mann, 2000). Although more difficult, it is also the method of choice for determining protein modifications. The posttranslational modifications, such as phosphorylation, glycosylation, and sulphation, provide extremely important information on protein function.

The second area is proteome-wide differential display of proteins under various growth conditions. Knowledge of where and when proteins are expressed is also essential for understanding their biological functions. Two-dimensional gel electrophoresis, followed by protein identification using MS, and protein microarrays can be used to monitor protein expression under different growth conditions (Pandey and Mann, 2000). By labeling protein samples with different isotopes, relatively quantitative data on gene expression can also be obtained by MS analysis.

The third area is protein–protein interactions. Systematic identification of protein interactions in complex molecular machines of life is very important in understanding how a given proteome works. Various approaches such as a two-hybrid system, phage display, ribosome display, protein arrays, and MS can be used to determine protein–protein interactions in a rapid, high-throughput manner (Pandey and Mann, 2000).

1.5.4 Metabolomics

Elucidation of gene function by analysis of the metabolome has several advantages. Like proteins, metabolites also constitute functional entities within the cell, and the total complement of metabolites varies with the physiological, developmental, or pathological state of a cell, tissue, organ, or organism. In contrast to mRNAs and proteins, the number of metabolites is far fewer than that of genes or gene products, and hence metabolite analysis is less complex than mRNA and protein analysis. For instance, *S. cerevisiae* has ~6,000 protein-coding genes but fewer than 600 low-molecular-weight intermediates (Rassmsdonk et al., 2001).

Two major approaches have been used to understand gene functionality via metabolic analyses (Delneri et al., 2001). The first approach is to identify the biochemical reactions catalyzed by enzymes encoded by genes of unknown function (Delneri et al., 2001). Proteins with novel biochemical activities were identified with this approach (Martzen et al., 1999). The second approach is functional analysis by coresponses. Since many genes may be involved in synthesizing or degrading a single metabolite, there is no direct relationship between metabolite and gene as there is for mRNAs and proteins. To overcome this difficulty, the coresponses-based functional analysis was proposed (Raamsdonk et al., 2001). Briefly, this approach uses the metabolic profiles of the genes of known function to elucidate the possible roles of an unknown gene. This can be achieved by comparing the differences in metabolic profiles between mutant strains defective in an uncharacterized gene and an experimentally characterized gene of known function. The metabolic profiles of a mutant strain carrying a deletion in an undefined gene will be similar to those of mutant strains defective in a gene of known function if their gene products act on the same functional domain of the cell. Otherwise, their metabolite concentrations will change in a different way or will not change at all (Raamsdonk et al., 2001). In general, comprehensive metabolite analysis can be achieved using MS or nuclear magnetic resonance (NMR) spectroscopy followed by sophisticated chemometric analysis of the resulting spectra. By quantifying the relative changes in the intracellular metabolites between mutant and wild-type cells, it is possible to determine the effects of mutations on cellular metabolism and hence on gene function. This approach is particularly useful in revealing phenotypes for gene deletions without measurable effects on cell growth (Oliver, 2002).

In contrast to proteomic analysis, the main difficulty with metabolomic analysis is conceptual, not technical (Delneri et al., 2001). It is technically feasible to simultaneously analyze several hundreds of metabolites, but linking metabolite analysis to functions of individual genes is sometimes difficult because many genes may be involved in synthesizing or degrading a single metabolite.

1.6 IMPORTANCE OF MICROBIAL FUNCTIONAL GENOMICS TO THE STUDY OF EUKARYOTES

Functional analysis of microbial genomes is not only crucial to understanding gene functions and regulatory networks in microorganisms themselves but also very important in defining gene functions and regulatory networks in higher organisms, such as plants, animals, and human. First, many genes involved in essential life processes (e.g., DNA replication, DNA repair, transcription, translation, central metabolism, signal transduction, and cell division) are conserved among microorganisms (i.e., prokaryotes and lower eukaryotes) and higher eukaryotes. Knowledge of the functions of genes and regulatory networks involved in basic life processes will provide important insights into the functions of their counterparts in higher eukaryotes. Many microorganisms such as *E. coli* and yeast can be used as surrogates in the functional analysis of eukaryotic genomes. No other eukaryote is as genetically and physiologically tractable as *S. cerevisiae*. The functions of genes from large or poorly studied eukaryotic genomes can be defined by complementing the deletion mutants of the increasingly well-defined yeast genome (Oliver, 2002). For instance, more than 40% of human-coding sequences are capable of complementing yeast mutations (Oliver, 2002). Therefore, trans-complementation of yeast mutations is a potentially powerful strategy for defining the functions of many human genes, and even more useful for functional analysis of other microbial eukaryotes such as fungal and protozoal pathogens. In addition, mitochondria and chloroplasts, which are of bacterial origin, are important components of eukaryotic life processes. Functional and evolutionary genomic analyses of microorganisms, therefore, could be important for understanding the function and evolution of these organelles, as well as gene function and regulatory networks in eukaryotic cellular processes. Finally, many microorganisms (e.g., symbionts, pathogens) have formed intimate relationships with higher eukaryotes. Functional genomic studies of these microorganisms will also facilitate the understanding of interactions and coevolution between microorganisms and their eukaryotic hosts.

1.7 SUMMARY

Biology is undergoing a revolution, and the nearly complete human genome sequence is the cornerstone of genome-based biology. This is the first time in history that we have access to the entire genomic content of living organisms. However, determination of entire genome sequences is only the first step in understanding the physiology and behavior of an organism. The next critical step is to elucidate the functions of these sequences, thus providing biochemical, physiological, evolutionary, and ecological meaning to nucleotide information.

We define *genomics* as the study that aims at a genome-level understanding of the molecular basis of the structure, function, and evolution of biological systems using whole-genome sequencing information and high-throughput genomic technologies. From the perspective of general system theory, genomics can be divided into structural genomics and functional genomics. The former is defined as the genome-wide structural study of genes and proteins, including genome mapping, sequencing, and organization as well as genome-wide protein structure characterization, whereas the latter refers to system-level understanding of gene functions and regulatory networks using global genome-wide approaches based on the information provided by structural genomics. To reflect the complementary functional investigations of genes at the levels of mRNA, proteins, and metabolites, functional genomics can be divided into transcriptomics, proteomics, and metabolomics. Various high-throughput approaches such as microarrays, mass spectrometry, phage display, two hybrid systems, and nuclear magnetic resonance are used to study the cellular dynamics of mRNAs, proteins, and metabolites. Genomics can also be classified into different subdisciplines from the perspectives of biochemistry, physiology, and ecology.

The ultimate goal of functional genomics is to use genome sequence information and related genomic technologies to link sequences with functions and phenotypes and to understand how biological systems at different levels function in nature. Achieving this goal means facing several challenges in defining gene function, identifying molecular machines, delineating genetic regulatory networks, and understanding biological systems beyond individual cells. The translation of the vast genomic information into metabolic function, physiology, growth, and behavioral potential through modeling and simulation is also a great challenge.

Functional analysis of microbial genomes is not only important to understanding gene function and regulatory networks in microorganisms themselves but also very important to defining gene function and regulatory networks in higher organisms, such as plants, animals, and human. Many microorganisms such as *E. coli* and yeast have been used as surrogates in the functional analysis of eukaryotic genomes, because they are more genetically and physiologically tractable.

FURTHER READING

Aebersold, R. and D. R. Goodlett. 2001. Mass spectrometry in proteomics. *Chem. Rev.* 101:269–295.

Brown, P. O. and D. Botstein. 1999. Exploring the new world of the genome with DNA microarrays. *Nat. Genet.* 21:33–37.

Eisenberg, D., E. M. Marcotte, I. Xenarios, and T. O. Yeates. 2000. Protein function in the post-genomic era. *Nature* 405:823–826.

Fiehn, O. 2002. Metabolomics—the link between genotypes and phenotypes. *Plant Mol. Biol.* 48:155–171.

Hieter, P. and M. Boguski. 1997. Functional genomics: it's all how you read it. *Science* 278:601–602.

Kitano, H. 2002. Systems biology: a brief overview. *Science* 295:1662–1664.

Oliver, S. G. 2002. Functional genomics: lessons from yeast. *Philos. Trans. R Soc. Lond. B Biol. Sci.* 357:17–23.

Pandey, A. and M. Mann. 2000. Proteomics to study genes and genomes. *Nature* 405:837–846.

Strauss, E. J. and S. Falkow. 1997. Microbial pathogenesis: genomics and beyond. *Science* 276:707–712.

Tennant, R. W. 2002. The national center for toxicogenomics: using new technologies to inform mechanistic toxicology. *Environ. Health Persp.* 110:A8–A10.

2

Microbial Diversity and Genomics

Konstantinos Konstantinidis and James M. Tiedje

2.1 INTRODUCTION

The microbial world is considered the largest unexplored reservoir of biodiversity on Earth, and because of this is considered the next frontier in biology. Genomics provides both the opportunity to more fully explore this unknown but also the concern that this level of complexity is too overwhelming to approach meaningfully at the genomic level. There is no doubt that some aspects of this diversity can be—indeed recently have been—discovered from genomic approaches, but the challenge is huge and the discovery process will be long and continuous and require some additional technological breakthroughs. This chapter focuses on providing the background for understanding the origin and development of such microbial diversity along with information on the importance and extent of this diversity to provide the context for its exploration at the genome level. The chapter also summarizes general trends from the 112 microbial genomes uncovered at the time of writing, which is beginning to reveal a more complete picture of genome adaptation to ecological niches.

2.2 BIOCHEMICAL DIVERSITY

Prokaryotic life emerged about 3.7 billion years ago, or about 2 billion years before eukaryotic life arose (Fig. 2.1). Thus, prokaryotic organisms have had a very long time to evolve, which accounts for the high biochemical and genetic diversity that characterizes this group. Life began based on lithotrophy in which energy is derived from the oxidation of reduced inorganic compounds by a chemical oxidant, a terminal electron acceptor. Although there is still considerable debate about where on Earth life began, for example, hydrothermal vents or surface soil, and the original electron donor and acceptor pair,

Microbial Functional Genomics, Edited by Jizhong Zhou, Dorothea K. Thompson, Ying Xu, and James M. Tiedje.
ISBN 0-471-07190-0   © 2004 John Wiley & Sons, Inc.

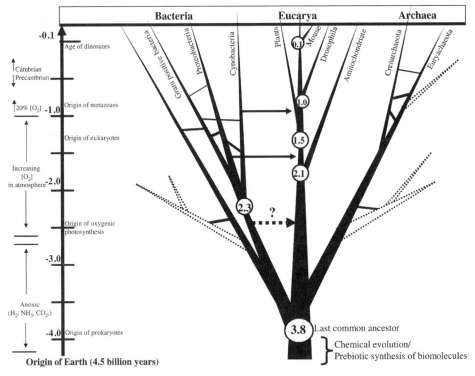

Figure 2.1. *The evolution of life.* Representative lineages of the three domains of life (solid branches) and the estimated time of their appearance (encircled numbers) are shown. Dashed branches represent (hypothetical) lineages that became extinct during the course of evolution. Gray lines connecting branches are used to illustrate the importance of lateral gene transfer and fusion events in the evolution of organisms. Two major fusion events are shown (solid arrows): the evolution of the mitochondria and the plant chloroplasts from a proteobacterium and a cyanobacterium symbiont, respectively. Hence, Eukaryotes are represented as the genomic hybrids of Archaebacteria and Eubacteria, although it is still uncertain whether the eukaryotic cell emerged from the fusion of an archaeon host and a eubacterial symbiont (dashed line with question mark). Cyanobacteria emerged from the main eubacterial line of descent around 2.3 billion years ago; the O_2 produced by the oxygenic photosynthesis carried out by Cyanobacteria is believed to have caused the extinction of many prokaryotic lineages and the adaptive evolution of most lineages currently known.

prokaryotic lithotrophy has evolved to such an extent that the currently known microorganisms can exploit a wide range of electron donors such as hydrogen, hydrogen sulfide, or ferrous iron and electron acceptors such as oxygen nitrate, sulfate, sulfur, ferric iron, protons, carbon dioxide, or a variety of oxidized metals.

Eukaryotes, on the other hand, can live using lithotrophy only when they are symbiotically associated with such prokaryotes. For example, the 2-m-long tubeworm *Riftia pachyptila* lives in the vicinity of seafloor hydrothermal vents and metabolizes hydrogen sulfide and carbon dioxide by means of sulfide-oxidizing, carbon dioxide-fixing bacterial symbionts (Cavanaugh et al., 1981; Lopez-Garcia et al., 2002). It is not necessary, however, to explore unusual (from our perspective) environments to encounter such fascinating symbiotic relationships. Several methanogens, which are members of Archaea, live intracellularly with eukaryotes and serve as their metabolic hydrogen sinks (Embley and Finlay, 1994).

Photosynthesis was the next autotrophic lifestyle to evolve, and this was a prokaryotic invention as well. Bacteria show diverse kinds of photosynthesis, with most bacterial photosynthesis being anaerobic and widely distributed among bacterial phyla. On the other hand, photosynthesis is rarely encountered among the cultured archaeal species and is probably restricted to species of *Halobacterium*. Oxygenic photosynthesis, the water-based photosynthetic mechanism that produces the powerful electron acceptor oxygen, arose only in the phylum-level lineage of Cyanobacteria. The time of oxygenic photosynthesis evolution is still debated, with most evidence supporting its origin about 2.5 to 3.0 billion years ago (Fig. 2.1) (Hedges, 2002). This process resulted in major changes in the chemistry and biology of Earth by opening new niches for the adaptive radiation of most prokaryotic lineages currently known, while causing the extinction of many other lineages. There is now conclusive genomic evidence that the eukaryotic photosynthetic machines, that is, the chloroplasts of plants, originated from a symbiotic event between a eukaryotic cell and cyanobacterium (Liester, 2003).

Organotrophy, in which reduced organic compounds are used for energy and carbon, is the most well-studied prokaryotic lifestyle probably because some of its members are very easy to grow and have become model organisms, for example, *Escherichia coli* and *Bacillus subtilis*. Fungi, protozoa, and animals are also organothrophs. Organothrophy is not the prevalent form of metabolism in the environment, however. Autotrophic metabolism and carbon dioxide fixation must necessarily contribute a greater biomass than the organothrophic metabolism that they support. Autotrophy is more widespread and diverse in prokaryotic than in eukaryotic organisms. Nonetheless, even in the case of organotrophy, the number of different metabolic processes carried out by Bacteria far exceeds those carried out by eukaryotes. The extent of bacterial metabolic and enzymatic diversity is such that it is believed a handful of microbial species can live on almost any carbon source available.

A unique characteristic of the Bacteria compared to the other two domains of life is that its species closely related by molecular criteria can display strikingly different carbon and energy metabolism. For instance, in the relatively closely related gamma-Proteobacteria subgroup, one can find very phenotypically different organisms such as the *E. coli* (organotroph), *Chromatium vinosum* (hydrogen sulfide-based phototroph), and the symbiont of *R. pachyptila*, the tubeworm (hydrogen sulfide-based symbiont). Furthermore, the ancient lithotrophic way of life is widely distributed phylogenetically, intermixed with organotrophic organisms. This pattern suggests that organotrophy arose many times from otherwise photosynthetic or lithotrophic organisms. Indeed, many bacterial species can switch between these modes of nutrition, carrying out photosynthesis in the light and lithotrophy or organotrophy in the dark.

The situation is even more profound when specific biochemical traits, for example, functional proteins, are considered. For instance, the ability to denitrify (making use of N oxides as terminal electron acceptors) occurs sporadically among the cultivated bacterial species of coherent SSU rRNA clusters (Zumft, 1997). SSU rRNA sequence information is commonly used for the construction of phylogenetic trees to infer the ancestry and relatedness of organisms. As apparent from the denitrification example, even organisms that are identical or cluster tightly by the SSU rRNA criterion may not share key physiological similarities. Furthermore, the functional genes involved in the denitrification pathway (e.g., nitrite reductase, nitrous oxide reductase) exhibit substantial sequence diversity in the cultivated representatives from a single gram of marine sediment or forest soil. Nonetheless, not all biochemical processes share the diversity revealed for the denitrification process. For example, the ammonia-oxidizing bacteria form a monophyletic group, which shows decreased sequence diversity in the ammonia monooxygenase gene (a key enzyme catalyzing this process) (Kowalchuk and Stephen, 2001). The lack of correspondence between metabolism and evolutionary relatedness is attributed to lateral gene transfer, large-scale symbiotic fusions, and the great ability of bacteria to evolve to exploit available ecological niches.

The other domain of Prokaryotes, the Archaea, shows considerably less metabolic diversity than the Bacteria. Many lithotrophic but comparatively few organotrophic representatives of Archaea have been cultivated. There are currently two Archaea kingdoms with cultivable representatives: the Euryarchaeota and the Crenarchaeota (Fig. 2.1). Both kingdoms rely on hydrogen as a main source of energy. In the former, the main electron acceptor is carbon dioxide, and the product, methane (methanogenesis). Isotopic analysis indicates that most of the methane encountered in the outer few kilometers of the earth's surface is the product of methanogenic Archaea. Thus, these Archaeal species must constitute an important component of the global biomass. The known Crenarchaeota, on the other hand, oxidize molecular hydrogen using a sulfur compound as the terminal electron acceptor. Most cultivated representatives of Crenarchaeota are thermophiles; some grow at the highest known temperatures for life, up to $113°C$ in the case of *Pyrolobus fumaris* (Stetter, 1999). Most of the Archaea metabolic diversity probably awaits discovery. Recent findings from culture-independent surveys for the presence of archaeal-specific molecular signatures in the environment suggest that archeal ecological significance and global distribution are much higher than represented in the currently cultured species (Barns et al., 1996; Bintrim et al., 1997; DeLong et al., 1999; DeLong and Pace, 2001). For instance, several phyla- and order-level lineages plus a new kingdom of Archaea, the Korarchaeota (Barns et al., 1996), have been proposed based on cloned SSU rRNA sequences from different environment sources. The lack of cultivable species representative of these lineages or information about the physiology of these species severely limits our ability to summarize archaeal metabolic diversity.

Another microbial lineage for which there is relatively little information about its metabolic breadth is the amitochondriate eukaryotes. This interesting microbial group appears to have spun off of the main eukaryotic line (the fung–plant–animal line) early in evolution (Cavalier-Smith, 1992; Hedges, 2002) (Fig. 2.1). However, the indisputable conclusion to be drawn from current knowledge on the metabolic and biochemical repertoire of the three domains of life is that the versatility of Bacteria makes the metabolic machineries of Archaea and Eukarya seem comparatively monotonous.

2.3 GENETIC DIVERSITY

2.3.1 The Unseen Majority

Although the previous discussion pointed out the immense metabolic diversity of the prokaryotic domains of life, what makes prokaryotic organisms and consequently the metabolic processes they carry out important to Earth is the huge number of prokaryotic cells and their ubiquity. The most recent estimates suggest the total number of prokaryotic cells on Earth to be 4 to 6×10^{30} and their cellular carbon to be 350 to 550 Pg (Whitman et al., 1998). Hence, prokaryotic carbon is 60 to 100% of the estimated carbon in terrestrial and marine plants! In addition, prokaryotes are presumably the largest pool of recyclable nitrogen and phosphorus for primary production since the nitrogen and phosphorus to carbon ratio is high for the prokaryotic cell. Most of the earth's prokaryotes are found in the open ocean and in soil, where the numbers of cells are in the order of 10^{29} to 10^{30} (Whitman et al., 1998). Furthermore, prokaryotes are believed to be abundant in environments where eukaryotic organisms are absent or very rare, such as the subsurface (e.g., below 8 m for the terrestrial environment and below 10 cm for ocean sediments). Estimates of the numbers of cells in the subsurface are huge (Hazen et al., 1991; Gold, 1992; Whitman et al., 1998), although there has been a limited sampling of these environments and thus uncertainty in the accuracy of these predictions. The activity of prokaryotes is substantial in marine and soil environments based on cell turnover times for these environments, which have been estimated at 6 to 25 days for the upper 200 m of ocean and 2.5 years for soil (Grey and Williams, 1971; Ducklow and Carlson, 1992; Whitman et al., 1998). However, prokaryotic activity in the subsurface is orders of magnitude lower, for example, turnover times of 1 to 2×10^3 years, than that in soil or the ocean water column (D'Hondt et al., 2002).

2.3.2 How Many Prokaryotic Species Are There?

While it is beyond question that prokaryotic cells are ubiquitous and far exceed any other type of life in numbers, the enumeration of prokaryotic species on Earth is far from being resolved. This is due to both fundamental problems regarding the definition of species as well as practical limitations in counting prokaryotic species.

The current species concept for prokaryotes (Rossello-Mora and Amann, 2001) despite being pragmatic, operational, and applicable (Stackebrandt et al., 2002) remains controversial (Brenner et al., 2000). The controversy stems from the prokaryotic species concept not being comparable to the eukaryotic one. Prokaryotes lack diagnostic morphological characteristics and are asexual organisms that exchange genetic material in unique and unusual ways compared to eukaryotes. In addition, it is not always feasible, due to technological limitations and/or poor understanding of the metabolic and physiological properties of prokaryotic cells, to define unique phenotypic characteristics that are required for a species description (Vandamme et al., 1996). For these reasons, none of the (at least) 22 species concepts described for eukaryotes is applicable to prokaryotes. This has led most microbial taxonomists to agree on a functional species definition for prokaryotes that is rooted in the degree of DNA–DNA reassociation. In this definition, two strains belong to the same species when their purified DNA molecules show at least 70% hybridization.

This definition of prokaryotic species does not translate well to eukaryotes. Application of the same definition to eukaryotes would lead to the inclusion of members of many taxonomic tribes in the same species (Sibley and Ahlquist, 1987). For example, all the primates (i.e., humans, orangutans, and gibbons) would then belong to the same species (Sibley et al., 1990). Furthermore, gorillas and orangutans would not be considered under threat because they would be viewed as the same species as humans, which are numerous and cosmopolitan. Thus, a simple comparison of the number of eukaryotic and prokaryotic species greatly underestimates prokaryotic diversity. Indeed, the prokaryotic species concept is probably comparable to that of an animal family or perhaps even an animal order.

The other obstacle in enumerating prokaryotic species is the fact that only a small fraction of the microbial community, typically about 1%, is cultivable. Furthermore, the habitats where prokaryotic species live are sometimes difficult to sample (e.g., deep ocean or subsurface) or too complex (e.g., soil or sediments) for an exhaustive count of prokaryotic species. This has led researchers to try to model the total number of prokaryotic species rather than exhaustively count them. In one such classical study, Torsvik and coworkers employed whole community DNA–DNA reassociation kinetics to estimate the total number of genome equivalents or species by considering the 70% DNA–DNA association cutoff as the definition of species (Torsvik et al., 1990). Based on this approach, 350 to 1,500 and 3,500 to 8,800 different prokaryotic species were found in Norwegian soil samples (Torsvik et al., 1998; Ovreas and Torsvik, 1998). Using the same method, Ovreas and colleagues found the prokaryotic diversity in aquatic environments to be orders of magnitude less than that in soil (Ovreas et al., 2001). Dykhuizen, using data from whole community DNA–DNA association between related communities, estimated that more than a billion (10^9) prokaryotic species exist in soil (Dykhuizen, 1998). Several reasons can explain the high soil microbial diversity, such as the high diversity of carbon resources; the rather stable, protective, even ancient environment; and what appears to be a high degree of spatial isolation that reduces competition, thereby maintaining less competitive members (Tiedje, 2000; Zhou et al., 2002; Treves et al., 2003).

Others have used clone libraries of the SSU rRNA gene from environmental samples to estimate prokaryotic diversity. The distribution of unique (representing different species) SSU rRNA gene sequences in these libraries was used for extrapolation to the total number of species in the environment. However, it is unlikely that such libraries are exhaustive, for example, species that appear in small numbers in the sample are likely absent in the clone libraries. To overcome this, Hughes and colleagues used statistical approaches that considered the proportion of species that were observed more than once in the clone library relative to those that were observed only once (Hughes et al., 2001). They concluded there is not sufficient information to assume that microbial populations follow a lognormal distribution in nature; that is, if species are assigned to log abundance classes, the distribution of species among these classes is normal. Therefore, they used a nonparametric approach based on mark and recapture methods and estimated about 500 species per gram in two grazed grassland soils. Assuming a lognormal distribution of species, Curtis and coworkers estimated 6,300 species for the same data set (Curtis et al., 2002). Extrapolating to a larger scale, they estimated the entire bacterial diversity in the oceans to be up to 2×10^6 species, while a ton of soil could contain 4×10^6 different species (Curtis et al., 2002).

There are several drawbacks to these approaches to estimated species richness. First, there is a limited sampling of environments and thus the extrapolation to a global scale may be insecure. Furthermore, it is uncertain how many different microbes exist in different environments, for example, different soils, or how much they vary in different geographic locations. In one study to address this question, Cho and Tiedje found that fluorescent *Pseudomonas*, a cosmopolitan heterotroph that is frequently recovered from soil, shows a high degree of endemicity, for example, genotypes recovered from distantly located sites show significant heterogeneity (Cho and Tiedje, 2000). If microbial populations have a high degree of endemicity, it greatly expands the earth's total microbial diversity. Second, the description of species based on SSU rRNA gene sequence is problematic mostly because the sequence of this molecule is too conserved to resolve species (Stackebrandt and Goebel, 1994). Third, the type of distribution that prokaryotic species follow remains controversial (Harte et al., 1999; Hughes et al., 2001) and hence the assumptions needed for modeling cannot be fulfilled. Although the accuracy of estimates of global microbial diversity is in question, there is no doubt that the number of prokaryotic species is large, most probably much larger than the most diverse eukaryotic phylum, the insects (with greater than 10^5 species). Currently, only 5,600 prokaryotic species are described, which appears to be less than the number of species in a few grams of soil!

2.4 THE CHALLENGE OF DESCRIBING PROKARYOTIC DIVERSITY

Culture-based approaches, despite their tremendous utility in basic microbiological research, cannot accurately describe naturally occurring prokaryotic assemblages simply because all such methods are incomplete. Most important, organisms that are recovered on standard microbiological media are rarely representative of the naturally occurring community, and for the majority of prokaryotic species no appropriate media and conditions for growth are available. The problem of studying naturally occurring pro-karyotic communities has been extensively discussed and reviewed (Staley and Konopka, 1985; Tiedje, 1995; Amman, 1995; Pace, 1997), and will not be further discussed here.

To deal with this problem, microbiologists have employed culture-independent approaches in studying naturally occurring prokaryotic assemblages. Culture-independent approaches most commonly target the cell's DNA because of its information content and discriminating capacity. Many of these methods were only developed during the last decade. The rest of this chapter describes these methods, their caveats, and particularly, how they have revolutionized our perception of prokaryotic diversity and the composition of naturally occurring prokaryotic communities.

2.4.1 Methods to Study Microbial Diversity

Methods for studying microbial diversity in the environment can be divided into those that target the total community (community composition, species abundance, etc.) and those that target the species or genes of interest. The former methods differ according to the level of resolution they can offer. Broad-scale analysis of community DNA is performed employing techniques such as DNA–DNA reassociation and shifts in guanine + cytosine (GC) content (Torsvik et al., 1990; Goodfellow and O'Donnell, 1993; Nuesslein and Tiedje, 1999), which provides information about the total genetic diversity and changes in structure of a given bacterial community, respectively. These techniques do not provide information about other diversity parameters, such as richness, evenness, and composition. Fingerprinting methods, based on either DNA or other biochemical molecules such as phospholipid fatty acids, give higher resolution and can provide detailed information about changes in the whole community structure. DNA fingerprinting methods typically employ the polymerase chain reaction (PCR) and are more popular due to their higher resolution and robustness. DNA fingerprinting methods include denaturing gradient gel electrophoresis (DGGE), amplified rDNA restriction analysis (ARDRA), terminal restriction fragment length polymorphism (T-RFLP), and ribosomal intergenic spacer analysis (RISA), or variations of them (Tiedje et al., 1999; Torsvik and Ovreas, 2002). Although these methods cannot be used as absolute estimations of species richness due to biases in the

PCR amplification step, their combination can reveal a great deal about the diversity of the prokaryotic community.

The characterization of particular species from the environment often involves cloning and sequencing of selected genes such as the SSU rRNA gene or key functional genes. Recently, the creation and sequencing of large insert genomic libraries from community DNA, such as the bacterial artificial chromosomes (BAC), have revolutionized the culture-independent study of prokaryotic diversity (Rondon et al., 2000). These so-called metagenomic libraries (see community genomics, Chapter 1) can link phylogenetic information with information on the potential function of uncultured community members as well as uncover interesting novel genes (Beja et al., 2000). Furthermore, metagenomic libraries can be employed for the whole-genome sequencing of uncultivated members of syntrophic associations or otherwise simple communities, giving new insight into interesting, yet unexplored, biological processes. A method based on integron-targeted PCR has also been developed to recover genes from environmental DNA without the necessity of knowing their sequences (Stokes et al., 2001). Lastly, the florescence in situ hybridization (FISH) technique should also be considered in the context of studying single prokaryotic species (DeLong et al., 1989; Amann et al., 1995). FISH enables the visualization of prokaryotic cells in situ by targeting functional genes in addition to the SSU rRNA gene and can reveal higher functional diversity than revealed by the more conserved SSU rRNA gene. FISH can be coupled to other techniques such as autoradiography or isotopic analysis to assess the activity of particular cells of the species under study (for an excellent review of FISH and its applications, see Gray and Head, 2001).

2.4.2 Limitations of Culture-independent Methods

Cultivation-independent molecular approaches, like other methodologies, have their own specific pitfalls and biases (Wintzingerode et al., 1997). Clone libraries are rarely sequenced to exhaustion and are biased by differential cell lysis, preferential DNA recovery from specific cell types, and several analytical complications introduced by the PCR step (Suzuki and Giovannoni, 1996; Wang and Wang, 1996; Wintzingerode et al., 1997; Poltz and Cavanaugh, 1998; Thompson et al., 2002b). In particular, Thompson and colleagues recently cautioned that the formation of heteroduplexes during the PCR step and the mismatch repair system of the *E. coli* vector during cloning may introduce artificial sequence diversity among highly identical sequences (Thompson et al., 2002b). Additionally, approaches that employ the selective amplification of genes from the environment or probes for the identification of species in the environment might give misleading results because they are necessarily limited to the currently available sequences in the database. However, as mentioned previously, only a small fraction of actual prokaryotic diversity has been described, and fewer than half of the deposited cultures have their SSU rRNA gene sequenced and deposited in central databases (Hugenholtz et al., 1998). One example is the recently described new phylum of Archaea that was isolated from a hydrothermal vent and has a SSU rRNA sequence so different from known groups that it could not be detected with the "universal" SSU rRNA probes (Huber et al., 2002).

Although there are certain biases in culture-independent studies, several reports exist in which researchers obtained similar results by using different techniques, each with different potential biases (Mullins et al., 1995; Beja et al., 2000). Furthermore, congruent conclusions have been drawn from independent microbial diversity studies conducted in similar environments. Hence, the biases in molecular methods may not be so severe, at least for some communities, to inappropriately skew knowledge of that community. However, knowledge of the technique and its limitations is essential for its successful and robust application.

2.4.3 Interesting Findings from Culture-independent Approaches

Novel Species Perhaps the most important contribution of culture-independent approaches to our knowledge of prokaryotic diversity is the discovery of many microbial taxa, ranging from new species to even new phyla! Bacteria now comprise more than 40 phyla (Pace, 1997; Hugenholtz et al., 1998; Zhou et al., 2003), compared to only 12 in 1987 (Woese, 1987), and at least 10 of the newly described phyla are represented only by environmental SSU rRNA sequences. Although there are currently about 70,000 SSU rRNA gene sequences in the Ribosomal Database Project (RDP) database, the description of new species is far from the saturation point, to judge from the high frequency with which novel species continue to be reported. Additionally, there are still a number of environmental niches that have been poorly investigated and these are likely to harbor novel taxa. For example, a nano-sized, hyperthermophilic, archaeal species from a novel phylum was recently reported. This unusual organism lives symbiotically with larger cell-sized archaeal species in hot submarine vents (Huber et al., 2002). The contribution of culture-independent approaches has been profound for the Archaea since so many members of this domain do not grow under the culture conditions used so far. Indeed, in a current assessment of archaeal diversity, most phyla- or order-level lineages are represented only by environmental SSU rRNA sequences (DeLong and Pace, 2001). The importance of the characterizing SSU rRNA sequences from the environment to enriching our knowledge of extant taxa is illustrated in Fig. 2.2 by the proportion of the SSU rRNA database populated by environmental (uncultivated) clones.

Do Isolated Strains Represent Ecologically Important Species? Another major contribution of the culture-independent techniques was the assessment of which species are ecologically important in terms of population sizes and activity. The vast majority of cultivable strains from any environment are members of only four bacterial phyla, and often of only specific species within these phyla: Proteobacteria (*Escherichia*, *Pseudomonas*), Firmicutes (*Bacillus*, *Streptococcus*, *Staphylococcus*), Actinobacteria (*Mycobacterium*), and Bacteroidetes (*Flexibacter*, *Cytophaga*, *Bacteroides* group). For instance, out of the 65,872 SSU rRNA gene sequences in the RDP database as of April 2003 (Cole et al., 2003), 81.7% are sequences from these four phyla (Fig. 2.2). This percentage is even higher for sequences of strains in the world's major culture collections. An injudicious interpretation of these results would be that species from these phyla are dominating the earth. However, culture-independent molecular surveys show that phylogenetic groups other than the four phyla above dominate terrestrial environments (Liesack et al., 1997; Hugenholtz et al., 1998). For instance, more than 75% of the SSU rRNA gene sequences in clone libraries from soil belong to members of the following taxa: the class alpha-Proteobacteria, and the phyla Actinobacteria, Acidobacteria, and Verrucomicrobia (Hugenholtz et al., 1998). Members of the phylum Acidobacteria appear to be numerically dominant, forming up to 52% of SSU rRNA gene sequences in

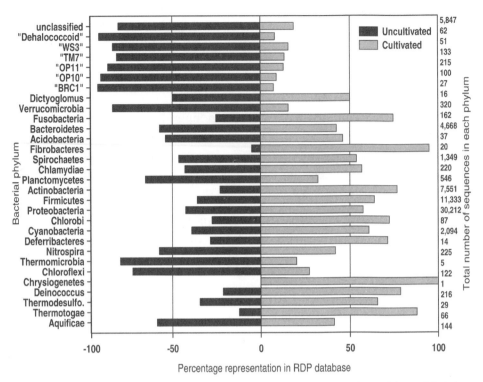

Figure 2.2. *Distribution of uncultivated vs. cultivated SSU rRNA sequences for each bacterial phylum.* Data were collected from 65,872 sequences in the RDP database (http://rdp.cme.msu.edu/html/), as of April 2003. (The classification is based on an annotation from GenBank and was provided courtesy of Ryan Farris and James R. Cole of RDP.)

clone libraries, and active members of most soils in several other independent studies (Kuske et al., 1997; Nogales et al., 1999; Felske et al., 2000). Yet until very recently, only one named species had been reported from soil, and this species, *Acidobacterium capsulatum*, was isolated from an acid mine drainage site (Kishimoto and Tano, 1987). Similar to the Acidobacteria case, the Verrucomicrobia phylum appears to be widespread in soil, although most of its cultured members are from aquatic sources (Hugenholtz et al., 1998). The remaining groups, for example, the phylum Actinobacteria and the alpha division of Proteobacteria, despite having large numbers of cultivated and described species, also contain large clades represented only by cloned SSU rRNA gene sequences in these libraries.

Much of the inconsistency between culture-based and culture-independent studies in terrestrial environments is probably attributable to the complex nature and the high prokaryotic diversity of soil. However, similar results were obtained in other less complex environments. In a case study, Suzuki and coworkers compared SSU rRNA gene sequences from an environmental library constructed from a marine sample with the SSU rRNA gene sequences of isolates recovered from the same sample (Suzuki et al., 1997). They found a limited number of sequences shared in the two collections, illustrating the principle that culturing is not representative of even simple communities. In a similar case, Beja and colleagues did not retrieve any SSU rRNA gene sequences belonging to the

Erythrobacter species (α-4 subclass of Proteobacteria) (Beja et al., 2002), which was suggested as a key member of the photosynthetic bacterial community in the ocean based on the high frequency with which this species was recovered from marine samples (Kobler et al., 2001).

Archaeal Diversity and Distribution   Archaea were originally believed to be a rather constrained group of prokaryotes that thrive in extreme environments. The application of culture-independent methods, however, revealed that archaeal diversity and distribution on Earth are much higher than initially anticipated. Furthermore, archaeal specific sequences are now commonly retrieved from environments that are far from extreme, such as soil (Bintrim et al., 1997), groundwater (Takai et al., 2001), or oceans (DeLong et al., 1999), disproving the notion that Archaea are exclusively extremophiles. For example, archaeal species are now believed to be present in oceans, where they can represent up to 20% of the total microbial community (DeLong et al., 1999; DeLong and Pace, 2001); this is a habitat where they were not thought to exist 10 years ago. Furthermore, several lines of evidence, for example, FISH microscopy targeting ribosomal RNAs (DeLong et al., 1999), seasonable fluctuation in the abundance of the archaeal clones (Murray et al., 1998), the detection of archaeal specific biomarkers in the phytoplankton (DeLong et al., 1998), suggest that the archaeal population in the ocean is active, dynamic, and probably a major component of community function. Lastly, archaeal species have recently been determined to be a significant member of the prokaryotic community in waters from deep South African gold mines (Watanabe et al., 2002). The biological role and significance of archaeal species in the environment are the subject of ongoing research.

Linking Species to Environmental Function   Linking species to function and deciphering the biological role and importance of uncultivated species are the current great challenges for microbial ecologists. Community clone libraries have offered new perspectives toward this goal. The clones from these libraries are expressed in well-studied vectors or the clone sequences are analyzed via bioinformatics to suggest functional information. Although there are several limitations in cloning, for example, vector compatibility and biases in library construction, this approach has garnered widespread interest because of the high probability of identifying novel functions and genes. In perhaps the most important example, Beja and coworkers described a rhodopsin-type protein from an uncultivated marine bacterium, a class of enzymes never observed before in the bacterial domain. These sequences, when expressed in *E. coli*, suggested a light-driven proton pump, that is, a new type of photosynthesis (Beja et al., 2000). Rhodopsin sequences were then successively found in many oceanic surface samples, suggesting that this type of photosynthesis is widespread in nature (Beja et al., 2001). This approach has also been exploited to identify phenotypes of biotechnological interest (instead of studying the ecology of natural populations), such as antibiotic production (Cottrell et al., 1999; Henne et al., 2000), industrially important biocalysist α-amylase (Richardson at et al., 2002), etc.

2.5 DIVERSITY OF MICROBIAL GENOMES AND WHOLE-GENOME SEQUENCING

Although its primary goal was a better understanding of the physiology and metabolism of a species, whole-genome sequencing has revolutionized the study of other major microbiology disciplines, including functional and genetic diversity. For instance, it has revealed that bacterial genomes are more fluid than originally expected, for example, mobile elements and lateral gene transfer events play a major role in the evolution and shaping of genome space (discussed in Chapter 4). This fluidity, nonetheless, is directly connected to diversity because it suggests that prokaryotes have great potential to diversify. The rest of the chapter discusses the recent insights into microbial diversity that are derived from whole-genome sequencing projects.

Prokaryotic genome sequencing projects have grown rapidly (Fig. 2.3) following the improvements in sequencing technology, capacity, and cost reduction such that over 115 genomes have been classified as of the second quarter of 2003 and more that 200 other projects are underway. The genomes sequenced so far are biased toward organisms with smaller genomes, often from strains living in simpler, resource-rich environments such as endocellular parasites or pathogens (Fig. 2.3). The currently sequenced microbial genomes

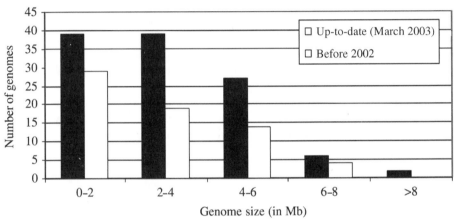

Panel A.

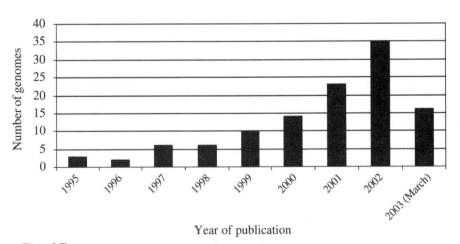

Panel B.

Figure 2.3. Genome size distribution of the fully sequenced prokaryotic genomes (panel A) and the number of published prokaryotic genomes per year (panel B). (Data were retrieved from NCBI and included all 112 prokaryotic genomes available as of March 2003.)

TABLE 2.1 Currently Sequenced Microbial Genomes (as of May 2003)

Species	Genome size	Total ORFs	G + C Content (%)
Aeropyrum permix	1.67	1,840	56.3
Agrobacterium tumefaciens	5.6	5,299	59.0
Agrobacterium tumefaciens (U. Wash.)	5.6	5,402	59.0
Aquifex aeolicus	1.6	1,560	43.5
Archaeoglobus fulgidus	2.18	2,420	48.6
Bacillus anthracis	5.1	5,311	35.4
Bacillus cereus	5.41	5,255	35.3
Bacillus halodurans	4.2	4,066	43.7
Bacillus subtilis	4.2	4,112	43.5
Bacteroides thetaiotaomicron	6.26	4,778	42.8
Bifidobacterium longum	2.26	1,729	60.1
Borrelia burgdorferi	0.9	1,638	28.2
Bradyrhizobium japonicum	9.11	8,317	64.1
Brucella melitensis	3.3	3,198	57.2
Brucella suis	3.28	3,264	57.3
Buchnera aphidicola	0.62	504	25.3
Buchnera sp.	0.71	574	26.4
Campylobacter jejuni	1.64	1,634	30.5
Caulobacter crescentus	4.01	3,737	67.2
Chlamydia muridarum	1.07	916	40.3
Chlamydia trachomatis	1.05	895	41.3
Chlamydophila caviae	1.17	1,005	39.2
Chlamydophila pneumoniae AR39	1.23	1,112	40.6
Chlamydophila pneumoniae CWL029	1.23	1,054	40.6
Chlamydophila pneumoniae J138	1.22	1,069	40.6
Chlorobium tepidum	2.16	2,252	56.5
Clostridium acetobutylicum	4.1	3,848	30.9
Clostridium perfringens	3.1	2,723	28.5
Clostridium tetani	2.8	2,373	28.7
Corynebacterium efficiens	3.15	2,950	63.1
Corynebacterium glutamicum	3.3	3,040	53.8
Coxiella bruneti	2	2,009	42.7
Deinococcus radiodurans	3.28	3,182	66.6
Enterococcus faecalis	3.35	3,113	37.5
Escherichia coli K12	4.6	4,279	50.8
Escherichia coli O157:H	5.5	5,361	50.5
Escherichia coli O157:H7 EDL933	5.6	5,324	50.4
Escherichia coil CFT	5.23	5,379	50.5
Fusobacterium nucleatum	2.17	2,067	27.2
Haemophilus influenzae	1.83	1,714	38.2
Halobacterium sp. NRC-1	2.57	2,622	65.9
Helicobacter pylori 26695	1.66	1,576	38.9
Helicobacter pylori J99	1.64	1,491	39.2
Lactobacillus plantarum	3.31	3,009	44.5
Lactococcus lactis	2.36	2,267	35.3
Leptospira interrogans	4.69	4,727	35.0
Listeria innocua	3.01	3,043	37.4
Listeria monocytogenes	2.94	2,846	38.0
Mesorhizobium loti	7.59	7,275	62.5
Methanococcus jannaschii	1.74	1,785	49.5
Methanopyrus kandleri AV19	1.7	1,687	31.3
Methanosarcina acetivorans	5.75	4,540	61.2

(*Table continued*)

Table 1   *Continued*

Species	Genome size	Total ORFs	G + C Content (%)
Methanosarcina mazei	4.1	3,371	42.7
Methanothermobacter thermautotrophicus	1.75	1,873	41.5
Mycobacterium leprae	3.26	1,605	57.8
Mycobacterium tuberculosis CDC1551	4.4	4,187	65.6
Mycobacterium tuberculosis H37Rv	4.4	3,927	65.6
Mycoplasma genitalium	0.58	484	31.7
Mycoplasma penetrans	1.36	1,037	25.7
Mycoplasma pneumoniae	0.81	689	40.0
Mycoplasma pulmonis	0.96	782	26.6
Neisseria meningitidis MC58	2.27	2,079	51.5
Neisseria meningitidis Z2491	2.18	2,065	51.8
Nitrosomonas europaea	2.81	2,461	50.7
Nostoc sp.	7.2	6,129	41.3
Oceanobacillus iheyensis	3.63	3,496	35.7
Pasteurella multocida	2.4	2,015	40.4
Pseudomonas aeruginosa	6.3	5,567	66.6
Pseudomonas syringae	6.4	5,471	58.4
Pseudomonas putida	6.18	5,350	61.5
Pyrobaculum aerophilum	2.2	2,605	51.4
Pyrococcus abyssi	1.76	1,769	44.7
Pyrococcus furiosus	1.9	2,065	40.8
Pyrococcus horikoshii	1.8	1,801	41.9
Ralstonia solanacearum	5.8	5,116	67.0
Rickettsia conorii	1.27	1,374	32.4
Rickettsia prowazekii	1.1	835	29.0
S.enterica ser. *Typhi*	4.8	4,767	51.9
Salmonella typhimurium LT2	4.95	4,553	52.2
Shewanella oneidensis	5.03	4,472	45.9
Shigella flexneri	4.61	4,180	50.9
Sinorhizobium meliloti	6.7	6,205	62.2
Staphylococcus aureus Mu50	2.9	2,748	32.8
Staphylococcus aureus MW2	2.8	2,632	32.8
Staphylococcus aureus N315	2.81	2,625	32.8
Staphylococcus epidermis	2.5	2,419	32.1
Streptococcus agalactiae	2.2	2,124	35.6
Streptococcus mutans	2.04	1,960	36.8
Streptococcus pneumoniae R6	2.03	2,043	39.7
Streptococcus pneumoniae TIGR4	2.2	2,094	39.7
Streptococcus pyogenes	1.85	1,697	38.5
Streptococcus pyogenes MGAS8232	1.9	1,845	38.5
Streptomyces avermitilis	9.03	7,575	70.7
Strepromyces coelicolor	8.67	7,512	70.6
Sulfolobus solfataricus	2.99	2,977	35.8
Sulfolobus tokodaii	2.7	2,826	32.8
Synechocystis sp. PCC 6803	3.57	3,167	47.7
Thermoanaerobacter tengcongensis	2.7	2,588	37.6
Thermoplasma acidophilum	1.56	1,482	46.0
Thermoplasma volcanium	1.58	1,499	39.9
Thermosynechococcus elongatus	2.6	2,475	53.9
Thermotoga maritima	1.85	1,858	46.2
Treponema pallidum	1.14	1,036	52.8
Tropheryma whipplei	0.93	783	46.3

Table 1 *Continued*

Species	Genome size	Total ORFs	G + C Content (%)
Ureaplasma urealyticum	0.75	614	25.5
Vibrio cholerae	4	3,835	47.5
Vibrio parahaemolyticus	5.18	4,537	45.4
Vibrio vulnificus	5.13	4,832	46.7
Wigglesworthia brevipalpis	0.7	654	22.5
Xanthomonas campestris	5.08	4,181	65.1
Xanthomonas pv. *citri*	5.17	4,312	64.8
Xylella fastidiosa	2.68	2,832	52.6
Xylella fastidiosa pv. *temecula*	2.52	2,036	51.8
Yersinia pestis	4.65	4,083	47.6
Yersinia pestis pv. *kim*	4.6	4,090	47.6
Average	**3.2**	**2,986.7**	**45.4**

along with their genome size, number of ORFs, and percentage of G + C content are summarized in Table 2.1. This set of genomic information is now large enough to reveal some major trends in and impressions about prokaryotic genomes and is consistent with the very high microbial diversity discussed above.

2.5.1 Genomic Diversity within Species

Whole-genome sequencing has revealed much higher genetic diversity within species than originally anticipated. An example is the *E. coli* where whole-genome sequences of four strains are now available. Comparative genomic analysis of these sequences reveals that the pathogenic O157 strain has a genome 1 Mb larger than that for strain K12, and about 25% of its genes are not conserved in the strain K12 genome (Perna et al., 2001; Hayashi et al., 2001). If one considers that prior to whole-genome sequencing, strains of the same species were believed to harbor minimum gene content differences because they only rarely could be differentiated based on phenotypic characteristics, the genetic heterogeneity revealed between *E. coli* strains was surprisingly high for its time (only 2 years ago!). Furthermore, the annotation of the strain-specific gene set offered novel insight into the pathogenic lifestyle of strain O157 compared to the innocuous K12, confirming that the previous culture-based approaches had exploited only a fraction of the metabolic repertoire of the best-studied pathogen. Most of strain O157's specific genes are now believed to have been acquired through lateral transfer events, based on atypical sequence characteristics (Lawrence and Ochman, 1998) and the enrichment of mobile elements such as phage, prophages, and insertion sequences in the strain O157-specific gene set (Perna et al., 2001; Hayashi et al., 2001). These findings also suggest that an environmental and benign strain could evolve relatively easily to a devastating pathogen since only about 4.5 million years (rather short in evolutionary time) has elapsed since the last common ancestor between the two strains (Reid et al., 2000).

The availability of two additional genomic sequences of *E. coli* strains (strains CFT073 and O157-EDL933) reveals further surprises. Only about 3,000 genes are shared among the four *E. coli* genomes (Welch et al., 2002), compared to about 4,000 genes shared between O157 and K12 strains (Perna et al., 2001). The 3,000 genes conserved in all *E. coli* strains show, however, a remarkable synteny (interrupted by strain-specific islands), suggesting a vertically transmitted backbone gene set for the species (Welch et al., 2002).

In summary, the genomic sequencing of *E. coli* strains has revealed not only an enormous genetic diversity at the subspecies level, but also the presence of very different selection forces that has led to the accumulation or deletion of genetic material. This subspecies diversity appears not to be unusual since preliminary evidence suggests that *Streptococcus pneumoniae*, based on comparative microarray hybridizations (Hakenbeck et al., 2001), and *Burkholderia cepacia*, based on genome size estimations (Lessie et al., 1996), seem also to have high diversity.

On the other hand, species such as *Mycobacterium tuberculosis* do not appear to share the genetic diversity observed in *E. coli*. Based on both comparative analysis of the sequenced strains (Fleischmann et al., 2002) and comparative microarray hybridization analysis of several strains (Berh et al., 1999), *M. tuberculosis* strains are unlikely to be more than 1 to 2% different in terms of gene content, although the current analysis might be biased due to the study of exclusively clinical isolates. These findings raise another fundamental issue as well. The current species definition based on 70% DNA–DNA association values poorly correlates with gene content differences within species, as is apparent from the comparisons among the *E. coli* and *M. tuberculosis* strains.

2.5.2 Genome Structure and Its Relation to the Ecological Niche

Genome sizes vary by more than an order of magnitude among the known prokaryotes (e.g., 0.5–10 Mb). However, the genome size distribution does not appear to be random, for example, it correlates with the ecological niche of the organisms. The smaller genomes are found in endocellular parasites or symbionts (0.5–1.2 Mb), because these organisms occupied a very narrow niche and have undergone reductive evolution (discussed in Chapter 4). For instance, the endosymbiont of aphids, *Buchnera* sp., has a genome size of only 650 Kb, compared to 4 Mb for its ancestors from whom *Buchnera* diverged 150 to 250 million years ago (Shigenobu et al., 2000). For free-living bacteria, genome size correlates with the species metabolism and the width of its ecological niche. Pathogenic species with a narrow range of hosts (or, more generally, species with a narrow ecological niche) also have small genomes, for example, *Helicobacter* sp. and *Streptococcus* sp. Anaerobic bacteria with a restricted metabolism, such as methanogens, typically have small genome sizes, ranging from 1.5 to 2.5 Mb. In contrast, aerobic organisms and opportunistic pathogens show higher diversity in genome sizes with some species such as *Pseudomonas* having genomes as large as 6 Mb. The largest genomes are found in species that have complex lifestyles, including myxobacteria and actinobacteria (8–9 Mb).

Some exceptions to these generalizations are also encountered. For instance, the two recently sequenced methanogenic archaeal *Methanosarcina* species have larger genomes than the typical archaeal methanogens (e.g., 4–5 Mb). This species, however, forms a separate order in the archaeal domain, and its members can thrive in a broader range of environments and form complex multicellular structures, that are not found in other archaeal species (Galagan et al., 2002). All these observations lead to the conclusion that the interaction between an organism and its particular habitat(s), for example, resource availability and diversity, stable or fluctuating environmental conditions, selects the genome size of the species. Nonetheless, what controls the upper genome size in prokaryotes remains poorly understood. Several hypotheses exist, such as the decreased fidelity of replication in large genomes, or the energy cost to successfully control an excessive metabolic repertoire, but none has been experimentally proven.

The variation of genome sizes within species is believed to be rather limited (Shimkets, 1998), which has been supported by recent genomic sequence data and by pulse field gel electrophoresis analysis of genomes (Maule, 1998). Some of the better-documented exceptions to this are the *E. coli* and *Burkholderia cepacia* species mentioned above, where different strains can vary up to 25% or up to 50% in genome size, respectively (Bergthorsson and Ochman, 1998; Lessie et al., 1996). On the other hand, genome size can vary up to 3-fold for different species of the same genera! At one end of the spectrum are species such as *Borrelia* sp., whose chromosomes vary by less than 15 Kb in size (Casjens et al., 1995), whereas species such as spirochete *Treponema* sp. (MacDougall and Saint Girons, 1995; Walker et al., 1995), and *Mycoplasma* sp. (Barlev and Borchsenius, 1991) show a variation in genome sizes of up to 3- and 2.3-fold, respectively. Perhaps more typical are genera such as *Streptomyces* and *Rickettsia*, which vary from 6.4 to 8.2 Mb (Leblond et al., 1990) and 1.2 to 1.7 Mb (Frutos et al., 1989), respectively. It should be pointed out, however, that too few strains within species and within genera have been studied to give us a complete understanding of the natural variation in the size of prokaryotic genomes.

Although Bacteria are believed to have a single circular chromosome, an increasing number of exceptions to this are being identified (Casjens, 1998; Volff and Altenbuchner, 2000). For instance, the *Streptomyces*, *Borrelia* and *Agrobacterium* species, contrary to their close relatives, have been shown to possess linear chromosomes. Interestingly, each of this species has solved in a different way the problem of replicating the telomeric sequences of the linear chromosomes (Kobryn and Chaconas, 2001). Several other species such as *Klebsiella*, *Escherichia*, and *Thiobacillus* harbor linear plasmids. Linearity, at least in *Streptomyces* and *Borrelia*, is believed to enhance genomic plasticity, because linear chromosomes (or plasmids) are very unstable and undergo, at high frequency, amplifications and large deletions, often removing the telomeres. This was confirmed with the whole-genome sequencing of *S. coelicolor*, which showed that the secondary metabolite-related genes (*Streptomyces* is notorious for its secondary metabolites like antibiotics) are more frequently encountered in the arms of the chromosome than in its center; the center is biased toward housekeeping genes (Bentley et al., 2002). However, whether linearity offers a selective advantage and why it is phylogenetically constrained to a limited number of species remain unclear. Furthermore, linearity does not appear to be necessary for replication in nature. For instance, Lin and coworkers using an insertion vector successfully circularized the chromosome of *S. lividans*, although the circularized chromosomes are also unstable (Lin et al., 1993), and Ferdows and colleagues have found a naturally occurring circularized version of a *Borrelia* plasmid (Ferdows et al., 1996).

Several species have multiple rather than single chromosomes (differentiated from large plasmids by harboring housekeeping genes like ribosomal or tRNA genes). Many alpha-proteobacterial genera such as *Agrobacterium*, *Brucella*, *Rhizobium* and *Rhodobacter*, the beta-proteobacterial genera *Burkholderia* and *Ralstonia*, and some spirochetes like *Leptospira* have multiple (between two and four) chromosomes and, very frequently, one or more plasmids. In at least two, the *Brucella* and *Burkholderia*, the multiple chromosomes are a stable property of the genus. In the proteobacteria phyllum, the multiple chromosomes correlate with a free-living, opportunistic lifestyle, whereas species that are obligatorily associated with animal hosts or vectors contain no plasmids and, with a few exceptions, single chromosomes (Moreno, 1998). *Brucella*, *Bartonella*, *Rickettsia*, *Anaplasma*, and *Neisseria* organisms are among the latter species. Based on these observations, multiple chromosomes are assumed to confer increased genome plasticity and potential for diversification. However, Itaya and Tanaka have recently divided the single circular *Bacilus subtilis* chromosome without any apparent defect in the viability of the engineered cells (Itaya and Tanaka, 1997). Thus, the advantage (if any) of multiple chromosomes remains a mystery.

2.5.3   General Trends in Genome Functional Content

In almost every genome sequenced to date, there has been a constant percentage (about 20–30%) of open reading frames (ORFs) that show no homology to any known protein (i.e., they encode hypothetical proteins) (Fraser et al., 2000). Although it has been suggested that the majority of these are noncoding DNA based on *in silico* analysis (Ochman, 2002; Mira et al., 2002), more recent evidence suggests that at least a portion of them code for functional proteins. Proteomic analysis of *Deinococcus radiodurans*, known for its extraordinary radiation resistance, revealed that at least part of these genes is translated to proteins (Liu et al., 2003). Furthermore, genes of unknown function are present, although in a smaller percentage, in the highly reduced genomes of symbiotic species. The significant number of "functionally unknown" genes in every genome suggests that novel processes are still likely to function in every prokaryotic cell and await characterization. Alternatively, some of these genes might function in well-studied cell processes, but their sequences have diverged too much to resemble any of the known annotated sequences.

While it is clear from the above discussion that there is considerable functional and sequence diversity among prokaryotic species, whole-genome sequencing has also revealed some universal functional trends as well. For instance, the genomes of endosymbionts have preferentially lost genes involved in metabolism, biosynthesis, and regulation while retaining most of the informational genes compared to their free-living relatives (Andersson et al., 1998; Mira et al., 2001; Moran, 2002). Interestingly, although there is a strong deletion bias toward the former major functional categories in the symbiotic genomes, the specific pathways lost appear to be lineage-specific; for example, *Buchnera* sp., an obligate symbiont of aphids, contrary to other endosymbionts, retains the genes for the biosynthesis of all amino acids (Shigenobu et al., 2000). On the other hand, genes involved in metabolism, regulation, and secondary metabolite biosynthesis are overamplified in the genomes of large genome-sized, free-living species (Fig. 2.4). These properties probably allow the latter species to thrive in diverse ecological niches and fluctuating environmental conditions. The genomic expansion appears to take place via two major processes, the gene duplication (Jordan et al., 2001) and the lateral gene transfer (genome evolution is discussed in detail in Chapter 4).

2.5.4   Biases in the Collection of Sequenced Species: A Limit to Understanding

The current collection of sequenced species is rather limited (compared to the extant species richness), and there are several issues that should be pursued in the future for a comprehensive understanding of prokaryotic genetic and functional diversity. For example, several major phylogenetic lineages remain under or overrepresented with sequenced representatives. For instance, 43 (38.2%) of the 112 completely sequenced strains as of April 2003 belong to the phylum Proetobacteria. At the same time, some of the most dominant phyla in nature (see Section 2.4.3) still have a limited number of sequenced representatives. For example, the Acidobacteria have no sequenced species and Bacteroides has one. Archaea have 16 completely sequenced species, but this collection is limited to methanogenic and thermophilic species and does not include mesophilic species that are widespread in the ocean and soil environments.

Another limitation of the current collection of sequenced species is that the collection is heavily biased toward clinical representatives. About 70% of the bacterial strains fully

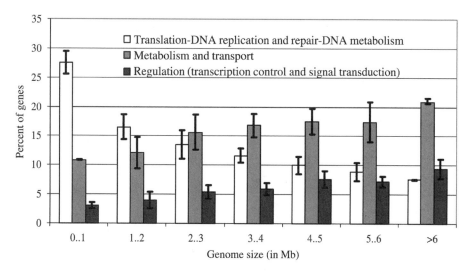

Figure 2.4. *Correlation between cellular processes and genome size for prokaryotic genomes.* Bars represent the percent of genes (averages from 81 genomes) that are functioning in specific cellular processes (graph legend). It appears that larger genomes preferentially accumulate metabolic, transport, and regulatory genes as opposed to informational ones, judging from the inverse pattern for these classes with genome size. Error bars represent the standard deviation from the mean except for the first and last genome size classes, where error bars represent the data range due to a small number of genomes in these classes (two and three genomes, respectively).

sequenced are of clinical importance. A representative example is the Actinobacteria phylum, a dominant group in soil based on culture-independent methods, which has nine sequenced species but all of them of clinical origin. Furthermore, our own analysis using the cluster of orthologous groups (COGs) database (Tatusov et al., 1997) showed that the Joint Genome Institute's collection of 23 exclusively environmental and partially sequenced prokaryotic genomes had a smaller percentage of genes, on average, assignable to the COGs database compared to the fully sequenced species (51.4% vs. 65% for the closed set). These results may indicate that this genome set samples more of the uncharacterized genes in nature.

2.6 SUMMARY

Our microbial world is the product of 3.7 years of evolution, or in other words, 85% of prokaryote history occurred before Pangea broke apart! Hence, microbes have explored a multitude of niches and been through a full range of climatic conditions due to continental drift and global climate cycles. The result is tremendous exploration of biochemistry and huge sequence diversity. Compared to higher eukaryotes, a major difference in microbes at the gene sequence level is the diversity of sequence for the same gene function. This is particularly important in ecology since it is this diversity of sequence that provides the different substrate range, kinetic, regulatory, and stability properties that are so important to the global recycling of elements that sustain life on Earth.

Much of the microbial world remains uncultured, but through extracting and analyzing DNA from the environment, we know that this world is much larger than we knew before. However, we are still limited in our knowledge since we know little more than we can infer from its string of A,G,C, T's. We now have molecular tools that have begun to reveal ecological patterns and function of species. We also know which taxa or functional genes dominate in particular environments; this provides some clue as to which organisms might be more important to target for genome sequencing. Although the vast microbial world remains uncultured, progress in culturing the previously unculturable has been made and some surprising new breakthroughs in biology have been reported as a result. This trend, aided by molecular tools, will no doubt continue and is a high priority.

The trend for sequencing microbial genomes continues to accelerate and will no doubt continue. With over 100 completely sequenced prokaryote genomes, some trends are emerging. Gene content and genome size do show patterns that are consistent with their environmental niche. Prokaryote genomes vary from 0.6 to over 10 Mb, and those with large genomes occupy environments where versatility is needed and the opposite is the case for the small genome-sized species. Genomes also vary in whether they are circular or linear and with regard to the number of replicons. This variation also seems to relate to the organism's ability to adapt to its environment. Finally, some species show considerable genome variability within a species, whereas others may not. This genomic difference at the subspecies level raises important questions about the validity of the current prokaryotic species concept.

While we do have sequence information from a significant number of prokaryotic genomes, it is biased to the clinically important strains, and to those with small genomes, for example, < 3 Mb. This may skew our current knowledge of prokaryotic genomics. In particular, it limits our genomics-based understanding of the large terrestrial microbial world that has many species with typically large genomes.

FURTHER READING

Brenner, D. J., J. T. Staley, and N. R. Krieg. 2000. Classification of prokaryotic organisms and the concept of speciation. In *Bergey's Manual of Systematic Bacteriology*, 2nd ed. D. Boone, R. Castenholz, and G. Garrity (eds.). Springer-Verlag, New York.

Casjens, S. 1998. The diverse and dynamic structure of bacterial genomes. *Annu. Rev. Genet.* 32:339–377.

DeLong, E. F. and N. R. Pace. 2001. Environmental diversity of bacteria and archaea. *Syst. Biol.* 50:470–478.

Gray, N. D. and I. M. Head. 2001. Linking genetic identity and function in communities of uncultured bacteria. *Environ. Microbiol.* 3:481–492.

Hedges, S. B. 2002. The origin and evolution of model organisms. *Nat. Rev. Genet.* 3:838–849.

Hugenholtz, P., B. M. Goebel, and N. R. Pace. 1998. Impact of culture-independent studies on the emerging phylogenetic view of bacterial diversity. *J. Bacteriol.* 180:4765–4774.

Pace, N. R. A molecular view of microbial diversity and the biosphere. 1997. *Science* 276:734–740.

Shimkets, L. J. 1998. Structures and sizes of genomes of the Archaea and Bacteria. In *Bacterial Genomes. Physical Structure and Analysis*. F. J. Bruijn, J. R. Lupski, and G. M. Weinstock (eds.). Chapman and Hall, New York.

Tiedje, J. M. 2000. 20 years since Dunedin: the past and the future of microbial ecology. In *Microbial Biosystems: New Frontiers. Proceedings of the 8th International Symposium on Microbial Ecology*. C. Bell, M. Brylinsky, and P. Johnson-Green (eds.). Atlantic Canada Society for Microbial Ecology, Halifax, Canada.

Torsvik, V. and L. Ovreas. 2002. Microbial diversity and function in soil: from genes to ecosystems. *Curr. Opin. Microbiol.* 5:240–245.

3

Computational Genome Annotation

Ying Xu

3.1 INTRODUCTION

The DNA sequence of a genome encodes the entire functionality of a living organism. Knowing the biological complexity exhibited by even the simplest living organism, we can imagine the enormous amount of information that must be encoded in a genome's DNA, which consists of four types of letters, A, C, G, and T. A complete genome sequence could have millions (for a typical microbial genome) to billions (for the human genome) of base pairs of {A, C, G, T}. What information is encoded in such a long contiguous string of A's, C's, G's, and T's? Where is the information located in the sequence? What encoded information is identifiable directly from the DNA sequence using existing computational methods? How should the identified information be presented? These are some of the basic issues that a *genome annotation* project should address. Currently, genome annotation attempts to identify functional elements in a genome sequence and to characterize their functional roles and possibly functional mechanisms. These functional elements may include protein- and RNA-encoding genes, transcriptional units and promoter/regulatory regions, and other identifiable signals such as simple and complex repeats or CpG islands. Identification of these functional elements provides a list of the basic parts (or building blocks) of the complex biological machinery that a genome encodes, and it represents the main focus of the current genome annotation efforts. In-depth functional inference of these functional elements, their functional mechanisms, and the biological networks they form may need additional information other than genome sequence alone (see Chapter 5).

There are two general approaches to the computational identification of functional elements in a genome sequence: (1) the ab initio approach and (2) the comparative approach. The ab initio approach predicts a functional element, based on its distinguished

Microbial Functional Genomics, Edited by Jizhong Zhou, Dorothea K. Thompson, Ying Xu, and James M. Tiedje. ISBN 0-471-07190-0 © 2004 John Wiley & Sons, Inc.

statistical features, and it is generally used to identify *novel* functional elements. The comparative approach identifies a functional element, based on its sequence similarity to previously known ones. This approach is becoming the preferred tool for identifying certain types of functional elements, as the number of known functional elements increases at an exponential rate. A computational identification algorithm/program could employ one or both approaches, as each may have its own strengths and weaknesses depending on the application.

Genome sequencing technology has advanced so much since the inception of the human genome project (Lander et al., 2001; Venter et al., 2001). Although it took scientists more than 15 years to decipher a single human genome, it now takes only days to sequence a microbial genome. We can realistically expect over 1000 genomes to be sequenced within the next few years. With this rate of sequence data production, fast and reliable genome annotation techniques will be essential to constructing a "road map," possibly coarse, incomplete, and not 100% accurate, on an otherwise incomprehensible string of letters to keep up with the sequence production rate. This initial road map should allow researchers to quickly focus on regions of a genome with interesting biological properties, directly relevant to their own research studies. It will also provide a bird's-eye view of a whole genome, that is, a list of basic building blocks in a genome, their spatial relationships and linear structures.

3.2 PREDICTION OF PROTEIN-CODING GENES

The single largest set of functional elements in a genome consists of genes. As we know now, the human genome has more than 30,000 genes (Lander et al., 2001; Venter et al., 2001). A typical microbial genome may have a few thousand genes. Most of them encode proteins, whereas a small fraction of them encode RNA, for example, tRNA or rRNA. While only a small percentage of the human genome contains gene-coding regions (Lander et al., 2001), 75 to 90% of the DNA in a typical microbial genome consists of coding regions (Doolittle, 2002). Throughout this section, we use the word "gene" to refer to a protein-coding sequence.

Since the information contained within a gene sequence is processed in triplets (three base pairs) during translation, there are three possible ways to read or translate a DNA sequence from left to right. Figure 3.1A shows an example. The upper sequence of Fig. 3.1A can be read in triplets as ACG, TTA, GAT, etc.; or CGT, TAG, ATA, etc.; or GTT, AGA, TAA, etc. These three ways of reading and translating a DNA sequence constitute three *reading frames*. A DNA sequence can also be read and translated in its reverse strand, that is, going from right to left in the complementary strand (see the lower part of Fig. 3.1A). Hence, each DNA sequence has six reading frames. For each reading frame, a DNA sequence could have a number of *in-frame* stop codons. The sequence fragment between two stop codons of the same reading frame is called an *open reading frame* (ORF). A gene always resides inside an ORF with its stop codon being at the boundary of the ORF. Since an ORF contains no stop codon in the middle and each gene always ends with a stop codon, each ORF can have at most one gene. By definition, a gene, when read in its translation frame, should not contain any stop codon in the middle of its sequence. It should be noted that different genes may have different reading frames. Figure 3.1B shows the typical structure of a few genes in a segment of a microbial genome. Generally, genes do not overlap in the genome sequence, although there have been a few rare cases of overlapping genes reported in the literature (Fukuda et al., 1999).

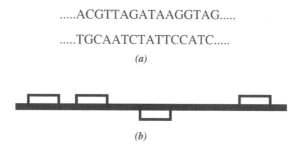

.....ACGTTAGATAAGGTAG.....

.....TGCAATCTATTCCATC.....

(a)

(b)

Figure 3.1 *(a) Segment of DNA sequence with its complementary strand. (b) The thick line represents a segment of genome sequence.* Each box above the line represents a gene in the forward strand and a box under the line represents a gene in the reverse strand.

It has been widely observed that most of the long ORFs contain genes in microbial genomes. For example, a majority of ORFs of at least 300 bases long contain genes (Li, 1997; Silke, 1997; Skovgaard et al., 2001). ORFs, particularly long ones, provide the initial list of gene candidates. The challenging issue is to determine if an ORF (part of it) has enough coding potential to be considered as a gene and where the first codon of the gene is, that is, to determine the translation start of the gene. A gene typically starts with an ATG. However, an ATG may not always be a translation start as an ORF could contain multiple ATGs. Determining which ATG is the correct start is not a trivial matter.

3.2.1 Evaluation of Coding Potential

Statistics-based methods for ab initio gene prediction are generally based on the observation that di-codons, or six-mers, may have different relative frequencies in coding regions and in noncoding regions. For example, the di-codon GACTGC, which encodes the amino acids aspartic acid (ASP) and cysteine (CYS), has a significantly higher relative frequency in noncoding regions than in coding regions in *Shewanella oneidensis*. Hence, we may predict that a region is noncoding if it is composed of mostly GACTGC. If most of the di-codons in a DNA segment have higher frequencies of occurrence in (known) coding regions than in (known) noncoding regions, we can probably expect that the region is more likely to be a coding region than a noncoding region. Otherwise, we will expect that it is noncoding. This is the basic idea used in most of the existing ab initio prediction programs for microbial genes.

This simple approach to evaluating the coding potential of a region can be implemented as follows. There are 4,096 different di-codons in a genome (there are $4^6 = 4,096$ possible combinations of forming a 6-mer using four letters). We can calculate the relative frequencies of each of the 4,096 di-codons in known coding and noncoding regions in a genome as follows. Count the total numbers of di-codons in known coding and noncoding regions of a genome, respectively. Then for each di-codon X, we can count the total numbers of occurrences of X in coding and noncoding regions, respectively. The relative frequency of X in coding regions is the ratio between the number of occurrences of X and the total number of di-codons in coding regions; the relative frequency of X in noncoding regions is estimated in a similar fashion. Based on these relative frequencies, one can estimate the probability that a sequence region is coding or noncoding. The following two methods represent the simplest ways of performing such an estimation.

Preference Model   In the preference model (Claverie and Bougueleret, 1986; Claverie et al., 1990), each di-codon X has a *preference value* defined as log*(FC(X)/FN(X))*, where *FC(X)* is X's relative frequency in a coding region and *FN(X)* in a noncoding region. If X happens to have the same relative frequencies in coding and in noncoding regions, then its preference value is zero. It will have a positive preference value if X has a higher relative frequency in coding than in a noncoding region; otherwise, it will have a negative preference value. Clearly, the higher its absolute value, the stronger the preference of the di-codon. The overall preference value of a DNA segment is calculated as the sum of all preference values of the di-codons in the segment. Clearly, if a majority of the di-codons in a DNA segment have positive preferences, the overall preference value will tend to be positive. Otherwise, it will tend to be negative. A segment can be predicted to be a coding region if it has a positive preference value; otherwise, it is a noncoding region.

There are a number of variations of this simple preference model, including *reading frame-dependent* or *reading frame-independent* models or a combination of the two (Uberbacher and Mural, 1991). But the essence is basically the same: estimating the coding potential by comparing relative frequencies of di-codons in previously known coding and noncoding regions. Some researchers have tried to use 7-mer-, 8-mer-, or even 9-mer-based preference models for gene finding. Although using longer mer models should increase the prediction accuracy as they contain more nucleotide-dependent information, such models may not be generally applicable for microbial gene finding due to the limited data available. A typical microbial genome may have a few million base pairs and 10 to 25% of that is noncoding, while the total number of 9-mers is $4^9 = 266,144$. If we need, on average, at least 10 occurrences of each 9-mer in noncoding regions (or coding regions) to assure the quality of our frequency estimation, we would need at least ~2.6 million noncoding 9-mers in our training data, the data set used to estimate the di-codon frequencies. This is clearly not practical for microbial gene predictions. The same argument could be made for 7-mer- and 8-mer-based models. Hence, 6-mer-based models are preferred for most microbial gene-finding programs.

Preference models were widely used in early gene prediction programs, including GRAIL (Uberbacher and Mural, 1991; Xu et al., 1996b) and SORFIND (Hutchinson and Hayden, 1992). However, they have been gradually replaced by more sophisticated models in the current generation of gene-finding tools, for example, hidden Markov models (Burge and Karlin, 1997).

Markov Chain Model   One drawback of the preference model is that it implicitly assumes consecutive 6-mers or di-codons are independent, which is probably not true in general and hence may lead to suboptimal performances by a gene prediction program using such a model. A Markov chain model (Borodovsky et al., 1986; Borodovsky and McIninch, 1993) provides the capability of modeling dependence relationships among consecutive di-codons. The model uses the Baysian formula to estimate the probability of a region being coding or noncoding. The following notations are useful for explaining the model. We use $P(S = s_1, s_2, \ldots, s_k | \text{coding})$ and $P(S = s_1, s_2, \ldots, s_k | \text{noncoding})$ to represent the conditional probability of having a DNA segment $S = s_1, s_2, \ldots, s_k$ in a coding and a noncoding region, respectively. These *a priori* probabilities can be estimated from known coding and noncoding sequence data of the target genome. We can use the Baysian formula to estimate the posterior coding and noncoding probabilities, $P(\text{coding} | S = s_1, s_2, \ldots, s_k)$ and $P(\text{noncoding} | S = s_1, s_2, \ldots, s_k)$, for DNA segment

$S = s_1, s_2, \ldots, s_k$, as follows:

$$P(\text{coding}|S) = P(S|\text{coding})/(P(S|\text{coding})$$
$$+ P(S|\text{noncoding})P(\text{noncoding})/P(\text{coding}))$$

$P(\text{noncoding})/P(\text{coding})$ represents the estimated ratio of coding and noncoding base pairs in a genome. We know that in the human genome about 2 to 3% of the DNA sequence is coding region, so the ratio is probably about 97/3 or $\sim$32. This ratio is much smaller in a microbial genome, perhaps approximately 1/9 to 1/3. The *a priori* probability $P(S|\text{coding})$ can be estimated based on the assumed Markov chain property of a DNA sequence, that is, *the probability of having a particular nucleotide in a particular position depends only on the nucleotide types of its previous k positions*. If $k = 1$, such a model is called a first-order Markov chain or a second-order Markov chain model if $k = 2$, etc. The most commonly used Markov chain model for gene finding is a fifth-order model. Hence,

$$P(s_1, s_2, \ldots, s_k|\text{coding}) = P(s_1, s_2, s_3, s_4, s_5|\text{coding})$$
$$\times P(s_6|\text{coding}, s_1, s_2, s_3, s_4, s_5)$$
$$\times P(s_7|\text{coding}, s_2, s_3, s_4, s_5, s_6) \ldots$$
$$\times P(s_k|\text{coding}, s_{k-5}, s_{k-4}, s_{k-3}, s_{k-2}, s_{k-1})$$

A term like $P(s_1, s_2, s_3, s_4, s_5, |\text{coding})$ can be estimated as the relative frequency of 5-mer $s_1s_2s_3s_4s_5$ in coding regions. A term like $P(s_6|\text{coding}, s_1, s_2, s_3, s_4, s_5)$ can be estimated as the ratio between the numbers of occurrences of '$s_1s_2s_3s_4s_5s_6$' and '$s_1s_2s_3s_4s_5$' in coding regions. Similarly, $P(s_1, s_2, \ldots, s_k|\text{noncoding})$ can be calculated based on the Markov chain assumption. Using these precalculated probabilities, one can estimate the coding probability $P(\text{coding}|S)$ of a segment S, as above. Similarly, the posterior probability of $P(\text{noncoding}|S = s_1, s_2, \ldots, s_k)$ can be calculated as

$$P(\text{noncoding}|S) = P(S|\text{noncoding})/(P(S|\text{noncoding})$$
$$+ P(S|\text{coding})P(\text{coding})/P(\text{noncoding}))$$

We will predict that a DNA sequence segment S is coding if $P(\text{coding}|S) > P(\text{noncoding}|S)$; otherwise, it is predicted to be noncoding. The value of $P(\text{coding}|S)$ [similarly, $P(\text{noncoding}|S)$] is generally "bio-polar," that is, it is either very close to one or zero. The transition from a value of one to zero, or vice versa, is fairly short. This sharpness in transition makes a Markov chain model a good tool for accurately predicting the gene boundaries.

Similar to preference models, Markov chains of a higher order, for example, sixth-, seventh-, or eighth-order, have been rarely used for gene prediction. Although higher-order Markov chain models should, in general, provide more accurate predictions, their applications have been limited for microbial gene prediction, due to the limited amount of data available.

3.2.2   Identification of Translation Start

For microbial gene predictions, a challenging problem is to identify accurately a gene's translation start, that is, the location of the starting codon ATG. The difficulty arises from the fact that an ORF may contain multiple in-frame (in the same reading frame of the ORF) ATGs, which are in close proximity to each other.

A general observation has been that translation starts in genes of a genome may share somewhat similar (or *conserved*) sequence patterns around the start codon ATG. By capturing and utilizing such conserved sequence patterns, one can possibly predict new translation starts, based on previously known ones. One way to capture weakly conserved sequence patterns is through the application of a *weight matrix*. To do this, one needs a set of known translation starts in the genome under investigation. One can collect the flanking DNA sequence segments around the (known) translation starts and align them so that the translation starts, ATGs, are lined up, as shown in Fig. 3.2A. How large of a flanking region needs to be considered depends on the *information content* level (Schneider et al., 1986) (this essentially measures how askew the distribution of the four letters is) of an aligned column. One needs only to consider those flanking regions with an information content above some threshold. For translation starts, a weight matrix model can be defined as follows. Each aligned position is represented by one column in the weight matrix. The first element of each column is the relative frequency of letter A among all the letters appearing in the corresponding aligned position. Similarly, the second, third, and fourth elements are defined for C, G, and T. A weight matrix is essentially a simple way to describe the common or similar sequence characteristics of flanks of translation starts.

After a weight matrix is calculated, one can use this information to assess the possibility of an unknown ATG site being a start codon by measuring how well its flanking region matches this weight matrix. One simple way to do the matching is to add up the corresponding values in the matrix from each column. That is, if the left-most position of the flank is a C, we will use the score 0.30 as the matching score, or 0.10 if it happens to be

TCTGTATGGTTAG
ACGGTATGCGATC
CATCCATGCTACC
CATGCATGTCAGT
ACGTGATGCGCGT
CGAGCATGGTTGC
GAACAATGCGTTT
AACGTATGACTGT
GTGGCATGGTACA
ACTCTATGTCTGA

(a)

0.40	0.40	0.20	0.00	0.10	1.00	0.00	0.00	0.10	0.00	0.40	0.10	0.20
0.30	0.40	0.30	0.30	0.40	0.00	0.00	0.00	0.40	0.10	0.10	0.20	0.20
0.20	0.10	0.20	0.60	0.10	0.00	0.00	1.00	0.40	0.30	0.00	0.50	0.20
0.10	0.10	0.30	0.10	0.40	0.00	1.00	0.00	0.10	0.60	0.50	0.20	0.40

(b)

Figure 3.2 *Example of a weight matrix for translation starts ATG.* (*a*) A list of 10 aligned start codon ATGs and their flanking regions. (*b*) Weight matrix based on the aligned 10 sequences.

a T. More sophisticated models can be used to assess the match score (Burge and Karlin, 1997; Salzberg et al., 1998b).

An implicit assumption used in the weight matrix model is that letter distributions are independent of each other among neighboring positions. Although this simple model may turn out to be effective for translation start predictions in some cases, the underlying assumption may not be generally true. To deal with dependent relationships among neighboring positions, one can use a generalized weight matrix model. For example, one can examine the letter distribution at position i, under the condition that the $i - 1^{st}$ position is of a particular letter type: A, C, G, or T. Hence, instead of having four elements in each column, such a weight matrix will consist of 16 elements. Basically, we will have a different letter distribution for position i, for different letter types at position $i - 1$. If needed, one can go even further to consider two previous positions, $i - 1$ and $i - 2$, when calculating the letter distribution for position i. This will be a second-order Markov model, whereas the previous one is a first-order Markov chain model. In this case, each column of the weight matrix consists of $4 \times 4 \times 4 = 64$ elements. The same argument holds here as in using higher-order Markov chains for gene predictions; a higher-order model may give more accurate predictions but may not be applicable due to the limited amount of data available to "train" the model.

More sophisticated models may take into consideration the possibility that there could be different types of translation starts. That is to say, translation starts may fall into more than one class, so that each class of translation starts may share some (possibly weak) consensus while translation starts of different classes do not. Dividing translation starts into classes could help improve the signal–noise ratio in a weight matrix and hence help improve prediction accuracy. Techniques such as data clustering (Burge and Karlin, 1997) or unsupervised learning (Reese et al., 2000) are generally used to partition a set of translation starts into clusters sharing common or similar features. This kind of "divide-and-conquer" strategy has been widely used for the detection of other sequence signals in a DNA sequence, including splice junctions (Reese et al., 2000).

3.2.3 Ab initio Gene Prediction through Information Fusion

One common strategy for gene identification is to first identify all ORFs for each of the six reading frames of a given DNA sequence. A prediction program will generally have some length cutoff, say, at least 100 base pairs, for an ORF to be considered. Then for each in-frame ATG within each ORF, one can measure the coding potential of the region starting at the ATG and ending with the codon right before the in-frame stop codon of the ORF, using the models described in Section 3.2.1, in conjunction with a prediction of this ATG being a translation start. A prediction of "gene" will be made if both the ATG has a high translation-start score and the whole region has high coding potential. Considering the in-frame coding potential of the region to the left of the ATG could also help the prediction. The rationale is that a correct translation start should have strong coding potential to its right and low coding potential to its left (if we assume that we are looking at the forward strand of a DNA) (Fig. 3.3).

Though the coding potential of a region and the score of an ATG as a translation start are the main factors to consider when predicting a gene, a few other factors also need to be considered. These may include the following:

an ORF

Figure 3.3 *Schematic of an ORF structure.* The two vertical bars at the two ends represent the boundaries of the ORF. An ORF may have multiple in-frame ATGs. Deciding which ATG represents the true translation start is not a trivial matter.

Gene Length Distribution. If one examines the length distribution of all known genes in a microbial genome, one can immediately see that the length distribution is not uniform. The distribution generally has a shape similar to an exponential distribution (e.g., $x^n e^{-x}$ for some positive integer n) or a gamma distribution. It is asymmetric and has a heavy tail on the right side. What this means is that some gene lengths are more favored (by nature) than other lengths. A computational gene-finding program should take this information into consideration when making a prediction.

G + C Composition. It has long been observed that genes in regions with different G + C compositions (defined as the percentage of letters G and C among all four possible letters) have different di-codon frequencies. Gene predictions, using one set of di-codon relative frequencies for regions of different G + C compositions, may lead to incorrect predictions. One possible remedy is to use different di-codon frequency tables when predicting genes in regions with different G + C compositions (Uberbacher et al., 1996). A related approach is to use the G + C composition directly as a normalization factor in the computational gene prediction.

Regions of Repeats. A genomic sequence may contain simple or complex repeats (see Section 3.5). These repeat regions should not overlap any genes even though some of these regions may "exhibit" coding potentials as measured by a preference model or other models, by chance. Several reliable prediction software programs for identifying repeated regions are available (Smit, 1999; Jurka et al., 1996). These regions should be masked out before running a gene-finding program.

Different computational models have been used for gene predictions based on the information outlined above. These include neural networks (Uberbacher and Mural, 1991; Snyder and Stormo, 1993; Xu et al., 1996b), decision trees (Salzberg et al., 1998a), Markov chain models (Borodovsky and McIninch, 1993), discriminant analysis (Zhang, 1997), and hidden Markov models (Burge and Karlin, 1997). Although these approaches employ different formalisms, one common feature is that they all attempt to put different types of information into the same computational framework. We now use a neural network approach, as an example, to illustrate how all this information can be combined to construct a better gene prediction program.

The difficulty in combining information from different sources into one computational framework is that the statistical relationship among different types of information could be very complex. Neural networks (Bishop, 1995) provide a practical and effective way to deal with this issue. A neural network predicts a gene or nongene, based on what was used to "train" it. A nongene is a region in an ORF that does not overlap any coding regions of the same reading frame. One can train a neural network by "telling" it that set A contains only genes and set B contains only nongenes, and by "showing" it various features of sets A and B. By examining the common features of sets A and B, respectively, a neural network can "learn" what features separate the two sets. For example, we can collect the following information for each known gene and known nongene. We use a *region* to represent a gene or a nongene that starts with an ATG and ends with a stop codon in the following:

(a) Coding potentials C1 and C2 of the region and the flanking region to the left of this region's translation start (the length of a flank region we should choose depends on the specific problem, e.g., 20 base pairs), as measured by either the preference model or a Markov chain model (or both)

(b) The score T of the candidate translation start ATG, as measured by a weight matrix model

(c) The G + C composition G calculated within a window of, say, 500 base pairs centered around the center of the region

(d) The length of the region L

The set A consists of a list of vectors (C1, C2, T, G, L, 1) for each gene in the 'training' set, and similarly, the set B consists of a list of (C1, C2, T, G, L, 0) for each nongene. Zero and 1 are used to 'tell' the neural network that one set consists of all genes and the other set all nongenes. For real applications, people often consider partial genes, too. Those are regions that overlap both gene and nongene regions. For each such region, we can calculate the same values C1, C2, T, G, and L. For the last element, we use the percentage of this region overlapping a coding region of the same reading frame (Xu et al., 1994). So, if a region is close to completely overlapping a gene, the last element should be close to 1; at the other extreme, it should be close to zero.

By going through such a training set containing complete genes, nongenes, and partial genes a number of times, a neural network can learn a mathematical formula that takes the input vector, for example, (C1, C2, T, G, L), and generates a value close to the desired value between zero and 1, representing the degree of the region being a gene. On a conceptual level, one can think of training a neural network as finding the parameters of a linear or quadratic discriminant function in a regression model by optimally fitting the provided data, although on a technical level these two approaches have very little in common.

Back-propagation (Bishop, 1995) represents one of the most popular learning algorithms for neural networks. A neural network typically has a few input nodes, one for each input parameter, and one or a few output nodes, plus possibly one or two hidden layers with some hidden nodes. Figure 3.4 shows a simple neural network architecture with four input nodes, one output node, and a hidden layer with five hidden nodes. Nodes of a neural network are arranged in layers, with the input nodes in the bottom layer and the output nodes in the top layer. Any layer in between is called a hidden layer. For a typical gene-finding problem, one or two hidden layers should suffice. Nodes on neighboring layers are connected with edges. Each edge has a weight associated with it. The training procedure is basically completed by adjusting the edge weights to best fit the data.

There are a number of neural network training software options, both free-ware and commercial versions, that can be used to implement a gene-finding program (see URLs in Table 3.1). These include SNNS (Stuttgart Neural Network Simulator), NeuralWorks (a product of NeuralWare), and STATISTICA Neural Network (a product of STATISTICA). GRAIL represents the first gene-finding program that uses a neural network as its main prediction framework (Uberbacher and Murel, 1991).

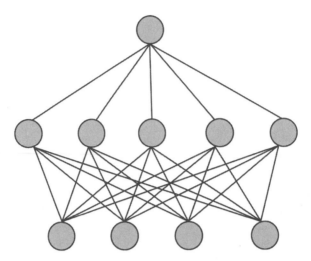

Figure 3.4 *Schematic of a neural network with four input nodes, one output node, and five hidden nodes.*

Currently, popular gene prediction tools for prokaryotic genomes include GeneMark (Borodovsky and McIninch, 1993) and Glimmer (Salzberg et al., 1998b). The best performing gene prediction programs can reach an accuracy level of $\sim 90\%$ (at the nucleotide level). Still these prediction programs may often predict the translation start incorrectly (Skovgaard et al., 2001). One common practice for genome annotation is to employ multiple gene prediction programs and use a voting scheme in making the final gene prediction.

TABLE 3.1 Web Servers for Genome Annotation

Program Name	Internet Address
SNNS	http://www-ra.informatik.uni-tuebingen.de/SNNS/
NeuralWorks	http://www.neuralware.com/
STATISTICA	http://www.statsoftinc.com/
GeneMark	http://opal.biology.gatech.edu/GeneMark/
Glimmer	http://www.tigr.org/software/glimmer/
GRAIL	http://compbio.ornl.gov/Grail-1.3/
Generation	http://compbio.ornl.gov/generation/
Genome Channel	http://compbio.ornl.gov/channel/
dbEST database	http://www.ncbi.nlm.nih.gov/dbEST/
GRAILEXP	http://compbio.ornl.gov/grailexp/
BLAST	http://www.ncbi.nlm.nih.gov/BLAST/
TRNAscan-SE	http://www.genetics.wustl.edu/eddy/tRNAscan-SE/
COG	http://www.ncbi.nlm.nih.gov/cgi-bin/COG/palox
Signal Scan	http://bimas.dcrt.nih.gov/molbio/signal/
TransTerm	http://www.tigr.org/software/transterm.html
BLAST	http://www.ncbi.nlm.nih.gov/BLAST/
FASTA	http://www.ebi.ac.uk/fasta33/
MegaBLAST	http://www.ncbi.nlm.nih.gov/blast/megablast.html
KEGG	http://www.genome.ad.jp/kegg/
MAGPIE	http://genomes.rockefeller.edu/magpie/
iPSORT	http://www.hypothesiscreator.net/iPSORT/
GO	http://protege.stanford.edu/ontologies/go/go.html
SignalP	http://www.cbs.dtu.dk/services/SignalP/

3.2.4 Gene Identification through Comparative Analysis

One could also identify genes in a genome through sequence comparisons without trying to study the statistical features common to coding regions. The basic assumption used in a comparative approach for gene finding is that a region having high sequence similarity with a known gene in the same or another genome is probably a gene. For a candidate gene region, one can run BLAST (Altschul et al., 1990) to query the GenBank database (Benson et al., 2002). If the queried sequence region exhibits high similarity to a known gene, the region could be predicted as a gene (see Chapter 5 for more details). The key advantage of such comparative approaches is that the prediction is generally more reliable than ab initio approaches, whereas their main drawback is that they apply only to genes with known homologues in GenBank. When annotating a new genome, one can first apply a comparative approach to find a subset of genes in the genome, and then use these genes and identified intergenic regions to build the di-codon tables needed by an ab initio method to find (hopefully) the rest of the genes in the genome.

Such a comparative approach can be and has been generalized in the following ways to make it more generally applicable: (1) sequence comparison against ESTs, and (2) gene prediction through the identification of conserved regions across multiple genomes.

EST-based Gene Predictions ESTs are expressed sequence tags. They are sequenced fragments of complete cDNA libraries of a genome. Most often, they are sequence fragments at the 5' or 3' ends of cDNAs. Although most of the EST projects have been for eukaryotic genomes, there are a few on-going projects attempting to produce EST sequences for microbial genomes (http://www.microbialgenome.org/links/parasites.html). The dbEST database (Boguski et al., 1993) represents the largest collection of ESTs; it currently has 12,845,578 entries (according to an August 2002 release). These EST sequences provide a rich source of information for gene predictions, particularly for genomes that are not closely related to any sequenced genomes.

Through BLASTing (Altschul et al., 1990, 1997) a whole new genome against the entire dbEST database, one can quickly identify a set of coding regions in the genome. Although a particular EST sequence may not be derived from the genome under investigation, the BLAST program can possibly find its homologous regions in this genome. Generally, these EST sequences only represent small portions of complete cDNA sequences. Hence, they cannot provide complete information about genes in a new genome. We can use the identified coding regions, based on the EST searches, to build the di-codon frequency tables of a new genome, and then apply such information to identify the rest of the coding regions.

ESTs can also be used directly in gene predictions, in conjunction with a neural network approach as outlined in Section 3.2.3. The basic idea is that if an EST matches well with a region in an ORF, one could infer that the ORF must have a gene in it. In addition, one could also infer that the left boundary of the gene must extend as far (to the left) as the left boundary of the region matching the EST. A generalized framework, of the kind outlined in Section 3.2.3, will be needed to handle such EST information. GRAIL-EXP represents one of the first gene-finding programs that fully utilize EST information (Xu and Uberbacher, 1997). ESTs are particularly helpful in accurately identifying the gene boundaries in GRAIL-EXP, when compared to prediction programs that do not use EST information.

Gene Prediction through Identifying Conserved Regions across Multiple Genomes It has been widely observed that conserved (long) regions across multiple genomes tend to be coding regions (Bafna and Huson, 2000; Batzoglou et al., 2000). Hence, by discovering such conserved regions, one could possibly identify novel genes, which may not have known homologs in GenBank nor have similar di-codon distributions to known genes in the target genome. This has been done mainly by aligning the whole-genome sequences of related genomes. Considering the amount of computing time needed, such approaches have been mainly applied to compare two genomes at a time.

There are a number of sequence alignment programs capable of doing alignments between two whole genomes. These programs can identify stretches of DNA with good alignment scores across two input genomes (as shown in Fig. 3.5). These aligned stretches do not need to have the same relative order in their respective genome. By going through such aligned DNA stretches, it is generally not difficult to identify which are probably coding regions. Note that a conserved region, discovered in this way, could also possibly be a regulatory region (see Section 3.4). A simple guideline for distinguishing a coding region from a regulatory region could be that coding regions are generally long DNA stretches sitting inside some ORFs, whereas conserved regulatory regions consist of interspersed short conserved regions. The following represent a few popular genome alignment programs available on the Internet.

(a) *megaBLAST* (Zhang et al., 2000). A member of the BLAST sequence alignment family, particularly designed for very long sequence comparisons. It employs a heuristic approach to speed up computational time and reduce the memory requirement by regular BLAST program.

(b) *SENSEI* (http://stateslab.wustl.edu/software/sensei/). Essentially follows the designing philosophy of BLAST for sequence alignment, that is, to first find short (size of 8) ungapped sequence matches and then extend them into longer gapped alignments. SENSI is probably more efficient computationally than BLAST.

(c) *MUMmer* (Delcher et al., 1999). Fast alignment program for long DNA sequences. It assumes that the sequences to be aligned are closely related, that is, that they have high sequence identity. The program achieves its high computational efficiency by utilizing a suffix trees data structure (Weiner, 1973), which is applicable only when two sequences are highly similar.

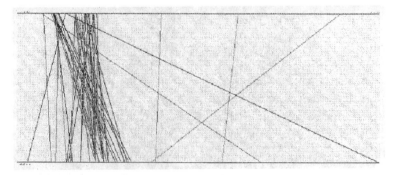

Figure 3.5 *Alignment result between two microbial genomes (segments) by PatternHunter (Ma et al., 2002).* The two parallel horizontal lines represent the two genomes. Each line connecting them represents two corresponding genes in the two genomes in their respective positions.

(d) *PatternHunter* (Ma et al., 2002). Represents a new class of fast sequence alignment program. It achieves the same level alignment performance as BLAST but runs significantly faster. Whereas BLAST attempts to find contiguous ungapped matches, PatternHunter uses sequence matches that do not have to be contiguous. This generalization has made the program more efficient, in terms of both computational time and memory requirement, than BLAST.

DIALIGN is one such program (Abdeddaim and Morgenstern, 2001), which predicts genes through genome-scale sequence comparison. In addition to serving as a parallel gene prediction approach to ab initio methods, comparative approaches can also be used as a way to validate gene predictions by ab initio methods. By providing additional and independent evidence, a comparative method can clearly help to increase the prediction confidence of genes with an ab initio approach.

3.2.5 Interpretation of Gene Prediction

Most of the existing gene prediction programs provide an indicator for its prediction reliability. For example, GRAIL (Uberbacher et al., 1996) provides one of the following descriptors: *marginal*, *intermediate*, or *strong*, to indicate the confidence level of a gene prediction. Such an indicator is typically estimated as follows. A prediction program usually provides a score, for example, the neural network score of GRAIL. One can tabulate all predictions by the program on a training data set and divide the predictions into bins based on the prediction scores. For example, all predicted genes with scores between zero and 0.1 are put into the first bin, all genes with scores between 0.1 and 0.2 in the second bin, etc. Then one can calculate the percentage of the correct predictions within each bin. For example, only 10% of the predictions in the first bin are correct, or 25% of the predictions in the second bin are correct. This number gives a reliability measurement of all predictions that fall into each bin. Different reliability thresholds may be applied for different applications and different purposes. If the goal is to design experiments for gene validation, one may want to consider using a high reliability threshold as we do not want to waste resources on things that are irrelevant. If the goal is to perform general screening for possible genes, one may want to lower the reliability threshold to a level that does not exclude many genes.

Pseudogenes It has been observed that some of the noncoding regions in microbial genomes are actually *pseudogenes*, segments that are still recognizable as having encoded proteins in the past, but which now contain stop codons and/or deletions that keep the gene from being expressed properly (Doolittle, 2002). Most of these pseudogenes have frameshifts due to deletions/insertions in the middle of the segment, making it hard for a regular gene prediction program to find them. However, they can be identified using a specialized coding-region detection program (Xu et al., 1996a) that takes into consideration possible frameshifts when calculating coding potentials. Some microbial genomes could have a significant number of pseudogenes. For example, *Mycobacterium leprae* has 1,100 predicted pseudogenes (Cole et al., 2001), while remarkably most of these pseudogenes are still fully functional in *M. tuberculosis*.

The identification of genes is only the first step in genome annotation. The ultimate goal is to understand what biological function each gene encodes and how it achieves its designed biological function. Existing genome annotation systems are generally content with some level of functional assignment of a gene without attempting to get into the detailed mechanistic issues. Functional assignment is now generally done based on the premise that genes having similar sequences have similar functions. Hence, if a gene encodes a protein that is a protease, we would assign a similar function to an unknown gene that has high sequence similarity, it having been determined by BLAST. Chapter 5 provides an in-depth discussion of this issue.

3.3 PREDICTION OF RNA-CODING GENES

In addition to proteins, a cell uses RNA molecules, also encoded by DNA, as functional elements. These RNA molecules are referred to as *functional* RNAs (Carter et al., 2001). They include tRNA (transfer RNA), rRNA (ribosomal RNA), sRNA (small RNA), srpRNA (signal recognition particle RNA), etc. Interestingly, RNA is the only biological polymer that serves as both a catalyst (like proteins) and as information storage molecules (like DNA). tRNAs work as adapter molecules that decode the genetic code. At one end of each tRNA is an anticodon that recognizes a particular codon, and at the other end of the molecule is the appropriate amino acid for that codon. A battery of enzymes, termed amino-acyl tRNA synthases, will add an amino acid to the growing protein chain, for each codon recognized by its corresponding tRNA molecule. rRNA molecules catalyze the synthesis of proteins through undergoing cleavages and dozens of nucleotide modifications before assembly with ribosomal proteins to form the mature ribosome. A typical microbial genome may have a few dozen to a few hundred RNA-coding genes.

Identification and functional assignments of RNA genes, just like protein genes, form an essential part of the genome annotation process. Generally, computational approaches for protein-gene prediction do not apply to RNA genes as the signals used for finding protein genes, including di-codon preference, $G + C$ composition bias, or translation starts/stops, generally are not present in RNA genes, at least not as strong signals. Traditionally, RNA gene prediction has been typically accomplished through sequence similarity searches, using programs like BLAST (Altschul et al., 1990). Some classes of RNA genes (e.g., some rRNA genes) are conserved across related genomes. A known RNA gene in genome X can be used to find its homologous genes in other genomes. Whereas this approach is useful for finding some classes of RNA genes, it is not generally useful for finding novel RNA genes.

A more general class of prediction methods for RNA genes relies on statistical features common to all or some large class of RNA genes. These methods are generally based on the following types of sequence signals. (1) RNA signals are frequently composed of a combination of sequence and structure motifs (Carter et al., 2001). Through identification of these sequence and structural motifs, one can design prediction programs for RNA genes. Since these motifs are generally specific to a particular class of RNA genes, for example, tRNA genes, most of the existing prediction programs are designed to recognize particular types of RNA genes. (2) An RNA gene consists of a set of secondary structures in its folded tertiary structure (see Fig. 3.6B). The dominating secondary structure types are *stems* (made up of a sequence of nucleotide pairs, A-U or C-G) and *loops* (see Fig. 3.6A). The secondary structures, particularly the stems, provide detectable signals for RNA gene recognition. Recognition of such signals generally requires a computer program to examine a large window of a DNA sequence for the detection of long-range relationships.

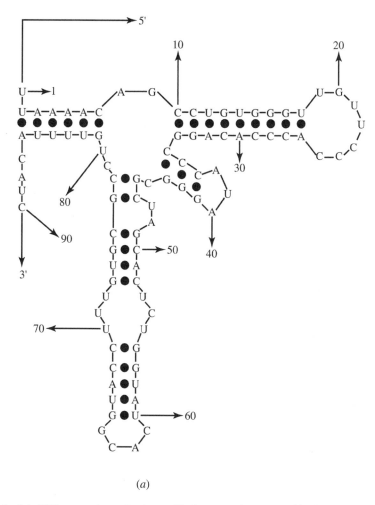

(a)

Figure 3.6 *(a) RNA secondary structure with loops and stems, with the numerical numbers representing the sequential positions of a nucleotide in the RNA sequence. (b) [next page] The tertiary structure of RNA 1gid (Woese et al., 1990).*

tRNAscan is one of the most popular prediction programs for tRNA genes (Fichant and Burks, 1991). It recognizes a tRNA gene through identification of conserved sequence and structural motifs commonly shared by known tRNA genes. It achieves high prediction fidelity by detecting sequence and structure motifs and utilizing RNA secondary structure information. Combining tRNAscan and another tRNA prediction algorithm formulated by Pavesi and colleagues (Pavesi et al., 1994), Lowe and Eddy (1997) developed a more sensitive prediction program called tRNAscan-SE, which can reach a prediction accuracy as high as 99% for tRNA genes, with a false positive rate at one false prediction per 15 gigabases.

Carter and coworkers (2001) have recently developed a computer program for the general purpose of predicting RNA genes. It involves examining all known RNA genes and finding the common features of these genes using machine-learning techniques,

(b)

Figure 3.6   *(continued)*

including support vector machines (Cristianini and Shawe-Taylor, 2000) and neural networks. The parameters used in this program to distinguish RNA genes and non-RNA genes include (1) the compositions of A, C, G, and U within a window of evaluation, (2) the frequencies of various sequence motifs that frequently occur in RNA tetraloops (Woese et al., 1990), including UNCG (Ennifar et al., 2000), GNRA (Jucker et al., 1996), CUYG (Woese et al., 1990; Moore, 1999), and AAR (Burge et al., 1992), and (3) free energy of folding for each presumed RNA sequence window (Zuker, 1989). The authors observed that these parameters provide a significant amount of information for distinguishing an RNA coding sequence from a non-RNA gene sequence. This program applies to both eukaryotic and prokaryotic RNA genes.

Until very recently, gene predictions have mainly focused on protein-coding genes. Only limited effort has been devoted to the identification of some special classes of RNA genes, for example, tRNA genes. Novel RNA genes, particularly small RNA genes, are being discovered on a regular basis through experimental methods, with little help from computational techniques. By applying general-purpose prediction programs like those described above, one could possibly identify novel RNA genes and even infer their functions.

3.4 IDENTIFICATION OF PROMOTERS

There are two main classes of functional information encoded in the genome sequence of every living organism (see Chapter 10 of Jiang et al., 2002). One class is the coding regions, which specify the structure and function of each gene product (protein or RNA molecules); the other class is the regulatory regions, which control and regulate when, where, and how genes are expressed. The promoter is the most important regulatory region; it regulates the very first step in gene expression, that is, mRNA transcription.

The transcription process is initiated when an RNA polymerase binds to a promoter. The RNA polymerase then moves along the DNA sequence using it as a template to produce an RNA molecule. It releases the DNA sequence and stops transcription when it encounters a terminator (see Section 3.5). A promoter is generally composed of a list of transcription factor (TF)-binding sites. The transcription factors (proteins) control the "on" and "off" of a gene and regulate the gene's expression level, through binding to these TF-binding sites. Often a regulatory event is controlled not by a single TF but by a collection of TFs that bind to the same promoter region of a gene. The RNA polymerase of a prokaryote typically consists of one type of various sigma factors (σ), each of which recognizes a specific subset of promoter sequences. For example, the *E. coli* genome is known to have the following sigma factors: σ^F (for flagellum genes), σ^S (for starvation stress response), σ^{32} (for heat shock response), σ^{54} (for nitrogen assimilation), and σ^{70} (the primary, or housekeeping, sigma). It is also known that the σ^{70} factor binds to two conserved sequence motifs at positions -10 and -35 (the transcription start is denoted as position $+1$; -10 denotes the tenth position upstream of position $+1$). The consensus sequences are TTGAGA at position -35 and TATAAT at position -10, respectively (von Hippel, 1992). σ^{54} is known to bind two consensus boxes at positions -12 and -24 (Korzheva et al., 2000). These consensus sequences placed at particular positions with certain distances between them are the basis for the computational prediction of promoter regions.

3.4.1 Promoter Prediction through Feature Recognition

The hidden Markov model (HMM) (Rabiner, 1989) represents a popular statistical tool for capturing and modeling long-range relationships in a biological sequence. This tool has been widely used to model various biological structures or processes, including gene structures in eukaryotic genomes (Burge and Karlin, 1997), protein family classification (Haussler et al., 1993), and RNA secondary structure prediction (Asai et al., 1993). A HMM typically consists of a number of states, each of which has certain probabilities to transit to other states. During each transition, some outputs are generated with certain probabilities. For a biological sequence analysis problem, like promoter identification, an output is often a nucleotide or an amino acid. By walking through a sequence of states, one can obtain a sequence of nucleotides or amino acids. All the probabilities associated with a HMM determine the most probable path of states that a given bio-sequence may walk through.

By collecting a set of known promoter sequences and a set of nonpromoter sequences, one can "train" a HMM by selecting all the probability values that will "generate" all the known promoter sequences with significantly higher probabilities than that of nonpromoter sequences. In essence, such a trained HMM can capture the key statistical features, including conserved sequence fragments and their spacing relationships in known

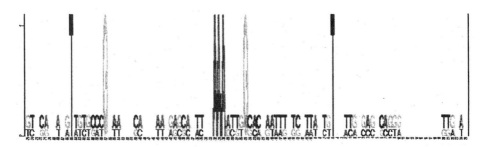

Figure 3.7 *Sequences recognized by the σ^{54} factor (Pedersen et al., 1996).* The height of each letter in a particular sequence position represents the relative frequency of that letter in that position in the aligned promoter regions from positions -75 to $+25$.

promoter sequences that distinguish promoter from nonpromoter sequences. Figure 3.7 shows an example of conserved sequence patterns and spacing relationships among these patterns that are recognizable by σ^{54}. Once such a HMM is trained, it can be used to detect if a given sequence fragment is a promoter sequence by calculating the probability that this sequence may be generated by this trained HMM. Note that one does not have to know in advance what features should be considered by an HMM, like the conserved promoter elements TTGAGA at position -35 and TATAAT at position -10 for σ^{70} binding σ^{70}. A HMM training program, like the expectation maximization (EM) program (Dempster et al., 1977), generally has the capability of picking up statistically significant sequence features and spacing features.

Pedersen and colleagues (1996) described a HMM for prokaryotic promoter recognition. The program was trained on a set of 248 promoter sequences recognized by RNA polymerase. Each sequence contains 75 base pairs upstream and 25 base pairs downstream of the transcription start. The program identifies a promoter by analyzing the features of a continuous block of sequence, as shown in Fig. 3.7.

There are also other types of promoter-recognition programs, which mainly focus on the identification of individual binding motifs. These programs attempt to find conserved sequence motifs across multiple DNA sequences that are from the upstream regions of genes suspected of being coregulated. If a region in a genomic sequence consists of multiple conserved sequence motifs, it may indicate that this is a promoter region. Below we list a few popular motif-finding programs.

CONSENSUS (Stormo and Hartzell, 1989) attempts to find conserved k-mers that occur in the majority of the input sequences using a heuristic approach (k could be any integer in some specified range, e.g., from $5-30$). The program examines one sequence after another to determine if the current sequence contains any k-mers that are similar to any k-mers of the previously examined sequences and gradually builds a consensus matrix, based on the identified conserved k-mers. The final consensus matrix provides information on whether the input sequences have any conserved k-mers.

MEME (Bailey and Elkan, 1994) searches for conserved k-mers from a set of input sequences, using an EM algorithm (Dempster et al., 1977) to estimate the maximum likelihood of the conserved k-mers. It is more general than CONSENSUS in the sense that it does not require a conserved k-mer to occur in most of the sequences and each sequence can have at most one copy of the same conserved k-mer.

Gibbs sampler (Lawrence et al., 1993) overcomes the problem that neither CONSENSUS nor MEME can guarantee finding the global optimality of the search problem they are designed to uncover—they generally stop at some local optima. Gibbs sampler employs a stochastic search algorithm, which is mathematically proven to converge with the global optima with probability 1 as the number of data samplings approaches infinity (or a very large number for any practical applications).

Other publicly available prediction programs for prokaryotic promoters include Signal Scan (Prestridge, 1991) and NNPP (Reese et al., 1996). Promoter prediction will be most effective when used in conjunction with gene prediction, as a promoter should sit right in front of one or a series of genes. Predictions of promoters and genes can validate each other, making the final prediction more reliable. A HMM for promoters can possibly be generalized to model a *promoter-gene* structure or the more general structure of *promoter-gene-gene-⋯-gene* (see Section 3.5).

3.5 OPERON IDENTIFICATION

Operons represent a basic organizational unit of genes, in the complex hierarchical structure of biological processes in a cell. They are mainly used to facilitate efficient implementation of transcriptional regulation. Genes in an operon are arranged in tandem in a genome and controlled by a common regulatory region consisting of a set of regulatory binding motifs (see Chapters 4 and 7 for a further discussion of this). Figure 3.8 shows schematically the structure of an operon having multiple genes with one common promoter and a terminator. The key information used for the computational identification of an operon may include (1) a predicted promoter region and a terminator, (2) a set of (predicted) genes arranged in tandem (with a short distance between two consecutive genes) on the same strand, and (3) functional information (possibly predicted) of the genes involved. A capability in accurately identifying operons will help to identify transcriptional regulatory networks in a cell.

Terminator Identification A terminator marks the end of an operon structure and provides a signal for the RNA polymerase to stop transcription. Two mechanisms of transcription termination and two classes of termination signals have been described in bacteria: rho-dependent and rho-independent (von Hippel, 1992). Both types of transcription terminators have recognizable sequence signatures. For example, in *E. coli*, the rho-independent termination signals consist of stable hairpins followed by U-rich regions. Three nucleic acid binding sites have been characterized in the transcription elongation complex: a double-stranded DNA binding site, an RNA–DNA hybrid binding site, and a single-stranded RNA binding site (Korzheva et al., 2000). By recognizing these sequence features, a number of computer programs for the recognition of prokaryotic

Figure 3.8 *Schematic of an operon structure, consisting of four genes and a promoter and a terminator.* Each box represents a gene.

transcription terminators have been developed (Lesnik et al., 2001). TransTerm (Ermolaeva et al., 2000, see Table 3.1) is a program that finds rho-independent transcription terminators in bacterial genomes. Each terminator found by the program is assigned a confidence value that provides an estimate of its probability of being a true terminator. Other programs include those developed by Brendel and Trifonov (1986) and d'Aubenton Carafa and coworkers (1990).

In addition to structural information such as promoters and terminators, another key piece of information useful for operon prediction is the fact that genes in operons tend to encode enzymes which catalyze successive reactions in metabolic pathways. Based on this phenomenon, Zheng and colleagues (2002) recently developed a computer program for operon prediction that achieves an impressive prediction accuracy. A large collection of predicted operons for microbial genomes are provided at http://genomics4.bu.edu/operons/, having been amassed by this program.

There are other prediction programs for operon structures, based on somewhat different types of information Craven and coworkers (2000) created a prediction program based on identified sequence patterns of known operons. One of the earliest and most fully characterized operons is the *lac* operon (Muller-Hill, 1996). Now, a large number of operon structures have been fully characterized, such as the *trp* operon (which controls the biosynthesis of tryptophan in the cell from the initial precursor chorismic acid) and the *mhp* operon (which catalyzes successive reactions in the phenylpropionate catabolic pathway). Using these known operons, one can collect statistics on (1) the intergenetic distance within an operon *vs.* between operons, and (2) the distribution of the number of genes that an operon may have, and possibly others. By combining these statistical distributions and the predictions of genes, promoters, and terminators, Craven and colleagues (2000) developed a machine-learning algorithm for the prediction of the detailed structure of an operon. Their program is capable of performing such predictions on a genomic scale.

3.6 FUNCTIONAL CATEGORIES OF GENES

A key step in genome annotation is to assign genes to functional classes after they are predicted. However, there is no standard set of functional categories for genes just yet, other than for a small subset of gene products, that is, EC classes for enzymes. As the number of microbial genomes being sequenced and annotated increases rapidly, we can expect that some standards for gene function classification will soon be established. Currently, functional assignments of genes are done in an ad hoc way. The microbial genome annotation projects sponsored by the Department of Energy (DOE) use the functional categories of KEGG (in the *Kyoto Encyclopedia of Genes and Genomes*) (Kanehisa et al., 2002) in their gene functional assignments at the low-resolution level. They essentially classify genes into functional categories defined by the metabolic and regulatory pathways of KEGG. For example, KEGG currently lists eight metabolic groups with about 120 metabolic pathways and four regulatory groups with a few dozen regulatory pathways (see Chapter 5 for more details). If a gene is identified as being involved in "amino acid metabolism" or a "cellular growth and death" pathway, its functional category will be labeled accordingly.

A typical example of the functional assignments of genes is provided in the following listing (available at http://genome.ornl.gov/microbial/mbar/26may01/kegg_fc.html). The genome is *Methanosarcina barkeri*, and all its genes are grouped into these functional categories at the low-resolution level:

Cellular processes
    Cell motility
Environmental information processing
    Membrane transport
    Signal transduction
Genetic information processing
    Replication and repair
    Sorting and degradation
    Transcription
    Translation
Metabolism
    Amino acid metabolism
    Carbohydrate metabolism
    Energy metabolism
    Lipid metabolism
    Metabolism of cofactors and vitamins
    Metabolism of complex carbohydrates
    Metabolism of complex lipids
    Metabolism of other amino acids
    Metabolism of other substances
    Nucleotide metabolism
Unassigned
    Nonenzymes

At a more detailed level, genes are assigned to different EC classes if they are predicted to be enzymes, or belong to some functional classes defined by existing functional classification schemes, such as Pfam (Bateman et al., 2002), for example. Figure 3.9 shows an example of detailed functional assignments of *Methanosarcina barkeri* genes in the "cell motility" pathway. In this example, three genes are assigned to EC classes (http:// genome.ornl.gov/microbial/mbar/26may01/kegg_fc.html). A detailed procedure for the functional assignments of genes, using various computational tools and databases, is provided in Chapter 5.

Contig	Gene	Subject Gene ID	Subject Gene Name	Subject Gene Desc
Contig1954	4227	eco:b1888	cheA	chemotaxis protein cheA [EC:2.7.3.-] [SP:CHEA_ECOLI]
Contig1918	3032	eco:b1884	cheR, cheX	chemotaxis protein methyltransferase [EC:2.1.1.80] [SP:CHER_ECOLI]
Contig1955	4281	eco:b3544	dppA	dipeptide transport system substrate-binding protein [SP:DPPA_ECOLI]
Contig1954	4231	eco:b1885	tap	methyl-accepting chemotaxis protein IV (MCP-IV) (dipeptide chemoreceptor protein) [SP:MCP4_ECOLI]
Contig1954	4226	eco:b1883	cheB	protein-glutamate methylesterase [EC:3.1.1.61] [SP:CHEB_ECOLI]
Contig1954	4232	eco:b1887	cheW	purine-binding chemotaxis protein [SP:CHEW_ECOLI]

Figure 3.9 *Detailed functional assignments of genes in the "cell motility" pathway.*

3.7 CHARACTERIZATION OF OTHER FEATURES IN A GENOME

Whereas genes and promoters represent the two most important classes of information directly extractable from a genomic sequence, other sequence signals could also provide highly important information for a full understanding of a genome. In a typical genome annotation process, the following signals and sequence features are additionally predicted and described.

G + C Composition   It is generally the case that the density of genes in a region correlates with the region's $G + C$ composition (the ratio of G's and C's within a sequence window). Typically, in a genome, higher $G + C$ compositions imply higher gene densities. $G + C$ composition is often calculated within a sliding window of a fixed size, say, 1,000 or 2,000. The calculated value is assigned to the center base pair of the window. By plotting the $G + C$ compositions along the sequence axis using a gray value (the higher the $G + C$ composition, the lighter a plot), one should be able to see the banding patterns of a whole genome.

CpG Islands   CpG islands are stretches of DNA with a higher frequency of CpG dinucleotides when compared with the entire genome (Bird, 1987). CpG islands are believed to preferentially occur at the transcriptional starts of genes. Whereas a CpG island is used to describe a region with a high CG dinucleotide ratio, there is no formal definition of how high that ratio should be. One commonly used threshold is 0.6. In annotating the sequenced human genome, 0.8 was used as the threshold to identify CpG islands (Venter et al., 2001). Because of the observed relationship between CpG islands and transcription starts, CpG islands can be used as one piece of evidence for predicting promoters and transcription starts.

Genomic Repeats   Repetitive sequences (*repeats*) are ubiquitous in both prokaryotic and eukaryotic genomes. Some of these repeats have known biological functions, whereas the functions for a significant portion of them are not clear. For example, transposons perform a function by allowing mobile elements to move around a genome. Identification of the repetitive elements in a genome represents an important component of a genome annotation process. Various computer programs exist for identifying different types of repeats. The following are few examples:

(a) *Tandem Repeat Identification*. The computer programs developed by Karlin and colleagues (Leung et al., 1991) and Benson (1999) can find simple tandem repeats through exact and approximate string matching.

(b) *RepeatMasker* (Smit, 1999). Has a database of all known repeats, for example, ALU or line repeats. It identifies repeat elements in a new DNA sequence by matching all the repeat sequences in its database against the DNA sequence.

(c) *RepeatFinder*   (http://www.proweb.org/proweb/Tools/selfblast.html). Finds repeated patterns, identified through either exact or approximate match, using a clustering technique. It represents a more general approach than previous methods.

Gene density   This is another feature that can be applied at a genome scale. It provides the number of genes per fixed length of genomic sequence. By putting this distribution together with the G + C composition plot, one can easily see the correlation between the two in a genome.

3.8 GENOME-SCALE GENE MAPPING

When combining genome alignments, as discussed in Section 3.2.4, with gene predictions, one can build a "complete" map of how genes from one genome compare to genes in related genomes. Such a map can help us to understand the following issues central a genome:

Genes Unique to a Genome   It is estimated that typically 20 to 30% of genes in a genome are unique to it. By analyzing the correspondence maps of genes between one genome and other genomes, one could possibly identify these genes. Further functional annotation of these genes (see Chapter 5) could provide clues to understanding a genome's unique characteristics.

Genome Rearrangement   A systematic study of how genes' locations in one genome sequence differ from the locations of their corresponding genes in related genomes can help us understand how genomes have been rearranged. Quantitative studies of genome rearrangement (Sankoff, 1999) could provide an effective measure for evolutionary distances among different genomes. Researchers have used the following quantitative measures to estimate genomes' evolutionary distances:

Reversal Distance. It measures how many reversal operations, defined from (a, b) to (b, a), it needs to transform one ordered list of genes $a_1, a_2, \ldots, a_n$ to another list $b_1, b_2, \ldots, b_n$, where $b_1, b_2, \ldots, b_n$ is a permutation of $a_1, a_2, \ldots, a_n$.

Transposition Distance. A transposition operation moves a block of genes from one position to another. The transposition distance between two sequences of genes is the minimum number of transposition moves that it takes to transform one sequence to another.

Using these two distances or their combinations, one can rigorously study the distance between two genomes. This will provide another useful distance measure besides the traditional gene-based distance measures between genomes.

3.9 EXISTING GENOME ANNOTATION SYSTEMS

Although there is not yet a set of established "standards," the following information was recommended for presentation in a policy paper of the American Society of Microbiology on an annotated microbial genome (available at http://www.asmusa.org/pasrc/microbialgenome.htm).

Annotation entails assembling information of several distinctive types, starting with refined DNA sequence data but extending beyond that level to varying degrees of complexity. For example, completed DNA sequence information may be segmented into . . . specific types of "product," such as proteins, transfer RNAs (tRNAs), and phage sequences. A particular gene of the first type may be annotated in such terms as its protein coding region, its transcript, promoter region, and so forth. At a higher level of annotation, a protein that is encoded by a particular gene may be annotated in terms of its physical attributes, such as molecular weight, membrane spanning regions, structural domains, or three-dimensional structure. Moreover, annotation at the level of comparative biology may include information linking a particular protein from a specific microorganism to similar proteins from other organisms or to members of similar protein families. Genes may also be annotated at a functional level, in terms of their respective roles in cellular metabolism, a particular systematic enzyme number (EC) designation, protein-protein interactions, and expression profiles.

There are a number of existing annotation systems for microbial genomes that have provided a basic framework for extracting and representing the information described above.

Genome Channel   Genome Channel is at the front end of a suite of genome annotation tools built and deployed at DOE's Oak Ridge National Laboratory (Table 3.1). It has been the key annotation tool for DOE's microbial genome annotation projects. Although Genome Channel has been used mainly by the DOE annotation team, its annotation results are publicly available. It presents its annotation results in the following main categories:

Modeled Genes. Gene models are represented in FASTA format with various prediction details, including sequential positions, methods used for prediction, BLAST hits, etc. Also presented are their translations to protein sequences. Figure 3.10 shows a screenshot of predicted gene models of Genome Channel.

Functional Assignments of Genes. Gene models are assigned to different functional classes, using different classification methods. These include KEGG pathways, Pfam families, EC classes, and COG groups (Tatusov et al., 1997).

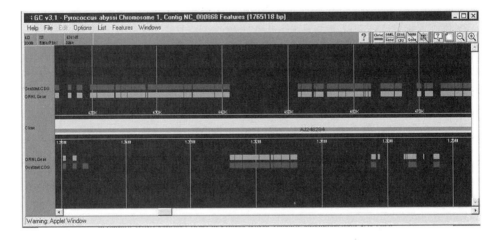

Figure 3.10 *Screenshot of Genome Channel.* It displays gene predictions of a DNA fragment of *Pyrococcus abyssi* on both strands. Each box represents a predicted gene. Boxes of different colors represent genes predicted by different programs. By superimposing them, one can easily observe the prediction consensus.

RNA Genes. Currently, only tRNA and rRNA genes (16S and 23S) are predicted and listed in the annotation results.

Repeats. Genome Channel provides a list of all the identified repeat sequences in a genome, using existing repeat finding programs, as discussed in Section 3.7.

General Sequence Features. General features in a DNA sequence, such as local GC composition, GC and AT skew, and others, are also presented as part of the genome annotation results.

MAGPIE (Multipurpose Automated Genome Project Investigation Environment) MAGPIE is an annotation environment particularly designed for microbial genomes at Rockefeller University (Table 3.1). It provides an environment for each annotation effort by supporting bookkeeping features for the project. MAGPIE generates a set of tables that contain the genomic features associated with a particular region of the genome. A unique feature of MAGPIE is that it provides an extensive set of graphical overviews of the analysis results. These graphical viewers include the following:

(a) Project page with various project statistics

(b) Contig page that provides an overview of the whole genome

(c) Status page on results reported from different tools

(d) ORF page that displays all the predicted genes

(e) Repeat page for displaying various identified repeats

(f) Search page for displaying search results by BLAST and FASTA

(g) tRNA page for displaying identified tRNA genes

(h) Metabolic pathway page for displaying the assigned metabolic pathways of identified ORFs

Using such an organizational framework and the display capabilities, one could run his or her own prediction tools, and display the results and organize the data through MAGPIE.

GenDB GenDB (Goesmann et al., 2002) is an open source annotation tool for microbial genomes, a unique feature among the existing genome annotation systems. It represents an effort to build a genome annotation system that enables a user with programming skills to adapt the system and extend it as required. It provides various generic viewing capabilities and supports the customization of these viewing tools for individual users. It currently runs the following prediction tools:

(1) BLAST and FASTA (Pearson, 1990) for sequence similarity searches

(2) Interpro BLOCKS (Henikoff and Henikoff, 1992) and Pfam for protein motif identification

(3) SignalP (Nielsen et al., 1997) for the identification of signal peptides

(4) TMmn for the identification of transmembrane regions in a predicted gene

(5) iPSORT (Bannai et al., 2002) for subcellular localization prediction

(6) KEGG (Kanehisa et al., 2002) for pathway identification

(7) GO for ontology identification

3.10 SUMMARY

Large-scale annotation efforts require handling massive amounts of genomic data using automated pipelines, with a need to combine diverse sources of data and methods. As the annotation technologies begin to mature, we can expect that some annotation standards will soon be established, which should guide the future development of annotation tools. Capabilities for providing prediction reliability indicators, for example, the p-value of BLAST search results, will be a key to the full automation of an annotation process. Good visualization capability will continue to be a key measure of the usability of an annotation framework. In addition, capabilities in supporting interactive operations between annotation systems and human annotators will be highly important as the integration of human expertise to assess the validity and authenticity of all computational results could go a long way in improving the quality of genome annotation.

FURTHER READING

Altschul, S. F., T. L. Madden, A. A. Schäffer, and J. Zhang, et al. 1997. Gapped BLAST and PSI-BLAST: a new generation of protein database search programs. *Nucl. Acids Res.* 25:3389–3402.

Burge, C. and S. Karlin. 1997. Prediction of complete gene structures in human genomic DNA. *J. Mol. Biol.*, 268:78–94.

Carter, R. J., I. Dubchak, and S. R. Holbrook. 2001. A computational approach to identify genes for functional RNAs in genomic sequences. *Nucl. Acids Res.* 29(19):3928–3938.

Delcher, A. L., S. Kasif, R. D. Fleischmann, and J. Peterson, et al. 1999. Alignment of whole genomes. *Nucl. Acids Res.* 27:2369–2376.

Doolittle, R. F., 2002. Biodiversity: microbial genomes multiply. *Nature* 416:697–700.

Jiang, T., Y. Xu, and M. Q. Zhang. 2002. *Current Topics in Computational Molecular Biology.* MIT Press, Cambridge, Mass.

Kanehisa, M., S. Goto, Kawashima, and A. Nakaya. 2002. The KEGG databases at GenomeNet. *Nucl. Acids Res.* 30:42–46.

Lawrence, C. E., S. F. Altschul, M. S. Boguski, and J. S. Liu, et al. 1993. Detecting subtle sequence signals: a Gibbs sampler strategy for multiple alignment. *Science* 262:208–214.

Tatusov, R. L., E. V. Koonin, and D. J. Lipman. 1997. A genomic perspective on protein families. *Science* 278:631–637.

Zheng, Y., J. D. Szustakowski, L. Fortnow, and R. J. Roberts, et al. 2002. Computational identification of operons in microbial genomes. *Genome Res.* 12(8):1221–1230.

4

Microbial Evolution from a Genomics Perspective

Jizhong Zhou and Dorothea K. Thompson

4.1 INTRODUCTION

Understanding the evolutionary (i.e., phylogenetic) relationships between extant organisms has been a central topic in biology but a great challenge to biologists for many centuries (Graham et al., 2000; Gupta, 1998). Classification of organisms based on phenotypic characteristics is of limited use in understanding the evolutionary relationships between organisms and does not necessarily reflect their evolutionary descent (Graham et al., 2000). Thus, much of modern evolutionary studies are focused on molecule-based phylogenetics (Doolittle, 1999a). Molecular evolution is generally investigated at the level of single individual genes, such as ribosomal rRNA genes, but the phylogenetic relationships of species derived from comparisons of single genes are rarely consistent with each other due to complicated evolutionary processes and relationships. The availability of whole-genome sequences from many organisms provides us with an opportunity to obtain the most objective and comprehensive descriptions of evolutionary relationships among different organisms. Recent comparative studies based on whole-genome sequences have generated some fundamentally different and exciting insights into the evolution of biological systems. The discovery that horizontal gene transfer and lineage-specific gene loss play major roles in the evolution of prokaryotic genomes resulted in a major shift in our understanding of the evolution of life (Aravind et al., 2000; Koonin et al., 2001). These evolutionary processes largely account for the extreme diversity of microbial genomes. Such results have changed the traditional view of evolutionary biologists that mutations within individual genes are the major source of phenotypic adaptive variations through natural selection (Ochman and Moran, 2001).

Microbial Functional Genomics, Edited by Jizhong Zhou, Dorothea K. Thompson, Ying Xu, and James M. Tiedje.
ISBN 0-471-07190-0 © 2004 John Wiley & Sons, Inc.

This chapter will provide an overview of the most recent insights into molecular evolution that have emerged from whole-genome sequence comparisons. The first section will discuss the identification of orthologous genes, followed by a presentation of recent genomics-based perspectives on several central issues of molecular evolution such as molecular clock, horizontal gene transfer (HGT), gene duplication, gene loss, and the universal tree of life. This chapter will also introduce the concept of minimal genomes and briefly discuss recent insights into the evolution of lifestyle and mitochondrial genomes.

4.2 IDENTIFICATION OF ORTHOLOGOUS GENES

Due to complicated evolutionary processes and history, the identification of orthologs is critical to constructing reliable phylogenetic trees. In this section, we will describe how to identify orthologous genes from a knowledge of genomic sequence.

Definitions of Orthologs and Paralogs   Once the sequence for an entire genome is available, the molecular and cellular functions for individual open reading frames (ORFs) or genes are generally first assigned by searching the databases for homologous sequences (see Chapter 3). However, homology is not function, and hence sequence comparison plays a limited role (Strauss and Falkow, 1997). The functional assignment of genes by similarity comparison can be misleading or incorrect due to the complicated evolutionary and structure-function relationships among different genes (Strauss and Falkow, 1997). It is important to differentiate two types of homologous genes, orthologs and paralogs, when homology is used to define gene functions and to study the evolutionary relationships of genes and organisms.

Orthologous genes, or orthologs, are defined as homologous genes that are vertically descended from a common ancestor and encode proteins with the same functions in different species (Koonin et al., 1996; Huynen and Bork, 1998) (Fig. 4.1). Since the homologous relationship among the orthologous genes results from speciation, the history of these genes reflects the history of the species. In contrast, paralogous genes (i.e., paralogs) are referred to as homologous genes that evolved through duplications and may encode proteins with similar but not identical functions. Since homology is the result of duplication rather than speciation, the history of genes does not reflect the history of species, and thus paralogs are not useful for determining phylogenetic relationships between different species.

Detection of Orthologous Genes   Under ideal conditions, one can expect that the orthologous genes between two genomes will have the highest pairwise identity and have bifurcated relatively recently compared to paralogous genes, which arose from duplication before speciation. Thus, to identify orthologous genes, the most straightforward approach is to obtain pairwise similarities for all the genes in genomes, and a pair of sequences with the highest degree of identity can be considered as orthologous genes.

In many cases, it is difficult to determine orthology unequivocally based solely on sequence identity; thus, other auxiliary information will be used. For example, one auxiliary method is based on gene order (Huynen and Bork, 1998). If the sequences neighboring the putative orthologs of two genomes are also orthologous, then the pair of sequences with the highest degree of identity is most likely orthologous genes. Since little conservation of the gene order is observed when the divergence of their orthologous genes

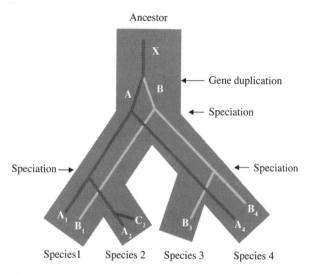

Figure 4.1 *Illustration of the concepts of orthologous and paralogous genes.* Genes A and B were derived from a duplication of gene X. During evolution, gene A retained the original function of gene X, whereas gene B acquired a slightly different function, as indicated by the color changes in the figure. The ancestor diverged into four species after several events of speciation. Notice that gene B was lost in species 2, while gene A was lost in species 3. Gene A was also duplicated again in species 2 after the speciation of species 2 from species 1, and gene C could have a different function from gene A. In this example, genes a_1, a_2, and a_4 and b_1, b_2, and b_3 are orthologous. Genes A, B, and C in various species are homologous, but they have paralogous relationships.

approaches the level of 50% identity of amino acid sequences (Huynen and Bork, 1998), this method is helpful in assessing orthology only for relatively recently speciated genomes.

An additional strategy for assessing orthology is to compare the pair of genes with those from a third genome (Huynen et al., 1997). If two genes from different genomes show the highest degree of identity to each other as well as to a single gene from a third genome, then this gene pair most likely corresponds to orthologs (Tatusov et al., 1997). For example, suppose that genes a, b, and c are homologous and from different genomes A, B, and C. If the pairwise similarity between these genes, a-b, a-c, and b-c, all have the highest sequence similarity among similar pairs from the corresponding genomes, then a, b, and c are considered to be orthologous genes.

Challenges to Identifying Orthology Due to complex evolutionary relationships, identifying orthology based on sequence identity is a great challenge. Application of the sequence identity approach to assessing orthology is complicated by the following processes. First, the identification of orthologs is hampered by sequence divergence. For orthologous genes with rapid evolution or from genomes that are distantly related such as Bacteria and Archaea, their sequences could be highly divergent. The sequence similarity between two orthologs could be very low and not higher than the similarity between nonhomologous sequences. As a result, the homologous sequences are rarely recognizable (Huynen et al., 1997).

The second process that restricts the identification of orthologous genes is gene duplication coupled with gene loss. If different paralogs of an ancient gene that was duplicated before speciation are lost in two genomes, the remaining homologous sequences will have the highest degree of sequence identity and hence will be considered orthologous although they are, in fact, paralogous (e.g., see genes A_1 and B_3 in Fig. 4.1). Also, it is very difficult to distinguish orthologous sequences from genes resulting from ancient duplications.

The third process that hampers the identification of orthologous genes is HGT. As mentioned before, nonorthologous genes can be transferred horizontally. The original orthologous genes in the recipient strain can be displaced by the transfer of unrelated or distantly nonorthologous genes. As a result, orthologs may be detectable or classified as paralogs (Huynen et al., 1997).

The fourth factor that complicates the identification of orthologous genes is the modular structure in multidomain proteins. Multidomain proteins may have orthology at the level of single domains but not at the level of the whole protein. Thus, it is possible that nonorthologous proteins may have orthologous domains. As a consequence, it will be difficult to assess the orthology of such proteins.

Although orthology was originally defined as a one-to-one relationship between proteins, in practice, one-to-many and many-to-many orthologous relationships exist because of the complex nature of evolutionary processes. There is no single, simple, and perfect approach to assessing orthology. Determining orthology is dependent on the methods used, the genomes compared, and the questions asked, and hence orthology is generally methodologically defined (Huynen et al., 1997).

Databases for Classifying Orthologous Genes Appropriately identifying orthologous relationships between genes among different genomes is critical to functional and evolutionary studies. Thus, the database, Clusters of Orthologous Groups (COGs) of proteins, has been designed for classifying genes from completely sequenced genomes (Tatusov et al., 2001; also available at http://www.ncbi.nlm.nih.gov/COG). If any group of at least three genes from distant genomes are more similar to each other than they are to any other proteins from the same genome, this group of genes is considered to belong to an orthologous family (Tatusov et al., 2001). The COGs database contains information that reflects one-to-one, one-to-many, and many-to-many orthologous relationships and will be useful for predicting functions of individual genes or gene sets and studying their evolutionary relationships.

4.3   GENOME PERSPECTIVES ON MOLECULAR CLOCK

4.3.1   Historical Overview

Notion of Molecular Clock Most changes in DNA sequences occur randomly by mutations, probably with more or less equal probability in most of the regions of the genome. Genetic mutations that produce deleterious effects on the organism will be eliminated by natural selection. Genetic mutations that have no effect on protein functions or that are advantageous to the organism will inevitably accumulate over time. After two species diverge, their shared DNA sequences will also diverge, and the genetic divergence of the two species will increase as time passes. Therefore, the sequence divergence rates (i.e., molecular evolution rates) of genes or proteins can be used to reconstruct evolutionary relationships of living species.

Amino acid sequence-based comparative studies in the early 1960s observed that the number of amino acid differences in a given protein appeared to be proportional to the time elapsed since the divergence of the organisms being compared (Zuckerkandl and Pauling, 1962, 1965). Such early empirical observations led to the notion of a molecular clock proposed by Zuckerkandal and Pauling (1965), which assumes that the rate of amino acid or nucleotide substitution is roughly constant among diverse lineages. It is also assumed that the underlying molecular mechanism for such rate constancy is that many amino acid and nucleotide sequence substitutions may have little or no effect on protein functions, and most of the substitutions that occur during evolution are this type of neutral mutation.

Theoretical Basis of Molecular Clock The neutral theory of molecular evolution provides a mathematical description of the molecular clock hypothesis. This theory assumes that new alleles in a species result from mutations. If alternative alleles are neutral with respect to natural selection, their frequencies will change only by random genetic drift (Kimura, 1983). Therefore, rates of allelic substitution will be stochastically constant. Based on this theory, it is expected that a universally valid and exact molecular clock would exist under the conditions that, for a given molecule, the mutation rates for neutral alleles are exactly equal among all organisms at all times (Kimura, 1983).

Under this theory, the difference in evolution rates among genes and proteins is due to the different magnitudes of selective constraints. Since many proteins differ in their functional constraints, there would be many molecular clocks that tick at different rates but they will time the same evolutionary events (Ayala, 1997). While the molecular clocks that tick at rapid rates are useful for studying the evolutionary relationships of recently diverged organisms, the molecular clocks that evolve slowly are appropriate for investigating ancient evolutionary events.

It should be noted that the molecular clock postulated by the neutral theory is not a metronomic clock that measures time exactly. The molecular clock is stochastic and behaves like a Poisson process. Although some variation occurs, the probability of the rate change is constant. A stochastic clock will be fairly accurate over appropriately long periods (Ayala, 1997). The molecular clock hypothesis has prompted much research and yielded important results ever since its inception and has also been controversial for several decades (Britten, 1986; Easteal et al., 1995; Li et al., 1996; Ayala, 1997; Nei et al., 2001). The property of rate constancy has been used extensively to reconstruct phylogenies, particularly when the fossil records are missing or incomplete (Ochman and Wilson, 1987; Wilson et al., 1987). However, it is apparent that the clock ticked differently not only among different lineages, genes or proteins, but also among different parts or sites of the genes or proteins (Britten, 1986; Wu and Li, 1985; Bousquet et al., 1992; Martin and Palumbi, 1993; Ohta, 1995; Li, 1993; Ayala, 1997; Gibbs et al., 1998), despite some empirical studies, which observed that proteins (genes) or parts of the proteins evolved constantly among some lineages.

Based on the neutral theory, a ratio of the variance to the mean of the Poisson process equal to 1 is expected. The results from many empirical tests showed that such a ratio is significantly greater than 1 for half of the proteins examined (Gillespie, 1991). As a result, several alternative models of the neutral theory were proposed to account for the excess variance of the molecular clock (Ayala, 1997). For instance, many mutations may have very slightly deleterious rather than strictly neutral effects on protein functions (Ohta, 1987) and DNA polymerases could differ in their ability to correct errors, and hence the mutation rate would change (Gillespie, 1991). In addition, the generation time is very different among different organisms; the larger the number of replication cycles, the greater the number of errors that will occur, and thus the molecular clock in organisms with short generation times will tick faster than that in organisms with long generation times (Ayala, 1997).

4.3.2 Current Genomic View on Molecular Evolutionary Clock

No Evidence for a Universal Molecular Clock   Many factors that determine substitution rates differ dramatically among different types of organisms such as bacteria, plants, and animals (Gibbs et al., 1998). One factor is that the fidelity of DNA replication could be different among different species due to differences in the effectiveness of DNA repair. Population size and generation time are also dramatically different among different species. A population with a short generation time will need less time to fix new mutations, whereas a large population will need more time to fix new mutations. In addition, the function and functional constraints may change over evolutionary time. Therefore, strictly speaking, no gene or protein would evolve constantly for a long evolutionary time among all lineages (Nei et al., 2001).

Evidence suggests that some genes and proteins evolve at approximately equal rates across animal taxa. However, no theoretical basis exists for equal substitution rates in prokaryotes and eukaryotes, or even among all bacterial lineages (Ochman et al., 1999). With the exploration of whole-genome sequences, many comparative studies support the early conclusion that a global constant molecular clock does not exist (Nei et al., 2001; Ochman et al., 1999; Grishin et al., 2000).

Although it appears that no universal molecular clock exists, some results do suggest that certain categories of sequences evolved constantly among some lineages (Ochman et al., 1999). For instance, similar substitution rates for synonymous sites and small subunit (SSU) rRNA genes were observed between *Buchnera* lineages and *Escherichia coli/ Salmonella enterica* lineages. This suggests that the absolute rates of some sequences could be constant among some bacterial lineages.

Possible Constant Mean Evolution Rates of a Group of Genes   Despite mounting theoretical and empirical evidence against the universal rates of sequence evolution among bacterial groups and other organisms, it is still commonly assumed that molecular evolution is sufficiently regular over time and across lineages, and thus it is assumed that a molecular clock can be useful for testing phylogenetic hypotheses or timing evolutionary events (Ayala, 1997). Although the rates of sequence divergence vary across time and among different lineages for individual genes, it is believed that the mean rates of sequence divergence for a group of genes may be constant or nearly constant (Doolittle et al., 1996; Wray et al., 1996). Although many factors affect the sequence-diverging processes, the aggregate process would be stochastic. Therefore, with a large data set, the anomalies should cancel out, and the mean rates of sequence divergence would be more homogeneous (Doolittle et al., 1996). Such mean rates of sequence divergence for a group of genes can be used to reconstruct evolutionary history (Doolittle et al., 1996; Wray et al., 1996). With the availability of complete genome sequences for many organisms, it is possible to test hypotheses of mean rate constancy. The distributions of substitution rates among proteins encoded in 19 complete genomes, including different prokaryotes, a yeast and worm, were analyzed recently by Grishin and coworkers (2000). The interprotein substitution rate distributions inferred from the genome-to-genome comparisons were similar to each other, suggesting that a generalized version of molecular clock hypothesis may be valid on a genomic scale.

Due to complicated evolutionary processes and different structural and functional constraints, great caution should be taken when drawing conclusions from molecular clocks (Ayala, 1997). First, the results from molecular clock studies are generally not decisive and thus should be used cautiously and weighed against other available evidence. Molecular clocks from several genes should be used together whenever feasible, in particular when molecular clocks are used to infer important evolutionary events (Wray et al., 1996). This is because statistical and other biases of evolutionary rates will tend to cancel each other as the number of genes or proteins increases. In addition, synonymous but not nonsynonymous nucleotide substitutions should be used to examine the molecular clock whenever possible. However, synonymous substitution cannot be used to infer evolutionary events over long time periods due to mutation saturation (Ayala, 1997).

4.3.3 Timing Genome Divergence

Approaches to Calibrating Molecular Evolution Rates in Prokaryotes To date evolutionary events using molecular data, the molecular chronometers must be calibrated using the divergence times for particular lineages. The calibration of molecular clocks in animals and plants is generally easier because of the relatively reliable fossil records. The estimation of reliable calibration points for dating the branching events in bacteria is extremely difficult because of the lack of reliable fossil records for microorganisms. The following three approaches can be used to calibrate molecular chronometers in prokaryotes:

(i) *Calibration Based on Ecological Events at Known Times.* One approach to calibrating evolution rates is to associate origins of bacterial taxa with ecological events that occurred at known times in the geological past and assign upper and lower boundaries to the age of each lineage. Ochman and Wilson (1987) used 10 calibration points, including oxygen appearance, diversification of *Cyanobacteria*, appearance of photosynthetic eukaryotes, and appearance of land plants and mammals to construct a time scale for bacterial evolution. Based on these calibration dates, a relationship between time and bacterial RNA divergence was obtained and used to estimate divergence times between other branches. The absolute rates of SSU rRNA genes ($\sim 1\%$ per 50 million years) were fairly uniform across bacterial species and very similar to those observed in eukaryotes. Based on this absolute rate, it was estimated that *E. coli* and *S. enterica* serovar *Typhimurium* diverged between 120 million and 160 million years ago.

(ii) *Calibration Based on Host Fossil Record.* The aphid endosymbionts, *Buchnera* in the *Proteobacteria*, live within specialized cells of aphids and are essential to the growth and reproduction of their host (Moran et al., 1993). The endosymbionts are inherited cytoplasmically by aphid offspring. It is estimated that the symbiotic relationship between aphid and *Buchnera* has been maintained for over 100 million years. During this time period, the bacterial and aphid lineages were cospecialized. Thus, the absolute rates of evolution of *Buchnera* lineages can be calibrated with the fossil-based dates of divergence of their aphid hosts. With this approach, it was estimated that the rate of SSU rRNA genes was about 1 to 2% per 50 million years in *Buchnera*, which is nearly twice that in their free-living relatives. The higher evolutionary rates in these endosymbionts are presumably due to their small population sizes (Ochman et al., 1999).

(iii) *Calibration Based on Eukaryotic Molecular Clock.* If a universal molecule diverges at similar rates in all living organisms, the rates of evolution in bacteria can be estimated based on the evolution rates of eukaryotic organisms with reliable fossil records (Ochman et al., 1999; Doolittle et al., 1996). Doolittle and colleagues (1996) estimated the evolutionary distances for different lineages within and among the major kingdoms based on the amino acid sequences of 57 enzymes from 15 different biological groups. The distance data were then calibrated based on the known divergence times drawn from the fossil record of vertebrates to obtain absolute evolution rates. Finally, the calibrated evolutionary rates were applied to all other distantly related groups to extrapolate their divergence times (Doolittle et al., 1996; Feng et al., 1997).

Assumptions for Estimating Genome Divergence Times   To estimate the species divergence time based on the evolutionary distances of protein sequences, it is generally required that (1) the proteins compared are of vertical descent; and (2) the proteins behave as a molecular clock among different lineages. Thus, it is critical to use genes or proteins that have evolved at a constant rate.

As we discussed previously, no genes or proteins would evolve at a constant rate across all lineages over a long period of evolutionary time. However, it is still possible to obtain rough estimates of species divergence times even if the evolutionary rate is not strictly constant (Nei et al., 2001). If a gene does not evolve constantly in some lineages, these lineages can be eliminated from estimating species divergence times (Nei et al., 2001). In addition, the accuracy of the estimated divergence times will increase as the number of genes used increases. The estimated divergence times are reasonably good if a large number of protein sequences are used (Kumar and Hedges, 1998; Feng et al., 1997; Nei et al., 2001).

Divergence Times of Archaea, Bacteria, and Eucarya   The estimation of the divergence times between Archaea, Bacteria, and Eucarya based on protein sequences has received great attention and also been the subject of controversy. Although numerous studies have been conducted to determine their divergence times (Doolittle et al., 1996; Ochman and Wilson, 1987; Wilson et al., 1987; Wray et al., 1996; Feng et al., 1997; Nei et al., 2001), the divergence times of major groups of organisms, especially for prokaryotes, have remained elusive. From the microfossil evidence, it is generally believed that eukaryotes and bacteria diverged 3.5 billion years ago (Schopf, 1993). Doolittle and coworkers (1996) showed that the divergence time between prokaryotes and eukaryotes was 2 billion years ago. With more sequences (64 enzyme sets) from several complete genomes and improved methods, Feng and colleagues (1997) reevaluated the divergence times of Archaea, Bacteria, and Eucarya. Six different, well-established divergence times based on the vertebrate fossil record were used to calibrate the evolutionary rates. The results suggested that Archaea and Bacteria might have diverged between 3 and 4 billion years ago.

Nei and coworkers (2001) obtained 104 putative orthologous sequences from the databases and calculated the evolutionary distances using an improved statistical method. The divergence time between Bacteria and Eucarya was estimated to be about 3 billion years, which is younger than the age of microfossils (Schopf, 1993). Such differences could be due to (1) large standard errors in the estimate; (2) horizontal gene transfer, which might have occurred between the ancestors of current bacteria and eukaryotes; and (3) the possibility that the microfossils fail to represent the ancestors of current bacteria and/or eukaryotes.

It should be noted that the estimated divergence times of different lineages also depend on the choice of calibration dates (Feng et al., 1997; Nei et al., 2001). Great caution is needed in estimating ancient divergence times. Using recent calibration dates to estimate ancient divergence times is more error-prone than to estimate recent divergence times due to complicated evolutionary processes and methodological difficulty in obtaining appropriate sequence alignments (Nei et al., 2001). Using multiple calibration points will help increase estimation accuracy, and it may be best to select several calibration points from both within and outside the groups examined (Lee, 1999).

It is believed that the availability of complete genome sequences from many different organisms makes it possible to obtain more reliable phylogenetic trees and estimates of divergence times. Once reliable phylogenetic trees are available, a reasonably good evolutionary history of different organisms can be reconstructed at the molecular level. By comparing the molecular-based evolutionary history with paleontological data, more reliable estimates of genome divergence times among different lineages and a united view of the tree of life can be developed (Nei et al., 2001).

4.4 GENOME PERSPECTIVES ON HORIZONTAL GENE TRANSFER

4.4.1 Historical Overview of Horizontal Gene Transfer

Horizontal gene transfer (HGT) or lateral gene transfer is an evolutionary phenomenon that involves the occurrence of genetic exchanges between different evolutionary lineages. The existence of HGT was established many years ago. As early as 1944, Avery and colleagues (1944) demonstrated that DNA can be absorbed by microorganisms from environments. However, the significance of HGT for bacterial evolution was not recognized until the 1950s, when worldwide antibiotic resistance patterns emerged.

Although the existence of HGT was convincingly established, the extent of HGT has been the subject of a lengthy debate (Syvanen, 1994; Koonin et al., 2001). This is mainly because HGT challenges the traditional tree-based view of the evolution of life, and it is very difficult to prove without ambiguity (Pennisi, 1998, 1999; Koonin et al., 2001). Because of sequence limitation, early studies attempting to understand the extent of horizontally transferred sequences in a genome were restricted to the very few microorganisms for which sufficient sequence information was available from individual genes. With this approach, 10 to 16% of the *E. coli* chromosome was estimated to have resulted from HGT (Lawrence and Ochman, 1997; Medigue et al., 1991). This clearly suggested that HGTs could not be dismissed as rare, insignificant evolutionary events.

The availability of whole genomic sequences from many diverse groups of bacteria provides great insight into microbial genome evolution. Comparative genome sequence analysis indicated that dramatic differences in genome sequences were observed among bacteria within the same lineages, such as between *E. coli* and *Haemophilus* (Tatusov et al., 1996) and between *E. coli* 0157:H7 and *E. coli* K12 (Perna et al., 2001). Such differences could not be explained by vertical descent. Although differential gene loss can explain such genome differences to some extent, HGT appears to be another major evolutionary process that shapes genome diversity, structure, and evolution (Koonin et al., 2001). In addition, massive HGT apparently occurred between archaeal and bacterial genomes (Nelson et al., 1999; Eisen, 2000). For instance, significant portions of *Thermotoga maritima* and *Aquifex aeolicus* genomes are of archaeal origins. Although comparative genomic studies indicate that HGT is a common evolutionary event, whether it is a major evolutionary force is still highly debatable (Kyrpides and Olsen, 1999; Logsdon and Faguy, 1999), because HGT may change the general view of the evolution of life, and there are some uncertainties in determining whether HGT is associated with phylogenetic tree topologies, unequal evolution rates in different lineages, and complicated life evolutionary processes. In the following sections, we will briefly present approaches to identifying HGT and the mechanisms involved in HGT, and then summarize the most recent findings on HGT from a genomics perspective.

4.4.2 Identification of HGT

Generally, it is very difficult to identify and prove HGT, because the evidence given to support HGT can be explained by other mechanisms. There are two general approaches to detecting HGT processes. One strategy is to experimentally determine the gene transfer process by directly observing the transferred genetic materials from a donor strain to a recipient strain. Such studies are more convincing and straightforward for establishing HGT events. However, historical HGT events (i.e., genetic transfers that have already occurred) rarely can be detected by the direct experimental approach, and evidence for their occurrence must derive from other sources. The other approach is to infer historical HGT processes from sequence analyses and comparisons across different organisms as described below.

Phylogenetic Analysis One of the most effective and widely used approaches to identifying HGT processes is to generate, and then compare the phylogenetic trees for every gene in every genome (Eisen, 2000; Huynen and Bork, 1998; Ochman et al., 2000; Nelson et al., 1999). The underlying bases for this approach are the following. First, DNA fragments introduced through HGT are limited to the descendents of the recipient strains and absent from closely related taxa, and thus a scattered phylogenetic distribution will be observed. Second, the transferred DNA fragments will be restricted to certain species and such species-specific regions may display unusually high levels of DNA or protein similarities to genes from very distantly related taxa inferred by other criteria. For example, if a bacterial gene is clustered in a well-supported tree with eukaryotic homologs rather than with the homologs from other bacteria, HGT from eukaryotes to bacteria most likely occurred (Koonin et al., 2001). However, phylogenetic analysis usually does not generate such clear-cut solutions in most suspected cases of HGT (Koonin et al., 2001).

One of the problems with this method is that it is generally difficult to distinguish orthologous sequences, HGT, and ancient gene duplications (Huynen and Bork, 1998). Also, the reconstruction of robust phylogenetic trees is sometimes difficult, because phylogenetic methods are prone to a variety of artifacts such as convergence and long-branch attraction (Moreira and Philippe, 2000; Koonin et al., 2001). Phylogenetic tree topology is a good indicator of probable evolutionary processes only when the critical nodes of the tree are statistically robust as indicated by bootstrap analysis (Brown, 1994; Efron et al., 1996). In many cases of HGT, the tree topology is not strongly supported statistically and thus phylogenetic analysis is useless for verifying the putative HGT events.

In addition, phylogenetic analysis is very time-consuming and labor-intensive and critically depends on correct sequence alignments. It is generally difficult, however, to obtain correct alignments when the compared sequences are quite divergent. Thus, it is difficult to automate phylogenetic analysis without compromising data quality.

Best Sequence Match Detection An alternative approach to generating phylogenetic trees for all genes is to determine the sequence similarity of genes between genomes (Eisen, 2000; Koonin et al., 2001). HGT events could have occurred when a gene sequence from a particular organism displays the strongest similarity to a homolog from distant taxa. This approach is generally accomplished by employing the following steps. First, all the protein sequences from a bacterial genome are compared with the entire protein database using a sequence search program such as BLAST. Then, the best-match detected homologs are classified based on their taxonomic origin, and the percentages of detected homologs are calculated for each taxon. By comparing the taxonomic distribution of the detected homologs, putative gene transfer events can be assigned. For instance, if a certain fraction of the bacterial protein sequences displays the greatest similarity to eukaryotic homologs rather than to homologs from other bacteria, gene transfer from eukaryotes to bacteria could have occurred. This approach was used to establish the extensive HGTs between thermophilic Bacteria and Archaea. Over 20% of the genes from the bacterium *Thermotoga maritima* were most similar to the genes from Archaea rather than those from Bacteria (Nelson et al., 1999).

Although the best-match approach can rapidly provide information on possible HGT events, it is of very limited use, because sequence similarity is not necessarily correlated to evolutionary relationship. As an example, the high degree of sequence similarity between thermophilic Archaea and Bacteria can be explained by the rapid evolutionary rates of the genes or gene loss in mesophilic bacteria. Generally, the genes in mesophiles evolved faster than those in thermophiles. If the genes in mesophiles evolved sufficiently faster than those in thermophilic Archaea and Bacteria, the genes in thermophilic bacteria will show less similarity to those in mesophiles than in Archaea. If the genes in mesophiles are lost, the genes from thermophilic Bacteria will definitely show higher similarity to those of Archaea. Thus, evidence derived from this approach should be considered preliminary, and this method must be used together with other approaches for detecting HGT.

Nucleotide Compositional Analysis Bacterial species exhibit a wide range of variation in their overall GC content, but genes within a genome are fairly similar in terms of base composition, codon usage patterns, and di- and tri-nucleotide frequencies (Karlin et al., 1998). As a result, DNA fragments obtained through HGT will retain sequence characteristics of the donor genome, and thus, they can be used to detect possible HGT events (Lawrence and Ochman, 1997, 1998). Species relationships can be determined by clustering the relative abundance of di-nucleotides and codon usage patterns (Karlin et al., 1998).

The advantage of this approach is that it only requires genome sequence from one species. However, unusual base compositions can be caused by other factors, such as selection and mutational bias, rather than by HGT (Eisen, 2000). In addition, HGT among similar species cannot be detected and insights into the evolutionary history of the genes cannot be obtained with this approach.

Gene Organizational Analysis   Gene order is rarely conserved in Bacteria and Archaea, because extensive gene shuffling occurred during their evolution (Huynen and Bork, 1998). Studies showed that it is extremely unlikely that three or more genes exist in the same order in distant genomes unless these genes form an operon (Wolf et al., 2001). Also, it appears that each operon emerges only once during its evolution and is maintained by natural selection (Lawrence, 1997, 1999). Therefore, if a gene order in distant evolutionary lineages is the same and it is not observed in other more closely related lineages, the horizontal transfer of these genes could have occurred. This approach has been successfully used to detect HGT in *Methanobacterium thermoautotrophium* and *Synechocystis* (Huynen and Bork, 1998).

Sequence Vestige Detection   In addition to information on the sequences and organization of the genes themselves, the regions adjacent to the putative horizontally transferred genes generally contain vestiges of the sequences involved in gene transfer (Ochman et al., 2000). The sequence vestiges include remnants of transposable elements, transfer origins of plasmids, or known attachment sites of phage integrases. The occurrence of such sequences further attests to putative HGT events. HGT is a complicated historical process, and hence it is very difficult to prove. Thus, all indications for HGT should remain probabilistic. To maximize the likelihood of these events being identified correctly, all the methods discussed above should be used in a complementary manner to obtain multiple lines of evidence. Also, it should be noted that HGT between species also occurs via homologous recombination (Ochman et al., 2000). Due to a mismatch correction in the recipient genome, it is more difficult to detect HGT events through homologous sequence analysis. In addition, it is more challenging to unequivocally determine the donor and recipient organisms in each case (Koonin et al., 2001). It is generally assumed that the taxon is the most likely source if all or the majority of its members have the probable genes, whereas the distant taxon is most likely the recipient if only a few of its members have the probable transferred genes. However, in many common cases of HGT, the direction cannot be determined with any confidence (Koonin et al., 2001).

4.4.3   Mechanisms Underlying HGT

The acquisition of new functional properties through HGT requires the delivery of donor DNA into a recipient cell. This transfer is accomplished through three mechanisms: transformation, transduction, and conjugation.

Transformation   Transformation is the process whereby competent bacterial cells take up naked DNA from the environment. Certain naturally competent bacterial species, such as *Neisseria gonorrhoeae* and *H. influenzae*, can accept foreign DNA throughout their entire life cycle, whereas other organisms, such as *Bacillus subtilis* and *Streptococcus pneumoniae*, are competent only during certain physiological stages. With transformation, DNA can be potentially exchanged between distantly related species.

Transduction   Transduction is another mechanism by which new genetic material is introduced into a bacterium. This process is mediated by a bacteriophage that has replicated within a microorganism different from the recipient organism and has packaged random DNA fragments of the donor chromosome (generalized transduction) or the DNA sequences adjacent to the phage attachment site (specialized transduction) (Ochman et al., 2000). Those bacteria capable of being transduced express cell surface receptors that are recognized by the bacteriophage. Phages are widely distributed in the environment, but the range of the microorganisms that can be transduced is determined by the cell surface receptors recognized by the bacteriophage.

Conjugation HGT can also occur via conjugation, which involves physical contact between the donor and recipient cells. The conjugative transfer of DNA from a donor to a recipient strain is generally mediated by a self-transmissible or mobilizable plasmid (Ochman et al., 2000). Conjugation can transfer DNA fragments between very distantly related bacteria such as Bacterial, Archaeal, and Eucaryal domains.

Once new DNA is introduced into a recipient cell's cytoplasm by one of the aforementioned mechanisms, a successful gene transfer will occur if the introduced foreign DNA persists as an episome or is integrated into the bacterial genome by other processes, including homologous recombination, integration via bacteriophage integrases, or illegitimate recombination by mobile element transposases. Recent studies based on whole-genome sequence comparison imply that virtually any sequence can be transferred among distantly related microorganisms by these mechanisms. However, since the bacterial genome is relatively small, it appears that the rates of gene transfer and/or the rate of maintenance of the introduced new sequences are very low (Ochman et al., 2000). For *E. coli*, it was estimated that the rate of successfully introduced DNA fragments is about 16 kb per million years (Lawrence and Ochman, 1998).

4.4.4 Types of Genes Subjected to HGT

Complexity Hypothesis Genome sequence comparison reveals two functionally distinct gene clusters: informational genes and operational genes (Rivera et al., 1998). The former includes genes with cellular functions involved in information processing such as transcription, translation, replication, the homologs of ATPases and GTPases, and most tRNA synthetases. The operational genes function in basic cell maintenance processes, for example, amino acid synthesis, biosynthesis of cofactors, cell envelope proteins, energy metabolism, intermediary metabolism, fatty acid and phospholipid biosynthesis, nucleotide synthesis, and gene regulation. Generally, informational genes, particularly the translational and transcriptional apparatuses, are large, complex systems with complicated interactions, whereas most operational genes are members of small assemblies of a few gene products.

Phylogenetic analyses of many genes indicate that the informational genes interacting with many other genes are less prone to horizontal transfer than operational genes, which have fewer interactions. This leads to the proposal of the complexity hypothesis, that is, the complexity of the gene interactions is a significant factor restricting successful HGT, because the products of the genes with complex interactions will function less easily in a foreign cytoplasm (Jain et al., 1999).

It is certainly reasonable to expect that interactiveness will affect exchangeability, but the situation could be more complicated. For example, genes with fewer interactions could more easily acquire idiosyncratic structure through coevolution with their partners. Such structure could be incompatible with integration into the homologous cellular complex in a distant species (Doolittle, 2000). In contrast, proteins or RNAs with complex interactions might change very little in structure, and thus, they may be better for retaining functional capability in foreign cellular environments. It is also possible that the complexity of interaction is not critical for the HGT process. For instance, many informational genes are required for cell survival and hence the chance to be replaced with a homologous gene is less compared to operational genes.

Traits and Genes Acquired via HGT

Antibiotic Resistance Genes   Antibiotic resistance genes are generally associated with highly mobile genetic elements such as plasmids and transposable elements and thus the antibiotic resistance genes are readily transferred among different taxa (Ochman et al., 2000). Antibiotic resistance can also be transferred by integrons (i.e., gene expression elements containing promotorless genes) through changing promoterless nonfunctional genes to functional genes (Ochman et al., 2000). The transfer of antibiotic resistance genes is a very common phenomenon, because the transferred antibiotic resistance genes allow microorganisms to expand their ecological niches.

Virulent Factors   HGT appears to play a major role in acquiring pathogenicity islands in many pathogenic bacteria (Hacker et al., 1997). Pathogenicity islands are chromosomally encoded regions that generally contain large clusters of virulence genes and can be incorporated into a benign organism, which then becomes a pathogen (see Chapter 12). Their flanking sequences generally contain short direct repeats originating from mobile genetic elements, and their open reading frames within the islands have homology to bacteriophage integrases. Thus, the pathogenicity islands can be transferred by phage and other mobile genetic elements (Cheetham and Katz, 1995; Lindsay et al., 1998; Karaolis et al., 1999; Ochman et al., 2000).

Metabolic Traits   Acquiring new metabolic traits through HGT can allow a microorganism to rapidly explore a new environment. It appears that HGT has made significant contributions to the metabolic diversification of prokaryotes by mobilizing metabolic traits (Ochman et al., 2000). For instance, *E. coli* obtained the ability to use the milk sugar lactose by acquiring the *lac* operon, which enabled the bacterium to expand to a new niche, the mammalian colon.

Informational Genes   Based on the complexity hypothesis, informational genes, such as those involved in the translation machinery, are less prone to HGT. However, several exceptions are observed based on whole-genome sequence analysis. For instance, phylogenetic analysis revealed several probable HGT events for the ribosomal protein RpS14, which is required for assembling ribosomal 30S subunits and involved in the central process of peptide elongation (Brochier et al., 2000). Aminoacyl-tRNA synthetases are central components of the translation machinery. Phylogenetic analysis also indicated that several gene transfers from eukaryotes to bacteria occurred for various aminoacyl-tRNA synthetase genes (Wolf et al., 1999; Koonin et al., 2001). These observations appear to contradict the complexity hypothesis (Doolittle, 1999b).

4.4.5 Classification and Scope of HGT

Classification of HGT HGTs can be divided into at least three groups based on the relationships of the horizontally acquired genes and the homologous genes in the recipient strains, and the fates and effects of the transferred genes (Koonin et al., 2001).

(i) *Acquisition of a Novel Gene.* A gene is transferred from a donor organism to a recipient organism that does not have the gene(s) homologous to the transferred genes. In this case, new capabilities could be acquired by the recipient strain. Using the best-match approach, Koonin and coworkers (2001) analyzed HGT events with proteins from 9 archaeal and 22 bacterial complete genomes. The percentages of acquisition of new genes were estimated to be 0.2 to 7.2% when archaea or bacteria were used as a reference taxon for comparison. The fractions of acquisition of new genes varied substantially within the range of 0.4 to 19.8% among different bacterial lineages when different bacterial groups were used as reference taxa for comparison.

(ii) *Acquisition of a Paralog.* A gene is transferred from a donor to a recipient lineage that has gene(s) homologous to the transferred genes. The transferred genes have similar but not identical functions. The homologous genes in the recipient strains can be replaced by the acquired paralogous gene through a phenomenon termed nonorthologous gene replacement. In this replacement, a gene coding for a protein responsible for a particular function is replaced with a nonorthologous (distantly related or unrelated) but functionally analogous gene. Phylogenetic analysis of the candidate horizontally transferred genes from *Methanobacterium thermoautotrophicum*, *Aquifex aeolicus*, *Haemophilus influenzae*, and *Vibrio cholerae* indicated that the fraction of paralog acquisition is less than 1% with strong statistical confidence (Koonin et al., 2001). This number might be an underestimate, because the category of the majority of the putative transferred genes could not be confidently assigned due to the complexity of the tree topology.

(iii) *Acquisition of a Phylogenetically Distant Ortholog.* A gene is transferred from a donor to a recipient lineage that has orthologous homologs to the transferred genes, but they are phylogenetically distantly related. The ancestral orthologous gene in the recipient strain can be replaced by the acquired distantly related orthologous gene, thus eliminating the ancestral gene from the recipient strain. This phenomenon is referred to as xenologous gene replacement (Gogarten, 1994; Koonin et al., 2001). Phylogenetic analysis of four genomes (*M. thermoautotrophicum*, *A. aeolicus*, *H. influenzae*, and *V. cholerae*) indicated that the proportion of acquisition of phylogenetically distantly related orthologs varied considerably among them, within the range of 0.4 to 2% (Koonin et al., 2001). Significant differences in the relative contributions of xenologous gene replacement and paralog acquisition were also observed among these genomes. While xenologous gene replacement appears to be prevalent in the hyperthermophile *A. aeolicus*, paralog acquisition occurred more frequently in the pathogen *V. cholerae*.

Scope of HGT The availability of whole genomic sequences from many diverse groups of bacteria provides a great opportunity to measure and compare HGT. A wide range of variation in HGTs was observed among different evolutionary lineages.

Interdomain Gene Transfers in Prokaryotic Genomes   Comparisons of completely sequenced genomes confirm that a large number of genes were transferred between Bacteria and Archaea. The hyperthermophilic bacteria *Thermotoga maritima* and *Aquifex aeolicus*, for example, each have a large number of genes that are most similar to their homologs in thermophilic archaea. About 25% of the 1,877 ORFs from *T. maritima* and 16% of the 1,512 ORFs from *A. aeolicus* have the highest similarity to archaeal proteins (Nelson et al., 1999; Aravind et al., 1998). In contrast, the proportions of the genes that are most similar to archaeal genes are much lower in mesophiles, such as *E. coli, Bacillus subtilis*, and *Deinococcus radiodurans*.

HGT from Eukaryotes to Prokaryotes   The horizontal transfer of genes from eukaryotes to prokaryotes is potentially of particular interest, because it may play an important role in bacterial pathogenicity (Ochman et al., 2000; see Chapter 12). It was estimated that the proportion of these types of HGTs is relatively small (less than 1%) based on the best-match approach (Koonin et al., 2001). It should be noted that such numbers could be an underestimate, because some transferred eukaryotic genes may not show highly significant similarity to their eukaryotic ancestors. Also, the acquisition of eukaryotic genes in pathogens appears to be related to their host. For instance, an apparent excess of animal genes and plant genes was observed in the animal pathogen *Pseudomonas aeruginosa* and in the plant pathogen *Xylella fastidiosa*, repsectively.

HGT in Different Bacterial Lineages   Sequenced genomes from different bacterial lineages show a wide range of variation in HGT events. A wide range of variations of HGTs were observed among different bacterial lineages (Koonin et al., 2001; Ochman et al., 2000). There appeared to be a correlation between genome size and the number of horizontally transferred genes present in a genome. For instance, virtually no or little HGT occurred in bacterial genomes of small size, such as those in *Rickettsia prowazekii, Borrelia burgdorferi*, and *Mycoplasma genitalium*, whereas up to 18% of genes in *E. coli* and *Synechocystis* PCC6803 appeared to have arisen from HGTs. However, the proportion of putative transferred genes is strikingly high in *Treponema pallidum* ($\sim$33%) and *Borrelia burgdorferi* ($\sim$30%), although their genome sizes are small (1.14 and 1.44 Mb). The proportion of the transferred genes could be overestimated due to differential gene loss in the two spirochetes with different lifestyles (Subramanian et al., 2000; Koonin et al., 2001). Also, in some bacterial species, a substantial fraction of horizontally transferred genes can be attributed to plasmid-, phage-, or transposon-related sequences (Ochman et al., 2000). A substantial proportion of acquired DNA in *E. coli, Synechocystis* PCC6803, and *Helicobacter pylori* is physically associated with mobile DNA elements, which probably served as a vehicle for integrating the transferred genes into bacterial chromosomes. In addition, sequence analysis indicated that many of the small genetic elements (e.g., small chromosomes, megaplasmids, and plasmids) in bacteria originated separately from the main chromosomes (Eisen, 2000). The small genetic elements in *D. radiodurans, V. cholerae*, and *X. fastidiosa* could have arisen through HGT, although the source of these elements is not clear.

Gene Transfers from Bacteria to Eukaryotes   HGTs from bacteria to eukaryotes are well documented, but in most cases, they involved gene transfers from organellar genomes rather than from nuclear genomes (Martin, 1999). Recently, analysis of the rough draft of the human genome sequence suggested that 223 bacterial genes were horizontally transferred into the human genome during the period of vertebrate evolution (International Human Genome Sequencing Consortium, 2001). However, whether these genes were directly transferred from bacteria to vertebrates remains uncertain. After careful reexamination of the human genome sequence along with the genome sequences available from all other organisms, about 40 genes were found to be shared exclusively by humans and bacteria; furthermore, these genes might have originated from an HGT event (Salzberg et al., 2001). However, gene loss combined with sample size effects and evolutionary rate variation provides an alternative and more biologically plausible explanation for the existence of genes shared by humans and prokaryotes but missing in nonvertebrates (Salzberg et al., 2001). In another similar study, Stanhope and colleagues (2001) found that 28 of the most putative HGT genes appear to be derived from ancient eukaryotes, and thus they cannot serve as examples of direct HGT from bacteria to vertebrates.

Apoptosis (i.e., programmed cell death) is one of the central cellular processes in multicellular eukaryotes, and it is important in development, stress responses, aging, and disease. By analyzing whole-genome sequences from humans and other organisms, a considerable portion of apoptotic protein domains in human was found in *Actinomycetes* and *Cyanobacteria*, suggesting that HGT could play a major role in the early evolution of apoptosis. There is no doubt that HGT has played an important role in microbial evolution. However, it should be noted that most of the proposed HGT involves only one type of evidence, such as unexpected sequence similarity and unusual sequence composition. Since other factors may also cause unusual sequence similarity and composition, other lines of evidence are needed to support the probable HGTs. Also, in most cases, nothing is known about the donor lineage and possible occurrence time of the probable gene transfers. Thus, all indications for HGT are putative and great caution needs to be taken when claiming a horizontal gene transfer event.

4.4.6 Evolutionary Impact of HGT

Impact of Acquired DNA on Recipient Strains   The potential advantages that HGT offers to the recipient strain are as follows (Doolittle, 2000). First, the newly acquired genes may impart novel biosynthetic or degradative capabilities on the recipient strain. These capabilities may allow recipient strains to exploit new ecological niches. The acquired genes may also provide a recipient with resistance to antibiotics or other toxic agents, thus permitting the recipient strain to expand its ecological niche and proliferate in the presence of toxic compounds. In addition, the new genes may encode a protein whose kinetic, biochemical, or biophysical characteristics enable the organism to better adapt to changing environmental conditions and thus enhance the fitness of recipient strains in competitive environments.

Although HGT offers several advantages, it also brings to the recipient organism several problems (Ochman et al., 2000). First, in order to take full advantage of the acquired genes, the recipient must appropriately regulate the expression of the transferred genes in coordination with the rest of the genome. Second, the bacterial genome size is relatively small and constant. Thus, the recipient strains must eliminate more expendable sequences to accommodate the transferred genes. Not all transferred DNA sequences are useful to the recipient strain, so the recipient must therefore delete the bulk of acquired useless DNA. Finally, this process requires substantial intragenome recombination.

Impact on Genomic Operon Structure   One of the key organizational differences that distinguish prokaryotic genomes from those of eukaryotes is that several genes which collectively provide for a single metabolic function are typically cotranscribed from a single promoter. Such gene structure is referred to as an operon. The obvious advantage of clustering genes into operons is that the genes encoding a single metabolic function are coordinately expressed and regulated, and the coordinated regulation will provide selective benefits to the individual (Lawrence, 1997). One interesting question is how the operon structure is formed. Operon organization could be formed through genomic rearrangement processes such as deletions, inversions, and translocations. To form an operon, previously unlinked genes must be precisely juxtaposed to allow cotranscription. However, genomic rearrangement has some potential difficulties in forming such precise operon structure. Although deleting the intervening DNA sequences is the easiest way to juxtapose two genes, it is impossible when the intervening DNA encodes essential functions. While inversions and translocations can bring some genes closer together, they also disrupt existing gene clusters. HGT has been proposed as an important mechanism in the organization of operons (Lawrence, 1997). The selfish operon model postulates that HGT allows gradual and efficient formation of gene clusters, and that the physical proximity will enhance their fitness. Any gene that is not under selection in a recipient strain is subjected to deletion after a chromosomal segment is transferred. Since transferred chromosomal fragments are recent acquisitions, those intervening genes are not important to cell survival. Consequently, deleting such intervening genes would not be detrimental. Therefore, horizontally transferred DNA can permit the juxtaposition of genes encoding single metabolic functions in a gradual and stepwise manner, and eventually these genes will be tightly linked together as an operon and cotranscribed from a single promoter, which could be provided at the site of integration. In turn, the formed operon will facilitate gene transfer among different taxa.

Impact on Microbial Speciation   HGT is a key venue for bacterial diversification and evolution by resorting existing capabilities. Although mutational processes are important in gene divergence and evolution, the roles of HGT in bacterial speciation cannot be ignored, especially in developing novel capabilities to exploit novel resources. HGT can rapidly confer novel functions to the recipient strains, and thus it can enable the rapid, effective, and competitive exploration of new ecological niches. In this manner, HGT may provide sufficient ecological differentiation for speciation. In contrast, the novel functions arising from mutational processes are unlikely to allow rapid and effective exploration of new resources, because the mutational processes are slow and inefficient (Lawrence, 1997). For instance, it seems that no phenotype used to distinguish *E. coli* and *Salmonella* is derived from the differentiation of ancestral genes via point mutations. Conversely, all the described differences in phenotypes such as lactose, citrate and propanedial utilization, and indole production appear to have arisen from gene gain and loss processes.

Although HGT plays important roles in microbial speciation, such interspecific transfer and recombination also blur species boundaries. In this manner, HGT acts to counter the process of speciation. As a result, not any one gene can reflect the evolutionary history of an organism.

4.5 GENOMIC PERSPECTIVES ON GENE DUPLICATION, GENE LOSS, AND OTHER EVOLUTIONARY PROCESSES

4.5.1 Gene and Genome Duplication

Gene duplication is a major force in genome evolution. The evolutionary importance of gene duplications originally proposed by Ohno (1970) is now universally accepted (Wagner, 2001). It is believed that duplicated genes often experience relaxed evolutionary constraints after duplication. This promotes functional diversification and biochemical innovation of the duplicated genes through mutations and genetic recombination. It is also believed that duplicated genes with redundant functions can protect an organism against deleterious mutations.

Despite the obvious significance of gene and genome duplications, many central issues concerning the rate, fate, and evolutionary consequences of gene/genome duplications could not be adequately addressed previously due to limited genomic sequence information. The availability of whole-genome sequence information from many organisms provides a great opportunity to further our understanding of gene/genome duplication processes. In this section, we will briefly summarize the recent findings and genomic insights into gene/genome duplication processes.

Types of Gene Duplications There are three major types of duplication: (1) duplication of the entire genome; (2) duplication of a single chromosome or part of a chromosome; and (3) duplication of a single gene or group of genes. The number of genes will increase rapidly if the entire genome is duplicated. Genome duplication can occur as a result of an error during meiosis in which diploid gametes are produced instead of haploid gametes. The fusion of the two diploid gametes leads to the formation of an autopolyploid, which is a tetraploid cell whose nucleus contains four copies of each chromosome. Since each chromosome still has a homologous partner to form a bivalent during meiosis, autopolyploids are generally viable. Autopolyploidy is very common in plants but less common in animals and is an important mechanism in plant speciation.

A single chromosome or part of a chromosome can also be duplicated. Based on our current knowledge of modern organisms, duplication of a single chromosome can be lethal or result in genetic diseases, probably due to an imbalance in gene products and disruption of the cellular biochemistry. Thus, chromosome duplication is not the major cause of expansions in gene number during the evolution of life.

Duplications of genes and groups of genes occur frequently in all genomes. Genes and groups of genes can be duplicated through the processes of unequal crossing-over, unequal sister chromatid exchanges, DNA amplification, or replication slippage (Brown, 1999).

Detection of Genome Duplication Generally, the duplication of individual genes or groups of genes can be identified by sequence-based phylogenetic analysis. Many previous studies indicate that sequence-based phylogenetic analysis is very powerful and effective for identifying duplications of individual genes. However, whole-genome duplication is not discerned as easily by examining modern genomes, and it may be difficult to obtain evidence for whole-genome duplication. It is expected that many of the extra gene copies resulting from genome duplication became pseudogenes and are not recognizable in DNA sequences. Even if duplicated genes that are retained can be identified, it is difficult to distinguish whether they resulted from genome duplication or simply from the duplication of individual genes.

Two approaches can be used to identify genome duplications: the map- and tree-based approaches. The map-based approach to identify ancient genome duplication involves examining the gene order and chromosomal locations of duplicated genes. Duplicated sets of genes with the same order of genes and the same relative positions in the chromosome are good indicators of genome duplication (Wolfe and Shields, 1997; Wolfe, 2001). Phylogenetic trees can also be used to identify ancient genome duplication. Based on the molecular clock hypothesis, the date of a genome duplication can be calculated. If whole-genome duplication occurred, all duplicated gene pairs on duplicated chromosomal segments and all duplicated segments in a genome should have the same age (Wolfe, 2001).

Evolutionary Fates of Duplicated Genes   Duplications of genes, chromosomal segments, or entire genomes provide a primary source of raw material necessary for evolutionary transitions such as the evolution of multicellularity, bilateral symmetry, and the evolution of vertebrates (Taylor et al., 2001). Duplications do not directly lead to gene expansion and gene differentiation, because the initial outcome of the duplication event is that an organism has extra copies of different genes. The potential for gene expansion and differentiation exists because the extra copies of the genes are not essential for functioning of the cell, and hence, mutations can be accumulated without harming the viability of the organism. Theoretically, there are four possible evolutionary fates for the duplicated genes (Lynch and Conery, 2000). (1) *Nonfunctionalization*: One copy of each duplicated gene is suppressed, either by physical deletion or by accumulating point mutations until it becomes a pseudogene. (2) *Neo-functionalization*: Random mutations cause one copy of the duplicate genes to diverge functionally and the diverged copy may acquire a new beneficial function, while the other copy retains the original function. (3) *Sub-functionalization*: Random mutations may cause both copies to diverge functionally to the point at which their total functional capacity is reduced to the level of the single copy of ancestral genes. (4) *Equifunctionalization*: Both copies of the duplicated genes may persist in the genome with perfect or near-perfect sequence identity and cause a higher level of the gene product to be produced. The combinations and iterations of these processes could lead to multigene families containing up to hundreds of more or less similar genes with similar or divergent functions.

Gene duplication is a common phenomenon in the evolution of life. One of the central issues is how frequently duplicated genes evolve new functions or become silenced as pseudogenes. Since gene duplicates are believed to be initially functionally redundant and the vast majority of mutations affecting fitness are deleterious, it is commonly believed that the vast majority of duplicate genes will become pseudogenes through loss-of-function mutations (Lynch and Force, 2000). Since the rate of mutation to null alleles is on the order of 10^{-6} per generation, it is expected that such nonfunctionalization of the duplicated genes can occur within a few million generations (Lynch and Force, 2000; Lynch and Conery, 2000).

Another interesting question that remains to be answered is why the duplicated genes are retained if they are initially functionally redundant. There are two possible evolutionary reasons. First, there can be selection for increased level of gene expression and divergence of gene functions (Seoighe and Wolfe, 1998). Both copies of the genes can persist in the genome with perfect or near-perfect sequence identity if a higher level of the gene product(s) increases the fitness of the organism. Second, both copies can be maintained by functional divergence through acquiring a new function (neo-functionalization) or retaining only a subset of the functions of the ancestral gene (subfunctionalization). For example, many genes involved in development have multiple, independently mutable functions related to timing and tissue specificity of expression. If different subfunctions are essential for survival, growth, and/or reproduction, the two duplicated gene copies with different subfunctions will be maintained indefinitely by natural selection (Lynch and Force, 2000).

Genomic Perspectives on Gene Duplication Duplications of genes and groups of genes have occurred frequently in all genomes. It appears that substantial portions of genes arose from duplications. But in the absence of genome-scale data, it is difficult to estimate the rate of gene duplication. Lynch and Conery (2000) analyzed duplicate genes in three completely sequenced eukaryotic genomes (yeast, *Saccharomyces cerevisiae*; fly, *Drosophila melanogaster*; and worm, *Caenorhabditis elegans*), and in three other partially sequenced genomes (human, *Homo sapiens*; mouse, *Mus musculus*; and plant, *Arabidopsis thaliana*). Very high variable duplication rates were observed, ranging from 0.002 (fruit fly) to 0.02 (worm) with an average of 0.01 per gene, per 1 million years. The proportion of duplicated genes was estimated to be about 8, 10, and 20% for the fly, yeast, and worm genomes, respectively (Lynch and Conery, 2000), 15% for human (Li et al., 2001), and up to 25% for *Arabidopsis* (Wolfe, 2001). Lynch and Conery also estimated that the half-life of the duplicate genes is about 3 to 7 million years, and more than 90% of the duplicated genes disappear before 50 million years have elapsed. Their results strongly support the consensus of earlier studies that the vast majority of duplicated genes will be lost after duplication (Wagner, 2001).

Jordan and coworkers (2001) assessed the contribution of gene duplication to genome evolution in Archaea and Bacteria by examining the clusters of related paralogous genes that could have expanded subsequent to the diversification of the major prokaryotic lineages. A total of 21 completely sequenced archaeal and bacterial genomes were analyzed, and each genome represents a unique lineage at the genus level. Similar to the observation of the proportion of duplicated genes in eukaryotes, a substantial fraction (5–33%) of the coding capacities in the genomes examined resulted from lineage-specific expansion. A strong positive correlation was also observed between genome size and the number of recently duplicated genes. It appears that the expanded gene families made substantial contributions to the genomic determinants of phenotypic differences between bacterial lineages (Jordan et al., 2001).

Although the importance of individual gene duplication has been recognized for some time, the evolutionary importance of whole-genome duplication is controversial. The availability of whole-genome sequences provides the possibility of rigorously examining the evidence for whole-genome duplication. One of the most comprehensive comparisons was performed with the yeast (*S. cerevisiae*) genome (Wolfe and Shields, 1997). Whole-genome sequence comparison identified 55 duplicated gene sets with a total of 376 pairs of genes, and these duplicated regions covered half of the genome. This strongly suggested that whole-genome duplication had occurred. The large duplicated chromosomal regions in *S. cerevisiae* and the limited gene order information from related species (Wolfe and Shields, 1997; Mewes et al., 1997; Coissac et al., 1997; Keogh et al., 1998) suggested that *S. cerevisiae* is a degenerate tetraploid. By comparing the evolutionary rates of the duplicated genes, it was estimated that the whole-genome duplication event occurred about

100 million years ago. However, it is not clear whether the tetraploid resulted from genome duplication within a single species (autopolyploidy) or from hybridization between two closely related species (allopolyploidy).

4.5.2 Genomic Perspectives on Gene Loss

Gene Loss as an Engine of Evolutionary Change   Since the vast majority of mutations are deleterious, loss of function is the most likely outcome when a novel selection acts on a population. It is expected that the loss of functions will occur much more often than the gain of new functions or new regulatory systems (Olson, 1999). The mutated genes exhibiting loss of function can persist in the genome and be reverted for adaptive changes if the selective environment shifts again. Olson (1999) hypothesized that gene loss plays a major role during the evolution of life and could serve as an engine of evolutionary changes.

It is understandable that acquiring genes by HGT and duplication could lead to the evolution of new traits, but it is less obvious that loss of gene functions could serve as a means of bacterial adaptation. The "less-is-more" hypothesis is supported by several direct lines of evidence. For example, loss of chemokine receptor genes increases the resistance of human cells to the infection of AIDS and *Plasmodium vivax*, because these genes are essential for entry of the pathogens into target cells (Olson, 1999). Another example is that the loss of the *omp*T gene, which encodes a surface protease, is critical to the development of *Shigella* virulent strains. Recent comparative genomics studies discussed below also support the view that gene loss is a major motif of molecular evolution.

Gene Loss in Prokaryotes   In contrast to eukaryotes, prokaryotes have a narrow range of variation in genome size (see Chapter 2) and have few nonfunctional sequences. The variation of genome size in prokaryotes is mainly due to gene number differences. Although prokaryotes increase their DNA content through HGT as well as gene duplication, as discussed above, HGT is the primary route for obtaining new genes. One interesting question is why bacterial genomes are so compact and not ever-expanding if HGT is an ongoing process. The obvious explanation is that the structural evolution of chromosomes is biased toward DNA loss rather than DNA insertion (Mira et al., 2001).

If natural selection is not strong enough to maintain them, genes could be lost by large deletions, which remove one or more genes in a single event. The existence of such deletional bias can be evaluated by examining the occurrence of known pseudogenes and their functional counterparts in different taxa. This is because pseudogenes do not express a cellular function and therefore are not subject to selective pressure. By comparing the known pseudogenes across 42 completely sequenced prokaryotes, Mira and colleagues (2001) found that deletions were more frequent than insertions and have a much greater effect on genome size. Gene deletions were also found to be considerably high among different natural isolates of *E. coli* and among different *Mycobacterium tuberculosis* strains (Ochman and Jones, 2000; Kato-Maeda et al., 2001). In addition, Ochman and Moran (2001) determined that a large set of intracellular symbionts and pathogens have undergone more massive gene loss due to a lack of effective selection for maintaining genes. Since the host presents a constant environment rich in metabolic intermediates, many genes involved in biosynthetic pathways are superfluous and hence eliminated through mutational bias favoring deletions. Genome sequence comparisons indicated that many apparent beneficial genes such as those involved in DNA repair, translation, and transcription were lost in the intracellular symbionts and pathogens due to the inefficiency of natural selection and small population size (Ochman and Moran, 2001).

Genes can also be lost by mutational inactivation and subsequent gene erosion (Mira et al., 2001). The existence of such deletional bias can be evaluated by comparing the length of intergenic spacers and related gene organization (Mira et al., 2001). The intergenic spacers between *E. coli* and *Buchnera* can be classified into two categories: *ancient spacers* and *amended spacers*. While ancient spacers are those flanked by the same genes in *E. coli* and *Buchnera* on both ends, amended spacers are those where one or more genes are missing from *Buchnera* but have flanking genes with the same order as in *E. coli*. Although the sizes of ancient spacers are very similar between two species, the amended spacers are more than three times longer than ancient spacers. The most probable explanation for the larger size of amended spacers is that they represent highly eroded pseudogenes, and the additional sequences arose from the residue of ancestral genes (Mira et al., 2001).

Gene Loss in Eukaryotes   Sequence comparisons of *Saccharomyces cerevisiae* to other fungi, *Schizosaccharomyces pombe* and *Neurospora crassa*, indicate that *S. cerevisiae* appears to have lost many gene families (Aravind et al., 2000; Braun et al., 2000). After the divergence of *S. cerevisiae* from *S. pombe*, *S. cerevisiae* lost about 300 genes, which account for about 5% of the total genes in the modern yeast genome. About 300 genes have diverged far beyond expectation in the *S. cerevisae* lineage (Aravind et al., 2000).

N. crassa is considerably more complex in terms of morphology and development than *S. cerevisiae*. Sequence comparison indicates the loss of specific genes in *S. cerevisiae*, some of which were involved in basic cellular processes such as translation, ion homeostasis, cytoskeleton, and some metabolic enzymes (Braun et al., 2000). Sequence analysis also revealed that substantial portions of *N. crassa* genes are "orphan" genes without any clear homologs in any species.

Impact of Gene Duplication and Loss on Species Speciation   As mentioned previously, gene duplications provide the raw material for evolutionary changes. It is believed that gene loss or silencing of duplicated genes is more important to the evolution of species diversity than the evolution of new functions in duplicated genes (Lynch and Conery, 2001). Since the loss of gene copies in two isolated populations could result in rapid accumulation of genetic differences, gene duplications and subsequent gene loss might have a significant role in speciation (Lynch and Conery, 2001; Taylor et al., 2001). The process by which different duplicated genes in different populations are lost is referred to as divergent resolution (Taylor et al., 2001). Many studies have shown that divergent resolution is an important process in the speciation of animals such as fish (Taylor et al., 2001). With ten to hundreds of duplicated genes present, divergent resolution would cause the passive buildup of reproductive isolation.

It is believed that gene loss through large deletions and/or mutational inactivation and subsequent gene erosion have also played significant roles in the evolution of genome size of prokaryotes. Mira and coworkers (2001) suggest a simple model for the evolution of prokaryotic genome size by considering several opposing forces (Fig. 4.2). While the processes of HGT and gene duplication increase DNA content through selection for gene functions, the processes of gene deletion and mutational inactivation and subsequent gene erosion remove DNA through mutation and random genetic drift. Due to these opposing forces, the genome size of prokaryotes remains relatively small and constant. However, not

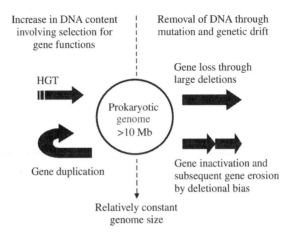

Figure 4.2 *The importance of evolutionary forces in the evolution of genome size in prokaryotes* (Mira et al., 2001). While the genome size increases through gene duplications and horizontal gene transfer, it decreases via gene deletions and gene inactivation followed by gene erosion. As a consequence, the prokaryotic genomes remain relatively small and constant.

all these processes exist at the same rates. Thus, some prokaryotic genomes may increase in size, whereas others may decrease in size.

4.5.3   Genomic Perspectives on Other Evolutionary Processes

Concerted Evolution   Multigene families are groups of genes of identical or similar sequences descended by duplication of some ancestral gene. They occur in all living organisms and usually encode related or identical functions. The members of a multigene family may be clustered together or dispersed on different chromosomes or different regions of a chromosome. Some multigene families such as rRNA genes are composed of genes with identical or near-identical sequences. It is believed that the multiple copies of identical genes are required for rapid synthesis of the gene product at certain stages in the cell cycle.

As we mentioned previously, the rate of mutation in null alleles is relatively high ($\sim 10^{-6}$ per generation), and duplicated gene sequences can be eroded quickly. The identical or near-identical sequences of a multigene family suggest that some mechanisms must exist to prevent individual copies from accumulating mutations and therefore diverging away from the functional sequences. It is most likely that the members of a multigene family coevolve in parallel to maintain their sequence homogeneity, that is, concerted evolution (Elder and Turner, 1995; Liao, 2000).

Three models can account for sequence homogenization of multiple gene copies of a multigene family: unequal crossing-over, gene conversion, and gene amplification (Liao, 2000). The most likely mechanism for achieving concerted evolution is gene conversion, which is a nonreciprocal recombination process whereby a sequence becomes identical by replacing one copy of a gene with all or part of the sequence of a second copy of the gene. Therefore, multiple events of gene conversion could maintain the sequence identity among individual members of a multigene family.

Ribosomal RNA genes (rDNA) are the most widely known example of a multigene family. Liao (2000) compared multiple paralogous rRNA gene sequences from 12 prokaryotic organisms whose complete genome sequences are available. The results indicate that the genic sequences of rRNA genes undergo much slower divergence than their flanking sequences. Putative sequence conversion tracts were identified throughout the entire length of each individual rRNA gene and their immediate flanking sequences. The results suggested that gene conversion initiated within the genic regions of the rRNA genes is likely to play a major role in concerted evolution of dispersed rRNA gene families.

Convergent Evolution   Convergent evolution is a phenomenon in which the same character state (e.g., gene function, catalytic enzyme mechanisms, structural motifs, and sequences) appears and evolves independently in separate evolutionary lineages. Functional and mechanistic convergence is common in evolution, whereas sequence convergence has never been proven (Doolittle, 1994; Karlin and Brocchieri, 2000). Galperin and colleagues (1998) used three completely sequenced prokaryotic and eukaryotic genomes to systematically evaluate analogous enzymes that catalyze the same reaction but have unrelated enzyme structure. A total of 105 enzymes were found to belong to the same enzyme classification categories but no sequence similarity was detected. Thirty-four of the enzymes had different structural folds. There are many analogous enzymes that are different in bacteria compared to eukaryotes and hence could be good potential targets for developing new antibacterial agents.

4.6   UNIVERSAL TREE OF LIFE

4.6.1   Establishment of a Universal Tree of Life

Historical Perspectives   Understanding the evolutionary relationships between living organisms has been a great challenge for centuries (Gupta, 1998). The interest in classifying all living things to reflect the natural order is an ancient one. Even before Darwin, biologists (and philosophers) believed that God or some other eternal principle created a natural order of living organisms. After Darwin, more than a century ago, biologists believed that all modern species descended from a very limited number of common ancestors, which themselves evolved from fewer progenitors back to the beginning of life. Therefore, a single genealogical tree could be reconstructed to describe the evolutionary relationships of living and extinct organisms. Most contemporary biologists believed that the last common ancestor lived approximately 3.5 to 3.8 billion years ago (Doolittle, 2000).

Before the 1960s, the phylogenetic classification of living organisms was dependent on morphological, physiological, and behavioral traits. Based on such phenotypical studies, scientists concluded that there were two basic types of living organisms, prokaryotes and eukaryotes, depending on the structure of the cells that composed them. The main difference between these two kinds of living organisms is that eukaryotes possess a true nucleus—a membrane-bound organelle housing the chromosomes—whereas prokaryotes do not have a nucleus. For complex eukaryotic organisms, it is possible to reasonably infer their genealogical relationships; however, for prokaryotes, this approach has very limited use because they do not have a complex morphology. Therefore, in the mid-1960s,

Zuckerkandl and Pauling (1965) proposed a novel revolutionary strategy to infer phylogenetic relationships based on genes and proteins. This marked the beginning of the field of molecular evolution.

During its early period (1965–1977), molecular phylogenetics was dependent on the sequences of such proteins as ferredoxins and cytochromes, because nucleic acid sequences could not be determined at that time. Protein-based phylogenetic studies confirmed and then expanded the family tree of well-studied groups such as vertebrates and demonstrated the general usefulness of molecular phylogeny. Molecular phylogenetic studies also lent strong support to the hypothesis that mitochondria and chloroplasts are descendants of alpha-proteobacteria and cyanobacteria, respectively (Doolittle, 2000).

In the late 1970s, Woese and his collaborators turned their attention to the small subunit ribosomal RNA (SSU rRNA) genes. SSU rRNAs have several advantages over ferredoxins and cytochromes as a universal molecular chronometer (Woese, 2000; Doolittle, 2000). First, it is a molecule that exists in all eukaryotes and prokaryotes. Thus, it can be used to compare the phylogenetic relationships among all living organisms. Also, this molecule has both fast- and slow-evolving mosaic domains and thus can be used to analyze the phylogenetic relationships of both closely and distantly related organisms. In addition, this molecule has essential cellular functions and interacts with more than 100 other coevolved rRNAs and proteins. As a result, it is highly resistant to horizontal gene transfer. On the basis of cluster dendrograms generated from the data of RNase T_1 oligonucleotide catalogs of rRNAs, Woese and colleagues showed that various prokaryotic and eukaryotic organisms fell into three distinct groups (Fox et al., 1980; Woese and Fox, 1977). While one group contained all eukaryotic organisms, the second group consisted of all known bacteria, which were referred to as eubacteria, and included gram-negative and gram-positive bacteria, and cyanobacteria. The third group was called archaebacteria (now Archaea) and was composed of a variety of unusual prokaryotes such as methanogens, thermophiles, and halophiles (Woese and Fox, 1977). Based on the similarity of coefficients from oligonucleotide catalogs, the archaebacteria were not more closely related to bacteria than to eukaryotes. By integrating this observation with other unique biochemical characteristics such as membrane lipids, cell walls, and RNA polymerase, Woese and colleagues proposed that archaebacteria were totally separate from other bacteria and therefore that prokaryotes consisted of two distinct monophyletic groups: archaebacteria and eubacteria (Woese and Fox, 1977).

Formal Proposal of the Three-domain Universal Tree   Molecular comparisons, in particular rRNA sequence-based phylogenies, showed that archaebacterial rRNAs were significantly different from bacteria, suggesting that this distinct group of prokaryotes should have the same taxonomic status as eukaryotes and eubacteria. In 1990, Woese and coworkers formally proposed replacing the bipartite view of life with a tripartite scheme. This proposal assigned each of the three groups, archaebacteria, eubacteria, and eukaryotes, to a Domain status, which is the highest taxonomic level. These three groups were renamed as Bacteria (eubacteria), Archaea (archaebacteria), and Eucarya (eukaryotes) (Fig. 4.3). Each domain contains three or more kingdoms (Woese et al., 1990a). In this universal tree, the length of the archaeal branch is significantly shorter than those of the bacterial and eubacterial branches. Archaea appear to be more similar to Bacteria or Eucarya than Bacteria are to Eucarya. Also in this universal tree, Archaea share a common ancestor (i.e., they are monophyletic). Following the work of Woese and colleagues (1990a), many more rRNA sequences were accumulated and now more than

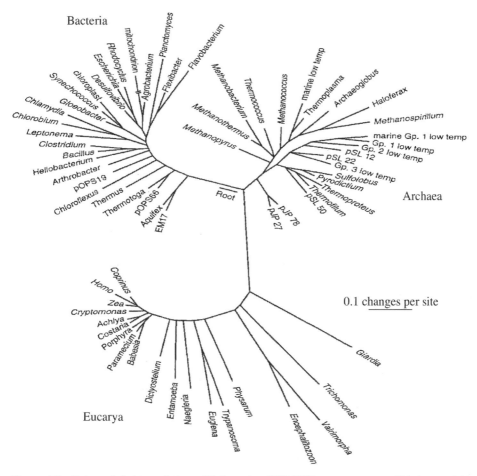

Figure 4.3 *Universal phylogenetic tree of life based on SSU rRNA gene sequences.* This tree contains 64 sequences representative of all known phylogenetic domains and was constructed using maximum likelihood methods. (Reprinted with permission from N.R Pace, A molecular view of microbial diversity and the biosphere, *Science*, 1997, Vol. 276, pp. 734–740. Copyright (1997) American Association for the Advancement of Science.)

10,000 rRNA sequences are available from both prokaryotes and eukaryotes. Phylogenetic analysis based on such massive sequence data support the new tripartite scheme (Olsen et al., 1994). The three-domain universal tree is widely accepted as the current paradigm of the evolution of life in the field of evolutionary biology (Edgell and Doolittle, 1997; Gray, 1996; Morell, 1996; Pace, 1997).

Many phylogenetic studies and phenotypic characteristics support the notion of the three-domain universal tree (Doolittle, 1999a; Brown and Doolittle, 1997). Many trees based on proteins such as RNA polymerases, ATPases, elongation factors, and the ileu-tRNA synthetases support the tripartite division (Forterre, 1997). Archaea, Bacteria, and Eucarya also each have some unique characteristics. For example, all Archaea have a unique lipid composition. The cell membranes of Archaea are mainly composed of diphtanylglycerol diether or dibiphytanyldiglycerol tetraether or both, whereas the cell membranes of Bacteria and Eucarya consist of diacylglycerol-derived lipids. The cell wall components also differ among the three domains of life. Only bacterial cell walls contain peptidoglycan and only eukaryotes have tubulin- and actin-based cytoskeletons. In addition, although neither Bacteria nor Archaea have a membrane-bound nucleus, the replication, transcription, and translation systems in Archaea are more similar to those in eukaryotes than to those in Bacteria.

Although the three-domain classification of living organisms is widely accepted, the concepts have been challenged by some researchers (Cavalier-Smith, 1992; Margulis and Guerrero, 1991; Mayr, 1990, Rivera and Lake, 1992; Gupta, 1998). For instance, Archaea were hypothesized to be paraphyletic rather than monophyletic (Rivera and Lake, 1992). Rivera and Lake proposed a new dichotomy of the living organisms in which halophilic, methanogenic, and some thermophilic archaea were grouped with Bacteria, whereas the other thermophiles were clustered with Eucarya. In addition, the relationships among the three domains appear to vary among different protein markers (Forterre, 1997). For instance, Bacteria are more closely related to Eucarya based on malate dehydrogenase, while Bacteria more closely resemble Archaea based on citrate synthases and the cell division protein FtsZ.

Since 1995, full genome sequences have become available for more than 60 prokaryotic and eukaryotic organisms. It is expected that the new information will support the hypothesis of the tripartite division of living organisms. However, information from whole-genome sequences has so far proved to be confusing rather than enlightening (Pennisi, 1998; Forterre, 1997). In the following section, we will critically review the current findings and genomic insights on the tree of life.

Rooting the Universal Tree   Once the three-domain classification of life was accepted, one natural and interesting question is which of the two structurally primitive groups (Bacteria or Archaea) gave rise to the first eukaryotic cell. To answer this question, appropriately rooting the universal tree is important. However, the universal tree derived from rRNA data is rootless because no outgroup sequence is available.

There are three possibilities to root the universal tree: (1) Bacteria diverged first from a lineage leading to Archaea and eukaryotes; (2) Archaea evolved first from a lineage producing Bacteria and eukaryotes; or (3) Eukaryotes emerged fully from the lineage leading to Bacteria and Archaea. It is impossible to root the universal tree based on a single gene because no organism can serve as an outgroup. This difficulty has been circumvented by reciprocally rooting trees based on paralogous genes, which arose by duplication prior to the divergence of prokaryotes and eukaryotes (Iwabe et al., 1989; Gogarten et al., 1989). For example, assume a gene was duplicated in the last common ancestor (cenancestor) so that all extant organisms have both copies of this gene. If the two duplicated genes still share some sequence similarity, then one of the two sequences can be used as an outgroup sequence for constructing rooted trees (Fig. 4.4). Reciprocally rooted gene trees can be constructed by using each of the duplicated genes as an outgroup (Brown and Doolittle, 1997).

The DNA sequences from two independent sets of anciently duplicated genes, that is, the genes encoding the elongation factors (EFs) Tu(1α) and G(2) and α and β subunits of membrane ATPases, were initially used to root the universal tree of life. The composite trees that were generated based on these genes showed that Bacteria diverged first from a lineage leading to Archaea and eukaryotes (Iwabe et al., 1989; Gogarten et al., 1989) (Fig. 4.5A). The sisterhood of Archaea and Eucarya was validated by rooting the universal tree

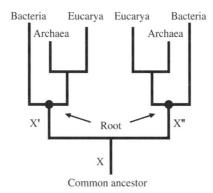

Figure 4.4 *Conceptual illustration for rooting universal tree using paralogous genes.* Assume that gene X was duplicated to give rise to X′ and X″ in the common ancestor, and both paralogous genes are present in all extant organisms with significant recognized homology. The duplicated genes can then be used as an outgroup to reciprocally root trees of paralogous genes. The relationships of Archaea to Bacteria and Eucarya illustrated in this figure are based on the results from the elongation factors (EFs) Tu(1α) and G(2) and α and β subunits of membrane ATPases.

based on the paralogous sequences of isoleucyl-tRNA and valyl-tRNA synthetase genes (Brown and Doolittle, 1995), and by reciprocal rooting based on the two repeats of an internally duplicated sequence of a carbamoylphosphate synthetase subunit (Lawson et al., 1996).

Since each set of anciently duplicated genes has its particular drawbacks (Brown and Doolittle, 1997), there are still doubts about the rooting of the universal tree. Most rooting

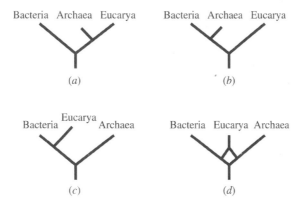

Figure 4.5 *Alternative rootings of the universal trees based on various individual genes (Brown and Doolittle, 1997).* (*a*) Archaea is more closely related to Eucarya. This tree topology is supported by the phylogenies of the following proteins: arginosuccinate synthetase, aspartyl-tRNA synthetase, ATP α and β subunits, DNA polymerase B, elongation factors (EFs) Tu(1α) and G(2), isoleucyl-tRNA synthetase, RNA polymerase subunit A and B, tryptophanyl-tRNA synthetase, tyrosyl-tRNA synthetase, carbamoylphosphate synthetase subunit, and the signal recognition particle SRP54. (*b*) Eucarya diverged first from Bacteria and Archaea. This tree topology is supported by the phylogenies based on the following genes: ALADH, citrate synthetase, glutamate dehydrogenase II, glutamine synthetase I, gyrase B, and heat shock protein 70. (*c*) Archaea diverged first from Bacteria and Eucarya. This tree topology is supported by the phylogenies of enolase and glyceraldehydes-3-phosphate dehydrogenase (GAPDH). (*d*) Eucaryotes originated from a fusion of two primitive ancient Bacteria and Archaea.

studies have been criticized due to unrecognized paralogy, horizontal gene transfer, unequal rates of evolution, and/or mutational saturation (Hilario and Gogarten, 1993; Forterre, 1997; Forterre and Philippe, 1999). Because of unequal evolutionary rates, the bacterial rooting obtained with elongation factors does not appear to be reliable (Forterre and Philippe, 1999). More recently, the two ancient duplicated genes encoding components of the protein-targeting machinery were used to root the universal tree (Gribaldo and Cammarano, 1998). One protein is a key subunit of the signal recognition particle (SRP), termed SRP54 for Eucarya and Ffh for Bacteria. The other paralogous protein is involved in the recognition and binding of the ribosome SRP nascent polypeptide complex, termed SRα for Eucarya and Ftsy for Bacteria. The root of the reciprocal trees is firmly placed between Bacteria and Archaea/Eucarya, thus providing strong support for the notion that Bacteria diverged first from a lineage leading to Archaea and eukaryotes. In contrast to metabolic enzymes but similar to EF-Tu(α)/EF-G(2), the SRP and SRP receptor-based phylogenies are unlikely to suffer from ambiguities due to horizontal gene transfer, as might be the case for ATPase subunits (Hilario and Gogarten, 1993; Brown and Doolittle, 1997). However, similar to the elongation factor, SRP may also be subjected to the problem of unequal evolutionary rates, and hence the bacterial rooting could be a methodological artifact (Forterre and Philippe, 1999).

By using the midpoint method, which simply places the root somewhere in the center of the tree, various alternative rootings of the universal tree were obtained based on different individual genes (Figs. 4.5B through D). One alternative hypothesis that was first proposed in the 1970s (Reanney, 1974; Darnell, 1978) is that the universal tree of life has its root in the eukaryotic branch (Fig. 4.5B) (Forterre and Philippe, 1999). Forterre and Philippe (1999) suggest that the universal ancestor is eukaryoticlike and many prokaryotic features were derived by simplification of the complex universal ancestor through gene loss and nonorthologous displacement. During evolution, simplification appears to be prevalent in most prokaryotic lineages, while the reverse is true in eukaryotic lineages. Another alternative "chimeric" or "fusion" hypothesis has also been proposed suggesting that eukaryotes arose from a merging of two primitive prokaryotes (Fig. 4.5D) (Zillig, 1991; Zillig et al., 1989; Lake and Rivera, 1994). The phylogenetic trees based on several proteins such as acetyl-coenzyme A synthetase and potolyase support this hypothesis (Brown and Doolittle, 1997).

Although the rooting of the universal tree is still controversial, the hypothesis that Bacteria diverged first from a lineage leading to Archaea and eukaryotes is generally favored, because the bacterial rooting explains the existence of eukaryotic features in Archaea, it strengthens the dogma that eukaryotes emerged from prokaryotes, and it strongly supports the hot-origin-of-life hypothesis (Forterre, 1997).

4.6.2 Challenges and Current View of the Universal Tree

As more whole-genome sequence data become available, there is less general agreement about the universal tree. Methodological problems and HGT challenge the concept of the universal tree as described below.

Methodological Artifacts More critical phylogenetic analyses based on both rRNA and protein genes indicate that the deep branchings and rooting of the universal tree are unreliable due to methodological artifacts originating from the differences in evolutionary rates and mutational saturation within molecules and between lineages (Forterre and

Philippe, 1999; Doolittle, 1999a). The artifacts include mutational saturation, '*long-branch distraction*,' and '*among-site rate variation*'. Since mutationally saturated sequences are maximally diverged, further mutations are most likely to make the sequences more similar rather than more diverse. Long-branch attraction exists when the rates of sequence changes are substantially different between lineages even in the case where the mutation is not saturated. Lineages with higher sequence change rates are artifactually clustered to each other and with outgroups when using all routine phylogenetic methods except the maximum likelihood method. The substitution rate also varies among different sites in a gene. In this situation, long-branch attraction also occurs even with the maximum likelihood method (Yang, 1996). The best example for illustrating the artifacts associated with methodologies is the uncertainty of the phylogeny of microsporidia based on different genes. Microsporidia are anaerobic protists. Based on cytological features, they had been classified as primitive eukaryotes, called archezoa, which diverged early during eukaryotic evolution before mitochondria acquisition. While they were positioned at the deepest branch in eukaryote phylogeny based on SSU rRNA genes, microsporidia were positioned with fungi or within fungi based on the largest subunit of RNA polymerase gene (Hirt et al., 1999).

Challenges of HGT on the Conceptual Basis of Universal Tree   If the universal tree is correct, bacterial genes should be found only in eukaryotes but not in any archaea. Also, the genes derived from bacteria in eukaryotes should be only those in mitochondria or chloroplast, and those genes would be involved only in respiration and photosynthesis but not in cellular processes, because the genes for the cellular processes would have come from the ancestral archaeon. However, these expectations do not hold true. As discussed previously, many Archaea have genes usually found in Bacteria. Modern eukaryotic molecular phylogenetic studies also indicate that many enzymes involved in cellular processes are of bacterial origin but not of arachaeal ancestry as the universal tree would imply (Brown and Doolittle, 1997; Rivera et al., 1998; Doolittle, 1999b). HGT is the most likely logical explanation for such observations.

Strictly speaking, precise phylogeny cannot be represented by a tree with any HGT event, and the tree will be meaningless if extensive HGT events occur (Koonin et al., 2000; Doolittle, 1999b). As we have discussed, extensive HGT events have been observed among different lineages of living organisms. This certainly leads to questions regarding the conceptual basis of the universal tree of life and phylogenetic classification.

Current Views of the Universal Tree   Although HGT occurred to a great extent among different lineages of living organisms, many researchers are trying to preserve the conceptual framework of the universal tree by treating HGT as just a nuisance in phylogenetic classification or as unessential to phylogenetic processes, and thus, HGT may not be a threat to the conceptual basis of phylogenetic classification (Doolittle, 1999a). Some of these views are discussed below.

Core of Nontransferable Genes   As we have discussed, the complexity hypothesis predicts that the genes involved in information processing, such as DNA replication, transcription, and translation, are less likely to be transferred, because these genes are fundamental to the cells and they are involved in complex cellular machinery. Thus, these genes could be used to reliably track the evolution of cellular or organismal lineages (Doolittle, 1999a).

There are several kinds of evidence against this view. First, cells do not contain a mechanism to identify fundamental genes or which genes are involved in operational and informational processes. Also, some cellular machinery involved in operational processes is more complex than some transcription/translation components. In addition, the horizontal transfer of some informational genes such as RNA polymerase, elongation factors, aminoacyl tRNA synthetases, and rRNA has been observed (Doolittle, 1999a; Asai et al., 1999; Ueda et al., 1999). Thus, the notion of the nontransferable genes has not been proven and may be unprovable.

Majority Rule View   The majority rule view states that although the rate of HGT is unexpectedly high, the tree topologies based on most genes will be the same as that based on the rRNA gene (Doolittle, 1999a).

Genetic Annealing Model   In this model, the universal ancestor is a diverse community of cells rather than a discrete entity (Woese, 1998). HGT occurred pervasively among the primitive cells of early evolution to create complex genetic systems. After the divergence of Bacteria, Archaea, and Eucarya, the extent of HGT decreased and the cellular genetic system began to anneal or crystallize and became refractory to HGT (Woese, 1998; Doolittle, 1999a). Thus, the universal tree based on rRNA is still a valid representation of organismal genealogy (Woese, 2000).

4.6.3   Genome-based Phylogenetic Analysis

Recognizing archaebacteria as a life form distinct from eubacteria and eukaryotes and establishing a three-domain model of evolution are considered the most significant developments in biology and have had a tremendous impact on our understanding of evolutionary relationships among different living organisms. The establishment of the three-domain universal tree is primarily based on SSU rRNA gene phylogeny. However, one key question is whether the three-domain tree is a true representation of relationships between organisms. In recent years, substantial new information based on a large number of genes and proteins, especially whole-genome sequences from many prokaryotic and eukaryotic organisms, has accumulated. With such information, it is now possible to critically assess whether the three-domain model represents the evolutionary relationships among living organisms or whether different types of relationships occur. Genome-based phylogenetic analyses with different methods indicate that the effects of horizontal gene transfer are not sufficient to wash out the phylogenetic signal, at least at the three-domain level, although they are important during evolution. In the following section, we will summarize the different approaches used in recent phylogenetic studies based on complete genome sequences.

Genome Phylogeny Based on Large Combined Protein Sequences

Phylogenetic relationships of species derived from comparisons of single genes are rarely consistent with each other because of HGT, unrecognized paralogy, and highly variable rates of evolution (Huynen and Bork, 1998). The availability of complete genome sequences from many species allows the construction of a phylogeny that is less sensitive to such inconsistencies and more representative of whole genomes than are single-gene trees. One approach to constructing whole-genome sequence-based phylogeny is to combine large protein sequence data sets together. It is expected that the combined data will increase phylogenetic accuracy by amplifying phylogenetic signals, dispersing noise, and increasing resolving power when the signal is masked by the phenomenon of convergent evolution (Brown et al., 2001). Phylogenetic analysis of combined protein sequence data sets represents an important approach in the exploration of gene sequences to address evolutionary questions.

Brown and coworkers (2001) compared the open reading frames from whole genomic sequences of 45 species of Bacteria, Archaea, and Eucarya. A total of 23 conserved proteins among all species were identified to be suitable for constructing universal trees. These proteins are mostly involved in translation, transcription, and DNA replication. Phylogenetic trees were constructed with various phylogenetic methods such as maximum likelihood, maximum parsimony, and neighbor-joining methods, and all gave highly congruent tree topologies. Although individual proteins are variable in their support of domain integrity, the combined data set of 14 proteins yields universal trees that are highly consistent with SSU rRNA trees and strongly support the separate monophyly of domains as well as the early evolution of thermophilic Bacteria. These results indicated that genomes have retained sufficient phylogenetic signals for the reconstruction of robust universal trees although HGT has probably played a critical role in genome evolution (Brown et al., 2001).

Genome Phylogeny Based on Gene Content   The second approach to minimizing the potential HGT interference is to construct the phylogenetic tree based on gene content rather than sequence identity (Snel et al., 1999). The similarity between two species is defined as the fraction of the number of genes common between genomes divided by the total number of the genes in the smallest one of the two genomes. Such similarity is then used to evaluate the evolutionary relationships among different species. The phylogenetic patterns based on gene content are the result of differential gene acquisition and loss along different evolutionary lineages (Snel et al., 1999).

Using this approach, Snel and colleagues (1999) compared the protein sequences encoded by 13 completely sequenced genomes with each other. The number of genes shared between the genomes was estimated based on an operational definition of orthology (Huynen and Bork, 1998), and a distance tree was constructed by the neighbor-joining distance method. The phylogenetic result showed that the tree topology based on gene content is highly consistent with the SSU rRNA gene tree. Similar results were also obtained by Tekaia and coworkers (1999) based on both gene content and overall gene similarities. The high bootstrap values indicated that gene content still carries strong phylogenetic signatures. The fact that gene content carries a strong phylogenetic signature implies that HGT plays only a limited role in determining the gene content of genomes (Snel et al., 1999).

Genome Phylogeny Based on Presence and Absence of Protein Families   The third approach to minimizing the potential HGT interference and methodological problems is to construct the phylogenetic tree based on the presence and absence of protein families (Fitz-Gibbon and House, 1999). This approach is analogous to using the distribution of morphological characteristics observed in a group of organisms for determining their phylogenetic relationships. But in contrast to most morphological characteristics, amino acid sequences are less likely to be highly similar unless they are from a common ancestor. This may eliminate the problem associated with convergent evolution.

Fitz-Gibbon and House (1999) analyzed complete genome sequences from 11 free-living microorganisms, and constructed their phylogenetic relationships based on the observed presence and absence of families of protein-coding sequences. The phylogenetic tree obtained was very similar to the SSU rRNA gene tree. These results indicate that there is a strong phylogenetic signal within the genomes that reflects the evolutionary histories of the organisms despite HGT, gene duplication, and loss. These results also suggest that phylogenetic analysis based on the presence and absence of protein families could be an important method for studying the relationships among living organisms, especially for distantly related organisms.

Genome Phylogeny Based on the Distributions of Evolutionary Rates   Unlike the molecular clock hypothesis that assumes the substitution rate of a gene does not change with time, Grishin and colleagues (2000) assumed that the absolute substitution rates may change with time in a correlated fashion and thus the ratio of any two rates is constant. In this case, the distribution of relative evolutionary rates, that is, each rate divided by the mean rate, will remain constant. By comparing the sequences from 19 complete genomes, Grishin and coworkers (2000) observed that the relative rate distribution among proteins is similar to each other and can be described by a single distribution with a long exponential shoulder. A rooted whole-genome phylogenetic tree was obtained based on the scaling parameter of the distribution of evolutionary rates. The resulting tree topology is largely consistent with that of global SSU rRNA-based trees.

Genome Phylogeny Based on Sequence Signatures of Indels

Sequence Signatures   Signature sequences of proteins are generally defined as specific amino acid sequence changes, such as amino acid substitutions, and specific insertions and deletions observed in all members of one or more taxa but not in other taxa (Rivera and Lake, 1992; Gupta and Singh, 1992; Gupta, 1998). The signatures were flanked by regions conserved among all the sequences examined. The presence of the conserved flanking regions ensures that the observed sequence changes are not the result of inappropriate alignment or sequence errors (Gupta, 1998).

Specific insertions or deletions, that is, indels, are most commonly used as sequence signatures in evolutionary studies for the following two reasons (Gupta, 1998). First, a conserved indel can be used to classify species. When a conserved indel flanked by conserved sequence regions is observed at precisely the same position in homologous sequences from different species, it is most likely that the indel was introduced only once during evolution and then vertically passed on to all descendants. Thus, such an indel bears evolutionary relationships among those species examined. As a result, based on the presence of an indel, the species with or without the conserved indel can be classified into a distinct group. A well-defined indel can also be used as a milestone for tracking evolutionary events and as a phylogenetic marker. For instance, if an indel was introduced to gram-negative bacteria after the speciation of gram-positive and gram-negative bacteria, all gram-positive bacteria and other bacteria that diverged before the speciation of gram-positive and gram-negative bacteria would lack this indel. Thus, this indel could be used to track the event of gram-positive and gram-negative speciation and as a phylogenetic marker for gram-negative bacteria.

There are two potential problems in using conserved indels for evolutionary analysis (Gupta, 1998). First, it is possible that the observed indels were introduced to a certain lineage multiple times due to similar functional constraints and selection pressure rather than to a common ancestry. Second, the presence of a conserved indel in particular groups of living organisms could also result from HGT. In practice, it is difficult to determine whether a conserved indel resulted from common ancestry or the two possible reasons cited above, and to obtain high confidence, other auxiliary information is needed.

Two types of auxiliary information can be used to assess the common ancestry of a conserved indel. The most important auxiliary information is obtained by examining cell structure and physiology. If the inferred conserved signature is consistent with the structural and physiological characteristics of the organisms, the inference of the signature is most likely correct. Otherwise, it may be better to ignore the inferred conserved signatures. Another source of auxiliary information results from assessing species distribution. If a shared indel exists in all known members of a given taxa, it is more likely that the indel was of common ancestry. In contrast, if only certain members of a particular taxa contain a shared indel and the species possessing the indel have no obvious structural and physiological relationships, this indel is most likely a result of independent evolution or HGT (Gupta, 1998).

The signature-based approach is advantageous to traditional phylogenetic analysis, because the evolutionary relationships inferred based on an indel will be less likely affected by differences in evolutionary rate. The differences in evolutionary rates may result in dramatic differences in the evolutionary tree using traditional phylogenetic analysis. Traditional phylogenetic analysis also appears to be sensitive to sequence errors resulting from polymerase chain reaction amplification, sequencing, and sequence contamination. Such sequence errors may change the branching orders of species in phylogenetic trees (Gupta, 1998). However, the signature approach is less likely affected by such sequencing errors. It is unlikely that a conserved indel of a defined length and sequence at a precise position within a conserved region is a result of sequence errors. In addition, the evolutionary relationships among species determined by signature sequences are based on the presence and absence of a given signature. Thus, in contrast to traditional phylogenetic analysis, the interpretation of the relationship is unambiguous. However, signature sequence analysis has two major limitations (Gupta, 1998). One is that this approach is not applicable to many proteins because they do not contain useful sequence signatures. The other is that a given signature can only be used to distinguish and establish the evolutionary relationships between the two groups of species and cannot provide evolutionary relationships among all species in a tree.

Signature Sequence-based Phylogeny Signature sequence analysis of different proteins revealed evolutionary relationships among living organisms that were distinctly different from the three-domain universal tree (Gupta, 1998). First, signature sequence analysis showed that archaebacteria are polyphyletic and more closely related to gram-positive bacteria. Second, signature sequence analysis also suggested that prokaryotes are not divided into archaebacteria and eubacteria, but rather two different groups: monodermata, which include all archaebacteria and gram-positive bacteria, and didermata, which include all true gram-negative bacteria. Whereas *Monodermata* are bounded by a single cell membrane, *didermata* have both inner and outer membranes. Molecular sequence data also suggested that diderm prokaryotes were derived from monoderm prokaryotes. In addition, signature sequence analysis revealed that the ancestral eukaryotic cell was derived from the fusion and integration between a thermoacidophilic archaebacterium and a gram-negative bacterium. The results based on signature sequence analysis provide an alternative view of the evolutionary relationships among existing organisms.

In summary, most of the genome-based phylogenetic studies support the notion of a three-domain structure of life, although this classification is still questioned by several authors. Each domain has unique phenotypic characteristics in terms of molecular structure and composition, physiology, and morphology. The apparent phenotypic and genetic coherence of each domain implies that the three-domain notion is operational and does reflect some ancient fundamental event in the evolutionary history of life (Forterre and Philippe, 1999; Wolf et al., 2002).

4.7 MINIMAL GENOMES

The number of genes in living organisms vary dramatically, ranging from as few as 480 in the pathogenic bacterium, *Mycoplasma genitalium*, to $\sim 30,000$ in multicellular eukaryotes. One interesting scientific question concerns the minimum number of genes necessary to support cellular life. The concept of the minimal gene set refers to the smallest possible groups of genes necessary to support a functioning cellular life form under the most favorable conditions with the availability of all essential nutrients and the absence of environmental stress (Koonin, 2000). Defining the minimal gene set is important to our understanding of the basics of cell functioning and evolution.

Based on the features of modern cells, the simplest form of a living cell should theoretically possess certain characteristics (Koonin, 2000). First, the organism should have an integral cell membrane and some minimal transport systems for material exchanges with the outside environment. It should be capable of importing most of the necessary metabolites, and thus the majority of metabolic enzymes can be dispensed. In addition, it should be incapable of taking up extracellular proteins, and thus all housekeeping genes involved in transcription and translation are needed. Finally, unlike nonliving entities, the cell should be capable of replicating and thus all the genes involved in replication are necessary. Genes that play a role in the above processes should be good candidates for inclusion in the minimal gene set.

Genes shared by multiple genomes are most likely to be essential and therefore are potential members of the minimal gene set. A minimal gene set was constructed by comparing whole-genome sequences from *H. influenzae* and *M. genitalium*. These two genomes appear to be particularly suitable for constructing a minimal gene set for the following reasons: First, *H. influenzae* and *M. genitalium* belong to different distantly related parasitic bacterial groups in phylogeny and are probably separated from their last common ancestor by at least 1.5 billion years. In addition, gene losses have taken place independently in both genomes, especially in *M. genitalium* that has the smallest genome sequenced. Since they have evolved independently, the common genes shared between these two genomes should be good candidates for constructing a minimal gene set. By comparing the sequences between these two genomes, a minimal gene set with ~ 250 genes was derived (Mushegian and Koonin, 1996). This minimal gene set included nearly complete systems for transcription, translation, and replication, and a minimum number of genes for repair machinery, signal transduction, and other cellular components (Koonin, 2000).

To test whether the *M. genitalium* genome already approximates a minimal genome or encodes the primary functions necessary to support cellular life, the global transposon-mediated knockout mutagenesis of *M. genitalium* and *M. pneumoniae* was performed (Hutchison et al., 1999). *Mycoplasma pneumoniae* is the closest known relative of *M. genitalium* and has a genome size that is 236 kb larger than *M. genitalium*. Whole-genome sequence comparison revealed that *M. pneumoniae* (680 ORFs) includes essentially all of the subset of the *M. genitalium* genome protein-coding genes (480 ORFs). The gene order is highly conserved between these two organisms, but substantial sequence differences were observed between these two genomes with an average of 65% amino acid sequence identity. Based on mutant variability, a minimal gene set of 265 genes was obtained. This is very close to the estimate obtained using a comparative genomics approach (Mushegian and Koonin, 1996). However, 38 of the 250 genes identified by sequence comparison have proved to be nonessential based on the mutagenesis study. These results imply that an evolutionarily conserved gene is not necessarily essential under all conditions (Koonin, 2000). One of the main problems for the mutagenesis approach is potential functional redundancy. The existence of paralogous genes or functionally equivalent genes will complicate the interpretation of genome-wide mutagenesis results (Peterson and Fraser, 2001).

The availability of whole-genome sequences from different microorganisms permits the reexamination of the minimal gene set more inclusively. Koonin (2000) assessed the original version of the minimal gene set in a fairly comprehensive manner by comparing whole-genome sequences from 21 microorganisms. While 32% of the genes in the original minimal gene set were universal, 30% were conserved in bacteria. Although 38% of them showed a less consistent phylogenetic distribution, about half of these genes were missing in only one or two species, suggesting that they are also highly conserved even though they are ubiquitous. The genes with scattered phylogenetic distribution were primarily involved in metabolic functions. These results suggested that the original minimal gene set is reasonably inclusive (Koonin, 2000).

By considering all the results from phylogenetic analysis, genetic mutagenesis, and biochemical inference, it is suggested that a cell could be supported by a bare-bones set of ~150 genes encoding the basal systems for transcription, translation, and replication (Koonin, 2000). However, it should be noted that although the minimal gene concept is scientifically interesting, it is meaningless without explicitly defining the growth conditions. A cell with the bare-bones minimal gene set could never be constructed in practice although possible in theory (Koonin, 2000).

4.8 GENOMIC INSIGHTS INTO LIFESTYLE EVOLUTION

Microorganisms form complex intimate relationships with a variety of multicellular organisms. Based on the effects on multicellular hosts, the bacteria that live only in eukaryotic cells and tissues can be divided into two basic types: mutualists (i.e., symbionts) and parasites (i.e., pathogens). The former is referred to as bacteria that increase the fitness of the individual host, whereas the latter is defined as those that decrease the fitness of the infected host (Moran and Wernegreen, 2000). Despite differences in the ultimate outcomes (i.e, harmful, benign, or beneficial), the basic biological processes needed to successfully infect a host for both pathogens and symbionts are very similar, typically including contacting and entering the host body, growing and replicating using the nutrients from host tissue, avoiding host defenses, and finally exiting and infecting new hosts (Ochman and Moran, 2001). Therefore, both pathogens and symbionts share some commonality in terms of genomic structure, genetic regulation, and evolution (Hentschel et al., 2000; Ochman and Moran, 2001). In the following section, we will provide a brief overview of the most recent understanding of the lifestyle evolution of pathogens and symbionts from a genomics perspective.

HGT and Gene Loss: The Major Evolutionary Forces of Pathogenesis and Symbiosis

HGT appears to play a major role in converting an organism into a successful pathogen or symbiont (Ochman and Moran, 2001). The most notable example is *E. coli* O157:H7, which is a worldwide threat to public health and causes severe haemorrhagic colitis (Perna et al., 2001). Genome sequence comparisons of the pathogenic *E. coli* O157:H7 and the nonpathogenic *E. coli* K-12 revealed the existence of massive gene transfer in O157:H7. A total of 1,387 new genes were found in O157:H7 that encode candidate virulence factors, alternative metabolic capacities, several prophages, and other new functions. It appears that the ancestral strain of *E. coli* O157:H7 was not pathogenic before the acquisition of these genes, but it was apparently preadapted to become pathogenic upon acquiring virulence determinants (Ochman and Moran, 2001). For instance, the strain already has many features necessary to cope with the new environments in animal hosts, such as mechanisms to counteract host defenses.

Gene loss also appears to play a critical role in the evolution of the interactions between pathogens or symbionts and their hosts (Ochman and Moran, 2001). Many of the obligate pathogens or symbionts have smaller genomes (see Table 2.1 in Chapter 2). Phylogenetic analysis indicated that the ancestors of the obligate pathogens and symbionts had a much larger genome size, suggesting that massive genome reduction occurred in these microorganisms (Ochman and Moran, 2001). The most notable examples are the pathogen *Mycoplasma genitalium* and the aphid symbiont *Buchnera* sp. APS. Whole-genome sequence comparisons revealed that a great number of genes such as those involved in biosynthesis are lost in pathogens and symbionts as an adaptive response to living within a host. The loss of genes involved in biosynthesis also reflects a lack of effective selection for maintaining genes in these microorganisms. Since the host provides a constant environment rich in metabolic intermediates, many molecules do not need to be synthesized (Gil et al., 2002) and the genes are therefore eliminated via mutational bias toward deletions (Mira et al., 2001).

Irreversible Pathogenic or Symbiotic Lifestyle

It has been hypothesized that the pathogenic and symbiotic interactions with a host are labile and thus the lifestyle of pathogens and symbionts is interchangeable (Herre et al., 1999; Moran and Wernegreen, 2000). It has also been proposed that symbionts arose as attenuated pathogens (Corsaro et al., 1999). If frequent transitions between pathogenic and symbiotic lifestyles occur, a mosaic of interaction patterns within clades of animal-associated bacteria would be observed (Moran and Wernegreen, 2000). However, phylogenetic analysis of animal-associated bacteria does not support this hypothesis. In contrast to theoretical predictions, no mosaic phylogenetic distribution patterns within deeply branching taxonomic clades were observed for the animal-associated bacteria. This suggests that the pathogenic or symbiotic interaction is not labile.

Recent genome sequence analysis provides the genetic basis supporting the irreversible specialization of obligate pathogens and symbionts, and thus pathogenesis and symbiosis most likely could not be switched. Due to the dramatic genome reduction, it is also less likely that specialized pathogens and symbionts can reacquire deleted genes, and thus they cannot revert to a free-living lifestyle independent of hosts (Ochman and Moran, 2001). Although both pathogens and symbionts share some commonalities of genetic changes in responding to host environments, such as dramatic reduction of genome size, low GC content, fast sequence evolution, and frequent gene rearrangement (Moran and Wernegreen, 2000), some fundamental differences are observed in their genome content. First, specialized mutualistic symbionts are capable of circumventing host defenses through various mechanisms. For instance, the aphid endosymbiont *Buchnera* lost virtually all the genes encoding surface molecules potentially involved in host defenses, whereas pathogens with reduced genomes, such as *Borrelia*, *Chlamydia*, *Mycoplasma*, *Rickettsia*, and *Treponema*, still have significant numbers of genes that are involved in cellular interactions and antigenic mechanisms (Ochman and Moran, 2001). Second, pathogenic bacteria with small genomes have also lost genes for intermediate metabolism and biosynthetic pathways, but a symbiont still possesses biosynthetic genes that are more directly beneficial to the host rather than to itself (Ochman and Moran, 2001). For instance, about 10% of the *Buchnera* genomes encode protein products for synthesizing amino acids that are unable to be produced by the host. The existence of an arsenal involved in biosynthetic pathways beneficial to the host in *Buchnera* strongly suggests that it did not evolve from a small genome pathogen (Moran and Wernegreen, 2000). It is possible that *Buchnera* evolved from a parasitic ancestor by reacquiring genes for amino acid biosynthesis, and hence became mutualistic. However, sequence comparison indicates that these biosynthetic pathways were not derived from an HGT event (Moran and Wernegreen, 2000).

4.9 GENOME PERSPECTIVE OF MITOCHONDRIAL EVOLUTION

Mitochondria are the energy-generating organelle of eukaryotes and encode a limited number of RNAs and proteins. The proteins encoded in mitochondria are largely components of respiratory complexes I through V of the electron transport chain (Burger et al., 1996). Mitochondria also encode rRNAs and tRNAs important in the organellar translation system. Although mitochondrial genomes contain some proteins essential to mitochondrial biogenesis and functions, many other proteins are encoded by nuclear genes, synthesized in the cytoplasm, and transported into the organelle. Due to their functional importance in energy generation, mitochondrial genomes have been extensively studied in terms of their structure, diversity, and evolution. In addition because of the high rates of nucleotide substitution and maternal inheritance, which is not subjected to recombination, mitochondrial DNA has been extensively used in population genetic studies.

Variations Among Mitochondrial Genomes Despite their universal functions, mitochondrial genomes exhibit remarkable diversity. Similar to a typical bacterial genome, many mitochondrial genomes have a circular structure, but some of them exist in a linear form (Gray et al., 1998a). The sizes of mitochondrial genomes vary dramatically, ranging from a ~ 6 kb in the human malaria parasite, *Plasmodium falciparum*, to more than 200 kb in land plants. The mitochondrial DNA (mtDNA) of *Arabidoposis thaliana* has a genome about 370 kb, which is the largest mitochondrial genome sequenced so far. Most mtDNAs have similar AT contents, ranging from 56 to 86% (Gray et al., 1998a).

Unlike typical bacterial genomes, the proportions of coding sequences vary significantly among mitochondrial genomes. For instance, more than 80% of *A. thaliana* mtDNA do not encode any functional product, whereas most protist mtDNAs typically contain 10% noncoding sequences (Gray et al., 1998a, 1999). Gene numbers encoded also vary significantly among different mitochondrial genomes (Gray et al., 1999). While only three protein genes are encoded by *Plasmodium* mtDNA, 97 are identified in the mtDNA of a heterotrophic flagellated protozoan, *Reclinomonas americana*. The *R. americana* genome contains 18 unique proteins that are not observed in other sequenced mitochondrial genomes (Gray et al., 1999; Kurland and Andersson, 2000). Another contrast to bacterial genomes is that the gene order in mitochondria is poorly conserved.

Mitochondrial genomes have evolved very differently among various eukaryotic lineages. Based on currently available mitochondrial genome sequences, mitochondria can be classified into two distinct categories: ancestral and derived (Gray et al., 1999). Ancestral mitochondrial genomes refer to those having clear bacterial vestiges such as the presence of many extra genes, rRNA genes, nearly complete tRNA genes, lack of introns, and use of a standard genetic code. Derived mitochondrial genomes are those having extensive genome reduction and gene loss, accelerated sequence evolution rates, and using nonstandard codons. Animal and most fungal mitochondrial genomes belong to this category (Gray et al., 1999). Currently, it is still not clear why and how mitochondrial genomes have evolved differently among different eukaryotes. With more mitochondrial genome sequences becoming available, a mechanistic understanding of the evolutionary processes of mitochondria should begin to emerge (Gray et al., 1999).

Nucleotide substitution rates are significantly different among mitochondrial genomes. By comparing whole mitochondrial genome sequences from six closely related vertebrates, Pesole and coworkers (1999) showed that high intragenomic variations in nucleotide substitution rates were observed, which is strongly dependent on the region being considered. For instance, the D-loop central domain and rRNA and tRNA genes evolved much more slowly than the two peripheral D-loop region domains. While the synonymous substitution rate is fairly uniform over the genomes, the nonsynonymous substitution rate is quite different among different genes. Also, the nonsynonymous nucleotide substitution rates of mtDNA are comparable to those of nuclear genes.

Origin of Mitochondria The hypothesis of an endosymbiotic origin of the mitochondrion arose in the nineteenth century and was precisely proposed by Margulis (1970). Before 1998, it was commonly believed that the endosymbiont was an alpha-proteobacterium such as *Paracoccus* and the host was an archaeon (Doolittle, 1998; Gray, 1992). Phylogenetic analysis based on the sequences of cytochrome c oxidase and cytochrome b indicated that mitochondria diverged from bacteria between 1.5 and 2.0 billion years ago, suggesting that the oxidative respiratory system in eukaryotes was very ancient (Kurland and Andersson, 2000).

Recently, members of the rickettsial subdivision of alpha-proteobacteria such as *Rickettsia*, *Anaplasma*, and *Ehrlichia* are thought to be among the closest known eubacterial relatives of mitochondria and the possible descendants of the endosymbiotic ancestor to mitochondria (Gray et al., 1999; Kurland and Andersson, 2000). These organisms are obligate intracellular parasites and are also thought to be the descendants of a free-living alpha-proteobacterium. Although the complete genome sequences of *Ricksettsia prowazekii* and *Bartonella henselae* are closely related to mitochondrial genomes (Kurland and Andersson, 2000), there is no evidence to show that the mitochondrial genomes evolved directly from an already reduced rickettsial-like genome (Gray et al., 1999).

Phylogenetic analyses based on both rRNA and protein sequences indicated that the mitochondrial genome was monophyletic. All mitochondrial genomes were descendants of a common protomitochondrial ancestor (Gray et al., 1999). In addition, the mitochondrial *rps* and *rpl* genes encoding ribosomal proteins were clustered and organized similarly to those in *E. coli* genomes, and the gene deletions in the *rps-rpl* cluster appeared to be conserved among mitochondrial *E. coli* and *R. prowazekii* genomes. These observations strongly suggest that mitochondria originated only once in evolution (Gray et al., 1999).

Genome Reduction   Two evolutionary modes appear to be responsible for the reduction of the coding capacity of mitochondrial genomes. One reductive mode is the loss of nonessential genes. Phylogenetic reconstructions using coding sequences from mitochondrial and bacterial genomes indicated that the common ancestor of *Rickettsia*, *Bartonella*, and mitochondria was a free-living bacterium, probably with a genome size larger than *Bartonella*. The relative small size of the extant mitochondrial genomes suggests that extensive gene loss occurred during the evolution of the protomitochondrial genome after endosymbiosis (Kurland and Andersson, 2000). For instance, due to the endosymbiotic lifestyle, operational genes, such as those involved in amino acid biosynthesis, nucleotide biosynthesis, cofactor biosynthesis, fatty acid and phospholipid metabolism, energy and intermediary metabolism, cell envelope synthesis and cell division, as well as most informational genes, were redundant and consequently deleted during the evolution of mitochondria. Such loss appears to have occurred at the early stage of mitochondrial genome evolution (Gray et al., 1999). In addition, the gene content of mtDNA varies substantially among different mitochondrial lineages, suggesting that differential gene loss also occurred after the divergence from protomitochondrial genome (Gray et al., 1999).

The other reductive mode involves the transfer of essential genes from the mitochondrial genomes to the nuclear genome. Most of the genes needed by mitochondria are found in the nuclear genome rather than in the mitochondrial genome. Sequence comparison revealed that some of them originated from the proteobacterial ancestor, whereas a considerable fraction arose within eukaryotic genomes (Andersson and Kurland, 1999). For instance, the yeast nucleus encodes about 400 proteins required by mitochondria and half of them appear to be of bacterial origin (Kurland and Andersson, 2000). Berg and Kurland (2000) argued that the higher mutation rates in mitochondria appeared to be the primary driving force for the transfer of mitochondrial genes to the nucleus.

Model for Mitochondrial Evolution and Origins of Eukaryotic Cells   The standard model of the endosymbiotic theory hypothesizes that the symbiosis was formed by exchanging ATP produced aerobically by the symbiont for organics produced by the host. This theory also assumes that the host had an anaerobic heterotrophic type of metabolism, having characteristics similar to those of the eukaryotic nucleocytoplasm (Margulis, 1981). But it is not clear whether the host was an archaebacterium or a full-fledged eukaryote with a nucleus (Gray et al., 1999). Various formulations of the endosymbiosis theory were proposed, but all of them imply that the nuclear genome was formed first and derived from the host, followed by a subsequent separate endosymbiotic event leading to mitochondria. After endosymbiosis, the nuclear genes involved in mitochondrial biogenesis and function were subsequently transferred from the mitochondria to the nucleus (Gray et al., 1999).

This traditional serial endosymbiosis-based view has been challenged by several recent observations. First, whole-genome sequence comparison showed that nuclear genomes contain informational genes largely from archaea and operational genes primarily from bacteria. Also, the number of genes with bacterial origins in the nuclear genome is much greater than that which could be transferred from a mitochondrial genome, and some of them are not involved in mitochondrial biogenesis and function. These observations suggest that the nuclear genome appears to be an evolutionary chimera derived from both archaeal and bacterial progenitors rather than from the anaerobic archaebacterium host, as suggested by the standard model of endosymbiosis theory (Gray et al., 1999).

A group of eukaryotes, collectively called Archezoa, lack mitochondria but have a nucleus and phagocytose. Phylogenetic analysis based on SSU rRNA genes suggests that Archezoa diverged away from the main eukaryotic lineages before the advent of mitochondria (Sogin, 1991; Gray et al., 1999). This led to the proposal that the anaerobic host involved in endosymbiosis was a primitive eukaryote. However, this view has also been challenged by several recent findings. First, the genes encoding typical mitochondrial proteins have been observed in the nuclear genomes of Archezoa. In some situations, mitochondrial proteins have been found in the hydrogenosome, an organelle capable of producing hydrogen and ATP anaerobically. These observations suggest that mitochondria were once present in Archezoa and were subsequently lost. Since they have similar functions, the hydrogenosome and the mitochondrion could be descended from the same alpha-proteobacterium ancestor (Gray et al., 1999). Second, Archezoa appear to have been misplaced in earlier phylogenetic trees due to methodological artifacts (Gray et al., 1999). Reappraisal of the sequences clustered microsporidia, a group of Archezoa, with fungi. Thus, Archezoa appear not to be primitive eukaryotes, and hence the proposal that the anaerobic host involved in endosymbiosis was a primitive eukaryote is also questionable.

The observations that the nuclear genome is chimeric in nature and that Archezoa are probably not primitive eukarytoes suggest that the serial endosymbiosis model is questionable. It seems that the major eukaryotic lineages, including those without mitochondria, diverged more or less simultaneously and the origin of mitochondria was much earlier than the time that alpha-proteobacteria separated from the rest of the bacterial lineages (Sogin, 1997; Gray et al., 1999). To accommodate these recent findings, various alternative hypotheses have been proposed (Gray et al., 1999). Based on the comparative biochemistry of energy metabolism, Martin and Muller (1998) proposed that the eukarytoic nucleus and mitochondrion were simultaneously created by the fusion of an anaerobic, strictly hydrogen-requiring methanogenic Archaebacterium (the host) with an alpha-proteobacterium (the symbiont) that was able to respire but generated hydrogen as a waste product of anaerobic heterotrophic metabolism. During the fusion event, it is possible that the same alpha-proteobacterium led to mitochondria-contributed genes involved in the cellular operation of the eukaryotic nucleus. This hypothesis is referred to as the *hydrogen hypothesis*. It can explain the chimeric origin of the nucleus and allows the possibility of the simultaneous origin of the ancestor of eukaryotic cells and its mitochondrion.

4.10 SUMMARY

Understanding the phylogenetic relationships between extant organisms has been a great challenge to biologists for many centuries. Because of the complicated evolutionary processes, phylogenetic relationships of species based on comparisons of single genes are rarely consistent with each other. With the availability of whole-genome sequences for many organisms, it is hoped that comprehensive descriptions of evolutionary relationships among different organisms can be obtained. Recent comparative genomic studies have revealed some fundamentally different and exciting insights into various evolutionary processes of biological systems.

Since most changes in DNA sequences occur randomly by small mutations, the genetic divergence of the two species will increase as time passes. The molecular clock hypothesis assumes that the rate of amino acid or nucleotide substitution is roughly constant among diverse lineages. Current studies with many genomic sequences showed that there are no genes or proteins that have evolved constantly among all lineages. But it is possible that the mean evolution rates of a group of genes are more or less constant. As a result, the evolutionary rate can be used to date the divergence of different genomes. Using this approach, we determine the divergence time between Bacteria and Eucarya to be about 3 billion years, which is younger than the age of microfossils.

One of the most exciting discoveries revealed by comparative genomics is that horizontal gene transfer (HGT) plays a major role in the evolution of prokaryotic genomes. This represents a major shift in our understanding of the evolution of life. Genome sequence analysis indicates that HGT is a common evolutionary event, but whether it is a major evolutionary force is still highly controversial. Various approaches can be used to identify horizontal HGT events, but it is very difficult to identify and prove them because the evidence supporting HGT can be explained by other forces. It has been observed that informational genes are less prone to HGT than operational genes. This observation leads to the hypothesis that the complexity of gene interactions is a significant factor which restricts successful HGT. In addition, a wide range of variations in HGTs was observed among different evolutionary lineages. It is believed that HGT is an important mechanism for forming operon organization and a key venue for bacterial speciation and evolution by resorting existing capabilities. However, interspecific gene transfer and recombination blur species boundaries, and thus HGT may also facilitate the process against speciation.

Gene duplication is another major force in genome evolution, because the duplication of genes and groups of genes has occurred frequently in all genomes through processes of unequal crossing-over, DNA amplification, replication slippage, etc. Genome comparison indicates that gene duplication rates vary from 0.002 (fruit fly) to 0.02 (worm) with an average of 0.01 per gene per million years. The proportion of duplicated genes ranged from 5 to 33% in various prokaryotes, yeast, *Arabidopsis*, worm, fly, and human. The duplicated genes could have different evolutionary fates through four functionalization processes: nonfunctionalization, neo-functionalization, subfunctionalization, and equifunctionalization.

Another exciting discovery revealed by comparative genomics is that lineage-specific gene loss also plays a key role during the evolution of life. Genome sequence comparison revealed that genes could be lost by large deletion and/or by mutational inactivation, followed by gene erosion. It is believed that gene loss or silencing of duplicated genes is more important to the evolution of species diversity than the evolution of new functions in duplicated genes. The narrow range of variation in genome size and lack of nonfunctional sequences in prokaryotes suggest that prokaryotic genome evolution is biased toward DNA loss rather than DNA insertion.

Ribosomal gene-based phylogenetic analysis revealed that life could be classified into three distinct domains: Archaea, Bacteria, and Eucarya. Various duplicated genes were used to root the universal tree of life and the results from many genes support the notion that Bacteria diverged first from a lineage leading to Archaea and Eucarya. The establishment of the three-domain universal tree has been marked as the most significant development in microbiology and has had a tremendous impact on understanding evolutionary relationships among different living organisms. Most of the genome-based phylogenetic studies support the notion of three domains. However, the universal tree concept is challenged by both methodological problems and HGT processes. Genome-based phylogenetic analyses with different methods indicated that the effects of HGT were not sufficient to wash out the phylogenetic signal, at least at the three-domain levels. Also, each domain had unique phenotypic characteristics in terms of molecular structure and composition, physiology, and morphology. The apparent phenotypic and genetic coherence of each domain implied that the three-domain notion is operational and does reflect some ancient fundamental events in the evolutionary history of life.

One interesting scientific question concerns what is the minimum number of genes necessary to support cellular life. Defining the minimal gene set is important to our understanding of basic cell functioning and evolution. Results from comparative genomic analysis, genetic mutagenesis, and biochemical inference suggested that a cell could be supported by a core set of ~ 150 genes encoding the basal systems for transcription, translation, and replication, but a cell with the minimal core gene set could never be constructed in practice although this is possible in theory.

Microorganisms form complex intimate mutualistic or parasitic relationships with a variety of multicellular organisms. Despite differences in the ultimate outcomes, both pathogens and symbionts share some commonalities in terms of genomic structure, genetic regulation, and evolution. Comparative genome sequence analysis indicates that HGT and gene loss are the major evolutionary forces of pathogenesis and symbiosis. Due to dramatic genome reduction, however, obligate pathogens and symbionts are irreversibly specialized and thus pathogenesis and symbiosis most likely could not be switched.

Mitochondria are the energy-generating organelles of eukaryotes and encode a limited number of RNAs and proteins. Despite their universally conserved functions, mitochondrial genomes exhibit remarkable diversity. The sizes of mitochondrial genomes vary dramatically, ranging from ~ 6 kb in the human malaria parasite, *Plasmodium falciparum*, to more than 200 kb in land plants. Genome sequence analysis suggested that the members of the rickettsial subdivision of the alpha-proteobacterium such as *Rickettsia*, *Anaplasma*, and *Ehrlichia* were thought to be among the closest known eubacterial relatives of mitochondria and were possible descendants of the endosymbiotic ancestor to mitochondria. The observations that the nuclear genome has a chimeric nature and that the archezoa are probably not primitive eukaryotes suggested that the traditional serial endosymbiosis model for mitochondrial evolution is questionable. A recent theory, the hydrogen hypothesis, states that the eukaryotic nucleus and mitochondrion were simultaneously created by the fusion of an anaerobic, strictly hydrogen-requiring methanogenic Archaebacterium (the host) with an alpha-proteobacterium (the symbiont) that was able to respire but generated hydrogen as a waste product of anaerobic heterotrophic metabolism.

FURTHER READING

Ayala, F. J. 1997. Vagaries of the molecular clock. *Proc. Natl. Acad. Sci. USA* 94:7776–7783.

Brown, J. R. and W. F. Doolittle. 1997. Archaea and the prokaryote-to-eukaryote transition. *Microbiol. Mol. Biol. Rev.* 61:456–502.

Brown, J. R., C. J. Douady, M. J. Italia, W. E. Marshall, and M. J. Stanhope. 2001. Universal trees based on large combined protein sequence data sets. *Nat. Genet.* 28:281–285.

Doolittle, W. F. 1999a. Phylogenetic classification and the universal tree. *Science* 284:2124–2129.

Doolittle, W. F. 1999b. Lateral genomics. *Trends Cell. Biol.* 9:M5–M8.

Forterre, P. and H. Philippe. 1999. Where is the root of the universal tree of life? *Bioessays* 21:871–879.

Gray, M. W., G. Burger, and B. F. Lang. 1999. Mitochondrial evolution. *Science* 283:1476–1481.

Gupta, R. S. 1998. Protein phylogenies and signature sequences: a reappraisal of evolutionary relationships among archaebacteria, eubacteria, and eukaryotes. *Microbiol. Mol. Biol. Rev.* 62:1435–1491.

Jain, R., M. C. Rivera, and J. A. Lake. 1999. Horizontal gene transfer among genomes: the complexity hypothesis. *Proc. Natl. Acad. Sci. USA* 96:3801–3806.

Koonin, E. V., L. Aravind, and A. S. Kondrashov. 2000. The impact of comparative genomics on our understanding of evolution. *Cell* 101:573–576.

Koonin, E. V., K. S. Makarova, and L. Aravind. 2001. Horizontal gene transfer in prokaryotes: quantification and classification. *Ann. Rev. Microbiol.* 55:709–742.

Kurland, C. G. and S. G. Andersson. 2000. Origin and evolution of the mitochondrial proteome. *Microbiol. Mol. Biol. Rev.* 64:786–820.

Lawrence, J. 1999. Selfish operons: the evolutionary impact of gene clustering in prokaryotes and eukaryotes. *Curr. Opin. Genet. Dev.* 9:642–648.

Lynch, M. and J. S. Conery. 2000. The evolutionary fate and consequences of duplicate genes. *Science* 290:1151–1155.

Mira, A., H. Ochman, and N. A. Moran. 2001. Deletional bias and the evolution of bacterial genomes. *Trends. Genet.* 17:589–596.

Moran, N. A. and J. J. Wernegreen. 2000. Lifestyle evolution in symbiotic bacteria: insights from genomics. *Trends Ecol. Evol.* 15:321–326.

Ochman, H., J. G. Lawrence, and E. A. Groisman. 2000. Lateral gene transfer and the nature of bacterial innovation. *Nature* 405:299–304.

Ochman, H. and N. A. Moran. 2001. Genes lost and genes found: evolution of bacterial pathogenesis and symbiosis. *Science* 292:1096–1099.

Pace, N. R. 1997. A molecular view of microbial diversity and the biosphere. *Science* 276:734–740.

Salzberg, S. L., O. White, J. Peterson, and J. A. Eisen. 2001. Microbial genes in the human genome: lateral transfer or gene loss? *Science* 292:1903–1906.

5

Computational Methods for Functional Prediction of Genes

Ying Xu

5.1 INTRODUCTION

The past decade has witnessed a gradual paradigm shift in the functional studies of genes as a result of worldwide sequencing and bioinformatic efforts. Traditionally, studies of genes typically start with the specific goal of understanding a particular biological function or process, for example, sporulation in budding yeast or electron transport in *Shewanella oneidensis*. Searching for genes involved in a specific biological function is often implemented through mutagenesis studies (see Chapter 8), which attempt to pinpoint the DNA sequence segments in a cell's chromosome that may be relevant to the function of interest. Techniques like positional cloning are often used to determine the precise boundaries of the candidate genes. These genes are identified and studied because of their possible relevance to a particular biological function.

The availability of the large quantity of genomic sequence, produced by the human and other genome sequencing projects (Lander et al., 2001; Yu et al., 2002b), has spurred widespread computational studies of gene prediction directly from genomic sequence. Using statistical analysis and pattern recognition techniques, a number of computational methods have been developed for gene predictions (Fickett and Tung, 1992; Uberbacher et al., 1996; Burge and Karlin, 1997; Salzberg et al., 1998a,b). Two types of information are generally used in these function prediction techniques: (1) statistical features that are common among previously known protein-coding genes but not shared by noncoding genomic sequences, and/or (2) full-length cDNA sequences and/or expression sequence tags (ESTs) (Boguski et al., 1993; also see Chapter 3 for more details). The reliability of

Microbial Functional Genomics, Edited by Jizhong Zhou, Dorothea K. Thompson, Ying Xu, and James M. Tiedje.
ISBN 0-471-07190-0 © 2004 John Wiley & Sons, Inc.

these computational methods for gene function prediction has reached such a high level that they have become the dominating technique for gene identification. The prediction of the 30,000+ genes in the human genome is one such successful example (Lander et al., 2001; Venter et al., 2001), although debates are still on-going about what percentage of protein-coding genes have been identified in the human genome. Currently, gene prediction accuracies for microbial genomes have generally reached the level of 90% or higher (Salzberg et al., 1998b), while eukaryotic gene predictions still have some room for improvement (Guigo, 1999).

One challenging problem is that we now have a vast majority of genes identified by computational tools for each sequenced genome but we have few clues about their biological functions. Can computational techniques help us to derive functional information about the novel genes? In this chapter, we present computational methods for the functional inference of genes, based on information derived from their sequences and structures (of their encoded proteins) and their relationships to other genes.

5.2   METHODS FOR GENE FUNCTION INFERENCE

5.2.1   Gene Functions at Different Levels

The biological function of a gene is typically realized through the protein encoded in its DNA sequence (with the exception of RNA-coding genes; see Chapter 3 for a more detailed discussion). Hence, the functional characterization of a gene can generally be carried out through functional analysis of the protein it encodes. There are a number of different classification systems for protein functions (see Section 5.2.2). Generally, a protein's function can be described at three levels: molecular, cellular, and phenotypic, representing, respectively, (1) its activity with respect to interactions with other molecules, (2) its role in a cellular system, and (3) its influence on the properties of an organism as a whole (Bork et al., 1998). In this chapter, we focus mainly on the characterization and analyses of protein functions at the molecular or cellular levels.

For either molecular- or cellular-level predictions, functional information can be provided at different levels of resolution or detail. At a low-resolution level, we may say that protein X is an enzyme or a DNA-binding protein. At a medium-resolution level, protein X may be described as a protease (one of the six classes of enzymes) or may function as a transcriptional repressor when it binds to DNA. At a high-resolution level, we may know that protein X is a protease with trypsin specificity or it is a *lac* repressor (Moult and Melamud, 2000). As we will discuss later in this chapter, different levels of function are computationally derivable for different proteins, depending on the amount of information available on their "related" proteins.

5.2.2   Searching for Clues to Gene Function

The most basic rule in predicting gene function using a computational approach is "guilt by association." The general idea is that if two genes or their encoded proteins are related evolutionarily, biophysically, or biochemically, then they may express similar functions or be functionally linked. These evolutionary or bio-physical/chemical relationships could possibly be detected through analyses of genomic sequences, sequence or structural motifs, protein structures, protein–protein interaction data, or other experimental data. This section provides a high-level roadmap for the commonly used prediction paradigms for the functional inference of individual genes. More detailed discussions of applying these prediction tools will be given in the following sections.

Scientists have accumulated a great amount of information about the biological functions of many genes in the past. Most of these genes, along with their known functions and DNA sequences, have been stored in several public databases. These databases can be roughly divided into the following categories: (1) protein sequence databases, (2) protein structure databases, and (3) protein family databases.

1. *Protein Sequence Databases.* GenBank represents probably the most comprehensive database of genes (Benson et al., 2002). It contains virtually all known DNA sequences and sequenced complete genomes. GenBank provides detailed information on the locations of all known genes in a DNA sequence. Swiss-prot is a database of all known protein sequences along with their annotated functions (Bairoch and Apweiler, 2000). As of August 2002, it contained 112,894 protein sequences. TrEMBL contains all protein sequences translated from gene sequences that are not in the Swiss-prot database (O'Donovan et al., 1999). Protein Identification Resource (PIR; Barker et al., 1999) is a database of nonredundant protein and nucleic acid sequences, along with bibliographic and annotated information for each protein. The *nr* database consists of nonredundant protein sequences from all the above protein databases plus translated protein sequences from GenBank.

2. *Protein Structure Databases.* PDB is a database containing virtually all experimentally solved protein structures, both X-ray crystallographic and NMR structures (Westbrook et al., 2000). As of August 2002, PDB contained 18,691 structures. DALI is a database of all unique structural domains of proteins (Holm and Sander, 1996), and its structural domains are organized in a hierarchical structure, based on their functional relationships. Protein-Nucleic Acid Complex Database is a database containing the structural data of protein-nucleic acid complexes (available from http://www.rtc.riken.go.jp/jouhou/3dinsight/complexdb.html). These data are classified according to the recognition motif of proteins and DNA forms involved in the complex.

3. *Protein Family Databases.* Pfam is probably the most widely used protein family database (Bateman et al., 1999, 2002). It currently consists of over 4,000 protein families. PRODOM is a database for homologous protein sequence domains (Corpet et al., 2000), that has been mainly used for the prediction of gene domain boundaries. Structural Classification of Proteins (SCOP) is a structure-based family classification database for proteins; it classifies proteins into a hierarchy of protein families (Murzin et al., 1995). A similar structure-based protein family database is the Class, Architecture, Topology and Homologous (CATH) superfamily (Orengo et al., 1998). ENZYME is an enzyme classification database (Bairoch, 1993).

These databases provide the basic source of information for the computational inference of protein functions. When a novel gene is identified or predicted in a particular genome, some level of function could possibly be assigned to this gene by discovering evolutionary, biochemical, or biophysical relationships between the gene and genes/proteins in these databases. The following subsections summarize a few key approaches to computation-based functional inference.

Homology Detection through Sequence Comparison Homologous genes from closely related organisms can be generally detected by comparison of their DNA sequences. Since the publication of the classical Smith–Waterman algorithm (Smith and Waterman, 1982), numerous sequence comparison algorithms have been developed. These algorithms include the popular BLAST program (Altschul et al., 1990), FASTA program (Pearson, 1990), and more recent and sensitive algorithms such as PSI-BLAST (Altschul et al., 1997) and SAM (see Table 5.2), which use information derived from multiple sequence alignments. As homologous genes generally share some level of common functionality (Murzin et al., 1995), low-resolution functional inference can be achieved through the detection of homologs in the protein databases. Functional inference at higher resolution may depend on our capability to distinguish orthologs from paralogs among the homologous genes.

Detection of Low Homology through Sequence-Structure Comparison In the most general situation, sequence-based methods may not be sensitive enough to detect genes with low homology from distantly related organisms. Sequence-structure alignment represents a more general and sensitive method for detecting remote homologs. A particular class of sequence-structure comparison methods is called *protein threading* (Bowie et al., 1991). A threading method detects a homolog by using information of both sequence similarity and structural information such as the physicochemical fit of a protein sequence with a three-dimensional structural template, and hence represents a more powerful class of tools for homology detection. The superiority of threading-based methods over pure sequence-based methods for remote homology detection has been rigorously demonstrated in the recent CASP protein structure prediction contest (CASP4, 2001).

Identification of Functional Motifs in Sequence It is well known that a significant portion of a protein structure may function only as a scaffold for its biologically functional parts. Hence, proteins with different overall structural folds could have similar functions as long as their functional sites (or active sites) are structurally similar. These "conserved" functional sites could possibly be detected through sequence comparisons or structural analyses. A number of sequence- or structure-based computational techniques have been developed for the detection of particular active sites of proteins, even though these proteins may not have a detectable sequence or structural homology. These tools include PRINTS (Attwood et al., 2002), PROSITE (Bucher and Bairoch, 1994; Falquet et al., 2002), and BLOCKS (Henikoff et al., 1999, 2000).

Nonhomologous Approaches *Phylogenetic profiling* (Pellegrini et al., 1999) and *gene fusion* (Marcotte et al., 1999) are two recently developed nonhomologous techniques for the functional inference of genes. The phylogenetic profiling approach predicts functionally related genes, based on the assumption that proteins functioning together in a pathway or structural complex are likely to evolve in a correlated fashion. During evolution, all such functionally linked proteins tend to be either preserved or eliminated altogether in a new species. Hence, by identifying two genes that have the same profile of presence or absence over a large set of genomes, one can predict that they may be functionally linked. The gene fusion approach is based on the observation that some interacting proteins have homologs in another organism fused into a single protein chain. By identifying proteins that exist as one single chain in one genome and exist as separate entities in other genomes, one can predict the possible interaction between two proteins.

5.3 FROM GENE SEQUENCE TO FUNCTION

The basic premise in sequence-based functional inference is that genes with high sequence similarity tend to be homologous, and hence generally have similar or related functions. A basic functional-inference paradigm typically involves two components: (1) a database of genes with known functions, often organized as a set of functional families (Bateman et al., 2002), and (2) a capability for assessing sequence similarities, typically through sequence comparison or alignment (Smith and Waterman, 1982; Altschul et al., 1990; Pearson, 1990). At a low resolution, the function of a novel gene is predicted to have the same function as its homologous genes. More detailed analyses may distinguish orthologs from paralogs among homologous genes. By definition, *orthologs* are genes in different genomes that have evolved from a common ancestral gene, whereas *paralogs* are genes related by duplication within a genome (Tatusov et al., 1997; see Chapter 4 for a further discussion). Generally, orthologs retain the same function during the course of evolution, while paralogs could have deviated somewhat from their original functions. The prediction of orthologs and paralogs among homologous genes represents a highly challenging and unsolved problem. For low-resolution analyses, people may be content with knowing if two genes are homologous.

5.3.1 Hierarchies of Protein Families

There are numerous classification schemes that serve as the foundation for the functional inference of genes. The concept of a *gene family* is generally used to represent a group of genes that are homologous. Operationally, gene families are in general defined in terms of sequence similarities or the similarities of their protein structures. Family classification is often done based on identifying genes/proteins having sequence or structure similarities within certain similarity cutoffs. Apparently, the detailed selection of similarity cutoffs may not be unique, which may translate into somewhat different classification results by different computer programs.

Pfam (Bateman et al., 2002) represents one of the most popular family classification schemes and databases. The latest version of Pfam consists of over 4,100 gene families (according to information released August 2002) that cover 69% of the Swissprot + TrEMBL gene sequences. Pfam builds its gene families based on the criterion that protein sequences of the same family have multiple sequence alignments of statistical significance. That is, when aligning the protein sequences of the same family, a significant portion of the aligned residues share the same or similar physicochemical features. Figure 5.1 shows a multiple sequence alignment (partial) among 24 members of the "envelope glycoprotein GP120" family from different species. The following description of this family is provided by the Pfam database: "The entry of HIV requires interaction of viral GP120, an envelope glycoprotein with human T-cell surface glycoprotein CD4 and a chemokine receptor on the cell surface. Proteins belonging to this family are found in HIV types 1 and 2, and Simian Immunodeficiency virus (SIV)." Note that residues of an aligned column (Fig. 5.1) are generally of the same or related amino acid types. If a novel gene matches well with the overall "profile" of the amino acid distribution of this family, identified through sequence comparison, one could conclude that this gene is homologous to this gene family and hence possibly has the same or similar functions.

For each gene family, Pfam provides not only the multiple sequence alignments but also various other types of information and annotation results. They include (1) functional

```
ENV_SIVM1/24-528 QYVTVFYGVPAWRNATIPLFCATKNR.......DTWGTTQCLPDNDDYSELALN.VTESFDAWE..NTVTEQAIEDVWQ:
ENV_HV2NZ/24-502 QFVTVFYGIPAWRNASIPLFCATKNR.......DTWGTIQCLPDNDDYQEITLN.VTEAFDAWN..NTVTEQAVEDVWN:
ENV_HV2G1/23-502 QYVTVFYGVPVWRNASIPLFCATKNR.......DTWGIIQCKPDNDDYQEITLN.VTEAFDAWD..NTVTEQAVEDVWS:
ENV_HV2D1/24-501 QYVTVFYGIPAWRNASIPLFCATKNR.......DTWGTIQCLPDNDDYQEITLN.VTEAFDAWD..NTVTEQAIEDVWR:
ENV_HV2CA/25-512 QYVTVFYGVPAWKNASIPLFCATKNR.......DTWGTIQCLPDNDDYQEIPLN.VTEAFDAWD..NTITEQAIEDVWN:
ENV_HV2BE/24-510 QYVTVFYGIPAWKNASIPLFCATKNR.......DTWGTIQCLPDNDDYQEIILN.VTEAFDAWN..NTVTEQAVEDVWH:
ENV_HV2D2/24-513 QYVTVFYGIPAWRNATVPLICATTNR.......DTWGTVQCLPDNGDYTEIRLN.ITEAFDAWD..NTVTQQAVDDVWR:
ENV_SIVAI/22-522 LYVTVFYGIPVWKNSTVQAFCMTPNT.......NMWATTNCIPDDHDNTEVPLN.ITEAFEAWD..NPLVKQAESNIHL:
ENV_SIVA1/24-538 QWITVFYGVPVWKNSSVQAFCMTPTT.......RLWATTNCIPDDHDYTEVPLN.ITEPFEAWADRNPLVAQAGSNIHL:
ENV_SIVCZ/33-496 LWVTVYYGVPVWHDADPVLFCASDAKAHSTEAHNIWATQACVPTDPSPQEVFLPNVIESFNMWK..NNMVDQMHEDIIS:
ENV_HV1ZH/33-511 LWVTVYYGVPVWKDAETTLFCASDAKAYDTEKHNVWATHACVPTDPNPQELSLGNVTEKFDMWK..NNMVEQMHEDVIS:
ENV_HV1W1/33-510 LWVTVYYGVPVWKEATTTLFCASDAKAYSTEAHKVWATHACVPTNPNPQEVVLENVTENFNMWK..NNMVEQMHEDIIS:
ENV_HV1J3/33-523 LWVTVYYGVPVWKEAATTLFCASDAKAYDTEVHNVWATHACVPTDPNPQEVVLENVTEKFNMWK..NNMVEQMHEDIIS:
ENV_HV1B1/34-511 LWVTVYYGVPVWKEATTTLFCASDAKAYDTEVHNVWATHACVPTDPNPQEVVLVNVTENFNMWK..NDMVEQMHEDIIS:
ENV_HV1A2/33-509 LWVTVYYGVPVWKEATTTLFCASDAKAYDTEVHNVWATHACVPTDPNPQEVVLGNVTENFNMWK..NNMVEQMQEDIIS:
ENV_HV1RH/33-519 LWVTVYYGVPVWKEATTTLFCASEAKAYKTEVHNVWAKHACVPTDPNPQEVLLENVTENFNMWK..NNMVEQMHEDIIS:
ENV_HV1BN/34-507 LWVTVYYGVPVWKEANTTLFCASDAKAYDTEIHNVWATHACVPTDPNPQELVMGNVTENFNMWK..NDMVEQMHEDIIS:
ENV_HV1OY/33-509 LWVTVYYGVPVWKEATTTLFCASDARAYATEFHNVWATHACVPTDPNPQEVVLGNVTENFDMWK..NNMVEQMQEDIIS:
ENV_HV1C4/35-522 LWVTVYYGVPVWKEATTTLFCASDAKAYDTEAHNVWATHACVPTNPNPQEVVLENVTENFNMWK..NNMVEQMHEDIIS:
ENV_HV1Z8/33-518 LWVTVYYGVPVWKEATTTLECASDAKSYEPEAHNIWATHACVPTDPNPRIEIEMENVTENFNMWK..NNMVEQMHEDIIS:
ENV_HV1EL/33-508 LWVTVYYGVPVWKEATTTLFCASDAKSYETEAHNIWATHACVPTDPNPQEIALENVTENFNMWK..NNMVEQMHEDIIS:
ENV_HV1ND/33-501 LWVTVYYGVPIWKEATTTLFCASDAKAYKKEAHNIWATHACVPTDPNPQEIELENVTENFNMWK..NNMVEQMHEDIIS:
ENV_HV1MA/33-513 LWVTVYYGVPVWKEATTTLFCASDAKSYETEVHNIWATHACVPTDPNPQEIELENVTEGFNMWK..NNMVEQMHEDIIS:
ENV_SIVGB/47-569 QYVTVFYGVPVWKEAKTHLICATDNS.......SLWVTTNCIPSLPDYDEVEIPDIKENFTGLIRENQIVYQAWHAMGS:
```

Figure 5.1 *Each row represents a protein sequence of the "envelope glycoprotein GP120" family.* The first column provides the name of each individual protein in this family and the starting and ending positions of a protein sequence being aligned here. The second column displays the multiple sequence alignment among 24 members of this family.

annotation of the family, (2) phylogenetic tree of the family, (3) sequence domains, and (4) links to other databases like SCOP and PDB. In addition, Pfam provides a powerful search engine that allows a user to search the database, using protein or gene sequences, keywords, or taxonomy. By searching the database with a query gene sequence, Pfam could provide a great amount of information about the gene, possibly including its biological functions, related protein structures, its family structure, and links to the relevant literature. Such a "one-stop-shop" search environment is clearly becoming a popular paradigm for many bioinformatic search engines.

Other Sequence-based Family Classification Schemes   There are a few family classification databases based on conserved sequence motifs or *fingerprints*. The basic idea is to classify genes into families based on the functional motifs they share or do not share. Such family classification schemes include PROSITE (Falquet et al., 2002), PRINTS (Attwood et al., 2002), and BLOCKS (Henikoff et al., 1999, 2000). Unlike Pfam, these classification schemes generally do not rely on sequence similarities of a whole gene or a significant portion of a whole gene with other genes. Section 5.4 provides a more detailed discussion of these databases.

One issue concerning the sequence-based family classification schemes is that they provide only one level of classification. In other words, two genes either belong to the same family or not, without any additional structure built over all the gene families, which can be used to relate one gene family to another. Clearly, a hierarchy over the gene families will provide additional information for gene functional assignments. Such a hierarchy currently exists only for a subset of all known gene/protein families, that is, those with known protein structures.

SCOP Database and Other Structure-based Family Classifications   A number of (protein) structure-based family classification schemes and databases have been developed. The SCOP database (Murzin et al., 1995) is one such database. Using additional information provided by the protein structures, SCOP provides a hierarchy among protein families. The hierarchy consists of three levels of classification: families, superfamilies, and fold families.

SCOP's definition of a family is a group of homologous proteins that typically have unambiguous sequence alignments, that is, correspondence between residues from different protein sequences is well defined. A group of gene families form a *superfamily* if their similar structural features and similar functions suggest that these proteins evolved from the same origin. The determination of a superfamily relies on structural and functional comparisons. Proteins from different superfamilies form a *fold family* if they have the same secondary structures in similar three-dimensional (or 3D) topological arrangements. Proteins of the same fold family only indicate that they have similar three-dimensional structures but not necessarily similar functions. By identifying a gene belonging to a particular class in this hierarchy, some level of function could be derived. For example, if a gene is determined to belong to a particular family, functions at a high-resolution level could possibly be assigned. When a gene is determined to belong to a particular superfamily (but not a particular family), some low-resolution functional inference can be made, based on the commonality or similarities of the functions among the members of this superfamily. Even when only able to identify that a protein is part of a fold family, one could possibly narrow down its possible functions to a small list. Currently, the SCOP database records 1,827 families, 1,073 superfamilies, and 686 fold families (according to a March 2002 release).

One example of a SCOP hierarchy is given here. *Immunoglobulin-like beta-sandwich* represents one fold family, and all proteins in this fold family have the structural feature of forming a sandwich with seven β strands in two sheets and one Greek key. This fold family consists of the following 15 superfamilies, which further consist of a total of 22 families. The numbers given in parentheses in the following list indicate the number of families each superfamily has. By identifying a protein belonging to this fold family, one can narrow down its possible function to one of the 15 possible functional classes. If one can further identify that the protein belongs to the immunoglobulin superfamily, we can predict if the protein's function has the same or similar function of immunoglobulin. Figure 5.2 shows two protein structures from the superfamilies immunoglobulin and Cu, Zn superoxide dismutase-like, respectively. We can see that although the two proteins have similar structures, they do not share a common biological function.

1. Immunoglobulin (6)
2. Fibronectin type III (1)
3. PKD domain (1)
4. Beta-galactosidase/glucuronidase domain (1)
5. Transglutaminase, two C-terminal domains (1)
6. Cadherin (1)
7. Actinoxanthin-like (1)
8. Cu, Zn superoxide dismutase-like (1)
9. CBD9-like (2)
10. Clathrin adaptor appendage domain (1)
11. Integrin domains (1)
12. PapD-like (2)
13. Purple acid phosphatase, N-terminal domain (1)
14. Superoxide reductase-like (1), and
15. Invasin/intimin cell-adhesion fragments (1)

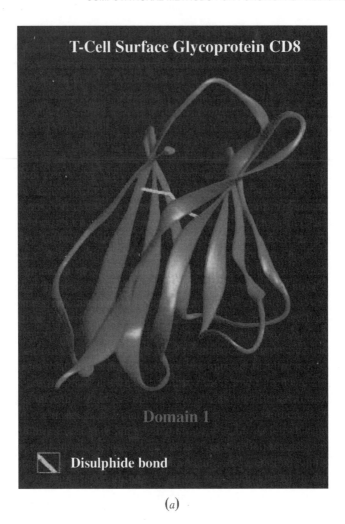

(a)

Figure 5.2   *See color insert. Three-dimensional structures of two proteins from (a) immunoglobulin and (b) Cu, Zn superoxide dismutase-like superfamilies, respectively.* Although they possess similar structures, their biological functions are unrelated.

There are a few other structure-based family classification schemes, including CATH (Orengo et al., 1997) and FSSP (Holm and Sander, 1996). The main difference between SCOP and CATH or FSSP is that SCOP's classification is mainly done based on manual inspections, while CATH and FSSP's classifications are done computationally. Whereas SCOP's classifications might be biologically more meaningful, its main drawback is that the manual process is too slow to keep up with the rate of protein structure submissions to PDB, and hence many new protein structures have not been included in SCOP in a timely fashion.

Enzyme Classification   Enzymes are among the most studied class of proteins. A great amount of information has been accumulated on enzymes and their molecular

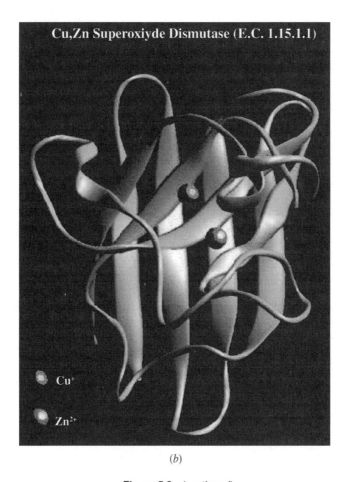

Figure 5.2 *(continued)*

functions and cellular functional roles in the past few dozen years. The ENZYME database (Bairoch, 1993) is a classification database of enzymes in Swiss-prot. It classifies all enzymes into six classes: *oxidoreductases, transferases, hydrolases, lyases, isomerases,* and *ligases,* which are further divided into over 200 subclasses of enzymatic functions organized as a functional hierarchy. By identifying the enzymatic class/subclass to which a protein may belong, a significant amount of functional information could be derived about the protein. Figure 5.3 shows the hierarchy of the isomerase class and the functional descriptions of each of its subclasses. Currently, the ENZYME database consists of 3,982 entries (according to an August 2002 release).

In addition to these general-purpose classification databases, there are also various specialized protein function databases, such as the Protein Disease Database (PDD) (Westbrook et al., 2000) or the enzyme active site database PROCAT (Wallace et al., 1996). By searching different functional databases and identifying the family/class that a gene/protein belongs to, one can derive some level of gene function. The main search tools include sequence-based and structure-based comparison methods.

```
5. -. -.-    Isomerases.
5. 1. -.-    Racemases and epimerases.
5. 1. 1.-     Acting on amino acids and derivatives.
5. 1. 2.-     Acting on hydroxy acids and derivatives.
5. 1. 3.-     Acting on carbohydrates and derivatives.
5. 1.99.-     Acting on other compounds.
5. 2. -.-    Cis-trans-isomerases.
5. 3. -.-    Intramolecular oxidoreductases.
5. 3. 1.-     Interconverting aldoses and ketoses.
5. 3. 2.-     Interconverting keto- and enol- groups.
5. 3. 3.-     Transposing C==C bonds.
5. 3. 4.-     Transposing S-S bonds.
5. 3.99.-     Other intramolecular oxidoreductases.
5. 4. -.-    Intramolecular transferases (mutases).
5. 4. 1.-     Transferring acyl groups.
5. 4. 2.-     Phosphotransferases (phosphomutases).
5. 4. 3.-     Transferring amino groups.
5. 4.99.-     Transferring other groups.
5. 5.  .-    Intramolecular lyases.
5.99. -.-    Other isomerases.
```

Figure 5.3 *An example of ENZYME hierarchy (partial).* The first column is the enzyme classification (EC) number. Each enzyme is represented as a four-digit number (separated by '.'), and the second column describes the functionality of an enzyme or enzymatic class. Enzymes with the same first-digit number in the EC numbers are from the same enzyme class (one of the six). For example, all isomerases start with 5. Enzymes from the same (next level) subclass all have the same first two digits. For example, all EC numbers of *cis-trans*-isomerases start with 5.2.

5.3.2 Searching Family Trees

Pairwise sequence comparison (or alignment) is currently the most popular technique for identifying or detecting homologous proteins. If two genes or their protein sequences have a "good" alignment, it probably means that the two genes are homologous. The basis for measuring the *goodness* of a sequence alignment is a mutation probability matrix (Gonnet et al., 1992). A mutation (or substitution) matrix is a 20×20 matrix, with each of its values representing the probability that one amino acid type can mutate to another amino acid type during evolution. PAM (Gonnet et al., 1992) and BLOSUM (Henikoff and Henikoff, 1992) matrices represent the two most commonly used mutation matrices. The goodness of a sequence alignment is defined as the overall probability that one sequence can mutate to another one, normalized with respect to the probability that two random sequences can mutate to each other. A common practice is that if this goodness score for a pairwise sequence alignment is above some threshold, the two proteins are considered as homologous.

The first rigorous algorithm for biological sequence alignment was developed by Needleman and Wunsch (1970). The algorithm finds an alignment between two sequences that globally optimizes the alignment score, defined in terms of the total mutation probability and the penalty for having alignment gaps. The algorithm employs an algorithmic technique called *dynamic programming* to achieve the global optimality. Since the publication of the paper describing this method, dynamic programming has become the main algorithmic technique for solving sequence alignment problems. Temple Smith and Michael Waterman were responsible for the popular applications of the sequence alignment technique for homology detection, the now classic Smith–Waterman algorithm (Smith and Waterman, 1982). It has improved both the sensitivity of homology detection and the computational efficiency of the Needleman–Wunsch algorithm. The algorithm has been the standard in the field since its publication in 1982.

Numerous sequence alignment algorithms using different design philosophies and implementation techniques have been developed and reported in the past 20 years, they offer improved sensitivity and computational efficiency over the classical algorithms such as Smith–Waterman. BLAST and FASTA are among the most popular alignment tools.

Homology Detection Using BLAST and FASTA Two basic problems gradually become apparent with rigorous sequence alignment algorithms such as Smith–Waterman when they are used for the detection of remote homologs and for long sequence comparisons (e.g., comparing a protein sequence against a whole sequence database). First, Smith–Waterman's alignment score does not depend only on the quality of a sequence alignment, but also on the "quantity" of an alignment, that is, the number of aligned positions. An alignment score of 1,000 may mean different things for proteins with different lengths and different amino acid compositions. A score with a well-defined meaning is clearly needed, particularly for remote homology detection. One such score is the *statistical significance score* of an alignment, originally proposed by Karlin and Altschul (1990) and later implemented in the BLAST search program. The second problem is that the Smith–Waterman algorithm is fast enough to compare two gene/ protein sequences, but it is clearly too slow for searching or comparing one sequence against a whole database with possibly tens of thousands or even millions of sequences. Both these issues are addressed in the next generation of search algorithms, including BLAST and FASTA.

Basic Local Alignment Search Tool (BLAST) was originally developed in 1990 (Altschul et al., 1990). It detects sequence homology by finding a series of ungapped sequence-fragment alignments and then assembling the ungapped alignments into longer sequence alignments. The ungapped alignment scheme allows BLAST to utilize a computational procedure different than dynamic programming; this has made the program run substantially faster. For each alignment, BLAST provides a *p-value*, which essentially represents the probability of such an alignment (and its quality) between two random sequences. For example, if the p-value of an alignment is e^{-10}, this essentially means that the probability of having such an alignment and the quality by chance is extremely small, that is, e^{-10}. Hence, one can infer that the obtained alignment is most probably due to homology. Generally, the lower the p-value of an alignment, the higher the probability that two sequences are homologous. There are some practical guidelines for setting the p-value threshold of BLAST to identify homologs, say, $< e^{-5}$. However, this is very problem-dependent. For example, if the goal is to make a protein structure prediction based on its homologous structure (Sali and Blundell, 1993), one may want to set a very low p-value threshold, say, e^{-10}, whereas a higher p-value, say, e^{-1}, may be used if the goal is to detect any remote homolog a protein might have.

Actually, BLAST is not a single computer program but rather represents a suite of alignment tools. This set of tools includes BLASTp, BLASTx, tBLASTn, megaBLAST, and PSI-BLAST. Each of these programs is designed for a particular range of applications. For example, megaBLAST is designed to run very fast so it can be used to compare two whole genomes directly. PSI-BLAST is a version of BLAST that is specifically designed to detect remote homology (Altschul et al., 1997). Although BLAST is very useful for the detection of sequence homology within a family class (defined by SCOP), PSI-BLAST has proven to be more effective in detecting remote homologs of the same superfamily (Murzin et al., 1995). When using PSI-BLAST to search for remote homologs against a database like Swiss-prot or PDB, the program first identifies all sequences in the database that have statistically significant alignments with the query sequence, that is, alignments with p-values lower than some specified threshold. Then it uses each of these identified sequences as a query sequence to search the database again. It uses such an iterative procedure to gradually expend the list of "related" sequences by applying the rule that "a friend's friend is a friend." Clearly, this expansion process has to be controlled carefully; otherwise, every sequence in the database could become "relevant" very quickly after a few iterations.

FASTA (Pearson and Lipman, 1988; Pearson, 1990) represents another popular sequence alignment tool for homology detection, built on sound statistical theory, like BLAST. When comparing two sequences, FASTA first identifies all identity groups existing in the two sequences and then extends them to local ungapped sequence alignments; in the final step, it links all these local alignments to form a "global" alignment between two sequences. On the level of designing principles, FASTA and BLAST both adopted similar ideas of finding ungapped alignment blocks first and then extending them into longer sequence alignments. However, their implementation techniques are significantly different, which has made them two unique searching methods. FASTA provides an *E-value* for each of its sequence alignments, which measures the statistical significance of the alignment. The E-value is calculated based on a model that is somewhat simpler than the model used for p-value calculation in BLAST. Because of its popularity, FASTA's data input format has become the standard in the industry. Go to http:// fasta.bioch.virginia.edu/for more information about this program and its server.

To determine if a novel gene has any homologs with known functions, one can run BLAST or FASTA against several databases, say, Swiss-prot + TrEMBL. Functions of any identified homologous genes can be predicted to be the functions of the novel gene. Alternatively, one can run the search tools against some specialized databases, such as PDD (available at http://www.lecb.ncifcrf.gov/PDD/), to determine if this gene may be involved in some known diseases.

Multiple Sequence Alignment   Whereas pairwise sequence alignment methods can help detect homologous genes and provide hints about a gene's function, they generally cannot provide detailed information about the possible mechanisms underlying a biological function. By aligning multiple gene/protein sequences from the same family, one can possibly derive such information. For example, a protein's functional sites are generally well conserved throughout evolution in order to preserve the functionality of the protein, while other positions may mutate to different amino acid types. These conserved regions or residues are generally not detectable through pairwise sequence alignment; they become apparent only when alignments exist among multiple members of the same protein family (or superfamily).

A multiple sequence alignment is just like a pairwise alignment except that it involves more than two sequences. Figure 5.4 shows one alignment of 16 sequences from the same protein family, "inhibitor of apoptosis domain." From this alignment, one can see that certain positions are highly conserved, while others are not. For example, the first aligned position consists of Arg(R) only; the same can be said about the fourth aligned position from the right—all residues are Cys(C). Based on such an analysis, one can possibly infer

```
09J849/133-201  RKQSFPK.........--FKTSRTH-.-YrDNRETLAKNG.FYHYGKKFEI...........--FRCSSCKFV
Q99GY5/89-157   RRDSFRQ......yKKAKSYFKNS.--.--LDLLAQNG.FYYYGVKTEV...........--RCAYCLLV
Q9YMP8/85-153   RAASFRA.........--FKAGCGK-.-YgSDANALAACG.FYNGRCRRA...........--QCSRCGMV
056307/85-152   RKQSFSS.........--FKWARRQFkSHnKLIDMLSRRG.FICFGKKARL...........--FCVGCKVV
IAP2 NPVOP/85-150  RKRSFAS.........--FKWARRQFgSRaREVDMLSRRG.FYCVGK--RL...........--FCAGCKVV
Q65368/85-150   RKRSFAS.........--FKWARRQFgSRaREVDMLSRRG.FYCVGK--RL...........--FCAGCKVV
902435/85-152   RKKSFTS.........--FKKSRRQFaSQsVVVDMLARRG.FYYFGKAGHL...........--RCSGCHIV
IAP2 NPVAC/85-152  RKKSFTS.........--FKSSRRQFaSQsVVVDMLARRG.FYYFGKAGHL...........--RCSGCHIV
BIR1 SCHPO/25-99  RLDTFCK.......KKWPRAKPT-.--.--PETIAIVG.FYNPISESNse.......erlDNVTCYMCTKS
BIR5 MOUSE/18-88  RIATFKN.........--WPFLEDCA.--.CTPERMAEAG.FIHCPTENEP...........DLAQCFFCFKE
BIR5 RAT/18-88   FIYTFKN.........--WPFLEDCS.--.CTPERMAEAG.FIHCPTENEP...........DLAQCFFCFKE
Q9GLN5/18-88    RISTFKN.........--WPFLEGCA.--.CTPERMAEAG.FIHCPTENEP...........DLAQCFFCFKE
BIR5 HUMAN/18-88  RISTFKN.........--WPFLEGCA.--.CTPERMAEAG.FIHCPTENEP...........DLAQCFFCFKE
Q9BVZ4/18-81    RISTFKN.........--WPFLEGCA.--.CTPERMAEAG.FIHCPTENEP...........DLAQCFFCFKE
Q9DDK0/20-90    RAATFRN.........--WPFTEGCA.--.CTPERMAAAG.FVHCPSENSP...........DVKQCFFCLKE
Q9VEM2/31-101   RVESVKS.........--WPFPETAS.--.CSISKMAEAG.FYWTGTKREN...........DTATCFVCGKT
```

Figure 5.4 *See color insert. Multiple sequence alignment among some members of the Inhibitor of Apoptosis domain family (Swiss-prot code: PF00653). The first column shows the names of proteins from different organisms of this family. The second column gives the multiple sequence alignments. The aligned residues under different colored shading illustrate the level of conservation of a particular aligned position. For example, the first aligned position consists of only Arg(R). Another three completely conserved positions are Gly(G) with orange shading and the Cyc(C) with pink shading.*

that these conserved residues are probably involved in the functional sites of the protein. Further functional mechanisms could possibly be derived, based on our general knowledge of the biochemistry of amino acid types that tend to predominate in active sites.

There are a number of computer programs/servers for multiple sequence alignments on the Internet, including Clustal W (Higgins et al., 1994), PILEUP (Genetics Computer Group, 1994), MALI (Vingron and Argos, 1989), and PIMA (Smith and Smith, 1992). Typically, these programs can handle up to a few dozen protein sequences, and their computational time goes up very quickly as more sequences are considered. Rigorously solving a multiple sequence alignment problem is computationally intractable, even for problems with 10 sequences (Jiang et al., 2002). Hence, a multiple sequence alignment problem is often solved using heuristic approaches. One common heuristic approach is to solve a series of pairwise sequence alignment problems and then "massage" these pairwise sequence alignments into a consistent multiple sequence alignment (see Chapter 4 in Jiang et al., 2002).

In addition to being useful for the detection of possible functional sites, multiple sequence alignments also provide a more sensible way of detecting remote homology than pairwise sequence alignments. Using aligned multiple sequences from the same protein family, one can calculate the sequence *profile* of the family. A sequence profile is basically the distribution of amino acid type for each aligned position. It is generally represented as a 20-element vector denoting the percentage of each amino acid type in a particular aligned position. Using such a profile to represent each protein family, homology identification becomes a matter of fitting a query sequence with a profile rather than aligning the sequence with each member of the family. It has been clearly demonstrated that the profile approach is more sensitive in remote homology detection than the traditional sequence alignment approach (Altschul et al., 1997), as such methods directly use common features of a protein family in a sequence comparison. There are a few computer programs specifically designed for the detection of remote homology based on such profile ideas. SAM is one such program (Karplus et al., 1998).

5.3.3 Orthologous vs. Paralogous Genes

The capability to distinguish orthologs from paralogs among identified homologous genes would allow functional inference at a higher resolution level. Short of a deeper understanding of the distinguishing features of orthologous vs. paralogous genes, existing methods generally use sequence similarity levels to make ortholog/paralog predictions.

A popular technique for ortholog/paralog predictions is based on the idea of *clusters of orthologous genes* (COGs) (Tatusov et al., 1997). A simple definition of a COG is that it "consists of individual orthologous genes or orthologous groups of paralogs from three or more phylogenetic lineages" (Tatusov et al., 1997).

One simple approach to the prediction of orthologous genes is based on the assumption that two genes, A and B, from two different genomes are orthologous if they have the highest sequence similarity among all A's (or all B's) in alignments against genes from B's (or A's) genome. However, this simple approach may not work well if two genomes are only remotely related as noted by the authors of the original COG paper. The more reliable approach proposed by the same authors is to use multiple genomes. The idea is that for any genes A, B, and C from three different genomes, if A-B, B-C, and A-C all have the highest sequence similarity among similar pairs from the corresponding genomes, then A, B, and C are considered to belong to the same COG. Clearly, this is a generalization of the simple idea outlined above; it can be generalized further to include more genomes to make the prediction more reliable. A computer server has been implemented, based on this concept, for the identification of COGs at http://www.ncbi.nlm.nih.gov/COG/. By applying this prediction capability to 43 sequenced genomes, over 3,300 COGs have been predicted and organized as a searchable database.

The most useful applications of such a COG database is to identify orthologous genes of a newly sequenced genome and to make functional inferences, at a relatively high-resolution level, of the identified orthologous genes. In their paper, Tatusov and colleagues (1997) identified 51% of the genes in the sequenced genome *Helicobacter pylori* to be orthologous genes to some of the known COGs and made functional predictions for them. One strategy to apply the capability of COG predictions is to perform high-resolution functional inference when a gene is identified to be an ortholog of genes with known functions and do low-resolution inference otherwise.

5.3.4 Genes with Multiple Domains

Some percentage of genes in a (microbial) genome might have multiple domains, where each domain could have its own biological function and exist independently of other domains of the gene. Some statistics suggest that the average size of a protein domain is about 150 amino acids. Hence, for genes significantly longer than this, one may want to consider the possibility of multiple-domain genes. It is often the case that one domain of a gene could appear as a domain in another gene. If we examine the protein structure of a multidomain gene, we can see that each domain forms a compact and semi-independent structure within the whole protein structure (Xu et al., 2000). Generally, the connecting regions between two domains in a protein structure form a bottleneck-looking conformation (see Fig. 5.5). Typically, a multidomain protein (gene) can accomplish a complex function through combination of different but related functions by each individual domain encoded by the gene.

The functional inference of a multiple-domain gene generally involves a two-step process: (1) identification of the boundaries of each domain and (2) functional inference of each domain. PRODOM is a commonly used computer program for the identification of protein domains (Corpet et al., 2000). It identifies domains through multiple sequence alignments of genes to detect long segments of sequences that exist in different protein

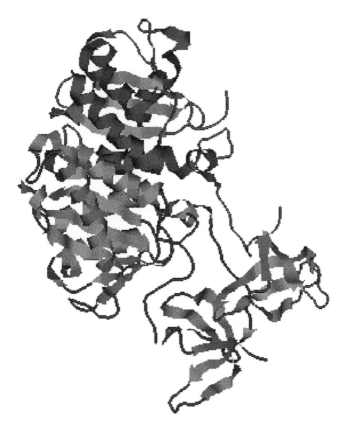

Figure 5.5 *See color insert. The X-ray crystallographic structure of Rous Sarcoma Virus Integrase (Yang et al., 2000).* The protein has two domains (the upper and the lower parts of the image), a conserved catalytic domain (the upper part) and a DNA-binding domain (the lower part).

families and then to cluster these segments, shared by different gene families, into clusters of domains. PRODOM keeps a database of all its previously identified domains. For a novel gene, one can search the PRODOM database to predict its domain boundaries, based on matches with the PRODOM domain sequences. Currently, PRODOM consists of 305,465 domain families (as of a September 2001 release). Figure 5.6 shows the distribution of the number of domains a gene may have for all genes in the Swiss-prot + TrEMBL database. From this illustration, we can see that multiple-domain genes are not rare at all.

There are also protein structure-based approaches for the identification of domains, including DALI (Holm and Sander, 1996), SCOP (Murzin et al., 1995), and DamainParser (Xu et al., 2000). They divide a protein structure into domains based on the structural features of the protein. These structure-based methods are generally more accurate than sequence-based approaches as they have additional information, that is, the coordinates of the three-dimensional protein structure. However, these methods apply only to genes with solved protein structures or structural homologs.

By the same token, a protein sequence may consist of other types of functional "fragments," which should be dealt with separately in the functional inference of a gene. For example, a protein may contain a signal peptide, typically about 12 amino acids long.

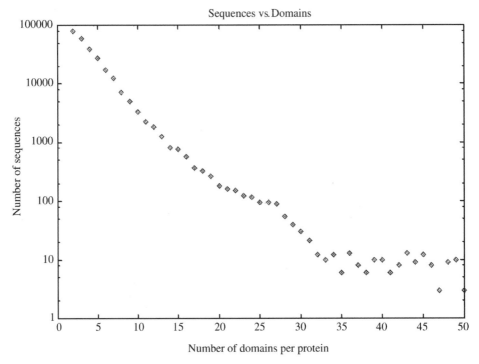

Figure 5.6 *Distribution of the number of domains per gene.* The *x*-axis is the number of domains, and the *y*-axis the number of genes in Swiss-prot + TrEMBL with a particular number of domains. (Available at http://prodes.toulouse.inra.fr/prodom/doc/prodom_stat2001.3.html.)

SignalP is a computer server that can identify whether a protein sequence has a signal peptide (Nielsen et al., 1997). Such an application should be made before the functional inference of a gene.

5.4 IDENTIFICATION OF SEQUENCE MOTIFS

Functional inference of unknown genes can also be accomplished through the identification of functional motifs. Genes sharing common functional motifs generally have the same or similar biochemical functions. Over the years, scientists have identified many functional motifs in gene sequences. These motifs, along with their functional annotations, have been stored in a number of public databases, including PRINTS, PROSITE, and BLOCKS. Through database searches for known functional motifs in a novel gene, one could possibly infer the function of a gene. Motif-based functional inference complements sequence-based approaches, such as BLAST, and extends the scope of genes that could be assigned functions, as these methods generally do not rely on high sequence similarity for the whole gene.

PROSITE (Falquet et al., 2002) is a database of protein families, whose classifications are based on the identification of common sequence motifs or patterns, of biological significance, existing in gene sequences. The following is an example of a conserved sequence motif, shared by proteins of the WW-domain family, where 'x' is a wild card and the numbers represent the numbers of sequence positions covered by 'x'.

```
W-x(9,11)-[VFY]-[FYW]-x(6,7)-[GSTNE]-[GSTQCR]-[FYW]-x(2)-P
```

PROSITE currently consists of 1,500 + sequence patterns (as of an August 2002 release). Some functional information associated with each of these patterns is provided in the database.

PRINTS (Attwood et al., 2002) is a compendium of protein fingerprints. A fingerprint is a group of conserved motifs used to characterize a protein family. Fingerprints can encode protein folds and functionality more flexibly than one single motif. The following gives two fingerprint groups of PRINTS with their names:

```
RIVELIYIDIVGLAQFK and GLGFPYEGPAPLEAIANGCAFL (name of the
fingerprint: RABGDIREP), and FLNPKFNPPKSSKN and
LQRINAFIEKQDF (name of the fingerprint: TREKCHANNEL)
```

These fingerprint groups are associated with certain PRINTS families, about which some level of functional information is known. Currently, PRINTS contains over 10,000 fingerprints (according to an August 2002 release).

BLOCKS (Henikoff et al., 1999, 2000) is a database of multiply aligned ungapped sequence segments, called *blocks*. Each block represents a highly conserved region of homologous proteins. By identifying matches between a novel gene and some blocks in the BLOCKS database, one can possibly infer homology relationships and further biological functions. BLOCKS currently consists of over 11,000 blocks from 2,600 + gene groups with known functions (as of an August 2002 release).

Searches for a particular sequence motif against a novel gene sequence can be done using a simple sequence comparison algorithm such as Smith–Waterman. In cases where no gaps are allowed within a motif, a simple string matching program may even suffice.

5.5 STRUCTURE-BASED FUNCTION PREDICTION

The central idea of structure-based functional inference is to first identify structural homologs with known functions and then infer the function of a novel gene based on the functions of the homologs, similar to sequence-based approaches. The main difference is that it uses structural information, rather than sequence information alone, to derive functional information. Structure-based and sequence-based approaches complement each other in the following sense. Sequence-based approaches are more generally applicable since they do not require that a query gene possess a homologous protein structure, and many genes may not have homologous protein structures. Structure-based approaches are generally more sensitive in homology detection when applicable, and protein structures could provide more detailed functional information than sequence-based methods can. The following is an example illustrating how protein structures can provide more detailed functional information. Proteins of the chymotrypsin class of serine proteases have a wide range of phenotypic and cellular functions, ranging from food digestion and sperm processing to regulation of the immune response, but they all have closely related molecular functions, that is, proteases with a common catalytic mechanism (Moult and

Melamud, 2000). The different phenotypic and cellular roles are the result of differences in the specificity of peptides cleaved (Shultz and Schmier, 1979), which is apparent only with the tertiary structures of these proteins. The general guideline for using these two approaches for functional inference is to use the structure-based approach whenever applicable; otherwise, try sequence-based methods.

In performing structure-based functional inference, one important question to ask is how generally applicable the structure-based methods are to novel genes? On the surface, they do not seem to be very generally applicable, as we know that the combined database of Swiss-prot and TrEMBL (as of an August 2002 release) has $\sim 800,000$ protein sequences, and only $\sim 18,000$ of them have known protein structures (according to a PDB release of August 2002). However, if we look at this issue more carefully, we find that probably a significant percentage of the Swiss-prot + TrEMBL genes are homologous to some of the protein structures in PDB. This is based on PDB submission statistics and some theoretical studies on the protein fold space. Examining structure submissions to PDB in the past three years, we see that only $\sim 10\%$ of these proteins have novel structural folds. This seems to suggest that a significant percentage of the structural folds in nature are probably already solved and stored in PDB. Theoretical studies have further suggested that the number of unique structural folds in nature is probably quite small (Wang, 1998; Portugaly and Linial, 2000; Wolf et al., 2000), possibly in the range of 1,000 to 2,000. Indications from genome-scale structure predictions and the recent CASP experiments (CASP3, 1999; CASP4, 2001) also support the widely held belief that we have probably already seen at least 50 to 70% of all the structural folds in nature in the current PDB database (Montelione and Anderson, 1999).

Now, the key issue is how to identify the structural homologs of a novel gene in the PDB database and to make an accurate structure prediction based on the identified structural homology. Protein fold recognition and structure predictions represent highly challenging problems. One key computational technique for tackling such problems is *protein threading*.

5.5.1 Protein Fold Recognition through Protein Threading

Protein threading is a computational technique for aligning a protein sequence onto a protein structure (Bowie et al., 1991; Godzik et al., 1992; Jones et al., 1992); it was pioneered by Eisenberg and coworkers (Bowie et al., 1991). As it uses both sequence-level and structure-level information for homology detection, it is generally more sensitive than sequence-based approaches. This fact has been well established in the recent CASP protein-structure prediction experiments (CASP3, 1999; CASP4, 2001).

Protein threading makes a structural fold prediction from the protein sequence by recognizing a nativelike fold of the query protein (if there is any) in a protein structure database, such as PDB. It aligns (or *threads*) each amino acid of a query protein sequence onto consecutive positions in a protein structure in such a way to optimize some potential energy function. Typically, a threading energy function is statistics-based, that is, derived from known protein structures. It often includes the following terms: (1) the overall fitness of each amino acid to its assigned structural environment (typically defined in terms of solvent accessibility and secondary structure), (2) the overall contact preference between two amino acids assigned to nearby structural positions, (3) sequence similarity between the query and template sequences, and (4) the total penalty for having alignment gaps. Figure 5.7 illustrates the basic idea of protein threading.

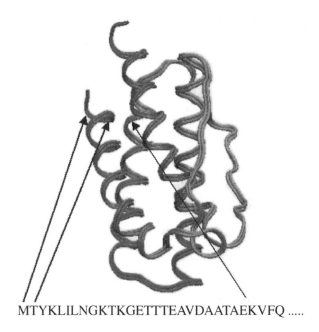

MTYKLILNGKTKGETTTEAVDAATAEKVFQ .....

Figure 5.7 *See color insert. Threading a sequence onto a protein structure.* The threading process involves placing each amino acid (possibly skipping some amino acids) onto structural positions (denoted by arrows) with their sequential order maintained. The arrows indicate that a Thr(T) is placed in a helical position, which is close to the structural position where a Glu(E) is placed.

For a query sequence, a threading program will try to thread it onto each of the structures in the structure database. Then it will determine which threading(s) is statistically significant. Each structural fold with a significant threading result will be predicted to be a structural homolog of the query protein, and the placed positions of the query sequence are predicted to be the query's (backbone) structure.

Numerous computer programs have been developed for protein fold recognition and structure prediction, based on the threading technique. They include 1-2-3D (Alexandrov et al., 1996), Threader (Jones et al., 1992), FUGUE (Shi et al., 2001), and PROSPECT (Xu and Xu, 2000; Kim et al., 2002). These and other similar programs have played a vital role in the functional inference of novel genes with no apparent homologs in PDB, as detected by PSI-BLAST. Table 5.1 shows a comparison (Kim et al., 2002) of homology recognition by different threading programs vs. PSI-BLAST, on a large data set consisting of 600 query and template sequences (Lindahl and Elofsson, 2001). As we can see, for genes with structural homologs from the same family (as defined in SCOP), PSI-BLAST's fold recognition rate is comparable with that of threading-based methods. However, as we move to proteins with only distant or very distant structural homologs (superfamily and fold family, respectively), threading-based approaches clearly perform significantly better.

There have been a number of studies that use threading-based methods to perform functional assignments of genes on a whole-genome scale (Di Gennaro et al., 2001). Generally, for a newly sequenced microbial genome, about 50 to 70% of the genes may have some level of functional assignments (go to http://genome.ornl.gov/microbial/), using a combination of all available techniques including both sequence- and structure-based methods.

TABLE 5.1 Comparison between Threading-based Approaches with PSI-BLAST and Its Variations (Kim et al., 2002) on the Lindahl and Elofsson Benchmark Set

Pairs at top 1/top 5 (%) Method	Family only		Superfamily only		Fold only	
	Top 1	Top 5	Top 1	Top 5	Top 1	Top 5
PROSPECT	84.1	88.2	52.6	64.8	27.7	50.3
FUGUE	82.2	85.8	41.9	53.2	12.5	26.8
PSI-BLAST	71.2	72.3	27.4	27.9	4.0	4.7
HMMER-PSIBLAST	67.7	73.5	20.7	31.3	4.4	14.6
SAMT98-PSIBLAST	70.1	75.4	28.3	38.9	3.4	18.7
BLASTLINK	74.6	78.9	29.3	40.6	6.9	16.5
SSEARCH	68.6	75.7	20.7	32.5	5.6	15.6
THREADER	49.2	58.9	10.8	24.7	14.6	37.7

Threading-based programs include PROSPECT, FUGUE, and THREADER. Top 1 means the sequencefold alignment with the highest score, and Top 5 means among the top five highest scoring sequence fold alignments. The numerical value at each entry, say, 88.2% by PROSPECT, means that in 88.2% of the cases, the correct homologous fold is among the top five alignments by PROSPECT for proteins in the "Family" category.

5.5.2 From Structure to Function

Functions can be derived at different resolution levels from a predicted structural model, depending on the model's accuracy. Knowing only the correct structural fold of a gene (the lowest resolution level of a threading prediction), one can make some functional inference based on the family or superfamily that the fold belongs to, just like what we have done with sequence-based methods. Even if we can identify only the fold family a gene belongs to, we can still derive some useful information. For example, the protein structure of the *ycaG* gene was found to have a fold similar to that of the amidohydrolase family (Colovos et al., 1998). Inspection of the structure revealed that the same catalytic apparatus was present in *ycaG* (Moult and Melamud, 2000). Such information is clearly valuable when nothing else is known about the protein.

When an accurate structure model of a novel protein is predicted, more functional information could be derived through analyses of the structural features of the model, in addition to copying the functional annotations of its homologs. Most structure prediction programs provide some type of prediction reliability indicator. For example, PROSPECT gives a reliability score for each prediction, that is, *z-score* (Xu and Xu, 2002; Kim et al., 2002). A strong correlation exists between this reliability score and the quality of a predicted structure. Utilizing scores like this, a user can roughly estimate the accuracy of a predicted structure. For example, when PROSPECT's reliability score (z-score) is above 20, we can generally expect that the predicted structure is of high accuracy. Through analysis of the structural features of a predicted structure, one could derive the following information:

1. *Ligand-binding Site.* In a systematic study, Laskowski and colleagues (1996) found that in 83% of the protein structures, the largest cleft is a known primary ligand-binding site, and in 9% of the cases, the second largest cleft is a known primary ligand-binding site. Hence, by identifying the large clefts in a protein structure, one could predict ligand-binding sites. When conducting this search in conjunction with analysis of conserved residues among all sequences from the same protein family, one could even possibly derive which residues are involved in ligand binding.

2. *Macromolecular Binding Site.* The prediction of protein–DNA, protein–RNA, or protein–protein binding sites represents a more challenging problem than protein–ligand binding. However, recent studies have indicated that there is a strong correlation between macromolecular binding sites and disordered regions in a protein (see Section 5.5.3). By identifying disordered regions in a protein, in conjunction with analyses of specific patterns in a protein structure, like nonpolar interactions, one could make predictions of possible macromolecular binding sites. Again, this analysis will be most effective when carried out in conjunction with the analysis and prediction of conserved sequence motifs/residues.

3. *General Functional Sites.* Through analysis of the geometric features of all known functional sites of all known protein structures, one can build a database of the functional sites with their corresponding geometric descriptors. SITE is such a database (Zhang et al., 1999). By matching each of these geometric descriptors against a new protein structure, one can possibly derive the functional sites of a protein and further the biochemical functions of these sites.

4. *Enzyme Catalytic Mechanisms.* If a target protein is an enzyme, we can generally expect that some mechanistic information could be inferred, knowing the large collection of known enzymes, along with their sequence patterns as represented in the PROSITE database and their known functional mechanisms. This information has been accumulated over many years of research.

5. *Posttranslational Modifications.* The identification of potential posttranslational modifications can help us understand the functional mechanism of a protein involved in a particular biological process. There are a number of computer programs that can be used for such predictions. These include RESID, NetPhos, and Sulfinator (all the websites are provided in Table 5.2).

Another level of functional inference could possibly be achieved when we have very accurate structure models of a query protein. That is done by studying the dynamic behavior of a protein as it could provide mechanistic information about the biological function of a protein. The key prerequisites for such studies are high-resolution structure modeling and high-performance computing. Such studies can be carried out through molecular dynamics (MD) simulation (Head-Gordon and Brooks, 1991; Nelson et al., 1996) of proteins or interacting proteins. Through such simulations, one could derive how a biological function is implemented at the atomic level.

In a MD simulation, Newton's equations of motion governing the microscopic evolution of a many-body system are solved numerically under the appropriate environmental conditions. In principle, it can provide a complete time evolution of each individual atom in a given system, from which dynamic and energetic information about the system may be obtained. Using such a simulation capability, one can "observe" how an ion channel opens up and closes, or how the mercuric ion reductase MerA reduces ionic mercury Hg(2) to elemental mercury Hg(0) (Fox and Walsh, 1983; Engst and Miller, 1998). Clearly, this type of simulation capability will be extremely useful in understanding the mechanisms of biological processes. However, in general, the current computing capability is too slow to support MD simulations long enough to meaningfully interpret biological functions. It is well known that the timescale classical MD simulation can reach is on the order of nanoseconds (10^{-9}), while most biologically interesting phenomena occur in a much longer time, for example, microseconds (10^{-6}) to milliseconds (10^{-3}). The encouraging news is that a recent simulation study by Peter Kollman and colleagues

TABLE 5.2   Web Server for Functional Inference of Genes

Program Name	Internet
GenBank	http://www.ncbi.nlm.nih.gov/Genbank/
Swiss-prot	http://www.expasy.ch/sprot/
nr	http://www.ncbi.nlm.nih.gov/BLAST
PIR	http://pir.georgetown.edu/
TrEMBL	http://www.expasy.ch/sprot/sprot-top.html
PDB	http://www.rcsb.org/pdb/
BLAST	http://www.ncbi.nlm.nih.gov/BLAST/
FASTA	http://www.ebi.ac.uk/fasta33/
Pfam	http://pfam.wustl.edu/faq.shtml
PROSITE	http://www.expasy.ch/prosite/
PRINTS	http://www.bioinf.man.ac.uk/dbbrowser/index/biblio.html
BLOCKS	http://www.blocks.fhcrc.org/
SCOP	http://scop.mrc-lmb.cam.ac.uk/scop/
CATH	http://www.biochem.ucl.ac.uk/bsm/cath_new/index.html
FSSP	http://www2.ebi.ac.uk/dali/fssp/fssp.html
ENZYME	http://us.expasy.org/enzyme/enzyme_details.html
MegaBLAST	http://www.ncbi.nlm.nih.gov/blast/megablast.html
FASTA	http://www.ebi.ac.uk/fasta33/
	http://fasta.bioch.virginia.edu/
PDD	http://www-lecb.ncifcrf.gov/PDD/
CLUSTAL W	http://www.ebi.ac.uk/clustalw/#
PIMA	http://bioweb.pasteur.fr/seqanal/interfaces/pima-simple.html
SAM	http://www.cse.ucsc.edu/research/compbio/sam.html
COG	http://www.ncbi.nlm.nih.gov/cgi-bin/COG/palox
PRODOM	http://prodes.toulouse.inra.fr/prodom/doc/prodom.html
DomainParser	http://compbio.ornl.gov/structure/domainparser/
SignalP	http://www.cbs.dtu.dk/services/SignalP/
SOSUI	http://sosui.proteome.bio.tuat.ac.jp/sosuiframe0.html
THREADER	http://www.hgmp.mrc.ac.uk/Registered/Option/threader.html
FUGUE	http://www-cryst.bioc.cam.ac.uk/$\sim$fugue/prfsearch.html
PROSPECT	http://compbio.ornl.gov/structure/prospect/
RESID	http://pir.georgetown.edu/pirwww/dbinfo/resid.html
NetPhos	http://www.cbs.dtu.dk/services/NetPhos/
Sulfinator	http://us.expasy.org/tools/sulfinator/
PONDR	http://www.pondr.com/background.html
DIP	http://dip.doe-mbi.ucla.edu/
KEGG	http://www.genome.ad.jp/kegg/
SPAD	http://www.grt.kyushu-u.ac.jp/eny-doc/
CSNDB	http://geo.nihs.go.jp/csndb/

(Duan and Kollman, 1998) has demonstrated the possibility of simulation at the timescale of microseconds. We can expect that as the computing speed of supercomputers increases at an exponential rate, we are probably not too far away from general-purpose MD simulations at the microsecond level or even longer.

5.5.3   Disordered vs. Ordered Regions in Proteins

Disorder represents a conformational state in which a protein or protein segment does not have a well-defined tertiary structural conformation, that is, it forms a dynamic molten globule or random coil, and hence is partially or completely unfolded (Dunker et al., 2002). People have long noticed the existence of such disordered regions when solving a protein structure. Recent studies have indicated that disordered regions are often associated with molecular recognition domains, protein folding inhibitors, flexible linkers, etc. (Dunker et al., 2002). These studies have also shown that disorder is a general phenomenon. It is estimated that more than 30% of eukaryotic proteins have disordered regions of at least 50 consecutive residues in length, and proteins from eukaryotes apparently have more intrinsically disordered regions than those of either bacteria or archaea (Dunker and Obradovic, 2001). The general belief is that it is the flexibility of such a disordered region which facilitates the binding and interactions between a protein and other macromolecules. Hence, identification of disordered regions can provide direct evidence for possible binding sites or docking interfaces between a protein and other macromolecules.

Disordered regions can be predicted, with a good level of accuracy, from a protein sequence just like an ordered structure can be predicted from its sequence. PONDR is one of the publicly available computer servers (Romero et al., 2001) for the prediction of disordered regions from a protein sequence. It makes disorder predictions, mainly based on the composition of amino acid types and their hydropathy, within a fixed window (of size 9) that slides through a protein sequence. The computational method it uses is a neural network classifier. Its current prediction accuracy (as of a June 2002 release) falls in the range of 50 to 80%. Using a capability such as PONDR, one can infer regions that are possibly associated with protein binding or interaction. Such information, when used in conjunction with other predictions of binding sites or conserved regions/residues or protein docking, could provide a powerful tool for the investigation of protein complexes, that is, how a group of proteins interact with each other to accomplish a biological function at a systems level.

5.6 NONHOMOLOGOUS APPROACHES TO FUNCTIONAL INFERENCE

Homology is the foundation of both sequence-based and structure-based approaches for functional inference, as they both derive functional information through the detection of homologous genes that have been previously studied and have known functions. Although powerful, homology-based approaches have their limitations. First, they do not apply to truly novel genes, or genes without functional homologs. These genes could account for at least 30% of the total predicted genes in a newly sequenced genome (go to http:// genome.ornl.gov/microbial/). Second, they generally cannot provide information about direct functional links among nonhomologous genes, which are essential for the study of biological functions at a systems (e.g., cellular or phenotypic) level.

Phylogenetic profiling (Pellegrini et al., 1999) represents a method that infers gene functions, not based on copying functions of identified homologs, but rather it derives functions based on the assumption that proteins functioning together in a pathway or structural complex are likely to evolve in a correlated fashion. That is, during evolution, such functionally linked proteins tend to be either preserved or eliminated altogether in a new species. By identifying which genes have the same presence/absence profile over a large set of genomes, Eisenberg and colleagues have developed the phylogenetic profiling approach (Pellegrini et al., 1999). The basic idea is as follows. By examining a set of genomes (say, 100), one can encode each gene as a binary string, for example,

101100100111 ..., which represents this gene as being present in genome no. 1 (so the first bit is 1), absent from genome no. 2 (so the second bit is 0), etc. Eisenberg and colleagues argue that if genes have identical or similar encoded binary strings, then they will have a good chance of appearing in the same biological pathway or the same protein complex, and hence they are functionally linked at the systems level.

In a recent study that applied this prediction technique to the yeast genome, Marcotte and coworkers (2000) found that 361 nucleus-encoded mitochondrial proteins could be identified at 50% accuracy with 58% coverage. From these numbers and the proportion of conserved mitochondrial genes, it can be inferred that approximately 630 genes, or 10% of the nuclear genome, is devoted to mitochondrial function. Based on this and related work, a database of protein interactions called the Database of Interacting Proteins (DIP) (Xenarios et al., 2000) has been developed. The database currently contains $\sim 6,800$ proteins and $\sim 17,700$ protein–protein interactions covering 110 genomes (as of an August 2002 release).

Gene fusion represents another nonhomologous method for functional inference, also developed by Eisenberg's group (Marcotte et al., 1999a). The method infers functionally linked genes based on the observation that some pairs of interacting proteins have homologs in another organism fused into a single protein chain. Through application of this method, the authors predicted 6,809 such putative protein–protein interactions in *Escherichia coli* and 45,502 in yeast. Many members of these pairs were confirmed as functionally related. Some proteins have links to several other proteins; these coupled links appear to represent functional interactions such as protein complexes or pathways. Experimentally confirmed interacting pairs have been documented in the DIP database.

We can expect that these types of nonhomologous approaches will become a key technique for the functional inference of genes and role assignments of these genes to biological pathways as more and more (microbial) genomes are sequenced.

5.7 FUNCTIONAL INFERENCE AT A SYSTEMS LEVEL

Functional assignments of individual genes represent only the first step in functional inference at a systems level. Discussions in the previous sections have mainly focused on the function of a gene at the molecular level. As we know, a gene's function, at the cellular and phenotypic levels, may differ in different biological processes where it participates. Understanding the detailed functional role of a gene in a biological pathway (signaling, regulatory, or metabolic pathway) represents the next step toward the full characterization of a gene's function in a living organism. The Kyoto Encyclopedia of Genes and Genomes (KEGG) database provides a nice framework for such studies.

The KEGG project started in 1995 (Ogata et al., 1996), its primary objective being to computerize the current knowledge of molecular interactions, namely, metabolic pathways, regulatory pathways, and molecular assemblies. Now KEGG (Kanehisa et al., 2002) represents probably the most comprehensive database of biological pathways. Among its five subdatabases (GENE catalogs, MOLECULAR catalogs, ORTHOLOG groups, GENOME maps, and PATHWAY maps), the PATHWAY subdatabase is the most unique because of its comprehensiveness. PATHWAY currently contains all known metabolic pathways (about 120 as of August 2002) and a few dozen regulatory pathways (this number increases rapidly). Its metabolic pathways are divided into 11 metabolic groups:

1. Carbohydrate metabolism
2. Energy metabolism
3. Lipid metabolism
4. Nucleotide metabolism
5. Amino acid metabolism
6. Metabolism of other amino acids
7. Metabolism of complex carbohydrates
8. Metabolism of complex lipids
9. Metabolism of cofactors and vitamins
10. Biosynthesis of secondary metabolites
11. Biodegradation of xenobiotics

Each of the KEGG metabolic pathways has a detailed diagram showing how all the functional units are linked together to accomplish a particular metabolic process, as shown in Fig. 5.8. These functional units, represented as boxes, are generally proteins that are capable of performing the needed biological functionality (specified by the EC number inside each box). It should be noted that each functional unit can be implemented by different proteins in different organisms as long as they belong to the same EC (sub)class. For example, scientists have identified that the protein performing box 2.8.2.- (Fig. 5.8) is SULT1C2 in human (Sakakibara et al., 1998), whereas no protein has been identified for the same functional unit in *E. coli*. For this particular cysteine metabolism pathway, 11 functional units in *E. coli* have their proteins identified, compared to only 9 in a human.

When carrying out the functional annotation of novel genes, KEGG provides a set of templates for the functional assignments of genes. Using the sequence-based or structure-based methods described in this chapter, one can search for genes that may fall into a particular EC class, for example, 4.3.1.17. Identified genes will form the initial candidate list for that particular box. Further analyses on protein–protein interactions (see Section 5.6) may help to rule out certain false candidates, as they probably will not directly interact with the proteins associated with its neighboring boxes in the pathway. Experimental procedures should generally be followed to validate the computational predictions. Clearly, such computational predictions could speed up the investigation process of metabolic pathways, particularly in a newly sequenced organism.

KEGG's signaling and regulatory pathways are currently divided into the following four categories (Kanehisa et al., 2002):

1. Generic information processing
    1.1. Transcription
    1.2. Translation
    1.3. Sorting and degradation
    1.4. Replication and repair
2. Environmental information processing
    2.1. Membrane transport
    2.2. Signal transduction
    2.3. Ligand–receptor interaction
3. Cellular process
    3.1. Cell motility

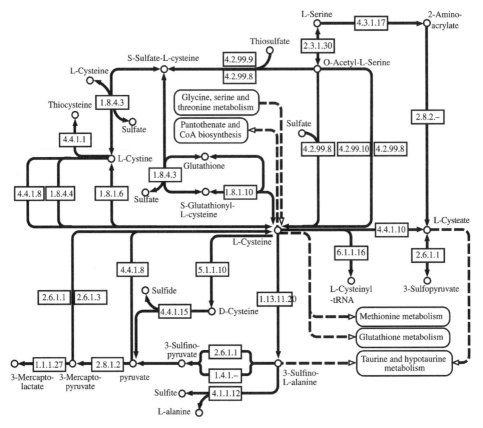

Figure 5.8 *Generic pathway of cysteine metabolism as part of the amino acid metabolism group.* Different organisms may have different variations of this pathway. Each box in the pathway model represents a functional unit, protein, or protein complex performing a specified functionality, for example, producing sulfate under certain conditions. (Available at http://www.genome.ad.jp/kegg/.)

   3.2. Cell growth and death
   3.3. Cell communication
   3.4. Development
   3.5. Behavior
   4. Human diseases

Due to the complexity of these signaling and regulatory pathways and the difficulty in deciphering them, the levels of their descriptions in KEGG are not nearly as detailed as the metabolic pathways. Figure 5.9 shows a typical regulatory pathway model in KEGG. Such a pathway model, even at this level of detail, could provide highly useful guidance in constructing regulatory pathways and assigning genes to them in a newly sequenced genome, in a fashion similar to what is described above for metabolic pathways.

In addition to KEGG, there are a few other public pathway databases. These include SPAD (an integrated database for genetic information and signaling transduction systems) and CSNDB (a database for signaling pathways in human cells). These databases can be

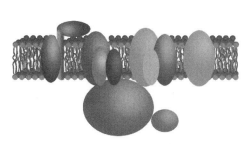

Sec-dependent pathway

| SecB | SecA | SecY | SecE | SecG |
| SecD | SecF | YajC | YidC | |

Signal peptidase

SPase I   SPase II

SRP (signal recognition particle)-dependent pathway   RNA

| SRP9 | SRP72 | SRP19 | SRPR | 4.5S |
| SRP14 | SRP68 | SRP54 | | |

Tat (twin-arginine translocation) system

TatA   TatB   TatC

Figure 5.9 *Schematic of a generic pathway for protein export in the sorting and degradation pathway group.* (Available at http://www.genome.ad.jp/kegg/pathway.)

used in a similar fashion to KEGG, in the functional annotation of novel genes and functional assignments to biological pathways.

It should be noted that all the biological pathways described/stored in these databases are experimentally validated, and they represent the results of many years of research by many scientists. In the past few years, we have witnessed a surge in the computational construction and modeling of biological pathways (Friedman et al., 2000; Jamshidi et al., 2001; Kato et al., 2000; Shmulevich et al., 2002), thanks to powerful emerging technologies such as microarray chips and two-hybrid systems. These modeling studies attempt to build pathway models that are consistent with information extracted from large-scale microarray gene expression data (DeRisi et al., 1997; Chu et al., 1998; Zhu et al., 2000a) and two-hybrid data for protein–protein interactions (Fields and Song, 1989; Uetz et al., 2000). Although such computational techniques for biological pathway construction are still in their infancy, we can expect that they will play an increasingly important role in functional inference at a systems level, as the high-throughput technologies like microarray chips mature (see Chapter 6).

5.8 SUMMARY

Computational methods for the functional inference of genes have played a significant role in genome annotation and are starting to play an indispensable role in selecting initial gene candidates in the systematic investigations of biological pathways. They allow functional inference based on the general principle of "guilt by association." The four main classes of inference techniques include (1) sequence-based methods, (2) sequence motif-based methods, (3) structure-based methods, and (4) nonhomologous methods. The foundation of all these inference techniques is the large collection of genes with experimentally verified functions and structures, compiled through many years of experimental biology work. Through identifying relationships between the novel genes and these known genes, these methods can infer gene functions. Predicted gene functions could guide further experimental studies, such as mutagenesis and proteomic studies, to validate and refine hypothesized pathway models. We expect that as the computational technologies for functional inference mature and as more high-throughput genomic and proteomic data become available, iterative computation-experiment processes will become standard protocol for functional studies in biology.

FURTHER READING

Altschul, S. F., T. L. Madden, A. A. Schäffer, and J. Zhang, et al. 1997. Gapped BLAST and PSI-BLAST: a new generation of protein database search programs. *Nucl. Acid Res.* 25:3389–3402.

Attwood, T. K., M. Blythe, D. R. Flower, and A. Gaulton, et al. 2002. PRINTS and PRINTS-S shed light on protein ancestry. *Nucl. Acid Res.* 30(1):239–241.

Bork, P., T. Dandekar, Y. Diaz-Lazcoz, and F. Eisenhaber, et al. 1998. Predicting functions: from gene to genomes and back. *J. Mol. Biol.* 283:707–725.

Bowie, J. U., R. Luthy, and D. Eisenberg. 1991. A method to identify protein sequences that fold into a known three-dimensional structure. *Science* 253:164–170.

Burge, C. and S. Karlin. 1997. Prediction of complete gene structures in human genomic DNA. *J. Mol. Biol.* 268(1):78–94.

Holm. L. and C. Sander. 1996. Mapping the protein universe. *Science* 273:595–602.

Jiang, T., Y. Xu, and M. Zhang (eds.). 2002. *Current Topics in Computational Molecular Biology.* MIT Press, Cambridge, Mass.

Kanehisa, M., S. Goto, S. Kawashima, and A. Nakaya 2002. The KEGG databases at GenomeNet. *Nucleic Acids Res.* 30:42–46.

Marcotte, E. M., I. Xenarios, A. M. van Der Bliek, and D. Eisenberg. 2000. Localizing proteins in the cell from their phylogenetic profiles. *Proc. Natl. Acad. Sci. USA* 97(22):12115–12520.

Uberbacher, E. C., Y. Xu, and R. Mural. 1996. Discovering and understanding genes in human DNA sequence using GRAIL. *Meth. Enzymol.* 266:259–281.

6

DNA Microarray Technology

Jizhong Zhou and Dorothea K. Thompson

6.1 INTRODUCTION

Microarrays (or microchips) are a powerful genomic technology developed a few years ago; in 1998 they were listed by the news and editorial staffs of the journal *Science* as one of the top ten breakthrough technologies. Similar to the situation in which microprocessors have accelerated computation, microarray-based genomic technologies have revolutionized the genetic analysis of biological systems. Microarray technology represents a powerful new tool that allows researchers to view the living cell under various physiological states from a comprehensive and dynamic molecular perspective. The widespread, routine use of such genomic technologies will shed light on a wide range of important research questions: from how cells grow, differentiate, and evolve to the medical challenges of pathogenesis, antibiotic resistance, and cancer, from agricultural issues of seed breeding and pesticide resistance to the biotechnological challenges of drug discovery and the remediation of environmental contamination.

Although there is some skepticism concerning this technology, microarray-based hybridization assays have generated considerable interest in the past few years and have become the standard technology for high-throughput analysis of biological systems. Some researchers consider microarray-based studies to be expensive and nonhypothesis-driven descriptive research, that is, fishing experiments (Brenner, 1999; Mir, 2000). This controversy is typical for any new field of research. DNA or oligonucleotide arrays have been successfully used to monitor messenger RNA (mRNA or transcript) abundance levels of differentially expressed genes under different cell growth conditions or in response to environmental perturbations or genetic mutations (Lockhart et al., 1996; DeRisi et al., 1997; Schena et al., 1996; Richmond et al., 1999; Ye et al., 2000; Thompson et al., 2002; Liu et al., 2003; Wodicka et al., 1997) and to detect specific mutations in DNA sequences (Hacia, 1999; Broude et al., 2001), but performing these experiments in a manner that produces accurate and reliable results presents unique technical challenges.

Microbial Functional Genomics, Edited by Jizhong Zhou, Dorothea K. Thompson, Ying Xu, and James M. Tiedje.
ISBN 0-471-07190-0   © 2004 John Wiley & Sons, Inc.

This chapter will review the technical underpinnings of microarray-based hybridization assays, the advantages and disadvantages of different types of array technologies, and the potential problems and challenges in microarray-based hybridization analysis. We present basic microarray concepts and advantages, a description of the various technologies used for microarray fabrication (Fig. 6.1), microarray hybridization and detection, and image processing. Finally, the basis and principles of using microarrays for monitoring gene expression will be discussed. Since glass-based DNA microarrays are the microarray technology of choice for many basic research laboratories, our discussion of the array technology focuses on this type of microarray. An effort is also made to cover additional types of microarrays and their associated analytical methods as comprehensively as possible. It should be noted that the goal of this chapter is to provide an in-depth description of the basis and the principles of microarray-based technologies rather than an exhaustive review of the current microarray technology. For technical details, we refer readers to the following recent books related to microarray-based technology: Schena (2002); Schena and Davis (2000) Bowtell and Sambrook (2002); Grigorenko (2002); Brownstein and Khodursky (2003); Causton et al. (2003).

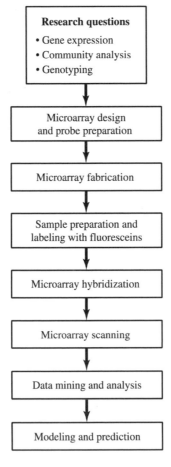

Figure 6.1 Flowchart of a microarray experiment.

6.2 TYPES OF MICROARRAYS AND ADVANTAGES

6.2.1 Concepts, Principles, and History

Microarrays are orderly miniaturized arrays containing large sets of DNA sequences that have been attached to a solid substrate using automated equipment such that each spot (element) corresponds to unique DNA (Schena et al., 1998; Schena, 2002). These arrays are also sometimes referred to as microchips, biochips, DNA chips, or gene chips. To avoid confusion with computer microchips, we will use the term microarray.

Microarray assays are based on the hybridization of a single-stranded molecule labeled with a fluorescent tag, or fluorescein, to a complementary molecule attached to a solid support, such as glass. In principle and practice, microarrays are extensions of conventional membrane-based hybridization that have been used for decades to detect and characterize nucleic acids in biological samples. In microarray assays, an unknown sample is hybridized to an ordered array of immobilized DNA molecules of known sequence to produce a specific hybridization pattern that can be analyzed or compared to a given standard. The fluorescein-labeled DNA strand in solution is generally called the target, whereas the DNA strand immobilized on the microarray surface is referred to as the probe. Because the sequence of the immobilized molecule is usually known, it is used to "probe" or investigate the target molecule in solution. This terminology is the opposite of the convention that originated with Southern blot hybridization, in which target molecules located on the membrane are interrogated by solution-phase probes.

The basic idea of microarrays was first proposed in the late 1980s. Augenlicht and his colleagues provided one of the first descriptions of DNA microarrays for simultaneously monitoring the expression level of thousands of human genes on nitrocellulose with radioactive labeling (Augenlicht et al., 1987, 1991). They spotted 4,000 complementary DNA (cDNA) sequences on nitrocellulose and used them to analyze differences in gene expression patterns among different types and stages of colon tumors. At the same time, four groups independently developed the concept of determining a DNA sequence by hybridization to a comprehensive set of oligonucleotides, that is, sequencing by hybridization or SBH (Bains and Smith, 1988; Drmanac et al., 1989; Khrapko et al., 1989; Southern et al., 1992). Although the concept of SBH is extremely elegant, it has several inherent problems, such as the presence of repeated sequences and the imperfect specificity of hybridization. These problems have limited the use of SBH for sequence determination. Such challenges have motivated most researchers to shift the emphasis to more immediately addressable applications, such as profiling gene expression.

By the mid-1990s, the reverse dot-blot scheme for monitoring genome-wide gene expression was recast by several different groups. Both DNA fragments and synthetic oligonucleotides were arrayed on various substrates, including nylon membranes, plastic, and glass (Lockhart et al., 1996; Schenea et al., 1995). All of them depended on sequence-specific hybridization between the arrayed DNA and the labeled nucleic acids from cellular mRNA. Later studies in yeast, with both DNA and oligonucleotide microarrays, clearly indicated that microarrays were powerful tools for monitoring gene expression (DeRisi et al., 1997; Wodicka et al., 1997).

Microarray-based genomic technology has greatly benefited from parallel advances in other fields. Without such advancements, the development of high-density microarrays and the various applications that we see today would not be possible (Schena and Davis, 2000; Eisen and Brown, 1999). First, many genome projects have produced large-scale sequence information for microarray-based analysis. Second, technical advances in robotics have made it possible to fabricate high-density microarrays in a very small area. Finally, recent advances in fluorescent labeling and detection offer significant advantages in speed, data quality, and user safety for microarray-based assays.

6.2.2 Microarray Types and Their Advantages

Types of Microarrays   Microarrays can be divided into two major formats based on the type of immobilized probe:

(i) *DNA Microarrays.* Constructed with DNA fragments typically generated with the polymerase chain reaction (PCR) (Schena et al., 1995; DeRisi et al., 1997; Marshall and Hodgson, 1998), DNA microarrays have certain advantages over oligonucleotide microarrays, especially for monitoring gene expression patterns. While oligonucleotide microarrays are limited to array elements of low sequence complexity, the specificity of hybridization for a complex probe is improved with DNA microarrays using DNA fragments substantially longer than oligonucleotides (Shalon et al., 1996). Oligonucleotide synthesis also requires prior sequence knowledge, but this is not the case for DNA arrays. Nucleic acids of virtually any length, composition, or origin can be arrayed (Shalon et al., 1996). However, oligonucleotide-based microarrays have the advantage of minimizing the potentially confounding effects of occasional cross-hybridization (Wodicka et al., 1997) and are uniquely suited for detecting genetic mutations and polymorphisms. Since oligonucleotide probes can be directly synthesized, the handling and tracking of oligonucleotide microarrays are easier than for DNA microarrays, which generally require PCR amplification. Amplifying all probes with desired minimum DNA concentration for printing is labor-intensive and time-consuming.

(ii) *Oligonucleotide Microarrays.* These are constructed with short (10- to 40-mer) or longer (up to 100-mer) oligonucleotide sequences that are designed to be complementary to specific coding regions of interest. There are two general types of oligonucleotide microarrays based on the strategy used for oligonucleotide immobilization and fabrication: (1) direct parallel synthesis on solid substrates by light-directed or photoactivatable chemistries (Lipshultz et al., 1999; Pease et al., 1994) or standard phosphoramidite chemistries (Southern et al., 1994); and (2) chemical attachment of prefabricated oligonucleotides to solid supports (Khrapko et al., 1989; Beattie et al., 1992, 1995; Eggers et al., 1994; Fotin et al., 1998; Guschin et al., 1997a,b; Khrapko et al., 1989; Lamture et al., 1994; Rehman et al., 1999; Rogers et al., 1999). Each strategy for oligonucleotide immobilization has specific advantages and disadvantages (Schena et al., 1998; Hoheisel, 1997).

There are two major advantages to the direct synthesis approach. First, the photoprotected versions of the four DNA bases allow microarrays to be manufactured directly from sequence databases, thereby removing the uncertain and burdensome aspects of sample handling and tracking. Second, the use of synthetic reagents minimizes microarray-to-microarray variations by ensuring a high degree of precision in each coupling cycle. The photolithographic approach, however, requires the use of photomasks, which direct light to specific areas on the array for localized chemical synthesis; these photomasks are very expensive and time-consuming to design and build. The yield and length of synthesized oligonucleotides are also subject to wide variation and uncertainty, which could lead to unpredictable effects on hybridization across the microarray. For the attachment of presynthesized probes, the concentrations and length of each oligonucleotide on the array can be controlled prior to immobilization. Standard synthesis chemistry is also well established for many nucleotide derivatives for which no light-inducible monomer equivalents are available. In addition, the postsynthesis approach is less complicated and can be customized according to the needs of the laboratory. However, the critical drawback of the postsynthesis approach is still the need for the external synthesis and storage of different oligonucleotides prior to array fabrication.

Advantages of Microarrays Microarrays offer a number of advantages over conventional nucleic acid-based approaches:

(i) *High-throughput and Parallel Analysis.* The attachment surface of nonporous substrates allows thousands to hundreds of thousands of array elements or probes to be uniformly deposited on a very small surface area. As a result, gene expression can be monitored at the genomic level, or many constituents of a microbial community can be simultaneously assessed in a single experiment using the same microarray. This is very important for studying gene expression at the genome-wide level, because the large amount of expression data generated from a single microarray experiment allows researchers to begin to develop a comprehensive, integrated view of a cellular system.

(ii) *High Sensitivity.* High sensitivity can be achieved in probe-target hybridization, because microarray hybridization uses a very small volume of probe and the target nucleic acid is restricted to a small area (Shalon et al., 1996; Guschin et al., 1997a). This feature enables high sample concentrations and rapid hybridization kinetics.

(iii) *Differential Display.* Different target samples can be labeled with different fluorescent tags and then hybridized in parallel to the same microarray, allowing the simultaneous analysis of two or more biological samples in a single assay. Multicolor hybridization detection minimizes variations resulting from inconsistent experimental conditions and allows the direct and quantitative comparison of target sequence abundance among different biological samples (Shalon et al., 1996; Ramsay, 1998).

(iv) *Low Background Signal Noise.* Nonspecific binding to a nonporous surface is very low; as a result, organic and fluorescent compounds that attach to microarrays during fabrication and use can be rapidly removed. This results in significantly less background than is typically encountered with porous membranes (Shalon et al., 1996).

(v) *Real-time Data Analysis.* Once the microarrays are constructed, hybridization and detection are relatively simple and rapid, allowing real-time data analysis in field-scale heterogeneous environments.

(vi) *Automation.* Microarray technology is amenable to automation and therefore has the potential of being cost-effective compared to conventional detection methods (Shalon et al., 1996).

6.3   MICROARRAY FABRICATION

6.3.1   Microarray Fabrication Substrates and Modification

Substrates   The substrate used for printing microarrays has a large impact on the overall microarray experiment. Poor surface treatment may result in poor DNA attachment to the slide, and nonuniform surfaces will cause variations in the amount of attached DNA. Residual substances on the surface from various steps in microarray experiments can also lead to high background fluorescence. In addition, the detection sensitivity could be decreased substantially by using materials with high intrinsic fluorescence. The selection of appropriate substrates for microarray experiments is therefore of critical importance.

A variety of substrates have been used for printing microarrays. They fall into two basic categories: porous and nonporous. Overall, nonporous substrates are preferred for many microarray experiments, because the unique physical and chemical characteristics of glass slides (e.g., little diffusion, low intrinsic fluorescence) allow miniaturization and the use of fluorescent labeling and detection, which are the most critical issues for large-scale genomic analysis. The miniaturized microarray format coupled with fluorescent detection represents a fundamental revolution in biological analysis (Schena and Davis, 2000).

Nonporous Substrate   At present, nonporous solid surfaces such as glass and polypropylene are the most commonly used substrates for microarray fabrication. Glass slides are most widely used, because they are inexpensive, have good physical characteristics, and are easily modified for nucleic acid attachment and synthesis (Southern, 2001).

Nonporous substrates have several advantages over porous substrates (Schena and Davis, 2000). First, they allow small amounts of biochemical molecules to be deposited at precise, predefined locations on the surface with little diffusion. Consequently, miniaturization and hence high-density microarray fabrication are possible. Second, nucleic acids are placed on the surface of the nonporous substrate. As a result, hybridization between target and probe molecules occurs at a much faster rate than on porous substrates because molecules do not have to diffuse in and out of the pores. In addition, there is no steric inhibition of hybridization due to confinement in the pores of the membrane (Southern, 2001). Third, with coverslips, a nonporous surface allows the use of small sample volumes, which then enables high probe concentrations, rapid hybridization kinetics, and high sensitivity. Fourth, a nonporous substrate does not absorb reagents or samples and excess labeled materials can be easily removed. This speeds up the procedure, improves reproducibility, and reduces background. Fifth, a nonporous substrate has low intrinsic fluorescence and thus allows the use of fluorescence detection. Finally, a solid substrate offers a homogeneous attachment surface. The inherent uniform flatness of a solid surface allows true parallel analysis.

The main disadvantage of using nonporous substrates such as glass is their susceptibility to dust and other airborne particle contamination. The quality of slides may also be rather low. Poor surface modification of slides is one of the main causes of the poor performance of microarrays (Southern, 2001). In addition, because of the planar surface, the capacity for immobilization is limited and consequently the sensitivity of the assay is relatively low compared to porous substrates (Afanassiev et al., 2000).

Porous Substrates   Porous substrates such as nitrocellulose and nylon membranes have also been used for microarray fabrication (Englert, 2000). The principal advantage of membranes is that large volumes and concentrations of samples can be immobilized on a small area because the pores of the substrates provide a larger total surface for binding. Consequently, relatively higher sensitivity and a better dynamic range for quantitative comparison can be achieved. Also, the deposited samples will immediately distribute into the membrane by capillary flow and homogeneous spots can be easily obtained. In addition, membrane-based microarrays can be reused several times (Beier and Hoheisel, 1999).

A disadvantage of the porous substrates is that the boundaries and shapes of the spots are poorly defined, and the membranes swell in solvent and shrink and distort when dried. Such fragility and flexibility make it difficult to precisely locate spot position during spotting and image analysis. In addition, many membranes have high intrinsic fluorescence and thus higher background noise. Finally, because the spot sizes on a membrane cannot be reduced to a level comparable with glass slides or other nonporous substrates, much more DNA is required for such microarray production (Beier and Hoheisel, 1999).

Flow-through Channel   To increase the binding capacity, porous silicon dioxide or channel array glass have been used for microarray supports (Steel et al., 2000). The probe solution will flow through the microchannels and be immobilized on the walls of the microchannels. This support has several theoretical advantages. First, the binding capacity and hence sensitivity and dynamic range are much greater than those for other methods because the surface area per unit of cross section increases on the order of 100-fold. Second, compared to a planar surface, the mass transport of reactants is potentially enhanced, and consequently, the rate of hybridization increases and less experimental time is required. In addition, more uniform probe deposition and higher array densities can be achieved due to the improved wetting properties of microporous materials. Finally, the samples can be concentrated by slowly flowing the samples through microchannels, and hence this substrate can be used to analyze dilute samples.

The disadvantage of this technology is that the current detection system developed for two-dimensional geometries is not appropriate for the flow-through channel platform. New detection systems, such as a charge couple device (CCD)-based scanner, need to be developed. It may also be very difficult to ensure that a sample solution is uniformly distributed among different microchannels across the entire microarray. Nonuniform flow will cause the nonhomogeneous performance of microarray hybridization.

Attachment Strategies   The appropriate attachment of nucleic acids to an array surface is very important for microarray analysis. For reliable microarray hybridization, the attachment chemistry must meet several criteria. The nucleic acids must be tightly bound to the microarray surface and the surface-bound molecules must be accessible for hybridization. In addition, the attachment chemistry must be reproducible. Both ionic interaction and covalent bonding are used for attaching nucleic acids to solid surfaces, depending on the size of nucleic acid molecules.

Electrostatic Interaction   Long DNA fragments (on the order of several hundred bases) will attach to the glass surface through ionic interaction between the negatively charged phosphodiester backbone and the positively charged slide surface (Fig. 6.2). Some studies have shown that oligonucleotides larger than 70 bp can also be bound to glass

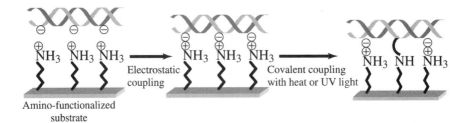

Figure 6.2 *Attachment of nucleic acids to solid surfaces through electrostatic interactions.* The microarray substrates contain primary amine groups (NH_3^+) attached covalently to the glass surface. The amines carry a positive charge at neutral pH, which permits attachment of native DNA through the formation of ionic bonds with the negatively charged phosphate backbone. Covalent attachment of the DNA to the surface can also be further achieved by treatment with UV light or heat.

surfaces through ionic interaction (Hughes et al., 2000). Generally, amine- or lysine-coated slides are used for adsorbing DNA to glass slides. Since amines carry a positive charge at neutral pH, they allow the attachment of native DNA through the formation of ionic bonds with the negatively charged phosphate backbone. Electrostatic attachment can be supplemented by treating the fabricated slides with ultraviolet (UV) light or heat. This induces free radical-based coupling between thymidine residues on the DNA and carbons on the alkyl amine. The combination of electrostatic bonding and nonspecific covalent attachment links native DNA to the substrate surface in a stable manner.

Attachment through ionic interaction is less expensive and more versatile than covalent bonding (Worley et al., 2000). However, the attached DNA molecules are susceptible to removal from the surface under high salt concentrations and/or high-temperature conditions. Therefore, covalent bonding methods are preferred.

Covalent Bonding   DNA can be covalently attached to glass surfaces using different attachment chemistries (Fig. 6.3). Although long DNA molecules can be attached

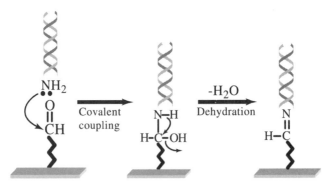

Figure 6.3 *Attachment of nucleic acids to solid surfaces through covalent bonding.* The microarray substrates contain primary aldehyde groups attached covalently to the glass surface. Primary amino linkers (NH_2) on the DNA attack the aldehyde groups to form covalent bonds. Such attachment is stabilized by a dehydration reaction in low humidity, which leads to the formation of Schiff base covalent bonds.

covalently to a microarray surface with several methods, the immobilization of aminated DNA to an aldehyde-coated slide is preferred (Zammatteo et al., 2000).

Because oligonucleotides are short, covalent bonding is generally required for the attachment of oligonucleotides to a glass surface. Usually, one end of the oligonucleotide molecule is fixed covalently onto a solid surface using a variety of methods. The attachment of biomolecules to a solid phase presents some problems that are unique to homogeneous solutions. Because the bound probe is not free to diffuse, a lower reaction rate is expected. In addition, target molecules in solution may not be able to effectively interact with the bound probes due to steric hindrance from the solid support and the other bound probes that are very close to each other (Shchepinov et al., 1997).

To minimize steric interference, there should be some distance between the oligonucleotide probes and the support. This is accomplished using additional molecules as linkers or spacers to tether the probe to the support. The desired linker must meet several criteria (Guo et al., 1994). First, the linkage must be chemically stable under the hybridization conditions used and must be sufficiently long to minimize steric interference. Also, the linker should be hydrophilic enough to be freely soluble in aqueous solution. In addition, there should be no nonspecific binding of the linker to the support. Shchepinov and coworkers (1997) showed that the optimal linker for both 5′ and 3′ immobilized oligonucleotides should have a low negative charge density and length of 30 to 60 atoms.

A linker is generally coupled to the probe during oligonucleotide synthesis. Various types of linkers have been used, such as poly-dT (Guo et al., 1994; Rehman et al., 1999; Afanassiev et al., 2000), polycarbon atoms (Afanassiev et al., 2000), and oligodeoxyribonucleotides with a hairpin stem-loop structure (Zhao et al., 2001). The length of spacers has a significant effect on hybridization. While virtually no hybridization signal was observed for poly-dT spacers (15 bp), about a 20-fold enhancement of hybridization was obtained for a poly-dT spacer of 15 nucleotides (Guo et al., 1994). The effect of linkers composed of multiple carbon atoms (e.g., C_{36}, C_{18}, C_{12}, and C_6) on hybridizations has also been examined (Afanassiev et al., 2000). Overall, signal intensity improved with an increase in the number of carbon atoms in the linker. The space between the probe and support can also be accomplished by chemically modifying the slide surface (Guo et al., 1994; Beier and Hoheisel, 1999).

Oligonucleotides can be immobilized onto solid supports through homo-biofunctional or hetero-biofunctional cross-linkers. For example, amino-modified oligonucleotides can be covalently attached to glass surfaces containing amine functional groups through homo-biofunctional cross-linkers (Guo et al., 1994), and to glass surfaces containing aldehyde and epoxide through heterobiofunctional cross-linkers (Schena et al., 1996; Lamture et al., 1994). Thiol- or disulfide-modified oligonucleotides can be attached to the glass surface containing an amine functional group via heterobiofunctional cross-linkers (Chrisey et al., 1996). A heterobiofunctional cross-linker is generally preferred over a homo-biofunctional cross-linker to prevent surface-to-surface linkages and probe-to-probe linkages as opposed to the desired surface-probe linkages (Steel et al., 2000). For using a hetero-biofunctional cross-linker, the probe should have a different modification chemistry from the array surface.

Surface Modification Microscope slides made from low-fluorescence glass that are used for microarray construction must be extremely clean and coated with chemicals before use. The glass surface must have a suitable functional group for the attachment of target DNA molecules because target DNA will not inherently bind to untreated glass. A hydrophobic surface is also essential for achieving high-density spots, because spotted hydrophilic samples will spread less on a hydrophobic surface than on an untreated hydrophilic glass surface (Rose, 2000).

The quality of slide coating ultimately impacts the quality of the microarray data. A poor surface coating can result in poor probe retention. For spot-to-spot consistency, the coating must be uniform and homogeneous without bare patches. In addition, the coating must be able to resist harsh conditions, such as boiling, baking, and soaking. Finally, the coating must be nonfluorescent. The following approaches are often used for glass surface modification.

Silanization   Silanes are most commonly used for slide surface modification to provide organic functional groups for covalently attaching biomolecules (Shriver-Lake, 1998). Glass slides can be modified to contain surface hydroxyls that can react with methoxy or ethoxy residues of a silane molecule. Many different commercially available silanes contain various functional reactive groups such as amino, epoxide, carboxylic acid, and aldehyde, which are suitable for covalent bonding with appropriately modified biomolecules (Schena, 2002). The most commonly used surface for microarray analysis contains reactive amine and aldehyde groups (Schena, 2002), which allow attaching biomolecules to glass slides by electrostatic interactions or covalent bonding (Fig. 6.2). Silanization can be accomplished by immersing the slides into a silane-containing solution or by vapor deposition (Steel et al., 2000; Worley et al., 2000). Vapor-phase coating is best able to deposit a monolayer of silane on the slide surface (Chrisey et al., 1996; Worley et al., 2000). Silanized slides are commercially available from a number of companies.

Dendrimeric Linker Coating   To increase binding capacity, a more elaborate chemistry was developed for synthesizing dendrimeric linkers on silanized glass slides to allow the covalent attachment of aminated DNA molecules (Beier and Hoheisel, 1999). Such a linker system multiplies the coupling sites by introducing additional reactive groups through branched linker molecules. There are several advantages to this linker system. First, it allows the covalent immobilization of both the presynthesized nucleotides and the in situ synthesis of oligonucleotides on glass slides. As a result, the loading capacity can increase by 10-fold. It has no nonspecific attachment of hybridization probes and low fluorescent background. In addition, covalent bonding is stable and the microarrays can be reused many times. Finally, since bonding occurs only through a terminal group of the attached molecules, no negative effects on hybridization efficiency were observed.

Gel Coating   To combine the advantages of porous and nonporous substrates, several attachment approaches that use polyacrylamide and agarose for surface modification have been developed (Zlatanova and Mirzabekov, 2001; Afanassiev et al., 2000). Polyacrylamide gel elements are affixed to the glass surface. The size of the pads varies from $10 \times 10 \times 5$ μm to $100 \times 100 \times 20$ μm, with volumes ranging from picoliters to nanoliters. Since each gel pad is surrounded by a hydrophobic surface that prevents solution diffusion among the elements, each one can function independently. The probe molecules are immobilized on gel pads using robotic means. The use of polyacrylamide gel pads as an immobilization support offers significant advantages over the use of probes attached to solid supports (Drobyshev et al., 1999). Three-dimensional immobilization of probes in gel pads may provide higher capacity and a more homogeneous environment than heterophase immobilization on glass, leading to higher sensitivity and a faster rate of hybridization (Vasiliskov et al., 1999). However, like nylon membrane-based supports, gel pads may cause higher background (Beier and Hoheisel, 1999). Only particular sizes of molecules can diffuse into the gel, so the probes and targets must be the appropriate size or fragmented (Englert, 2000). The gel pad attachment method can be inconvenient because it requires the activation of gels and probes with labile reactive chemicals (Rehman et al., 1999). A more flexible attachment method uses copolymerization of 5'-terminal modified oligonucleotides with acrylamide monomers (Rehman et al., 1999). The advantages of this method are easy probe preparation with standard DNA synthesis chemistry, easy immobilization without involving highly reactive and unstable chemical cross-linking agents, and efficient and specific attachment.

Agarose can be used for coating glass slides for probe attachment (Afanassiev et al., 2000). The agarose film is activated to produce reactive sites that allow the covalent immobilization of molecules with amino groups. Compared to a glass-based planar surface, agarose film has a higher binding capacity. Also, agarose does not interfere with fluorescent detection. In contrast to acrylamide gels (Zlatanova and Mirzabekov, 2001) and dendrimeric branched systems (Beier and Hoheisel, 1999), this method does not require complex preparation technology.

Nitrocellulose Coating   Glass slides coated with a proprietary nitrocellulose-based polymer have been examined as an immobilization support (Stillman and Tonkinson, 2000). The nitrocellulose-based polymer can bind DNA and proteins noncovalently but irreversibly. This support has better spot-to-spot consistency and much higher binding capacity and better dynamic range than traditional modified glass slides. It is suitable for fluorescent detection due to its relatively low light scatter. Although it has a higher fluorescent background, the background-subtracted signal is significantly higher on this support than on the glass support coated with polylysine. However, the ability of miniaturizing the array dimensions in such slides remains to be examined.

6.3.2 Arraying Technology

A critical step in using microarrays is the fabrication procedure, which involves attaching or printing the DNA probes on the array (Fig. 6.1). The microarray format is compatible with many advanced fabrication technologies. The most widely used printing technologies are photolithography, mechanical microspotting, and ink-jet ejection. Each technology has advantages and disadvantages for microarray fabrication (Schena et al., 1998; Schena and Davis, 2000). All three technologies allow the manufacture of microarrays with sufficient density for genetic mutation detection and gene expression applications. The key considerations in selecting a fabrication technology include microarray density and design, biochemical composition and versatility, reproducibility, throughput, density, and cost. Because of its versatility, affordability, and wide applications, microspotting may become the microarray technology of choice for the basic research laboratory. Thus, our discussion on array technology in this chapter will primarily focus on microspotting technology.

Light-directed Synthesis   In photolithography, oligonucleotides are synthesized in situ on a solid surface in a predefined spatial pattern by using a combination of chemistry and photolithographic methods borrowed from the semiconductor industry (Fodor et al., 1991; McGall and Fidanza, 2001) (Fig. 6.4). This approach has been developed by Affymetrix. First, a glass or fused silica substrate is covalently modified with a silane

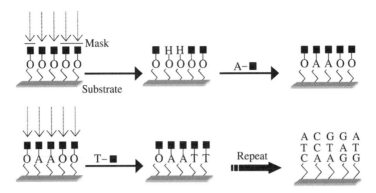

Figure 6.4 *In situ light-directed oligonucleotide probe array synthesis.* The solid surface contains linkers with a photolabile protecting group symbolized by the solid square (e.g., MeNPOC). MeNPOC is resistant to many chemical reagents, but it can be removed selectively by shining UV light for a short time. When MeNPOC is removed, the deprotected region on the surface can form chemical bonds with DNA bases containing a MeNPOC photoprotecting group at its 5' hydroxyl position. In this illustration, light is directed through a photolithographic mask to specific areas of the array surface, which are activated for chemical coupling. The first chemical building block A containing a photolabile protecting group X is then attached. Next, light is directed to a different region of the array surface through a new mask. The second chemical building block T containing a photolabile protecting group X is added and the process is repeated until the desired product is obtained.

reagent to obtain a surface containing reactive amine groups, which are then modified with a specific photoprotecting group, methylnitropoperonyloxycarbonyl (MeNPOC). Then the specific regions of the surface are activated through exposure to light, and a single base is added to the hydroxyl groups of these exposed surface regions using a standard phosphoramidite DNA synthesis method. The process of photodeprotection and nucleotide addition is iterated until the desired sequences are obtained. Typically, the probes on the arrays are 20 to 25 bp in length. Since the average stepwise efficiency of oligonucleotide synthesis ranges from 90 to 95%, the proportion of the full-length sequences for 20-mer probes is approximately 10%. However, this should have a relatively minor effect on the performance of microarray hybridization because of the high absolute amount of full-length probes on the support (McGall and Fidanza, 2000).

Another emerging light-directed synthesis approach for constructing high-throughput oligonucleotide arrays is to use a digital light processor (DLP), that is, a micromirror (Singh-Gasson et al., 1999; Nuwaysir et al., 2002). This maskless array synthesizer (MAS) technology uses DLP to create "virtual" masks and to direct a UV light beam to discrete locations on a glass substrate for DNA synthesis. Similar to the Affymetrix photolithographic approach, MAS is capable of constructing high-density microarrays of any desired nucleotide sequence. In contrast, MAS does not require photomasks, which are very expensive and time-consuming to manufacture. The MAS technology makes photolithography much more flexible and user-friendly (Nuwyasir et al., 2002).

Photolithographic parallel synthesis offers a very efficient approach to fabricating high-density arrays. The maximum achievable density is ultimately dependent on the spatial resolution of the photolithographic process. Due to its steric and/or electrostatic repulsive effects, an optimum probe density for maximum hybridization signal should exist. Affymetrix microarrays currently contain $\sim$250,000 oligonucleotides in an area of 1 cm^2.

One of the main advantages of this approach is that microarrays of extremely high density can be constructed (Ramsay, 1998). However, photolithography can only be used with oligonucleotides, not with DNA.

Contact Printing The most commonly used microarray fabrication technology is mechanical microspotting, which uses an array of pins, tweezers, or capillaries to deliver picoliter volumes of premade biochemical reagents (e.g., oligonucleotides, cDNA, genomic DNA, antibodies, and small molecules) to a solid surface. Currently, more than 1,000 individual cDNA molecules can be deposited in an area of 1 cm^2 (Rose, 2000). The advantages of microspotting are ease of implementation, low cost, and versatility. One disadvantage of this method is that each sample to be arrayed must be prepared, purified, and stored prior to microarray fabrication. In addition, microspotting rarely produces the densities that can be achieved with photolithography.

Microarray printing is accomplished by direct contact with the surface using a computer-controlled robot (Eisen and Brown, 1999). Pin printing can be achieved with the following pin technologies.

Solid Pins Solid pins have flat or concave tips. Since the volume of sample with both types of solid pins is relatively small, generally only one microarray can be printed with a single sampling load. Consequently, the overall printing process is slow. This printing technology is therefore well suited for constructing low-density microarrays. Due to the large pin surface-to-sample volume ratio of the solid pins, sample evaporation is a significant problem. Under standard laboratory conditions, about half of a 250-pl volume is lost in 1 sec (Mace et al., 2000). To minimize evaporation, a highly humid environment is necessary. However, high humidity may prevent the sample from drying sufficiently on the slide, and this may cause sample migration or spreading.

Split Pins Split pins have a fine slot at the end of the pin for sample holding (Fig. 6.5). When the slit pin is dipped into the sample solution, the sample is loaded into the slot, which generally holds 0.2 to 1.0 μl of sample solution. A small volume of sample (0.5–2.5 nl) is deposited on the microarrays by tapping the pins onto the slide surface with sufficient force (Rose, 2000) or touching the pins lightly on the surface like an ink stamp (Martinsky and Haje, 2000). Tapping the pins on the substrate surface is not recommended, because it may cause bulk transfer of the sample from the pins, deformation of the pin tip, or may fracture the surface coating (Martinsky and Haje, 2000; Mace et al., 2000). The split pin from the company TeleChem International is manufactured using digital control, and hence there are virtually no mechanical differences among pins (Martinsky and Haje, 2000). The TeleChem pins therefore provide very high printing consistency under conditions of good sample preparation, proper motion control, and a homogeneous printing substrate. Because the split pins hold a larger sample volume than other pins, more than one microarray can be printed from a single sampling event. Although split pin technology has been successfully used, one of its drawbacks is that dust, particulates, evaporated buffer crystals, and/or other contaminants can clog the pin slot.

Pin and Ring The pin-and-ring method is a variation of the pin-based printing process. The sample is taken by dipping the ring into the sample well, and then a small volume of sample solution is deposited onto the surface of the slide by pushing the sample in the ring with solid pins (Mace et al., 2000). Different-sized rings can be selected to hold 0.5 to

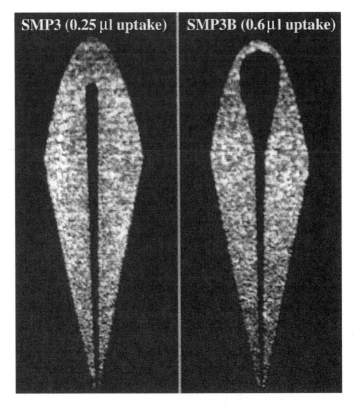

Figure 6.5 *Microspotting pins from TeleChem with an uptake channel, which holds liquid volume of 0.25 or 0.6 μl. Direct contact between the liquid and a hydrophobic solid surface deposits a drop of sample, typically 0.5 to 2.5 nl. Reprinted with permission from Tele Chem International, Inc. http:// www.arrayit.com/Products/Printing/Stealth/stealth.html.*

3.0 μl of sample. Many different spot sizes can be obtained by using pins with different diameters. In addition, sample evaporation is alleviated by minimizing fluid exposure through ring configuration.

The pin-and-ring arraying technology has several advantages. First, since the pin is used only for spotting and the sample fluid captured by the ring is relatively large, the deposition of samples among different slides is accomplished in an identical manner for each cycle. As a result, the microarrays fabricated with the ring-and-pin technology are highly consistent and reproducible (Mace et al., 2000). Second, since the ring is not subjected to any acceleration such as tapping and the ratio of ring to pin diameter is large, the ring geometry is capable of handling a wide variety of volumes and fluid viscosity. Unlike split pin printing, ring-and-pin printing is resistant to dust, particulate matter, high-viscosity fluids, debris, buffer or salts, and other materials. Finally, the ring and pin can deposit samples on soft substrates such as agar, gels, and membranes. One disadvantage of this technology is that the majority of sample captured by the ring is not used for spotting and is lost through washing prior to the next cycle of sampling.

Noncontact Ink-jet Printing Ink-jet ejection technologies provide another way to manufacture microarrays. The sample is taken from the source plate, and a droplet of sample is ejected from the print head onto the surface of substrate. Similar to microspotting approaches, ink-jet ejection allows the spotting of virtually any biological molecule of interest, including cDNA, genomic DNA, antibodies, and small molecules. In contrast to microspotting, ink jets have the advantage of avoiding direct surface contact. Ink-jet ejection, however, cannot manufacture microarrays as dense as those prepared by photolithography or microspotting approaches.

Currently, two types of noncontact ink-jet print technologies, piezoelectric pumps and syringe-solenoid, are used for printing microarrays.

Piezoelectric Pumps   This printing technology utilizes a piezoelectric crystal, which contacts with a glass capillary containing the sample liquid (Rose, 2000; Mace et al., 2000; Englert, 2000). When the crystal is biased with a voltage and is deformed, the capillary is squeezed and a small volume (0.05–10 nl) of fluid is ejected through the tip from the reservoir. The main advantages of piezoelectric printing technology are an extremely fast dispensing rate (on the order of several thousand drops per second), very small print volumes, and consistency of droplet size.

The piezoelectric printing technology is still in the early stages of development. The main problem with this technology is clogging by air bubbles and particulates. This makes the system less reliable. The void volume of sample solution contained in the capillary is also very large (100–500 µl) and not recoverable. In addition, it is difficult to change samples.

Syringe-solenoid Printing Technology   This technology uses a syringe pump and a microsolenoid valve for dispensing samples (Rose, 2000). The sample is taken by syringe and sample droplets are ejected by pressure onto the surface through the microsolenoid valve. The printing volume ranges from 4 to 100 nl. The main advantages of this technology are reliability and low cost. However, it is not suitable for fabricating high-density microarrays because of the large printing volume and spot size.

6.3.3 Critical Issues for Microarray Fabrication

This section will highlight some practical issues that are important for microarray fabrication: microarray density, reproducibility, storage time, contamination, and printing quality.

Microarray Density   Microarray density is one of the most important parameters for microarray fabrication. The number of spots that can be fabricated on a slide will depend on spot size and pin configuration.

Spot Size   The size of the DNA spot deposited on the slide determines the microarray density and directly depends on the volume of sample deposited on the surface. The volume deposited per spot generally ranges from 50 to 500 pl, with the possible extremes of 10 pl to 10 nl (Mace et al., 2000). Several factors affect the volume deposited. First, the surface properties (e.g., surface energy) of slides and pins, and the sample solution characteristics (e.g., viscosity) have significant effects on the volume of sample deposited and the spread area of the deposited sample on the slide surface. For printing higher-density microarrays, a hydrophobic glass surface (e.g., aldehyde slide) is preferred, because spotted hydrophilic samples will spread less on a hydrophobic surface than on a hydrophilic one. Second, since the pin contact surface area determines the initial contact between the sample and slide, the spot size increases as the pin contact surface area increases. In addition, pin velocity has a great effect on spot size. The loading sample volume for a split pin (e.g., ChipMaker and Stealth from TeleChem) typically ranges from 0.2 to 0.6 μl. Thus, if the pins tap the surface at high speed (> 20 mm/sec), a large sample volume may be forced out of the pin and large spots will be produced (Rose, 2000).

Pin Configuration   For printing, pins are mounted in a print head, which can hold up to 64 pins. The distance between pins on the print head is 4.5 mm and matches the well spacing of a 384-well plate. DNA samples are first taken from 96-well or 384-well source plates by dipping the pins into the sample wells with either a single pin or multiple pins, and then deposited on the slide surface by generally touching the pins to the surface. Fabricating arrays with a single pin is most straightforward, but it is time-consuming. The main advantage of single-pin printing is that the DNA samples on the source plates can be directly mapped onto an array with the same sample orders as those in the source plate. This will make sample tracking and posthybridization analysis easier. Also, no pin-to-pin variations are introduced. If we use a single pin and 250-μm spot-to-spot spacing, more than 20,000 spots can be deposited on a printing area of 22×72 mm. Since the printing process is much faster with multiple pins than a single pin, multiple pins are generally used for printing high-density microarrays. To print with multiple pins, the pins are dipped into sample wells of a 384-well plate and then touched to the slide surface simultaneously to create separate spots at a 4.5-mm spacing in the first round. Later rounds of printing are achieved by spotting with a predefined spot-to-spot offset distance from the previous location. Each pin deposits samples in a subgrid. Since some areas within each subgrid might not be completely filled with spots due to the restriction of the layout, the density of microarrays will generally decrease as the number of pins used increases. Printing microarrays with multiple pins requires more planning for the array layout and sophisticated sample tracking and deconvolution in data analysis. Also, using multiple pins for printing introduces pin-to-pin variations.

Reproducibility   Reproducibility is one of the most critical requirements for microarray fabrication. For reliable and reproducible data, the uniformity of individual spots across the entire array is very important for simplifying image analysis and enhancing the accuracy of signal detection. Several factors will affect the uniformity of spots, including array substrate, pins, printing buffer, and environmental controls. As we mentioned earlier, nonhomogeneous surface treatment will cause variations in the amount of attached DNA.

Significant variations could be caused by pin characteristics due to the mechanical difference in pin geometry, pin age, and sample solutions. Movement of the pin across the surface in the XY direction may cause the tip to bend (Rose, 2000). Tapping the pins on the surface may result in deformation of the pin tips. Also, dragging the pin tip across the surface may cause clogging of the pin sample channel. Therefore, great care is needed in handling pins, even though they are robust. Pins should be cleaned with an ultrasonic bath after each printing (Rose, 2000).

Environmental conditions have significant effects on spot uniformity and size (Hegde et al., 2000). Humidity control is necessary to prevent sample evaporation from source plates and the pin channel during the printing process. Sample evaporation can cause changes in DNA concentration and viscosity. As a result, the deposited DNA will be changed. Reducing evaporation will allow the spotted volumes of DNA to bind at equal rates across the entire surface. As a result, DNA spots of high homogeneity will be obtained (Diehl et al., 2001). Generally, the relative humidity is controlled between 65 to 75% (Rose, 2000). The choice of printing buffer is critical for obtaining spots of high homogeneity. Spot homogeneity as well as binding efficiency is often poor with the widely used buffer, saline sodium citrate. Using printing buffer containing 1.5 M betaine improves spot homogeneity as well as binding efficiencies because betaine increases the viscosity of the solution and reduces the evaporation rate (Diehl et al., 2001). More uniform spots can also be obtained with printing buffer containing 50% dimethyl sulfoxide (DMSO) (Hegde et al., 2000; Wu et al., 2001).

Storage Time   Another very important practical issue is the shelf life of microarrays. The maximum time that microarrays can be stored is still unknown. The shelf life could depend on the coating chemistry of the slide and the storage conditions. Unprocessed microarrays can be stored in a dessicator for many months without the deterioration of performance (Worley et al., 2000).

Contamination   To produce high-quality microarrays, contamination from airborne dust and impurities on slide surfaces must be minimized or eliminated during array fabrication. Dust and particulate matter can settle on the slide surface and cause printing inaccuracies and difficulty in processing array image. Enclosing the array device in a humidity chamber can minimize dust contamination.

Sample carry-over from using a pin for taking many different samples is another source of contamination that will complicate hybridization results. Efficient cleaning of the pins is therefore required for the printing process. Generally, the pins can be cleaned by dipping them into distilled water or detergent and then sucking out the wash solution from the pin channel with a vacuum. Repeating this process three times is generally sufficient to eliminate sample carry-over. However, caution should still be taken for potential sample cross-contamination.

Sample cross-contamination during sample preparation and handling is another major concern. For making microarrays, plasmids containing the desired cDNA clones are generally extracted from bacterial cultures and the desired genes are amplified from the plasmid DNA. Recent studies showed that up to 30% of clones contained the wrong cDNA (Knight, 2001). This is most likely due to bacterial contamination and handling errors during sample preparation; great care must be taken to eliminate or minimize such errors. Errors in the public sequence database could also lead to the failure of microarray-based detection. For instance, some mouse sequences in the public databases correspond to the wrong strand of the DNA double helix. As a result, the designed oligonucleotides could not detect their target mRNAs (Knight, 2001).

Evaluation of Printing Quality   After printing, it is important to assess the quality of the arrayed slides prior to hybridization in terms of the surface quality, integrity, and homogeneity of each DNA spot; the amount of deposited DNA; and consistency among replicated spots. It is common that DNA deposition with some pins is poor. Staining prior to hybridization is therefore important to identify potential problems during the fabrication process. Microarrays can be stained with various fluorescent dyes, such as PicoGreen, SYBR Green II, and Topo Green (Battaglia et al., 2000), followed by fluorescence scanning. Such information is critical to data interpretation and an important step in microarray quality control.

6.4 MICROARRAY HYBRIDIZATION AND DETECTION

6.4.1 Probe Design, Target Preparation, and Quality

Probe Design and Synthesis   One of the difficulties associated with microarray analysis is probe specificity. Probe design and synthesis are critical steps in generating high-quality microarrays for analysis. Three types of probes are used for microarray fabrication: PCR products, cDNA clones, and oligonucleotides. For microarray fabrication, individual open reading frames (ORFs) can be amplified with gene-specific primers. However, because cross-hybridization among homologous genes is a potential problem, full-length genes cannot be used for microarray construction. Therefore, gene-specific fragments should be amplified using gene-specific primers. Several computer programs are available for automatically identifying a DNA fragment specific ($<75\%$ identical in sequence) to each ORF by comparing the target gene with all other genes in a genome (Xu et al., 2002). Once the specific fragments are identified, more than one set of good primers can be obtained based on the identified unique fragments using the PCR primer design program Primer 3 (developed at the Whitehead Institute). The optimal forward and reverse primers are generally selected based on the following considerations. First, for genes shorter than 1,000 bp, the PCR-amplified unique fragments should be as long as possible. For genes longer than 1,000 bp, the optimal amplified fragments should be within the range of 500 to 1,200 bp. Second, each primer should have 20 to 28 oligonucleotides. In addition, to simplify PCR amplification in 96-well plates, most of the primer sets should have annealing temperatures of approximately 65°C. If the desired target annealing temperature cannot be obtained, a lower annealing temperature may be used. In the case that no specific fragments are identified for some homologous genes, fragments with higher than 75% sequence identity will be selected and appropriate primers can be designed based on these fragments. However, hybridization signals for these genes should be carefully interpreted during microarray data analysis. The designed primers are generally synthesized commercially. One of the great practical problems is that PCR product yields vary considerably among different genes and some primers may fail to yield PCR products. This may result in significant variation in the DNAs deposited on the slide surface.

The cDNA clone-based probes are generally derived from whole genes or fragments of genes that are amplified from clone libraries using vector-specific primers. The size of clone probes generally ranges from a few hundred to a few thousand base pairs. Generally, since vector-specific primers are used for amplifying the cloned inserts, clone-based probes cannot be specifically designed for regions of low homology to other genes. Also, as we mentioned earlier, a substantial portion of clones may contain the wrong cDNA due to bacterial contamination or mishandling (Knight, 2001).

Oligonucleotide probes are different from other probes in that they can be deposited by printing or synthesized in situ on a solid surface. Specific oligonucleotide probes can be designed based on gene sequences. Generally, the sizes of the oligonucleotide probes are shorter than 25 bp, and several different oligonucleotide probes are used per gene for high-density oligonucleotide microarrays. For discriminating mispriming, a probe is designed deliberately to experience a single mismatch at the central position (Lockhart et al., 1996; Warrington et al., 2000). Recently, the utility and performance of oligonucleotide microarrays containing 50- to 70-mer oligonucleotide probes were evaluated (Kane et al., 2000). The results indicate that these oligonucleotide microarrays can be used as a specific and sensitive tool for monitoring gene expression.

Characteristics of Fluorescence Fluorescent compounds absorb light of a certain wavelength, followed by exciting electrons within the compound to a higher-energy state. From this state, photons are emitted at a specific length to produce fluorescence light. The wavelength for fluorescence emission is often longer than that for fluorescence excitation. Generally, the excitation light used should produce reasonably efficient excitation at least at the 50-to-70% level, and the excitation wavelength should not be very close to the emission peak (Schermer, 1999).

Many microarray experiments for differentiating gene expression use two common fluoresceins, cyanine 3 (Cy3, green) and cyanine 5 (Cy5, red). The excitation peaks for Cy3 and Cy5 occur at wavelengths of 550 and 649 nm, respectively, and the emission peaks for Cy3 and Cy5 occur at wavelengths of 570 and 670 nm. Many other fluorescent dyes, such as Alexa analogs, are also available for microarray assays (Schena and Davis, 2000).

Fluorescent labeling and detection use fluorescent DNA bases that absorb and emit light at distinct and separable wavelengths. The inherent low level of intrinsic fluorescence of glass and other microarray substrates permits the use of fluorescent labeling and detection schemes, which enable a great leap in terms of speed, data quality, and user safety (Eisen and Brown, 1999). Fluorescein-based detection allows the simultaneous analysis of two or more biological samples in a single assay and hence permits the direct and quantitative comparison of target sequence abundance among different biological samples. In addition, fluorescence-based assays make it possible to utilize advanced data acquisition technologies such as confocal scanners and CCD cameras.

Target Labeling and Quality Target labeling is another critical step in successful microarray-based experimentation. There are many methods available for labeling nucleic acids for microarray hybridization that can be classified into two categories: direct and indirect labeling.

Direct Labeling In the direct labeling approach, fluorescent tags are directly incorporated into the nucleic acid target mixture before hybridization by enzymatic synthesis in the presence of either labeled nucleotides (e.g., Cy3- or Cy5-dCTP) or PCR primers (Fig. 6.6). The most common method is to label the target mRNA or total RNA using reverse transcriptase. In a reverse transcription reaction, one of the nucleotides contains fluorescently labeled nucleotides that will be incorporated into the transcribed cDNA during first-strand cDNA synthesis. Reverse transcription requires a primer for initiation, which can be random hexamers, oligo(dT), or gene-specific primers. Since prokaryotic mRNA has no poly(A) tail, random hexamers are generally used for reverse transcription. In this case, total RNA is transcribed, and hence a greater degree of background fluorescence intensity can occur. Although gene-specific primers reduce such background levels by copying gene-specific fragments, it requires reverse transcription with hundreds or thousands of primers.

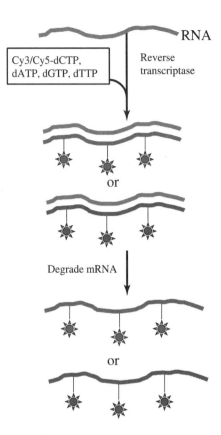

Figure 6.6 *Direct incorporation of fluorescent dyes into target sample through reverse transcription.*

A variation of the direct labeling approach uses the amplification of mRNA up to 1,000- to 10,000-fold by T7 polymerase to obtain antisense mRNA (aRNA). The aRNA is then reverse-transcribed to obtain labeled cDNA (Salunga et al., 1999). One of the advantages of the T7 polymerase-based amplification method over other methods is that because amplification is a linear process, all mRNAs are almost equally amplified. Another advantage is that mRNA can be easily labeled with reverse transcriptase, which incorporates fluorescent tags much more readily than DNA polymerase.

One of the problems with the reverse transcriptase-based labeling approach is that nucleotides tagged with different dyes are differentially and nonuniformly incorporated into cDNA. This could be due to structural differences between the fluorescent dyes, although the basis of this effect is poorly understood. To solve this problem, a two-step approach was proposed. The FairPlay™ system developed by Stratagene Corporation uses a two-step chemical coupling method to fluorescently label cDNA. First, an amino allyl-dNTP is uniformly and efficiently incorporated into cDNA by reverse transcriptase because the amino allyl-dNTP does not exhibit steric hindrance (Fig. 6.7). An amine-reactive cyanine is then chemically coupled to the amino-modified cDNA. The main advantage of such an approach is that this system efficiently produces uniformly labeled cDNA without any dye bias. As a result, this system is highly sensitive (having a 5-fold increase in sensitivity compared to direct dye incorporation method), requires less RNA,

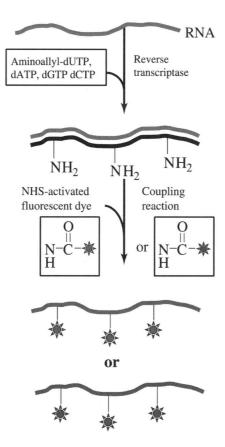

Figure 6.7 *Incorporation of fluorescent dyes into target samples through reverse transcription in the presence of amino-allyl-dUTP, followed by chemical coupling with fluorescent dyes.*

and allows the detection of low abundance genes. Also, any labeling bias resulting from fluorescent dye incorporation appears to be negligible and thus the dual labeling experimental approach is not needed.

Indirect Labeling   The indirect labeling approach is different from the direct approach in that the fluorescence is introduced into the procedure after hybridization. Briefly, epitopes are incorporated into the target samples during cDNA synthesis. After hybridization with the epitope-tagged target samples, the microarray is incubated with a protein that binds the epitopes to provide fluorescent signals. The most common indirect labeling method uses a biotin epitope and a fluorescent streptavidin-phycoerythrin conjugate (Warrington et al., 2000). The biotinylated nucleotides are incorporated into cDNA by reverse transcription and hybridized with microarrays. After hybridization, the array is stained with a streptavidin–phycoerythrin conjugate, which binds to biotin tags and emits fluorescent light when excited with a laser.

The second indirect labeling approach is known as tyramide signal amplification (TSA) (Adler et al., 2000). This approach uses biotin and dinitrophenol (DNP) epitopes as well as streptavidin and antibody conjugates linked to horseradish peroxidase (HRP). In the presence of hydrogen peroxide, HRP catalyzes the deposition of Cy3- or Cy5-tyramide compounds on the microarray surface. By this method, DNP- or biotin-dCTP analog is first incorporated into cDNA, and then the epitope-tagged cDNA is hybridized with the microarray. After hybridization, the microarray is incubated with anti-DNP-HRP and Cy3-tyramide is deposited on the microarray surface, followed by incubation with streptavidin-HRP and the deposition of Cy5-tyramide (Adler et al., 2000). The main advantage of this approach is that it can provide 10- to 100-fold signal amplification over the direct labeling approach. Thus, this approach can be used for monitoring the gene expression of rare transcripts or for analyzing samples prepared from small numbers of cells. The main disadvantage of this method is that it is generally less precise for comparative analysis due to the differences resulting from the variations of labeling efficiencies and protein binding affinities (Schena and Davis, 2000). Also, the signal intensity is only semiquantitative because of the involvement of enzymatic signal amplification (Alder et al., 2000).

The third indirect approach is to use DNA dendrimer technology (Stears et al., 2000) (Fig. 6.8). Dendrimers are stable, spherical complexes of partially double-stranded oligonucleotides with a determined number of free ends. The free ends are tagged with fluorescent dyes, Cy3 or Cy5. In this technology, the cDNA is first synthesized by reverse transcriptase with primers containing specific capture sequences that can bind the

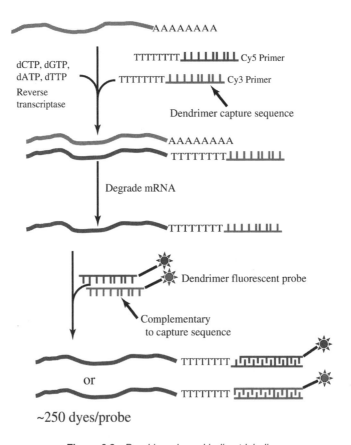

Figure 6.8 *Dendrimer-based indirect labeling.*

dendrimers tagged with Cy3 or Cy5 through sequence complementarity. The synthesized cDNAs are then hybridized to microarrays, and the bound cDNAs on the microarrays are detected by incubating the arrays with fluorescent dye-tagged dendrimers. The dendrimer detection approach is highly sensitive and requires up to 16-fold less RNA for probe synthesis. Also, since the fluorescent dye is attached to the free ends of dendrimers, signal intensity is independent of probe size and composition. In addition, this detection system has a high signal-to-background ratio and can be used for multiple channel detection on a single microarray.

6.4.2 Hybridization

After microarray fabrication, the most important issue in microarray-based analysis is probe–target hybridization. Conceptually, microarray hybridization and detection are quite similar to the traditional membrane-based hybridization (Eisen and Brown, 1999). Before hybridization, the free functional groups (e.g., amine) on the slide should be blocked or inactivated to eliminate nonspecific binding. Nonspecific binding causes high background and depletion of probes. In addition, the unbound DNA on the slides can be washed away during the prehybridization process. Removal of unbound DNA in prehybridization is important, because any DNA that washes from the surface during hybridization competes with DNA bound to the slide. Since the rate of hybridization in solution is much faster than that on surfaces, the presence of unbound probe DNA can lead to a dramatic decrease in the measured signals obtained from microarrays.

After prehybridization, the microarray is hybridized with fluorescently labeled target DNA or RNA for a period of time. The unbound material is washed away. Regardless of the hybridization format, the hybridization solution should be mixed well so that the labeled targets are evenly distributed on the array surface to obtain optimal target–probe interactions across the entire microarray. The wash solutions should be uniformly distributed to eliminate unbound probes, remove nonspecific hybridization, and minimize background signal.

6.4.3 Detection

The detection of microarray hybridization signals is another critical step in microarray-based studies. The confocal scanning microscope (used more commonly) and CCD camera have been used successfully for microarray detection, and many such devices are commercially available (Hegde et al., 2000). Generally, a confocal scanner uses the laser excitation of a small region of the glass slide (~ 100 μm^2), and the entire array image is acquired by moving the glass slide, the confocal lens, or both across the slide in two directions (Shermer, 1999). The fluorescence emitted from the hybridized target molecule is gathered with an objective lens and converted to an electrical signal with a photomultiplier tube (PMT) or an equivalent detector. The main drawback of a confocal scanning microscope is that each excitation wavelength must have its own laser, which can be expensive. The confocal scanning microscope is also very sensitive to any nonuniformity of the glass slide surface.

The CCD camera exploits many of the same principles as a confocal scanner, but the CCD camera utilizes substantially different excitation and detection technologies (Shermer, 1999). CCD systems typically use broad-band xenon bulb technology and spectral filtration (Basarsky et al., 2000). The key advantage of the CCD camera-based imaging systems is that they allow the simultaneous acquisition of relatively large images of a slide (1 cm^2) and hence do not require moving stages and optics, which reduces cost and simplifies instrument design. However, since the CCD camera does not move the optics or stages, several images need to be captured from different fields of the microarray and then combined to represent the entire information on the slide. In addition, it is difficult to effectively separate excitation and emission light in the spectral filtration process because the most commonly used fluoresceins have similar excitation and emission maxima.

6.4.4 Critical Issues in Hybridization and Detection

This section will highlight some important issues related to microarray hybridization and detection from a practical point of view.

Probe DNA Retention and Quantitative Hybridization In solution-based hybridization, the hybridization signal intensity depends on both target and probe DNA concentrations. In microarray-based gene expression studies, however, it is assumed that the concentrations of all probe DNAs deposited on the microarrays are much higher than the mRNA concentrations in the fluorescently labeled target samples. As a result, hybridization signal intensity is dependent only on the mRNA concentration in the target samples rather than on the arrayed probe DNA concentration. Therefore, many factors that cause probe deposition variations will have negligible effects on hybridization signal intensity.

For accurately quantifying gene expression, it is essential to ensure that the arrayed DNA probes are in excess relative to the labeled target cDNAs. Generally, a DNA concentration of 100 to 200 ng/μl is used for spotting. This corresponds to 100 to 200 pg/ spot for 1-nl deposition. The retention is about 20 to 30% on silanized glass surfaces (Worley et al., 2000). Thus, after boiling and hybridization, this corresponds to approximately 20 to 60 pg of double-stranded DNA present in each spot for binding. Studies indicated that the arrayed DNA appears to be in excess for all the protein-coding genes in *E. coli* (Worley et al., 2000). However, probe DNA retention depends on slide surface, coating chemistry, postprocessing, hybridization, and washing conditions. Therefore, to ensure accurate quantitative results for highly expressed genes, it is important to understand how much spotted DNA can be actually retained after hybridization for each type of slide and for the hybridization conditions used.

Whether the probe DNA on the slide surface is in excess also depends on the amount of target sample used. Typically, 10 to 20 μg of total cellular RNA is used for monitoring gene expression in prokaryotes. For monitoring rare transcripts, higher RNA concentrations (e.g., 50 μg) are generally used. In this case, the DNA probes on the slide for the abundant transcripts may not be in excess relative to the target samples. As a result, the hybridization may not be quantitative. It is important to select the appropriate amount of RNA to ensure that the microarray signal is within the range of linear response for the system being used.

Target Labeling and Availability RNA quality is very important for obtaining high-quality microarray hybridization results. Impurities in RNA preparations could have an adverse effect on both labeling efficiency and the stability of the fluorescent dyes. Thus, the RNA must be free of contaminants such as polysaccharides, proteins, and DNA. Many commercial RNA purification kits are available for producing RNA with sufficient purity for microarray studies. Unincorporated nucleotides must also be removed in order to reduce background noise. In addition, both Cy3 and Cy5 are sensitive to light, and thus great caution must be taken to minimize exposure to light during labeling, hybridization, washing, and scanning.

Labeling is the most critical step for obtaining high-quality microarray data. The most often encountered experimental problem is that microarray hybridization signal varies greatly from time to time. In many cases, poor hybridization signal results from poor dye incorporation. Very low dye incorporation (<1 dye molecule/100 nucleotides) yields unacceptably low hybridization signals. However, studies showed that very high dye incorporation (e.g., >1 dye molecule/20 nucleotides) is not desirable, because high-dye incorporation significantly destabilizes the hybridization duplex (Worley et al., 2000). Thus, it is important to measure dye incorporation efficiency prior to hybridization. The specific activity of dye incorporation can be determined by measuring the absorbance at wavelengths of 260 and 550 nm for Cy3 or 650 nm for Cy5. A suitable labeling reaction should have an 8 to 15 A_{260}/A_{550} ratio for Cy3 and a 10 to 20 ratio A_{260}/A_{650} for Cy5. Another problem encountered routinely is the quality of fluorescent dyes. The labeling efficiency and hybridization can vary significantly from batch to batch, especially for Cy5. Fresh reagents are very important in achieving highly sensitive detection (Wu et al., 2001).

Microarray hybridization is generally performed in the absence of mixing. Since the diffusion coefficient is very small for large labeled target DNA molecules, the probe on each arrayed spot is, in effect, hybridizing with its labeled counterpart from its immediate or nearly immediate local environment (Worley et al., 2000). Thus, the target solution should be mixed well and uniformly distributed over the microarray surface area. Otherwise, the availability of the labeled target molecules to the arrayed spots could be significantly different across the microarray surface. Labeled target molecules may be depleted in some area, while they may be abundant in other areas. As a result, significant differences in signal intensity could be observed. Nonuniform hybridization is a common problem associated with microarray experiments. To detect and assess this problem, it is important to keep replicate spots well separated on a slide.

Spatial Resolution and Cross-talk The spatial resolution of microarray detection systems is usually expressed in terms of pixel size or the physical "bin" in which a single datum is acquired. The spatial resolution for many commercial systems usually ranges from 5 to 20 μm. The selection of spatial resolution depends on spot size. A general guideline is that the pixel dimension should be less than one-tenth of the diameter of the smallest microarray dot on the array. For example, microarrays containing 100-μm spots require fluorescent detectors with a spatial resolution of 10-μm pixel size.

Cross-talk refers to the phenomenon in which an emission signal from one channel is detected in another channel. This results in an elevated, erroneous fluorescence reading in the other channel. Cross-talk is more likely to emanate from a shorter wavelength channel into a longer wavelength channel than vice versa. For example, the fluorescence intensity from the Cy3 channel can contaminate the Cy5 channel but not vice versa. Cross-talk is the most common potential concern for the simultaneous scanning approach, which acquires images from two channels at the same time (Basarsky et al., 2000). For gene expression experiments, the maximum cross-talk should be kept to less than 0.1%. The most common and cost-effective way to minimize cross-talk is to use emission filters that reject light outside the desired wavelengths. Cross-talk can also be minimized by selecting fluorescent dyes and lasers with sufficient differences in wavelength (Schermer, 1999).

Photobleaching and Scanning Parameters   Light is emitted from a fluorescent dye when it is illuminated by excitation light. Generally, the emitted fluorescence is directly proportional to the power of the excitation light. Therefore, to increase sensitivity, higher power of excitation light is preferred. However, if the excitation light is excessive, the incoming photons can damage the dyes and reduce the fluorescent signals during successive scans. More powerful light sources and/or longer laser exposure time can lead to significant photobleaching. Generally, photobleaching should be less than 1% per scanning.

Although Cy3 (0.15) has a lower quantum yield than Cy5 (0.28), Cy3 is more efficiently incorporated into cDNA during reverse transcription. Such dye characteristics can cause variations in the signal intensity obtained for these two dyes in reverse labeling experiments. The signal should be balanced during scanning by using a higher PMT setting for the dye with a weaker signal to allow the detection of more spots with low signal intensity.

Different dyes have considerable differences in their photostabilities. For example, Cy5 is more sensitive to photobleaching than Cy3. The differences in photostability among different dyes could potentially cause problems when multiple dyes are used in experiments because the ratios measured can lead to significant quantitative errors. To minimize photobleaching, the Cy5 channel is always scanned first, followed by the Cy3 channel.

6.5   MICROARRAY IMAGE PROCESSING

6.5.1   Data Acquisition

The fundamental aim of image processing is to measure the signal intensity of arrayed spots and then quantify gene expression levels based on the signal intensities for each spot. Therefore, the spots on the array image must be correctly identified.

Microarray images comprise arrays of spots arranged in grids. An ideal microarray image for easy spot detection should have the following properties: (1) The location of spots should be centered on the intersections between the row and column lines; (2) the spot size and shape should be circular and homogeneous; (3) the location of the grids on the images should be fixed; (4) the slides should have no dust and other contaminants; and (5) the background intensity should be very low and uniform across the entire image. However, in practice, it is difficult to obtain such ideal images. First, spot position variation occurs because of mechanical limitations in the arraying process, including inaccuracies in robotic systems, the printing apparatus, and the platform for holding slides. Second, the shape and size of the spots may fluctuate considerably across the array because of variations in the size of the droplets of DNA solution, DNA and salt concentration in the printing solution, and slide surface properties. In addition, contamination from airborne dust and impurities on the slide surface is a significant problem for processing array images. To obtain accurate measurements of hybridization signals, all these potential problems should be taken into consideration.

Many methods are available for resolving the spot location errors, spot size and shape irregularities, and contamination problems to accurately estimate spot intensities (Zhou et al., 2000). A variety of commercial and free software, such as ImaGene[TM] from BioDiscovery (Los Angeles, Calif.), QuantArray[TM] from GSI Lumonics, and the software on Axon GenePix[TM] systems (Bassett et al., 1999), can be used for microarray image processing. Typically, a user-defined gridding pattern is overlaid on the image and the gridding circles can be automatically defined and adjusted based on the spot size and shape. The areas defined by patterns of circles are then used for spot intensity quantification.

The data are extracted and generally expressed as the total (the sum of the intensity values of all pixels in the signal region), mean (the average intensity of pixels), and median (the signal intensity of the median pixel). The total intensity is not the best measurement because it is sensitive to variations in the amount of DNA deposited on the surface and the presence of contamination (Zhou et al., 2000). Because the mean intensity measurement reduces variations in the amount of DNA deposited on the spot, it is probably the best measurement when using an advanced image processing system permitting accurate segmentation of contaminated pixels. However, the mean measurement is vulnerable to outliers (Petrov et al., 2002), The median is a better choice than the mean if the image processing software is not good enough to correctly identify signal, background, and contaminated pixels. The median intensity value is very stable, and it is close to the mean if the distribution profile of pixels is uni-modal. The median is equal to the mean when the distribution is symmetric. In addition, an alternative to the median measurement is to use a trimmed mean (the mean of the pixel intensity after a certain percentage of pixels are removed from both tails of the distribution).

Some comparative studies indicate that the choice of measurements depends on the segmentation techniques used. The mean is the best measurement if the combined and trimmed segmentation techniques are used, whereas the median will be best without trimming (Petrov et al., 2002).

6.5.2 Assessment of Spot Quality and Reliability, and Background Subtraction

Because of the inherently high variation associated with array fabrication, hybridization, and image processing, the intensity data for some spots may not be reliable. Thus, the first step in data processing is to assess the quality of spots and remove unreliable, poor spots or problematic slides prior to data analysis. Also, in many cases, because of slide quality, background, and contamination, the quality of data could vary significantly among different slides (Tseng et al., 2001). Finally, some outlying spots (outliers) could occur due to experimental errors and should be removed prior to data analysis (Heyer et al., 1999). Without assessing the reliability of the data, conclusions drawn from the analyses of such data could be misleading.

Identification of Poor Slides Due to the complicated nature of microarray experiments, it is important to evaluate slides as a whole prior to rigorous data analysis and to eliminate unreliable hybridization signals. Two measures can be used to assess the overall slide quality if replicate spots are deposited on the slide (Worley et al., 2000): (1) the average coefficient of variation (CV) of replicates in the spot pairs; and (2) the R^2 value of the regression line from a scatter plot of duplicate spots. Although there is no general consensus for rejecting slides, slides are generally accepted if the average CV is less than 50%.

If there are no replicate spots on microarrays, the slide quality can be assessed by determining the number of spots that are of poor quality. Generally, microarray experiments should be repeated if more than 30% of the spots on the microarray are flagged as poor spots.

Identification of Poor-quality Spots   There are no rigorously defined rules for identifying poor spots from a biological and statistical perspective. The spot quality and integrity are generally assessed based on the following criteria:

(i) *Spot Size and Shape.* Spots with excessively large or small diameters should be discarded. Discarding such low-quality spots significantly improved the reliability of data (Zhou et al., 2000).

(ii) *Spot Homogeneity.* The distribution of pixels within the spots can be used to assess spot homogeneity. Generally, spots with less than a certain percentage (e.g., 55–60%) of all pixels having intensities greater than average background intensities (Khodursky et al., 2000) or one standard deviation (SD) above local background are flagged as poor-quality spots (Murray et al., 2001).

(iii) *Spot Intensity.* Spots with signals not significantly above background should be identified using various standards. For example, spots with median or mean signals less than one to three times the standard deviation above background in both channels (Chen et al., 1997; Hegde et al., 2000; Basarsky et al., 2000) are flagged as poor-quality spots. Also, spots whose signal is not at least 2.5 times higher than the background signal in both channels are excluded (Evertsz et al., 2000).

Another way to define poor spots is based on the signal-to-noise ratio (SNR), which is often defined as the ratio of the difference between a signal and background divided by the standard deviation of background intensity (Verdnik et al., 2002). This ratio indicates how well one can resolve a true signal from instrumental noises. A commonly used criterion for the minimum signal that can be accurately determined is a SNR equal to 3. Below 3, the signal could not be accurately quantified, and thus these spots can be treated as poor spots.

The commercially available software, ImaGene from Biodiscovery, is able to flag poor spots automatically. Spots identified as poor quality are not included in the data analysis (see Chapter 7). Although the criteria for defining poor spots are based on subjective thresholds rather than statistically robust tests, they take into account the major factors affecting the quality of data and are likely to be very effective in reducing the amount of noise.

Removal of Outlying Spots   Outliers are extreme values in a distribution of replicates. Outlying spots could be caused by uncorrected image artifacts such as dust or by factors undetectable by image analysis such as cross-hybridization. Outliers significantly affect the estimation of expression values and its associated random errors. Thus, the removal of outlying spots is an important step for predata analysis. However, distinguishing outliers from differentially expressed genes is very challenging because there is not a general definition of outliers. In the following subsections, we briefly describe several commonly used methods for outlier identification.

Simple Threshold Cutoff   If a gene's CV (i.e., standard deviation divided by the mean intensity among replicate spots) is greater than a certain threshold (Murray et al., 2001), for example, 30 to 50%, this gene is not included in the data analysis.

Intensity-dependent Threshold Cutoff by Windowing Procedure (Tseng et al., 2001)   The CV values for individual genes are plotted against the average signal intensity of the two channels $[(Cy3 + Cy5)/2]$. For each gene, a windowing subset is constructed by selecting a certain number of genes (e.g., 50) whose mean intensities are closest to this gene. If the CV of this gene is within the top certain percentage (e.g., 10%) among genes in its windowing subset, then data on this gene are regarded as unreliable, and hence all replicate data for this gene are discarded. However, in many cases, it is possible that some of the replicate spot data for this gene are reliable. To salvage the information for this gene, the most outlying spots can be eliminated, and the CV of the intensity ratios for the remaining spots associated with this gene can be recalculated. If the CV is significantly reduced below the threshold level, the data for the remaining spots can be used in subsequent analysis. The CV values can also be utilized for assessing the quality of different slides and different experiments (Tseng et al., 2001).

Removal of Outliers with Jackknife Procedure   Heyer and colleagues (1999) proposed using a jackknife correlation to remove outliers for the time series of microarray expression data. First, the correlation coefficient is calculated for each pair of genes with all the time series of data points. Then the data at one time point are deleted and the correlation coefficient is recalculated, respectively, for each pair of genes with all the time series of data points but one. The jackknife correlation is the minimum of the correlation coefficients obtained and can then be used for further cluster analysis. Jackknife correlation is robust and insensitive to single outliers. Applying jackknife correlation reduces false positives, while capturing the shape of an expression pattern. Iyer and coworkers (1999) showed that the genes showing similar expression patterns generally had a jackknife correlation of 0.7 or higher.

Identification of Outliers Based on Pooled Error Methods   Several methods are used for the statistical detection of outliers, but they are generally less adequate for typical microarray studies due to the small number of replicates (Nadon et al., 2001). The random error estimation for each gene based on a small number of replicates is imprecise, which makes statistical tests insensitive. As a result, many replicate spots may be falsely identified as outliers or many true outliers may not be identified (Nadon and Shoemaker, 2002). In addition, potential violation of the normality assumption makes inferences of outliers and gene differential expression less reliable (Nadon et al., 2001). To circumvent these difficulties, the pooled error method assumes that all probes or probes of similar intensities within a specific study have the same true random error. Thus, the variance estimates can be pooled together across many genes and their precisions can be greatly improved. It is assumed that the standardized residuals have a normal distribution if the pooled error model is correct. Under these assumptions, the existence of outliers will cause the distribution of the entire data set to deviate from normality. Removal of spots with large residuals will improve the normality of the entire data set. Generally, outliers are identified in an iterative fashion. First, spots with large absolute residuals are removed from the data set, data are examined for normality, and the residuals are calculated again. The process is iterated until the index asymptotes approach a stable value, which indicates that the further removal of data values would not improve the normality of the distribution of the remaining data set. To facilitate array-based statistical analysis, a commercially available software, ArrayStat™, was developed (Nadon et al., 2001). In this software package, outliers are detected automatically. The pooled error method provides a better and more sensitive method for outlier detection, and can be used appropriately for microarray experiments having as few as two replicates.

Background Subtraction   The second step in microarray data processing is background subtraction. Background subtraction is necessary to distinguish signals from noise and allows the comparison of specific signals. There are two approaches to background subtraction. The first approach is to take a signal from blank areas on the array and use this for subtraction. The problem with this approach is that background varies across the array, and thus the background noise among spots is significantly different. The second approach is to use a local background for the area around each spot for background determination. Local sampling of the background is generally used to specify a threshold that the true signal must exceed. By doing this, it is possible to detect weak signals and extract an average density above the background for each array element (Chee et al., 1996).

After removing poor slides, poor-quality spots, outliers, and background, the microarray data are ready for further normalization and data analysis (see Chapter 7).

6.6   USING MICROARRAYS TO MONITOR GENE EXPRESSION

Microarrays have been widely used to quantify and compare gene expression in a high-throughput fashion. In the following section, we will briefly review: the use of microarrays to monitor gene expression, fundamental basis, general approaches, and performance, as well as key issues related to designing microarray experiments. The scientific findings and new insights obtained with microarray analysis are described in detail in Chapters 11 and 12.

6.6.1   General Approaches to Revealing Differences in Gene Expression

Information about when and where genes are turned on or off and changes in their expression levels under different environmental conditions is valuable for understanding gene function and regulation. Three comparative approaches have been used for the display of differential gene expression. In one approach, cells of interest are cultured under different physiological conditions, and the differences in mRNA abundance between the test and reference samples are compared using high-density microarrays. This is the most straightforward and widely used approach for identifying patterns in gene expression associated with various physiological states (DeRisi et al., 1997; Tao et al., 1999; Ye et al., 2000; Beliaev et al., 2002a,b). In another approach, cells of interest are grown under a specific physiological condition and then harvested at different times during growth. Changes in mRNA levels are revealed using microarrays. Information on the temporal dynamics of gene expression is very useful in understanding when genes are turned on or off and how genes interact with each other (DeRisi et al., 1997; Liu et al., 2003). Finally, comparisons of gene expression patterns between wild-type and mutant cells can be conducted. The expression of genes in response to changing environmental conditions can be very complicated, and often the expression profiles of many genes are altered as a result. This presents a great challenge to understanding the underlying regulatory mechanisms. The most effective approach to defining the contributions of individual regulatory genes in a complex metabolic process is to use DNA microarrays to identify genes whose expression is affected by mutations in putative regulatory genes (DeRisi et al., 1997; Thompson et al., 2002; Beliaev et al., 2002a).

The basic scheme for microarray-based gene expression studies is as follows (Fig. 6.9). In a typical microarray experiment for monitoring gene expression, gene-specific PCR

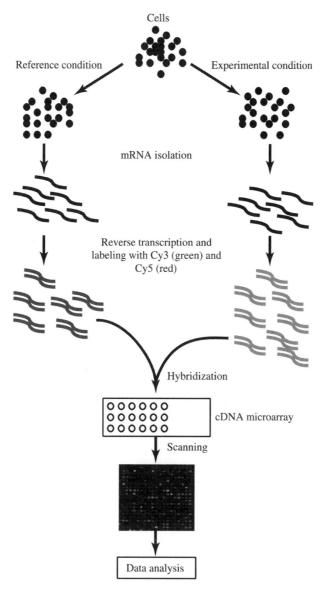

Figure 6.9 *See color insert. General approach to using microarrays for monitoring gene expression.*

primers are designed based on whole-genome sequence information and synthesized. Gene-specific fragments are then amplified with specific primers, purified, and arrayed on solid substrates. Once the microarrays are ready, total cellular RNA isolated from bacterial cells grown under two different conditions is fluorescently labeled with different dyes (Cy3 or Cy5) via the enzyme reverse transcriptase. The microarray is then simultaneously hybridized with fluorescently tagged cDNA from the test and reference samples. The signal intensity of each fluorescent dye on the array is then measured with a confocal laser scanning microscope or CCD camera. The quantitative ratio of red (Cy5) to green (Cy3) signal for each spot reflects the relative abundance of that particular gene in the two experimental samples. With appropriate controls, the intensity can be converted into biologically relevant outputs (e.g., the number of transcripts per cell). A series of samples can be compared with each other through separate cohybridizations with a common reference sample, and the data can be analyzed with various statistical methods. A detailed discussion of the technical aspects of microarray experiments for monitoring gene expression can be found in the review by Eisen and Brown (1999).

6.6.2 Specificity, Sensitivity, Reproducibility and Quantitation of Microarray-based Detection for Monitoring Gene Expression

Specificity, sensitivity, and quantitation are the three major concerns for any detection technology. Microarrays were initially used to measure gene expression levels in *Arabidopsis thaliana* (Schenea et al., 1995). The microarrays of 48 duplicate cDNA sequences were constructed and simultaneously hybridized with a mixed set of fluorescently labeled probes that also contained cDNA from rat and yeast as negative controls. There was no detectable response from the rat and yeast controls, suggesting that microarray hybridization was very specific. DeRisi and coworkers (1997) showed that microarray hybridization could differentiate homologous gene sequences that were less than 75% similar, and oligonucleotide microarrays can, in theory, distinguish single-base-pair differences among probes (Hacia, 1999; Wang et al., 1998).

The detection limit, or sensitivity, can be evaluated by spiking labeled samples with a known amount of labeled control mRNA. The detection limit of DNA microarrays used in human genome research allows monitoring of transcripts that represent 1:500,000 (w/w) of the total mRNA. A similar detection limit was obtained using oligonucleotide microarrays (Lockhart et al., 1996). Such a detection limit is higher than that obtained with conventional Northern hybridization methods. For yeast, mRNA transcripts present at a level less than one molecule in 100,000 can be detected, which is equivalent to one copy per 20 yeast cells (Johnston, 1998). This result suggests that microarray hybridization is very sensitive for detecting gene expression.

Microarray hybridization has inherent high variability as a result of variations in array fabrication, labeling, target concentration, and scanning. The quantitative aspects of microarray hybridization have not been well established. Comparison of microarray hybridization results with previously known results suggested that microarray hybridization appears to be quantitative enough for detecting differences in gene expression patterns under various conditions (Lockhart et al., 1996; DeRisi et al., 1997; Schena et al., 1996; Thompson et al., 2002a). DNA microarrays were used to measure differences in DNA copy number in breast tumors (Pinkel et al., 1998; Pollack et al., 1999). Single-copy deletions or additions can be detected (Pollack et al., 1999), suggesting that microarray-based detection appears to be quantitative.

6.6.3 Microarray Experimental Design for Monitoring Gene Expression

Microarray experiments generate massive data sets and the challenge is how to analyze and interpret them in a rapid meaningful way. To improve the efficiency and reliability of the experimental data, careful experimental design is needed; otherwise, the collected data may fail to answer the research question of interest or lead to a biased, inadequate interpretation of the experimental results (Yang and Speed, 2002).

The main objective of experimental design is to make the data analysis and interpretation as simple and powerful as possible. For a competitive microarray hybridization experiment in which two fluorescent dyes are used, the most important experimental design issue is to determine how the mRNAs are labeled and which mRNAs are hybridized together on the same slide (Yang and Speed, 2002). In most experiments, several designs can be devised. An appropriate choice of design will depend on the research questions addressed, numbers of comparisons, number of slides available for hybridization, the amount of mRNAs available, and the cost. Various design schemes were described in great detail by Yang and Speed (2002) and several designs could be devised for a particular microarray experiment. We refer readers to the review by Yang and Speed (2002) for a further understanding of the experimental design issues. The microarray experimental design scheme can be classified into the following three categories (Fig. 6.10). (1) Reference design: In this scheme, all the treatment samples are labeled with one dye and hybridized respectively with a common reference sample labeled with another dye (Fig. 6.10A). This is an indirect design that is widely used in gene expression studies and is especially suitable when the amount of mRNAs from treatment samples is limited and when many treatment samples are compared. The design allows for easy analysis and interpretation without using sophisticated statistical tools. However, the average variances for this design are considerably higher than those for other designs. Since it is straightforward, the reference design is used much more often than other designs. (2) All-pairs design: In this design, all the treatment samples are labeled with different fluorescent dyes and directly hybridized together in pair-wise fashion (Fig. 6.10B). The main advantage of the all-pairs design is that more precise comparisons among different treatment samples can be obtained. However, this design is unlikely to be feasible and desirable when a large number of comparisons are made due to the constraints on mRNA quantity and cost. (3) Loop design: In the loop design, all the treatments are successively connected as a loop (Fig. 6.10C) (Kerr and Churchill, 2001a,b). Using the same number of microarrays as the reference design, the loop design obtains twice as much data on the treatments of interest. The loop design requires far fewer slides than the all-pairs design. However, long paths between some pairs of treatment samples are needed in a larger loop, and thus some comparisons are much less precise than others (Yang and Speed, 2002). Another practical problem with this design is that each sample should be labeled with both red and green dyes, which doubles the number of labeling reactions. In addition, the failure of microarray hybridization in one sample will affect the analysis of other samples in the loop.

6.7 SUMMARY

Microarrays are a recently developed, widely used genomic technology that have had a great impact on many areas of research, including medical science, agriculture, industry,

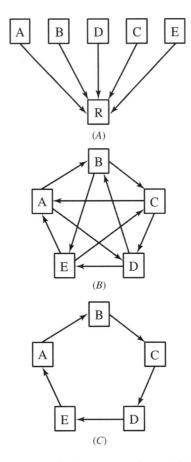

Figure 6.10 *Illustrations of basic types of microarray experimental design schemes with five treatment samples.* By convention, the sample labeled with green dye (Cy3) is placed at the tail, while the sample labeled with red dye (Cy5) is placed at the head of the arrow. (A) Reference design: The five treatment samples (A–E) are labeled with one dye and hybridized respectively with the common reference sample R, which is labeled with the other dye. Altogether five hybridizations are needed. (B) All-pair design: Each sample is labeled twice with Cy5 and twice with Cy3. Altogether 10 pair-wise hybridizations are needed. (C) Loop design: Each sample is labeled once with Cy5 and once with Cy3. Altogether five successive pair hybridizations are needed.

and environmental studies. There are two basic types of microarrays: DNA microarrays containing DNA fragments typically generated with the polymerase chain reaction, and oligonucleotide microarrays containing oligonucleotide sequences. Oligonucleotide microarrays can be further divided into two general types based on fabrication strategies. One is the direct parallel synthesis of oligonucleotides on solid substrates by light-directed or photoactivatable chemistries, or standard phosphoramidite chemistries. The other is the chemical attachment of prefabricated oligonucleotides to solid supports.

Microarray fabrication is one of the key steps of microarray technology. The substrate used for printing microarrays has a significant impact on the overall microarray experiment. A variety of substrates have been used for microarray fabrication; a nonporous solid surface, such as glass, is the most commonly used, because this surface has unique physical and chemical characteristics that allow miniaturization and the use of fluorescent labeling and detection. The miniaturized microarray format coupled with fluorescent detection represents a fundamental revolution in biological analysis. Nucleic acids can be attached to the modified glass slide surface by electrostatic interaction or covalent bonding. A variety of approaches have been used to modify glass slides, including silanization, dendrimeric linker coating, gel coating, and nitrocellulose coating. Photolithography, mechanical microspotting, and ink-jet ejection are the most widely used printing technologies, but because of its versatility, affordability, and wide applications, microspotting may become the microarray technology of choice for any basic research laboratory. Microarray density, reproducibility, storage time, contamination, and printing quality are very important issues for consideration in microarray fabrication.

Probe design and synthesis are critical steps in generating high-quality microarrays for analysis. Generally, DNA fragments with less than 75% sequence identity are used as probes on microarrays. Many microarray experiments for differentiating gene expression use two fluoresceins, such as Cy3 and Cy5. The fluorescent dyes can be directly or indirectly incorporated into target molecules. The most important issue in microarray-based analysis is the successful performance of target hybridization. To obtain gene-specific detection, microarray hybridization is generally performed under highly stringent conditions (i.e., 65°C). The detection of microarray hybridization signals is also important in microarray-based studies. Both the confocal scanning microscope and coupled charge device (CCD) camera have been successfully used for microarray detection, each with advantages and disadvantages.

Image processing is an important step for microarray-based analysis. Since microarray hybridization has inherent variations, rigorous image processing prior to data analysis is needed. Poor spots and outliers must be removed prior to data analysis. The spot quality and integrity are generally assessed based on the spot's size, shape, homogeneity, and intensity. Various methods have been used to remove outliers, although no general definition of outlying genes exists.

Whole-genome microarray is a powerful approach to studying gene expression. The gene expression can be differentially displayed under different physiological conditions, different time points, and different genetic backgrounds. Many previous studies indicated that microarray-based gene hybridization is specific and sensitive enough for answering many biological questions. However, great caution is needed in microarray experimental design to improve the efficiency and reliability of the obtained microarray hybridization data.

FURTHER READING

Bowtell, D. and J. Sambrook. 2002. *A Molecular Cloning Manual: DNA Microarrays.* Cold Spring Harbor Laboratory Press, Plain View, N.Y.

Brownstein, M. J. and A. B. Khodursky. 2003. *Functional Genomics: Methods and Protocols.* Humana Press, Totowa, N.J.

Causton, H., J. Quackenbush, and A. Brazma. 2003. *Microarray Gene Expression Data Analysis: A Beginner's Guide.* Blackwell Publishing, Malden, Mass.

DeRisi, J. L., V. R. Iyer, and P. O. Brown. 1997. Exploring the metabolic and genetic control of gene expression on a genomic scale. *Science* 278:680–686.

Eisen, M. and P. Brown. 1999. DNA microarrays for analysis of gene expression. *Method. Enzymol.* 303:179–205.

Grigorenko, E. V. 2002. *DNA Arrays: Technologies and Experimental Strategies.* CRC Press, Boca Raton, Fla.

Nadon, R. and J. Shoemaker. 2002. Statistical issues with microarrays: processing and analysis. *Trends Genet.* 18(5):265–271.

Nuwaysir, E. F., W. Huang, T. J. Albert, and J. Singh, et al. 2002. Gene expression analysis using oligonucleotide arrays produced by maskless photolithography. *Genome Res.* 12:1749–1755.

Schena, M. 2002. *Microarray Analysis.* John Wiley & Sons, New York.

Schena, M. and R. W. Davis. 2000. Technology standards for microarray research, In M. Schena (ed.), *Microarray Biochip Technology*, pp. 1–18. Eaton Publishing, Natick, Mass.

Southern, E. M. 2001. DNA microarrays. History and overview. *Meth. Mol. Biol.* 170:1–15.

7

Microarray Gene Expression Data Analysis

Ying Xu

7.1 INTRODUCTION

One of the greatest challenges in functional genomics, as noted by Eric Lander (1996), is to "decipher the logical circuitry controlling entire developmental or response pathways" in a cell. Tackling such an enormously complex problem in a systematic manner has not been possible until very recently, due to the emergence of microarray-based genomic technologies. Microarrays allow the simultaneous observations of expression levels of many or all genes in a cell. Through the rational design of gene expression experiments under different conditions, one can possibly derive how genes function together within a cellular network and how they regulate each other's activities (e.g., by turning genes "on" and "off," or changing their activity levels). The (technical) task is perhaps akin to reverse engineering a microprocessor based on recordings from each register (Lander, 1996).

Modeling the internal structure of a complex machinery based on its observed external "behavior," for example, patterns of flashing lights or expression profiles, is not a new science. During World War II, scientists used computational techniques to decode encrypted messages by modeling the internal structures of an enemy's message encoders, based on previously observed relationships between their military maneuverings and intercepted encoded messages. A simple example of such a modeling problem is provided in Fig. 7.1. The goal here is to determine the internal control mechanism of a black box, which has two inputs (with values on and off represented as 1 and 0, respectively) and one output. For this simple example, our observations have been that whenever one of the input values is 1, the output of the black box is always 1; otherwise, it is 0. The simplest model that is consistent with these observations is an OR logic gate, which can be rigorously derived using mathematical modeling techniques. More complicated logic circuitry, for example, a computer CPU or a memory board, can be similarly derived (or "designed"),

Microbial Functional Genomics, Edited by Jizhong Zhou, Dorothea K. Thompson, Ying Xu, and James M. Tiedje. ISBN 0-471-07190-0 © 2004 John Wiley & Sons, Inc.

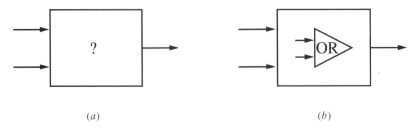

Figure 7.1 *(a) Black box with two inputs and one output. (b) Possible model for the black box of (a), based on observed data.*

based on specified input–output relationships, possibly represented as differential equations. There are numerous mathematical frameworks for dealing with this type of modeling problem, including control theory (Luenberger, 1979), information theory (Pierce, 1980), and coding theory (Vertibi and Omura, 1979). Apparently, the control mechanism of a living cell with tens of thousands of states or possibly even more is going to be much more complex than a message encoder or a computer CPU. A significant amount of information, represented as input–output relationships, will be needed to model such a complex machinery. Note that a simple model consisting of 100 binary states could require up to 2^{100} input–output data pairs for an accurate modeling work.

Microarray technology provides a powerful tool that enables researchers to observe simultaneously mRNA expression levels of thousands of genes. By examining the expression data, one can observe directly which genes are differentially expressed under a particular experimental condition, thus indicating possible functional connections between the inputs (experimental conditions) and certain components (differentially expressed genes) of the gene network. From this data, one could also derive which genes are coexpressed—possibly indicating genes working closely together in a cell—and even possibly infer causality relationships among certain genes, providing hints about how genes might interact together to form networks. A number of recent studies have suggested the feasibility of deriving or interpreting local gene networks (metabolic or regulatory networks) related to particular biological functions, using large-scale gene expression data (DeRisi et al., 1997; Chu et al., 1998; Iyer et al., 1999). These include investigations of transcriptional mechanisms of sporulation in budding yeast (Chu et al., 1998) and the response of human fibroblasts to serum (Iyer et al., 1999). Although the reported results are highly encouraging, it should be noted that these studies have been carried out mainly on well-studied and generally well-understood systems. It could represent a totally different level of difficulty to investigate novel gene networks using gene expression data. As an analogy, such an endeavor would be like trying to model a complex electronic circuitry when having only a small number of observations on its input–output relationships: not enough data to make a sensible model. Additional information from genomic sequence or other sources could be useful in helping to narrow down the modeling possibilities. For example, in a bacterial cell, genes involved in the same pathway are generally organized into one or a few *operons*, a sequence of genes arranged in tandem that share a common promoter region (Stephanopoulos et al., 1998; Chapter 3). Coexpressed genes that are also organized in the same operon should suggest strongly that they belong to the same local gene network. Other types of information that could help in deciphering a gene network may include protein–protein interaction data generated using two-hybrid systems (Fields and Song, 1989; Uetz et al., 2000), known pathways and networks (possibly from other genomes) like those stored in the KEGG database (Ogata et al., 1996; Kanehisa et al., 2002), and functional assignments of individual genes (see Chapter 5).

This chapter addresses issues related to (1) preprocessing of observed microarray gene expression data, (2) analysis and clustering of microarray data, and (3) applications of microarray data for gene network inference.

7.2 NORMALIZATION OF MICROARRAY GENE EXPRESSION DATA

As discussed in Chapter 6, microarrays are used to determine gene expression profiles under designed experimental conditions. Ideally, the expression level of a gene should be "accurately" reflected by the hybridization intensity level captured by the microarray scanner at each spot on the slide. Under this premise, the observed data could be used to infer the internal structure of a gene network, possibly in a fashion similar to that of deciphering the internal structure of a message encoder. However, the reality is that a high level of background noise is often associated with microarray data. Many technical factors, in addition to the biological ones we intend to capture, could contribute to the observed data, leading to systematic errors (or variations) to the readings of gene expression levels. Identification of the sources of such systematic errors and making the needed corrections represent the first step in the process of interpreting the observed microarray data in a meaningful way. This step is generally referred to as *data normalization*.

7.2.1 Sources of Systematic Errors

Numerous technical factors exist that could contribute to the creation of systematic errors or variations in microarray experiments (Schuchhardt et al., 2000; Tseng et al., 2001). They may include the following.

Variations within a Slide   Studies have shown that substantial variations can exist for multiple copies of the same gene on the same slide (Dudoit et al., 2001). Differences in pin geometry, slide homogeneity, hybridization, and target fixation could all contribute to these variations. Different pins may have slight differences in their lengths, in the opening of the tips, or in other surface properties. All these could lead to differences in the amount of target DNA transfer, possibly leading to systematic variations in microarray signal intensity. The amount of deposited target DNA could also vary even for the same printing pin (Dudoit et al., 2001). Also, the fraction of target DNA that is chemically fixed on slides is unknown, which could vary considerably in different areas of a slide. In addition, for various reasons, the labeled targets may be distributed unevenly over the slide and/or the hybridization reaction may occur unequally in different parts of the slide. Some areas of a slide may be contaminated, thus giving a high background intensity. The influence of these factors on signal intensity measurement within a slide is generally referred to as *spatial effect*.

Variation Across Slides   Differences in surface properties, slide fabrication, hybridization, and imaging could lead to systematic variations in hybridization signals among different slides. The amount of probe DNA immobilized on a slide during array printing and probe fixation could be significantly different among different slides due to various factors, including differences in slide surface properties and sample evaporation during printing. The amount of cDNA added to the slides also could be different, especially when different RNA preparations are used, and/or when the local environment and hybridization conditions such as temperature, buffer pH, target concentrations, incubation, and washing time in each hybridization chamber are different. In addition, background noise and the local curvature of the surface among different slides may have a significant impact on scanner readings especially for confocal scanners, which are sensitive to the location of the focal point. The influence of these factors on measuring signal intensity is referred to as *slide effect*. Tseng and coworkers (2001) demonstrated that such effects could be significant.

Variation Caused by Probe Labeling   The two commonly used dyes Cy3 and Cy5 may not be equally incorporated into DNA molecules through reverse transcriptase and DNA polymerase. Generally, Cy3 can be incorporated into DNA more efficiently than Cy5 with the same amount of RNA under the same preparation. In addition, the two dyes have different quantum efficiencies that are detected by the scanner with different efficiencies. While the detection limit of Cy5 with the scanner is lower than Cy3, Cy5 is more sensitive to photobleaching. The influence of these factors on intensity measurements is referred to as *label effect*.

The use of two dye labels may also introduce gene–label interactions. First, fluorescent labeling may vary systematically, depending on nucleotide composition of the target sequences. Compared to Cy5-dCTP, Cy3-dCTP may be preferentially incorporated into specific sequences; the same could be true for Cy5-dCTP for other sequences. Also, the length of Cy3- and Cy5-labeled cDNA by reverse transcription with random priming could be significantly different from sequence to sequence. The longer labeled cDNA could potentially lead to higher intensity for certain sequences. If such interactions occur, certain sequences may always show higher intensity in one channel than in the other channel even under nondifferential conditions.

Variation in Growth Conditions and mRNA Preparation   In a comparative microarray experiment, two RNA samples from different treatments are labeled with different fluorescent dyes. Because of the differences in genetic identity (e.g., wild-type vs. mutant strains) and environmental growth conditions, cell biomass and mRNA abundance could vary significantly among different cultures. Also, the RNA purity could be very different from sample to sample, which could lead to different labeling and hybridization efficiencies. In addition, sensitivity to mRNA degradation could be significantly different with different preparations. All these factors affecting signal intensity are referred to as *sample effects*.

7.2.2 Experimental Design to Minimize Systematic Variations

Although microarray experiments can have high variations, measures can be taken, during the design phase of experiments, to minimize systematic variations. These measures may include the following.

Minimizing Spatial Effects   To minimize spatial effects, multiple spots for a gene or control DNA should be fabricated on the microarrays. For control sequences, various concentrations of sequences should be spotted on arrays. Multiple spots of genes or control

DNAs within the same slide are very useful for identifying contaminated spots, spots having high background noise, and problematic slides in each experiment (Tseng et al., 2001). Also, to minimize spatial effects, normalization can be performed for each sector of a microarray based on all the genes in that sector of the array. Since DNAs in different sectors are deposited by different pins, such normalization is effective in eliminating pin-to-pin variations (Yang et al., 2001; Finkelstein et al., 2001; Dudoit et al., 2001). By comparing the normalization results for different genes and control sequences among different sectors, one should be able to assess and minimize the systematic variations associated with slide surface properties.

Minimizing Labeling Effects and Gene–Label Interactions To eliminate systematic variations in probe labeling and gene–label interactions, a reverse labeling experimental design is recommended (Kerr and Churchill, 2001b; Marton et al., 1998; Tseng et al., 2001). In this design, two aliquots of the two RNA samples (A and B, e.g.) are labeled with Cy3 and Cy5, separately, and then hybridized with two microarrays. The hybridization solution for the first microarray consists of Cy3-labeled sample A and Cy5-labeled sample B, whereas the labeling for the second microarray hybridization is reversed. Then the signal intensity for each microarray is normalized based on all genes (or a selected subset) on the microarray using different normalization approaches (see below). After normalization, the signal intensities from both channels for each sample are averaged and the intensity ratios of the two samples calculated based on the averaged signals. The reverse labeling experimental design is effective in eliminating labeling effects and gene–label interaction (Tseng et al., 2001).

Minimizing Slide Effects and Sample Effects In a typical comparative study, multiple replicated treatments, say, three, under each condition are used, and mRNAs from two different conditions are fluorescently labeled and cohybridized to the same single or multiple slides. By using such an experimental design, the signal intensity is compounded with both slide and sample effects. It will be difficult to eliminate the systematic variation resulting from slide and sample effects based on all arrayed genes for normalization. In this situation, one should identify a sufficient number of nondifferentially expressed genes on each slide and use them to construct a normalization curve, because the expression level of the nondifferentially expressed genes is expected to remain constant under the experimental conditions tested.

7.2.3 Selection of Reference Points for Data Normalization

The ultimate goal of data normalization is to detect and correct systematic errors or variations caused by technical factors, as discussed in Section 7.2.1. A key step in implementing such a normalization procedure is to select a set of reference points, through comparison against which systematic errors could be detected and even possibly be corrected. The following are three commonly used strategies for selecting such reference points:

(i) *Using All Genes on the Microarray.* Under a particular experimental condition, only a small percentage of the genes in a cell is expected to be differentially expressed. Thus, the remaining genes should have constant expression levels between two channels so they can be used to calibrate spatial effects (Dudoit et al., 2001), slide effects, and label effects (Tseng et al., 2001). The prerequisite for using almost all genes on the array as reference points is that only a small fraction of the genes are differentially expressed and the numbers of down- and up-regulated genes are approximately equal.

(ii) *Using Constantly Expressed Housekeeping Genes.* The housekeeping genes that are believed to be constantly expressed across a variety of conditions can be used for normalization (Duggan et al., 1999). Although it could be difficult to identify a set of housekeeping genes that do not change their expression levels significantly under all conditions, it may be possible to identify small sets of housekeeping genes for particular experimental conditions.

(iii) *Controls.* A third approach is through the use of spiked controls or a titration series of control sequences. In the spiked control method, DNA sequences from organisms different from the one being studied are printed on the array and then the mRNAs of the control sequences are mixed with the two different mRNA samples of the same amount. These spotted controls should have equal Cy5- and Cy3-derived intensities and thus they can be used for normalization. One limitation of this approach is that the composition of the control sequences could be considerably different from the target sequences and they therefore may not be representative of the genes of interest. Another limitation is that it is sometimes very difficult to know how much mRNA to spike, because there are always varying amounts of rRNA and tRNAs present and the degree of RNA degradation varies from sample to sample. In the titration series method, a series of concentrations of the control sequences are printed on the arrays. These control spots are expected to have equal Cy5- and Cy3-derived intensities across a range of concentrations. Genomic DNA can be used in the titration method, because it should have a consistent expression level across various conditions.

7.2.4 Normalization Methods

By comparing the expression levels of reference genes under different conditions, one could possibly detect systematic variations. One simple technique for achieving this is through visualizing the two-dimensional (2D) *scatter plot* (see Fig. 7.2), in which signal intensities under two different conditions are plotted on a two-dimensional plane with the two intensity values representing the x-axis and y-axis coordinates, respectively. If one uses reference genes, one expects that the overall intensity values of these genes should be approximately symmetric to the x- and y-axes, under two different conditions. Deviation from this symmetry generally indicates systematic errors. The same can be said about the overall intensity values of the two channels from the same microarray slide. To "correct" systematic errors, one generally needs to model the relationship between the correct data and the erroneous data. These models could be either *linear* or *nonlinear* models, depending on the complexity of such a relationship.

Linear models have often been used to model *multiplicative errors* (the magnitude of the error is proportional to the data value, e.g., the larger the data value, the larger the error) and *additive errors* (the magnitude of the error is independent of the data value itself). In data sets with multiplicative errors, an intensity value x will be read as $x + \epsilon x$, where ϵx is the error whose quantity increases as x. Errors caused by different mRNA concentrations in solution or by hybridization time fall into this category. By plotting the intensity levels of genes under two different experimental conditions in a scatter plot like

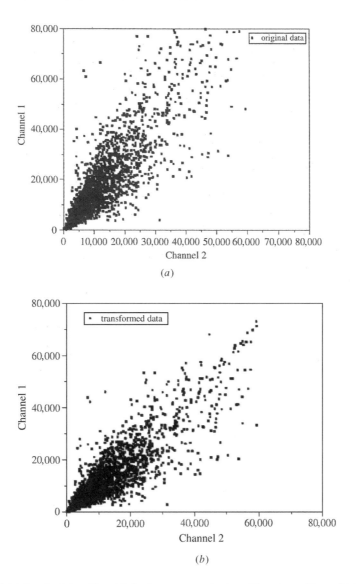

Figure 7.2 *(a) Intensity data with multiplicative errors under two different conditions. (b) "Corrected" intensity data.*

Fig. 7.2A, one could easily see if any systematic error exists. When no multiplicative errors exist, the orientation of the *first principal component* of the data set should coincide with the line of $y = x$, where the first principal component can be loosely interpreted as the orientation along which the data set has the widest spread. A brief description of principal component analysis is given in Section 7.3.2. If the orientation of the first principal component coincides with $y = ax$ with $a \neq 1$, we have then detected a multiplicative error, as shown in Fig. 7.2A.

The correction of such an error is a very nontrivial matter since the observed data do not contain any information about the underlying model of the errors. One simple solution could be simply multiplying all intensity values of x by a so that the first principal component of the adjusted data set will approximately coincide with the line $y = x$. By doing so, we essentially assume that the errors come from the x data set. Similarly, we can divide all the intensity values of y by a, assuming that the errors are all in the y data set. One drawback of this strategy is that modifying the intensity values by a significant factor could have negative affects on our analysis of differentially expressed genes under different biological conditions (see Section 7.3). The goal here is to compare a series of expression data for the same gene under different conditions or at different time points, and to decide how the gene's expression levels change as a function of time or biological conditions. "Artificially" adjusting the expression levels clearly will not make this analysis any easier. One way to minimize the effects of error correction is to adjust the intensity values on both x and y data sets simultaneously. For example, we can multiply the x intensities all by $\sqrt{a}$ and divide all y values by $\sqrt{a}$. By doing so, we have achieved both error corrections and "minimized" the effects of incorrect adjustments. Figure 7.2B shows an example of "corrected" intensity data.

Another type of linear error is the additive error, which changes the original intensity level by adding a quantity to it, and this quantity is independent of the intensity value. Errors caused by the background intensity of a microarray slide fall into this category. Detection of such errors can be done as follows. One can examine the two sets of intensity values, using a scatter plot as we have done above. If the first principal component of the plot has the same orientation of line $y = x$ but with a nonzero y-intersect β, we have then detected an additive error. To correct the error, one could either subtract β from all the y values or add β to all x values. Alternatively, one could subtract $\beta/2$ from the y values and add $\beta/2$ to the x values, similar to what we have done with multiplicative errors. The same argument that we used above for the multiplicative errors is applicable.

For a general linear error, a linear transformation could be derived through analysis of the intensity data of a reference gene set. For example, if one believes that reference genes should have a constant expression level under different conditions but finds that the observed data indicate otherwise, a linear correction factor could be derived through simple mathematical analysis, for example, linear regression. Then this linear transformation can be applied to the whole gene set for the "correction" of systematic errors. The above discussions on additive and multiplicative errors are special cases of a general linear error.

In the more complicated cases where linear models may not be adequate to capture the complex nature of certain systematic errors, nonlinear models could prove to be more effective. We now outline two nonlinear error models.

Trimmed Geometric Mean (TGM) Model   In this model, signal intensities from each channel are log-transformed (see Section 7.3), and 5% of the extreme values (minimum and maximum) are discarded. The log-trimmed geometric mean (TGM) and the standard deviation of the log-trimmed means are calculated. The normalized value for a gene is obtained by dividing the difference between log intensity and log-trimmed means by the standard deviation of the log-trimmed means. The normalized values are then converted back from log space to the normal values. This nonlinear method was initially described by Morrison and colleagues (N. Morrison et al., Nature Genetics Microarray Meeting, Scottsdale, Ariz, 1999), and has been widely used since then (Murray et al., 2001; Thompson et al., 2001; Beliaev et al., 2002a,b).

Intensity-dependent Nonlinear Normalization Method It is often the case that the dye bias is dependent on spot density (Yang et al. 2001; Dudoit et al., 2001; Tseng et al., 2001). Thus, an intensity-dependent normalization method may be preferable. Yang and coworkers (2001) proposed an intensity-dependent nonlinear normalization method. In this method, the normalized intensity ratio is set to be the difference between the actual log intensity ratio and the intensity ratio estimated based on the Lowess function (Venables and Ripley, 1994). For a detailed discussion of this method, we refer the reader to Yang et al. (2001) and Dudoit et al. (2001).

7.3 DATA ANALYSIS

Before statistical analyses of microarray gene expression data for information extraction, some preprocessing of the collected intensity data is generally needed. This may include data transformation for the purpose of easier mathematical handling of the data, and identification/removal of the possible outliers of a data set.

7.3.1 Data Transformation

Prior to statistically analyzing the microarray data, one key question that should be answered is whether the data meet the underlying assumptions of the statistical analysis techniques to be used. A common requirement for statistical techniques is that the data should follow a normal distribution and have homogeneous variance. If the assumptions are not met approximately, the statistical analysis and test will be invalid. The data could be transformed to meet these requirements.

There are many different approaches to data transformation, among which the most commonly used in the microarray field is to take the logarithm of the expression data, mainly for the following reasons. First, the variation of log-transformed intensities and log-transformed ratios of intensities is less dependent on absolute magnitude. Log transformation could equalize variability in the wildly varying microarray raw data. Second, log transformation could even out highly skewed distributions and thus bring the data closer to a normal distribution. Studies have shown that log transformation is very effective in bringing the microarray data approximately to normal distribution and is suitable for microarray-based gene expression data.

7.3.2 Principal Component Analysis

Principal component analysis (PCA) is a multivariate technique for examining relationships among a group of data points in Euclidean space. It has been widely used in the analysis of gene expression data for various purposes, including the identification of outliers in a data set, reduction of dimensionality, etc. Considering its wide applications, we provide a detailed description of a PCA procedure here.

The basic idea of k principal component analysis is to find an orthogonal transformation of multidimensional data points in a k-dimensional space that would maximize the scattering of projections of the data points in the new space. Each principal component is a linear combination of the original data points. The principal components have the following properties:

(1) Linear correlation between any two different principal components is zero.

(2) The jth principal component has the largest variance among any unit-length linear combination of the original data points, which is not correlated with the first $j - 1$ principal components.

(3) The j-dimensional linear subspace spanned by the first j principal components gives the best possible fit to the data points, measured by the sum of squared distances between the data points and the subspace.

These properties make PCA a good tool for representing a data set at different levels of resolution. Let $\{x_{j1}, x_{x2}, \ldots, x_{jk}\}, j = 1, 2, \ldots, n$, be a set of n data points in k-dimensional Euclidean space. We use

$$E(t) = \Sigma_{i=1}^{n} x_{it}/n$$

to represent the average of the tth component of all the data points and

$$U = \{\Sigma_{i=1}^{n}(x_{is} - E(s))(x_{it} - E(t))\}_{s,t=1,2,\ldots,k}$$

to denote the covariance matrix of the data set. The first principal component is given by the coefficients $\{c_{11}, c_{12}, \ldots, c_{1k}\}$, of a linear combination $y_{1t} = \Sigma_{i=1}^{n} c_{1i} x_{ti}$, which maximizes the variance of $\{y_{11}, y_{12}, \ldots, y_{1n}\}$ under the condition $\Sigma_{i=1}^{n} c_{1i}^2 = 1$. Eigenvectors of the matrix U with the highest eigenvalue give these coefficients, that is, the first principal component. Generally, the eigenvectors with the jth highest eigenvalue gives the jth principal component.

PCA is often used in the early phase of a data analysis project as it provides only low-resolution information, for example, overall shape of a data set or its geometric orientation, and do so in an efficient manner.

7.4 IDENTIFICATION OF DIFFERENTIALLY EXPRESSED GENES

A basic application of the microarray technology is to detect genes whose expression levels are affected by a particular biological condition. Through systematic analyses of such data under different biological conditions, one could possibly derive mechanistic information of certain biological processes or the organizational information of gene networks. Typically, the change in a gene's expression level, under two different conditions, is measured either by the ratio or the difference between the two expression levels. One popular measure is the *fold-change* (Gray et al., 1998b; Baldi and Long, 2001). This technique measures if the ratio between two expression levels (possibly averaged over a set of replicates) of the same gene is above or below a specified threshold. Often in the literature, we see a 3- or 5-fold-change cutoff used to determine if a substantial change has taken place, and hence infer that the gene is affected by a particular biological condition. Such a simple fold-change analysis probably represents the most prevalent approach currently used in biological labs. The approach has proved to be practically useful for genes with significant expression-level changes. However, the detection of genes with subtle changes in their expression levels may require more sophisticated analysis tools.

When analyzing microarray data, one needs to understand what contributes to the observed data. In Section 7.2, we discussed technical factors that could affect the observed intensity levels and approaches to correct the errors caused by these factors so that the "normalized" data will more accurately reflect the underlying activities of gene expression. To interpret the (normalized) data, we need to further understand that mRNA hybridization is not a deterministic process. The number of mRNA molecules hybridized to a particular slide spot is stochastic in nature, even under the same controlled physical and biological conditions. That is, the expression level of a particular gene under a particular condition is generally not a fixed value, but rather follows a probabilistic distribution over a set of possible values. Hence for different times, somewhat different numbers of mRNA molecules might be hybridized to the same chip spot. To effectively estimate if a gene's expression level has changed under different experimental conditions, one needs to perform multiple measures from repeated microarray experiments under the same experimental conditions. It is the repetition that can give us the confidence in our estimated changes of expression levels. The same argument applies to other random reading errors of chip data caused by various technical factors.

One simple way to use multiple measurements of gene expression levels is to compare the averaged intensity values of genes and then to assess if there is a significant change in each gene's expression level. Comparisons of the averaged values should clearly increase the accuracy of our assessment of "differentially expressed" genes. Such an assessment can be done using well-established statistical analysis tools. The *t-test* is a classical statistical method for accepting or rejecting a hypothesis with a rigorously defined confidence level. The basic idea behind the *t*-test can be explained as follows. It is well known from statistical theory (Bury, 1975) that the average value over a large number of random variables approximately follows a normal distribution, and the difference between two such average values for two different sets of variables follows a distribution called the *Student* distribution. Assessing a hypothesis that the two normal distributions are statistically different can be done by using the Student distribution table, which provides a confidence value for accepting such a hypothesis. An application procedure for a *t*-test is as follows.

Let $E_a^1, E_a^2, \ldots, E_a^k$ and $E_b^1, E_b^2, \ldots, E_b^k$, represent two sets of k repeated measurements of intensities of the same gene under conditions A and B, respectively. By completing the following statistical analysis, one can either accept or not accept the hypothesis that the underlying expression levels reflected by these two sets of intensity data are intrinsically different. Let μ_a and σ_a be the estimated mean and standard deviation of $E_a^1, E_a^2, \ldots, E_a^k$, and μ_b and σ_b be the estimated mean and standard deviation of $E_b^1, E_b^2, \ldots, E_b^k$. We use σ to represent $(\sigma_a + \sigma_b)/2$. From known results of statistics theory, we know that $\sqrt{(k/2)}(\mu_a \mu_b)/\sigma$ obeys the Student distribution (Fisher and Yates, 1963), particularly $\sqrt{(k/2)}(\mu_a - \mu_b)/\sigma \sim St(2k-2)$, where $St(n)$ represents the Student distribution with n degrees of freedom. This implies that if $|\mu_a - \mu_b| > \sigma\sqrt{(2/k)}t_\alpha(2k-2)$, we can say, with α level of confidence, that the two gene expression levels are intrinsically different, that is, they are differentially expressed. A confidence level at α means a false positive rate of $(1 - \alpha)/2$. α is a real number between zero and 1, and $t_\alpha(2k-2)$ represents the percentile value for Student distribution t_α. t_α is typically presented in the form of a table (which can be found in any mathematical handbook), and its values can be obtained through a simple table lookup. Table 7.1 shows a small portion of such a table for the purposes of illustration.

We now use an example to explain how this table can be used to estimate the confidence level to determine if two expression levels reflected by intensity values $E_a^1, E_a^2, \ldots, E_a^k$ and

TABLE 7.1 Partial Table of Student Distribution

n	$t_{0.995}$	$t_{0.990}$	$t_{0.975}$	$t_{0.950}$	$t_{0.900}$	$t_{0.800}$
1	63.66	31.82	12.71	6.31	3.08	1.38
2	9.92	6.96	4.30	2.92	1.89	1.06
3	5.84	4.54	3.18	2.35	1.64	0.98
4	4.60	3.75	2.78	2.13	1.53	0.94
5	4.03	3.36	2.57	2.02	1.48	0.92
6	3.71	3.14	2.45	1.94	1.44	0.91
7	3.50	3.00	2.36	1.90	1.42	0.90
8	3.36	2.90	2.31	1.86	1.40	0.89
...	...	...	...	...	...	...

Each row (starting from row 2) represents Student distribution values for a particular degree of freedom, and each column (starting from column 2) represents the Student distribution values with a particular confidence value, say, $t_{0.900}$. A t_α means a confidence level α, for example, $t_{0.900}$ means a confidence level of 90%.

$E_b^1, E_b^2, \ldots, E_b^k$ are intrinsically different. Assume that we have two replicates for each data point, so $k = 2$; and we set our confidence level at $\alpha = 0.80$, that is, the false positive rate is $(1 - \alpha)/2 = 10\%$. Hence, if $|\mu_a - \mu_b| > \sigma * 1 * 1.06 = 1.06\sigma$, for the estimated values μ_a, μ_b, and σ, then we can say that the gene is differentially expressed at confidence level 0.80, that is, the probability of making a false positive prediction is 10%. It is worth noting that with more measures for each gene, we shall have a higher confidence in making such a prediction. For example, if $k = 4$ (i.e., we have four replicates for each slide spot) and we have the same estimation result $|\mu_a - \mu_b| > 1.06\sigma$, our confidence level will increase to between 0.90 to 0.95 [note that if $t_\alpha(2k - 2) = t_\alpha(6) = 1.06/0.71 = 1.49$, α should be between 0.90 and 0.95—see Table 7.1; and hence, the false positive rate is between 0.05 to 0.025]. The basis for making a call that the gene is differentially expressed under two different conditions is that their mean values μ_a and μ_b are different. *The larger the difference they have, the smaller number of measurements we need to make such a call with the same level of confidence, or the higher level of confidence we will have for the same number of measurements.* It should be noted that this analysis provides a confidence value only for a call that a gene is differentially expressed, which is typically what we need to address. It does not provide any confidence assessment if the call is that the gene is not differentially expressed.

Apparently, this analysis can only help us assess if a gene's expression levels change, go up or down, but does not provide any information about the amount of change under different conditions. For many applications, it is generally adequate to simply know that a gene is differentially expressed. Such information can be used to guide further experimental or computational experiments. To estimate the actual level of changes, more sophisticated statistical analysis tools will be needed. For this, we refer the reader to Claverie (1999).

The idea of using a t-test to determine if a gene is differentially expressed can be extended from one data point to multiple data points, collected either from time-series data or under multiple experimental conditions. Suppose we have m expression data points for each gene. For each data point t, we have a confidence level p_t to state that the gene is differentially expressed. Then our total confidence level for making a statement based on all m data points can be defined as the minimum value of $\{p_1, p_2, \ldots, p_m\}$.

One practical problem with using a t-test is that biologists may be reluctant to run multiple replicates for each gene spot due to economical considerations, while the number of replicates needed for an effective application of a *t*-test may possibly be high (e.g., > 3 or 4) in cases where the intensity levels fluctuate significantly among the replicates. A generalized version of the *t*-test, called the Baysian probabilistic-model-based regularized *t*-test (Baldi and Long, 2001), is generally considered to need a fewer number of replicates to achieve the same level of confidence for the determination of differentially expressed genes. This method has been applied to identify global expression profiles in *E. coli* K12 (Long et al., 2001). The results showed that this approach identified a set of differentially expressed genes, which seem to be biologically more meaningful than the results obtained by a regular *t*-test when assessed based on known knowledge of these genes. For additional information on detecting differentially expressed genes, we refer the reader to Nadon and Shoemaker (2002).

7.5 IDENTIFICATION OF COEXPRESSED GENES

By examining changes in gene expression levels over a period of time or under different experimental conditions, one could possibly detect correlations among the expression profiles for some genes (expression levels as a function of time or experimental condition). Figure 7.3 shows the expression levels of three genes under six different conditions (or at six different time points). The expression profiles of genes denoted with square and black circle symbols appear to be correlated as they share a similar pattern of going up or down as time/condition changes, while the diamond-denoted gene seems to have a different pattern of change. Correlations among expression profiles could exist in a less apparent manner. For example, two genes' expression profiles may be correlated in a time-delayed fashion, for example, one gene's expression level at time t may be correlated with another gene's expression level at time $t + 1$. Generally, two genes are considered to have correlated expression profiles if their expression profiles are not independent from a statistical point of view. The general belief is that genes with related biological functions or working in a common biological process often have correlated expression patterns while

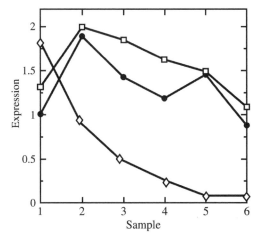

Figure 7.3 *Schematic of expression profiles of three genes (represented by square, closed circle, and diamond symbols) under six different conditions or over six time points.*

unrelated genes generally do not. By discovering genes with correlated expression profiles, we could possibly derive which genes may have related functions, or may exist in the same biological pathway. More sophisticated analysis could even suggest a causal relationship between genes in a common genetic pathway or network.

Data clustering represents one popular approach to identifying coexpressed genes, that is, genes with correlated expression profiles. The basic idea of data clustering is to partition a data set into nonoverlapping subsets (or *clusters*) such that data points of the same cluster are "highly" correlated, whereas data points from different clusters are not. Figure 7.4 shows a set of 68 gene expression profiles from the budding yeast *S. cerevisae* (Eisen et al., 1998), where each row represents one gene and each column represents a time point or an experimental condition. A green spot represents a gene's expression level going down and a red spot represents a gene's expression level going up, with a black square indicating no change, when comparing the expression levels between the test and the control conditions. These 68 genes fall into four categories: (1) protein degradation, (2) glycolysis, (3) protein synthesis, and (4) chromatin. These four classes of genes are separated using gray lines (Fig. 7.4). Genes from each of the four categories display similar expression patterns, while genes from different categories have significantly different expression profiles.

Since the publication of the seminal work by Pat Brown and colleagues on a systematic study of metabolic and genetic control networks through the analyses of gene expression data (DeRisi et al., 1997), a number of similar studies have been carried out. These include studies of the transcriptional program of sporulation in budding yeast (Chu et al., 1998) and the transcription program on the response of human fibrolasts to serum (Iyer et al., 1999), all using clustering techniques to identify functionally related genes or genes involved in a particular biological process. We can expect that many more similar studies

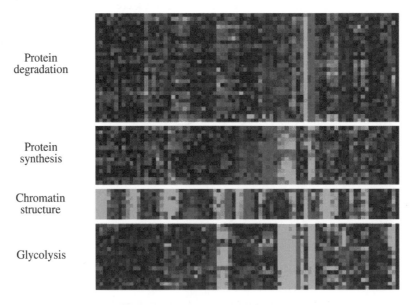

Figure 7.4 *Four classes of genes with distinct gene expression profiles for each class of genes.* The data set is from Eisen et al. (1998), and the clustering is done using EXCAVATOR (Xu et al., 2001, 2002).

will follow, and computational clustering techniques will play a vital role in such systematic studies.

7.5.1 Basics of Gene Expression Data Clustering

This subsection provides a short description of the basic ideas of data clustering. Two major classes of clustering techniques will be introduced. They are (1) the classical data clustering approach and (2) data cluster extraction from a noisy background. For different applications with data having different characteristics, different procedures should be employed in order to achieve the best clustering performance.

A data clustering program typically has three key ingredients: (1) a *distance measure* between data points, (2) an *objective function* that defines the quality of clustering results, and (3) a *clustering algorithm*, which should, in general, be capable of finding the globally optimal clustering result, measured by the specified objective function.

Distance Measures In a gene expression data clustering problem, a distance measure between two gene expression profiles could be defined as the Euclidean distance between the two profiles. For example, if $(E_a^1, E_a^2, \ldots, E_a^n)$ and $(E_b^1, E_b^2, \ldots, E_b^n)$ are two profiles with E_x^i representing the expression level of gene x (a or b) at time point i, their Euclidean distance is $\sqrt{(\Sigma_{i=1}^n (E_a^i - E_b^i)^2)}$.

The distance measure could also be defined in terms of the linear correlation coefficient of the two profiles, that is, as $1 - CC(E_a, E_b)$ with $CC(E_a, E_b)$ being the correlation coefficient between E_a and E_b. Another related distance measure is defined as the cosine of the angle between the two vectors $(E_a^1, E_a^2, \ldots, E_a^n)$ and $(E_b^1, E_b^2, \ldots, E_b^n)$. For different applications, different distance measures may be more effective. Most of the existing gene expression data analysis software (see Section 7.4.2) provides the option of different distance measures and allows a user to choose a particular one for a specific application.

Whereas these simple distance measures are probably adequate for detecting apparent coexpressed genes, more sophisticated distance measures may be needed for detecting the more subtly coexpressed genes. These may include cases where genes have correlated profiles in a time-delayed fashion, or coregulated genes have changes in their expression levels in the opposite directions (some go up, while others go down). Different distance definitions will result in different clustering results. Selection of the "right" distance measure for a particular clustering problem could be based on our general understanding of the data set and also possibly on a trial-and-error process. An ideal clustering software should provide a user with the capability to specify his/her own distance measure so different distance measures can be tried and evaluated easily.

Objective Functions There are two types of clustering problems relevant to gene expression data analyses. The first type is the classical clustering problem, one in which we partition a data set into clusters with data points of the same cluster having small distances and data points of different clusters having larger distances (see Fig. 7.5A). The second type deals with identifying data clusters from a noisy background as illustrated in Fig. 7.5B, where a data cluster is considered as a group of data points that form regions with higher densities than their neighboring regions. Figure 7.5B has three such dense clusters sitting in the middle of a noisy background. The objective here is not to partition the whole data into clusters, but rather to identify and extract the dense clusters out of the noisy background.

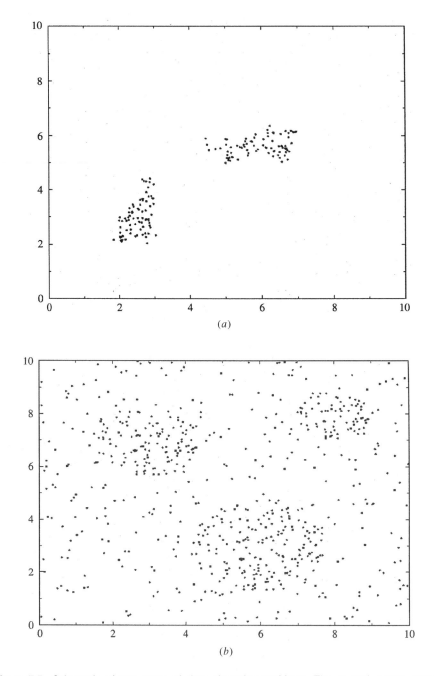

Figure 7.5 *Schematic of two types of data clustering problems.* The *x*- and *y*-axes represent expression levels at two different time points (or under two different conditions). (*a*) Data set with two apparent clusters. (*b*) Three data clusters in a noisy background.

For the first type of clustering problem, a popular objective function is defined as the sum of the distances between data points of each cluster and the center of the cluster. Whereas this objective function captures our intuition about data clustering, developing a rigorous computer algorithm for finding the globally optimal solution represents a highly challenging problem. Objective functions for the second type of clustering problem are more complex, which we will discuss in Section 7.5.3.

Algorithms   Clustering algorithms are used to find the best possible clustering results, as defined by the specified objective function. The following subsection describes three popular clustering approaches. Throughout this subsection, we will use the averaged intensity value if a data set consists of multiple replicates for each gene.

7.5.2   Clustering of Gene Expression Data

There are a number of popular clustering techniques for gene expression data. They include *K*-means clustering (Herwig et al., 1999), hierarchical clustering (Eisen et al., 1998; Wen et al., 1998), and self-organizing maps (Tamayo et al., 1999). We now outline the basic idea of each of these three clustering techniques.

K-means Clustering   *K*-means clustering represents one of the simplest clustering methods. The objective of a *K*-means clustering algorithm is to divide a whole data set into *K* clusters so that the total distance between each data point and its cluster's centroid (or mean) is minimized. Intuitively, this will require partitioning the data set into groups so that the data points of each group are closely centered around the cluster's centroid. The following provides the basic procedure for a *K*-means algorithm:

Step 1. Make initial guesses of the *K* centroids (or means) $m_1, m_2, \ldots, m_K$ of *K* clusters to be found (one common practice is to make the initial guesses of centroids spread out in space).

Step 2. Do the following operations until there is no change in the centroids m_1, $m_2, \ldots, m_K$.

  Step 2.1. Assign each data point to the cluster whose distance to its centroids is the smallest.

  Step 2.2. For each cluster, calculate its centroid and replace its current centroid by the new one.

Clearly, *K*-means is an iterative algorithm: It iterates until it converges to a stable set of centroids. It is generally difficult to assess theoretically how fast this procedure will converge. However, often in real applications, a *K*-means algorithm converges quickly to some local minima. A key advantage of a *K*-means algorithm is that it is very easy to implement as a computer code, and it generally produces good clustering results when the underlying clusters do not overlap with each other, and the data set is fairly clean, that is, it does not have a significant amount of noise. One main drawback to this approach is that its clustering quality heavily depends on how good the selection of the initial centroids is. It could take a significant amount of trial and error before finding the "right" set of initial centroids. Another drawback is that the user has to provide the number of clusters *K* into which he or she may want to partition a data set. A user often has to try different values of

K and then assesses the clustering quality for each K value before making a decision on the "correct" K.

From a mathematical point of view, a K-means algorithm generally does not guarantee the best possible clustering result defined by an objective function as outlined above. Hence, it is classified as a heuristic clustering approach, like many other clustering algorithms.

The following is a list of a few available computer software programs for gene expression and other data clustering, based on (or including) the K-means algorithm. Each of these programs (and those listed in the following subsections) can be accessed through the Internet (see the URLs provided in Table 7.2).

(1) *GeneSpring* gene expression data analysis system is commercial software for the manipulation, analysis, and visualization of gene expression data; its analysis capability consists of a number of clustering techniques, including K-means.

(2) *Spotfire DecisionSite* is commercial software for the interactive visualization system for a variety of data mining applications, including microarray analysis. Its analysis capabilities include K-means-based clustering analysis.

(3) *Expression Profiler* is freeware for the analysis and clustering of gene expression and sequence data. Its analysis capabilities include data clustering using the K-means approach.

Hierarchical Clustering   Hierarchical clustering has been widely used in gene expression data analyses, possibly influenced by the work of M. Eisen and coworkers (1998). It represents another simple clustering technique. The objective of this clustering paradigm is to provide a hierarchical view of a clustering problem at different levels of resolution. At the highest resolution, every data point forms a cluster by itself. At the lowest resolution, the whole data set forms one cluster. In between, each cluster at a particular level is formed by merging the two closest clusters at a higher (resolution) level. Figure 7.6 illustrates this idea. Figure 7.6A is a set of expression data with two time points (or under two experimental conditions), and the tree structure in Fig. 7.6B represents the

TABLE 7.2 Web Servers for Microarray Gene Expression Data Analysis and Applications

Program Name	Internet Address
KEGG	http://star.scl.kyoto-u.ac.jp/kegg/
GeneSpring	http://www.silicongenetics.com
Spotfire DecisionSite	http://www.spotfire.com
Expression Profiler	http://ep.ebi.ac.uk/EP/
Cluster and Treeview	http://rana.lbl.gov/EisenSoftware.htm
DNA-Chip Analyzer	http://www.biostat.harvard.edu/complab/dchip/
R Packages	http://www.stat.uni-muenchen.de/ ~ strimmer/rexpress.html
GenoMax	http://www.informaxinc.com/solutions/genomax/gene_expression.html
EPCLUST	http://ep.ebi.ac.uk/EP/EPCLUST/
EXCAVATOR	http://compbio.ornl.gov/structure/excavator/
Yeast database	http://cmgm.stanford.edu/pbrown/explore/
GXD	http://www.informatics.jax.org/mgihome/GXD/aboutGXD.shtml
GEO	http://www.ncbi.nlm.nih.gov/geo/

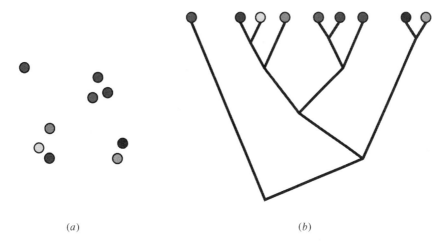

(a)   (b)

Figure 7.6 *Schematic of hierarchical data clustering.* (*a*) Set of data points in two-dimensional space. (*b*) Representation of a clustering tree of the data set.

result of a hierarchical clustering. Each tree node represents a cluster at some resolution level. The whole data set, represented by the root of the tree, forms one cluster at the lowest resolution, which is formed by merging two clusters (one cluster containing the leftmost green node in B and the other one containing the rest of the data points). The larger cluster of the two is formed by merging two clusters at an even higher resolution, one with six data points and one with two data points (blue and light blue nodes), etc. With this clustering tree structure, one can select different thresholds on the intercluster distance to obtain clustering results at different resolution levels. For example, if we want to acquire a clustering result with its intercluster distance being at least D, then all we have to do is cut the tree edges with distances larger than D. Note that a tree edge has the corresponding intercluster distance, defined as the shortest distance between data points of two clusters. Each connected piece, after the cutting, represents one cluster for this specified D.

A hierarchical clustering approach employs a procedure like the following for constructing a clustering tree as outlined above:

> *Step 1.* Label each data point as a cluster and assign them as tree leafs (the top row of Fig. 7.6(*b*).
>
> *Step 2.* Do the following operation until there is only one cluster left:
>> *Step 2.1.* Merge two clusters with the shortest intercluster distance into one and replace the two clusters by it; and connect the corresponding tree nodes to a new tree node representing the new cluster.

Like K-means clustering, hierarchical clustering is easy to implement and can often produce good clustering results when the "structure" of clusters is not very complicated, and the data set is clean. Its main drawback is its lack of robustness as noise could have significant negative effects on the clustering results. This is so because the hierarchical clustering method mainly uses local information to accomplish data clustering. When noise affects the local features of a data set, it could have a major impact on the overall clustering result.

The following is a list of hierarchical clustering software for gene expression data analysis:

(1) *Cluster and Treeview* is freeware for gene expression data analysis developed by M. Eisen. Its data clustering capability is based on a hierarchical clustering approach.

(2) *DNA-Chip Analyzer* is another freeware for gene expression data analysis. Hierarchical data clustering is one of the main data analysis tools employed in the package.

(3) *R Packages* for gene expression analysis is freeware that consists of a set of various tools for gene expression data analysis. Among numerous clustering and statistical analysis tools, hierarchical clustering is one of the tools employed.

Self-organizing Maps Self-organizing maps (SOM) represent a class of neural networks often used for data clustering (Kohonen, 1982a,b). On the conceptual level, an SOM approach has a similar objective to that of a *K*-means approach, that is, it tries to identify a good representative (e.g., centroid) for a group of nearby (similar) data points and to group data points around these representatives. However, on a technical level, its approach is quite different from that of *K*-means clustering. It uses a neural network to represent the data set and trains the "weights" of the neural net to capture data clusters. Its clustering procedure involves a sequence of weight-adjusting operations on the neural network when reviewing the entire data set. Without getting into the detailed mathematical procedure of training a self-organizing map, we use Fig. 7.7 as an example to illustrate the essence of this procedure. Figure 7.7A represents a data set (with each data point represented as a small black dot) divided into three clusters as marked. An SOM initially is made of a grid of (neural network) nodes represented by green dots, with each node having eight neighbors (except the boundary nodes). When a data point is processed, the SOM finds its closest neural network node and pulls the node a "little" closer to the data point. It does this also to the node's neighbors but with lesser forces (how much closer a neighbor will be pulled toward the direction of the data point depends on its distance and other SOM

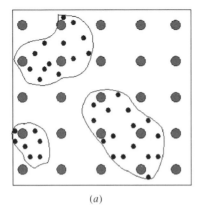
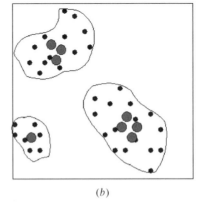

(a)   (b)

Figure 7.7 *Schematic of SOM training.* (*a*) Representation of a two-dimensional data set with a two-dimensional grid of neurons in its initial configuration. (*b*) Representation of a partially trained SOM.

parameters). The nodes will continue to be pulled toward the direction of their nearby data point as it is being processed. As we can imagine, the overall effect of all such pullings will be that nodes close to or within a cluster will be pulled toward the center of the cluster. Eventually, these nodes will be merged into one and other nodes (nodes that are not close to any data points) will disappear due to the lack of activities. Figure 7.7A and 7.7B schematically illustrate the starting configuration of such a grid and a close-to-end configuration after all the pullings.

Recent studies have shown that SOMs could perform significantly better than the classical *K*-means algorithm and hierarchical clustering, although mathematically analyzing an SOM's performance presents a very challenging problem. Like *K*-means clustering, SOMs conduct data clustering based on very local information, that is, which data points are close to a particular point (e.g., centroid). Information about regional densities or homogeneity is mostly ignored. Hence, such clustering techniques may not do well when a data cluster's boundary shape is far from being a sphere, or when a data set has a great amount of noise, like a typical set of gene expression data.

The following is a list of a few SOM-based data clustering software programs, in addition to the software listed above, which also include SOM as part of their clustering capabilities:

(1) *GenoMax* gene expression analysis software is commercial software for gene expression data analysis and visualization. Its data clustering capabilities consist of numerous clustering algorithms, including self-organizing maps.

(2) *EPCLUST* is part of the Expression Profiler package. It employs SOM as one of its clustering techniques.

(3) *GENECLUSTER* is freeware developed at MIT/the Whitehead Institute. It uses self-organizing maps as its data clustering techniques for gene expression data analysis.

7.5.3 Cluster Identification from Noisy Background

When examining gene expression data, one may find that it is often the case that not every gene expression profile is closely related to the profile of any other genes. So the general picture for gene expression data is something like this: Some gene expression profiles are closely related with each other and form relatively dense data clusters, while the profiles of other genes are scattered all around, constituting a general "noisy" background, inside of which the clusters sit. Figure 7.5B provides a schematic of such a picture. Data clustering by simply dividing the data set into groups does not make much sense for such a problem, as clustered data points only represent a portion of the whole data set and the rest of the data points are not clustered. The problem we are interested in solving here is *to identify and extract the clustered data points from a noisy background.* None of the clustering techniques outlined above is adequate for this more general clustering problem.

We now describe a computational framework that can be used to effectively deal with such a problem. The framework is built on a graph-theoretic concept called *minimum spanning trees* (Xu et al., 2001, 2002). Some simple concepts from graph theory are needed here to explain the basic idea of the algorithm. Since our goal is to identify clustered data points from a noisy background, we need to arrive at a clear definition of a *data cluster.* Intuitively, a data cluster is a group of data points that are close to each other and occupy a region in space that has higher density than its neighboring regions. A rigorous definition of a data cluster, which captures the essence of this intuition, can be found in Xu et al. (2002).

For a set D of gene expression profiles and a distance measure ρ, we define a graph $G(D) = (V, E)$ as follows. The vertex set V contains all data points, that is, each data point is defined as a vertex. The edge set E contains all pairs of data points, that is, each pair of data points forms an edge, and the edge distance is defined as the distance ρ between the two data points. A *spanning tree* T of $G(D)$ is a connected subgraph of $G(D)$ such that (1) T contains every vertex of $G(D)$ and (2) T does not contain any cycle. A *minimum spanning tree* (MST) is a spanning tree with the minimum total edge distance (Aho et al., 1974). Any connected component of a MST is called a *subtree* of the MST. MSTs have long been used for pattern classifications in the field of image processing and pattern recognition (Duda and Hart, 1973; Gonzalez and Wintz, 1987). It represents a powerful tool for capturing the essential cluster structures in an unstructured data set.

There are numerous computational methods for constructing a MST from a data set. Prim's algorithm represents one of the classical methods (Prim, 1957; Aho et al., 1974). The basic idea of the algorithm can be outlined as follows: *The initial solution is a singleton set containing an arbitrary vertex; the current partial solution is repeatedly expanded by adding the vertex (not in the current solution) that has the shortest edge to a vertex in the current solution, along with the edge, until all vertices are in the current solution.*

Figure 7.8 shows an example of building a MST from a data set. Figure 7.8A is a data set in two-dimensional space (representing gene expression profiles under two experimental conditions), and Fig. 7.8B is a minimum spanning tree of the data set built by Prim's algorithm. The label attached to each vertex represents the selection order by Prim's algorithm, that is, 1 means the vertex selected first, 2 the vertex selected second, etc. Figure 7.8C illustrates the distances of the MST edges in the selection order by Prim's algorithm. This list of edges, along with their distances, is called a *linear representation* of the two-dimensional data set of Fig. 7.8A. An interesting property of this simple linear representation is that though simple, it captures all the essential information of a data set needed for data clustering, as stated below.

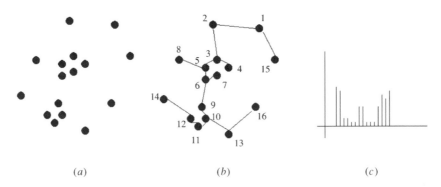

(a) (b) (c)

Figure 7.8 *Schematic of Prim's algorithm.* (*a*) Set of data points in two-dimensional space representing a set of gene expression data with two time points (or under two experimental conditions). (*b*) Minimum spanning tree constructed using Prim's algorithm. (*c*) Plot of edge distances of the MST in the selection order of Prim's algorithm. The *x*-axis represents the selection order 1, 2, 3, . . . , and the *y*-axis the edge distance.

Olman and coworkers (2002) have rigorously proved the following: For any data set D (of any finite dimension) and any subset C of D, C is a data cluster of D *if and only if* (1) C's elements form one consecutive segment in D's linear representation, and (2) all edge distances in this segment are smaller than the distances of the two edges right outside of the segment.

We refer the reader to Olman and colleagues (2002) for the technical details of this fundamental result on data clustering. What this mathematical result implies is that finding a data cluster from a noisy background can be rigorously solved by searching through a linear sequence (a linear representation of the data set) and finding *valleys* (segments with small edge distances) in this linear sequence. Now we can go back to check the plot of Fig. 7.8C. We can see two valleys in the linear representation plot, which correspond precisely to the two clusters in the noisy data set.

A nice feature of this MST-based approach is that it rigorously transforms a multiple dimensional (could be of any dimension, theoretically speaking) data clustering problem to a string search problem. In addition, the two-dimensional plot of a linear representation of a data set allows a user to visualize a high-dimensional data set to "see" if the data set contains any substantial data clusters. Hence, it also provides a useful visualization tool for gene expression data analysis.

This capability of extracting data clusters from a noisy data set has been incorporated into the EXCAVATOR gene expression data analysis package (Xu et al., 2001, 2002), which we now introduce.

7.5.4 EXCAVATOR: A Software for Gene Expression Data Analysis

EXCAVATOR is computer software (Xu et al., 2002) that implements the clustering techniques described in Section 6.5.3, plus a few additional MST-based clustering techniques. We use this software as an example to illustrate how MST-based clustering techniques can be used for gene expression data analysis. Like other gene expression data analysis software packages, EXCAVATOR has the standard capabilities for gene expression data processing and clustering, including using different distance measures, different clustering algorithms, and different ways of displaying clustering and related results. In addition, EXACVATOR has a number of unique capabilities, which we now outline.

Data Clustering vs. Cluster Identification Using the MST framework, EXCAVATOR supports both the regular data clustering, that is, partitioning a data set into groups of similar elements, and cluster identification as we discussed above. Solving a data clustering problem under the MST framework can be done optimally and rigorously, unlike virtually all other data clustering software packages (Xu et al., 2002). As established by Xu and coworkers (2002), a data clustering problem is equivalent to partitioning an MST representing the data set. Clearly, partitioning a tree is a much simpler problem than partitioning a multidimensional data set. A typical objective function could be to partition a tree T into k subtrees $\{T_1, \ldots, T_k\}$, such that the following function:

$$\Sigma_{i=1}^{k} \Sigma_{d \in Ti} \rho(c(T_i), d)$$

is minimized, where $c(T_i)$ represents the center (or centroid) of the subtree T_i and $\rho(\ )$ represents the distance between two data points. Intuitively, the problem is to divide the tree into k subtrees so that data points of each subtree are as close to the center of the subtree as possible. Xu and colleagues (2002) provide a detailed algorithm for rigorously and efficiently solving such a problem. Figure 7.4 shows the clustering result of a data set of 68 genes by EXCAVATOR's MST-based clustering algorithm.

If a data set is not as clean as the example shown in Fig. 7.4, EXCAVATOR provides the option of allowing the user to apply the cluster identification algorithm described in Section 7.5.3. Figure 7.9 illustrates an application example that uses a set of simulated data to demonstrate the effectiveness of such a cluster identification algorithm to extract a data cluster from a very noisy background.

Automatic Determination of the Number of Clusters   When applying a clustering algorithm for gene expression data analysis, it is often the case that the user does not know how many clusters a data set may have. EXCAVATOR provides a mechanism that can assist a user in making such a decision. The basic idea is as follows: When EXCAVATOR partitions a data set into clusters by partitioning its minimum spanning tree, it estimates the "homogeneity" of the edge distances within each cluster. Clearly, the more homogeneous, the better the quality of the data clustering. One can measure such homogeneity using one numerical value (for details, see Xu et al., 2002). Figure 7.10 shows a plot of this measurement for the data set of Fig. 7.4. As one can see, the quality improves as the number of clusters increases, but then the improvement becomes minimal as the number goes beyond a particular value. This means that when a cluster is overcut, although homogeneity improves, such improvement becomes negligent. Finding the transition point where significant improvement becomes minimal improvement provides a candidate for the actual number of clusters in a data set. 4 is the transition point for this particular example. That is why four clusters are selected for the example of Fig. 7.4. For each clustering application, EXCAVATOR plots a figure like the one in Fig. 7.10. EXCAVATOR provides an option that allows the user to specify the number of clusters based on this plot, or the program will select the number automatically based on where the transition point is.

Data-constrained Clustering   When clustering gene expression data, one may have specific information about the genes under investigation. Such information could help improve the quality of an ab initio data clustering procedure as we have described above. EXCAVATOR allows a user to specify if any genes should or should not belong to the same cluster, based on the user's specific knowledge, and finds the optimal clustering that is consistent with the specified constraints. This feature is implemented as follows in the program. If data points are found to belong to the same cluster, the algorithm marks the whole MST path connecting the two points as "cannot be cut" when doing the clustering. So every data point on this path will be assigned to the same cluster of these two points. A similar process is devised for two genes that should belong to different clusters. EXCAVATOR provides a graphics interface to facilitate easy incorporation of such constraints into the clustering process.

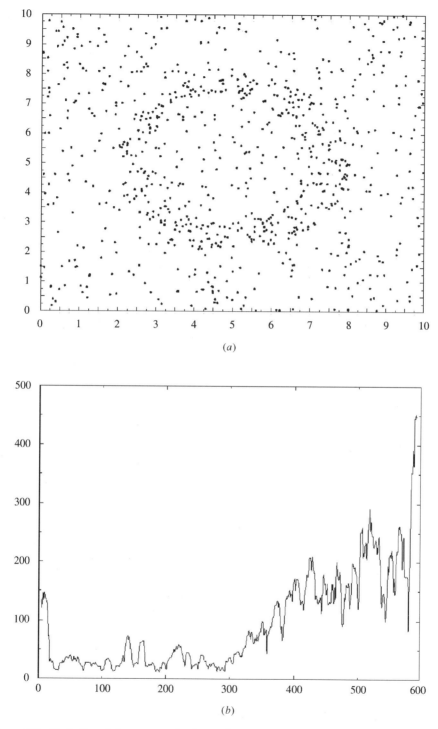

Figure 7.9   *Illustration of cluster extraction from noisy data.* (*a*) Ring-shaped cluster in the middle of a noisy background. (*b*) Linear representation of the data set based on Prim's algorithm. The large "valley," which represents the ring-shaped data cluster, is underlined using a red line. (*c*) [next page] Extracted data cluster corresponding to the valley.

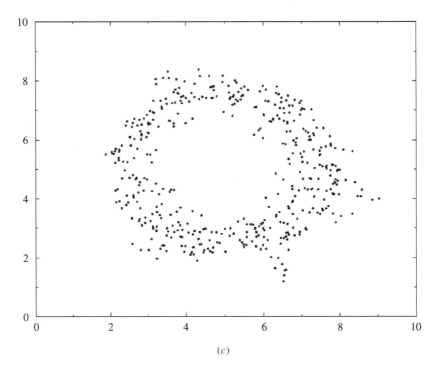

(*c*)

Figure 7.9   (*continued*)

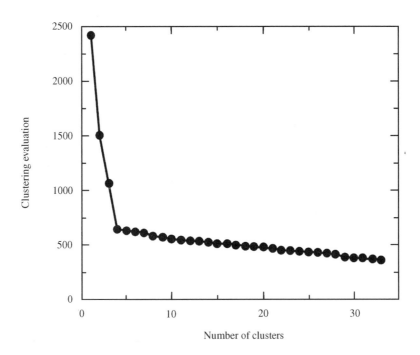

Figure 7.10   *Illustration of clustering quality vs. the number of clusters.* The *x*-axis is the number of clusters, and the *y*-axis the inverse of clustering quality.

7.5.5   Discovering Subtypes through Data Clustering

Gene clustering, as described in Sections 7.5.2 and 7.5.3, is mainly used to identify genes working in the same biological machinery or process. The same type of clustering technique could also be applied to clustering of experimental conditions, based on similarities of the same sets of genes that are expressed or not expressed (or expression levels going up or down) under these conditions. Such analyses could be used to detect all genes that are expressed or not expressed under particular experimental conditions. Scientists have applied such analyses to discover disease-related human genes through studies of expression data of genes from normal and diseased cells or tissues. The investigation of adult lymphoid malignancy is one such example (Alizadeh et al., 2000).

7.6   APPLICATIONS OF GENE EXPRESSION DATA ANALYSIS FOR PATHWAY INFERENCE

Though powerful as a tool for probing the internal structures of a biological process and/or organization in a cell, microarray technology alone may not be able to provide a sufficient amount of information/constraints for constructing a unique pathway model, at least not in its current stage of development. It is often the case that many (possibly significantly) different models could all be consistent with the observed expression data, leading to many possible interpretations of the observed data. Additional information should generally be helpful in narrowing down the possibilities of the model structures of a target pathway under investigation. This information could derive from either other types of high-throughput genomic/proteomic data or from specific knowledge about a particular biological process or organization. The following is a listing of a few such examples that could be used for pathway construction, in conjunction with microarray gene expression data.

Operons/regulons in a Genomic Sequence   An operon is a series of genes arranged in tandem that share a common promoter region, and a regulon is a network of operons that are regulated by a common regulatory protein and its effector(s) (Stephanopoulos et al., 1998). Genes predicted to be both coexpressed and of the same operon/regulon provide good candidates for genes belonging to the same gene network. A number of computational methods have been developed for the prediction of operons/regulons from genomic sequences (Craven et al., 2000; Terai et al., 2001; Ermolaeva et al., 2001).

Protein–protein Interactions   The prediction of protein–protein interactions provides another important piece of information about which genes may be wired together and how they are wired in a gene network. There are numerous ways of predicting the interaction of proteins. These may include (1) predictions based on high-throughput experimental data, which could be highly noisy, like two-hybrid systems (Fields and Song, 1989; Uetz et al., 2000), and (2) predictions based on bioinformatic analyses of genomic sequence data like the gene fusion method (Marcotte et al., 1999a) or the approach based on phylogenetic profiles (Pellegrini et al., 1999).

Functional Assignment of Genes   Knowing the biological functions of genes can help to establish more detailed relationships among genes that are predicted to belong to the same gene network. Such an inference of detailed gene relationships may require general knowledge about biochemistry pathways and specific knowledge about a particular biological process under investigation, for example, electron transporter pathway. It is estimated that 50 to 70% of genes in a microbial genome can have some level of functional assignments, using bioinformatics tools (see Chapter 5 for details).

Data like these could provide structural constraints when building a model for a gene network. Then mathematical modeling techniques can be used to build pathway models that are consistent with these structural constraints and the observed gene expression data. Numerous modeling methods have been proposed for such purposes; they include Boolean networks (Shmulevich et al., 2002), Bayesian networks (Friedman et al., 2000), differential equations (Jamshidi et al., 2001; Kato et al., 2000), and steady-state models (Kyoda et al., 2000).

7.6.1   Data-constrained Pathway Construction

We have witnessed a major surge in computational studies of inference and the construction of regulatory pathways in the past few years. The major goals of these studies are (1) to derive causality relationships among *relevant* genes (e.g., which genes may turn on/off other genes) that are consistent with experimental data, particularly large-scale microarray gene expression data; and (2) to infer the detailed wiring map by which genes are wired with other genes, which are consistent with observed experimental data. There are generally two classes of models that have been used to model regulatory pathways: (1) qualitative models and (2) quantitative models.

Qualitative Models   Boolean networks (Liang et al., 1998; Akutsu et al., 1999, 2000a, b) probably represent the most popular class of qualitative models. In a Boolean network model, a gene has two possible states: *on* or *off* (or up- or down-regulated). Each gene's state is controlled by a set of other genes, and their functional relationship is represented as a Boolean function. For example, gene A's state S_A may be represented as a Boolean function of genes B, C, and D's states S_B, S_C, S_D as $S_A = S_B \lor S_C \land S_D$. "V" represents an "or" operation and " $\land$ " an "and" operation. What this Boolean expression means is that as long as either S_B or ($S_C \land S_D$) is on, then S_A will be on. For ($S_C \land S_D$) to be on, both S_C and S_D have to be on. A Boolean network is essentially a wiring diagram connecting all the relevant genes, and the functional relationships among connected genes are represented as Boolean functions.

To apply Boolean networks to model a genetic network, we need to convert the observed expression data of each gene to one of the three values: up, down, or "unchanged." Genes with unchanged expression levels will not be considered in the pathway modeling process. Mathematical procedures will then be applied to *identify* Boolean networks that are most consistent with the up/down expression patterns among the relevant genes (Akutsu et al., 1999). A key advantage of Boolean networks or their variants, for example, qualitative networks (Akutsu et al., 2000c), is that they allow a modeling procedure to use the most reliable part of the observed gene expression data, that is, expression levels going up or down. As we know, the detailed numerical values of the observed expression data could be difficult to use as basis for computational network modeling, knowing many less understood technical and biological factors that could contribute to the observed intensity data. On the other hand, Boolean networks may be too simple to capture all the essential information hidden in the gene expression data.

Quantitative Models Qualitative models, like Boolean networks, should generally be adequate if our goal is to build static (or time-less) models for biological pathways. Such models are clearly inadequate if the goal is to study the dynamic behavior of a biological pathway, for example, conditions of reaching an equilibrium state in a negative feedback regulation process or modeling the stability region of an oscillation process, which are needed for the study of any dynamical system. There are numerous mathematical models for studying such dynamic behaviors. The work on generalized qualitative circuit models with feedback loops, by Thieffry and Thomas (1995, 1999), has shown these models to be effective in generating multistationarity (i.e., alternative states of gene expression) using one positive regulatory circuit and in generating stable oscillatory behavior using one negative circuit (Thieffry and Thomas, 1999).

Differential equations have long been a popular tool for modeling electronic circuits. They are now becoming one of the popular methods for modeling regulatory networks (Chen et al., 1999; Solomen et al., 2000). Using gene expression data, one could possibly formulate kinetic (differential) equations of regulatory networks. Studies have indicated that a relatively small set of temporal gene expression will be "adequate" to construct kinetic equations, which are capable of modeling various dynamical behaviors of regulatory networks, including the feedback of transcriptional regulation, stable and semistable systems, time-delayed regulation (Chen et al., 1999). Although such modeling work is still in its infancy, the initial modeling results are clearly encouraging.

7.7 SUMMARY

A great deal of information could be revealed about genes through the rationale design of microarray gene expression experiments and sensible data interpretation of gene expression patterns. Some level of functional information of unknown genes could possibly be derived from their gene expression data, based on the premise of "guilt by association" (see Chapter 5). More detailed role assignments of genes to a particular gene network are possible when additional information from other sources, like genomic or proteomic data, is available. A number of major efforts in building public databases for gene expression data are underway, to facilitate the functional inference of genes and gene networks in a systematic manner. These databases include the yeast *Saccharomyces cerevasiae* located at Stanford University, the mouse genes expression database GXD, and the Gene Expression Omnibus database GEO at NCBI. Information derived from these general-purpose databases could help a researcher to derive a rough picture of a particular biological machinery/process under investigation; target-specific gene expression experiments may be needed to collect more detailed and specific information on a selected set of genes to fill the gaps in a gene network construction. Bioinformatics will play a significant role in systematic studies of biological machineries, based on high-throughput experimental data, including gene expression data.

FURTHER READING

Akutsu, T., S. Miyano, and S. Kuhara. 2002. Infering qualitative relations in genetic networks and metabolic patwhays. *Bioinformatics* 16:727–734.

Claverie, M. 1999. Computational methods for the identification of differential and coordinated gene expression. *Human Molec. Gen.* 8(10):1821–1832.

DeRisi, J. L., V. R. Iyer, and P. O. Brown. 1997. Exploring the metabolic and genetic control of gene expression on a genome scale. *Science* 278:680–686.

Eisen, M. B., P. T. Spellman, P. O. Brown, and D. Botstein. 1998. Cluster analysis and display of genome-wide expression patterns. *Proc. Natl. Acad. Sci. USA* 95:14863–14868.

Fields, S., and O. Song. 1989. A novel genetic system to detect protein-protein interactions. *Nature* 340:245–246.

Jamshidi, N., J. S. Edwards, T. Fahland, and G. M. Church, et al. 2001. Dynamic simulation of the human red blood cell metabolic network. *Bioinformatics* 17:286–287.

Michal, G. 1999. *Biochemical Pathways*. John Wiley & Sons, New York.

Nadon, R., and J. Shoemaker. 2002. Statistical issues with microarrays: processing and analysis. *Trends Genet.* 18(5):265–271.

Stephanopoulos, G. N., A. A. Aristidou, and J. Nielsen. 1998. *Metabolic Engineering: Principles and Methodologies*. Academic Press, San Diego, Calif.

Wen, X., S. Fuhrman, G. S. Carr, and S. Smith, et al. 1998. Large-scale temporal gene expression mapping of central nervous system development. *Proc. Natl. Acad. Sci. USA* 95:334–339.

8

Mutagenesis as a Genomic Tool for Studying Gene Function

Alexander S. Beliaev

8.1 INTRODUCTION

As the exploration of prokaryotic organisms moves into the postsequencing era, there is an increasing need for experimental approaches that will facilitate the functional analysis of bacterial genomes. Currently, the characterization of many microorganisms of medical, environmental, and biotechnological importance is limited by the lack of adequate genetic tools. Although new technologies, such as DNA microarrays and mass spectrometry-assisted protein identification, have been or are being developed to address this need, mutagenesis remains one of the most powerful techniques for uncovering gene function.

Understanding the function of a particular gene within the context of a complex biological system is a multistep process that requires an integrative systematic approach. Initially, information regarding the function of an open reading frame product is derived from comparative analysis of the nucleotide sequence using bioinformatic tools (see Chapter 3 for an in-depth discussion). However, not all predicted genes within a genome can be assigned a putative function based on homology. Moreover, sequence homology does not automatically translate into function, and the functional assignment of genes by similarity comparison can be misleading or incorrect due to the complicated evolutionary and structure–function relationships among different genes (Strauss and Falkow, 1997). The second step in defining gene function involves the measurement of gene and protein expression patterns within the complex cellular environment using expression profiling techniques such as DNA microarrays or mass spectrometry-assisted proteomics (Lockhart and Winzeler, 2000; Pandey and Mann, 2000). Using these techniques, one can gather a

Microbial Functional Genomics, Edited by Jizhong Zhou, Dorothea K. Thompson, Ying Xu, and James M. Tiedje.
ISBN 0-471-07190-0   © 2004 John Wiley & Sons, Inc.

large amount of information; however, much of the data generated are correlational, and conclusions regarding gene function are often inferred (Dean, 2001). The third, perhaps most crucial, step in functional analysis involves system perturbation where the gene in question is inactivated, and the resulting changes are determined in order to confirm or refute the proposed hypothesis. Functional assignments are made by comparing the wild-type phenotype to that of the mutant, in which the wild-type gene is altered or replaced by a mutated counterpart.

Most of the detectable mutations are associated with discrete changes in the phenotype of an organism. In bacteria, mutagenesis is a relatively straightforward process as they are haploid organisms where any single mutation results in altered expression or impaired function, which more than often translates into a detectable phenotype. There are a variety of ways to introduce a mutation, including chemical, physical, and genetic techniques. Although chemical and physical agents can be used to generate massive libraries of mutants, their utility for large-scale genomic analysis is hampered by the lack of easily selectable markers. The goal of this chapter is not to provide an exhaustive review of the wide variety of mutagenesis approaches, but to provide an in-depth description of those strategies that have proved particularly useful in the functional analysis of bacterial genomes. Therefore, in this chapter we will discuss the application of three major strategies for genome-wide mutagenesis: transposon insertion, gene disruption by allelic exchange, and expression inhibition using antisense RNA molecules. We will focus on the applications, modifications, advantages, and disadvantages as well as provide some case studies for each of these three methods.

8.2 TRANSPOSON MUTAGENESIS

As noted above, the central approach to studying gene function and regulatory networks in any organism involves the generation and analysis of comprehensive collections of mutations encompassing the entire genome. With the increasing availability of complete genome sequences, the potential for transposon mutagenesis has grown significantly. The randomness with which transposons insert into the host DNA allows one to generate large collections of mutants that can be analyzed en masse for the loss or impairment of a particular function or process. A number of transposition systems have been applied for genome-scale functional analysis, with a particular emphasis placed on the identification and characterization of genes important for organism survival. In this section, we will discuss several methods that utilize insertional transposon mutagenesis to conduct comprehensive genetic screens and identify genes essential for survival under defined conditions.

8.2.1 Overview of Transposition in Bacteria

Classification of Transposable Elements Transposable elements (TEs) are mobile DNA sequences that are able to translocate as discrete units to other locations within the genome. TEs are common occurrences in bacterial genomes. Since their discovery in the 1950s (McClintock, 1952), over 500 TEs from 159 prokaryotic species have been identified and described and a number of classifications have been proposed (Craig, 1996; Mahillon and Chandler, 1998). Transposable elements are diverse in size, structure, insertion specificity, and transposition mechanisms, with two major groups distinguished in bacteria: insertion sequences (IS) and transposons (Tn). Insertion sequences, the simplest of transposable elements, are represented by short DNA fragments that range anywhere from 700 to 1,300 bp in length and contain two 9- to 40-bp copies of terminally inverted nucleotide repeats. These inverted repeats flank a transposition-related gene, called transposase, which encodes a special DNA-binding protein that mediates the transposition. In contrast to IS elements, the structure of transposons is more complex, as they contain auxiliary genes unrelated to transposition that code for antibiotic and heavy-metal resistance or catabolic functions. Many transposons, such as Tn*5*, Tn*9*, and Tn*10*, are organized as "compound" molecules, in which the auxiliary genes are flanked by two copies of IS elements located in a direct or inverse orientation. Another group of TEs, known as "complex" transposons, features a single IS element flanked by short (30–40 bp) indirect repeats with the genes for drug resistance and transposition encoded in the middle. It should be noted that temperate integrating mutator phages, such as lambda (λ) and mu (μ), are also considered as TEs and form a separate group of transposable elements, although their structure and gene repertoire are more complex than those of a typical transposon (for reviews, see Thompson and Landy, 1989; Pato, 1989).

The majority of prokaryotic transposons are separated into two large classes according to whether they become duplicated during transposition. To date, two major mechanisms for transposition, conservative and replicative, have been described. The conservative (or nonreplicative) mechanism, characteristic for Tn*5* and Tn*10* families, involves an excision of the transposon sequence from the donor molecule and subsequent insertion at the target site without duplication. The first step of this process involves a double-strand cleavage of the donor molecule that exposes the 5'-phosphate and the 3'-hydroxyl terminus of the transposon molecule. Following the excision, the 3'-ends of the transposon are transferred and covalently joined to the 5'-phosphate overhangs at the target site. Upon completion of the strand-transfer step, replication takes place to repair the gaps between the 5'-end of the newly inserted transposon and the 3'-end of the target. During nonreplicative transposition, DNA synthesis is limited only to the short single-stranded regions and does not extend through the whole transposon (for a review, see Berg, 1989). In contrast, replicative transposition is not confined only to the strand breaks and involves duplication of the transposon sequences. The first step involves generation of single-stranded 3'-end nicks, which are introduced precisely at the ends of the transposon sequence, while the second strand is never cleaved. The 3'-OH groups are then transferred and linked to the 5'-phosphate groups of the target site forming a complex structure, or *Shapiro intermediate*, that is converted into a cointegrate by replication. It is important to note that during this process both strands of the donor molecule remain attached to the transposon sequences, which are duplicated during DNA synthesis. Similar to nonreplicative transposition, the free 3'-hydroxyl groups serve as primers for replication, which is not confined to the target sequence but continues in both directions throughout the transposon molecule. The resulting cointegrate, representing a fusion between two transposon copies and the donor molecule, is subsequently resolved into a donor molecule and the target sequence carrying a new insertion (reviewed in Sherrat, 1989).

Insertion Specificity and Effects on Distal Gene Expression Compared to other processes leading to genomic rearrangements, transposition does not rely on the similarity between the donor and recipient DNA sites. Transposons can insert into multiple sites within the chromosome, although their target selection is generally not random and differs significantly from element to element, with most elements displaying different degrees of selectivity, stringency, and patterns of target site selection (Mahillon and Chandler, 1998). Whereas some transposons will preferentially insert into certain sites, others will avoid specific regions on the chromosome. Transposon insertion patterns depend on both the frequency with which a particular site is used as a target and the detectability of the insertion. Several mechanisms for site selection have been described implicating the DNA-binding properties of the element-encoded transposases (Craig, 1997). A target site can be selected either through direct interaction of the element-encoded transposase with the target DNA sequence or through interaction between the transposase and other element- and/or host-encoded accessory proteins. In both cases, the accessibility of a particular target site is influenced by its spatial positioning relative to the donor molecule as well as by the global and structural features of the target DNA region (Craig, 1997).

Upon integration into the target site, transposons and insertion elements can affect the transcription of genes located in the vicinity of the insertion. Most often the transcriptional effects are due to terminator signals or polarity caused by long untranslated stretches of mRNA when the insertion inactivates a gene (or genes) located in the beginning of an operon (Galas and Chandler, 1989). The degree of polarity usually depends on the transposon size, presence of internal transcription terminators, presence of promoter elements within the transposon or within the target operon, position of the transposon insertion relative to the target gene, and the possibility of forming new promoter sequences at the insertion junctions (Reynolds et al., 1981). In addition, chromosomal transposon insertions have been found to turn on the transcription of cryptic genes. The latter phenomenon is primarily explained by conformational and sequence changes in the adjacent DNA that may result in increased gene expression (Berg, 1989). The effects of transposon insertion on the expression of distal genes have been used to obtain preliminary information about operon organization and to determine the presence of internal transcription initiation sites (Berg et al., 1979; Berg and Shaw, 1981).

Transposon Delivery Systems   For in vivo transposon integration into the target replicon and isolation of transposition events, a variety of delivery vehicles have been developed. These include suicide phages and plasmids that are unable to replicate within the host strain, but possess mobilization ability and a broad host range of transfer.

Most phage-based vectors used for transposon delivery carry integration and replication defects such as the absence of a phage attachment site or nonsense mutations in replication genes that prevent the potentially lethal phage DNA from propagating within the host. Some phage suicide vectors can be utilized to deliver transposons into heterologous hosts, although this requires additional genetic modification of the host organism. The recipient strains are sensitized to phage infection by introducing genes encoding the appropriate phage receptor. By this approach, suicide derivatives of λ and P1 have been used to deliver transposons into the genomes of *Klebsiella pneumonia*, *Vibrio cholera*, *Pseudomonas aeruginosa*, and several other organisms (Belas et al., 1984; DeVries et al., 1984; Harkki and Pelva, 1984; Kuner and Kaiser, 1981; Ludwig, 1987). Generally, use of phage delivery vehicles is restricted by the host specificity range and cannot be efficiently adapted for distantly related organisms that are not sensitive to bacteriophage infection. In contrast, plasmid vectors are more versatile with respect to transfer ability and can be used in a broader range of hosts. In theory, any transposon-harboring conjugative plasmid that does not replicate in the recipient strain can be used as a delivery vehicle. A number of strategies have been used to develop suicide vector systems, including the construction of conditionally defective origins of replication, utilization of plasmids incapable of replication in the recipient strain, and insertion of counterselectable markers on the plasmid (see Section 8.2). In general, the choice of a delivery vehicle largely depends on the properties of the recipient strain and on the transposition target.

8.2.2 Transposons as Tools for Mutagenesis

Because transposons cause sequence rearrangements (e.g., insertions, deletions, and inversions), their utility has been proven extremely useful in a variety of molecular biology applications. For example, transposons have been used for gene inactivation, gene delivery, sequencing, isolation of selectable markers, and other applications (for a review, see Berg and Berg, 1996). A variety of endogenous and heterologous transposon systems have been designed and adapted for the inactivation of both chromosomal and plasmid-borne genes. The choice of the transposon system for mutagenesis largely depends on the properties of the target DNA (i.e., chromosome, phage, or plasmid), as well as the properties of the transposable element itself (i.e., resistance marker, target specificity, and transposition mechanism). An important consideration for selecting the appropriate technique is the availability of genetic systems for the host organism. Transposon mutagenesis has been adapted using in vivo and in vitro approaches in a number of gram-positive and gram-negative organisms.

In vivo Mutagenesis For in vivo mutagenesis, the transposon carried on the appropriate suicide delivery vector is introduced by transformation or conjugation into the host organism. The in vivo mutagenesis can be either direct or indirect, depending on whether the transposon molecule is introduced directly into the organism of interest or the mutagenesis of the target DNA is conducted in an appropriate host (e.g., *E. coli*), respectively. The latter approach involves utilization of a suicide shuttle vector that is capable of replicating in the intermediary host but is not maintained in the organism of interest. Appropriate protocols are used to efficiently saturate the cloned DNA fragment with transposon insertions, and then the mutated clones are transformed into the host of origin. Following transposon integration into the target site and loss of the suicide vector, the mutants are selected by plating the organism of interest on a medium containing the resistance marker expressed by the transposon.

The major advantage of in vivo mutagenesis is that the target organism does not have to be naturally competent. Transposon molecules on a suicide donor plasmid can be introduced into the host organism using different transformation methods including electroporation or conjugation. However, in vivo systems face a number of technical limitations. For example, the transposon must be introduced into the host on a suicide vector, the transposase must be expressed in the target host, and the transposase usually is expressed in subsequent generations, resulting in potential insertion instability (Goryshin and Reznikoff, 1998).

In vitro Mutagenesis To bypass the inherent difficulty of using plasmid-mediated transfer of transposon sequences, a number of in vitro transposition systems have been designed, allowing generation of high levels of mutations. The in vitro approach is based on the ability of purified transposases to catalyze strand-transfer reactions between linear DNA molecules in a cell-free environment (Haniford and Chaconas, 1992; Gary et al., 1996; Goryshin et al., 2000). In this method, target DNA from the organism of interest is incubated with transposon molecules in the presence of purified transposase, producing a population of target DNA molecules, each containing the transposable element at a different position. The target DNA may be carried by a suicide plasmid, cosmid, or bacterial artificial chromosome (BAC) vector depending on one's experimental goals. The resulting insertions are screened using an appropriate mapping technique (e.g., PCR, Southern blotting, restriction digest). The mutated DNA fragments are introduced into the host by transformation, followed by homologous recombination between the interrupted mutant and wild-type alleles.

One of the great advantages of in vitro-based methods is the ability to reach high-saturation levels of mutagenesis, which allows one to conduct analyses of the target locus on either large or small scales. The availability of cloned DNA targets provides the advantage of prioritizing different loci based on the researcher's goals. The distinct disadvantage of the in vitro approach is the prerequisite for preliminary information on the target sequence. As with many in vitro approaches, this method is quite labor-intensive, especially when working with restriction maps, thus limiting the number of mutants that can be examined. Within the past several years, however, the increasing amount of sequence information has resulted in the development of different in vitro transposition systems (Colegio et al., 2001; Gehring et al., 2000; Guo and Mekalanos, 2001; Griffin et al., 1999; Sun et al., 2000; Wong and Mekalanos, 2000).

An approach designated Transposome™, which combines the advantages of in vivo and in vitro systems, has been developed utilizing the Tn5 transposition system (Goryshin et al., 2000). This method involves the in vitro formation of released transposon–transposase complexes (transposomes) followed by introduction of the complexes into the target cell of choice by transformation. Briefly, the Tn5 transposon and transposase complex is formed at the early stages of in vitro transposition when the transposase binds to the transposon molecule for excision. At this stage, there are no cations or target DNA available to activate the transposition. The latter is triggered by the internal salt concentration of the host when the Transposome™ complex is introduced into the target organism. Although this technology is dependent on the availability of a transformation system, it overcomes the host barrier posed by in vivo transposition and the need for homologous recombination (Hamer et al., 2001a).

Advantages of Transposon Mutagenesis
Transposon mutagenesis offers several advantages over other techniques as a result of the simplicity with which mutations are generated, insertion stability, as well as the presence of unique genetic markers that enable mapping of transposon insertions.

(i) Compared to various chemical and physical mutagens, transposon mutagenesis results in the integration of foreign DNA into the target gene. The reversion frequency of transposon-induced mutations is usually several orders of magnitude lower than that of the single-base mutations. The excision of transposons from the target site usually occurs at low frequencies, with revertants occurring at rates lower than 10^{-8} (Berg and Berg, 1996).

(ii) Stability issues have been further addressed by using minitransposon derivatives lacking transposase genes. To prevent secondary transposition events, the transposition-related functions are provided by a *cis*-acting transposase located on the suicide donor molecule, which is lost following the transposition. Since the majority of transposons encode for antibiotic resistance or other distinctive catabolic functions, their insertion not only inactivates the target gene, but also confers an antibiotic or other type of resistance to the mutant. Mutant cells containing transposon insertions are selected from the wild-type cells by positive selection on a medium that contains the resistance marker encoded by the transposon.

(iii) The selection marker of the transposon can be used to identify the size of the fragment of DNA that contains the transposon. Upon fractionation of the mutant genomic DNA, the region carrying the transposon insertion can be cloned and selected against the marker, mapped using the polymerase chain reaction (PCR), or used as a probe to identify the wild-type copy of the gene in a genomic library of the wild-type bacterium. Since the ends of most transposons are unique, the interrupted DNA region flanking the Tn insertion can be identified by sequencing outward into the adjoining chromosomal DNA using oligonucleotide primers complementary to the transposon ends.

These features make transposable elements powerful tools for insertional gene inactivation, notable for the scale of throughput and ease of analysis they offer. As a result, transposon-based systems provide an excellent method for genome-wide mutagenesis.

8.2.3 Transposon-based Approaches for Identification of Essential Genes

Historically, gene essentiality has been determined using random libraries of conditionally lethal mutations generated by chemical mutagenesis. This approach has been used successfully to isolate large numbers of temperature-sensitive yeast and bacterial mutants capable of growth on rich complex medium at permissive but not elevated temperatures (Kaback et al., 1984; Schmid et al., 1989). As with any chemical mutagen, mapping of the mutated loci in these temperature-sensitive mutants is generally carried out using complementation and cloning, a strategy that poses a major setback for any organism with an undeveloped genetic system. Even for species with available genetic vectors, the complementation and mapping of several hundred mutations require great effort and certainly cannot be utilized in a high-throughput fashion. Moreover, analysis of temperature-sensitive mutations suggests that random screens for mutants can result in "jackpots" created by repeated isolation of the same mutant classes, presumably because such gene products are particularly easy to mutate to temperature-sensitive alleles (Kaback et al., 1984; Judson and Mekalanos, 2000b).

Transposon mutagenesis is an alternative method for identifying the essentiality of a given gene. The availability of many bacterial genome sequences has spurred the development of various transposon-based approaches to detect and isolate essential genes. As outlined by Judson and Mekalanos (2000b), there are two general strategies for the identification of essential genes: (1) the *negative approach*, which isolates regions of the genome that are not essential, while presuming that everything else *is* essential; and (2) the *positive approach*, which identifies essential genes by generating conditional mutations and observing the lethal phenotype under defined conditions. The utility of each approach varies depending on the properties of the target organism as well as the experimental goals. In the following sections, we will briefly describe the techniques used for both negative and positive identification of essential genes and outline the major advantages and disadvantages of each method.

In vivo Global Transposon Mutagenesis The major issue with the identification of essential genes lies in the lethality of the resulting mutation. A way to circumvent this problem is to define regions within the genome that are dispensable by introducing a statistically significant number of mutations in nonessential genes. The in vivo negative selection technique can be used to generate massive libraries of transposon-induced mutants that can survive under defined conditions (Fig. 8.1). Following transformation into the host, transposon molecules are immobilized within the cells and selected by an appropriate screening approach. Most often the screening method involves positive selection for an antibiotic marker encoded by the transposon linked to the loss or impairment of a certain physiological function. The transposon junctions in each mutant are determined using either sequencing or PCR-based techniques, and each insertion is mapped to a precise position within the genome.

An excellent demonstration of the in vivo approach for high-throughput analysis of gene essentiality involved the determination of the minimal genome in *Mycoplasma* (Hutchison et al., 1999). In this study, global saturating mutagenesis was applied to *M. genitalium* and *M. pneumonia* to identify nonessential genes in both genomes while defining the minimal genome repertoire of these bacteria required for viability under laboratory growth conditions. Although this particular in vivo approach may not be feasible for many other microorganisms, the two *Mycoplasma* species appear to be ideal candidates due to the small size of their respective genomes [580 kb for *M. genitalium* and 816 kM for *M. pneumonia* (Fraser et al., 1995; Himmelreich et al., 1996)]. The global mutagenesis in *M. genitalium* and *M. pneumonia* was carried out using transposon Tn*4001*, originally isolated from *Staphylococcus aureus* (Knudtson and Minion, 1993). The transposon, carried on suicide plasmid pISM 2062 (Knudtson and Minion, 1993), was introduced into the host by electroporation, and the resulting mutants were selected for gentamycin resistance encoded by Tn*4001*. Prior to antibiotic selection, both cultures were

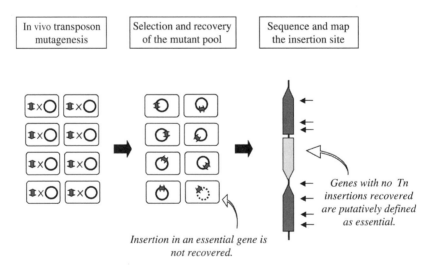

| In vivo transposon mutagenesis | Selection and recovery of the mutant pool | Sequence and map the insertion site |

Genes with no Tn insertions recovered are putatively defined as essential.

Insertion in an essential gene is not recovered.

Figure 8.1 *Global transposon mutagenesis.* The whole bacterial chromosome is a target for transposon mutagenesis. A large number of viable insertions are analyzed by sequencing. This method requires many insertions to be sequenced before statistically significant conclusions can be drawn. [Adapted with permission from N. Judson and J.J. Mekalanos, Transposon-based approaches to identify essential bacterial genes, *Trends in Microbiology*, vol. 8, 521–526. Copyright (2000) Elsevier Science.]

divided into eight separate populations, with each population resulting in 200 to 1,000 viable mycoplasmas harboring independent transposon insertions. To map the sites interrupted by the trasposon insertions, total genomic DNA was isolated from all the mutants, digested with endonuclease and circularized. Transposon junctions were amplified using inverse PCR with two primers specific for the ends of Tn*4001*, cloned into pUC18 and subsequently sequenced. The obtained junction sequences were aligned with the appropriate *Mycoplasma* genomic sequence to determine the exact location of the transposon-induced mutation. These junctions defined 1,354 distinct sites of insertion that were not lethal, while identifying up to 350 potential essential genes in the genome of *M. genitalium* (Hutchison et al., 1999).

The in vivo transposition approach has been further improved by the utilization of high-density microarrays used for identification of genes required for growth under defined conditions (Sassetti et al., 2001). In this modification of the in vivo technique, designated TraSH for transposon site hybridization, a large library of *Mycobacterium bovis* mutants was generated using a mariner-based *Himar1* transposon and efficient phage transduction system. The obtained mutants were compared for growth on rich and minimal medium. Genomic DNA was extracted from both pools represented by the surviving colonies grown under both conditions and used to identify the insertion sites by producing labeled RNA probes complementary to the chromosomal sequences flanking each transposon. These probes were hybridized to an *M. bovis* microarray containing fragments from each predicted ORF in the genome. By comparing hybridization signals derived from both mutant pools, a set of genes required for the growth of *M. bovis* on minimal but not rich medium was determined that consisted of genes of known and unknown functions (Sassetti et al., 2001).

Although the in vivo methods are robust and can be applied to studying any sequenced organism that has a developed genetic system available, they may not be appropriate for most research as it requires a large number of transposon insertions before statistically significant conclusions regarding gene essentiality can be made. Moreover, not every transposon insertion within a gene results in a null mutation, as many insertions near the 3′-end may remove only the nonessential COOH-terminal part of the protein. The presence of outward-facing promoters in the transposon could also drive the transcription of the entire ORF when the transposon is inserted near the 5′-end. Therefore, the major disadvantage of the in vivo negative approach lies in the inherent target specificity and efficiency of in vivo transposition. Due to the presence of "hot" and "cold" spots in every genome, it may be extremely difficult to obtain a high density of mutations for certain regions. Unless saturation is approached, one cannot determine those regions on a chromosome that are essential; however, reaching saturation before any conclusions are drawn may not be possible. Therefore, this technique can be utilized for the preliminary prediction of gene essentiality, and any results generated by it need to be further verified.

Genomic Analysis and Mapping by in vitro Transposition An approach that allows greater Tn insertion density for the negative selection of essential genes was developed using in vitro transposition (Akerley et al., 1998; Reich et al., 1999). This method, designated genomic analysis and mapping by in vitro transposition (GAMBIT), is particularly useful for studying gene essentiality in naturally competent organisms that are able to take up the resulting mutagenized fragments. It involves saturating mutagenesis of large DNA fragments, which are subsequently integrated into the host genome to generate a comprehensive library of transposon mutants (Fig. 8.2). Briefly, PCR-amplified

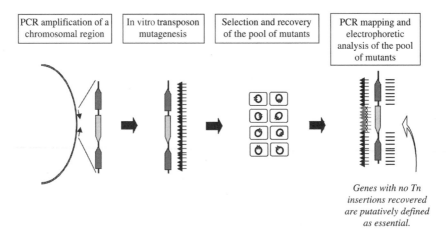

Figure 8.2 *Negative selection of essential genes using PCR-based GAMBIT technique.* A specific region of the bacterial chromosome is amplified by PCR and subjected to in vitro mutagenesis with a transposon containing a drug-resistance marker. The small horizontal arrows signify some of the possible locations of insertions in the PCR product. Genes are designated by vertical black (non essential) or gray (essential) arrows. Insertions that allow viable colonies to form are pooled and analyzed by PCR. Upon analysis, those regions that do not allow insertions (crossed out arrows) show up as empty regions on the agarose gel and are presumed to define essential genes. [Adapted with permission from N. Judson and J.J. Mekalanos, Transposon-based approaches to identify essential bacterial genes, *Trends in Microbiology*, vol. 8, 521–526. Copyright (2000) Elsevier Science.]

fragments of the chromosome are subjected to in vitro mutagenesis, resulting in a population of recombinant DNA molecules that contain multiple mutations in each ORF. Following transformation into the host organism, the mutated alleles integrate into the chromosome by homologous recombination and generate mutations. The resulting mutants are screened for gene function en masse, while DNA samples are isolated from the library before and after selective screening. Mutations represented in each sample are then analyzed by genomic footprinting (Singh et al., 1997), which involves PCR amplifications using a transposon-specific primer and an ORF-specific primer. This method is designed so that a mutation at a particular location is visualized as a band of discrete size by conventional gel electrophoresis. For the whole library, the genomic footprinting method results in a ladder of bands, each one representing a specific transposon insertion. Since mutations in essential genes required for the selected function fail the selection, GAMBIT allows the visualization of regions, or windows, where no transposon insertions are found. Alignment of these windows with the organism's genome sequence identifies putative ORFs that are essential for growth under defined conditions (Judson and Mekalanos, 2001b).

The genomic analysis and mapping by in vitro transposition were adapted and utilized to conduct genome-wide identification of ORFs essential for the growth and survival of the human pathogen *Haemophilus influenzae* (Akerley et al., 2002). In this study, 366 fragments (∼10 kb in length) representing the 1.83-Mb *H. influenzae* genome were amplified by PCR. For replication and verification purposes, the amplified segments were designed so that an average overlap of 5 kb exists between adjacent fragments. These fragments were subjected to in vitro saturating mutagenesis using a *mariner-based* minitransposon and hyperactive *Himar1* transposase (Lampe et al., 1999). The resulting library, which consisted of 366 individual pools of mutants, was analyzed by genetic footprinting and mapped to determine the transposon insertion sites. Based on the obtained results, it was estimated that 478 ORFs (38% of the total coding capacity) of *H. influenzae* are critical for growth and survival on rich medium (Akerley et al., 2002). Of these, 259 ORFs encode proteins of unknown function. Comparison of the *H. influenzae* essential gene repertoire with other complete genomes identified the presence of essential gene homologs in *E. coli, M. genitalium, Mycobacterium tuberculosis,* and *Pseudomonas aeruginosa* (Akerley et al., 2002).

As mentioned above, a major advantage of GAMBIT over in vivo global mutagenesis is the increased insertion frequency that makes the saturation of a particular genome region relatively easy. The high degree of saturation substantially improves the statistical significance of the conclusions regarding the essentiality of a given gene. A particularly attractive feature of GAMBIT is that analysis can be performed on any region of the chromosome, making this method flexible in terms of experimental scale. Although natural competency is a prerequisite for this technique, an approach utilizing suicide vectors as delivery vehicles has been proposed and adapted for efficient genomic footprinting in filamentous fungi (Hamer et al., 2001b). The primary disadvantages of GAMBIT concern its labor intensity and the large number of primers required for mapping.

Mariner-based TnAraOut Transposon System The major restriction of all transposon-based approaches is that their use is limited to characterizing nonessential genes since the insertional inactivation of an essential gene would yield a lethal phenotype. Transposons creating transcriptional fusions overcome this problem. In contrast to the negative strategy, the 'positive approach' identifies essential genes by generating conditional mutations and observing lethality under defined conditions. For this, a transposon-localized inducible promoter is used to replace the gene's natural promoter using in vivo transposition mutagenesis. Insertion of a transposon carrying an outward-facing promoter into the promoter/operator region of the target gene creates a transcriptional fusion, where the inducible promoter replaces the function of the natural one (Fig. 8.3) (Chow and Berg, 1988; Rappleye and Roth, 1997; Judson and Mekalanos, 2000b).

Recently, this approach was used to identify genes essential for the growth or survival of *Vibrio cholerae* on rich medium (Judson and Mekalanos, 2000a). A mariner-based transposon, designated TnAraOut, carrying an arabinose-inducible promoter (pBAD) was used to generate 16 insertional mutants that displayed an arabinose-dependent growth phenotype. Five of these mutants displayed lethal phenotypes on solid medium in the absence of arabinose. Analysis of the chromosomal regions flanking these transposon insertions identified four known essential genes (*gyrB, proRS, ileRS,* and *aspRS*), while one TnAraOut insertion occurred upstream of a hypothetical gene that was not shown previously to be essential. The other eleven mutants that grew in the absence of arabinose, however, showed significantly reduced growth rates manifested by smaller colonies or reduced cell density. The majority of these mutants revealed TnAraOut insertions upstream of genes encoding metabolic or biosynthetic enzymes that are not essential but their expression is required for optimal growth on rich medium (Judson and Mekalanos, 2000a).

It is important to emphasize that the conditions used to test for gene essentiality may substantially affect the experimental results. Initially, it was shown that some of the TnAraOut mutants which did not display complete arabinose dependence for growth on rich medium may be, in fact, more arabinose-dependent under other conditions (Judson

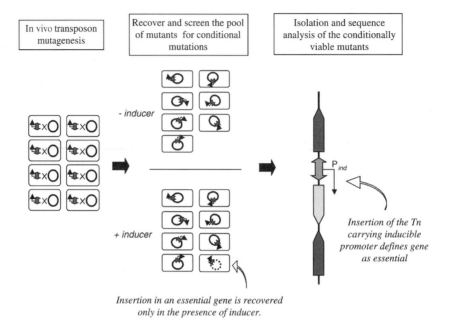

Figure 8.3 *The positive approach to identify essential genes.* Transposition with a transposon containing an outward-facing inducible promoter at one edge in the presence of the inducer results in many possible transposon insertions. The horizontal arrows signify possible insertion locations on the bacterial chromosome. Screening identifies insertions that disrupt the promoter region of an essential gene (gray arrow). The strain generated by such an insertion is dependent on the inducer for viability. The insertional junction is sequenced, allowing the identification of the downstream essential gene. [Adapted with permission from N. Judson and J.J. Mekalanos, Transposon-based approaches to identify essential bacterial genes, *Trends in Microbiology*, vol. 8, 521–526. Copyright (2000) Elsevier Science.]

and Mekalanos, 2000a). This was clearly demonstrated using a strain with an arabinose-inducible *tpi* gene that encodes a key glycolytic enzyme triose phosphate isomerase. Although the TnAraOut-*tpi* mutant was able to grow on LB in the absence of arabinose, it could not form colonies on a minimal M9 medium containing glycerol as the sole source of carbon and energy. Therefore, this approach has the potential to identify genes of unknown functions that are not essential for growth on rich medium but could be essential under other conditions (Judson and Mekalanos, 2001a). The use of the TnAraOut system is simple, cost-efficient, and can be adapted to a variety of bacteria that are permeable to arabinose and possess regulatory proteins of the AraC family of regulators.

In general, the positive approach for identification of essential genes has several advantages over other methods. First, every essential gene identified by this method is a gene of interest, whereas genes that are not essential can be tested further under different growth conditions. No additional strain or plasmid construction is required for this, as there is a built-in assay for testing of biological function. Second, this approach can identify essential genes that contain nonessential 5′-regions which are permissive for transposon insertions, as the presence of an inducible promoter will still generate a conditional phenotype. However, there are several drawbacks to the positive approach. An inducible promoter might not provide enough expression to overcome the inactivation of the natural promoter or, conversely, the basal expression levels from an inducible promoter might be too high to allow the identification of essential genes for which only minute amounts of gene product are required. Also, as with any method involving in vivo transposition, saturating mutagenesis of a genome is difficult to achieve; the insertion frequency varies significantly depending on the gene, and it might be difficult or impossible to inactivate smaller genes using this method (Judson and Mekalanos, 2000b).

8.2.4   Signature-tagged Mutagenesis for Studying Bacterial Pathogenicity

Although in vivo transposon mutagenesis has been the method of choice for many molecular geneticists, this technique remains restricted to individual mutant analysis and appears to be impractical for some high-throughput applications. Many limitations of the conventional approach have been circumvented by the development of a new transposon mutagenesis technique, termed signature-tagged mutagenesis (STM), which utilizes a unique DNA sequence to tag each individual transposon molecule. This approach was originally developed to investigate bacterial pathogenesis in vivo (Hensel et al., 1995; see Chapter 12) and has primarily been used to identify a variety of virulence factors from a wide range of plant, animal, and human pathogens (for reviews, see Mecsas, 2002; Shea et al., 2000). It should be noted, however, that signature-tagged mutagenesis can be applied to any microbial genome where in vivo negative selection is required. The key to STM is the generation of a collection of mutants, each one modified by the incorporation of a transposon molecule carrying a unique sequence. The tags are short DNA segments that contain a variable central region flanked by invariant arms to facilitate PCR amplification and isolation, and labeling of the unique sequences. In principle, each individual mutant can be distinguished from the pool of other mutants by the presence of a unique sequence integrated into its genome.

STM Procedures   The application of STM generally involves a two-step process. During the first step, a pool of transposon molecules carrying unique sequences is used to generate a library of tagged mutants. In the original method (Hensel et al., 1995), a complex mixture of double-stranded DNA tags carried on transposon miniTn5Km2 was used to generate a library of random insertional mutants (Fig. 8.4). Each tag consisted of a unique 40-bp sequence flanked by a common arm of 20 bp at both ends. The variable region of each tag ($[NK]_{20}$, where $N = A, C, G, T$ and $K = G, T$) was designed so that the same sequence occurs once in 2×10^{17} molecules. In addition, the variable regions did not contain any *Hind*III, *Kpn*I, *Pst*I, or *Sal*I restriction sites, which were used for subsequent cloning of the resulting tags into the suicide vector. The double-stranded tags were incorporated into pUTminiTn5Km2 by ligation and maintained in *E. coli* as a pool of transposons tagged by random signature sequences. The transposons were subsequently transferred into the recipient host strain by conjugation to generate a bank of mutants, each carrying a random miniTn5 insertion. The obtained mutant strains were stored individually in 96-well microtiter plates and used to prepare dot blots (Hensel et al., 1995).

The second step of signature-tagged mutagenesis involves in vivo screening of the library. In the original method (Hensel et al., 1995), the suitability of tags was determined prior to the use of PCR amplification, labeling, and hybridization to blots of DNA probes prepared from mutants. Mutants containing tags that yielded clear signals on autoradiograms were reassembled into new 96-well pools and subjected to a selective process that involved the infection of an animal (Hensel et al., 1995). Upon recovery of the mutants following the selective screen, PCR amplification was used to prepare labeled probes representing the tags found in the preselection population, input pool, and those

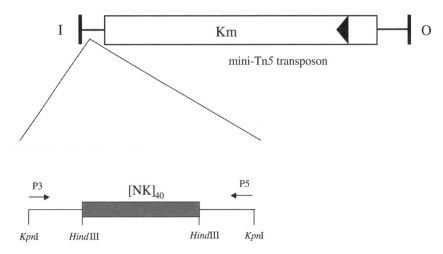

Figure 8.4 *Design of random signature-tagged transposons.* A complex mixture of double-stranded DNA signature tags was generated using PCR. Each tag comprises a different sequence of 40 bp ([NK]$_{20}$, where N = A, C, G, or T; K = G or T) flanked by arms of 20 bp unique to all tags. For generating a signature transposon library, the tags are amplified with primers P3 and P5, digested with KpnI, and cloned into the Tn5Km2 minitransposon. The transposon is carried on a suicide vector that is maintained in the appropriate host.

recovered after the screen, the output pool. Hybridization of the labeled probes from both input and output pools to dot blots of arrayed mutants and identification of tags that were lost in the output pools allowed the isolation of strains that were unable to survive during the infection process (Fig. 8.5). These mutants were recovered from the original microtiter plates, and the nucleotide sequence of DNA flanking the transposon insertion was determined by sequencing (Hensel et al., 1995).

As shown above, STM relies on the ability of the mutant to propagate in vivo as a mixed population and can identify only nonvirulent mutants that cannot be *trans*-complemented by other virulent strains present in the same inoculum (Chiang et al., 1999). Application of STM to a bacterial pathogen for the first time requires that a number of parameters be considered in order to obtain reproducible identification results.

Common Considerations for Application of Signature-tagged Mutagenesis There are several crucial steps involved in the generation and screening of signature-tagged mutant libraries. Here, we summarize the major issues associated with STM and list possible solutions that are used to improve the performance of this method.

Generation of Mutant Libraries One of the critical steps in STM depends on the high efficiency of the procedure chosen to generate random tag insertions in the genome. This is not easily achievable, as mutations in essential genes will result in a lethal phenotype, and therefore, such mutants will not be viable (Lehoux and Levesque, 2000). More often, however, the difficulties associated with STM are due to the lack of a suitable in vivo transposition system. To improve the efficiency of insertional mutagenesis, several modifications to the standard in vivo procedure have been proposed. In an effort to identify

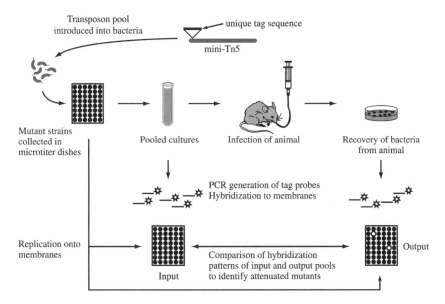

Figure 8.5 *Schematic representation of the original STM strategy.* For this method, a library of insertional mutants resulting from transposition events is pooled into microtiter plates. These pools, each containing 96 different sequence-tagged mutants, are screened for attenuated virulence in an appropriate infection model. Bacteria are recovered from the infected host on a solid medium, and the DNA is isolated from each pool of infection. The tags present in these DNA pools are amplified and labeled, and colony blots are probed and compared to the hybridization patterns obtained from the inoculum. [With permission, from the *Annual Review of Microbiology*, Volume 53 © 1999 by Annual Reviews www.annualreviews.org.]

virulence genes from *Streptococcus pneumonia*, a library of signature-tagged mutants was constructed by insertion-duplication mutagenesis using short random genomic DNA fragments inserted into a suicide plasmid vector carrying the molecular tag (Polissi et al., 1998). Because the in vivo transposition was shown not to produce random mutations in *Streptococcus* mutants (Hallet et al., 1994), the transposon molecule was replaced by randomly selected 400- to 600-bp chromosomal fragments, which facilitated homologous recombination between the tag-carrying molecule and the host chromosome. Another modification of the STM procedure involves utilization of shuttle vectors where the in vivo transposition is carried out in *E. coli*, and the resulting plasmids are subsequently transformed into the target organism. Shuttle mutagenesis using signature-tagged transposons was employed to generate a library of *Neisseria meningitidis* random mutants, which allowed in vivo screening in a bacterium where it is normally impossible to use STM (Claus et al., 1998).

Hybridization Specificity   In the method developed by Hensel and coworkers (1995), the specificity of DNA tags was tested by hybridization prior to use in an animal infection model. To avoid the prescreening process and to increase the sensitivity and specificity of the detection method, several modifications were made to the original procedure that involved selection of 96 tagged transposons before mutagenesis (Mei et al., 1997). The major criteria used for tag selection are based on efficiency of amplification and labeling as well as on high hybridization specificity. Each tagged transposon from the selected set was then used to generate 96 separate pools of mutants. Since the tag identity for each pool of mutants is known, the prescreening of mutant strains for the suitability of the tags is no longer required. Moreover, utilization of highly specific preselected tags allowed the hybridization analysis to be carried out by plasmid or tag DNA dot blots rather than colony blots, thus increasing the sensitivity of the assay (Chiang et al., 1999; Mei et al., 1997). In another modification of the STM procedure, screening by hybridization was replaced by PCR detection with a set of reusable tags (Lehoux et al., 1999). For this, 12 specific tags designed for optimal PCR amplification were generated and used to construct 12 libraries of mutants, which were arrayed in 96-well microtiter plates. In a defined library, each mutant had the same tag but presumably inserted at a different location in the bacterial chromosome: As an STM working scheme, one mutant from each library was picked to form 96 pools of 12 uniquely tagged mutants that were subsequently used for the in vivo screening and identification of attenuated virulence genes (Lehoux et al., 2002).

Pool Complexity   As the number of different mutant strains within the pool increases, so does the probability that virulent mutants will not be recovered in sufficient numbers to yield hybridization signals, which can lead to false identification of attenuated mutants. The pool complexity was shown to be inversely proportional to the hybridization signal intensities, therefore limiting the number of mutants that can be used in each pool (Chiang et al., 1999). As shown for *S. typhimurium*, strong and reproducible hybridization signals were obtained for pools of 96 different mutants, but not for pools of 192 (Hensel et al., 1995). The pool complexity, however, may vary for different pathogen–host systems. To obtain reproducible results in studies of *V. cholerae* colonization factors, the pool complexity was reduced to 48 strains (Chiang and Mekalanos, 1999).

Inoculum Dose and Administration   The amount of bacteria used for inoculation and routes used to introduce bacterial inocula can markedly affect the ratios of different mutants in the output pool. These events are primarily due to the selection process that occurs as the infection proceeds. Studies of *S. typhimurium* typhoid infection using an inoculum size of 10^4 cells represented by a pool of 96 different mutants produced inconsistent results (Chiang et al., 1999), whereas an inoculum of 10^5 cells yielded reproducible hybridization patterns and an attenuated virulence frequency of $\sim 4\%$ (Hensel et al., 1995). In contrast, high inoculum doses can overwhelm the animal's immune defenses, thus resulting in the growth and survival of attenuated mutants (Chiang et al., 1999). In addition, an intraperitoneal administration of 10^5 cells represented by a pool of 96 *S. typhimurium* mutants yields reproducible hybridization signals for the majority of strains recovered from the spleens of infected animals. At the same time, after oral administration of the same inoculum, only a small percentage of the mutants are found in the spleens, while the identity of the recovered strains varies from animal to animal (Chiang et al., 1999).

Duration of Infection   Another issue affecting the results of STM studies concerns the postinoculation time point at which mutant pools are recovered for hybridization screening. Short incubation times can result in insufficient attenuation where virulent strains do not have time to sufficiently outgrow the attenuated mutants. On the other hand, long incubation times can cause some virulent strains to outgrow other virulent strains in a nonspecific manner. These parameters are obviously interrelated and must be optimized empirically for each pathogen–host interaction to obtain reproducible hybridization patterns with tags recovered from at least two animals infected with the same pool of mutants (Chiang et al., 1999).

Integration of STM Approach with Microarray Technology As discussed above, an increasing number of STM studies have yielded a variety of modifications to the original method. The basic protocol was improved further by combining STM with high-density microarrays. In a recent study of *Yersinia pseudotuberculosis* murine infection model, signature tagging was utilized in conjunction with DNA microarrays (Karlyshev et al., 2001). Initial experiments utilizing libraries of randomly generated tags resulted in inconsistent hybridization signals in the input pools of *Y. pseudotuberculosis* mutants. Sequencing analysis of the tags derived from different clones revealed that variability in signal intensities was caused by deletions in the variable regions as well as different G + C content of the tags (Karlyshev et al., 2001). To overcome these problems, a set of 96 transposons was constructed using preselected tags derived from the *Saccharomyces cerevisiae* whole-genome mutagenesis project (Winzeler et al., 1999). A total of 192 sequences were selected based on similarity in melting temperatures, lack of secondary structures, and a maximum of 75% sequence identity among all tags. To improve the method's robustness and to facilitate data interpretation, each transposon was labeled with two tags using three-step PCR amplification with nested primers containing signature sequences (Karlyshev et al., 2001). To verify that hybridization signals carrying identical signature sequences were similar in strength, the tags were amplified from three different transposon molecules using biotinylated primers complementary to the conserved regions and hybridized to a high-density oligonucleotide microarray containing approximately 4,000 different 20-mer sequences. Although minor clonal variations were observed, the three initial sets yielded reproducible results of very similar hybridization intensities.

For in vivo screening experiments, a master library of *Y. pseudotuberculosis* signature-tagged strains was used to assemble the input pool of 603 transposon mutants. Upon recovery of the mutants, the tags from each pool were amplified and labeled with biotinylated primers and hybridized to oligonucleotide microarrays. Comparison of the signal intensities derived from both input and output pools identified 31 mutants that were attenuated, including a phospholipase mutant whose relative abundance was significantly reduced in the output pool. Sequence analysis of the transposon insertions identified 26 genes that were present in both *Y. pestis* and *Y. pseudotuberculosis*, while five genes were unique to *Y. pseudotuberculosis*. The putative virulence genes identified in this study included those involved in lipopolysaccharide biosynthesis, adhesion, phospholipase activity, iron assimilation, and gene regulation (Karlyshev et al., 2001).

In summary, this study identified 31 potential genes out of 603 mutant strains, a rate higher than those usually reported for the majority of STM studies. As suggested by the authors (Karlyshev et al., 2001), the increase in identification rate is primarily due to the modifications to the standard STM procedure. The combination of preselected sequences and high-density array hybridization technology offers improved performance, efficiency, and reliability by increasing the detection sensitivity and allowing quantitative analysis of data. Furthermore, utilization of double-tagged sequences to label each mutant increases the robustness of the method and data reliability. The hybridization patterns were found to be reproducible even with small amounts of inoculum or the number of colonies recovered in the output pool. As the probe sequences on the arrays are selected to have similar hybridization properties, the variability in signal intensities often observed with random sequences can be reduced, as one may select probes that are unique, allowing the analysis of highly complex environments such as mixed cultures, biofilms, and microcosms (Karlyshev et al., 2001).

8.3   TARGETED MUTAGENESIS THROUGH ALLELIC EXCHANGE

Allelic exchange through homologous recombination has been extensively utilized to introduce recombinant or mutated alleles into the genomes of both gram-negative and gram-positive prokaryotes as well as unicellular eukaryotes. Most often, gene inactivation is achieved by transforming the mutated allele into the host strain using nonreplicating or conditionally active plasmid vectors. Following homologous recombination and integration of the inactivated allele into the host genome, the mutants are identified using selectable or counterselectable markers, which confer resistance or sensitivity to a variety of drugs or chemical compounds. In the following section, we will review the principles and steps involved in gene inactivation through allelic exchange, and provide examples of large-scale functional genomic studies where this approach has been employed to determine gene function.

8.3.1   Suicide Vector Systems for Allelic Exchange

Understanding the regulation of plasmid replication and maintenance in vivo is an essential prerequisite for developing efficient methods of allele exchange. Suicide delivery systems, used as the cornerstone of allelic exchange mutagenesis, should possess several properties desirable for this type of gene inactivation. The suicide plasmid must be conditional for replication to allow selection for integration into the chromosome. This can be achieved by using a plasmid that is able to replicate autonomously only in permissive hosts or by using conditional replicons. Second, the suicide plasmid must carry a selectable marker (e.g., genes encoding antibiotic resistance) for subsequent selection procedures. Finally, the suicide vector should be transferable to a wide variety of organisms, in which transfer by conjugation is preferable in situations where other means of transfer such as transformation or electroporation are not efficient. The latter is often achieved by incorporating an origin of transfer from broad-host range plasmids such as RK2, RP4, or RP1 into the suicide vector (Guiney and Yakobson, 1983; Yakobson and Guiney, 1984; Furste et al., 1989; Parke, 1990; Jaworski and Clewell, 1995).

A number of approaches have been employed to develop delivery vehicles for allelic exchange. One of the easiest strategies for identifying a suitable suicide vector involves utilization of the host range specificity for replication. Examples include several *E. coli* plasmids of the ColE1 and p15A families that are unable to propagate in most nonenteric species (Parish and Stoker, 2000), while plasmids such as pUB110, pC194, and pHT1030, isolated from the gram-positive organisms *Bacillus subtillis* and *Staphylococcus aureus* (Keggins et al., 1978; Lofdahl et al., 1978; Arantes and Lereclus, 1991), do not replicate in *E. coli*. Similar approaches were developed using plasmids that require the presence of host-specific proteins for their replication (Hashimoto and Sekiguchi, 1976; Biek and Cohen, 1986; Leenhouts et al., 1991). It was shown that the presence of functional DNA polymerase I (PolA) and replication initiation protein DnaA is required for the replication of vectors pBR322 and pSC101 in *E. coli*, respectively (Hasunuma and Sekiguchi, 1977; Oka et al., 1980). Moreover, both plasmids become unstable in $recD^-$ strains when grown on minimal medium (Biek and Cohen, 1986). Requirement of host-specific replication factors was also employed for the construction of suicide delivery systems with temperature-sensitive origins of replication that become unstable during incubation at elevated temperatures. Strains containing temperature-sensitive PolA or RecD proteins cannot maintain ColE1-based plasmids at 42°C (Blum et al., 1989; Russell et al., 1989; Biek and Cohen, 1986). Examples of temperature-sensitive suicide vectors unable to propagate in replication-deficient strains include pSC101, pHSG415, and pWV01 plasmids, the utility of which was demonstrated in a number of gram-positive and gram-negative bacterial species (Leenhouts et al., 1996; White et al., 1999a; Wolska et al., 1999).

A robust suicide delivery system for targeted mutagenesis was developed utilizing plasmid vectors carrying the R6K origin of replication (*oriR6K*). Plasmid vectors containing the *oriR6K* can only be maintained in host strains genetically engineered to express the π protein and do not replicate in strains lacking the essential replicase-encoding *pir* gene. In *E. coli*, the π protein can be supplied in *trans* by a prophage (λ *pir*) that carries a cloned copy of the *pir* gene. Since the vast majority of gram-negative bacteria do not posses the *pir* gene, R6K *pir*-dependent suicide vectors are widely used for gene replacement mutagenesis (Kaniga et al., 1991; Skorupski and Taylor, 1996; Biswas et al., 1993; Kalogeraki and Winans, 1997; Alexeyev, 1999). In addition, these vectors contain an origin of transfer that allows conjugal transfer of the plasmids via a broad host range transfer system from appropriate donor strains to a wide range of recipient strains. Plasmid pGP704, a derivative of pBR322 that was constructed by replacing the deleted ColE1 origin of replication (*oriE1*) by a cloned fragment containing the origin of replication of plasmid R6K, is an example of such a system (Miller and Mekalanos, 1988).

Another approach used for selecting marker exchange events through homologous recombination is based on the incompatibility of certain plasmids. It was shown that superinfection with a plasmid from the same incompatibility group could be applied to displace the resident plasmid (Novick and Hoppensteadt, 1978). Although this approach is somewhat laborious and requires a series of screening steps, it proved to be useful for generating targeted gene disruptions in organisms that lack suitable selective markers or suicide vectors (Bringel et al., 1989; Fedorova and Highlander, 1997a,b).

The incorporation of counterselectable genes has been widely used for the construction of suicide vectors. Under appropriate growth conditions, a counterselectable gene promotes the death of the microorganisms harboring it (Stibitz, 1994). The most commonly used counterselectable markers include genes that confer sucrose (*sacB*), streptomycin (*rpsL*), or fusaric acid (*tetAR*) sensitivity (Dean, 1981; Maloy and Nunn, 1981; Gay et al., 1985). Transformants, which have integrated a suicide vector containing a counterselectable marker, retain a copy of the latter in the chromosome and therefore are eliminated in the presence of the counterselective compound. Counterselectable markers are often instrumental for the construction of knockout mutations or sequence modifications in microorganisms with poorly developed genetics systems. In addition, the potential of such markers has been demonstrated for other applications such as the construction of mutants, the isolation of insertion sequence elements, and the curing of endogenous plasmids (Reyrat et al., 1998).

8.3.2 Strategies Commonly Utilized for Targeted Mutagenesis by Allelic Exchange

In bacteria for which genetic systems are available, the elucidated genomic sequence allows generation of mutated alleles for subsequent knockout inactivation of the target genes. A number of strategies have been developed to transfer the modified allele into the chromosome using homologous recombination. Gene inactivation by allelic exchange can be achieved by integration of a synthetic DNA into the target region, replacement of the wild-type gene with a foreign DNA, and deletion of the target region.

Integration of Conditional Replicons by Single-cross-over Recombination: The Insertion–Duplication Method Suicide plasmids conditional for their replication can be used for insertional mutagenesis to generate defined duplications and null alleles within a target gene. For this, an internal fragment of the gene is generated either by PCR amplification or endonuclease digestion and cloned into a suicide vector of choice. The resulting construct is introduced into the host organism through conjugation, electroporation, or chemical transformation. Selection for a plasmid-encoded property, such as antibiotic resistance, results in the integration of the entire plasmid into the host chromosome via homologous recombination between the cloned internal fragment and the corresponding region of the recipient chromosome. The single cross-over will produce two truncated copies of the target gene flanking the plasmid insertion site (Miller and Mekalanos, 1988) (Fig. 8.6A).

Since single-cross-over events result in two truncated copies of the target gene, suicide vector integration by homologous recombination has been utilized to generate transcriptional and translational fusions while simultaneously disrupting the gene (Fig. 8.6B). A number of suicide plasmids have been specifically modified to accommodate this need by incorporating promoter-less reporter genes (*lacZ, phoA, galU, gfp*) directly

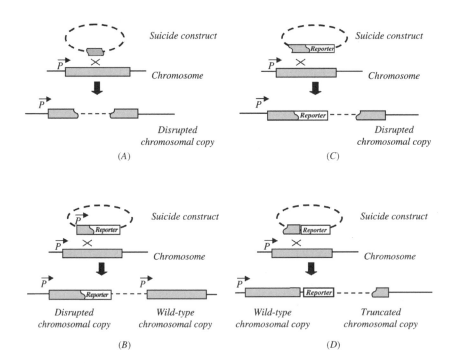

Figure 8.6 *Targeted gene inactivation and generation of transcriptional/translational fusions using suicide vector integration.* If gene disruption is the final objective, care must be taken to ensure that the resulting fragment(s) does not express functional protein.

downstream of the cloned gene fragment (Kalogeraki and Winans, 1997; Shalom et al., 2000; Lai et al., 2001). Upon vector integration into the chromosome through homologous recombination, a fusion between the 5'-end of the target gene and reporter gene is generated, followed by the inserted suicide vector carrying the second copy of the truncated target gene. In situations where the function of the target gene has to be retained, an intact 5'-end including the promoter can be cloned into the suicide vector. In this case, plasmid integration will result in a fusion between the target and reporter genes while a wild-type allele containing the native promoter sequence is downstream of the insertion site (Fig. 8.6C). Alternatively, a complete 3'-end of the target gene can be used to construct a *cis*-merodiploid strain, where a complete copy of the wild-type gene is followed by a reporter cassette and a truncated copy is located downstream of the vector sequence (Kalogeraki and Winans, 1997) (Fig. 8.6D).

Integration of suicide vectors through homologous recombination is an efficient method for site-specific mutagenesis in a variety of microorganisms. This approach is relatively easy to perform and compares favorably to most mutagenesis procedures. It is more direct in comparison to other techniques, since a single cross-over generates a knockout mutation of the wild-type gene. Integrative vectors can be used for random gene inactivation using total cloned fragments derived from total chromosomal DNA digests (Hoch, 1991). Utilization of suicide vectors for gene disruption, however, should be carefully reviewed based on the experimental goals, since this approach has several significant drawbacks. First, the insertion polarity, as well as the presence of any internal promoters carried by the vector, can substantially alter the expression levels of genes adjacent to the integration site. Second, formation of two mutant copies of the target gene can affect insertion stability and cause plasmid excision through homologous recombination. And finally, the physiological stress caused by the selective pressure (antibiotic, heavy metal, etc.) can affect cell fitness and result in phenotypes unrelated to the mutation.

Gene Replacement by Double-cross-over Recombination: The Deletion– Substitution Method To overcome the instability issues associated with gene disruption using integrative vectors, the wild-type allele carried on the suicide plasmid can be mutated by transposon insertion or by partial replacement with a selectable marker (Newland et al., 1985; Goldberg and Mekalanos, 1986) (Fig. 8.7). Regions flanking the mutated gene are retained to promote homologous recombination into the target genome. The resulting mutation in the target gene is usually constructed in a permissive host capable of maintaining the suicide vector and the construct carrying the disrupted allele is then transferred into the host organism. Putative mutants are isolated by direct selection and screening for double-cross-over recombinants that retain the selectable marker used to replace the wild-type allele but lose the vector-encoded drug resistance.

Several other variations of the deletion–replacement method have been developed using plasmids of the IncP incompatibility groups (Mekalanos et al., 1983; Kaper et al., 1984; Sar et al., 1990). Briefly, the wild-type allele is mutated using a cassette encoding drug resistance to the antibiotic kanamycin (Km). The mutated locus is subcloned into a mobilizable RP4 derivative, such as pLAFRII, which confers tetracycline resistance, and subsequently, is transferred by conjugation into the wild-type *Vibrio* recipient. Exconjugates are selected by screening for antibiotic resistance encoded by the vector (Tc), the drug resistance marker in the mutated allele (Km), and the recipient strain. Isolation of recombinants resulting from the exchange of the mutant for the wild-type allele through homologous recombination requires the elimination of the delivery plasmid

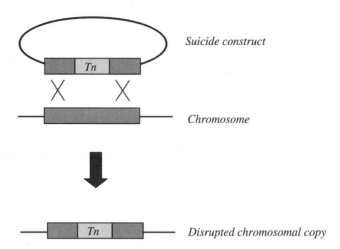

Figure 8.7 *Targeted gene disruption by double-cross-over.* The wild-type allele is replaced by a selectable marker or transposon insertion.

from the recipient. This is accomplished by introduction of a second incompatible "kick-out" plasmid, which also belongs to the incP group but encodes a different drug resistance marker, for example, gentamycin (Gm). Since two plasmids of the same incompatibility group cannot coexist in the same cell, selection of the 'kick-out' plasmid will result in the loss of the delivery vector. However, if both Gm and Km markers are selected, exconjugants arise that have lost the pLAFRII, and the mutated allele has integrated into the chromosome via homologous recombination.

A simple and highly effective method of allele replacement was developed for organisms for which high transformation and recombination rates can be achieved (Baudin et al., 1993; Datsenko and Wanner, 2000; Murphy et al., 2000; Yu et al., 2000). This technique permits gene replacement without prior cloning of the target gene and involves transformation of linear DNA substrates into the organism of interest. Mutated alleles are generated by PCR amplification using primers with 30- to 50-nucleotide extensions that are identical to regions adjacent to the target site and template vectors that carry an antibiotic resistance gene. When introduced into the host organism, the resulting PCR products can replace the targeted allele by homologous recombination. Following transformation, putative mutants are selected for marker-encoded drug resistance and screened using PCR amplification of the disrupted region. The utility of this approach for targeted gene replacement was first demonstrated in *Sacharomyces cerevisiae* by exploiting high rates of meiotic recombination in this naturally competent organism (Baudin et al., 1993). In contrast to yeast and a few naturally competent bacterial species, the transformation efficiency of most bacteria by linear DNA is extremely low due to the presence of intracellular exonucleases (Lorenz and Wackernagel, 1994). To circumvent this problem, recombination-proficient strains of *E. coli* were constructed in which the *recBCD* genes encoding exonuclease V have been replaced by the recombination functions of the bacteriophage λ (Murphy, 1998; Yu et al., 2000). Electroporation of the resulting *recBCD* :: λ strains with linear DNA gave transformation rates of 10^{-2}, with gene replacement frequencies ranging from 10^{-4} to 10^{-5} recombinants per survivor (Murphy et al., 2000). The gene inactivation approaches described here allow for generating stable, easily selectable mutations in a variety of microorganisms. Utilization of PCR-based gene disruption offers an advantage over other allele exchange techniques in which the gene of interest does not have to be cloned. A main disadvantage involves the decreased efficiency of double-cross-over recombination, requiring the need to screen greater numbers of mutants. In addition, polar effects are possible if there is a transcriptional terminator present in the cassette used for disruption.

Construction of Unmarked Deletions through Use of Counterselectable Markers

The construction of clean and unmarked mutations in bacteria, where a gene is replaced by an in vitro-modified allele, is a fundamental approach to understanding gene function at a molecular level as well as to defining structure–function relationships (Reyrat et al., 1998).

Because double-cross-over events that incorporate a gene from a plasmid into the chromosome are rare, it is not feasible to simply screen for such events if the cloned gene cannot be directly selected. In such cases, a two-step procedure is used instead (Fig. 8.8). First, the entire plasmid carrying a counterselectable marker is integrated into the chromosome by a single-cross-over event between the homologous genes, producing a chromosomal duplication. Second, the chromosomal duplication is segregated by homologous recombination between the flanking direct repeats, ultimately leaving either the wild-type copy or the mutant copy of the gene on the chromosome. Because the direct

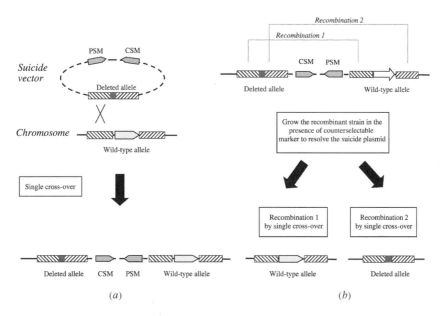

Figure 8.8 *Positive selection of allelic exchange mutants in a two-step selection strategy, using a counterselectable marker.* (a) During the first step, an intermolecular recombination leads to the integration of the suicide vector carrying the mutated (deleted) allele. The resulting merodiploid strains are isolated by positive selection (resistance function encoded by the positive selection marker, PSM). (b) The second step involves plasmid resolution through intermolecular recombination. Strains that have lost the vector during a second cross-over are selected by plating on counterselective medium (counterselection marker, CSM). Since two different recombination events can occur, the resolution of the suicide vector can lead to either a wild-type or mutant phenotype.

repeats are often short, the desired recombination event may be very rare. The marker-resistant phenotype of the plasmid provides direct selection for integration of the plasmid into the chromosome. Once the plasmid is integrated, it is possible to simply screen for segregants by testing for the loss of the plasmid resistance marker. However, because the segregants are rare, often a way of selecting for the loss of the integrated plasmid from the chromosome is needed. One way of selecting against the integrated plasmid is to utilize the *Bacillus subtilis sacB* gene, which confers sensitivity to sucrose (Gay et al., 1983). Expression of the *sacB* gene is toxic for gram-negative bacteria when grown in the presence of sucrose, providing a direct selection for loss of the plasmid. The resulting sucrose-resistant colonies are screened for the simultaneous loss of antibiotic resistance to ensure that the sucrose-resistant phenotype is due to loss of the integrated plasmid. If one of the two copies of the duplicated region has a mutation, a certain percentage of segregants will have lost the wild-type sequence and left the mutation in the chromosome. One factor that determines the percentage of segregants which retain the mutant allele is the position of the mutation within the region of homology. For example, if the deletion mutation is perfectly centered, about 50% of the segregants would be expected to retain the mutation in the chromosome.

Many modifications on this basic theme have been used in a variety of bacteria (Kaniga et al., 1991; Skorupski and Taylor, 1996; Fur and Voordouw, 1997). The ColE1 origin does not support replication in a variety of bacterial species and thus acts as a suicide vector when delivered from *E. coli* into such strains. The inability of ColE1 plasmids to replicate in nonenteric bacterial species has been used for allelic exchange in *Pseudomonas aeruginosa* (Schweizer, 1992). Another approach takes advantage of the recessive nature of streptomycin resistance conferred by chromosomal *rpsL* mutations. Plasmid pRTP1 is a ColE1 replicon carrying the wild-type *rpsL* gene in addition to *ampR*. Integration of the plasmid carrying the wild-type *rpsL* gene into a streptomycin-resistant recipient results in a merodiploid strain that is streptomycin-sensitive (Roberts et al., 1991). Subsequent selection for streptomycin resistance results in colonies that have lost the vector (like the segregation of *sacB* vectors described above).

Other Genome Engineering Approaches Utilizing Homologous Recombination

To extend the applicability of the homologous recombination approach for functional genomic studies, a number of high-efficiency site-specific recombination techniques have been developed for mutagenizing and modifying bacterial genomes. As discussed above, high-throughput mutagenesis methods utilizing recombination properties of the bacteriophage λ red system have been developed to permit efficient integration of linear DNA fragments into the bacterial chromosome (Datsenko and Wanner, 2000; Murphy et al., 2000; Yu et al., 2000). Further studies of the λ-mediated recombination demonstrated that in comparison to dsDNA integration, which requires the expression of phage-encoded Exo, Beta, and Gam proteins, the recombination between synthetic single-stranded oligonucleotides and the *E. coli* chromosomal DNA is solely dependent on the presence of the Beta protein (Ellis et al., 2002). Comparison of recombination levels between several sets of complementary oligonucleotides indicated that the strand preference correlates with the direction of DNA replication, where the ssDNA molecules that recombine more readily correspond in sequence to the lagging strand. It was proposed that the bacteriophage λ Beta protein promotes the incorporation of ssDNA molecules into the host chromosome by binding and annealing the oligonucleotide near the replication fork to generate a recombinant strand. Initial experiments using the *galK$_{tyr145am}$* mutant of *E. coli* suggested that Beta-mediated recombination can be used to modify genomic DNA with efficiencies of up to 6% recombinants among the transformed cells using 30- to 70-nucleotide-long oligonucleotides (Ellis et al., 2002). In addition to point mutations, recombination using ssDNA was also utilized to create deletions of a 3.3.-kb chloramphenicol resistance marker inserted at the *galK* amber site of the *E. coli galK$_{tyr145am}$* mutant. Comparison of recombination frequencies in these two experiments suggested that ssDNA recombination can create deletions as efficiently as single base changes. The ability to delete individual genes or chromosome segments makes the Beta-mediated oligonucleotide recombination extremely useful for in vivo chromosome modification and systematic gene deletion analysis of the *E. coli* genome. Since this type of recombination requires the function of a single protein, appropriate shuttle vectors containing the *bet* gene could be used to adapt the ssDNA mutagenesis method to a wide range of organisms (Ellis et al., 2002).

A different deletion approach combining the ability of transposons to introduce regions of portable homology with the Cre/*loxP* excision system has been developed for determining essential genes and minimizing the genome of *E. coli* (Yu et al., 2002a). In this method, two modified Tn*5* transposons carrying a *loxP* recombination site and encoding kanamycin- or chloramphenicol-resistance markers were used to generate pools of random mutants. Using phage P1 transduction, two selected mutations from each pool were brought in parallel, creating a single *Km^r-Cm^r* strain that contains two *loxP* sites in tandem. Subsequent expression of the Cre recombinase resulted in a deletion of the chromosomal region located between the two *loxP* sites. By using the described method, several deletion mutants missing large portions of the chromosome (59–117 kb) were generated. Some of the individual mutations were subsequently combined together resulting in a single *cumulative-deletion strain* that lacked 287 open reading frames spanning the total of 313 kb, but nevertheless displayed the wild-type phenotype under standard laboratory conditions. As suggested by Yu and coworkers (2002a), this combinatorial approach carries a number of advantages over standard allele exchange techniques as it is does not require the construction of targeting vectors for each deletion experiment and can be used for the construction of "minimized" bacterial genomes consisting only of genes essential for an organism's survival.

8.3.3 Application of Allele Exchange Approach in Functional Genomic Studies for Sequenced Microorganisms

Characterization of Unknown Genes in* E. coli *Using In-frame Precise Deletions Random mutagenesis used to identify gene function on a whole-genome scale is relatively rapid, but the subsequent matching of phenotypes to genes is slower. In addition, because of the operon structure in bacteria, traditional methods such as insertional mutagenesis run the risk of introducing polar effects on downstream genes or creating secondary mutations elsewhere in the genome. These limitations can be overcome by deleting each gene in the genome in a directed fashion through a targeted allelic exchange approach. A system of chromosomal mutagenesis involving PCR-based in-frame deletion was performed to identify the function of several putatively essential genes in *E. coli* (Link et al., 1997). The resulting PCR products carrying the deleted target were placed in the *E. coli* chromosome by using a gene replacement vector that contains a temperature-sensitive origin of replication and markers for positive and negative selection for chromosomal integration and excision. Two poorly understood genes, *hdeA* and *yjbJ*, encoding highly abundant proteins were selected as targets for this approach. When the system was used to replace chromosomal *hdeA* with insertional alleles, vastly different results that were dependent on the exact nature of the mutations were observed: Both essential and nonessential phenotypes were obtained (Link et al., 1997). In contrast, by using PCR-generated deletions, *hdeA* and *yjbJ* genes proved to be nonessential in both rich and glucose-minimal media. In competing experiments using isogenic strains, the strain with the insertional allele of *yjbJ* showed growth rates different from those of the strain with the deletion allele of *yjbJ*. These results illustrate that in-frame, unmarked deletions are among the most reliable types of mutations available for wild-type *E. coli*. Because these strains are isogenic with the exception of their deleted ORFs, they may be used in competition with one another to reveal phenotypes not apparent when cultured singly (Link et al., 1997).

Genome-wide Phenotypic Analysis of S. cerevisiae *Mutants Using Molecular Bar-coding Strategy*
A quantitative and highly parallel approach to analyzing gene replacement mutants was developed for the large-scale mutational analysis of the yeast genome, for which individual deletions or knockouts of each gene were constructed using a PCR-based strategy (Shoemaker et al., 1996; Winzeler et al., 1999). For each gene, a deletion cassette that contains a kanamycin resistance gene, two *molecular barcodes*, and yeast sequences homologous to the upstream and downstream flanking sequences of the target gene is amplified by PCR. Upon transformation into a diploid yeast strain, the deletion cassette replaces the coding sequences of one of the two chromosomal copies. The diploid strains are then sporulated, and haploid segregants, if viable, are recovered. A salient feature of the deletion cassette is the molecular barcodes. The molecular barcodes constitute 20-bp sequences that are unique to each deletion and allow the identification of each deletion strain within a pool of many strains (Shoemaker et al., 1996). The barcode can be amplified by PCR using common primer sequences that flank the barcode. A pool of yeast mutant strains thus can be analyzed in a single experiment by amplifying the barcodes and probing a DNA microarray containing sequences complementary to the barcodes. As an example of the utility of this approach, a pool of 558 homozygous diploid strains was grown in rich and minimal media (Shoemaker et al., 1996). Samples from the culture were taken at multiple time points, and the results from the microarrays were used to monitor the growth rate of each mutant in the population. After 60 cell doublings in rich medium or minimal medium, 15% of the strains exhibited some growth defects, and 10% of the strains severe growth defects. The advantage of this procedure is that a comprehensive collection of null mutants can be screened for a specific phenotype. However, there are several limitations. First, mutations in essential genes will not be represented in these haploid strains. This problem may be circumvented in some cases by the use of heterozygous mutant strains (see below). Second, only annotated ORFs are deleted; uncharacterized ORFs that are smaller than 300 bp (the limit used by the sequence annotators) are not represented in the collection (Mewes et al., 1997). Third, ~8% of the yeast deletion strains are aneuploid for a given chromosome region such that an extra copy of the gene may be retained by the cell (Hughes et al., 2000b). Finally, other mutations may exist as well; these secondary mutations probably exist in all mutant collections. Nevertheless, since this collection is available as either homozygous or heterozygous diploid strains and as haploid strains of both mating types, it is a great resource. For example, the homozygous diploid strain collection can be screened for diploid-specific processes such as sporulation or bipolar budding pattern (Vidan and Snyder, 2001). In the

past, these screens have proven difficult using classical approaches, since both copies of the gene need to be deleted. However, the haploid strain collection can be screened directly or used in synthetic lethal screens by deleting or mutating the gene of interest in all 6,000 strains. Despite the astronomical numbers, the transformations can be done quickly using a 96-well plate format (Ross-Macdonald et al., 1999).

8.4 GENE SILENCING USING ANTISENSE mRNA MOLECULES

Antisense technology is an effective approach to inhibiting the expression of specific genes for the validation of their putative functions. Antisense RNA (asRNA) silencing has been widely utilized in eukaryotic cells to alter gene expression through injection of synthetic oligonucleotides complementary to mRNA or by synthesis of antisense RNA from DNA cloned in an antisense orientation (Agrawal et al., 1997; Taylor and Dean, 1999; Crooke, 2000). The antisense RNA molecules have also shown great promise as a novel class of therapeutic agents in oncology, immunology, neurology, and virology (for a review, see Varga et al., 1999). Recently, the antisense technology has been adapted for high-throughput analysis of gene function and utilized to alter gene expression on a genome-wide scale in *Streptococcus aureus* and *Candida albicans* (Ji et al., 2001; DeBacker et al., 2001). Although not large in number, antisense studies hold great promise for the development of novel functional genomic tools. In the following section, we will describe the asRNA silencing methods, as well as focus on the utility of this technology for whole-genome analysis.

8.4.1 Antisense RNA regulation in vivo

Antisense RNAs are small, highly structured single-stranded molecules that act through sequence complementarity to inhibit target RNA (sense RNA) function. Naturally occurring antisense RNAs are between 35 and 150 nucleotides long and comprise between one and four stem loops (Brantl, 2002). These stem structures are important for metabolic stability and are often interrupted by bulges to prevent double-stranded RNase degradation as well as to facilitate melting upon binding to sense RNA (Hjalt and Wagner, 1995). Prior to 1981, when the first asRNA species were discovered in prokaryotes (Stougaard et al., 1981; Tomizawa et al., 1981), RNA had been exclusively recognized for its roles as messenger (mRNA) and as part of the protein translation apparatus (rRNA, tRNA). However, it quickly became apparent that in addition to converting DNA-encoded information into proteins, RNA molecules are involved in other cellular processes. An important aspect of RNA function in bacteria is antisense control, which plays a key role in the regulation of plasmid replication and maintenance, transposition, phage infection, transcriptional attenuation, and translation inhibition (Persson et al., 1990; del Solar et al., 1995; Le Chatelier et al., 1996; Jerome et al., 1999; Delihas and Forst, 2001; Masse and Gottesman, 2002; for reviews, see Brantl, 2002; Wagner et al., 2002a). The majority of asRNA-regulated systems described so far are found in prokaryotic species, whereas only a few have been shown to exist in eukaryotes. In prokaryotes, most asRNA control systems are associated with plasmids, phages, transposons, while fewer examples are identified for chromosomally encoded systems (Altuvia et al., 1997; Lease and Belfort, 2000; Wagner et al., 2002a).

Although mechanisms of antisense regulation are still not completely clear, it was postulated that asRNA molecules function by hybridizing with complementary mRNA transcripts. For the ability of asRNA to down-regulate gene expression, several mechanisms of action were proposed, including (1) translation blockage by antisense hybridization to target mRNAs; (2) translation initiation inhibition by occlusion of the ribosome binding site; (3) premature termination of mRNA transcription due to antisense binding to the genomic DNA template; (4) stimulation of rapid mRNA degradation by duplex-specific RNases; and (5) reduction of enzymatic activity by antisense binding to the target protein (Simons, 1988; Nordstrom and Wagner, 1994; Wagner and Simons, 1994; Wagner et al., 2002a). Based on these concepts of the asRNA regulatory process, a variety of strategies have been employed to artificially down-regulate levels of gene expression in bacteria and yeast (Coleman et al., 1984; Ellison et al., 1985; Engdahl et al., 1997; Kernodle et al., 1997; Olsson et al., 1997; Sturino and Klaenhammer, 2002). Antisense silencing was also utilized for metabolic engineering in *Clostridium acetobutylicum* (Desai and Papoutsakis, 1999) and *Desulfovibrio vulgaris* (van den Berg et al., 1991), and the results suggested that asRNA can be employed to affect primary metabolism in prokaryotes.

8.4.2 Antisense Approach to Large-scale Functional Genomic Studies

Although asRNA-mediated gene regulation has been studied in bacteria for nearly two decades, the antisense silencing approaches have not been routinely used to inhibit gene expression in bacteria. With the advent of microbial genomics, regulated antisense systems offer a comprehensive genomic approach to the identification and characterization of gene functions critical for bacterial growth and have been used successfully in several pathogenic microorganisms.

Genome-scale Antisense Silencing in S. aureus *Using a Random Antisense RNA Library*   The antisense RNA approach has been pioneered in *S. aureus*. Initially, an isogenic α-toxin (*hla*) staphylococcal strain was constructed by cloning a 600-bp fragment of the *hla* gene into an *E. coli–S. aureus* shuttle vector in an antisense orientation (Kernodle et al., 1997). The resulting isogenic strain produced 16-fold less α-toxin than the wild-type and displayed significant attenuation of lethal activity in a murine infection model, thus demonstrating the utility of antisense RNA technology for creating novel live-attenuated strains of bacteria. Further improvements to this technique have been made by incorporating the tetracycline (*tet*) transcription system to regulate asRNA expression (Ji et al., 1999; Yin and Ji, 2002). Earlier studies have shown that the *tet* repressor can be successfully utilized for the regulation of gene expression in a number of prokaryotic and eukaryotic species (Gossen and Bujard, 1992; Lutz and Bujard, 1997; Nagahashi et al., 1997; Stieger et al., 1999; Baron and Bujard, 2000; Wu et al., 2000). In constructing the *tet* regulatory system for *S. aureus*, the inducible *xyl/tet* promoter–operator fusion was inserted into an *E. coli–S. aureus* shuttle vector to generate plasmid pYJ335 (Ji et al., 1999). A 621-bp antisense fragment containing the 5'-region of the *hla* gene was cloned downstream of the pYJ335 *tet* promoter. Upon induction with anhydrotetracyclin, which is a less potent antibacterial analogue of tetracycline with excellent inducer properties, the resulting construct exhibited a 50- to 100-fold dose-dependent level of induction of *hla* asRNA, while the expression of the chromosomally derived *hla* gene was down-regulated by approximately 14-fold. Similarly, induction of the *hla* asRNA dramatically reduced α-toxin expression in two murine infection models, indicating that the constructed *tet* regulatory antisense system can effectively inhibit *S. aureus* gene expression both in vivo and in vitro (Ji et al., 1999).

Using the developed approach, several genome-wide studies have been conducted to identify a set of staphylococcal genes critical for growth on rich nutrient medium (Ji et al., 2001; Forsyth et al., 2002). In particular, a random genomic library containing 200- to 800-bp fragments was obtained by cloning sheared *S. aureus* chromosomal DNA into pYJ335 downstream of the *xyl/tet* promoter (Ji et al., 2001; Fig. 8.9). The resulting library was transformed into *S. aureus* and screened by replica plating colonies in the presence or absence of anhydrotetracyclin (ATc). Colonies that displayed growth defects in the presence of ATc but grew in the absence of the antibiotic were selected for further analysis. Out of 20,000 transformants screened in this manner, 600 colonies (3%) were identified as growth-defective or lethal after induction with ATc (Ji et al., 2001). These phenotypes were confirmed by restreaking for single colonies, purifying the plasmid DNA, transforming the plasmid DNA back into a *S. aureus* host, and retesting growth with or without ATc.

To further identify the specific cloned DNA fragments that resulted in loss of cell viability, the DNA inserts were reamplified by PCR with a set of vector-specific primers

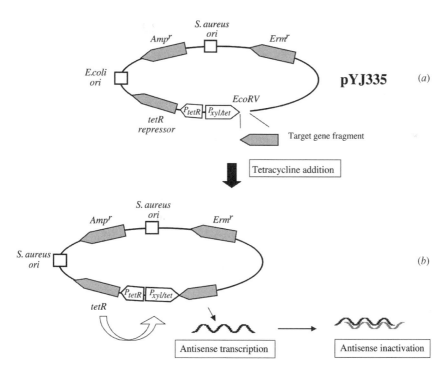

Figure 8.9 *Antisense mRNA inhibition using the Tc-inducible shuttle vector pYJ335 (Ji et al., 2001). (a) The origins of replication from pUC19 and pE194, respectively, allowing plasmid replication in E. coli and S. aureus hosts. A unique EcoRV site positioned downstream from the xyl/tet promoter is used for antisense cloning of the target gene fragment. (b) Upon addition of tetracycline or anhydrotetracycline, transcription from the xyl/tet promoter yields an antisense mRNA of the target gene. [Adapted with permission from D. Yin and Y. Ji, Genomic analysis using conditional phenotypes generated by antisense RNA, Current Opinion in Microbiology, vol. 5, 330–333. Copyright (2002) Elsevier Science.]*

and sequenced. Sequence analysis of the DNA inserts indicated that about 30% of the selected clones contained small, antisense-oriented fragments derived from different single ORF regions. The remaining clones contained sense-oriented ORF and non-ORF fragments, a mixture of sense- and antisense-oriented chimeric fragments, and sense and antisense fragments spanning multiple ORFs. The overall efficiency of this random insertion method to generate conditional growth-defective events during induction of antisense RNA was estimated at approximately 1%. This number was several orders of magnitude higher than that reported for other conditional growth selection procedures (Ji et al., 2001).

With this antisense silencing method, more than 150 critical staphylococcal genes were identified that exhibited lethal or growth-defective effects on antisense induction. About 40% of these genes were defined as homologs of known essential bacterial genes, and ~ 30% of the genes appeared to be homologs of bacterial genes with proposed functions. The remaining 30% of the genes represented apparently critical genes of unknown function. Within this group of critical genes, two distinct phenotypic classes of isogenic *S. aureus* strains were identified. The first class consisted of mutants that displayed a lethal phenotype after asRNA induction; the strains in the second class exhibited a slow- or reduced-growth phenotype, but did eventually give rise to definitive small colonies (Ji et al., 2001). As suggested by Ji and colleagues (2001), the latter may be due to suboptimal antisense inhibition resulting in residual expression of the essential gene. In addition, interpretation of essentiality on the basis of obtaining lethal phenotypes from those genes configured in polycistronic operons should be approached with caution, because the potential effects of defined sub-ORF antisense RNA induction on the expression of genes upstream or downstream of the mRNA region to which the antisense is made are unknown. In order to rule out the antisense polarity effect, direct monitoring of all gene products made from the polycistronic message is required (Ji et al., 2001).

Gene Suppression in Candida albicans *Using a Combination of Antisense Silencing and Promoter Interference* Another example of a genome-scale antisense inactivation approach was demonstrated by identifying genes critical for the growth of *C. albicans*, the major pathogen that causes human fungal infections (DeBacker et al., 2001). For this study, a gene suppression system was designed that uses a combination of antisense RNA silencing and promoter interference strategies (Fig. 8.10). The latter can occur by physical collision of RNA polymerase complexes elongating on opposite strands (Del Rosario et al., 1996), where the production of antisense RNA becomes a byproduct of transcription-level interaction (DeBacker et al., 2001). Genomic libraries of *C. albicans* containing antisense or complementary DNA fragments were cloned into two integrative vectors, pGAL1PNiST-1 and pGAL1PSiST-1, under the control of the *GAL1* promoter. Utilization of the *GAL1* promoter, which is induced in the presence of galactose and repressed by the addition of glucose, allowed dual control of the transcription of cloned fragments. The resulting antisense cDNA library was transformed back to *C. albicans*, and individual library clones were subsequently integrated into the genome by homologous recombination. Since the genome of *C. albicans* contains wild-type copies of *GAL1*, the recombination occurred either at the *GAL1* sites or the target gene loci. Depending on the integration site, the introduction of the constructed library into *C. albicans* and subsequent induction of the *GAL1* promoter led to gene inactivation either by the transcription of asRNAs alone or by a combination of antisense silencing and promoter interference (DeBacker et al., 2001). Out of 2,000 individual transformants screened for a promoter-induced diminished-growth phenotype, 198 displayed growth

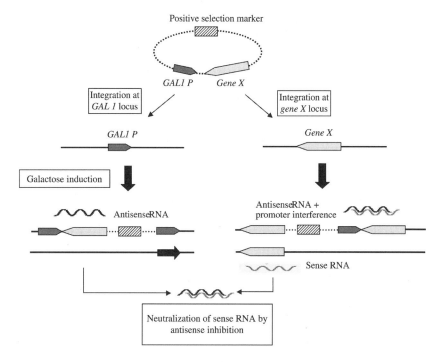

Figure 8.10 *Integration of antisense library plasmids into C. albicans genome (DeBacker et al., 2001).* The gene X antisense mRNA is produced from GAL1 promoter upon induction with galactose. Plasmid integration through homologous recombination can occur either at the GAL1 or the gene X genomic loci. In both cases, antisense RNA of gene X will be produced. Neutralization of a majority of sense RNA molecules transcribed from gene X occurs by hybridization with antisense RNA and leads to the translation of a reduced number of sense RNA molecules. Promoter interference may also occur when plasmids integrate at the gene X locus by "collision" of transcription from both the endogenous gene X and the GAL1 promoters, thus potentially contributing to a decrease in the total number of full-length sense RNAs. [Adapted with permission from D. Sanglard, Integrated antifungal drug discovery in *Candida albicans, Nature Biotechnology,* vol. 19, 212–213. Copyright (2001) Nature Publishing Group.]

defects in the presence of galactose and were selected for further analysis. Each DNA insert causing a galactose-dependent growth defect was amplified by PCR and subsequently sequenced. Sequence comparison of the identified PCR products using public and proprietary sequence databases resulted in the identification of 86 genes, half of which were of unknown function. Moreover, 33 of these genes did not have any homologs in the genome databases for other organisms. Although these findings are somewhat surprising, they have far-reaching implications for the drug-target discovery process in pathogenic organisms. As noted by Sanglard (2001), the identification of a considerable number of unknown genes essential for the growth of *C. albicans* (38% of the total number of identified genes) suggests that the genome of *S. cerevisiae* may not be the optimal source of basic information for the selection of candidate essential targets, as several pathogen-specific genes appear to be missing. On the other hand, several essential genes may not have been detected in this work because of the limited genome coverage of the antisense libraries as well as the absence of integrations at specific sites.

Advantages of Antisense Strategy for Functional Genomic Studies The antisense silencing strategy offers a comprehensive genome-wide approach to readily identifying genes essential for growth and viability, where each identified gene is maintained as one of a collection of conditional growth-defective/lethal isogenic strains (Ji et al., 2001). Antisense RNA inhibition can be utilized to conduct functional genomic studies in organisms for which existing en masse techniques of gene disruption are not applicable. It is particularly useful for studying gene function in diploid organisms as well as in haploid species where the target gene is present in multiple copies in the genome.

Utilization of regulated promoters to synthesize asRNA molecules can avoid the pitfalls of gene essentiality. In addition, regulated promoters allow for quantitative evaluation of the relevance of each gene product to cell viability, which can be obtained by selective titration of any gene product (Ji et al., 2001). More important, the titration of essential genes can also be achieved in vivo by the murine model of hematogenous pyelonephritis. This presents the prospect of examining the importance of a gene product when it is switched off after infection has been established (Ji et al., 2001; Yin and Ji, 2002).

The asRNA silencing method also provides geneticists with an approach to control, on a genome-wide scale, the expression of genes and to assess their essential nature or their role in specific phenotypes, such as those associated with dimorphism (i.e., the ability to change cell morphology) or drug resistance (DeBacker et al., 2001). The strategy is thus useful for identifying potential essential genes in unsequenced organisms since the construction of random libraries does not require prior sequence information. However, once an organism's genome has been completely deciphered, further selection of target genes can be made on the basis of their potential cellular role, the location of their translated products (e.g., localization in the membrane or cytosol), or their lack of equivalence in mammalian or human genomes (Sanglard, 2001). The availability of whole-genome sequence information can also significantly increase the efficiency of the method by specifically generating antisense fragments to each predicted ORF. It should be possible to test a genome-wide library of antisense-RNA-expressing strains at a subinhibitory concentration of antibiotic to identify strains showing increased sensitivity, thereby elucidating their targets (Ji et al., 2002).

8.5   SUMMARY

The large number of genome sequencing projects has led to the determination of the genetic code for many different species, and functional characterization of these sequences is underway. To facilitate the genomic analysis, the development of cognate mutagenesis methods is of crucial importance, where gene function can be elucidated methodically under defined conditions. In this chapter, we focused on the application of gene inactivation methods applicable for genome-scale analysis that include transposon mutagenesis, gene disruption through allelic exchange, and gene silencing using antisense RNA.

Traditionally, the function of a given ORF is defined empirically by testing for survival of the respective knock-out strain under a particular condition and identifying the phenotype of the resulting mutant. In that respect, the availability of entire genomic sequences makes reverse genetic analysis a logical approach. The construction of defined mutants by allelic replacement has proven to be a powerful method for determining gene function in numerous prokaryotes. On the other hand, the low throughput of the allele replacement techniques makes them very difficult and impractical to perform on a genome-wide scale. Moreover, it makes the identification of genes essential for viability nearly impossible. Comparative genomic studies suggest that approximately half of the *E. coli* ORFs might be needed for cell growth, while the rest have evolved as accessory genes or have been laterally transferred from other species (Lawrence and Ochman, 1998). To efficiently extract novel biological information pertinent to an organism's survival, high-throughput mutagenesis techniques are required. Specific methods including mRNA expression inhibition as well as saturating and signature-tagged mutagenesis were designed to identify genes essential for various cellular processes, such as energy metabolism, amino acid and protein biosynthesis, DNA repair, stress response, cell division, motility, and drug resistance. In addition, the genome-scale tagging of mutants has great potential impact for in vivo and population dynamics studies. The tagging and analysis of prokaryotes could not only be used for studying pathogenicity genes, but also to measure the survivability of mutants in other complex environments ranging from biofilms to anaerobic sediments and soils.

FURTHER READING

Akerley, B. J., E. J. Rubin, A. Camilli, and D. J. Lampe, et al. 1998. Systematic identification of essential genes by *in vitro* mariner mutagenesis. *Proc. Natl. Acad. Sci. USA* 95:8927–8932.

Chiang, S. L., J. J. Mekalanos, and D. W. Holden. 1999. *In vivo* genetic analysis of bacterial virulence. *Ann. Rev. Microbiol.* 53:129–154.

Datsenko, K. A. and B. L. Wanner. 2000. One-step inactivation of chromosomal genes in *Escherichia coli* K-12 using PCR products. *Proc. Natl. Acad. Sci. USA* 97:6640–6645.

De Backer, M. D., B. Nelissen, M. Logghe, and J. Viaene, et al. 2001. An antisense-based functional genomics approach for identification of genes critical for growth of *Candida albicans*. *Nat. Biotechnol.* 19:235–241.

Ellis, H. M., D. Yu, T. DiTizio, and D. L. Court. 2002. High efficiency mutagenesis, repair, and engineering of chromosomal DNA using single-stranded oligonucleotides. *Proc. Natl. Acad. Sci. USA* 98:6742–6746.

Forsyth, R. A., R. J. Haselbeck, K. L. Ohlsen, and R. T. Yamamoto, et al. 2002. A genome-wide strategy for the identification of essential genes in *Staphylococcus aureus*. *Mol. Microbiol.* 43:1387–1400.

Hensel, M., J. E. Shea, C. Gleeson, and M. D. Jones, et al. 1995. Simultaneous identification of bacterial virulence genes by negative selection. *Science* 269:400–403.

Hutchison, C. A., S. N. Peterson, S. R. Gill, and R. T. Cline, et al. 1999. Global transposon mutagenesis and a minimal *Mycoplasma genome*. *Science* 286:2165–2169.

Ji, Y., B. Zhang, S. F. Van Horn, and P. Warren, et al. 2001. Identification of critical staphylococcal genes using conditional phenotypes generated by antisense RNA. *Science* 293:2266–2269.

Judson, N. and J. J. Mekalanos. 2000a. TnAraOut, a transposon-based approach to identify and characterize essential bacterial genes. *Nat. Biotechnol.* 18:740–745.

Judson, N. and J. J. Mekalanos. 2000b. Transposon-based approaches to identify essential bacterial genes. *Trends Microbiol.* 8:521–526.

Link, A. J., D. Phillips, and G. M. Church. 1997. Methods for generating precise deletions and insertions in the genome of wild-type *Escherichia coli*: application to open reading frame characterization. *J. Bacteriol.* 179:6228–6237.

Reyrat, J. M., V. Pelicic, B. Gicquel, and R. Rappuoli. 1998. Counterselectable markers: untapped tools for bacterial genetics and pathogenesis. *Infect. Immun.* 66:4011–4017.

Ross-Macdonald, P., P. S. Coelho, T. Roemer, and S. Agarwal, et al. 1999. Large-scale analysis of the yeast genome by transposon tagging and gene disruption. *Nature* 402:413–418.

Shoemaker, D. D., D. A. Lashkari, D. Morris, and M. Mittmann, et al. 1996. Quantitative phenotypic analysis of yeast deletion mutants using a highly parallel molecular bar-coding strategy. *Nat. Genet.* 14:450–456.

Winzeler, E. A., D. D. Shoemaker, A. Astromoff, and H. Liang, et al. 1999. Functional characterization of the *S. cerevisiae* genome by gene deletion and parallel analysis. *Science* 285:901–906.

Yin, D. and Y. Ji. 2002. Genomic analysis using conditional phenotypes generated by antisense RNA. *Curr. Opin. Microbiol.* 5:330–333.

9

Mass Spectrometry

Nathan VerBerkmoes, Joshua Sharp, and Robert Hettich

9.1 INTRODUCTION

The recent technological revolution in instrumentation and computational power has enabled the development of sophisticated analytical technologies that now permit a detailed glimpse into the molecular machinery of the living cell. These important tools include electron microscopy, X-ray crystallography, nuclear magnetic resonance, mass spectrometry, and many others. The availability of these advanced analytical technologies at a time concomitant with revolutionary developments in molecular biology has enabled the investigation of biological structures and interactions that have previously been inaccessible.

Mass spectrometry (MS) is a rapidly advancing family of structural biology tools that have in common the measurement of molecular *ions* of intact and fragmented biomolecules. A common misconception about MS is that it only provides molecular mass information. In reality, MS is not only powerful for molecular mass measurement, but also provides ion manipulation capabilities for obtaining detailed structural information at the isomeric level (including differentiation of isomers in many cases). While MS is a well-established technique that historically has been important for small molecules, advances since the late 1980s have made MS applicable to large biomolecules such as proteins, nucleic acids, and their complexes. The key to forming gas phase ions from these larger molecules, a prerequisite for MS analysis, lies in using electrospray ionization (ESI) (Fenn et al., 1989) or matrix-assisted laser desorption/ionization (MALDI) (Hillenkamp et al., 1991; Nakanishi et al., 1994). With these ionization techniques, MS can be used as a new type of high-resolution "readout" for many existing biological and biochemical procedures, providing biological information in the form of molecular mass.

The completed sequencing of genomes for several organisms, including humans (Venter et al., 2001), has ushered in a new era of biology. The impact that this information

Microbial Functional Genomics, Edited by Jizhong Zhou, Dorothea K. Thompson, Ying Xu, and James M. Tiedje.
ISBN 0-471-07190-0   © 2004 John Wiley & Sons, Inc.

will have on a more complete understanding of the processes of life, including human health, is only now beginning to be realized. A natural extension of *genomics* (which is the study of the complete set of genes for an organism) research is the characterization of gene products, most of which are proteins. This has spawned the research area of *proteomics*, which is the study of the entire suite of proteins from a genome. Mass spectrometry has quickly become one of the leading technologies for proteome measurements, due to its inherent ability to identify proteins, including hypothetical species, at high mass accuracy, resolution, and throughput, even from complex mixtures (Peng and Gygi, 2001; Larsen and Roepstorff, 2000). For instance, the advent of modern MS techniques has greatly expanded the utility of the venerable proteolytic digest. Accurate MS measurement of the masses of the proteolytic peptides yields a so-called 'mass map' that can be used to identify a protein from published databases, and is a foundation technology for the rapidly expanding field of proteomics (Pandey and Mann, 2000). Furthermore, tandem mass spectrometry (MS/MS) enables full or partial sequencing of the proteolytic peptides, in some cases without prior separation (Yates, 2000).

The objective of this chapter is to illustrate how MS is becoming an essential tool for characterizing complex mixtures of proteins. The chapter will be divided into two major sections. The first section will provide a condensed description of the fundamentals of biological MS, discussing the major ionization methods, the different types of mass analyzers, how ion structure can be interrogated, and finally how proteins and peptides can be characterized. The second section will outline the status of MS for complex protein measurements such as proteomes, in particular detailing how bottom-up approaches (employing proteolytic digestion and peptide identifications) and top-down approaches (measurement of the intact proteins) can provide comprehensive proteome information.

9.2 FUNDAMENTALS OF MASS SPECTROMETRY

In order to understand the impact that MS can have on protein measurements, it is useful to define the basic operating principles. The goal of this section is not an exhaustive review of MS, which is given elsewhere (Dass, 2001), but rather to discuss the basic concepts of ionization, mass analyzer types, and biomolecular detection in sufficient detail to prepare the reader for a detailed discussion of complex protein characterization with this technique. These concepts will then be demonstrated for protein and peptide characterization.

9.2.1 Basic Components of Any Mass Spectrometer

Even though there are several different types of mass spectrometers, they all share three important fundamental components. These are:

(i) *Ion source*, where gas phase ions are generated from the sample.

(ii) *Ion analyzer*, where the charged particles are separated or sorted by their mass/charge (m/z) ratios.

(iii) *Ion detector*, where the abundances and mass/charge values of the ions are measured.

In almost all current instruments, a dedicated computer workstation controls all the aspects of this measurement, ranging from ion introduction to data analysis. The first component of the MS experiment is the ionization step, in which gas phase ions are generated in an ion source from a solution phase sample. This process is not trivial and historically has been the source of much frustration for large biomolecules. Early work in fast-atom bombardment mass spectrometry (FAB-MS) and plasma-desorption mass spectrometry (PD-MS) revealed some promise for biomolecules with molecular masses up to about 10 kDa. It was not until the advent of MALDI and ESI (which will be discussed in more detail below) that MS became a useful method for examining much larger species (i.e., molecular masses exceeding 100 kDa). The functions of the ion source of any mass spectrometer are 3-fold: (1) to efficiently and nondiscriminately generate gas phase ions from the analyte, (2) to avoid fragmenting the analyte in the ionization process, and (3) to inject the gas phase ions into the mass spectrometer.

Mass spectrometry would be a useless tool if the gas phase ions could not be partitioned or sorted in some meaningful way to disperse them for subsequent detection. Thus, the second critical component of any MS instrument is the ion analyzer. All mass spectrometers use electric and/or magnetic fields to alter the motion of the ion packet emerging from the source in such a way that ion dispersion occurs. This sorting process enables ions of similar mass/charge to be grouped together and separated either spatially or temporally from ions of dissimilar mass/charge. For example, in a time-of-flight mass spectrometer, a high-voltage direct current pulse is used to accelerate the ions out of the ion source and into an open evacuated flight tube. Because this voltage imparts the same kinetic energy to each ion with similar charge, all the ions have different velocities (i.e., $KE = \frac{1}{2} mv^2$). Thus, this process provides spatial resolution of the ions in that the lighter ions (i.e., those with lower mass/charge) travel faster and reach the ion detector before the heavier ions. By recording the ion signal as a function of arrival time, it is possible to obtain a time-of-flight mass spectrum.

The third and final component of all mass spectrometers is the detector. In every case, this process relies on the fact that charged particles provide a macroscopic electronic effect on a suitable detector. The most common detector employed is an electron multiplier. For this device, the ions impinge on a metal target. Electrons dislodged by this process are multiplied by a cascade process and give rise to an amplified signal each time an ion strikes the device. This analog signal is converted to a digital signal that can be recorded and manipulated by the computer workstation. A more detailed description of these three fundamental components will be provided in the following sections.

9.2.2 Ionization Methods

Although there are numerous ionization methods available for MS, two main ones dominate the study of large biomolecules: electrospray ionization and matrix-assisted laser desorption/ionization. The development of these methods has been reviewed in several recent articles (Vestal, 2001; Cech and Euki, 2001). The basis for the popularity of these two ionization methods is the ability of these techniques to provide gas phase ions from biomolecules whose molecular masses exceed 100 kDa. Until recently, MS was useful primarily for molecules with masses below 10 kDa; however, these two ionization methods have provided more than an order of magnitude in the accessible molecular mass, and thus have revolutionized MS for biological studies. Being able to directly examine compounds with molecular masses at least to 100 kDa has opened the door to peptide, protein, and even oligonucleotide studies, thus significantly engaging the attention of the biological community.

The principle ionization methods for protein investigations by MS are the following:

(i) *Electrospray Ionization (ESI or ES).* This technique utilizes a high-voltage needle (typically, about 4,000 V) to transfer preformed ions from solution phase into the gas phase. This resulting mass spectra usually consist of a range of multiply charged ions, such as $(M + nH)^{n+}$. For most ESI-MS experiments, a protein solution in the nM to μM concentration range would be prepared in water:acetonitrile ($\sim 50:50$ by volume), with about 0.1 to 1.0% acetic acid added to protonate the proteins in solution. Because the electrospray ionization process is quite sensitive to charge-carrying species, great care must be taken to minimize or exclude buffers, salts, carrier proteins, and other species that might interfere with measurement of the analyte of interest. If protein stability demands the inclusion of buffer, it is desirable to use a volatile buffer such as ammonium acetate (concentrations of $\sim 1-100$ mM). The online coupling of liquid separation techniques, such as HPLC, with ESI is relatively straightforward.

(ii) *Matrix-assisted Laser Desorption/ionization (MALDI).* This method is conducted by using a pulsed laser to desorb biomolecules that have been imbedded in a spectrally absorbing matrix compound (typically, a small organic acid). The resulting mass spectra consist primarily of singly charged species, such as $(M + H)^+$, although some higher charged species, especially doubly charged ions, are observed in some cases. Typical matrix compounds for proteins and peptides include sinapinic acid, alpha-cyano-4-hydroxycinnamic acid, and 2,5-dihydroxy-benzoic acid. MALDI is reasonably tolerant of the presence of salts, buffers, and other additives in the sample, and is often superior to ESI for "dirty" samples. The online coupling of liquid separation techniques, such as HPLC, with MALDI is fairly difficult.

9.2.3   Mass Analyzers

Although ionization is vital for transferring the solution phase analyte into the gas phase, it is critical to have a method to separate and detect those ions. This is the function of the mass analyzer. Because ion motion can be controlled by electric and magnetic fields, these fields are used to manipulate the ion motion for subsequent detection. There are several types of mass analyzers, as will be discussed below, but they all share the common attribute of altering ion motion for selective detection. When comparing different mass analyzers, it is useful to evaluate them based on the following figures of merit:

(i) *Mass Resolving Power.* The measure of how well adjacent peaks can be differentiated in the mass spectrum. This value is typically given as the peak full width at half maximum (FWHM) and is determined by dividing the ion m/z ratio by the peak width at 50% height.

(ii) *Mass Accuracy.* The comparison of the measured mass to the calculated mass. This value is typically given as error in either percentage or parts per millions (ppm).

(iii) *Mass Range.* The difference between the largest and smallest molecular mass that can be measured.

(iv) *Detection Limits.* The smallest amount of sample that can be measured with a signal/noise of at least $3:1$.

(v) *Dynamic Range.* The molar difference between the least abundant component and the most abundant component that can be detected in a single sample.

(vi) *Scan Speed.* The time that it takes for a given mass analyzer to record a complete mass spectrum.

(vii) *Tandem MS.* The capability of a given mass spectrometer to isolate and fragment a selected ion for structural interrogation.

The following mass analyzers are commonly employed for biological studies:

(i) *Linear Quadrupoles (Q).* The heart of this type of MS consists of four parallel metal rods, electronically isolated from each other, that are arranged to permit ion transmission in the longitudinal space between them. A combination of dc and ac radio-frequency (rf) voltages is placed on the rods to control ion motion. The mass/charge of the ions that are passed through the rods is proportional to the voltage applied to the rods (the ratio of rf to dc voltage is held constant). By scanning the rf voltage, ions of a given mass/charge ratio are sequentially allowed to pass through the rods and into the detector, allowing a mass spectrum to be acquired. Because ion detection is accomplished in a continuous mode, ESI is the most common ionization technique used for this instrument. Ions are detected by an electron multiplier placed at the end of the rods.

(ii) *Time of Flight (TOF).* This instrument consists of an ion source, a flight tube (about 1 m in length), and an ion collector. A high voltage is applied to the ion source, to accelerate the ions out of the source and into the flight tube. Because all ions of the same charge state are given the same kinetic energy, the lower mass/charge ions have higher velocities than the higher mass/charge species. Thus, ion spatial separation is achieved in the flight tube. The lightest ions strike the detector first, followed eventually by the heavier ions. By recording ion intensity as a function of time, it is possible to obtain a time-of-flight mass spectrum. A reflectron (or ion mirror) can be used to compensate for the range of kinetic energies resulting from the ionization process, and provides enhanced mass resolution. Both ESI and MALDI are employed for TOF, although MALDI is much more common.

(iii) *Sectors.* This instrument uses a combination of electrostatic and magnetic fields to guide an ion beam from the ionization source to an electron multiplier detector. The electric and magnetic fields serve to focus and resolve the ions spatially, and to transmit only one particular mass/charge species into the detector at one time. By scanning the magnetic and/or electric fields, a complete mass spectrum can be recorded. Because ion detection is usually accomplished in a continuous mode, ESI is the most common ionization source. MALDI is a pulsed technique and less suitable for interfacing to this type of instrument.

(iv) *Quadrupole Ion Traps (QIT).* Based on the linear quadrupole principles discussed above, it is possible to construct a three-dimensional QIT (Hao and March, 2001). This device has a ring electrode and two end caps. Unlike the analyzers discussed above, the QIT has the ability to trap ions spatially for a period of time. Ion trapping is achieved with a combination of dc and rf voltages applied to the ring electrode and end caps. This permits the ions to be stored in the enclosed volume of the trap for a period of time for further interrogation. Ion detection is accomplished by ramping the rf voltage to sequentially eject the ions from the trap into an electron multiplier. ESI is the most common source, although MALDI has been employed to some extent as well.

(v) *Fourier Transform Ion Cyclotron Resonance (FTICR).* This is one of the highest-performance types of mass spectrometers (Marshall et al., 1998; Marshall, 2001).

Ions are trapped in an evacuated cylindrical ion cell, which is situated inside a high-field superconducting magnet. By confining the ions electrically and magnetically, it is possible to store them for extended periods of time. Ion detection is achieved by measuring the rf cyclotron motion of the orbiting ion packet. Because the electronic frequencies can be measured so precisely, it is possible to obtain both high-resolution and high-mass accuracies. ESI is the most common source (Hendrickson and Emmett, 1999), although MALDI is used to a limited extent.

Table 9.1 compares performance features for these mass analyzers. There is not one specific mass analyzer that is completely universal for all applications. Often, the nature of the biological problem (and the financial resources available) will dictate the choice of instrument. It is not uncommon for different types of mass analyzers to be linked together, to overcome some of the individual limitations. For example, a hybrid quadrupole time-of-flight (QqTOF) instrument is becoming widely used for protein studies (Chernushevich et al., 2001). This instrument is designed to exploit the positive capabilities of each of the two mass analyzers and provides rapid mass measurement under enhanced resolution. It should be noted that some of these instruments are very straightforward for nontechnical experts to use, whereas the higher-performance instruments tend to require substantial technical training for operation.

9.2.4 Coupling Separation Methods with Mass Spectrometry

Although mass spectrometers have sufficient mass resolving power to be able to simultaneously differentiate very complex mixtures, numerous experimental parameters diminish the practical application of this capability. For example, consider the proteolytic digestion of a sample containing 100 different proteins. This process will generate an even more complex sample, which might contain between 2,000 to 5,000 peptides. Even though the higher-performance mass spectrometers discussed above have sufficient resolution to measure all these peptides in a single mass spectrum, the more easily ionized species will be overrepresented in abundance in the mass spectra, whereas the more poorly ionized species will be suppressed to the extent that they might not even be observable. This is

TABLE 9.1 Performance Factors for Different Mass Analyzers

Mass Analyzer	Most Common Ionization Mode	Ion Separation	Resolving Power (FWHM)	Mass Accuracy	Mass/charge Range
Quad.	ESI	Electronic band-pass filter	1,000–2,000	0.1 Da	500–3,000 Da
TOF	MALDI ESI	Flight time	500–1,000 (linear) 2,000–10,000 (reflect. TOF)	0.1 Da (linear) 0.001 Da (reflect. TOF)	500–1,000,000
Sector	ESI	Magnetic field (ion momentum)	5,000–100,000	0.0001 Da	1,000–15,000
QIT	ESI	rf/dc Voltage	1,000–2,000	0.1 Da	500–4,000
FTICR	ESI	rf Frequency	5,000–5,000,000	0.0001 Da	200–20,000

termed *ion suppression* and is readily observed in both ESI and MALDI mass spectra because of the competition for charge in the ionization process. Thus, even though two different peptides may be present at equimolar concentrations in the sample, it is quite possible that one species will be observed in high abundance and the other species not even detectable in the mass spectrum. One obvious solution to this dilemma is to couple some type of separation stage, such as chromatography, with the MS experiment. In order to minimize sample losses due to handling, the current state-of-the-art instrumentation often incorporates liquid chromatography directly online with the mass spectrometer. For example, high-performance liquid chromatography is readily interfaced with ESI-MS. The advantage of this setup is easily recognized; the online separation step, which is often based on reverse phase liquid chromatography, provides partial purification of the proteins or peptides, so that they enter the mass spectrometer at different times. This partially alleviates the ion suppression problems, as fewer species are measured at the same time. The practical details of how this is accomplished and the benefits it provides will be outlined in the second half of this chapter.

9.2.5 Ion Structural Characterization

Mass spectrometry has earned the reputation of being an excellent method for accurately measuring masses. However, the real power of this technology is its ability to manipulate the gas phase ions in order to study their reactivities and fragmentation pathways. These experiments provide detailed structural information. The two basic types of MS experiments to investigate structure are:

(i) *Ion Fragmentation Studies.* This is generically referred to as *tandem mass spectrometry* or *mass spectrometry-mass spectrometry (MS/MS)*. This can be achieved with collisional dissociation or photofragmentation.

(ii) *Ion Reaction Studies.* These are conducted by reacting selected gas phase ions with various reagents. For example, gas phase ions can be reacted with deuterium oxide to study hydrogen/deuterium exchange reactions to locate mobile hydrogens.

Collisional activated dissociation (CAD) is the most common type of ion fragmentation experiment. This procedure usually involves three key steps: ion isolation, ion dissociation, and fragment ion detection. Ion isolation is accomplished by removing all unwanted ions in the mass spectrometer except for a selected parent ion mass/charge ratio. This parent ion is accelerated translationally by applying appropriate voltages and then collided with a target gas such as helium, argon, or nitrogen. Upon impact, the translational energy of the ion is converted to internal energy, causing fragmentation. By controlling the amount of excitation energy, it is possible to induce either limited or extensive fragmentation. The resulting fragment ions are then measured and provide information about the structure and sequence of the parent ion. Alternatives to gas phase CAD include fragmenting the parent ions with photon, electron, or surface collisions.

Ion reactivity studies provide another route to unravel biomolecular structures. The most common experiment conducted for proteins is hydrogen/deuterium exchange reactions, which can be conducted either in the solution phase or in the gas phase inside the mass spectrometer. The primary goal of these experiments is to determine which hydrogens are immobile due to participation in higher-order biological structure (Smith,

1997b). For example, it is possible to investigate protein secondary structure such as alpha helices and beta sheets with this methodology. This technology is also being extended to the examination of protein–protein interactions by mapping which hydrogens are inaccessible in the protein complex relative to the individual monomers. The reader is referred to the following references for a summary of how mass spectrometry is becoming an important tool for studying biomolecular conformations: (Hernandez and Robinson (2001), Kaltashov and Eyles (2002), Bennett et al. (2000).

9.3 FUNDAMENTALS OF PROTEIN AND PEPTIDE MASS SPECTROMETRY

In the measurement of mass spectra, it is helpful to define some of the basic terms. They are:

  (i) *Average Molecular Mass.* The sum of the atomic masses of all the composite elements in a given molecule. Since average atomic masses are used in the calculation, the molecular mass is therefore calculated as an *average* mass.
  (ii) *Monoisotopic Molecular Mass.* The sum of the most abundant isotopes of all the composite elements in a given molecule.
  (iii) *Isotopic Packet.* The isotopic distribution observed resulting from the natural abundance of isotopes from the composite elements in the molecule.
  (iv) *Protonated Molecule.* The ion that results from the addition of one or more protons to the molecule. In MALDI-MS, the most abundant positive ion is often the singly charged $(M + H)^+$, whereas electrospray usually reveals a series of multiply charged $(M + nH)^{n+}$ ions.

9.3.1 Protein Measurements

One of the driving forces of MS for protein and peptide characterization is its ability to measure the molecular masses to a high degree of accuracy, often better than 0.01%. In MS investigations of small molecules, such as peptides with molecular masses less than 5,000 Da, it is convenient to consider the monoisotopic molecular mass, as this is usually the most abundant ion. In this case, the isotopic packet is fairly small (often only one to three measurable peaks). MS examinations of larger proteins reveal a substantially different scenario. For proteins with molecular masses in excess of 10,000 Da, the monoisotopic peak is virtually unmeasurable, and the most abundant peak is closer in mass to the average value. The isotopic packet is substantially larger in this case and includes abundant ions that range over several Da. These factors obviously affect the nature of the MS measurement and thus must be taken into account.

  In order to illustrate the capability of MS for protein measurements, consider the protein ubiquitin. This has become an almost universal standard for MS, as this protein is readily available in high purity, is remarkably stable, and produces excellent MS signals in either MALDI or ESI modes. Ubiquitin is a small protein consisting of 76 amino acids. This protein has an empirical formula of $C_{378}H_{629}N_{105}O_{118}S_1$, yielding a *calculated average molecular mass* of 8,564.8721 Da. In a low-resolution MS measurement, such as those obtained on the common linear MALDI-TOF instruments, the molecular ion region will appear as a broad peak with a mass centroid at the average mass value. The presence of salt, which is a very common contaminant from handling, in the sample will be reflected by the presence of additional peaks slightly heavier than the molecular mass. At moderate resolution, such as that obtainable on a sector MS, the broad molecular ion region begins to partially resolve into discernible peaks corresponding to the isotopic packet. Under high-resolution conditions, such as those possible with an FTICR-MS, the molecular ion region appears as a Gaussian distribution of well-resolved peaks separated by one Da intervals. These correspond to the isotopic signature due to the naturally occurring isotopes of carbon, hydrogen, nitrogen, oxygen, and sulfur in the molecule. The primary contributor to this packet is ^{13}C, which is 1.1% in natural abundance. The abundance of this isotopic contribution in the molecular ion of ubiquitin may be somewhat surprising, until one realizes that each protein molecule contains more than 350 atoms of carbon. Statistically, at this level, it is much more likely that one or more ^{13}C are present than that all the carbons are ^{12}C. The isotopic packet is about 10 Da wide, with an almost symmetric appearance. The monoisotopic peak at 8,559.6152 Da is almost undetectable, due to the low abundance of this species in the natural isotopic packet. The most abundant isotopic peak is calculated to be 8,564.6296 Da, which is close but not identical to the average mass of 8,564.8721 Da. It should be apparent that the mass resolution of the MS measurement has to be taken into account with respect to the expected masses to be measured. All these effects are further accentuated with larger proteins, which have a substantially wider isotopic packet (Horn et al., 2000c).

Figure 9.1A reveals the *measured* ES-FTICR mass spectrum of ubiquitin. Note that this figure shows a variety of multiply charged ions, which is very typical for electrospray ionization. Figure 9.1B illustrates the deconvoluted molecular mass spectrum, obtained by mathematically removing the multiple charges to yield the neutral protein species. *The measured mass of the most abundant isotope is 8,564.637 Da, which is 0.0074 Da (or 0.9 ppm) different from the calculated mass listed above.* It is this level of mass accuracy that has enabled MS to become an essential tool for biological studies.

If MS had only the ability to measure protein molecular masses accurately, this would be an important, but limited, capability. What drives MS beyond this level is the ability to manipulate and interrogate these ions, thereby providing structural information (Horn et al., 2000b; Loo et al., 1992; Little et al., 1994; Senko et al., 1994). For example, Fig. 9.2 illustrates the MS/MS of the $(M + 10H)^{10+}$ ion of ubiquitin in an ESI-FTICRMS experiment. Several observations can be drawn from the fragment ions observed in this mass spectrum. First, there are a variety of fragment ions ranging from the molecular mass down to small peptide species. Second, there are blank zones in which limited or no fragmentation is observed. Third, the fragment ions can be measured to high accuracy as well, providing definitive information about dissociation pathways. Even though complete sequence information is often not possible from such a measurement, an enormous amount of structural information can be gleaned from such a mass spectrum, as will be illustrated in the later sections of this chapter. In order to interpret the MS/MS and identify fragmentation sites within the protein, the following nomenclature is used. Proteins typically fragment by cleavage along the peptide backbone. Because the cleavage can occur at multiple sites, a systematic alphabetic code is used, as shown in Fig. 9.3. Fragment ions that retain the charge on the *N*-terminus end of the original protein are designated as *a*, *b*, or *c* ions, depending on the cleavage site. Fragment ions that retain the charge on the *C*-terminus end of the original protein are designated as *x*, *y*, or *z* ions, depending on their cleavage site (Roepstorff and Fohlman, 1984; Biemann, 1988). The most common fragment ions observed for proteins are usually *b*- and *y*-type ions. This is illustrated by

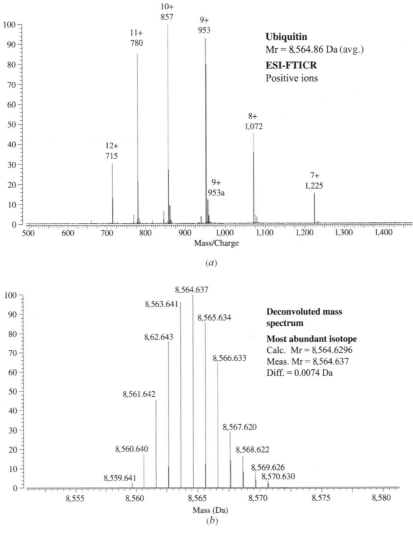

Figure 9.1 *Electrospray Fourier transform ion cyclotron resonance (ES-FTICR) mass spectra of the protein ubiquitin, illustrating the multiply charged ions observed in the positive ion mass spectra (a), and the deconvoluted view (b) showing the isotopic molecular region.*

the annotation in the inset for Fig. 9.2 for the MS/MS of ubiquitin. There are several web-based tools, such as those available at the Rockefeller PROWL website (http://prowl.rockefeller.edu), that can be used to predict the fragment ions from a given protein sequence.

9.3.2 Peptide Measurements

Even though accurate molecular mass and limited fragmentation are achievable for intact proteins, this information often is insufficient for the *unambiguous identification of the complete sequence.* One of the most logical solutions to this dilemma is to utilize solution phase proteolytic digestion procedures to cut the protein into characteristic peptides. The

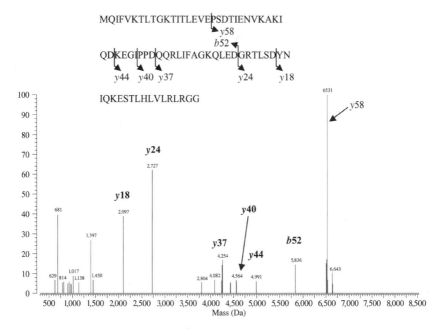

Figure 9.2 *Deconvoluted mass spectrum of the collisional dissociation (MS/MS) of the 10$^+$ charge state (with an m/z of 857) of ubiquitin.* Inset reveals the fragment ion identities and sequence locations.

resulting peptides can be measured under high-resolution conditions and fragmented with MS/MS techniques. Because peptides are smaller in molecular mass than intact proteins, they fragment more readily and provide more extensive sequence information. By combining this information with known details about the protein digestion process (i.e., the identities of the amino acids cleaved), it is possible to obtain the detailed information necessary to identify the protein. In fact, this information can be used to query protein databases for automated protein identification, as will be illustrated in the following sections.

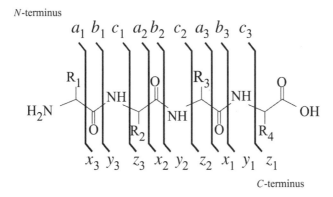

Figure 9.3 *Alphabetic code used to designate fragment ion types and locations from a generic peptide.* (Data from Roepstorff and Fohlman, 1984; Biemann, 1988.)

There are a variety of common proteases that can be used to cleave proteins into peptides. These range in cleavage specificity, providing control over the desired degree of protein fragmentation. These proteases are active under different solution phase conditions. The most common protease employed is trypsin, which is readily available in purified form and cuts proteins at the *C*-termini of arginine and lysine residues (provided that the adjacent residue is not proline). This produces fragment ions containing basic amino acids at the *C*-terminus (which is advantageous for positive ion MS measurements). A protease digestion experiment generally begins with protein denaturation and disulfide bond reduction to open up the protein for more efficient degradation. After the digestion is complete, it is essential to clean up the sample with reverse phase HPLC or solid phase extraction to remove the salts prior to MS measurements of the peptides. In order to minimize autodigestion products (generated by trypsin digesting itself), it is desirable to utilize sequencing grade modified trypsin, which is available from a variety of vendors as a solution phase reagent or immobilized on beads.

Because the digestion procedure often does not proceed to completion, it is useful to consider peptides that might result from missed cleavage sites. For example, we have used a tryptic digestion procedure to characterize an important yeast DNA checkpoint protein, Sml1p (Uchiki et al., 2002). Two important points were noted in this experiment: (1) The tryptic peptides were observed at dramatically different abundances in the mass spectra, which reflect the efficiencies of the digestion and cleanup procedures, as well as the common ion suppression effect, and (2) the sequence coverage was less than 100% (i.e., some of the predicted peptides are absent). The masses of the peptides observed are listed in Table 9.2, revealing how the measured peptide masses correlate with the predicted masses. For this particular case, 94% of protein sequence was determined from the peptide measurements. Any of these peptides can be interrogated with MS/MS experiments. For example, the MS/MS of the peptide with Mr = 1,865 Da (residues 1–16) yielded a variety of abundant *y*- and *b*-type fragment ions. Because the *y*-type ions were the most abundant series of ions, it was possible to identify an amino acid sequence tag of NSQDYFY in this peptide. This sequence tag information is very useful for not only confirming the identity of the peptide but also for determining its position within the

TABLE 9.2 Peptides Generated by Trypsin Digest of Sml1p-HisTag Protein

Measured Mass[a]	Tryptic Fragment	Residues	Calculated Mass[a]	Mass Difference
1,864.780 Da	1	1–16	1,864.779 Da	0.001 Da
1,260 (by MALDI-MS)	2	17–27	1,258.608	*1.392*
953.463	3	28–35	953.464	0.001
3,705.765	4–5	36–70	3,705.749	0.016
4,347.999	4–6	36–75	4,348.046	0.047
1,413.646	7	76–87	1,413.630	0.016
736.383	8	88–94	736.375	0.008
2132.007	7–8	76–94	2,131.995	0.012
810.373	9	95–101	810.369	0.004
1,068.468	10	102–110	1,068.454	0.014
1,860.809	9–10	95–110	1,860.813	0.004
(927)—not observed	11	111–117	927.396	

[a]Monoisotopic masses.

protein. Furthermore, it is possible to query the protein databases with sequence tag information, which is sufficient in many cases to identify a protein without any additional information.

9.4 MASS SPECTROMETRY FOR PROTEIN AND PROTEOME CHARACTERIZATION

The first section of this chapter outlined the basic concepts of MS for protein analysis. This section will discuss how that methodology can be applied to the analysis of individual proteins, protein complexes, and whole proteomes, with a focus on microbial proteomes. It should be noted that while research into MS-based proteomics has virtually exploded in the last 5 years, there are other methods to analyze whole proteomes and protein complexes, such as protein chips, antibodies, and yeast two-hybrid analyses (see Chapter 10). Although some of these other techniques may show great promise for the future, none have the current capabilities for rapid and molecular level analyses of protein mixtures that can be afforded by the MS-based techniques. For this reason, we will limit our discussion to techniques that are either directly or indirectly coupled with mass spectrometers for complex protein mixture analysis.

Proteome analyses, whether in simple microbes, yeast, or higher organisms, present a much greater challenge than the genomics sequencing efforts. *While the genome is static, the proteome is dynamic.* The genome generally contains a set number of copies of every gene; however, proteins in the proteome can be expressed in a wide concentration range, varying from only a few copies per cell for regulatory proteins to many thousands per cell for ribosomal subunits. The proteins derived from theses genes can take many different forms due to alternate transcript splicing (primarily observed in higher eukaryotes). Furthermore, proteins can be highly decorated with any number of posttranslational modifications (PTMs); more than 100 have been recorded (O'Donovan, 1999). These modifications can be static or dynamic, and may be present in multiple places on a protein. Finally, in the current state of proteomics technologies, there is no amplification technique for proteins similar to the polymerase chain reaction (PCR) that has become so important for oligonucleotide studies. Thus, the proteome of even simple microbes presents a much greater analytical challenge than the corresponding genome (or even transcriptome) analysis. Even in light of these difficulties, a complete understanding of microbes and microbial communities necessitates the development of analytical techniques for rapid and accurate analysis of whole proteomes and protein complexes. In order to achieve such a "systems biology approach," it is essential to interface proteomics, bioinformatics, genomics, transcriptomics, and metabolomics technologies.

9.4.1 Overview of Mass Spectrometry Approaches for Protein Studies

Protein analysis by MS-based methodologies can be broken down into three general areas. Although each is unique, they all can be addressed with similar MS technology.

Individual Protein Analysis This involves the analysis of purified proteins for quality control in structural or biochemical experiments, as a means of studying posttranslational processing, or in structural analysis of individual proteins by techniques such as H/D exchange (Dharmasiri and Smith, 1996), cross-linking (Young et al., 2000), and surface labeling (Bennett et al., 2000). While these methods of structural analysis by MS all show great promise for the future, a detailed discussion is beyond the scope of this chapter. Protein detection over a wide dynamic range is not much of an issue here, but sensitivity may be, for example, if the goal is to purify a low-copy number protein with a transient modification.

Protein Complex Analysis   This involves the analysis of purified protein complexes, which may also include the analysis of signaling pathways in which the protein complexes are generally more transient and difficult to analyze (Pandey et al., 2000). Protein complexes are typically purified by centrifugation/sucrose gradient techniques, although more recent approaches employ immunoprecipitation (Pandey et al., 2000) or tandem affinity purification tags (TAP) (Puig et al., 2001). The greatest challenge is the biochemical purification and sensitivity required in the analysis, since typically 100 to 1,000 ng of the complex can be prepared, although the preparation of protein complexes from microbes can often yield much greater quantities. Dynamic range in the MS detection is not as much of a problem, but may be important if a transient protein is associating with a large complex or if one is searching for PTMs on this complex. A key element of these measurements is the need to purify such complexes from cell lysates. This field is developing rapidly at present, as evidenced by attempts to characterize protein–protein interactions at the proteome level (Gavin et al., 2002; Ho et al., 2002; von Mering et al., 2002).

Whole-proteome Analysis   This typically refers to the analysis of whole cell lysates, organelle preparations, or crude fractions obtained by affinity chromatography or centrifugation, such as membrane preps, cleared serum, etc. The greatest challenge in this analysis is dynamic range, because medium- to high-abundance proteins mask most low-abundance proteins. Sensitivity is generally not as much of a problem for whole cell lysates due to large quantities of starting material, but may be an issue for organelle preparations or affinity purified/cleaned fractions. Whole-proteome analysis is very complex and can be broken down into four separate steps. The first step of sample preparation (i.e., cell growth or sample isolation) is followed by the second step of protein fractionation or separation. This leads to the third step of MS characterization, which is followed by the final step of computational analysis of the data (bioinformatics). For the analyses of protein complexes and whole proteomes, the most important aspect is the dynamic range of the measurement. The demands on MS can be relaxed somewhat by incorporating separation technologies (such as either offline or online fractionations of proteins/peptides) prior to mass spectrometric detection. These fractionations can be very crude or highly specific, depending on the nature of the application. Each of the four steps for proteome analysis will be discussed below.

Sample Preparation   The first step in obtaining a proteome sample entails common microbiology techniques of cell growth and isolation. For typical microbial proteome analysis, the cells can be cultured under a variety of growth conditions for comparative analysis. Depending on the species, 500 µL to 4 L of cell culture are more than sufficient for multiple analyses by any of the proteome methodologies that will be discussed below. The cells are typically harvested and washed with a buffer to remove excess media. The cells are then disrupted with various techniques such as sonication, bead beating, or

French press. The methodology for cell disruption will depend on the species type and kind of proteome analysis that will be applied.

Crude Fractionation Techniques

(i) *Centrifugation.* Cell extracts can be processed by centrifugation (varying speed and duration) to create different soluble and insoluble fractions. Sucrose gradients can be used for finer fractionation. In most cases, three lysate fractions are prepared: a crude soluble, a cleared soluble, and membrane fraction.

(ii) *Chemical Extraction.* Cell lysates or membrane fractions can be treated chemically with chloroform/methanol or acetone to remove lipids and small molecules from protein samples. This technique is very useful in cleaning up membranes, especially from photosynthetic organisms such as cyanobacteria and plants.

(iii) *Affinity Depletion.* Although this methodology is a form of chromatography, its primary purpose is to remove high-abundance proteins from a sample to enhance MS detection of more minor species. For example, the removal of albumin and immunoglobulins from serum is a common type of affinity depletion and enables the detection of other more minor proteins in this physiological media.

Chromatography Separation Techniques

(i) *Gel Filtration (or size exclusion).* In this technique, proteins are separated according to their sizes through a gel filtration column. This crude separation lacks resolving power and generally is used in the first step of fractionation.

(ii) *Affinity Chromatography.* This technique is a very powerful tool for analyzing low-abundance proteins and protein complexes. Basically, intact proteins or peptides are enriched over an affinity matrix by exploiting intrinsic properties (such as the presence of phosphate groups on protein) or chemically or biologically added labels.

(iii) *Ion Exchange Chromatography.* This technique separates peptides or proteins by their surface charge. Typical experiments employ cation exchange for peptides and anion exchange for proteins.

(iv) *Reverse Phase Chromatography.* This method relies on the separation of intact proteins or peptides according to their overall hydrophobicities. As such, it provides a high level of resolution and is typically the final separation method prior to MS analysis, due to the fact that its solvent requirements (a mixture of aqueous and organic phase with a small amount of organic acid) are directly compatible with ESI and MALDI.

Electrophoresis Separation Techniques

(i) *Capillary Electrophoresis.* This separation methodology is based on the differential electrophoretic mobility of peptides or proteins in the presence of an electric field gradient applied along a flowing liquid in a gel-filled capillary. Although this method originally showed great promise due to its speed, resolving power, and sensitivity, it has proven not to be very rugged and is difficult to couple online with MS. These factors render it very limited in general proteomics applications. It is possible that this methodology could be revitalized in microseparations on chips.

(ii) *Gel Electrophoresis.* This separation methodology is based on the differential electrophoretic mobility of peptides or proteins in the presence of an electric field gradient applied across a slab gel. This technique has several variations, but in proteomics commonly involves polyacrylamide gel electrophoresis (PAGE), in which isoelectric focusing and molecular mobility provide for a two-dimensional protein separation.

Mass Spectrometry Analysis Techniques   Currently, there are two major methods for analyzing proteins by MS. The *top-down* method involves measuring intact proteins as well as interrogating the MS/MS of these intact proteins. This method was first introduced with ESI-FTICR-MS (Little et al., 1994; Mortz et al., 1996; Kelleher et al., 1998) and expanded to ion traps with novel ion–ion reactions (McLuckey and Stephenson, 1998). In the *bottom-up*, or *shotgun* method, intact proteins are digested with a protease such as trypsin, Glu-C, or cyanogen bromide (CNBr), and the resulting peptide mixtures are analyzed by MS or MS/MS. It should be noted that in this definition it does not matter whether separations are performed on intact proteins or peptides; rather, the experiment type is defined by the species measured by the MS. Thus, 2D-PAGE followed by in-gel digestion and MS analysis is considered a bottom-up approach. The actual development of the bottom-up methodology cannot be traced to a single lab, but rather evolved from multiple labs using very different techniques including gel-based (Hess et al., 1993; Mortz et al., 1994; Shevchenko et al., 1996b; Wilm et al., 1996; Gatlin et al., 1998) and solution-based separations (Hunt et al., 1981, 1986; McCormack et al., 1997; Martin et al., 2000; Shen et al., 2001) followed by MS or MS/MS for protein identifications. These two general approaches can be summarized as follows:

(i) *Bottom-up Proteomics.* Complex protein mixtures (from cell lysate or protein complexes) are proteolytically digested (usually with trypsin), and the resulting peptide mixture is examined by MS. The MS data are used to query a peptide database from the specific organism to identify the protein components of the original mixture. This method is excellent for determining protein identities, but provides very limited information about the molecular form of the intact proteins.

(ii) *Top-down Proteomics.* Complex protein mixtures from cell lysates or protein complexes are examined directly by online or offline MS. No digest is conducted; rather, the intact proteins are measured with MS and MS/MS. This method provides fewer protein identities, but does give detailed information about protein processing (PTMs, truncation, mutations, signal peptides).

Both techniques have advantages and disadvantages and will be discussed in detail below with highlights of the current technologies used. Bottom-up proteomics is by far the more widely used method, mainly because it is much simpler to conduct and does not require high-performance MS instrumentation. The progress in the field of bottom-up proteomics has been staggering. Table 9.3 highlights some of the groundbreaking research over the last 5 years in large-scale proteome analysis. Each paper illustrates either a new technology in bottom-up proteomics or a major advance in protein characterization from the given organism. This list should not be considered all-inclusive or compared directly, since there were many differences in the number of growth conditions, the size of the experiments,

TABLE 9.3 Landmark Papers on Large-scale Proteome Analysis by Mass Spectrometry

Author	Year	Species	Growth/ Fractionations[a]	Seperation Methods	MS Methods	No. of ids[b]
Shevchenko et al.	1996	*Saccharomyces cerevisiae*	1/1	2D-PAGE	MALDI PMF/Nano-ES-MS/MS	150
Wasinger et al.	2000	*Mycoplasma genitalium*	1/1	2D-PAGE (4 zoom gels)	MALDI PMF	158
Langen et al.	2000	*Haemophilus influenzae*	1/1	2D PAGE (multiple zoom gels), multiple LC	MALDI PMF/AAC	502
Wasburn et al.	2001	*Saccharomyces cerevisiae*	1/3	MudPIT (Serial 2D SCX-RP)	Nano-ES-MS/MS	1,484
VerBerkmoes et al.	2002	*Shewanella oneidensis*	1/4	Serial 2D SCX-RP/1D RP(MMS)	Capp-ES-MS/MS	868
Lipton et al.	2002	*Deinococcus Radiodurans*	15/1	2D SCX-RP (MMS)	Nano-ES-MS (AMT)	1,910
Koller et al.	2002	*Oryza Sativa*	1/3	2D-PAGE-(RP)/MudPIT	Nano-ES-MS/MS	2,528
Mawuenyega et al.	2003	*Caenorhabditis elegans*	1/2	Switching 2D SAX-RP	Capp-ES-MS/MS	1,616
Peng et al.	2002	*Saccharomyces cerevisiae*	1/1	Off Line 2D SCX-RP	Nano-ES-MS/MS	1,504

[a]Number of different growth conditions and crude centrifuge fractionations.
[b]The reported no. of protein ids (varied in criteria and redundant vs. nonredundant).
PMF = peptide mass fingerprinting; PAGE = polyacrylamide gel electrophoresis; AAC = amino acid composition; MudPIT = multidimensional protein identification technology; AMT = accurate mass tag; MMS = multiple mass range scanning or gas phase fractionation; Capp = ES flow rates over 1 μl/min; Nano = ES flow rates less than 1 μl/min; SCX = strong cation exchange; RP = reverse phase; SAX = strong anion exchange.

and scoring of proteins identified. Rather, this list should be viewed as representing a trend in the increase in depth of proteome coverage and speed of analysis over the last few years. It has now become possible (if not routine) to measure ~ 1,000 proteins from a microbe under a given growth condition with a high degree of confidence in 1 to 3 days, depending on the technology used. Furthermore, if enough mass spectrometers are assembled, this analysis can be rapidly repeated for protein identification for an organism under a variety of different growth conditions.

On the other hand, top-down proteomics has moved along at a much slower pace. This is primarily due to the following factors:

(i) Liquid-based separations of intact proteins are more difficult than peptides.

(ii) MS and MS/MS analyses of intact proteins are more difficult to conduct and interpret than peptides.

(iii) The MS instruments capable of the adequate analysis of intact proteins from complex mixtures are either very expensive or not commercially available, and have not been designed for routine operation in most cases.

(iv) The algorithms to analyze MS/MS of intact proteins are not as well developed or commercially available.

Even with these experimental challenges, top-down proteomics provides a level of information that the bottom-up technique does not, which is the *intact state of the protein*. This is critical, as proteins function as intact molecular species, not as a combination of simple, small peptides. Thus, a full understanding of the intact state of proteins (PTMs, truncation, mutations, signal peptides) is necessary, suggesting that an integrated top-down, bottom-up proteomics method would be the most comprehensive. This approach will be discussed in more detail below.

Data Processing and Bioinformatics   One of the key hallmarks of MS-based proteomics is the large quantity of data that can be produced in short periods of time. Without the advancement of bioinformatics tools to process, sort, and compile these data sets, the field would not be able to move forward. Furthermore, the recent completion of large numbers of genomic sequences has been as much of a factor in the rapid evolution of the field of proteomics as have been the advances in MS. While the proteins from *unsequenced* microbes can be identified with MS by a methodology termed de novo sequencing, this methodology is limited in the speed and depth of analysis of the proteome. In typical de novo sequencing experiments, MS/MS spectra from enzymatically derived peptides are analyzed manually or with a computer algorithm to identify sequence tags (stretches of at least five to seven consecutive amino acids). These sequence tags then are used to blast against known microbial databases in an attempt to determine the nature of the proteins in the unsequenced organism. The most common technique is to analyze proteomes of microbes after their DNA sequence has been completed and annotated. These microbial DNA or protein databases can be used for the analysis of either top-down or bottom-up data sets. Currently, search algorithms for bottom-up data have evolved more rapidly and are more widely available than search algorithms for top-down data. The heart of the common search algorithms for bottom-up data is the comparison of peptide masses and/or MS/MS spectra directly against an *in silico* generated peptide list created from the DNA or protein sequences. The peptide identifications are then used to reconstruct the potential proteins that were in the original sample. This can be broken down into two main methodologies: peptide mass fingerprinting (PMF), which uses the masses of the peptides to query protein databases, and MS/MS spectral searching, which uses the fragmentation patterns of peptides as well as their intact masses for searching either protein or DNA databases.

The PMF technique is most widely used on simple mixtures of proteins (one to three proteins) derived from in-gel digest or solution-based digest of purified proteins. This is due to the confidence of protein identification by measuring a large number of peptides (four or more) in a single MS analysis from a given protein. The MS/MS spectral database-searching algorithms can be used on either in-gel digest or peptides generated from in-solution digest of entire proteomes. This is due to the fact that each peptide MS/MS spectrum is analyzed on its own merits. Of course, the more peptides that are identified from a protein, the more confidence that can be placed in its assignment. Because MS/MS spectra reveal predominantly both *y*-type and *b*-type ions (which are difficult to distinguish), the sequence of peptides is challenging to directly determine from the MS/MS spectra. Thus, all search algorithms in some fashion compare experimentally derived MS/MS spectra with *in silico* MS/MS spectra of peptides. Though there are numerous search algorithms for MS/MS spectra, two commercial products, MASCOT (Perkins et al., 1999) and SEQUEST (Eng et al., 1994), are the most widely used. MASCOT utilizes a probability-based algorithm for comparing MS/MS spectra with *in silico* MS/MS spectra, while SEQUEST uses a cross-correlation algorithm for this comparison. It is important to realize that no matter what methodology or search algorithm is used, the correct filtering processes for that algorithm must be applied in order to obtain confident protein identifications from large MS-based proteome data sets. The criteria for protein identification and lack of standardization are some of the most hotly debated subjects in the field of MS-based proteomics.

9.4.2 Bottom-up Mass Spectrometry Proteomics

There are two different approaches for conducting bottom-up MS proteomic measurements. The first, more traditional method employs conventional gel electrophoresis as the first step to separate and visualize proteins. Individual spots can be excised from the gel, digested with a suitable protease, and then characterized by MS. The second, more recent technique exploits the capabilities of high-resolution liquid chromatography (either in a one- or two-dimensional mode) as an online interface with MS. This method involves less sample handling and has the potential to be more comprehensive and to overcome some of the limitations of the gel electrophoresis approach. Each of these will be discussed in detail below.

Gel Electrophoresis with Mass Spectrometry   Gel electrophoresis has been the gold standard for protein separations over the last 25 years. Within the last 10 years, methodologies for coupling these powerful separation techniques with MS for protein identification have become available (Patterson and Aebersold, 1995; Shevchenko et al., 1996a; Gatlin et al., 1998). Furthermore, one-dimensional (1D) and two-dimensional polyacrylamide gel electrophoresis (2D-PAGE) followed by MS or MS/MS analysis have been the most common methods for analysis of microbial proteomes over the last 5 years (Shevchenko et al., 1996b; Sazuka et al., 1999; Wasinger et al., 2000; Langen et al., 2000; Nouwens, 2000; Molloy et al., 2000, 2001; Fulda et al., 2000; Grunenfelder et al., 2001; Hernychova et al., 2001; Bumann et al., 2001; Hermann et al., 2001; Wagner et al., 2002).

The steps in the analysis of proteins by PAGE followed by MS will be highlighted and discussed below. These methodologies have been reviewed in great depth (Jungblut and Thiede, 1997; Jensen et al., 1998; Pandey and Mann, 2000). Detailed protocols for the methodology may be easily found on the World Wide Web at http://www.expasy.org/ ch2d/protocols, http://donatello.ucsf.edu/ingel.html, and http://proteomics.uchsc.edu/ protocols/index.html.

The following discussion summarizes the key steps in a gel electrophoresis MS experiment.

Gel Electrophoresis   The constraints of this chapter will not allow for a detailed discussion of gel electrophoresis and gel staining for visualization. Rather, we direct the readers to a comprehensive reference on proteomics that mainly focuses on 2D-PAGE proteomics (Pennington and Dunn, 2001). This reference provides detailed explanations of all aspects of 2D-PAGE. We will focus on highlighting the most common PAGE methodologies used in proteomics today.

1D SDS-PAGE   This is an effective methodology for separations of intact proteins by molecular weight. This technique is commonly used as a first or second separation technique in protein analysis. In general, proteins are dissolved in $\sim 0.1\%$ sodium dodecylsulfate (SDS) buffer, which is a strong negatively charged detergent. SDS acts to completely coat and denature the proteins, thus imparting the same amount of charge to similarly sized proteins. This allows a mixture of proteins to be run through a gel matrix such as polyacrylamide by applying a voltage between two electrodes. The gel acts as a molecular sieve; larger proteins are retained on the gel and move more slowly than smaller proteins. A set of protein standards usually are run in the first lane of the gel, so that approximate protein molecular weights can be estimated. These molecular mass measurements cannot be determined to a very high level of accuracy or precision. Proteins are typically visualized by any number of stains and then examined by MS analysis (see below). This methodology is most commonly employed for purified, soluble proteins or protein complexes, although membrane proteins can be effectively analyzed as well. The major drawback of this technique is resolution; typically, only 50 to 100 proteins can adequately be resolved in a single gel.

2D-PAGE   For whole-proteome separations and analysis, 2D-PAGE followed by MS analysis is the gold standard. This is driven primarily by the high resolving power of 2D-PAGE, which can distinguish over 1,000 proteins in a single gel. The original methodology went through several modifications until it was shown to be useful for the analysis of proteins from whole cells (Scheele, 1975; Klose, 1975; Iborra and Buhler, 1976). In this methodology, proteins are separated in the first dimension by isoelectric focusing (IEF), which differentiates the proteins based on their isoelectric points. The IEF gel is then equilibrated in a buffer containing SDS, which is essential for proper separation in the second dimension. The IEF gel is then placed at a 90° angle with a SDS-PAGE gel, and the proteins are eluted from the IEF gel into the polyacrylamide gel, which then separates them into a second dimension by their molecular masses. As with 1D-PAGE, any number of stains can be used to visualize proteins in the gel after separation. The original methodology of 2D-PAGE had very poor reproducibility, due to a number of inherent problems in the isoelectric focusing. These problems were solved with the development of immobilized pH gradient (IPG) for the first dimension

(Bjellqvist et al., 1982; Gorg et al., 2000). IPG-IEF is now the most common method for first dimension separation of proteins in 2D-GE applications in proteomics. It is now possible to purchase precast gels for 2D-GE, thus increasing reproducibility and throughput. For increased dynamic range, *zoom gels* are typically used, which provide narrow pH ranges in the first dimension (Wildgruber et al., 2000). The sample is analyzed under a number of different zoom ranges to amplify low-abundance proteins for MS detection.

The 2D-PAGE methodology can also be coupled with MS detection. This methodology is most commonly used for whole-proteome analysis or comparative proteome analysis, as well as analysis of preparations from organelles, periplasms, or protein complexes.

Gel Staining and Visualization   The heart of comparative analysis for 2D-PAGE is gel staining, visualization, and comparison. The most important aspects of the staining process for these types of applications are sensitivity and compatibility with MS. Currently, silver stain and Coomassie blue stain are the most common visualization reagents. Silver stain has outstanding detection limits (spots containing 1–10 ng of protein can be observed) and takes about 1 to 2 hours for the entire process. It also has the advantage that no destaining step is needed. Coomassie is not quite as sensitive (with detection limits of 50–100 ng) and takes longer, generally 1 hour to stain and overnight to destain. Furthermore, with Coomassie stain, the gel pieces must be thoroughly washed before MS analysis, or adducts from the stain can occur on peptides. The development of colloidal Coomassie blue stain led to improved detection limits (8–10 ng), due to decreased background (Brush, 1998). Because of the complexity of the protein patterns, it is difficult for the human eye to compare two gels. Thus, 2D-PAGE gels usually are scanned by using a computer-assisted digital camera, and the picture is converted to a digital image. This allows for precise determination of spot location and intensity, and allows for computer-aided comparison of different gels. Protein spots are quantified by their visual intensities. This can then serve as a differential display of two different organism growth types or mutants. Very often in differential display analysis, only protein spots that show quantifiable change are excised, digested, and analyzed by MS, thus saving the time of analyzing every spot. Current 2D-PAGE software tools allow for integration of the entire analysis. Future directions in this area will include stains that are faster, more sensitive, simpler, and fully compatible with MS analysis. For example, development of the fluorescent SYPRO stains (Steinberg et al., 1996a,b; Valdes et al., 2000) has provided a sensitive and simple alternative to the silver and Coomassie blue stains.

In-gel Digestions   While other methodologies exist for spot identification, such as electroelution followed by Edman sequencing or blotting to membranes followed by MALDI-TOF-MS of membranes (Eckerskorn et al., 1997), we will focus on the most common methodology of *in-gel digestion* for MS. Initial concerns of protein analysis by MS following either Coomassie or silver stain due to permanent modification of peptides resulted in the delayed application of this technology. However, in 1996, Matthias Mann and coworkers silenced those fears when they showed convincingly that proteins separated on a 1D SDS-PAGE and stained with either Coomassie or silver stain could be digested in-gel, the resulting peptide mixture extracted from the gels, and then analyzed with MS, using either MALDI-TOF-MS or Nano-ESI-MS/MS (Shevchenko et al., 1996a). Multiple standards and fractions from yeast proteomes were analyzed, with a total of over 1,000 peptides measured. Their results indicated that neither Coomassie nor silver staining adversely affected protein identification by MS. It is important to note that they omitted the fixation/sensitization step with glutaraldehyde in the silver staining process, which is known to covalently modify proteins. Their final conclusion was that silver stain is overall a better choice for staining and subsequent in-gel digestion with MS analysis. The silver stain had a better limit of detection (it was nearly 100 times lower than that for Coomassie), was faster overall to process, and resulted in less background. Detailed protocols for in-gel digestions can be found in the literature (Shevchenko et al., 1996a) and on the World Wide Web at http://proteomics.uchsc.edu/protocols/index.html, http://www.bio.vu.nl/vakgroepen/mnb/proteomics/6.in-geldigestion.html, and http://www.hmc.psu.edu/core/Maldi/malditofprotocols.html.

Quality sample preparation is critical for these measurements. For example, inadvertent contamination of samples with human keratins from skin and hair from normal sample handling often is the biggest barrier to successful protein identification. Likewise, poor-quality reagents also are likely to present contamination problems, due to the high sensitivity of MS. It is therefore highly desirable to use sequencing grade trypsin (which has reduced chymotrypsin activity) for high-quality digestions. In large proteomics facilities, this entire process is automated with robots, which pick spots based on coordinates from the digital image, digest these spots in microtiter plates, extract peptides from the digested spots, and finally clean the peptide samples and deposit them on MALDI targets or into microtiter plates for nano-ES-MS/MS (see below). Membrane proteins, due to their hydrophobic characteristics, are generally harder to analyze by the typical in-gel digestion method than soluble proteins. An interesting variation on the procedure has recently been reported; it uses cyanogen bromide (CNBr) as well as trypsin and results in 100% sequence coverage of some standard membrane proteins (van Montfort et al., 2002).

MS Analysis and Database Searching   Analysis of gel-separated spots in proteomics can be accomplished by a variety of methodologies. The review listed above (Jungblut and Thiede, 1997) gives a very detailed schematic of all the possible modes of protein identification, and how they are related. We have simplified this to the three most common methodologies for protein identification after in-gel digestion, as illustrated in Fig. 9.4. The choice as to which of these three methodologies to use will depend on the instruments available, the throughput required, the necessary confidence of identification, and the complexity of the spot (very often "single" gel spots may contain up to 5–10 proteins). All three techniques have been demonstrated with low femtamole sensitivity, although the dynamic range for each is dramatically different. The three common techniques are highlighted below.

PEPTIDE MASS FINGERPRINTING (PMF)   This methodology is the most commonly used technique for analyzing proteins from PAGE gels. Briefly, the extracted peptides are spotted on a MALDI target plate and mixed with a suitable matrix compound (typically, alph-cyano-4-hydroxycinnamic acid, CHCA). The target is loaded into a MALDI-TOF mass analyzer, and each spot is analyzed by multiple laser shots. This provides a composite mass spectrum of the peptides in a given spot. Typically, all peptides are singly charged, and MALDI-TOF analyzers with reflectrons can provide 50 to 100 ppm mass accuracy on the peptides. Using commonly occurring autotryptic peptides as internal standards can increase mass accuracy. Querying a protein database with these

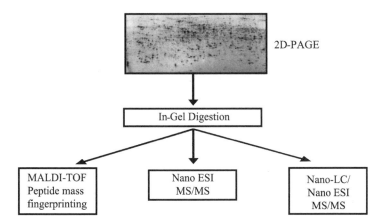

2D-PAGE

In-Gel Digestion

MALDI-TOF
Peptide mass
fingerprinting

Nano ESI
MS/MS

Nano-LC/
Nano ESI
MS/MS

Figure 9.4 *Flow diagram depicting the three most common mass spectrometry experiments that are conducted to identify peptides from the enzymatic digestions of 2D-PAGE gel spots.*

peptide masses for *in silico* generated tryptic peptides can identify the original protein. Generally, four to five peptides matching the protein with high-quality mass accuracy (50–100 ppm) are considered a good match. Search algorithms of these types are very easy to create in-house or can be found on the World Wide Web (go to www.matrixscience.com, http://prowl.rockefeller.edu/cgi-bin/Profound, http://us.expasy.org/tools/peptident.html).

The major advantages of this technique are speed and simplicity. The MALDI-TOF instruments are very easy to operate and can be automated easily. A single instrument can analyze $\sim$ 100 spots per day in an automated fashion. The major disadvantages of this technique are the limited accuracy in the peptide identifications and lack of dynamic range. If the gel spot contains three or more proteins or if in-gel digestion does not provide enough peptides per protein, identification becomes very difficult.

NANOSPRAY-MS/MS   This methodology was demonstrated to be an effective and robust approach for analyzing peptide mixtures generated from in-gel digestion (Wilm et al., 1996). In this technique, the peptide mixture is introduced into a small glass needle with a small tip (10 μm id). The needle is then mounted on a nanospray source and a high voltage is applied. A typical flow rate from the tip is 1 to 10 nL per minute, with ion detection usually accomplished by QIT, triple quadrupole, or QqTOF MS. The mass analyzer is programmed for a data-dependent MS/MS experiment, in which the most abundant peptides are interrogated. The measurement process is conducted until the sample is depleted, which is generally about 30 to 60 minutes under these conditions. This experiment provides both the intact peptide mass and MS/MS spectra, which can be used for querying protein databases with search algorithms such as SEQUEST and MASCOT, or for de novo sequencing. The major advantage of this technique is its ability to identify a protein from only a few peptides. Furthermore, this method is readily applicable to simple mixtures of 5 to 10 proteins. The major disadvantages of this technique are low throughput and difficulty of operation. In general, nanospray MS is more difficult to master and automate than MALDI-TOF-MS. This methodology has an intermediate dynamic range when compared with the other two methods.

NANOLC-NANOSPRAY-MS/MS   If a protein spot contains more than 5 to 10 proteins, or if large sequence coverage of each protein is desired (this is useful for identifying PTMs), then the first two methodologies are insufficient. For the best dynamic range measurement, NanoLC-Nanospray-MS/MS is the methodology of choice. While several variations exist for this technique, fritless nanospray columns provide the best sensitivity (Gatlin et al., 1998). In this technique, C18 particles are packed directly with a packing bomb into laser-pulled fused silica needles (generally $100-75$ μm id and $10-15$ cm in length, with $5-10$ μm id tips). The tip of the needle acts as the frit for the packing material. The sample can then be loaded offline with a loading bomb, or online via an injection loop. The integrated column/tip is placed in a nanospray source behind a microtee or a microcross. A high voltage is applied at this cross through a gold wire or rod. The effluent from a high-performance liquid chromatography (HPLC) system is then directed into the back of the cross from a transfer line. If the HPLC can provide low enough flow rates, then a split at the cross is not required, and a microtee is employed; otherwise, the flow must be split into a waste line (microcross). The flow through the nanocolumn is usually 100 to 200 nL per minute. This integrated column and nanospray tip is are generally incorporated into a QIT, triple quadrupole, or QqTOF MS system. The mass analyzer is again set up for a data-dependent MS/MS experiment. The HPLC will employ a gradient elution of 95% water to 95% acetonitrile over 30 minutes to 1 hour for the MS analysis. This experiment provides both the intact peptide masses and MS/MS spectra, which can be used for querying protein databases with search algorithms such as SEQUEST and MASCOT, or for de novo sequencing. The main advantage of this technique is enhanced dynamic range, due to the increased physical separation of peptides. Multiple proteins can be easily identified from a given spot with greater sequence coverage than the other two techniques. Again, the major disadvantage when compared with peptide mass fingerprinting by MALDI-TOF is low throughput and more complex operation (which is more difficult to automate).

Future of 2D-PAGE for Proteome Analysis   As stated earlier, 2D-PAGE followed by MS analysis has been established as the gold standard for proteome analysis, especially for microbial species. But within the last few years, there has been a noticeable migration away from this methodology toward pure liquid-based approaches, which will be described in the final section of this chapter. This movement is mainly due to the inherent weaknesses in the 2D-PAGE methodology. The advantages and disadvantages of 2D-PAGE as they relate to the liquid-based methods are highlighted below. Some of these inherent weaknesses are currently being addressed with new techniques; however, the fundamental fact remains that at the current time, the depth of analysis of whole proteomes by 2D-PAGE does not compare well with the emerging liquid-based methodologies. 2D-PAGE is still routinely used in many labs around the world and undoubtedly this will continue.

Advantages of 2D-PAGE for Proteome Analysis

(i)   Gold standard, widely used and understood.
(ii) Very high resolving power.
(iii) Sensitive staining methodologies available.
(iv) Commercial software is available for automated gel processing and quantitation.

Disadvantages of 2D-PAGE for Proteome Analysis

   (i) Poor reproducibility.
  (ii) Limited recovery of low-abundance proteins.
 (iii) Limited pI and MW ranges.
 (iv) Time-consuming.
  (v) Membrane proteins do not enter the second dimension effectively.
 (vi) Coupling with MS is an indirect process.
(vii) Intact protein analysis is very difficult.

Liquid Chromatography with Mass Spectrometry (gel-less method) The coupling of liquid chromatography with MS (LC-MS) is one of the most promising approaches to overcome some of the limitations of 2D-PAGE discussed above. The advent of ESI has provided a natural way to interface liquid chromatography directly to MS, since ESI involves dynamic introduction of a flowing liquid stream directly into a mass spectrometer. It is reasonable to propose connecting a liquid chromatography system to the electrospray source, so that the benefits of liquid-based separation can be combined with high-resolution molecular mass (and MS/MS) measurements. While some work has been done on the chromatography of intact proteins in conjunction with MS (see below), most of the effort has focused on the chromatography of enzymatically generated peptides in conjunction with MS, the so-called bottom-up or shotgun method for proteomics. This is primarily due to the fact that peptides are much easier to handle, separate, and analyze than intact proteins. It is somewhat counterintuitive that it is desirable to take a complex protein mixture and make it more complex by digesting the proteins into representative peptides. For example, each averaged size protein can generate ~20 peptide fragments. So, the proteolytic digestion of a sample containing 1,000 proteins will generate a new sample that contains ~20,000 peptides. Although this appears to be a poor choice, in practice liquid chromatography and MS of peptides are currently well developed and robust, even for very complex peptide mixtures. The advantages of coupling liquid chromatography with MS are obvious when one considers some of the necessities of proteomics, as highlighted below.

Dynamic Range The need for multiple dimensions of separation has become most apparent in the use of LC-MS for proteome analysis, due to the large dynamic range necessary for measuring a whole proteome. While 2D-PAGE offers a very high resolving power for proteins, this methodology is currently limited with respect to the types of proteins it can analyze and the number of quality identifications that can be made from any one gel. The coupling of multiple dimensions of chromatography with MS offers a solution to this problem. This can be easily noted from the fact that while ~1000 proteins have been visualized on a 2D-PAGE gel, no published reports of more than a few hundred proteins being identified from a single gel exist. It has now become routine to accurately identify 500 to 1,000 proteins from a single sample in 20 to 30 hours on a LC-MS/MS system operated in the automated mode.

Sensitivity While 2D-PAGE gels have very sensitive staining methodologies for *observing* spots, the ability to *identify* proteins from these spots by MS is not as successful. This is primarily due to the large sample losses in the in-gel digestion step. Pure LC-MS methodologies promise to be more sensitive, due to the reduced overall sample handling while keeping the sample in the liquid phase.

Quantitation   Differential analysis between two or more sample types is a primary need for successful proteome applications. One of the reasons that 2D-PAGE has remained the gold standard is that it currently is inherently better at quantitation than liquid-based methodologies, due to the fact that LC-MS suffers from matrix effects that make the comparison of run-to-run peak intensities very difficult. This is primarily due to the electrospray ionization methodology and not MS. Furthermore, the apparent accuracy of 2D-PAGE for quantitation became questionable when it was realized that many spots on any 2D-PAGE gels contained more than one protein. The use of stable isotopes for peptide labeling (see the quantitation discussion below) has proven that quantitation can be accomplished by LC-MS.

Protein Diversity   A major advantage of liquid-based methodologies in comparison with 2D-PAGE is the diversity of proteins that can be analyzed. Virtually any protein that can be subjected to either chemical or enzymatic digestion may be analyzed. This includes membrane proteins, proteins of high and low pI values, and proteins of high and low molecular mass.

Throughput   In the field of proteomics, one of the biggest concerns is sample throughput, including not only how fast samples are analyzed, but how well they may be characterized in a short period of time. This is just as important in bacterial proteomics as mammalian or plant proteomics. Currently, at least forty different microbes have been fully sequenced at the Joint Genome Institute (information available at http://www.jgi.doe.gov) and many more have been completed worldwide. Researchers need to be able to analyze the proteomes from these organisms under many different growth conditions and with many different mutants for a systems biology approach to be truly effective. For sample throughput, LC-MS has already been well developed in the pharmaceutical industry, where thousands of samples are processed by large numbers of mass spectrometers in hundreds of laboratories every year.

The trend toward liquid chromatography methods for proteome analysis is made clear by examining Table 9.3. Over the last 2 years, all large-scale proteome analyses have been accomplished with some form of LC-MS methodology. The analysis of *Oryza sativa* provides the best example for a direct comparison of 2D-PAGE analysis and LC/LC-MS/ MS analysis for whole proteomes (Koller et al., 2002). In this study, the rice plant was broken into three fractions: leaves, roots, and seed tissues. Proteins were isolated from each fraction and analyzed by 2D-PAGE, followed by automated Nano-LC/MS/MS or by multidimensional protein identification technology, or 'MudPIT' (Washburn et al., 2001). The analysis of all three fractions by 2D-PAGE-MS resulted in 556 nonredundant identifications, whereas the analysis of all three fractions by MudPIT resulted in 2,363 nonredundant identifications. Koller and coworkers made no mention of the length of time each analysis consumed, but we estimate the MudPIT analysis could have been accomplished in 5 days on a single mass spectrometer once the system was optimized. This throughput, as well as the enhanced dynamic range, is why the liquid-based methodologies are having the greatest impact on proteome analysis. Because of the current trend toward LC-MS for proteome analysis, we will direct the next section to detail some of the protocols used in LC-MS and some of the cutting-edge techniques such as MudPIT now being routinely used in proteome analysis. For further reading on the advancement of LC-MS in proteomics, see the excellent reviews by Peng and Gygi (2001), Mann et al. (2001), and Liu et al. (2002).

Methodologies For continuity, we will first outline the entire process of LC-MS/MS for shotgun proteomics in Figs. 9.5 and 9.6, as it could apply to microbial proteomes. Sample preparation, different versions of LC-MS procedures, and quantitation will then be explained in detail below. The shotgun proteomics technique begins with the enzymatic digestion of a microbial proteome sample and analysis of the resulting peptide mixture by automated LC-MS/MS or LC/LC-MS/MS in a data-dependent manner (Fig. 9.5 details this process). Bear in mind that the sample which is injected into the chromatographic system consists of a mixture of several thousands of distinct peptides. These peptides are separated physically over a period of time by their hydrophobicity or net charge, and are sequentially injected into the mass spectrometer and ionized by ESI. By using the mass spectrometer to record the overall ion intensity as a function of time, it is possible to obtain a base peak chromatogram (BPC) much like a UV chromatogram (Fig. 9.5, top left). During the entire chromatographic run, the mass spectrometer is oscillating between *full scan mode*, where it is acquiring m/z values of peptides entering the mass spectrometer at

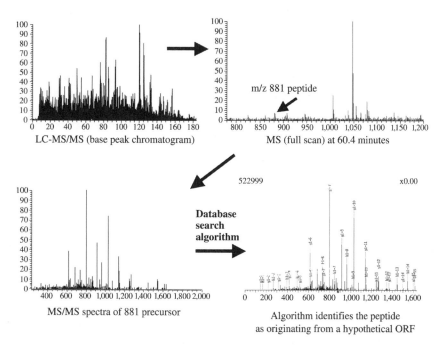

Figure 9.5 *Illustration of the shotgun proteomics technique for the microbe Shewanella oneidensis.* The entire cell lysate is digested with trypsin, cleaned up, and then injected directly onto a C18 reverse phase column interfaced online with a QIT mass spectrometer. The base peak chromatogram (upper left) reveals a complex mixture of peptides in this sample. A full scan mass spectrum at 60.4 minutes. (upper right) into the HPLC run reveals the presence of several peptides at this particular time slice. Even the minor peptides can be isolated and examined by MS/MS techniques (bottom left) to yield fragment ions that provide structural information (bottom right).

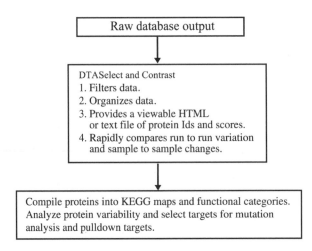

Figure 9.6 *Flow diagram depicting how the mass spectra from Fig. 9.5 are processed.* The raw data output is processed with the software algorithms DTAselect and Contrast (Tabb et al., 2002) to filter and compile the data into files of protein identifications and scores. This information can be used to generate biologically informative KEGG maps and functional category piecharts.

that time point (Fig. 9.5, top right), and subsequent *MS/MS mode*, which examines the fragmentation of the most abundant peptides (generally three to five) as they elute from the column (Fig. 9.5, bottom left). This latter mode is accomplished by the gas phase isolation of individual peptides, followed by collisional dissociation. The mass spectrometer records the fragment ions and mass of the precursor ion. To increase dynamic range, most methodologies employ some type of dynamic exclusion so that peptides that have already been fragmented are not fragmented again. The precursor masses and fragmentation patterns are then submitted to search algorithms such as MASCOT and SEQUEST, which can query thousands of MS/MS spectra against protein or nucleotide databases (Fig. 9.5, bottom right). The final stage of the process is illustrated in Fig. 9.6. Typically, a single LC-MS/MS experiment produces tens of thousands of MS/MS spectra. These spectra must be filtered and sorted in order to extract useful information from them. Filtering and sorting software such as *DTASelect* (Tabb et al., 2002) are used to extract and sort positive identifications, whereas the program *Contrast* (Tabb et al., 2002) is used to compare run-to-run variations and sample-to-sample changes. The protein identifications can then be compiled into KEGG maps and functional categories for the rapid viewing of metabolic and signaling pathways that are activated. This information allows targets to be designed for mutations, gene knockouts, and protein–protein interaction assays.

Sample Preparation, Digestion, and Cleanup Protocols   In general, the sample preparation for LC-MS is fairly simple and robust, but extremely important for the final outcome. If the protein sample is sufficiently concentrated (i.e., 1–5 mg per mL, which is a typical concentration of cell lysate from a microbe), it can be digested immediately. If the sample is dilute (<0.5–1.0 mg per mL), or contains interfering substances such as chlorophyll, lipids, or pigments, then acetone or chloroform/methanol precipitation may need to be employed. The sample is then denatured with 6 to 8 M guanidine or urea, and reduced with DTT or some other reducing agent at 60°C for 10 to 60 minutes. Iodoacetamide is commonly added at 0.1 M to alkylate cysteine residues. This step is performed at room temperature in the dark for 30 to 60 minutes. Care must be taken not to overalkylate the peptides, or nonspecific alkylation can occur. Some labs skip this step altogether, opting for a second harsh reduction after final digestion and directly before sample cleanup in a low pH solution. The denaturant concentration is lowered by dilution with Tris or bicarbonate buffer (with $1-10$ mM $CaCl_2$) and sequencing grade protease is added to the sample. The most common protease used is trypsin due to specificity, cost, and the fact that it produces a positively charged residue on the C-terminus (lysine or arginine), which aids ionization and peptide sequencing by MS/MS. The sample can be digested overnight, and fresh protease may be added in the morning to increase digestion. Some common variations of the digestion technique include the use of LysC protease along with trypsin (Link et al., 1999). LysC is active in up to 8 M urea and cuts at the C-terminal side of lysine residues (similar to trypsin) but not at arginine residues. Thus, the enzyme can be used under denaturing conditions, which increases overall sequence coverage when followed with the common trypsin digest. CNBr is another common chemical protease that can be used for membrane digestions since most hydrophobic domains lack adequate numbers of lysine or arginine residues for trypsin. Furthermore, the CNBr digestion is performed in 70% formic acid, which readily dissolves the membranes. This methodology has been shown to be effective for the analysis of whole yeast membranes (Washburn et al., 2001). The main disadvantage is the toxicity of CNBr; all reactions and sample handling should be performed in a fume hood. Another method for the digestion of membrane samples is a mixed aqueous-organic trypsin digestion. This methodology was demonstrated for the bacterium *Deinococcus radiodurans* (Blonder et al., 2002). Basically, the trypsin digestion is performed in a mixture of buffer and organic solvent such as methanol or acetonitrile. The organic phase allows for the solubilization of membrane proteins. After final digestion, the sample is desalted by solid phase extraction offline, or can be directly analyzed online with desalting utilizing either a trap cartridge or the analytical column itself. If sample quantity is not a concern, offline desalting by solid phase extraction is the most robust method for sample cleanup.

Types of LC-MS for Bottom-up Proteomics

ONLINE 1D LC-MS/MS The most simple bottom-up method for analyzing whole proteomes or protein complexes by LC-MS is the connection of a single reverse phase column directly to an electrospray mass spectrometer. In most cases, a high-performance liquid chromatography (HPLC) system is used to deliver a programmable gradient of high water to high organic over a C18, C8, or C4 reverse phase resin over a given length of time (usually $1-3$ hours). The solvents generally contain an organic acid such as formic or acetic acid ($0.01-0.1$%) to aid in ionization, but the standard HPLC ion-pairing reagent trifluoroacetic acid (TFA) should be avoided due to problems associated with electrospray (unless it is used at lower than 0.01%). The columns are typically 500 to 75 μm id, with lengths of 25 to 10 cm, and utilize flow rates of 10 to 0.100 μL per minute. The type of electrospray source used must match the flow rate of the column. For the higher-end flow rates, pneumatically assisted electrospray sources are used. For the lower-end flow rates, nanospray sources must be employed. The primary advantage of a lower flow rate system is an increase in sensitivity and resolution, whereas the main advantages of the higher flow rate system are robustness and ease of sample loading. The choice of column size, flow rate, and methodologies applied will depend on the application and amount of starting material available. The common mass spectrometers used for LC-MS applications in proteomics are QIT, triple quadrupoles, and QqTOF. Recently, online liquid chromatography in conjunction with high resolving FTICR-MS instruments has shown great promise and will be discussed separately below. A typical instrument setup for the high-throughput analysis of protein mixtures would include an autosampler to inject peptide sample onto the column, a high-performance pump to deliver the programmed gradient, and a reverse phase column, which can either be directly connected to an electrospray source through a low dead volume connection or the tip of the column can be directly pulled into a nanospray tip (this eliminates almost all dead volume) (Lee et al., 2002; Gatlin et al., 1998). The UV flow cell often used for detection in HPLC applications is generally omitted in MS-proteomics applications due to the large dead volume associated with its use and the lack of definitive information it provides. The mass spectrometer in most cases is operated in a data-dependent MS/MS mode, with computer control over the autosampler, HPLC and the mass spectrometer. The samples can be loaded onto the nanocolumns either offline with a pressure bomb or online with an autosampler. This avoids sample losses that may occur with an autosampler/sample loop, but will not be as high-throughput as online loading of sample. One of the biggest challenges in LC-MS applications for proteomics is rapidly loading a dilute sample onto a narrow bore column ($<$ 100 μm id), due to the low operational flow rates ($<$ 250 nL per minute). There has been much work in this area, and two methods have shown promise. The first involves loading the sample onto a small trap cartridge at a very high flow rate (van der Heeft et al., 1998; Devreese et al., 2001). This trap cartridge typically is very wide and short, in order to handle the high flow rate. The sample can be rapidly concentrated onto the cartridge and desalted. The trap is then back flushed on the resolving column, and a reverse phase gradient is run over both columns into the electrospray mass spectrometer. The advantages of this system include rapid sample loading and desalting, but this system can suffer from sample loss, as well as reduced resolution and sensitivity. The second method employs a vented precolumn of the same diameter as the resolving column (Licklider et al., 2002). This methodology may offer the same advantages as the previous method, without the loss of resolution and sensitivity.

Whole-proteome analysis by a single dimension of chromatography can be difficult due to the limited dynamic range of the system (this has led to the use of multidimensional systems, as discussed below). This is primarily a result of the slow scanning speeds of conventional mass spectrometers operating in data-dependent mode and the ionization suppression effects of some peptides over other peptides. At any given point in time in an LC-MS analysis of a whole bacterial proteome, hundreds of peptides can be simultaneously entering the mass spectrometer. Even the fastest mass spectrometers cannot isolate and fragment so many peptides in a short period of time. The use of multiple mass range scanning, or gas phase fractionation, for the analysis of whole bacterial proteomes has shown promise as a simple, cheap, and effective method for analyzing complex protein mixtures (Spahr et al., 2001; Davis et al., 2001; VerBerkmoes et al., 2002; Lipton et al., 2002). This method relies on injecting the same proteome sample onto an LC-MS system repeatedly, in which the mass spectrometer is programmed to scan a different narrow mass/charge range for each injection, thus allowing for more MS/MS spectra to be acquired from the sample. This methodology is not as sensitive as some of the methods described below, since the sample must be repeatedly analyzed (typically, 4–8 m/z ranges are examined, thus requiring that four to eight injections be made per sample), but this method is very simple to implement and is faster than the two-dimensional methods described below. Multiple mass/charge range scanning can now be used for the analysis of whole bacterial proteomes with the identification of 500 to 1,000 proteins per day, depending on the criteria used for identification, sample type, and the type of MS system employed. Furthermore, in microbial proteome analysis, sample quantity is usually not much of an issue, since 100 to 200 mg of total protein can easily be obtained from 4 to 5 g of cell paste.

MULTISTAGE LC-MS/MS   Due to the limited dynamic range offered by a single dimension of chromatography, methodologies are now being employed for coupling two or more dimensions of chromatography, either offline or online, in conjunction with ES-MS. While methodologies employing multidimensional chromatography have been commonly used in the separations field for many years, it has only been in the last few years that these methodologies were used in conjunction with MS for whole-proteome analysis. Recently, many different methods have been attempted to analyze whole proteomes in a relatively short period of time. One of the first methodologies described and probably the most widely cited is the previously mentioned multidimensional protein identification technology or MudPIT (Link et al., 1999; Washburn et al., 2001; Wolter et al., 2001). This technique employs a biphasic microcolumn integrated with the nanospray tip directly placed in front of an ion trap mass spectrometer. The first part of the column is packed with strong cation exchange (SCX) particles and the second part of the column with reverse phase (RP) C18 particles. A peptide mixture derived from the LysC/trypsin digestion of a whole proteome is directly loaded onto the back of the biphasic column with a pressure bomb. The column is placed in front of the mass spectrometer and an automated two-dimensional gradient is started by the HPLC. This gradient consists of an initial injection of low-concentration ammonium acetate by the HPLC onto the system, which moves a plug of peptides from the SCX material to the RP material. Peptides are then eluted from the RP material into the mass spectrometer with a typical, high-resolution RP gradient procedure. The system is then reequilibrated. This routine is repeated with increasing salt steps of ammonium acetate until all peptides have been run through the system (typically, 10–15 cycles or salt steps). The entire system is run in an automated mode, with computer control over the HPLC and mass spectrometer. The methodology was shown to be capable of analyzing a total of 1,484 proteins from three crude fractions from *Saccharomyces cerevisiae* (Washburn et al., 2001) in $\sim$ 4 to 5 days with a single mass spectrometer. The analysis revealed that the system was not biased toward any given group or type of protein. A variation of this methodology was recently reported and compared with the MudPIT technique (Peng et al., 2002). In this technique, the SCX column was not connected online with the RP column. Instead, peptides were loaded onto the SCX column and eluted with a linear gradient of increasing salt. Fractions were collected, concentrated, and then loaded onto the online vented column described above via an autosampler. This technique was again used to analyze the *S. cerevisiae* proteome, resulting in the identification of 1,504 proteins. It is difficult to compare the two techniques directly head-to-head, because of the variation in protein identification methodology, the amount and type of starting material, and the total length of time for analysis. The MudPIT technique appears to have the advantage of being more sensitive, since sample loss should be minimized in an online system; however, the offline method has the advantage of allowing the use of a much larger first SCX dimension and a true linear gradient instead of a step gradient. Variations on the two-dimensional methodology include the use of strong anion exchange instead of strong cation exchange in the first dimension (Mawuenyega et al., 2003), as well as the separation of intact proteins in the first dimension by strong anion exchange followed by tryptic digestion of the fractions and 1D-LC-MS/MS of the resulting peptide mixtures (VerBerkmoes et al., 2002).

LC FTICR-MS Within the last few years, a very promising new MS methodology for the analysis of whole proteomes has emerged. This technology involves the use of high-performance liquid chromatography in conjunction with electrospray ionization on high-performance FTICR mass spectrometers (Martin et al., 2000; Quenzer et al., 2001; Shen et al., 2001). The FTICR instruments provide better dynamic range, sensitivity, and mass accuracy over other more conventional mass spectrometers used in proteome research. These advantages, coupled with the high resolving power and sensitivity of nano HPLC columns, may provide access to low-copy number proteins in cells (one to five copies per cell). The current disadvantages of this technology include the high price of FTICR mass spectrometers, the expertise required to operate the instruments, and the difficulty of employing data-dependent MS/MS methods on the system. An intriguing experimental approach termed *accurate mass tags* (AMT) has been developed by Richard D. Smith and colleagues by using very high-pressure liquid chromatography (Tolley et al., 2001) in conjunction with an 11.4-Tesla FTICR mass spectrometer (Smith et al., 2002a,b; Lipton et al., 2002). This methodology promises to alleviate the need for the routine MS/MS of peptides. In this technique, peptides are separated by very high-pressure liquid chromatography (5,000–10,000 psi) on very long, narrow columns (85-cm length, 150 μm id) packed with C18 particles. The peptide mixtures are first analyzed by LC-MS on QIT-MS to obtain MS/MS data and retention times. Peptides that pass the minimum criteria (Xcorr > 2.0) are labeled *potential mass tags*. The same peptide mixtures are then analyzed on the high-magnetic field FTICR-MS by the same chromatography system, except in this case the FTICR-MS is only acquiring accurate mass/charge values rather than MS/MS. By the internal calibration of each mass spectrum with known peptides, mass accuracies to within 1 ppm can be obtained. Once a peptide has been identified on both systems within the criteria limits, it is labeled an accurate mass tag, and its elution time and mass are stored in a database. Thus, when the sample is analyzed again, the peptide mixtures can be screened only by the FTICR-MS system, without the need for time-consuming MS/MS acquisitions. The authors contend that the AMT databases can be used for rapidly screening many different growth states of a given organism, without the need for routine MS/MS. This methodology was recently used for the largest analysis of a proteome to date, in which >61% of the proteome of ionizing radiation-resistant bacterium *Deinocoocus radiodurans* was identified with the AMT approach by analyzing the microbe grown under 15 different growth conditions.

Quantitation The analysis of whole proteomes and protein complexes by MS can provide very useful *qualitative* information, but one of the most interesting areas of proteomics is the *quantitative* comparison between different growth conditions or mutants for a given organism. For example, the quantitative analysis of microbial proteomes has been dominated by 2D-PAGE followed by MS analysis, but this is shifting toward pure liquid-based methods due to the deficiencies in the 2D-PAGE methodology described above. This is primarily due to the fact that the pure liquid-based methods are inherently higher-throughput and are not biased against any protein type. There are difficulties with the liquid-based methods for quantitation, and technologies to address these are only now being developed and implemented. Due to matrix effects associated with both ESI and MALDI, direct comparisons of ion peak heights or area for given peptides or proteins eluting from LC columns into the MS should only be used as approximations for abundance levels, and are most likely not accurate for absolute quantitation. Recent developments in stable isotope labeling has allowed for the accurate, relative quantitation of proteins in two different samples, such as *E. coli* grown under high salt and low salt conditions. In these experiments, a given protein(s) can be compared with its counterpart from a different growth condition to obtain a relative expression level of up- or down-regulation, but an absolute level of protein expression is still very difficult to determine. Three main methodologies that employ stable isotopes for relative quantitation have developed over the last 4 years. Each has advantages and disadvantages, which are highlighted below.

ISOTOPE CODED AFFINITY TAGS (ICAT)   ICAT was originally developed in 1999 (Gygi et al., 1999) and has become commercially available through Applied Biosystems. The methodology has since been applied to the analysis of protein expression and comparison with microarray data in *S. cerevisiae* (Ideker et al., 2001; Griffin et al., 2002), as well as the analysis of human cell line HL-60 microsomal proteins (Han et al., 2001). This technique entails the use of an isotope encoded affinity tag. The proteins in a sample are mixed with the ICAT reagent, which specifically reacts with cysteine residues. The reagent has an isotopic label (either a "light" version containing either hydrogen atoms on the aliphatic chain or a "heavy" version containing eight deuterium atoms in the same location) and a biotin affinity tag for the isolation of cysteine-containing peptides. Thus, a tagged peptide can contain either the light or heavy version of the tag. This technique is applied to protein samples by labeling one sample with the light reagent and the other sample with the heavy reagent. The samples are then combined, digested with trypsin, and passed over an avidin column to enrich the cysteine-containing peptides. LC-MS methodologies described above can then be used to analyze the complex peptide samples to obtain peptide identifications as well as quantitative information by comparing the peak heights of the heavy and light versions of the same peptides. Accurate protein quantification should be obtained by averaging multiple labeled peptides from the same protein. The main advantage of this technique stems from the fact that the labeling is done directly after the cell lysates have been prepared, so any changes involving sample handling affect both samples equally. Also, the interrogation of only cysteine-containing peptides makes the peptide mixtures much simpler and easier to analyze. Finally, ICAT peptides can be easily sequenced by MS/MS and identified with MASCOT or SEQUEST. The software to analyze the large data sets is under development by the inventors of the technique, as well as many MS instrument companies. The major disadvantages of this technique are the commercial price of the reagent and the fact that the current version of the commercial reagent only labels cysteine residues. This latter point presents a serious problem for some bacterial species. The average number of cysteines per protein is much lower in bacterial species, as compared with some common eukaryotic species. Furthermore, a large percentage (50–60%) of proteins in bacterial species contain either zero, one, or two cysteine residues, as shown in Table 9.4, which either prevents quantitation, or requires that the quantitation be based on one or two data points.

^{18}O WATER LABELING   This methodology was recently introduced as an alternative to ICAT for accurate protein quantification (Yao et al., 2001). In this technique, one protein sample is digested with trypsin in the presence of ultrapure ^{18}O water, while the other sample is digested in normal water. The samples are then pooled and analyzed by LC-

TABLE 9.4 Comparison of Cysteine-containing Proteins in Different Organisms

S. oneidensis (bacterium)			*S. cerevisiae* (yeast)		
Average No. of Cysteine Residues per Protein = 3.159			Average No. of Cysteine Residues per Protein = 6.268		
Cysteines per protein	Proteins	%	Cysteines per protein	Proteins	%
0	1,000	19	0	627	9
1	950	18	1	663	9
2	830	16	2	668	10
3	646	12	3	645	9
Four or more	1,751	31	Four or more	4,332	62
Total proteins	**5,177**		Total proteins	**6,935**	
E. Coli (bacterium)			*A. Thaliana* (plant)		
Average No. of Cysteine Residues per Protein = 3.725			Average No. of Cysteine Residues per Protein = 7.892		
Cysteines per protein	Proteins	%	Cysteines per protein	Proteins	%
0	597	14	0	1,328	5
1	606	14	1	1,794	7
2	634	15	2	2,080	8
3	574	14	3	2,188	8
Four or more	1,829	43	Four or more	18,426	71
Total proteins	**4,240**		Total proteins	**25,816**	

MS/MS or MALDI methodologies on high-resolution mass spectrometers. The results clearly demonstrated that the carboxy termini of the tryptic fragments digested in ^{18}O water are fully labeled with ^{18}O, and this label is stable. Thus, all tryptic peptides from the H_2 ^{18}O sample have an increase of 4 daltons in mass (two incorporated oxygens on the *C*-terminus of each peptide). The peptides can then be quantitated by comparing peak areas of coeluting peptides separated by 4 daltons. This technique was also demonstrated to work with another common protease endoprotease Glu-C (Reynolds et al., 2002), thus adding to its versatility. The main advantage of this technique is that it works on any protein which can successfully be digested with either trypsin (K or R residues) or Glu-C (E or D residues). Furthermore, the technique has been shown to work successfully on phosphorylated, glycosylated, and disulfide-containing proteins (Reynolds et al., 2002). Elution times were also shown to be consistent with the labeled and unlabeled peptides (Reynolds et al., 2002). The main disadvantage of this technique is the price and availability of H_2 ^{18}O, as well as the need for a high-resolution mass spectrometer to analyze the peptides with such small mass differences.

NITROGEN LABELING In this methodology, the microbe of interest is grown under a defined media with either normal media (containing naturally occurring isotopic abundances) or isotopically enriched or depleted media (Oda et al., 1999; Paša-Tolić et al., 1999). The most common method is to grow the microbe in defined media without amino acids with only ammonium sulfate as a nitrogen source. The ammonium sulfate can then be either ammonium-^{15}N sulfate or normal ammonium sulfate. The microbe will incorporate the stable heavy isotope in its proteins. The normal and heavy-labeled samples can then be grown under the desired conditions, combined, lysed, and digested with a protease. The peptides will have heavy and light pairs that should elute at the same time in a LC-MS/MS experiment, and again the peptides can be quantified by comparing peak areas. This methodology has recently been employed for the largest quantitative proteome analysis to date of the yeast proteome by nitrogen labeling, followed by MudPIT analysis (Washburn et al., 2002). The major advantages of this technology are its low cost and ability to quantitate any type of protein that can be digested with either chemical or enzymatic methods. The samples are mixed immediately after growth so that any changes in sample preparation affect both samples in the same manner. The major disadvantage is that this technique may only be used for species whose growth conditions can be exquisitely controlled.

9.4.3 Top-down Mass Spectrometry Proteomics

Although proteolytic digestion and MS characterization (i.e., bottom-up proteomics) have become very powerful over the past 5 years, it is clear that this is an *indirect* protein identification technique, as the *intact* protein species are never measured directly, but rather only a fraction of the proteolytic peptides for any given protein are examined. This leads to some concern that subtle aspects of the protein, such as posttranslational modifications, might be missed by the bottom-up approach. This has prompted investigation into developing MS technology for the measurement of intact forms of proteins in complex mixtures. Although developing such technology may seem straightforward based on the extensive past work on characterizing purified protein samples, in fact, this approach turns out to be a formidable analytical challenge for proteomes due to at least three factors. First, the protein molecule masses can range from 5 to 200 kDa, requiring high-performance MS technology for accurate measurements. Second, the extreme heterogeneity of protein sequences gives rise to a significant ionization suppression effect when complex mixtures are examined. Thus, the proteins with the largest amount of surface charge will ionize most easily and be overrepresented in the mass spectrum relative to their abundance in the sample. Likewise, the proteins that are not easily ionized will be underrepresented. This factor suggests that some type of prefractionation, or online chromatography, will most likely need to be used for intact protein measurements. Third, the unambiguous identification of larger proteins is difficult, due to the isotopic packet that confounds accurate mass measurements and the inability to extensively fragment these proteins under tandem MS conditions to obtain complete sequence information. All three factors are much less of an issue for peptides, because of their lower molecular masses and more extensive fragmentation. However, research underway in several laboratories has shown remarkable progress in overcoming these challenges for the top-down approach. One particular factor worthy of note is that most of the developments of the top-down approach have focused on the experimental MS measurement technology. As a result, the bioinformatics component is much less developed for top-down data analysis.

This section addresses three important subtopics: sample preparation, molecular mass measurement, and structural interrogation. As will be obvious from the following discussion, these factors are somewhat different for the measurement of intact proteins relative to proteolytic peptides.

Sample Preparation   One of the challenges in isolating complex protein mixtures is keeping the proteins intact and soluble during the fractionation process. Because MS measurements do not require proteins to be in their active forms, it is often desirable to denature the entire complex mixture as early in the cleanup process as possible. While this usually inactivates cellular proteases, it often causes undesirable protein precipitation in the samples. For the bottom-up MS approach, it is advantageous to denature and digest the complex protein samples as early as possible in the cleanup process. Because only peptides are measured, protein stability is not an issue for this method. In contrast, protein stability is critical for the top-down MS approach. To enhance this, during the cellular lysing process, a protease inhibitor cocktail is often added to arrest protein degradation. The protease inhibitors, which are often small molecules, stabilize the protein samples, but need to be removed prior to MS characterization. This is often achieved by dialysis (conducted under cold conditions to minimize enzymatic activity) or reverse phase cleanup procedures. The samples are usually maintained at $-80°C$ until analysis.

Molecular Mass Measurement   The critical component for top-down proteomics by MS is measurement of the molecular masses of the intact proteins. The four important experimental aspects of this measurement are mass accuracy, mass resolution, dynamic range, and detection sensitivity. Because of the wide molecular mass range of possible proteins, some researchers have proposed the application of MALDI-TOF-MS technology. Although this is probably the best approach for very large species, the mass resolution and accuracy are very limited for this technique. For example, a protein with a molecular mass of 60 kDa can only be measured under the *best* TOF-MS conditions to about 0.02% (~ 12 Da). While this mass measurement is far superior to what is obtainable using gel electrophoresis, this value could still correspond to many proteins within a given database. A much higher level of mass accuracy would limit the number of possible proteins. This is the driving force for employing techniques such as ESI-FTICR-MS for intact protein measurements. Although there are still many experimental parameters to be worked out to make this a routine tool for intact proteome characterization, the technology does provide unprecedented capabilities for high-performance measurements. For example, the same protein with a molecular mass of 60 kDa could be measured with the FTICR-MS technique to about 0.0005% or 5 ppm (~ 0.3 Da). This level of accuracy has been demonstrated recently by ESI-FTICR-MS on the intact form of GroEL from *E. coli*, in which the *measured* most abundant isotopic mass of 57,196.431 Da compares very favorably with the analogous *calculated* value of 57,196.734 Da (an error of 5 ppm) (Hettich, unpublished results, 2002). The most straightforward top-down MS approach would attempt to identify the protein from only its molecular mass. For example, searching the SWISS-PROT protein database (available at http://us.expasy.org/) with a molecular mass of 57,196 $\pm$ 0.3 Da for *E. coli* yields only one possible match—the GroEL chaperonin protein. In fact, for this particular bacterium, this protein could be identified from the TOF-MS data as well, since this is the only protein within 57,196 $\pm$ 12 Da. For eukaryotic systems, however, the situation becomes more complicated. By searching the SWISS-PROT database for humans, we determine that there are five possible proteins falling within the mass window of the TOF-MS measurement (57,196 $\pm$ 12 Da). The higher-resolution mass measurement with the FTICR-MS technique would be sufficient to identify a unique protein from this batch and, in fact, would be adequate to resolve the cochlin precursor protein (COCH_HUMAN) at 57,193.15 Da from the major capsid protein L1 (VL1_HPV2A) at 57,193.61 Da. *Thus, high-resolution and accurate mass* *measurements of intact proteins are often sufficient information to identify many bacterial proteins, without further structural information.* This statement should be tempered with the knowledge that posttranslational modifications would alter the measured molecular masses and make it difficult to correlate the measured protein mass with the value predicted from the genome data. For this reason, it is best to integrate the measured molecular mass information with either structural data obtained by tandem MS or with data obtained by the bottom-up method on the same organism (VerBerkmoes et al., 2002). This latter approach will be discussed in Section 9.4.4.

High-resolution molecular mass measurements of intact proteins reveal the complex isotopic packet resulting from the combination of naturally occurring isotopes. This necessitates comparing the measured and calculated isotopic distributions to verify protein identification (Blank et al., 2002). In practical terms, the high-resolution molecular mass measurement is used to query a protein database for a given organism. The possible protein matches falling within the specified mass accuracy window are tabulated, and a calculated isotopic distribution is determined for each one (for FTICR-MS measurements, there are usually no more than three to four possible proteins within the 5–10 ppm range of the measured mass). For each putative protein, the calculated isotopic distribution and most abundant peaks are compared to the measured values for final protein determination.

Even with the high-resolution molecular mass measurements discussed above, the dynamic range and heterogeneity of intact proteins in these complex mixtures confound the MS measurements. The basic problem stems from the limited ability to simultaneously measure hundreds (or even thousands) of proteins in a single mixture. An obvious solution to this dilemma is to incorporate some aspect of protein fractionation, either offline or online, with the MS measurement. Although this increases the sample handling and thus possible contamination or sample losses, the MS measurement requirements are greatly relaxed. For example, offline anion exchange chromatography can be used to fractionate complex protein mixtures from crude cell lysates. Each fraction, which contains between 10 to 100 proteins, is more easily interrogated by MS. Figure 9.7 illustrates the ESI-FTICR mass spectrum of an anion exchange fraction from yeast. Note the appearance of at least six protein species. Each of the molecular ion regions can be expanded to illustrate the high-resolution mass measurements possible with this technique. Note that both phosphoglycerate mutase and phosphoglycerate kinase were identified at high accuracy (i.e., both less than 3 ppm error). By integrating these data with the bottom-up MS data from this fraction, it was confirmed that the molecular ion form of the mutase was missing the *N*-terminal methionine, whereas the molecular form of the kinase was not only missing the *N*-terminal methionine, but was also acetylated as well.

The most common protein fractionation approach has been to incorporate reverse phase liquid chromatography online with MS. This arrangement permits proteins to be physically separated by their hydrophobicity on the stationary phase of the column, and then eluted sequentially directly into the mass spectrometer. Although the entire measurement takes longer (usually about 1 hour for the LC-MS experiment), a much more extensive analysis of the complex protein mixture is possible. This approach has been demonstrated for the characterization of the chloroplast grana proteome (Gomez et al., 2002), and the yeast large ribosomal subunit (Lee et al., 2002), and resulted in not only protein identifications but also detection of posttranslational modified species. It is feasible to employ a multidimensional chromatographic approach for more enhanced protein fractionation. For example, a two-dimensional LC-MS experiment has been conducted on *S. cerevisiae* by

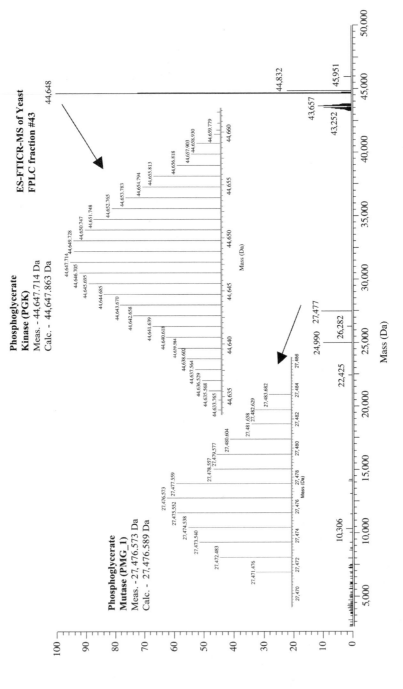

Figure 9.7 *ES-FTICR-MS of intact proteins from an anion exchange liquid chromatography fraction from yeast lysate. Note the appearance of at least six different protein species. The insets reveal how high-resolution measurements of the deconvoluted isotopic packets provide measured molecular masses that are in excellent agreement with calculated values.*

using a version of gel electrophoresis employing acid-labile surfactants, followed by reverse phase LC directly into an FTICR-MS (Meng et al., 2002).

There are several alternatives to online chromatography. One such approach involves a surface-enhanced laser desorption/ionization TOF-MS method (Merhant and Weinberger, 2000). For this method, a variety of chemical (hydrophobic, ionic, or mixed) or biochemical (antibody, DNA, enzyme, or receptor) surfaces are used to preferentially absorb selected protein species. This allows the fractionation to be fairly generic or highly specific, thereby selectively reducing the complexity of the protein sample. These surfaces can be incorporated into protein chips, providing a high-throughput sampling methodology for MALDI-TOF-MS. Another alternative to liquid chromatography focused on exploiting the demonstrated power of gel electrophoresis. As a modification of conventional 2-D PAGE, MS has been used to replace the size-based separation component of SDS-PAGE separation (Ogorzalek Loo et al., 2001). For this method, the proteins separated according to pI are then measured by MALDI-TOF-MS, with either postsource decay dissociation of intact proteins, or peptide mass mapping experiments. Such information can be used to construct virtual two-dimensional gels.

Structural Interrogation   To unambiguously verify the protein assignment by top-down MS, it is advantageous to acquire at least some structural information for the intact proteins (Mortz et al., 1996; Reid and McLuckey, 2002). As discussed earlier, this can be accomplished with a variety of tandem MS experiments, involving collisional dissociation, electron dissociation, or photodissociation. Although proteins usually fragment much less extensively than peptides, there is often sufficient fragment ion information to confirm or reject a possible protein identification from the accurate mass measurement. For example, the presence of only three or four fragment ions from a protein was found to be sufficient for a 99.8% probability of identifying the correct protein from a database of 5,000 bacterial protein forms (Meng et al., 2001). This methodology can be applied for proteins both with and without disulfide bonds (Nemeth-Cawley and Rouse, 2002). Electron capture dissociation shows promise for the most extensive fragmentation of intact proteins in a high-throughput manner (McLafferty et al., 2001; Horn et al., 2000a). This fragmentation process yields abundant c- and z-type ions, and usually provides extensive sequence coverage of proteins even up to 45 kDa in size (Ge et al., 2002). A combination of collisional dissociation and electron capture dissociation can be used to provide complementary information on intact proteins in bacterial proteomes (Demirev et al., 2001). For very large proteins (molecular masses exceeding 150 kDa), it may be advantageous to employ partial proteolytic digestion to make large peptides (5–50 kDa), and then characterize these species (Forbes et al., 2001). One of the most comprehensive techniques for top-down MS is a combination of capillary LC-MS with infrared multiphoton dissociation (IRMPD) (Li et al., 1999). IRMPD provides a rapid method of fragmenting the intact proteins, offering structural information on species whose accurate molecular masses have been measured.

9.4.4   Relating Mass Spectrometry Proteomic Data to Biological Information

In the long term, the value of proteome measurements will be judged not by the *extensive catalog lists* of proteins from an organism, but rather by the *biological information* it provides. Because of this, effort should be taken to make sure that the data output presentation and analyses are done in such a way to enable biological scientists to sort through the information in a meaningful fashion. At present, the procedures for achieving this goal are somewhat undefined, but are rapidly developing. Strong effort should continue to be placed on this "interface" region between analytical technology and molecular biology. If done successfully, it is very likely that developments in each respective area will fuel advances in the other area.

A few large-scale proteomic studies have been conducted on both prokaryotic and eukaryotic organisms, and are beginning to reveal detailed and somewhat surprising information about the proteins present. For example, a systematic proteomic study of rice leaf, root, and seed tissue has been completed, yielding the identification of 2,528 unique proteins (Koller et al., 2002). A comparative examination of the expressed proteins indicated that that enzymes involved in central metabolic pathways were present in all tissues, whereas metabolic specialization was supported by tissue-specific enzyme complements. The ADP-glucose pyrophosphorylase was cited as one specific example, in this case providing evidence for distinct regulatory mechanisms involved in the biosynthesis and breakdown of separate starch pools in different tissues. Furthermore, several allergenic proteins were identified in the seed sample, suggesting that proteomic measurements may have the potential to survey food samples for the presence of allergens.

A global proteomic analysis of the radiation-resistant microbe *D. radiodurans* has been reported (Lipton et al., 2002). More than 60% of the predicted proteome were identified in this study. Seventy-four of the 148 predicted stress-response proteins were identified, including the two classes of annotated proteins (catalase and superoxide dismutase) known to play a role in the detoxification process. Stable isotope labeling with ^{15}N media was employed to investigate the quantitative changes in protein expression as a function of radiation dosage. Initial studies revealed that the expression of the proteins RecA and DNA-directed RNA polymerase I was significantly induced during recovery from 17.5 kGy irradiation, in agreement with previous observations. In total, although it will take some time to work through the extensive proteomic data, it is clear already that such measurements have the potential to yield important information about the function of this unique organism.

The discussions in the previous sections of this chapter have indicated that both bottom-up and top-down MS proteomic approaches are powerful, but each still has some incompletely characterized issues. It appears that the most comprehensive proteome characterization by MS would probably be achieved by employing a combination of both methods. This would capitalize on the strength of the bottom-up approach for cataloging a large fraction of the proteins in the proteome, whereas the top-down approach would be essential for elucidating their molecular forms and the presence of posttranslational modifications (acetylation, phosphorylation, signal peptide truncation, etc.) and gene annotation start site errors. This type of approach is likely to yield the most detailed information for the biological characterization of proteomes.

We have recently developed and demonstrated such a comprehensive method for proteome analysis that integrates both the top-down and bottom-up approaches VerBerkmoes et al., 2002). Our technique was applied to the proteomic characterization of the gram-negative facultative anaerobe *Shewanella oneidensis* MR-1, which is of substantial interest for the bioremediation of metals in contaminated soils. Our integrated approach for *S. oneidensis* enabled the facile detection of such common posttranslational modifications as the loss of *N*-terminal methionine, signal peptide cleavages, as well as incorrect translational start sites. To present the data in a meaningful format, the proteins

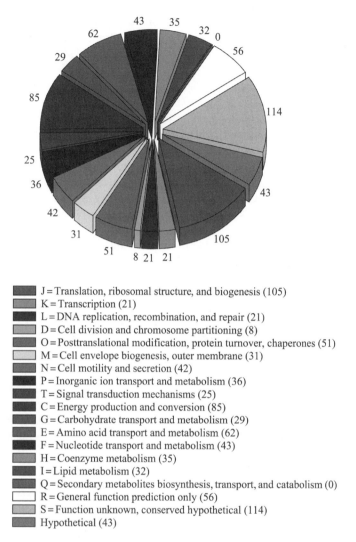

J = Translation, ribosomal structure, and biogenesis (105)
K = Transcription (21)
L = DNA replication, recombination, and repair (21)
D = Cell division and chromosome partitioning (8)
O = Posttranslational modification, protein turnover, chaperones (51)
M = Cell envelope biogenesis, outer membrane (31)
N = Cell motility and secretion (42)
P = Inorganic ion transport and metabolism (36)
T = Signal transduction mechanisms (25)
C = Energy production and conversion (85)
G = Carbohydrate transport and metabolism (29)
E = Amino acid transport and metabolism (62)
F = Nucleotide transport and metabolism (43)
H = Coenzyme metabolism (35)
I = Lipid metabolism (32)
Q = Secondary metabolites biosynthesis, transport, and catabolism (0)
R = General function prediction only (56)
S = Function unknown, conserved hypothetical (114)
Hypothetical (43)

Figure 9.8 *Functional category piechart of protein classes identified by shotgun proteomics LC-MS/ MS methodology for the microbe Shewanella oneidensis.* (Reprinted from VerBerkmoes et al., 2002.) Reprinted with permission from *J. Proteome Res.*, 2002, vol. 1, pp. 239–252. Copyright (2002) American Chemical Society.

identified in the study were sorted into functional categories using the clusters of orthologous groups of proteins (COG) classification system (Tatusov et al., 1997, 2001). This is summarized in the piechart of Fig. 9.8. Proteins from all functional categories except Q-secondary metabolite biosynthesis, metabolism, and transport were identified. Many chaperones, numerous peptidases, and a number of thiol-disulfide isomerases, thioredoxins, and peptidyl-prolyl *cis-trans* isomerases, which are all involved in protein folding or turnover, were found. The complete gluconeogenic pathway, tricarboxylic acid cycle, and electron transport system (ETS) were identified, although most of the membrane-embedded subunits of the ETS complexes were not found. Many of the basic pathways for shunting three and five carbon skeletons to various locations are detected, as are the pathways for purine and pyrimidine metabolism. Most of the enzymes from the amino acid biosynthetic pathways are absent, while many from the metabolic pathways are present. Interestingly, many of the enzymes involved in vitamin biosynthesis were detected.

Surprisingly, three of the five decaheme cytochromes and one of the two outer membrane proteins from a cluster thought to be involved in iron and heavy metal reduction under anaerobic conditions (Beliaev et al., 2001) were present. In addition, there were other cytochrome c-like proteins and cytochrome c maturation factors present. Considering the genes mentioned above, it is not surprising that many of the genes involved in heme biosynthesis are detected. A periplasmic binding protein whose signal peptide was putatively identified in the top-down approach is the first gene in an operon also containing an ABC transport ATPase and permease, and three genes involved in molybdopterin biosynthesis, one of which was also detected. In addition, only 135 bp separates the beginning of this operon from a two-component signal transduction system, both members of which are detected, being transcribed in the opposite direction. This is a good candidate for a bidirectional promoter. Finally, there is one 7-gene and one 3-gene operon, and three individual genes that are all also involved in molybdenum/tungstate transport and/or cofactor biosynthesis, of which at least one component was detected. All of this taken together with the detection of some oxidoreductases (nitrate reductase, formate dehydrogenase) that use a molybdo/tungstate-pterin cofactor makes an interesting system for future study. Finally, 43 purely hypothetical or *in silico* predicted genes with no BLAST hits were detected, thereby confirming their existence. This study demonstrates a novel integration of top-down and bottom-up proteomics analysis strategy for the characterization of the proteome of *S. oneidensis*. In several cases, this approach was critical for the identification of posttranslational modifications and an incorrect gene start site (for thioredoxin). Improved and/or more extensive separation procedures should help improve the sensitivity of both approaches, especially for the top-down method.

9.5 SUMMARY

The conventional approach to molecular biology has been to identify a particular protein or protein class, and then systematically purify and study the structure, function, and interactions of that protein in extensive detail. The advent of experimental techniques for characterizing very complex mixtures of proteins has enabled the emergence of *proteomics*, which now permits the simultaneous examination of virtually all the proteins expressed by an organism under a specific growth condition. This research direction represents a paradigm shift in molecular biology, in which one does not need to target a particular protein or protein class, but rather can monitor the entire protein complement of the genome in a single experiment. This provides a top-down view of all the proteins that are essential for an organism's life cycle, and is likely to provide new information that might not have been easily achieved with the traditional molecular biology hypothesis-driven approach. This mindset has given rise to a systems biology approach for studying the processes of life. Clearly, the most comprehensive picture of life would involve complete characterization of an organism's genome, transcriptome, proteome (including protein–protein interactions), and finally, metabolome. As recently as a few years ago, this kind of detailed information seemed unattainable. However, with the current level of experimental technology, including high-performance analytical instrumentation and sophisticated bioinformatics tools, this level of detail is becoming achievable, at least for simple prokaryotic organisms. If successful, such information would not only help us to better understand the mechanics of life, but would certainly be likely to favorably impact basic issues, such as human health.

Although enormous technological challenges must be addressed to achieve these goals, at the present time there appears to be no fundamental roadblocks to present such a scenario from becoming a reality. The technological hurdles, while substantial, do not appear to be insurmountable. The future of this field is likely to be marked by a replacement of the slower, labor-intensive gel electrophoresis technologies with higher-throughput, wider dynamic range gel-less methods, such as the multistage LC-MS/MS and AMT approaches discussed previously. Likewise, the top-down proteomics approach will continue to see rapid development, thus providing a powerful complementary means of interrogating intact protein structures and identifying modifications. The concomitant advances in computer hardware and software are already enabling complicated bioinformatics analyses of huge datasets. The driving force for success in this general area of defining the "molecular machinery of life" will continue, at least for the short term, to be the high-throughput, accurate, comprehensive characterization of all biological components and their interactions. There will certainly be a move toward increased automation to accomplish these goals. Recall that at the inception of the Human Genome Program, there was much skepticism in the scientific community that the entire human genome could be acquired within our lifetimes. In spite of this, the gauntlet was thrown down, and the technology and bioinformatics rose to meet the challenges. This remarkable achievement has paved the way for the entire field of proteomics, in which there appears to be an urgency to identify and characterize all the gene products that are predicted to be present based on the genome.

This is truly an exciting time to be involved in such scientific endeavors. The completion of the Human Genome Project has been compared to at least the equivalent of landing a human on the moon. As in that case, the development of genomics and proteomics has opened up an entire landscape of new research opportunities, and is very likely to take us to places that we have been unable to even comprehend previously.

FURTHER READING

Dass, C. 2001. *Principles and Practice of Biological Mass Spectrometry.* Wiley-Interscience, New York.

Dreger, M. 2003. Subcellular proteomics. *Mass. Spec. Rev.* 22:27–56.

Mann, M., R. C. Hendrickson, and A. Pandey, 2001. Analysis of proteins and proteomes by mass spectrometry. *Annu. Rev. Biochem.* 70:437–473.

10

Identification of Protein–Ligand Interactions

Timothy Palzkill

10.1 INTRODUCTION

Genome sequencing projects have identified large numbers of previously unknown genes. Even for well-studied model organisms, such as *Escherichia coli* and *Saccharomyces cerevisiae*, the specific function of approximately one-third of their genes is unknown. The challenge of understanding the function of each gene in the genome has led to the development of high-throughput experimental techniques that allow the large-scale mapping of protein–protein and protein–ligand interactions.

Protein–protein interactions play critical roles in the functioning of cells. Defining protein interactions on a genome-wide scale can establish networks of interacting proteins, which can provide important clues about the function of a gene product. For example, if a protein of unknown function interacts with a cluster of proteins whose function is known, it would suggest that the protein is involved in the same function (Schwikowski et al., 2000). Identifying protein–protein interactions can also provide new insights into the mechanism of a biological process by providing detailed knowledge of the binding partners within a complex. For example, even if a group of proteins is known to be involved in a biological process, one does not know how or if these proteins interact in complexes to perform the function. Protein–protein interaction mapping data can detail the protein interactions and thereby provide precise information on the organization of complexes.

This chapter provides an overview of the methods currently being used to establish protein–protein interactions on a genome-wide scale. The methods discussed include in vivo approaches such as the yeast two-hybrid system as well as in vitro approaches such as mass spectrometry, phage display, and protein arrays.

Microbial Functional Genomics, Edited by Jizhong Zhou, Dorothea K. Thompson, Ying Xu, and James M. Tiedje.
ISBN 0-471-07190-0   © 2004 John Wiley & Sons, Inc.

10.2 HIGH-THROUGHPUT CLONING OF OPEN READING FRAMES

Functional genomic studies aimed at determining protein function on a genome-wide scale are critically dependent on efficient cloning of the open reading frames identified by genome sequences. The open reading frames must be cloned into plasmid vectors that permit large-scale expression or facilitate functional analysis of the encoded proteins. Because all proteins are not expressed equally well from a single system, it is necessary to obtain constructs whereby the gene of interest is present in multiple protein expression systems. For example, it would be useful to test expression of a gene from various bacterial promoters to determine which system is optimal for expression. In addition, for functional analysis, it would be desirable to place the gene of interest into vectors used for phage display, two-hybrid analysis, or glutathione S-transferase (GST) fusions. If we use conventional cloning techniques that employ restriction endonucleases and DNA ligase, the time and cost of constructing such a set of vectors are prohibitive. However, alternate methods have been developed to facilitate rapid cloning of PCR products.

10.2.1 Bacteriophage λ att Recombination-based Cloning

The bacteriophage lambda (λ) site-specific recombination system has been used to create an efficient cloning system. Bacteriophage λ undergoes both a lytic and lysogenic cycle. In the lysogenic cycle, the phages do not multiply but, instead, their DNA integrates into the host chromosome (Ptashne, 1992). Lysogeny occurs when, immediately after infection, the λ DNA circularizes and the Int protein promotes the integration of the circular DNA into the chromosome. The Int protein catalyzes the site-specific recombination between an attachment sequence on the phage DNA (called *attP* for attachment phage) and an attachment sequence on the *E. coli* chromosome (called *attB* for attachment bacteria). An *E. coli* host protein, integration host factor (IHF), is also required for the integration of λ DNA (Landy, 1989). The integration reaction is highly sequence-specific and there is no net gain or loss of nucleotides. The site-specific recombination reaction is a type of nonhomologous recombination, because the *attB* and *attP* sites are mostly dissimilar. The sites have the 15-base pair (or bp) common core sequence of 5′-GCTTTTTTATACTAA. Because the region of homology is small, the reaction will not occur without the Int protein, which recognizes both the *attB* and *attP* sites. The sites resulting from recombination between *attB* and *attP* are called *attL* and *attR* (Fig. 10.1) (Landy, 1989).

Phage λ DNA will remain in the prophage state until the host cell DNA is damaged (Ptashne, 1992). DNA damage triggers expression of the λ *int* and *xis* genes, among others. The Int and Xis gene products catalyze the excision of the λ DNA from the bacterial chromosome. Excision results from recombination between the *attL* and *attR* sites to regenerate the *attB* and *attP* sites on the bacterial and phage chromosomes. The excision reaction is not simply the reverse of the integration reaction in that the Xis protein is required in addition to the Int and IHF proteins (Landy, 1989). Thus, providing the appropriate combination of proteins can control the directionality of the recombination reaction.

The λ recombination cloning system is a method whereby a DNA fragment flanked by *att* recombination sites can be combined in vitro with a vector that also contains recombination sites and incubated with integration proteins to transfer the DNA fragment into the vector (Hartley et al., 2000). This system has been termed recombinational cloning

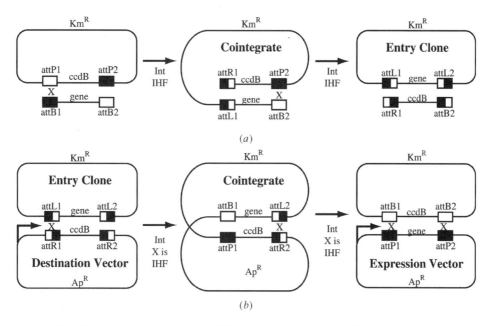

Figure 10.1 *Recombinational cloning (RC).* (*a*) Cloning of PCR products using the attB + attP > attL + attR reaction is catalyzed by the Int and IHF proteins. The result is an entry clone that can be used to create functional vectors. (*b*) Conversion of an entry clone to a functional vector using the attL + attR > attB + attP reaction is catalyzed by the Int, Xis, and IHF proteins. A wide variety of functional vectors can be constructed by using a destination vector that encodes the appropriate promoters and tags. [Figure adapted with permission from *Proteomics*, T. Palzkill, 2002, Kluwer Academic Publishers.]

(RC) (Hartley et al., 2000). The system can be used to directly clone PCR fragments into a vector in the absence of DNA ligase. The reaction used to insert PCR fragments is *attB* + *attP* > *attL* + *attR*, which is similar to insertion of the λ prophage and is catalyzed by the addition of the Int and IHF proteins. The substrates for the reaction are a PCR fragment containing an *attB* site at each end and a plasmid containing a selectable marker flanked by *attP* sites (Fig. 10.1). In contrast to λ integration, this reaction utilizes two *attB* sites and two *attP* sites. Furthermore, the *att* sites are mutated such that *attB1* will recombine with *attP1* but not with *attP2*. The engineered differences in *att* sites permit directional cloning of PCR products into the vector (Fig. 10.1) (Hartley et al., 2000).

The end result of cloning PCR fragments using RC is a vector containing a gene flanked by *attL* sites. This plasmid is termed an *entry clone*, because it can be used to generate a wide variety of functional vectors by an additional recombination reaction (Hartley et al., 2000). The reaction used for vector conversion is *attL* + *attR* > *attB* + *attP*, which is similar to the excision of the λ prophage by the Int, Xis, and IHF proteins (Landy, 1989). The gene within the entry clone is transferred to a destination vector that contains the desired transcriptional promoters and protein tags by incubating the two vectors with the Int, Xis, and IHF proteins. The reaction differs from excision of the λ chromosome, because the entry clone contains two *attL* sites and the destination vector contains two *attR* sites (Hartley et al., 2000). The *att* sites are mutated to ensure that recombination only occurs between *attL1* and *attR1* and between *attL2* and *attR2*. The recombination reaction proceeds through a cointegrate molecule that is resolved to create a destination vector containing the gene of interest with the desired promoter and tag sequences (Fig. 10.1).

The RC system has been used recently for several large-scale cloning projects. For example, the RC system is being used to clone genes from the worm *Caenorhabditis elegans* to create vectors for use in protein expression and the analysis of protein–protein interactions using the yeast two-hybrid system (Walhout et al., 2000). Recently, it has been reported that over 12,000 *C. elegans* genes have been cloned using this system (Reboul et al., 2003). In addition, greater than 100 human cDNAs have been cloned to create a set of entry clones using the RC system. These clones were then converted to green fluorescent protein (GFP) fusions using an appropriate destination vector, and the cellular localization of the proteins encoded by the cDNAs was determined (Simpson et al., 2000).

10.2.2 Topoisomerase-based Cloning

Vaccinia Virus Topoisomerase I-adapted Vectors   Another alternative to DNA ligase-based cloning is the use of topoisomerase to mediate cloning. This method exploits the ability of the vaccinia virus DNA topoisomerase I to both cleave and rejoin DNA strands with high sequence specificity (Shuman, 1992a,b). In the reaction, the enzyme recognizes the sequence 5′-CCCTT and cleaves at the final T, whereby a covalent adduct is formed between the 3′-phosphate of the cleaved strand and a tyrosine residue in the enzyme (Fig. 10.2). The covalent complex can combine with a heterologous acceptor DNA that has a 5′-hydroxyl tail complementary to the sequence on the covalent adduct to create a recombinant molecule (Shuman, 1994).

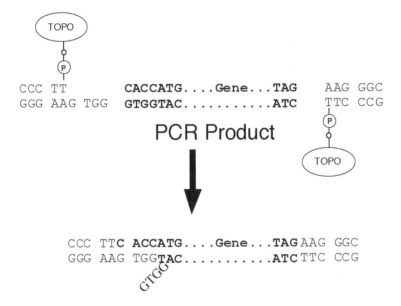

Figure 10.2 *Schematic of directional cloning of PCR products.* The sequence 5′-CACC is required at the 5′-end of the PCR product for directional topoisomerase-mediated cloning. In the example shown, the 5′-CACC sequence is appended immediately 5′ of the ATG start codon of the gene to be inserted. [Figure adapted with permission from *Proteomics*, T. Palzkill, 2002, Kluwer Academic Publishers.]

Topoisomerase I-based cloning exploits the reaction described above to join DNA fragments containing 5′-hydroxyl groups to acceptor plasmids, to which the vaccinia topoisomersase I enzyme is covalently attached. Because the joining reaction only occurs if the incoming DNA has a free 5′-hydroxyl, it is ideal for cloning PCR products, because these do not possess 5′-phosphates as the oligonucleotide primers used for amplification do not have 5′-phosphate groups (Shuman, 1994). The efficiency of this cloning method has recently been demonstrated by its use in the cloning of 6035 ORFs from *Saccharomyces cerevisiae* into a plasmid vector for protein expression in yeast as well as a vector for expression in mammalian cells (Heyman et al., 1999). One drawback of the original topoisomerase cloning method is that the PCR product can insert in either orientation. The cloning system has recently been improved to allow directional cloning of PCR products (Fig. 10.2). The only requirement for PCR primer design of this system is that the sequence 5′-CACC be included at the 5′-end of the PCR product. The 5′-CACC sequence is complementary to a sequence on the 5′-side of the plasmid-topoisomerase I adduct and, as such, controls the orientation of insertion of the PCR product (Fig. 10.2). This system was used to clone 99% of the ORFs from the genome *Treponema pallidum*, the causitive agent of syphilis (McKevitt et al., 2003). Thus, similar to the λ recombination system, topoisomerase-based reactions increase the efficiency of cloning to allow the systematic insertion of large numbers of ORFs into a plasmid vector.

10.2.3 *In vivo* Recombination-based Cloning in Yeast

Cloning by Transformation   The λ Int recombination system described above utilizes in vitro recombination followed by transformation of the products into *E. coli* cells. An alternative in vivo method that utilizes the highly efficient homologous recombination machinery of yeast has been used to clone greater than 99% of the *Saccharomyces cerevisiae* genes (∼6,000) (Uetz et al., 2000). This was accomplished using a two-step PCR procedure. Initially, a set of ∼6,000 primer pairs was used to amplify the ∼6,000 ORFs from the *S. cerevisiae* genome. Each forward primer contained a sequence unique to the ORF as well as a 22-base pair sequence at the 5′-end that was common to all the forward primers. Similarly, the reverse primer contained a sequence unique to each ORF as well as a 20-base pair sequence at the 5′-end that was common to all the reverse primers (Fig. 10.3). Each of the 6,000 ORFs was then PCR-amplified to generate the initial set of PCR products.

The second set of PCR reactions was performed using primers complementary to the 22-base pair and 20-base pair sequences appended on the initial forward and reverse primers. In addition to the complementary sequences, these forward and reverse primers contained an additional 50-base pair sequence homologous to sequences flanking a cloning site in the yeast vector to be used as the recipient plasmid for cloning (Hudson et al., 1997; Uetz et al., 2000). The resulting PCR products therefore contained a 70-base pair sequence at each end that was homologous to 70 base pairs on either side of the cloning site in the recipient vector.

Each of the ∼6,000 PCR products was then cotransformed into yeast along with the recipient vector that had been linearized using a restriction enzyme which digests the plasmid at the desired cloning site (Fig. 10.3). The 70 base pairs of a homologous flanking sequence on each end of the PCR products were sufficient for the yeast homologous recombination system to act on and insert the PCR DNA into the vector (Hudson et al., 1997; Ma et al., 1987).

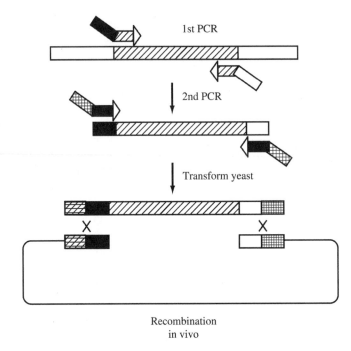

Figure 10.3 *Recombinational cloning in yeast.* Two successive PCR reactions are performed. Each set of primers contains 5′ flanking sequences that are eventually used for homologous recombination with vector sequences after transformation of yeast with the second PCR fragment and the linearized vector. [Figure adapted with permission from *Proteomics*, T. Palzkill, 2002, Kluwer Academic Publishers.]

10.2.4 Advantages and Disadvantages of Recombinational Cloning Systems

Each of the recombinational cloning methods is an efficient means of cloning PCR products into a desired set of vectors. The main advantage of topoisomerase-based cloning is the minimal requirement for extra sequences being appended to the 5′-end of the primers used for the amplification of individual ORFs. Efficient directional cloning of PCR products can be achieved with only four base pairs appended to the 5′-end of the forward primer and no additional nucleotides added to the reverse primer. In contrast, cloning of PCR products using the λ recombination reaction requires an additional 25 base pairs added to both the forward and reverse primers (Hartley et al., 2000).

The main advantage of cloning PCR products using yeast recombination is the relative ease of the procedure. The necessity of having >50 base pairs of sequence at each end of the PCR product that is homologous to sequence in the vector, however, is cumbersome in that the homology requirement necessitates the two sets of PCR reactions. Performing multiple PCR reactions increases the probability that the cloned gene will contain a mutation. In addition, the primers used for the initial PCR reaction contain 20 to 22 base pairs of sequence in addition to the sequence complementary to the individual ORF, which increases the cost of primer synthesis.

In summary, cloning based on recombination has greatly facilitated the high-throughput cloning of large sets of open reading frames. This, in turn, has enabled large-scale functional genomics experiments, such as those described below, to be implemented.

10.3   YEAST TWO-HYBRID SELECTION SYSTEM

The yeast two-hybrid system is a powerful tool for studying protein–protein interactions. It is a genetic method based on the modular properties of site-specific transcriptional activators (Fields and Song, 1989). Hybrid proteins composed of a DNA-binding domain fused with a protein X and a transcriptional activation domain fused with a protein Y are produced in yeast. If protein X and protein Y interact, it reconstitutes the transcription factor and leads to the expression of a reporter gene (Fig. 10.4). The DNA-binding domain fusion is frequently referred to as the "bait," while the activation domain fusion is the "prey."

Several versions of the yeast two-hybrid system are commonly used. Many of these systems make use of fusions to the DNA-binding domain of the Gal4 transcription factor. An alternate version utilizes a fusion of the *E. coli* LexA DNA-binding domain to the amino terminus of the protein of interest to create the bait. This system also requires a LexA-binding site placed upstream of the reporter gene. A prey construct consisting of a Gal4 activation domain or another acidic activation domain, such as B42 derived from *E. coli* sequences, can be used with either the Gal4 or LexA DNA-binding fusion bait constructs (Brent and Finley, 1997). A commonly used reporter for either of these baits is the *lacZ* gene from *E. coli*, which encodes β-galactosidase. A binding site for the Gal4 or LexA DNA-binding domain is placed upstream of the *lacZ* gene. Interaction of the proteins being tested recruits the activation domain to the DNA-binding domain and activates transcription of the *lacZ* gene (Fields and Song, 1989). Gene activation can be

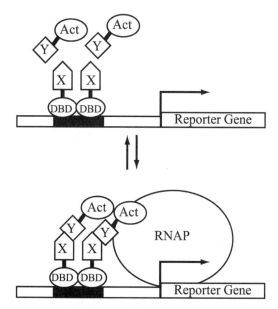

Figure 10.4 *Overview of yeast two-hybrid system.* Interaction of proteins X and Y upstream of a reporter gene leads to transcriptional activation. Protein X is part of a fusion protein that binds to a site on DNA upstream of the reporter gene by means of a DNA-binding domain (DBD). Protein Y is part of a fusion protein that contains a transcriptional activation domain (Act). Interaction of proteins X and Y places the activation domain in the vicinity of the reporter gene and stimulates its transcription. [Figure adapted with permission from *Proteomics*, T. Palzkill, 2002, Kluwer Academic Publishers.]

monitored, because β-galactosidase production causes a colony to turn blue on X-Gal indicator plates.

A challenge in the use of two-hybrid systems is the elimination of false positives. Such clones result from the activation of reporter gene transcription independent of specific binding between the bait and prey fusion proteins. For example, any bait protein that self-activates transcription will be a false positive. However, self-activating baits can be easily isolated and eliminated by screening the bait construct alone for reporter gene activation. A high background of false positives is also known to occur for certain reporter systems (James et al., 1996). To avoid this problem, recent versions of the two-hybrid system use multiple reporter genes with selectable or screenable phenotypes to monitor gene activation. In one version, the reporter genes include HIS3, ADE2, and *lacZ* (James et al., 1996). In addition, this system utilizes different Gal4-responsive promoters, with the HIS3 gene under the control of the GAL1 promoter, the ADE2 gene under GAL2 control, and the *lacZ* gene under the control of the GAL7 promoter. Protein–protein interactions are identified by the growth of yeast on agar plates lacking histidine. False positives are eliminated by scoring for the adenine and *lacZ* markers using colony color screens (James et al., 1996). The use of multiple reporter genes under the control of different promoters is reported to provide high levels of sensitivity with low background levels of false positives (Brent and Finley, 1997; James et al., 1996).

The most established use of the two-hybrid method has been to isolate new proteins from activator domain libraries that interact with LexA or Gal4 fusion baits. The activator domain libraries can consist of cDNA or fragmented genomic DNA inserted at the *C*-terminus of the activator domain. New interacting proteins are identified by introducing the activator domain library into the yeast strain containing the protein of interest fused to the DNA-binding domain and isolating colonies using the reporter systems described above (Bai and Elledge, 1997). This approach has been used by many laboratories and has resulted in the identification of many new interactions (Schwikowski et al., 2000).

10.3.1 Analysis of Genome-wide Protein–Protein Interactions in Yeast

High-throughput Yeast Two-hybrid Screens The availability of the complete genome sequence of *Saccharomyces cerevisiae* has motivated attempts to map protein–protein interactions on a genome-wide basis using the two-hybrid method and gene sets consisting of all the yeast ORFs amplified and cloned individually (Hudson et al., 1997; Uetz et al., 2000). Two types of two-hybrid experiments have been performed on a large scale. In the array method, yeast clones containing individual ORFs constructed as fusions to the Gal4 DNA-binding domain or activation domain are arrayed onto a grid, and the reciprocal fusions are screened individually against the array to identify interacting clones. In the comprehensive library screening method, a set of cloned ORFs are pooled to create a library of fusions and then individual ORF fusions are mated against the library to identify interacting clones.

In the first large-scale array experiment, a set of approximately 6,000 genes was individually cloned as fusions to the Gal4 activation domain (Hudson et al., 1997) and transformed into a two-hybrid reporter strain. Each strain was then inoculated into a well of a 384-well microassay plate. Sixteen such plates contained the $\sim$6,000 strains and constituted a living protein array (Uetz et al., 2000). A set of 192 yeast genes was then individually fused to the Gal4 DNA-binding domain and transformed into a reporter strain of the opposite mating type as the activation domain clone set. Each of the 192 bait strains was then mated to each of the ∼ 6,000 prey strains on the array to systematically screen for protein–protein interactions (Uetz et al., 2000) (Fig. 10.5). It was found that 87 of the 192 DNA-binding domain fusions participated in a protein–protein interaction, providing a total of 281 interacting pairs.

The initial comprehensive library screen was performed by collecting each of the 6,000 individually cloned Gal4 activation domain fusions into a single pool (Uetz et al., 2000). The ∼6,000 yeast genes were then individually fused to the Gal4 DNA-binding domain. The pooled activation domain fusions were mated individually to each of 6,000 unique Gal4 DNA-binding protein fusions (Fig. 10.5). Potential interactors were identified using reporter genes, and 12 clones from each mating that yielded interactors were sequenced to identify the activation domain fusion. A total of 817 ORFs were found to participate in 692 protein pairwise interactions (Uetz et al., 2000). It is of interest to note that 45% of the 192 proteins assayed were found to interact in the protein array experiment, whereas only 8% of the 5,345 potential ORFs in the high-throughput screen using the 6,000 pooled prey constructs were found to interact. Thus, the array screen, although of low throughput, generates a proportionally higher number of interactors. It is possible that pooling all the activation domain clones for the high-throughput experiment may select against interactions that involve cells with reduced growth rates or mating ability (Uetz et al.,

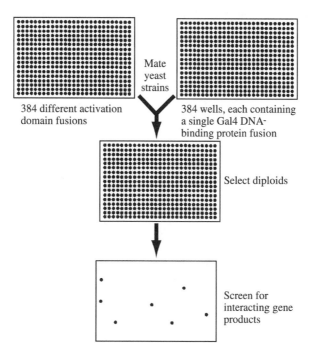

Figure 10.5 *High-throughput mating assay for two-hybrid protein interaction screening.* Yeast strains containing individual bait and prey clones are combined in a well and allowed to mate. Diploids are then selected and scored for a protein–protein interaction using the selection provided by the transcriptional reporter gene. [Figure adapted with permission from *Proteomics*, T. Palzkill, 2002, Kluwer Academic Publishers.]

2000). Regardless, the result does indicate that the set of interactors identified by the two-hybrid screen is critically dependent on how the method is implemented.

A high-throughput, comprehensive library screen has also been formulated by Ito and coworkers (2001), in which a DNA-binding domain fusion and an activator domain fusion were constructed for each of the $\sim$6,000 yeast ORFs (Fig. 10.6) (Ito et al., 2001). The DNA-binding domain and activator domain fusions were transformed into yeast reporter strains of the opposite mating type and pooled in sets of 96 ORF fusions per pool to create 62 pools each for the DNA-binding fusions and activation domain fusions. To examine all possible binding interactions, 3,844 (62 $\times$ 62) mating reactions were performed between the DNA-binding domain and activation domain pools. After mating, diploids containing interacting proteins were selected using multiple reporter genes, and inserts from both the DNA-binding and activation domain fusions were amplified by PCR and sequenced to determine the identity of the interacting clones. A total of 3,268 yeast proteins were found to participate in 4,549 pairwise interactions (Ito et al., 2001). When the authors considered only those interactions that were identified at least three times, a core group of 806 interactions among 797 proteins emerged (Ito et al., 2001).

A comparison of the interactions identified from the high-throughput library screens described above reveals a surprisingly small amount of overlap, with only about 20% of the interactions in common (Ito et al., 2001; Uetz et al., 2000). The lack of overlap suggests that the library screen experiments are not saturating. This is not surprising in that the use of pools combined with DNA sequencing of selected clones is unlikely to sample a significant fraction of the potential interactions. A screen of every possible ORF by ORF interaction would require testing $6,000 \times 6,000 = 3.6 \times 10^7$ pairwise combinations.

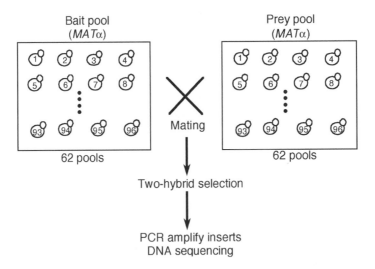

Figure 10.6 *Systematic mating of yeast two-hybrid bait and prey pools.* Each yeast ORF was cloned individually into both as a DNA-binding domain fusion (bait) and activation domain fusion (prey). The bait fusions were introduced into a MATa strain, and the prey fusions were introduced into a MATa strain. The bait and prey fusions were pooled in sets of 96 clones to generate a total of 62 pools of each. The pools were systematically mated (62 $\times$ 62) in a total of 3,844 crosses. Interacting clones were selected and the bait and prey inserts were PCR-amplified and sequenced to determine their identify. [Figure adapted with permission from *Proteomics*, T. Palzkill, 2002, Kluwer Academic Publishers.]

Given that many proteins are known to have multiple interaction partners, the number of actual interactions could be much larger.

Computationally Directed Two-hybrid Screens   Further indication that the *S. cerevisiae* genome-wide, two-hybrid library screens provide an underestimate of total interactions is provided by a computationally directed screen focused on identifying only those interactions mediated by a coiled coil protein motif (Newman et al., 2000). Coiled coils are a protein interaction motif consisting of two or more α helices that wrap around each other (Cohen and Parry, 1994). Sequences capable of forming coiled coils are characterized by a simple repeat pattern that has led to the development of accurate computer programs which identify coiled coil motifs from the primary amino acid sequence of proteins (Berger et al., 1995; Wolf et al., 1997). Use of such a program to identify coiled coils within yeast ORFs predicted approximately 300 proteins with two-stranded coiled coils and 250 proteins with three-stranded coiled coils encoded in the genome (Newman et al., 2000). Thus, approximately 1 in 11 yeast proteins are predicted to contain a coiled coil. Newman and colleagues (2000) examined interactions among 162 of the putative coiled coil regions using the two-hybrid system. A total of $162 \times 162 = 26{,}244$ pairwise tests identified 213 interactions involving 100 coiled coil motifs derived from 77 different proteins.

Strikingly, none of the interactions identified using the coiled coil directed approach were identified in the comprehensive two-hybrid experiments described above. This result points to a high frequency of false negatives in the comprehensive two-hybrid screens. This observation is similar to that described above in which a higher frequency of interactors was found when pairwise two-hybrid tests were performed vs. the use of libraries of activation domain clones (Uetz et al., 2000). In addition to the use of a pairwise screen, the coiled coil experiment may have also identified more interactors, because only the coiled coil regions of the yeast proteins were used in the screen. The use of full-length proteins in the two-hybrid screen could obscure interactions that are detected only when using fragments of proteins. For example, the full-length protein may contain an interaction site that is masked until an allosteric change in the protein reveals it (Hu, 2000). Taken together, these experiments indicate that estimates of the yeast interactome based solely on two-hybrid assays are underestimates. Thus, computationally directed screens that can approach saturation, such as the coiled coil motif screen, are likely to provide information not obtained in the comprehensive library screens.

Networks of Protein–Protein Interactions in Yeast   The results of genome-scale analyses of protein–protein interactions in yeast as determined by the two-hybrid method have been combined with interactions known from other biochemical methods to gain a more comprehensive view of interactions in the entire proteome. Schwikowski and coworkers (2000) analyzed 2,709 published interactions involving 2,039 yeast proteins, available public databases and from large-scale two-hybrid experiments. Surprisingly, they identified a single large network of 2,358 interactions among 1,548 proteins as well as several smaller networks (Schwikowski et al., 2000).

There are a number of interesting features associated with the network of interacting proteins. First, proteins of similar function tend to cluster together within the network. For example, 89% of all annotated proteins involved in chromatin structure are located in clusters with other proteins involved in chromatin structure (Schwikowski et al., 2000). In total, 63% of all interactions occur between proteins with a common functional assignment. Second, proteins with a common subcellular location tend to cluster together within the network (Schwikowski et al., 2000). Third, the network of interactions reveals sets of interactions that link cellular processes, thus revealing cross-talk between cellular compartments. For example, cell cycle control proteins exhibited the most cross-connections between cellular processes. Interactions were detected between proteins in the cell cycle control cluster and proteins involved in mitosis, protein degradation, mating response, DNA synthesis, transcription, and signal transduction as well as other processes (Schwikowski et al., 2000). These cross-connections reflect the central role of the cell cycle control proteins in regulating other cellular processes. Finally, the network provides information on the function of unknown proteins. The network approach to function prediction involves identifying the most common function among the interaction partners of a protein of unknown function and assuming that the protein of interest shares the same or a related function (Mayer and Hieter, 2000; Schwikowski et al., 2000). The limiting factor for assigning protein function based on interaction partners is a lack of knowledge about the function of proteins within a genome. For example, of the 554 proteins of unknown function within the large network identified by Schwikowki and colleagues (2000), only 69 had two or more partners of known function. As knowledge of the function of proteins within an organism improves, the number of inferences about protein function that can be made based on network analysis will also increase.

Organization of Protein Networks   The organization of a large network of protein interactions in yeast has been studied in detail (Jeong et al., 2001; Maslov and Sneppen, 2002). The networks that were analyzed consist of $> 1,500$ proteins connected by $> 2,000$ physical interactions. One goal of these studies was to determine if the architecture of the networks is best described by a uniform exponential topology, with proteins on average possessing the same number of links to other proteins, or by a heterogeneous scale-free topology, in which proteins exhibit widely different connectivities. The analysis of the probabilities of interactions indicated a highly heterogeneous scale-free network in which a few highly connected proteins play a critical role in mediating interactions among a large number of less connected proteins (Jeong et al., 2001). This type of network architecture, which follows a power law distribution, is common to other complex systems, including the Internet and metabolic networks (Barabasi and Albert, 1999; Jeong et al., 2000). Another feature of protein networks is that highly connected proteins are mostly connected to those with low connectivity (Maslov and Sneppen, 2002). This organization decreases the likelihood of cross-talk between different functional modules of the cell and increases the overall robustness of the network by localizing the effect of deleterious mutations (Maslov and Sneppen, 2002).

The architecture of the networks suggests that they tolerate random mutations, because the majority of these lesions would occur in proteins that are not highly connected to other proteins (Jeong et al., 2001). The network, however, is predicted to be highly vulnerable to mutations at the multiply connected node positions. These ideas were tested by the rank-ordering of all interacting proteins based on the number of links they exhibit, and correlating this with the phenotype effect of the deletion of a corresponding gene from the genome (Jeong et al., 2001). The correlation was aided by the large data set available from systematic gene disruption experiments performed on the yeast genome (Ross-Macdonald et al., 1999; Winzeler et al., 1999). It was found that the likelihood that removal of a protein will be lethal to a cell correlates with the number of interactions the protein has (Jeong et al., 2001). For example, proteins with five or fewer interactions constitute 93% of the yeast proteins for which gene disruption data are available, and yet only 21% of them are essential. In contrast, only 0.7% of the yeast proteins have more than 15 links, but the deletion of 62% of these proves lethal (Jeong et al., 2001). Thus, highly connected proteins that are central to the architecture of the network are much more likely to be essential than proteins that have few connections.

10.3.2   Genome-wide Yeast Two-hybrid Analysis of Other Organisms

Two-hybrid Analysis of Protein–Protein Interactions in Viral Systems
Comprehensive two-hybrid experiments have also been performed using ORFs from viral, bacterial, and animal genomes (Bartel et al., 1996; McCraith et al., 2000; Rain et al., 2001; Walhout et al., 2000). For the viral systems, the bacteriophage T7 and vaccinia virus genomes have been examined. The advantage of the viral genomes is that it is possible to systematically test all possible pairwise interactions. For example, a set of 266 potential ORFs from vacccinia virus was cloned as fusions to the Gal4 activation domain as well as the Gal4 DNA-binding domain (McCraith et al., 2000). Each of the potential 70,756 pairwise combinations of proteins was assayed by mating each DNA-binding domain fusion to an array of the 266 activation domain fusions. A total of only 37 protein–protein interactions were identified, of which 28 were previously unknown (McCraith et al., 2000). As McGraith and coworkers pointed out this is likely to be only a fraction of the interactions occurring during a viral infection. One reason for the low number of inter-actions could be the large number of vaccinia proteins that are membrane-associated. These proteins were expressed as full-length ORFs and are unlikely to reach the yeast nucleus to participate in two-hybrid interactions (McCraith et al., 2000). Expression of full-length ORFs may also mask certain interactions and lead to a high frequency of false negatives (Hu, 2000). The important message from these experiments is that, even if a screen is saturating for all pairwise combinations of protein–protein interactions, a significant proportion of interactions will not be identified using the two-hybrid screen exclusively.

Two-hybrid Analysis of Protein–Protein Interactions in Bacteria   The genome of the *Helicobacter pylori* bacterium is 2 Mb in size and encodes 1,742 ORFs. The comprehensive two-hybrid library screen performed with these ORFs differs from the yeast experiments described above in that the Gal4 activation domain library used consists of over 10 million random genomic fragments (Rain et al., 2001). Thus, the potential problem of full-size ORFs masking protein–protein interactions is reduced. A total of 261 ORFs were fused to the Gal4 DNA-binding domain to create a set of baits. These ORFs were selected to avoid hydrophobic proteins that may not target to the yeast nucleus (Rain et al., 2001). The activation domain library was mated to each of the 261 DNA-binding domain fusions to screen for interactions. A total of 1,200 interactions were identified, which connected nearly 50% of the genome. This approach seems a very efficient means of avoiding the problems associated with ORF by ORF pairwise screens. In addition, sequence data are accumulated from multiple fragments within each protein interactor identified from the activation domain library. Alignment of the fragments allows one to map the region of interaction within the protein (Rain et al., 2001).

Two-hybrid Analysis of Protein–Protein Interactions in **Caenorhabditis Elegans**   In contrast to the viral and bacterial systems, the *Caenorhabditis elegans* genome is large, encoding approximately 20,000 predicted ORFs. A comprehensive pairwise screen of all the ORFs would require 4.0×10^8 matings, which is not feasible with the present technology. However, a directed screen has been performed using 27 proteins known to be involved in vulval development fused to the Gal4 DNA-binding domain and a *C. elegans* cDNA library fused to the Gal4 activation domain (Walhout et al., 2000). The two-hybrid assay identified 148 interactions involving 124 different proteins. In an attempt to determine the biological relevance of the identified interactions, the authors searched for conserved interactions in orthologous proteins from other organisms (Walhout et al., 2000). Such conserved interactions were labeled *interlogs* and represent a useful method for increasing confidence in the validity of an interaction. In addition, Walhout and coworkers used a systematic clustering analysis to search for networks of interactions that form closed loops, reasoning that ORFs included in such loops have an increased likelihood of being involved in biologically significant interactions (Walhout et al., 2000).

It is apparent from the large amount of data from several organisms that the two-hybrid system is an efficient, automatable, and thus high-throughput method for identifying protein–protein interactions on a genome-wide basis. However, it is also apparent that a significant proportion of false positives and false negatives are inherent in the assay. In fact, it has been estimated based on the frequent linkage of functionally unrelated proteins and proteins from distinct cellular compartments that more than half of all current high-throughput interaction predictions are spurious (von Mering et al., 2002). Therefore, other methods for identifying protein–protein interactions are required both validate two-hybrid data and discover new interactions.

10.4 USE OF PHAGE DISPLAY TO DETECT PROTEIN–LIGAND INTERACTIONS

10.4.1 Display of Proteins on M13 Filamentous Phage

Phage display has emerged as a powerful tool for studying protein–protein and protein–ligand interactions (reviewed in Smith and Petrenko, 1997). The most common implementation of the method involves the fusion of peptides or proteins to a coat protein of a filamentous bacteriophage (Smith, 1985). The peptides or proteins are normally fused to the *N*-terminus of either the gene *III* or gene *VIII* phage proteins. The gene *III* protein is a minor coat protein (three to five copies per phage) located at the tip of the phage and is responsible for attaching the phage to the bacterial F pilus during the course of the normal infection process (Riechmann and Holliger, 1997). The gene *VIII* protein is the major coat protein that is present in 2,700 copies per phage particle (Rasched and Oberer, 1986). Because the gene encoding the fusion protein is packaged within the same phage particle, there is a direct link between the phenotype, that is, the ligand-binding characteristics of a displayed protein and the DNA sequence of the gene for the displayed protein. This permits large libraries of peptides of random amino acid sequence to be rapidly screened for desired ligand-binding properties (Fig. 10.7) (Smith, 1985). In addition, large collections of mutants of a displayed protein can be screened for variants with altered ligand-binding characteristics (Katz, 1997).

An important application of phage display has been to map protein–protein interaction sites. For example, random peptide libraries have been used to determine the epitopes of

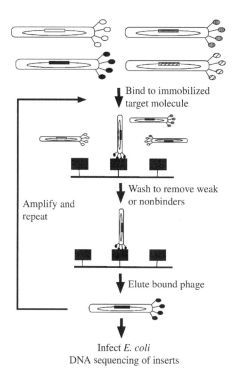

Bind to immobilized
target molecule

Wash to remove weak
or nonbinders

Amplify and
repeat

Elute bound phage

Infect *E. coli*
DNA sequencing of inserts

Figure 10.7 *Filamentous phage display.* Random sequence peptides or variants of proteins are fused to the gene III capsid protein of the M13 bacteriophage. The target ligand is immobilized in the solid phase. Phage displaying a protein that interacts with the target are enriched by affinity purification in a process called panning. After multiple rounds of panning, phage are used to infect *E. coli* and the identity of the selected inserts is determined by DNA sequencing. [Figure adapted with permission from *Proteomics*, T. Palzkill, 2002, Kluwer Academic Publishers.]

monoclonal and polyclonal antibodies (Felici et al., 1993; Folgori et al., 1994; Yao et al., 1995). Phages displaying peptides of random sequence are selected for their ability to bind antibodies, and the identity of the selected peptides is determined by DNA sequencing. The sequences are often found to match a region of the antigen protein of interest. A variation of this method has been to create a library of peptides from the protein of interest by randomly fragmenting the gene encoding that protein with DNAseI, and then shotgun cloning the small fragments of the gene into a phage display vector (Petersen et al., 1995). This method has the advantage that, if binding peptides are identified, they are certain to match a region of the protein of interest. This method has been used to map interactions other than antibody–antigen interactions. For example, the method was used to localize binding determinants within the β-lactamase inhibitory protein (BLIP) that are important for binding to the β-lactamase enzyme (Rudgers and Palzkill, 2001). In addition, it has been shown that in some cases the method can identify binding motifs with a conformational character (Williamson et al., 1998).

Shotgun cloning of fragmented genes into phage display vectors has been extended to encompass entire bacterial genomes (Jacobsson and Frykberg, 1995, 1996; Jacobsson et al., 1997). In test experiments, the total genomic DNA of *Staphylococcus aureus* was fragmented by sonication and cloned into a gene *III* and gene *VIII* phage display vector.

The potential of the method was demonstrated by selecting fragments of the gene-encoding protein A from the *S. aureus* genomic library by enriching for phage that bind to immobilized IgG protein (Jacobsson and Frykberg, 1995, 1996). This system has also been used to clone a fibrinogen-binding protein from *Staphylococcus epidermidis* (Nilsson et al., 1998), as well as surface proteins from group C streptococci that bind alpha (2)-macroglobulin, serum albumin, and IgG (Jacobsson et al., 1997). A limitation of this approach, however, is that the random fragments of DNA must be cloned between a signal sequence and the *N*-terminus of the gene *III*-encoding protein (g3p) or gene *VIII*-encoding protein (g8p). Therefore, only one in 18 clones ($3 \times 3 \times 2$) will contain a fusion that exists in the correct orientation and is in-frame with both the signal sequence and the phage coat protein. The standard phage display system is also not suited to the construction of cDNA libraries from eukaryotic organisms, because the presence of the stop codon at the end of the gene encoded in cDNA precludes fusion to the phage capsid protein. This problem is avoided by displaying proteins or fragments of proteins on the capsid of lytic bacteriophages.

10.4.2 Display of Proteins on the T7 Bacteriophage

The development of phage display systems that use lytic bacteriophage vectors, such as lambda (Maruyama et al., 1994; Santini et al., 1998), T4 (Ren et al., 1996), and T7 (Rosenberg et al., 1996), has provided an alternative that is independent of the *E. coli* secretion machinery. All these systems have the additional advantage that cloned proteins are fused to the *C*-termini of phage capsid proteins, which facilitates genomic and cDNA library constructions. The display of cDNA libraries on T7 phage has recently been used to identify interactions among signaling proteins in the EGF-receptor signaling pathway (Zozulya et al., 1999). In addition, the T7 system has been utilized to identify interactions between proteins and small molecules using a cDNA library (Sche et al., 1999). These studies suggest that phage display can be used for high-throughput protein–protein interaction studies.

An advantage of the genomic or cDNA phage display libraries is that the same library can be used to isolate binding partners for many different ligands. In addition, the ligand used for screening is not restricted to other proteins in that any molecule that can be immobilized may be used as a target for panning. This enables the library to be searched for proteins that bind DNA, RNA, proteins, carbohydrates, amino acids, or other small molecules. Peptide phage display libraries have even been used to probe the vasculature of living animals (Pasqualini and Ruoslahti, 1996). Such studies are performed by injecting a phage library into the circulation of a mouse, waiting a short time, harvesting the organ or tissue of interest, and using a homogenate to infect *E. coli* and amplify the phage that bound to this tissue (Pasqualini and Ruoslahti, 1996). This approach has been used to identify peptides that are home specifically to the vasculature of many different tissues and tumors (Arap et al., 2002; Pasqualini and Ruoslahti, 1996; Rajotte et al., 1998). These results suggest that the vasculature of most tissues displays markers selective for that tissue. It would be of interest to perform in vivo phage display experiments using genomic or cDNA libraries from pathogenic microbes to examine tissue targeting mediated by microbial ORFs.

The ability to identify binding proteins for many types of ligands will make phage display a useful tool for proteomics. Genome sequencing has identified many open reading frames for which no function has been assigned. Genomic phage display libraries could be used to classify open reading frames by binding function. For example, genomic DNA could be immobilized on beads, and the set of DNA-binding proteins could be selected from the genomic phage display library. By using different classes of ligands, it should be possible to build large categories of open reading frames based on binding properties.

10.4.3 Combining Yeast Two-hybrid and Phage Display Data

A significant limitation of both the yeast two-hybrid and phage display systems is that they generate a high proportion of false positive data. Because the techniques are quite different, that is, the yeast two-hybrid system is an in vivo assay while phage display is an in vitro assay, a combination of the two approaches may yield more useful biological data. The power of such an approach has recently been demonstrated for yeast proteins containing SH3 domains (Tong et al., 2002). SH3 domains are protein recognition modules that mediate many protein–protein interactions coordinating specific biochemical functions (Pawson and Scott, 1997). To identify proteins that interact with SH3 domains found in yeast, a total of 24 SH3 domains were expressed in a soluble form as glutathione S-transferase fusion proteins in *E. coli* (Tong et al., 2002). These proteins were used as targets for phage display to select SH3 domain ligands from a random amino acid sequence nonapeptide library. After multiple rounds of enrichment for peptide ligands, positive clones were sequenced, and consensus ligands were determined for 20 different SH3 domains (Tong et al., 2002). The consensus sequences were then used to search the yeast proteome for potential natural SH3 ligands. It was found that the proteins containing natural SH3 binding sites formed a network of interactions. To verify the findings, a second protein–protein interaction network was constructed by performing a series of two-hybrid screens with 18 different SH3 domain proteins as well as several proline-rich targets as bait (Tong et al., 2002). Finally, the common elements of the phage display and two-hybrid interaction networks were determined by finding the intersection of the data sets. A total of 59 interactions from the phage display network were also found in the two-hybrid network (Tong et al., 2002). Several of these interactions were verified in vivo using coimmune precipitation assays. The use of multiple sets of data obtained when utilizing different experimental methods is, therefore, a useful means of eliminating artifacts particular to a certain method.

10.5 DETECTING INTERACTIONS WITH PROTEIN FRAGMENT COMPLEMENTATION ASSAYS

10.5.1 Overview

Protein fragment complementation assays are based on an enzyme reassembly strategy, whereby a protein–protein interaction promotes the efficient refolding and complementation of enzyme fragments to restore an active enzyme. The approach was initially developed using the reconstitution of ubiquitin as a sensor for protein–protein interactions (Johnsson and Varshavsky, 1994). Ubiquitin is a protein consisting of 76 amino acids that is present in cells either free or covalently linked to other proteins. Ubiquitin fusions with other proteins are rapidly cleaved by ubiquitin-specific proteases, which recognize the folded conformation of ubiquitin. If ubiquitin is expressed as a fusion to a reporter protein, the cleavage reaction can be followed in vivo by release of the reporter (Johnsson and

Varshavsky, 1994). However, if a *C*-terminal fragment of ubiquitin is expressed as a fusion to a reporter and the *N*-terminal fragment of ubiquitin is expressed in the same cell, cleavage does not occur. If, on the other hand, proteins that do interact are fused to the *N*- and *C*-terminal fragments of ubiquitin, the full-sized ubiquitin is reconstituted and cleavage occurs (Fig. 10.8) (Johnsson and Varshavsky, 1994). Therefore, protein interactions can be tested in vivo by fusing them to the *N*- and *C*-terminal fragments of ubiquitin and assaying for release of the reporter protein (Fig. 10.8). As discussed below, the approach has been extended to other systems and now represents a viable alternative to two-hybrid and phage display methods for detecting protein–protein interactions in vivo.

10.5.2 Protein Fragment Complementation Using Dihydrofolate Reductase

The dihydrofolate reductase enzyme is involved in one-carbon metabolism and is required for the survival of prokaryotic and eukaryotic cells. The enzyme catalyzes the reduction of dihydrofolate to tetrahydrofolate, which is required for the biosynthesis of serine, methionine, purines, and thymidylate. The mouse dihydrofolate reductase (mDHFR) is a small (21 kD), monomeric enzyme that is highly homologous (with 29% sequence identify) to the *E. coli* enzyme (Pelletier et al., 1998). The three-dimensional structure of DHFR indicates that it is comprised of three structural fragments: F[1], F[2], and F[3], (Gegg et al., 1997).

E. coli DHFR is selectively inhibited by the antibiotic trimethoprim. The mDHFR enzyme, however, has 12,000-fold lower affinity for trimethoprim than does bacterial DHFR. Thus, *E. coli* cells expressing mDHFR are able to grow in the presence of trimethoprim. It has been demonstrated that expression of the F[1, 2] and F[3] domains of mDHFR separately as fusions to oligomerizing leucine zipper sequences results in trimethoprim-resistant *E. coli* cells due to reconstitution of the enzyme in vivo (Pelletier

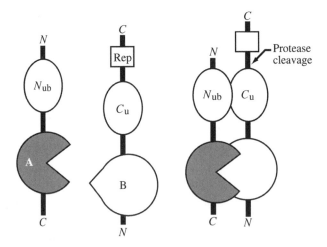

Figure 10.8 *Split ubiquitin as a sensor for protein–protein interactions.* Protein A is fused to the *N*-terminal domain, and protein B is fused to the *C*-terminal domain of ubiquitin. The interaction of A and B reconstitutes a full-sized, folded ubiquitin. Folded ubiquitin is recognized by a specific protease, and cleavage releases the reporter protein. [Figure adapted with permission from *Proteomics*, T. Palzkill, 2002, Kluwer Academic Publishers.]

et al., 1998). The reconstitution is due to the interaction of the leucine zipper sequences, which brings the F[1, 2] and F[3] fragments of mDHFR into close proximity, where they are able to fold into a single, functional enzyme. Potential protein–protein interactions can therefore be tested by fusing one of the proteins to the F[1, 2] domain and its putative binding partner to the F[3] domain and assaying for trimethoprim resistance in *E. coli* cells containing the fusion constructs. This system is analogous to the ubiquitin system described above.

The use of the mDHFR protein fragment complementation assay has recently been extended to mammalian cells (Remy and Michnick, 1999). Reconstitution of mDHFR activity in mammalian cell culture is monitored in DHFR-negative cells grown in the absence of nucleotides. An active mDHFR enzyme is required for the growth of these cells, because DHFR activity is required for the biosynthesis of purines and thymidylate (Remy and Michnick, 1999). A second approach to monitoring mDHFR reconstitution in cell culture involves a fluorescence assay based on the detection of fluorescein-methotrexate (fMTX) binding to reconstituted mDHFR in vivo. The basis for this assay is that when mDHFR is reassembled in cells, it binds with high affinity to fMTX in a 1:1 complex (Remy and Michnick, 1999). Bound fMTX is retained, while unbound fMTX is rapidly transported out of the cells. Therefore, the presence of reconstituted mDHFR can be monitored by fluorescence microscopy, FACS, or spectroscopy (Remy and Michnick, 1999).

The mDHFR protein complementation assay has been used to map a signal transduction network that controls the initiation of translation in eukaryotes (Remy and Michnick, 2001). A total of 35 different pairs of full-length protein pairs were analyzed, and 14 interactions were identified using the survival selection of cells grown in the absence of nucleotides. In addition, the use of the fMTX reagent in combination with fluorescence microscopy was used to localize the protein complex within cells (Remy and Michnick, 2001).The mDHFR protein complementation system possesses potential advantages compared to the yeast two-hybrid system for the detection of protein–protein interactions. For example, the mDHFR system can be used in *E. coli* or in mammalian cells (Pelletier et al., 1998; Remy and Michnick, 1999). When assaying interactions of mammalian proteins, it may be advantageous to work with a native system. In addition, the ability to localize interactions within cells using the fMTX reagent and fluorescence microscopy provides important information that is not available from the yeast two-hybrid system. Therefore, it is likely that this technology will find wide use for genome-scale proteomic studies in the future.

10.5.3   Monitoring Protein Interactions by Intracistronic β-Galactosidase Complementation

The β-galactosidase enzyme is widely used as a reporter of gene expression because of its ability to cleave the chromogenic substrate, 5-bromo-4-chloro-3-indoyl β-D-galactopyranoside (X-Gal), to yield a blue product. A protein–protein interaction assay has been developed based on the classical bacterial genetic phenomenon of intracistronic complementation (Mohler and Blau, 1996). In *E. coli*, deletions of either the *N*- or *C*-terminus of *lacZ* produce an enzyme that is inactive but can be complemented by coexpression of a second deletion mutant that contains the regions missing from the first mutant. Complementation occurs by assembly of the deleted fragments into a stable octameric protein that contains all the essential domains of the wild-type homotetramer

(Mohler and Blau, 1996). The *N*- and *C*-terminal domains present in the mutants involved in complementation are known as the α and ω regions, respectively. The interaction assay is based on the fact that when β-galactosidase fragments lacking the α domain ($\Delta\alpha$) and the ω domain ($\Delta\omega$) are coexpressed, complementation to create an active enzyme is very inefficient (Mohler and Blau, 1996). When the $\Delta\alpha$ and $\Delta\omega$ β-galactosidase mutants are fused to proteins that interact, however, the association of the $\Delta\alpha$ and $\Delta\omega$ fragments is favored and efficient complementation occurs (Mohler and Blau, 1996; Rossi et al., 2000). Therefore, potential protein–protein interactions can be tested by fusing the proteins of interest to the $\Delta\alpha$ and $\Delta\omega$ β-galactosidase deletion mutants and assaying for enzyme function using X-Gal (Mohler and Blau, 1996; Rossi et al., 2000).

The β-galactosidase complementation assay has also been adapted for use in mammalian cells (Rossi et al., 1997). The availability of fluorescent substrates for β-galactosidase allows for fluorescence microscopy and FACS analysis of mammalian cells expressing the fusion proteins of interest. Therefore, similar to the mDHFR system, β-galactosidase complementation assays may prove useful for genome-scale studies of protein–protein interactions in mammalian cells.

10.6 USE OF MASS SPECTROMETRY FOR PROTEIN–PROTEIN INTERACTION MAPPING

10.6.1 Overview

Rapid protein identification can be achieved by searching protein and nucleic acid databases directly with peptide mass data generated by mass spectrometry (Yates, 2000). The most common application of mass spectrometry to protein–protein interaction mapping has involved identifying the components of protein complexes. These experiments are performed using affinity-based methods to isolate entire multiprotein complexes (Fig. 10.9). This usually requires knowledge of the identity of at least one protein in the complex. This protein can then be tagged with an affinity handle, such as glutathione S-transferase, a poly-Histidine (His) repeat, or an epitope for an antibody such as the Flag tag. The tagged protein can then be overexpressed in cells and affinity purified under nondenaturing conditions such that the interaction partners copurify. The complex is then eluted and individual proteins are resolved by SDS-PAGE, the bands are cut from the gel and proteolyzed, and the exact mass of the peptides is determined by mass spectrometry. Database searching with the peptide mass data yields the identity of proteins from the complex (Pandey and Mann, 2000; Yates, 2000). There are numerous examples in the literature illustrating this general approach, including studies of the nuclear pore complex (Rout et al., 2000), the yeast Arp2/3 complex (Winter et al., 1997), TATA-binding-protein-associated factors (Grant et al., 1998), the yeast spindle-pole body complex (Wigge et al., 1998), spliceosome components (Neubauer et al., 1998) and proteins bound to the chaperonin GroEL (Houry et al., 1999).

10.6.2 Identification of Substrates for *E. coli* GroEL

The chaperonin GroEL, along with its cofactor GroES, is essential for the growth of *E. coli* (Fayet et al., 1989). The function of GroEL is to facilitate folding by limiting the side reaction of aggregation. GroEL is a homo-oligomer of 14 subunits that forms a cylindrical

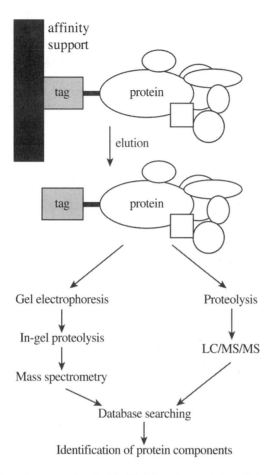

Figure 10.9 *Generic approaches to identify interacting proteins within complexes.* The complex is isolated from cells by affinity purification using a tag sequence attached to a protein known to be in the complex. Alternatively, the complex can be immunprecipitated with an antibody to one of the proteins in the complex. The proteins are resolved by polyacrylamide gel electrophoresis, proteolyzed, and the mass of the resulting peptides is determined by mass spectrometry. Alternatively, the proteins can be proteolyzed and the resulting peptides resolved by liquid chromatography. The peptide masses are then determined by mass spectrometry and used for database searching to identify the component proteins. [Figure adapted with permission from *Proteomics*, T. Palzkill, 2002, Kluwer Academic Publishers.]

structure with two large cavities. Substrate protein binds in the central cavity of the cylinder via hydrophobic surfaces exposed within the GroEL complex. GroES then binds the apical surface of the cylinder to trap the substrate in an enclosed cavity where aggregation is prevented. Approximately 10% of newly translated peptides in a cell are known to interact with GroEL, but what is not known is the identity of these substrate proteins.

The preferred substrates for GroEL were identified by the pulse-chase labeling of growing *E. coli* cells (Houry et al., 1999). At various times of chase, the GroEL-substrate complexes were isolated by immunoprecipitation with anti-GroEL antibodies. The precipitated proteins were resolved on two-dimensional polyacrylamide gels, and their identity was determined by digestion of the spots with trypsin, followed by peptide-mass fingerprint analysis using MALDI-TOF mass spectrometry (Houry et al., 1999). In this way, a total of 52 different proteins were identified as substrates for GroEL. These proteins were analyzed for a common structural motif that may form the basis of their interaction with GroEL. It was found that GroEL substrates preferentially contain several $\alpha\beta$ domains compared to other *E. coli* proteins (Houry et al., 1999). These experiments illustrate the power of the mass spectrometry approach to identifying protein–protein interactions.

10.6.3   Identification of Protein Complexes in *Saccharomyces cerevisiae*

Mass spectrometry has recently been used for systematic investigations of protein–protein interactions in yeast (Gavin et al., 2002; Ho et al., 2002). These studies are analogous to the systematic two-hybrid experiments described above, with the important difference being that the mass spectrometry studies were designed to study complexes of proteins, whereas the two-hybrid method examines binary interactions. In one approach, tandem-affinity purification (TAP) was used to purify 589 protein complexes (Gavin et al., 2002). The TAP method utilizes a combination of affinity tags for two consecutive affinity purifications to isolate the protein complexes. The complexes were then separated on one-dimensional SDS-PAGE and the individual proteins identified by MALDI-TOF mass spectrometry (Gavin et al., 2002). Of the 589 purified TAP tagged proteins, 78% presented associated proteins, indicating that the method efficiently purifies complexes. A total of 245 purified complexes corresponded to 98 known protein complexes from yeast, which serves to validate the method (Gavin et al., 2002). In addition, 242 purified assemblies represented 134 newly identified complexes. Finally, purifying several of the complexes twice and evaluating the associated proteins assessed the reproducibility of the experiments. These experiments indicated that the probability of detecting the same protein in two different purifications in approximately 70%. Therefore, 30% of the observations may be spurious.

A second systematic study of protein complexes in yeast utilized a single affinity tag on a bait protein to purify the complexes of interest (Ho et al., 2002). For these experiments, 725 proteins were tagged as bait and the associated complexes purified. The proteins in the assemblies were resolved on SDS-PAGE, excised from the gel, trypsinized, and identified using tandem mass spectrometry (Ho et al., 2002). This study identified 1,578 different interacting proteins representing 25% of the yeast proteome.

An important question is whether the data obtained from the high-throughput mass spectrometry experiments overlaps those obtained using the two-hybrid approach. The study making use of TAP tags produced data that overlapped with only 7% of the interactions seen in the two-hybrid assays (Gavin et al., 2002). Examination of the types of interactions detected by the two approaches suggests they are complementary. The two-hybrid method is not well suited for detecting complexes of proteins since it identifies binary interactions. However, the two-hybrid approach is useful for identifying pairwise interactions and interactions with low affinity.

10.7   PROTEIN ARRAYS FOR PROTEIN EXPRESSION PROFILING AND INTERACTIONS

Protein arrays are another recently emerged, promising, high-throughput technology for monitoring protein expression and interactions. A variety of formats for protein arrays are possible. For example, a set of antibodies can be gridded on a filter or glass slide and used to detect protein expression levels (Pandey and Mann, 2000). Another type of array consists of proteins from an organism arrayed directly onto a glass slide, nylon filter, or in microtiter wells (MacBeath and Schreiber, 2000). This format could be used to map protein–protein interactions or to associate a catalytic function with a protein.

The difficulty with protein arrays is that proteins do not behave as uniformly as nucleic acids. Protein function is dependent on a precise and fragile three-dimensional structure that may be difficult to maintain in an array format. In addition, the strength and stability of interactions between proteins are not nearly as standardized as nucleic acid hybridization. Each protein–protein interaction is unique and could assume a wide range of affinities. Currently, protein expression mapping is performed almost exclusively by two-dimensional electrophoresis and mass spectrometry. The development of protein arrays, however, could provide another powerful method to explore protein expression and protein–protein interactions on a genome-wide scale.

10.7.1 Antibody Arrays for Protein Expression Profiling

One of the goals of proteomics is to examine protein expression levels within and between tissues or organisms. Antibodies have historically been used to detect proteins using methods such as Western blotting and enzyme-linked immunosorbent assay (ELISA). It is cumbersome, however, to examine hundreds to thousands of proteins using these methods. An alternative is an antibody array whereby antibodies specific to each protein in the organism being examined are arrayed on a filter or glass slide. Protein expression mapping is then performed by obtaining a crude protein lysate of the tissue or organism of interest and labeling the proteins with a fluorescent tag. The protein mixture is then allowed to bind to the antibody array, so that proteins expressed in the tissue of interest can bind their cognate antibodies. After washing to eliminate nonspecific binders, bound protein is detected by a fluorescent tag attached to the protein in the original lysate (Fig. 10.10). This method is essentially a high-throughput ELISA experiment. This approach has the advantage that it is not necessary to fractionate the crude protein mixture before binding to the array. In addition, it may be possible to automate the procedure to examine multiple samples in parallel. The obvious limitation of this approach is the requirement for antibodies that are specific to each protein in the organism under study.

Obtaining antibodies by immunizing animals with purified proteins would be prohibitively difficult and expensive for large sets of proteins. An alternative strategy is to isolate specific antibodies using phage display libraries. The phage display of combinatorial antibody libraries has been extensively used to select monoclonal antibodies of a desired specificity without the use of conventional hybridoma technology (Rader and Barbas, 1997). The ability to isolate specific antibodies to a number of different proteins from a single phage display library lends itself to automation and, therefore, has the potential of providing antibodies for genome-scale projects.

10.7.2 Functional Analysis Using Peptide, Protein, and Small-molecule Arrays

One of the goals of protein chip technologies is to array all the proteins encoded within a genome for functional studies. Global analysis of protein function has historically been studied using library screens. For library screens, sets of related elements, such as a cDNA

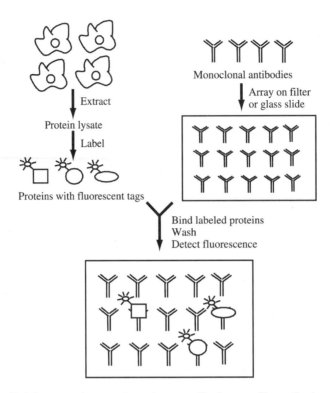

Figure 10.10 *Protein expression mapping using an antibody array.* The antibody array consists of monoclonal antibodies specific for a set of proteins in the organism of interest gridded onto a filter. To determine if a protein is expressed under the conditions being tested, a crude lysate is obtained, and the proteins within the lysate are labeled with a fluorescent tag. The lysate is applied to the filter, and the proteins are allowed to bind to the relevant antibody. Bound proteins are visualized via the fluorescent tag. [Figure adapted with permission from *Proteomics*, T. Palzkill, 2002, Kluwer Academic Publishers.]

expression library, are tested in batch for a desired biological property. For example, the screening of expressed proteins in λ-phage plaques has been performed for many years (Young and Davis, 1983). For these experiments, a cDNA library is inserted into a λ-phage vector such that the encoded protein is expressed as a fusion protein with the *E. coli* β-galactosidase protein. The phage library is used to infect *E. coli*, and λ-plaques are formed on agar plates. Each λ-plaque expresses a different fusion protein, and the plaques are probed for the presence of specific proteins using antibodies specific for those proteins (Young and Davis, 1983). When a positive plaque is identified, it is picked, and the DNA sequence of the insert is determined to identify the relevant gene. In this way, it is possible to clone genes based on a screen for an antibody–antigen interaction.

Protein arrays may offer advantages over libraries. The array format provides a precise, spatially oriented grid that allows side-by-side comparison of assay results for all the proteins on the array. The spatial arrangement also permits immediate identification of a clone that tests positive in an assay based on its location within the array. Therefore, less effort is required to identify the protein responsible for an interaction than with a library screen. A disadvantage of a protein array, however, is that fewer proteins can be efficiently arrayed and screened ($\sim 10^4$) vs. a library ($\sim 10^9$). It is also a technical challenge to purify the hundreds to thousands of individual protein species to be arrayed. Therefore, the number of elements that can be effectively assayed currently limits protein arrays.

Peptide Arrays   The earliest applications of the array format have been for peptide arrays. For example, the pin method for peptide synthesis involves the parallel synthesis of peptides in a 96-well microtiter plate format. Peptides are synthesized on amino-functionalized 4-mm polyethylene pins using a fluorenylmethoxycarbonyl (Fmoc) amino acid protection strategy (Geysen et al., 1984, 1986). For each cycle of synthesis, the pins are immersed in an appropriate amino acid solution. Because the peptides are synthesized on a solid support, they can be washed extensively between cycles of addition to increase the purity of the final preparation. This method was initially used for epitope-mapping of the VP1 coat protein of foot-and-mouth disease virus (Geysen et al., 1984, 1986). All 208 possible overlapping hexapeptides present in the 218-amino-acid VP1 sequence were synthesized and assayed for interaction with an antibody. In this way, an immunodominant region was identified and subsequently fine-mapped by introducing single-amino-acid substitutions into a peptide encompassing the region (Geysen et al., 1984). This was one of the first uses of high-throughput methods to define protein–protein interactions.

The SPOT synthesis method is another approach for simultaneous, parallel solid phase peptide synthesis. The method employs Fmoc protection chemistry, but uses hydroxyl groups present on cellulose membranes as the solid support for synthesis (Frank, 1992). When a small aliquot of liquid is dispensed onto a porous membrane, the liquid is absorbed and forms a circular spot. By utilizing a solvent of low volatility containing the appropriate reagents, each spot will form an open reactor for the stepwise addition of activated amino acids to a chemical group that is anchored to the membrane support (Frank, 1992). Arrays of spots providing suitable anchors for peptide assembly on the membranes can be generated through esterification of an αN-Fmoc-protected amino acid to available hydroxyl groups on the cellulose membrane, followed by Fmoc cleavage (Frank and Overwin, 1996). Fmoc-β-alanine is the most commonly used reagent for derivatizing the membrane. After deprotection, the free amino group of the β-alanine can be used as a platform for the synthesis of peptide arrays (Fig. 10.11) (Frank, 1992;

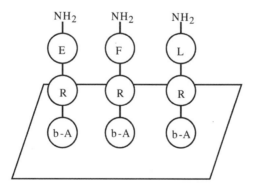

Figure 10.11   *Peptide array construction by SPOT synthesis.* β-alanine groups are covalently attached to the cellulose membrane that serves as a planar support. Peptide synthesis then proceeds using Fmoc chemistry with the β-alanine group as a starting point. The peptide is attached to the filter via its carboxy-terminus. [Figure adapted with permission from *Proteomics*, T. Palzkill, 2002, Kluwer Academic Publishers.]

Gausepohl et al., 1992). This method has been widely used for mapping antigen–antibody interactions as well as protein–DNA, protein–metal, and other protein–protein interactions (Reineke et al., 2001).

The most frequent application of SPOT synthesis has been in the preparation of peptide arrays for the identification of linear B-cell epitopes. If the antigenic protein is known, a set of overlapping peptides that encompass the entire sequence can be readily synthesized and assayed for binding of antibody (Reineke et al., 1999a). The individual residues critical for binding can then be determined by the SPOT synthesis of peptides containing amino acid substitutions. As an example, this approach was used to determine the binding epitopes of six monoclonal antibodies (MAbs) raised against listeriolysin, a toxin produced by the pathogenic bacterium *Listeria monocytogenes* (Darji et al., 1996). The binding sites for the MAbs were determined by using SPOT synthesis to create overlapping peptides that covered the entire amino acid sequence of the listeriolysin molecule. A total of 166 peptides, each 12 amino acids in length, with an offset of three amino acids, were synthesized as spots in the array format (Darji et al., 1996). Binding was detected by incubating a MAb with the immobilized peptides and detecting bound antibody with a peroxidase-conjugated secondary antibody. The cellulose membrane was reused several times after elution of bound antibody. In this way, a binding site of six to eight amino acids could be localized for each of the six MAbs (Darji et al., 1996). In addition, it was demonstrated that the binding of the MAbs to the membrane could be blocked by preincubation of the antibody with soluble peptide corresponding to the identified binding site for each MAb.

SPOT synthesis has also been used to create combinatorial peptide libraries for epitope mapping. For example, peptide epitopes recognized by the murine anti-p24 of HIV-1 monoclonal IgG2a antibody CB4-1 were identified using combinatorial libraries (Kramer et al., 1997). For these experiments, a complex positional scanning combinatorial peptide library $XXXX[B_1,B_2,B_3,X_1,X_2,X_3]XXXX$ consisting of 68,590 peptide mixtures was synthesized. This library was comprised of 10 sublibraries containing the different hexapeptides cores $[XXB_1B2B3X]$, $[XB_1XB2B3X]$, $[XB_1B_2XB3X]$, $[XB_1XXB_2B_3]$, $[XB_1XB_2XB_3]$, $[XB_1B_2XXB_3]$, $[B_1XXXB_2B_3]$, $[B_1XXB_2XB_3]$, $[B_1XB_2XXB_3]$, and $[B_1B_2XXXB_3]$ corresponding to all possible distance patterns of three defined positions (Kramer et al., 1997). Screening of the cellulose-bound positional scanning combinatorial library for CB4-1 binding resulted in the detection of 225 peptide mixtures out of a total of 68,590 spots. Deconvolution of the randomized positions of the selected peptide mixtures led to the identification of a peptide related to a sequence within p24, as well as three completely different peptides (Kramer et al., 1997). Determination of peptide-binding affinities and competition experiments revealed that all the peptides bound to the same region of the monoclonal antibody but with different affinities. In addition, substitution analysis of each peptide clearly showed that the molecular basis of antibody recognition is specific and unique for each identified peptide (Kramer et al., 1997). The substitution analysis was also used to define a *supertope* sequence based on the pattern of amino acid substitutions that were consistent with binding of the p24 monoclonal antibody. For example, one of the identified peptides had the sequence GATPEDLNQKLAGN, but systematic amino acid substitutions of the peptide using SPOT synthesis revealed the supertope sequence $XXXXX[DE]L[HKNR]XX[IL]XXX$, where X is any amino acid (Kramer et al., 1997). The supertope consensus sequence was used to search the SWISSPROT protein database, of which 5,517 matches were found. Several of the identified proteins were purified and shown to bind to the p24 monoclonal antibody. These experiments defined the poly-specificity of p24 MAb and demonstrated that SPOT synthesis can also be used to generate and search unbiased combinatorial random sequence libraries.

Mapping of discontinuous epitopes is more difficult because of the low affinity for antibody binding of peptides derived from separate regions of the antigen protein. An interesting application of SPOT synthesis has been to map a discontinuous binding site on interleukin-10 (IL-10) for a neutralizing anti-IL-10 antibody called CB/RS/1 (Reineke et al., 1999b). An overlapping peptide scan of the IL-10 sequence was performed using 15-mer peptides shifted by one amino acid. The peptide array was then probed using the CB/RS/1 antibody and bound antibody was detected using an antimouse IgG peroxidase-labeled antibody. The CB/RS/1 antibody was found to bind to peptides representing two regions of the protein that are distant in the primary sequence, but continuous on the folded structure (Reineke et al., 1999b). Amino acid residue positions within peptides that did bind the antibody were then systematically substituted and arrayed by spot synthesis. Binding of the CB/RS/1 antibody was again assayed to determine the positions within the peptides that are critical for antibody binding. A number of substitutions were also found that appeared to increase antibody binding. The two regions that exhibited antibody binding were linked to a single peptide that also incorporated the substitutions which increased binding. The single peptide was then systematically substituted with cysteine residues to identify disulfide bonds that would increase binding by lowering the conformational entropy of the peptide. The end result of these multiple spot synthesis experiments was a tight-binding 32-mer-peptide mimic of the discontinuous binding site between IL-10 and the CB/RS/1 antibody (Reineke et al., 1999b). This study is another demonstration of the power of the SPOT synthesis method for the rapid construction and testing of peptide arrays.

A number of other protein–protein interactions have been studied using SPOT synthesis. These include a number of interactions between proteins involved in signal transduction, such as PDZ domains (Schultz et al., 1998), SH3 domains (Cestra et al., 1999) and tumor necrosis factor receptor-associated factors (TRAFs) (Pullen et al., 1999). In addition, the method has been used to define the substrate specificity of the bacterial chaperone protein, SecB (Knoblauch et al., 1999). This study is of particular interest, because chaperone proteins exhibit relaxed substrate specificity and thereby bind to a large number of proteins. SecB assists translocation of precursor proteins across the bacterial cytoplasmic membrane by associating with newly synthesized precursors, either during or after translation, and thereby maintains them in a translocation competent state. However, the determinants of the substrate specificity of SecB are unclear (Knoblauch et al., 1999). To address this question, a total of 2,688 peptides derived from 23 different proteins were synthesized and screened for SecB binding (Knoblauch et al., 1999). This set of proteins contained known substrates of SecB as well as a number of proteins that are not secreted. The large data set allowed a robust statistical analysis of the substrate motif recognized by SecB. All screened peptides were grouped into four classes based on their affinity for SecB (high, medium, low, and no affinity) as determined by the quantitation of spot intensities (Knoblauch et al., 1999). Large differences existed between the amino acid distribution of SecB-binding and nonbinding peptides. High-affinity SecB binders were enriched for basic residues (Arg and Lys) and aromatic residues (Phe, Tyr, and Trp), whereas acidic residues (Asp and Glu) were strongly disfavored (Knoblauch et al., 1999). The results were used to identify a recognition motif, which can be used to accurately predict SecB-binding peptides. These experiments represent a step toward the use of SPOT synthesis for proteome-wide screens of binding specificity. SPOT arrays containing overlapping peptides for an entire bacterial proteome may soon be generated. Such an array would be a powerful tool for the study of antigen–antibody and protein–protein interactions at the level of the proteome.

Recently, SPOT synthesis has been extended to the generation of a protein array consisting of 837 different variants of the hYAP WW protein domain (Toepert et al., 2001). The WW domain was chosen as a model due to its short length of approximately 40 amino acids. The WW domain is found in proteins with diverse functions such as structural, regulatory, and signaling proteins in yeast, nematodes, and mammals (Chen et al., 1997a). The WW domains bind to short segments of proline-rich sequences to mediate protein–protein interactions. The structure of the WW domain is known, and several site-directed mutagenesis studies have previously been performed to identify the residues critical for its structure and function (Chen et al., 1997). Each of the 44 residues of the WW domain was systematically substituted by each of the 19 L-amino acids (Toepert et al., 2001). The results agreed well with earlier mutagenesis studies and could be rationalized based on the structure of the domain. One of the concerns of solid phase synthesis of a 44-amino-acid peptide is that a large percentage of the peptides within a spot will be incomplete products. Mass spectrometry was performed to verify the sizes of the synthesized peptides (Toepert et al., 2001). In addition, all possible single-amino-acid deletions were synthesized on an array and, in contrast to the array containing the full-sized variants, nearly all the deletions were inactive. These results suggest that the majority of the peptides within a spot are full-sized (Toepert et al., 2001). This study, coupled with the rapid advances in the chemical synthesis of proteins, indicates that SPOT synthesis may become a means of producing protein arrays using solid phase synthesis.

Although the SPOT synthesis method has great utility, the density of spots on the filters is not of the same order as arrays on DNA chips or glass slides (Brown and Botstein, 1999). It has been shown, however, that photolithography can be used to greatly increase the density of peptides on an array (Fodor, 1991). The method is similar to that used for the construction of high-density oligonucleotide arrays. Photolabile protecting groups are used for peptide synthesis. Growing peptides are selectively deprotected using masks that allow light to reach only those peptides to which an amino acid is to be added. With this technology, peptide densities up to 250,000 per cm^2 have been achieved (Fodor, 1991). The arrays described above are constructed by the synthesis of peptides directly onto a solid support. At present, SPOT synthesis is limited to approximately 30 to 40 amino acids due to difficulties with product purities and coupling efficiencies (Molina et al., 1996). Therefore, the construction of protein arrays by direct synthesis is not possible. Instead, proteins must be expressed and purified subsequent to use in an array. Recombinant proteins can often be expressed in organisms such as *E. coli* or *S. cerevisiae*, or from tissue culture. Alternatively, proteins can be produced by in vitro transcription and translation. The difficulty with any of these methods is obtaining properly folded protein for use on the array. Improper folding and aggregation are common problems for recombinant protein expression in heterologous systems. In addition, in vitro expression of proteins in the absence of specific chaperone proteins may lead to improperly folded structures. This problem is the major factor that distinguishes protein arrays from DNA or oligonucleotide arrays. Thus, progress in the development of protein arrays has been slower than that for DNA arrays. Nevertheless, several recent developments suggest protein arrays may be possible.

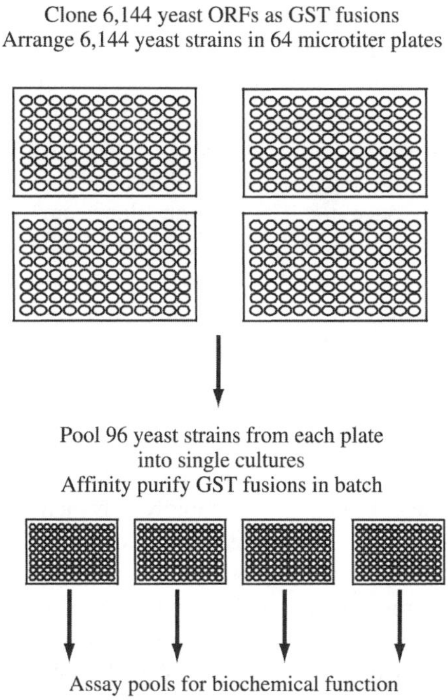

Clone 6,144 yeast ORFs as GST fusions
Arrange 6,144 yeast strains in 64 microtiter plates

Pool 96 yeast strains from each plate
into single cultures
Affinity purify GST fusions in batch

Assay pools for biochemical function

Figure 10.12 *Purification of protein from pooled yeast strains.* Each yeast ORF was cloned as a fusion to glutathione S-transferase in a protein expression vector to create 6,144 yeast strains. The individual strains were pooled in groups of 96 to create a set of 64 pools. Each pool was grown and the 96 fusion proteins were purified in batch. Each pool was then assayed for a biochemical function (Martzen et al., 1999). Pools positive for function were then deconvoluted using smaller pools consisting of strains from rows and columns of a 96-well plate. [Figure adapted with permission from *Proteomics*, T. Palzkill, 2002, Kluwer Academic Publishers.]

Arrays of Pooled Recombinant Proteins   The first obstacle in the development of a protein array is the large-scale expression and purification of proteins encoded by open reading frames of an organism. The feasibility of this approach has been demonstrated for *S. cerevisiae* (Martzen et al., 1999). For these experiments, an array consisting of 6,144 yeast strains was constructed, each containing a plasmid expressing a different GST–ORF fusion under the transcriptional control of the P_{CUP1} promoter. Because it would be prohibitively difficult to purify 6,144 individual proteins, the strains were collected in 64 pools consisting of 96 different GST fusion strains in each pool (Fig. 10.12). The GST fusion proteins from each of the pools were purified in batch by affinity chromatography using a glutathione agarose resin (Martzen et al., 1999). The pools were then used for biochemical assays of protein function. For example, an assay of the GST fusion pools demonstrated that each of the two previously known tRNA splicing activities from yeast were present only in the pools expected based on ORF number.

When a pool was identified as containing a biochemical function, the individual strain responsible for the activity was determined by preparing and assaying the GST fusion proteins from each of the 8 rows and 12 columns of strains from the appropriate microtiter plate (Martzen et al., 1999). Using this approach, biochemical assays assigned a putative function to the proteins encoded by three previously unknown genes. These included two cyclic phosphodiesterases involved in tRNA processing and a methyltransferase capable of modifying cytochrome c (Martzen et al., 1999). In principle, addressable, pooled GST fusion proteins could be used to identify proteins associated with any biochemical activity, if we assume that the fusion protein is soluble, folded, and functional. The method has the additional advantage that, once the GST fusion clones are constructed, it is a rapid technique. Martzen and coworkers state that only 2 weeks are required to purify the 64 pools, and the assays can be accomplished in a day (Martzen et al., 1999). In addition, the method is sensitive, because only 96 recombinant proteins are assayed at one time in contrast to the use of cell lysates, where thousands of proteins are present. This leads to a much higher concentration of each protein, which greatly facilitates the detection of a biochemical activity (Martzen et al., 1999).

The power of the pooled GST fusion protein approach may increase as new biochemical reagents and assays become available. The development of chemical probes for biological processes, termed chemical biology, is a rapidly advancing field. For example, the chemical synthesis of an active site-directed probe for identification of members of the serine hydrolase enzyme family has recently been described (Liu et al., 1999). The activity of the probe is based on the potent and irreversible inhibition of serine hydrolases by fluorophosphate (FP) derivatives such as diisopropyl fluorophosphate. The probe consists of a biotinylated long-chain fluorophosphonate, called FP-biotin (Liu et al., 1999). FP-biotin was tested on crude tissue extracts from various organs of the rat. These experiments showed that the reagent can react with numerous serine hydrolases in crude extracts and detect enzymes at subnanomolar concentrations (Liu et al., 1999). Clearly, reagents such as FP-biotin would work well with the pooled GST fusions where proteins exist at higher concentration than in crude extracts (Martzen et al., 1999). Other such chemical probes will likely be developed in the next few years.

Protein Microarrays   The use of pooled GST fusion proteins allows tests of biochemical activity, but does not lend itself to the identification of protein–protein or protein–small molecule interactions. For these purposes, it is necessary to immobilize proteins on a solid support, so that nonbinding molecules can be washed away. It is also necessary that the protein, once attached to the solid support, retain its folded conformation.

A method has been described for the attachment of proteins to glass slides at high spatial densities (MacBeath and Schreiber, 2000). These protein microarrays were constructed using a high-precision robot to deliver nanoliter volumes of samples to glass slides at a density of 1,600 spots per square centimeter. The proteins were attached covalently by pretreating the slides with an aldehyde-containing silane reagent (MacBeath and Schreiber, 2000). The aldehydes react readily with primary amines from lysine residues as well as the α-amine at the NH_2-terminus of the protein. Because lysines are usually present at multiple positions on the surface of proteins, the molecules attach in multiple orientations.

The high-density protein microarray was used to examine protein–protein interactions using three pairs of proteins that are known to interact: protein G and IgG; p50 and IkBa; and the FKBP12-rapamycin-binding domain (FRB) and FKBP12 (MacBeath and Schreiber, 2000). For each of these experiments, one of the binding partners was immobilized, while the other was labeled with a fluorescent tag and allowed to bind to the

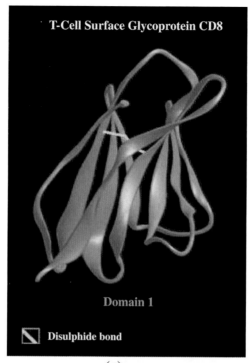

(a)

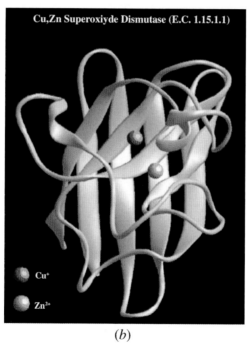

(b)

Figure 5.2 *See page 120 for caption.*

```
09J849/133-201    RKQSFPK........--FKTSRTH-.-YrDNRETLAKNG.FYHYGKKFEI.............---RCSSCKFV
Q99GY5/89-157     RRDSFRQ......yKKAKSYFKNS.--.--LDLLAQNG.FYYYGVKTEV.........---RCAYCLLV
Q9YMP8/85-153     RAASFRA........--FKAGCGK-.-YgSDANALAACG.FFYNGRCREA.........---QCSRCGMV
056307/85-152     RKQSFSS........--FKWARRQFkSHnKLADMLSRRG.FYCFGKKARL.........---RCVGCKVV
IAP2 NPVOP/85-150 RKRSFAS........--FKWARRQFgSRaREVDMLSRRG.FYCVGK--RL.........---RCAGCKVV
Q65368/85-150     RKRSFAS........--FKWARRQFgSRaREVDMLSRRG.FYCVGK--RL.........---RCAGCKVV
902435/85-152     RKKSFTS........--FKKSRRQFaSQsVVVDMLARRG.FYYFGKAGHL.........---RCSGCHIV
IAP2 NPVAC/85-152 RKKSFTS........--FKSSRRQFaSQsVVVDMLARRG.FYYFGKAGHL.........---RCSGCHIV
BIR1 SCHPO/25-99  RLDTFQK.......KKWPRAKPT-.--.--PETLATVG.FYYNPISESNse.......er1DNVTCYMCTKS
BIR5 MOUSE/18-88  RIATFKN.........--WPFLEDCA.--.CTPERMAEAG.FIHCPTENEP.........DLAQCFFCFKE
BIR5 RAT/18-88    FIYTFKN.........--WPFLEDCS.--.CTPERMAEAG.FIHCPTENEP.........DLAQCFFCFKE
Q9GLN5/18-88      RISTFKN.........--WPFLEGCA.--.CTPERMAAAG.FIHCPTENEP.........DLAQCFFCFKE
BIR5 HUMAN/18-88  RISTFKN.........--WPFLEGCA.--.CTPERMAEAG.FIHCPTENEP.........DLAQCFFCFKE
Q9BVZ4/18-81      RISTFKN.........--WPFLEGCA.--.CTPERMAEAG.FIHCPTENEP.........DLAQCFFCFKE
Q9DDK0/20-90      RAATFRN.........--WPFTEGCA.--.CTPERMAAAG.FVHCPSENSP.........DVXQCFFCLKE
Q9VEM2/31-101     RVESYKS.........--WPFPETAS.--.CSISKMAEAG.FYWTGTKREN.........DTATCFVCGKT
```

Figure 5.4 *See page 125 for caption.*

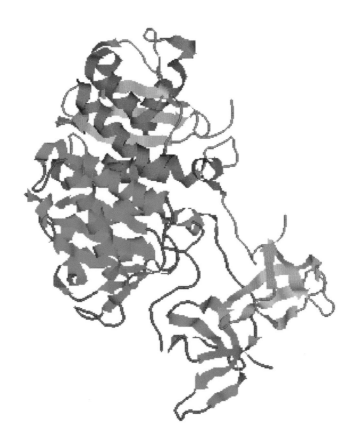

Figure 5.5 *See page 127 for caption.*

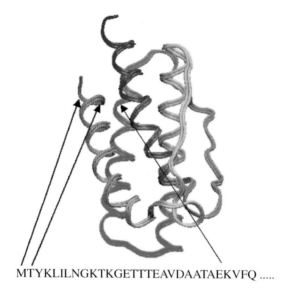

MTYKLILNGKTKGETTTEAVDAATAEKVFQ .....

Figure 5.7 *See page 131 for caption.*

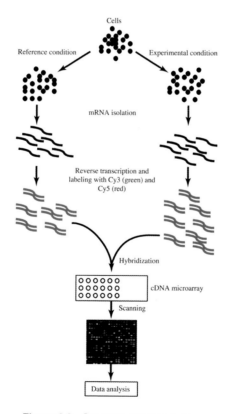

Figure 6.9 *See page 171 for caption.*

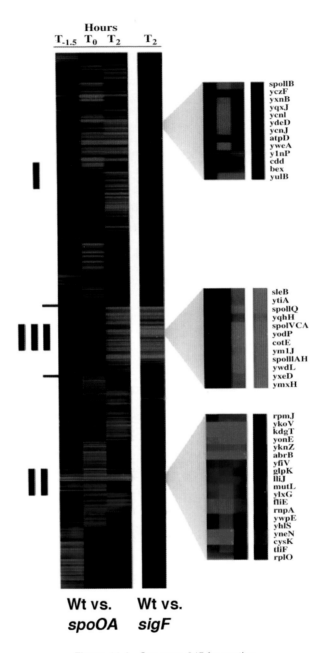

Figure 11.4 See page 345 for caption.

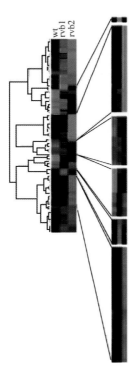

Figure 11.7   See page 359 for caption.

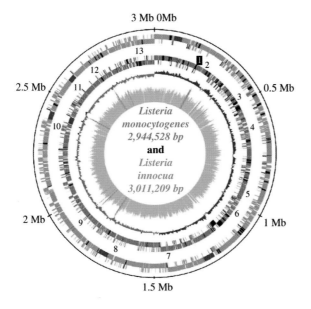

Figure 12.2   See page 397 for caption.

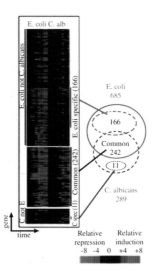

Figure 12.4 *See page 409 for caption.*

Figure 12.5 *See page 418 for caption.*

Oligonucleotide arrays
with probes perfectly
matching to the wild type
sequence. The sequences
for the probes illustrated
are identical except 1 bp.

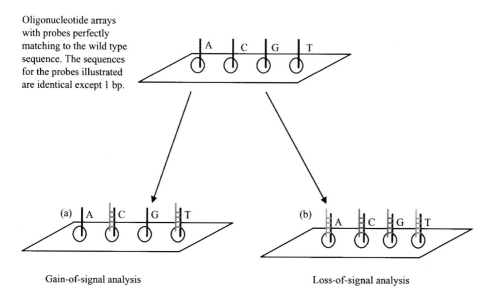

Gain-of-signal analysis   Loss-of-signal analysis

Figure 14.2 *See page 455 for caption.*

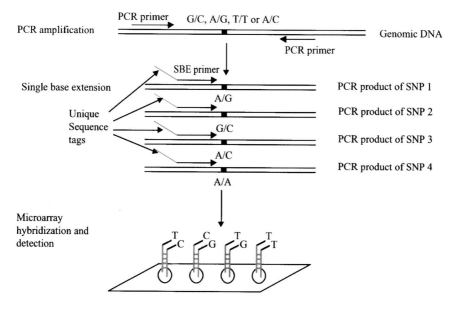

Figure 14.4 *See page 460 for caption.*

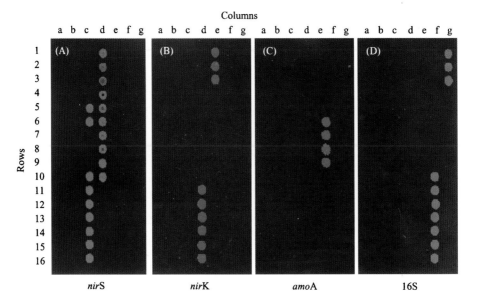

Figure 14.5 *See page 464 for caption.*

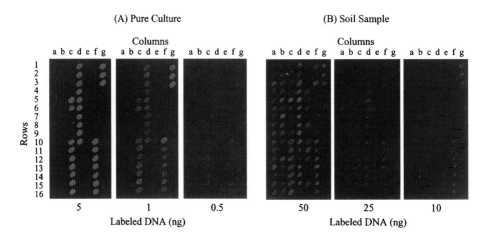

Figure 14.6 *See page 466 for caption.*

slide. Bound protein was detected by retention of the fluorescent tag on the slide after extensive washing (MacBeath and Schreiber, 2000). These experiments demonstrated that microarrays can be used to efficiently detect protein–protein interactions in extremely small sample volumes. The immobilized proteins were spotted at a concentration of 100 μg per mL. Because the binding reaction takes place in nanoliter volumes, the amount of the solution phase protein needed to detect binding is very small. For instance, specific binding could be detected for the FRB–FKBP12 interaction using picogram quantities of FKBP12 (MacBeath and Schreiber, 2000). MacBeath and Schreiber (2000) note that, because the concentration of solution phase protein necessary for binding is so low, it may be possible to label proteins with a fluorescent tag directly in a cell lysate and use the array to quantitate the amount of a specific protein within the lysate. Thus, the protein microarray could be used for protein expression mapping as well as for detecting protein–protein interactions (MacBeath and Schreiber, 2000). By using a similar approach, it was shown that the binding of small molecules to proteins on the microarray can be efficiently assayed. At this point, it is unclear what fraction of proteins will retain a folded structure when immobilized, but based on initial experiments, the protein microarray has great potential for high-throughput protein assays.

A protein microarray has also been used to study the specificity of protein–protein interactions (Newman and Keating, 2003). The bZIP transcription factors are an important class of eukaryotic DNA-binding protein in which the dimerization of coiled coil motifs plays an important role in binding specificity. There are a large number of bZIP coiled coil motifs encoded in the human genome, and the combinatorial binding specificity of these motifs was analyzed using a protein array. For these experiments, a leucine zipper array composed of 49 human bZIP coiled coil domains was constructed by expressing and purifying each of the domains and attaching the proteins to an aldehyde-coated glass slide (Newman and Keating, 2003). Each of the 49 proteins was then fluorescently labeled and used individually as a probe for binding to the array. In this way, a total of 49^2 interactions could rapidly be evaluated. An important feature of these experiments is that there was >90% agreement in reciprocal experiments, that is, a domain bound the same set of partners when it was used as the solution probe and when immobilized (Newman and Keating, 2003). This result suggests that artifacts due to protein unfolding on the solid support or other surface chemistry artifacts are not a major limitation for the approach. This method could be adapted to examine other binary sets of interactions within a protein family.

The use of protein microarrays on a proteome-wide scale has been demonstrated with the construction of an array containing 5,800 different yeast proteins representing 93.5% of the yeast proteome (Zhu et al., 2001). For these studies, a collection of 5,800 yeast ORFs were cloned into a yeast high-copy expression vector as fusions with glutathione S-transferase and polyhistidine (GST-HisX6) (Zhu et al., 2001). The proteins were expressed in yeast to facilitate proper folding and to incorporate any modifications. Following a 96-well format, 1,152 samples were purified at one time, and after all 5,800 proteins were purified, the samples were spotted at high density onto nickel-coated glass slides (Zhu et al., 2001). The protein microarray was tested for protein–protein interactions by probing the glass slide with biotinylated calmodulin in the presence of calcium. Calmodulin is a highly conserved calcium-binding protein involved in many calcium-regulated cellular processes and has many known partners (Hook and Means, 2001). These experiments identified six known calmodulin target proteins as well as 33 additional potential partners (Zhu et al., 2001). Because a precise tertiary structure is required for an interaction with calmodulin, these studies suggest that a large fraction of the proteins on the slide are folded into a functional state.

The yeast proteome microarray was also used to test for interactions that cannot be tested by in vivo approaches. Specifically, protein–lipid interactions were examined by screening for phospoinositide (PI)-binding proteins (Zhu et al., 2001). PIs are important constituents of the cell membrane and also serve as second messengers that regulate a wide range of cellular processes. The proteome microarray was probed with liposomes containing five different types of PIs as well as a liposome-containing phosphatidylcholine (PC). Each liposome also contained a biotinylated lipid, which was used for the detection of liposomes that were bound to proteins on the microarray (Zhu et al., 2001). The six liposomes identified a total of 150 different protein targets that produced signals significantly higher than the background (Zhu et al., 2001). Fifty-two of these proteins correspond to uncharacterized proteins. Of the remaining proteins, 45 are membrane-associated, including integral membrane proteins and proteins with lipid modifications (Zhu et al., 2001). In addition, a large percentage of the characterized proteins that bound to lipid are kinases. In order to verify the interactions detected on the array, several proteins were immobilized onto nitrocellulose filters at varying concentrations and were shown to bind lipid in a concentration-dependent manner (Zhu et al., 2001).

The yeast proteome chip studies illustrate a number of advantages that protein arrays have over other approaches for identifying protein–ligand interactions. For example, with the yeast two-hybrid approach, interactions are detected in the nucleus, which limits the number of interactions that can be detected. In contrast, because the protein array-binding experiments are performed in vitro, protein localization constraints are not an issue. An additional advantage of the in vitro detection of interactions is that many different types of ligands can be tested. Thus, interactions such as protein–lipid, protein–small molecule, or protein–nucleic acid interactions can be assayed.

Small-molecule Arrays   Genetics has been an important contributor to understanding biology. It relies on mutant alleles to gain insights into pathways of interest. In chemical genetics, small molecules instead of genetic mutations are used to modulate the functions of proteins conditionally and temporally, thereby allowing many biological processes to be explored (Mitchison, 1994). Examples of this approach include the use of colchicine to discover tubulin (Borisy and Taylor, 1967); the discovery of tetrodotoxin, which enabled the dissection of the action potential (Narahashi et al., 1964); and the identification of agonists of peroxisome-proliferator-activated receptor-γ, which facilitated understanding of the regulation of adipogenesis (Lehmann et al., 1995). The key elements to the development of chemical genetics as a systematic technique to study biology are the efficient synthesis of large, diverse collections of small molecules and facile methods to screen the small-molecule collections. Significant progress in the synthesis of diverse collections of small molecules has been achieved via diversity-oriented synthesis (Schreiber, 2000), solid phase purification (Merrifield, 1963), and the split-pool synthesis strategy (Furka et al., 1991). Recently, a technology platform has been described for split-pool, diversity-oriented synthesis using high-capacity macrobeads as individual microreactors (Blackwell et al., 2001). Each macrobead delivers approximately 5 mM of synthesized stock solution upon compound cleavage and resuspension in assay plates (Blackwell et al., 2001). This amount of product is sufficient for numerous biological assays.

The second key component for chemical genetics is an efficient screen for molecules affecting the process of interest. Both phenotypic and protein-binding assays have been successfully employed. For example, a combination of two phenotype-based screens, one based on a specific posttranslational modification and the other visualizing microtubes and chromatin, was used to identify compounds that affect mitosis (Mayer et al., 1999). One compound, named monastrol, arrested mammalian mitosis by interacting with the mitotic kinesin Eg5 (Mayer et al., 1999). Monastrol is the first compound to be identified that acts on mitosis without targeting tubulin.

Small-molecule screens based on protein-binding assays are of interest because of the potential to develop high-throughput screens. Protein-binding assays that allow proteins to be screened against large collections of small molecules would provide an effective means of identifying useful small molecules. Methods have been described for the creation of microarrays consisting of thiol-containing small molecules using a thiol capture reaction (MacBeath et al., 1999), as well as microarrays of alcohol-containing small molecules via activation of the surface of glass slides with thionyl chloride (Hergenrother et al., 2000). These methods were used to create a high-density microarray of 3,780 structurally complex 1,3-dioxane small molecules resulting from diversity-oriented synthesis (Kuruvilla et al., 2002). The utility of the microarray was demonstrated by probing it with fluorescently labeled yeast protein, Ure2p. This protein is the central repressor of genes involved in nitrogen metabolism in yeast. A compound that binds to Ure2p, named uretupamine, was identified and used to dissect a glucose-sensitive transcriptional pathway that acts downstream of Ure2p (Kuruvilla et al., 2002). New information was obtained, because uretupamine modulates only a subset of Ure2p functions and thus its effects are more specific than a knockout of the *URE2* gene. This study suggests that small-molecule microarrays provide a systematic method for acquiring small-molecule probes that can modulate different aspects of a protein's function, thus facilitating dissection of the biological process in which the protein participates.

The SPOT synthesis method, described above, has been used most commonly to synthesize membrane-bound peptides in an array format. However, the method has been extended to the efficient synthesis of arrays of small organic molecules on the membrane surface (Scharn et al., 2000). Parallel assembly of trisamino- and amino-oxy-1,3,5-triazines was achieved by applying the SPOT technique to both cellulose and polypropylene membranes. This was done using amines and phenolate ions as building blocks with a linker system that was cleavable with trifluoroacetic acid vapor (Scharn et al., 2000). A total of 8,000 cellulose-bound 1,3,5-triazines were synthesized and probed in parallel for binding to the antitransforming growth factor-α monoclonal antibody Tab2 in order to identify epitope mimics (Scharn et al., 2000). These studies suggest that SPOT synthesis is also an effective method for creating small-molecule arrays. Thus, small-molecule arrays are likely to be important tools for future proteomics studies to probe the function of proteins in biological processes.

Protein Chips and Mass Spectrometry The use of MALDI-TOF mass spectrometry in combination with a protein array has been described for the analysis of amyloid β-peptide variants secreted from tissue culture cells (Davies et al., 1999). Aggregated forms of the 4-kDa amyloid β-peptide form the senile plaques that are often found in the brain tissue of patients suffering from Alzheimer's disease (Selkoe, 1998). Numerous variants of the peptide have been identified in clinical samples and, therefore, an efficient method for identifying the variants is required. For this purpose, a protein chip was constructed whereby an antibody to the amyloid β-peptide was immobilized on the chip surface (Davies et al., 1999). Amyloid β-peptide variants secreted from cultured cells were captured from the media by placing 1 μL of media onto the chip surface. The chip was then washed, and bound peptide was eluted and analyzed by MALDI-TOF mass spectrometry (Davies et al., 1999). The high sensitivity and accuracy of mass spectrometry allowed the accurate identification of several amyloid β-peptide variants. In addition, a control bovine IgG was immobilized at a different position on the chip to show that the antibody to the amyloid peptide was responsible for capturing the variants from the media (Davies et al., 1999). This study demonstrated that the combination of a protein chip with mass spectrometry is an efficient means of identifying peptides with subtle differences in composition.

The protein chip-mass spectrometry approach has been expanded to include a number of immobilization platforms for molecules of interest. Thus, molecules can be captured by an affinity method, such as an antibody or using chip surfaces with other chromatographic properties such as anion exchange, cation exchange, metal affinity, and reverse phase (Fung et al., 2001). These chips can be used to reduce a complex mixture of proteins to sets of proteins with common properties that are then analyzed by mass spectrometry. In concept, this approach is similar to the liquid chromatography (LC) and tandem mass spectrometry approach described in Chapter 9 (Link et al., 1999). The goal of both approaches is to reduce the complexity of proteins to a number that can accurately be examined. The spectrum of proteins identified using a given chromatographic fractionation method represents a fingerprint of the cell state when the protein was extracted. For example, the protein chip method has been used to study protein expression profiles in normal and cancerous prostate samples to identify protein markers characteristic of a disease state (Wright et al., 2000). The protein chip-mass spectrometry platform is commercially available, and a number of applications have been published (reviewed in Fung et al., 2001).

Use of DNA Microarrays to Study Protein Function DNA microarrays are widely used to study gene expression by measuring mRNA levels (Brown and Botstein, 1999). An interesting alternative is to use these arrays to study the DNA-binding specificities of transcription factor proteins. A DNA microarray-based method has been developed to characterize sequence-specific DNA recognition by zinc-finger proteins (Bulyk et al., 2001). The zinc-finger transcription factors are among the best-understood families in terms of sequence-specific DNA binding. The Zif268 transcription factor from the mouse was used as a model system for these studies. The experiment involved placing the Zif268 protein on the surface of the M13 filamentous phage to isolate a number of variants with altered binding specificity using mutagenesis and phage display screening. The binding specificity of the wild-type and mutant Zif268 proteins was then determined using a DNA microarray.

The Zif268 protein contains three zinc fingers (F1, F2, F3), and each of these fingers interacts with a 3-base-pair sequence (Pavletich and Pabo, 1991). Amino acids within the F2 finger were mutagenized and screened by phage display to isolate binding variants. A DNA microarray was constructed that contained all 64 possible combinations of the 3-base-pair-binding region of the F2 zinc finger in addition to the flanking wild-type F1 and F3 recognition sequences (Bulyk et al., 2001) (Fig. 10.13). Phage displaying wild-type or mutant Zif268 protein were allowed to bind the DNA sequences on the array; nonbinders were washed away, and bound phage were detected with an anti-M13 antibody containing

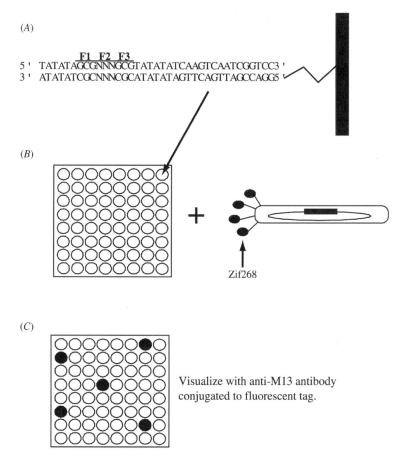

Figure 10.13 *Use of DNA microarray to define the binding specificity of transcription factors.* (A) Immobilized DNA fragment containing the Zif268-binding site. The binding site for the F2 finger is shown as NNN, where N represents any nucleotide. (B) A microarray was constructed that contained the sequence in (A) with all 64 combinations of the 3-base-pair F2-binding site, and phage displaying a variant Zif268 protein were allowed to bind to the array. (C) Bound phage were detected with an anti-M13 antibody. The position of bound phage defines the substrate specificity of the Zif268 protein variant (Bulyk et al., 2001). [Figure adapted with permission from *Proteomics*, T. Palzkill, 2002, Kluwer Academic Publishers.]

a fluorescent tag (Bulyk et al., 2001). Because the binding assay is highly parallel, it was possible to obtain a complete description of the binding specificity of each mutant in a single microarray experiment.

It should be possible to extend the DNA microarray-binding experiment for whole-genome analysis of transcription factor-binding sites. Bulyk and coworkers suggest that a microarray spotted with 12,000 1-kb sequences would span the entire *Saccharomyces cerevisiae* genome (Bulyk et al., 2001). Such an array could be used to characterize the sequence specificity of *S. cerevisiae* transcription factors. These experiments would be useful for predicting functions of previously uncharacterized transcription factors and for identifying new regulatory networks (Bulyk et al., 2001).

10.8   SURFACE PLASMON RESONANCE BIOSENSOR ANALYSIS

Surface plasmon resonance (SPR) biosensors have become an established method to measure molecular interactions. SPR biosensor experiments involve immobilizing one reactant on a surface and monitoring its interaction with another molecule in solution. SPR is an optical phenomenon used to measure the change in refractive index of the solvent near the surface that occurs during complex formation or dissociation (Jonsson et al., 1991) (Fig. 10.14). The SPR signal is expressed as resonance units (RU) and is proportional to the mass of the molecule in solution interacting with the immobilized ligand. Therefore, the standard experiment involves immobilizing the low-molecular-weight ligand and detecting the change in signal that occurs when the partner molecule is passed over the immobilization surface in a continuous flow (Schuck, 1997). One of the advantages of SPR is that binding reactions are monitored in real time without the need to label ligands. Hence, SPR can be used to study interactions between proteins, carbohydrates, nucleic acids, lipids, and small molecules.

10.8.1   Measuring Interactions of Biomolecules with SPR

Several hundred studies on macromolecular interactions using SPR biosensors have been published in a variety of fields (Rich and Myszka, 2000). Many of these studies are focused on detecting and quantitating protein–protein interactions. A typical experiment consists of covalently attaching one of the proteins to the sensor surface. A number of surfaces are commercially available for attachment, including carboxymethyl dextran, which can be derivatized to give a number of different functional groups to allow for a variety of immobilization chemistries (Schuck, 1997). Other surfaces include streptavidin for the capture of biotinylated molecules and a nickel chelation surface for the capture of His-tagged proteins (Rich and Myszka, 2000). Binding of soluble protein to the

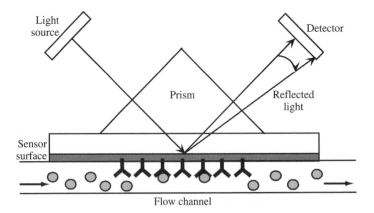

Figure 10.14   *Schematic diagram of surface plasmon resonance biosensor.* One of the binding partners is immobilized on the sensor surface. With the BIACORE instrument, the soluble molecule is allowed to flow over the immobilized molecule. Binding of the soluble molecule results in a change in the refractive index of the solvent near the surface of the sensor chip. The magnitude of the shift in refractive index is related quantitatively to the amount of the soluble molecule that is bound. [Figure adapted with permission from *Proteomics*, T. Palzkill, 2002, Kluwer Academic Publishers.]

immobilized protein gives the SPR signal in real time that can be used to monitor association kinetics. A buffer solution lacking protein is then allowed to flow over the complex to monitor dissociation kinetics. The kinetic data (k_a, k_d) can be used to obtain an equilibrium constant (K_D) (Morton and Myszka, 1998). Recent developments in SPR instrumentation may permit high-throughput analysis of protein–ligand interactions. For example, BIACORE has developed an instrument that can analyze samples in a 96-well plate format. Thus, SPR is likely to make an important contribution to genome-scale protein–protein interaction mapping.

New developments in immobilization surfaces have led to the use of SPR biosensors to monitor protein interactions with lipid surfaces and membrane-associated proteins. Commercially available (BIACORE) hydrophobic and lipophilic sensor surfaces have been designed to create stable membrane surfaces. It has been shown that the hydrophobic sensor surface can be used to form a lipid monolayer (Evans and MacKenzie, 1999). This monolayer surface can be used to monitor protein–lipid interactions. For example, a biosensor was used to examine the binding of Src homology 2 domains to phosphoinositides within phospholipid bilayers (Surdo et al., 1999). In addition, a lipophilic sensor surface can be used to capture liposomes and form a lipid bilayer resembling a biological membrane.

An interesting application of the lipid monolayer on the hydrophobic sensor surface has been the attachment of major histocompatibility complex (MHC) molecules to the surface in a specific orientation (Celia et al., 1999). This was accomplished by incorporating into liposomes a chemically modified lipid containing a nickel salt at the polar head group position (Celia et al., 1999). The liposomes containing the nickel lipids were used to coat the hydrophobic sensor surface to form a monolayer. The recombinant MHC molecule used for the experiments contained a poly-histidine tag in place of the transmembrane spanning region. It was demonstrated by electron microscopy that the histidine-tagged MHC molecules bind to the surface of the liposome. SPR measurements then demonstrated that the histidine-tagged MHC molecule binds specifically to the surface of the lipid bilayer (Celia et al., 1999). Furthermore, SPR experiments demonstrated that the MHC protein interacted with the monolayer in a specific orientation. This was inferred from the observed binding of an anti-$\alpha 1 \alpha 2$ antibody to the MHC molecules but the failure to detect binding of an anti-His-tag antibody (Celia et al., 1999). The failure of the anti-His-tag antibody to bind the MHC molecules was interpreted as steric hindrance due to the His-tag being associated with the nickel lipid at the monolayer surface. Finally, by incorporating a fluorescently labeled lipid into monolayers, Celia and coworkers were able to show that the lipids within the monolayers are laterally mobile (Celia et al., 1999). Thus, sophisticated SPR applications are being developed to simulate membrane protein interactions in vitro. Further advances in reconstituting membranes and membrane proteins on sensor surfaces will be an exciting contribution to proteomics studies. Nearly one-third of the open reading frames in a genome are thought to encode membrane-associated proteins. Therefore, the ability to assay and quantitate interactions of proteins within membranes will be crucial to proteomics studies whose goal is to obtain genome-wide protein–protein interaction maps.

10.8.2 Integration of SPR Biosensors with Mass Spectrometry

As described above, SPR biosensors can be used to detect and quantify protein–protein interactions without the need to label either of the binding partners. The method has also been used to detect interacting proteins from complex mixtures including cell lysates and conditioned media. This approach, known as *ligand fishing*, involves immobilizing a known protein as a functional hook to fish for unknown binding partners in complex mixtures (Lackman et al., 1996; Nelson et al., 2000). The difficulty with this approach is that, once a potential binding partner is identified in a complex mixture, the protein must be purified from the mixture using standard biochemical techniques before it can be identified by amino acid sequencing (Lackman et al., 1996). Coupling the biosensor with mass spectrometry provides a more direct route to identification of a binding partner. For this approach, the biosensor serves as a micropurification platform for mass spectrometry analysis and protein identification (Williams and Addona, 2000). The amount of protein recoverable from a SPR sensor surface is low (femtomoles), but this is a sufficient amount for identification using sensitive MALDI-TOF or tandem mass spectrometry methods (Nelson et al., 2000; Williams and Addona, 2000).

The potential of coupling SPR with mass spectrometry has been demonstrated using glutathione S-transferase (GST) and an anti-GST antibody as a model system (Nelson et al., 1997). For these experiments, anti-GST antibody was coupled to a carboxy-methyl dextran chip and free GST was injected and allowed to bind the antibody. The analysis was then stopped; the chip was removed from the machine and the area of the chip containing bound protein was coated with matrix material for MALDI analysis (Nelson et al., 1997). The chip was subsequently analyzed by MALDI-TOF mass spectrometry, and the resulting spectrum revealed a mass consistent with the mass of the GST protein. This method has been extended to include the proteolytic digestion of samples by using multiple flow cells on a single chip. For these experiments, antihuman interleukin alpha (anti-IL-1α) antibody was immobilized in flow cell 1 of the sensor chip, while pepsin was immobilized in flow cell 2 (Nelson et al., 2000). A solution containing IL-1α was routed to flow cell 1 where SPR measurements indicated that it was bound by the anti-IL-1α antibody. Following washing of nonspecifically bound proteins, the IL-1α was eluted from the surface and routed from flow cell 1 to flow cell 2. After allowing sufficient time for proteolytic digestion by the immobilized pepsin, MALDI-TOF mass spectrometry analysis was performed on the surface of flow cell 2. The resulting peptide masses clearly indicated the presence of IL-1α (Nelson et al., 2000).

The experiments described above indicate that technology is available to couple SPR with mass spectrometry. These methods should be useful for protein–protein interaction mapping. For example, immobilized proteins can be used as hooks for fishing binding partners from complex protein mixtures under native conditions. The coupling of techniques can lead not only to the rapid identification of interacting proteins, but will also provide information on the kinetic parameters of the interaction. This approach should serve as an excellent complement to the use of in vivo techniques such as the yeast two-hybrid system.

10.9 SUMMARY

The efficient, high-throughput detection of protein–protein and protein–ligand inter-actions has become an increasingly important aspect of proteomics studies. The most extensively used method for identifying binding partners has been the yeast two-hybrid system. This is an in vivo detection method that relies on the reconstitution of a functional transcriptional activator through the recruitment of a transcriptional activation domain to a

DNA-binding domain via interaction of the proteins being tested for interaction. The two-hybrid system has been used to systematically study interactions among yeast proteins and has identified extensive networks of interactions among thousands of proteins. Analysis indicates that these are highly heterogeneous scale-free networks in which a few highly connected proteins play a critical role in mediating interactions among a large number of less connected proteins. This type of network architecture is common to other complex systems including the Internet and metabolic networks. Complex networks have been defined in other organisms including viral, bacterial, and animal systems using the two-hybrid method. Phage display has also contributed to the identification of protein–ligand interactions. A particularly powerful approach has been the use phage display and two-hybrid data to reduce the number of false positive interactions. The inclusion of interaction data from protein fragment complementation studies is likely to further enhance protein interaction databases and provide an accurate picture of interactions on a proteome-wide scale.

Identifying protein–ligand interactions using protein chips is very challenging because of the need to retain the three-dimensional structure of a protein when it is immobilized on a solid support. To date, protein–protein interactions have been detected and dissected using both peptide and protein arrays. SPOT synthesis has been used extensively for the synthesis of peptide arrays, and these arrays have been used to identify antibody epitopes as well as to dissect binding sites for protein–protein interactions. Protein arrays have also been developed to study protein–protein and protein–ligand interactions. Recently, a protein array containing over 90% of the yeast proteome (5,800 proteins) was constructed and used to identify calmodulin-binding proteins and lipid-binding proteins. These experiments suggest that proteins are able to retain their tertiary structure when immobilized on glass slides, which bodes well for the widespread development of protein chips. Small-molecule arrays have also been developed to rapidly identify small-molecule probes for chemical genetics experiments. Small molecules are powerful tools for dissecting the role of individual proteins in complex biological processes. Protein and small-molecule arrays make use of fluorescently labeled molecules for the detection of interactions on a surface of a glass slide. A more detailed and quantitative method for the detection of interactions that does not utilize fluorescent labels is surface plasmon resonance. This method relies on the change in refractive index at a surface containing an immobilized target that occurs when a ligand binds the target. An advantage of this technique is that on rates, off rates, and equilibrium constants can be determined for the interacting pair of molecules. Thus, the technology for the high-throughput detection of protein–ligand interactions is developing very rapidly and is likely to become widely used. (Sections of this chapter are adapted with permission from *Proteomics*, T. Palzkill, 2002, Kluwer Academic Publishers.)

FURTHER READING

Ito, T. et al. 2000. Towards a protein–protein interaction map of the budding yeast: a comprehensive system to examine two-hybrid interactions in all possible combinations between the yeast proteins. *Proc. Natl. Acad. Sci. USA* 97:1143–1147.

Kuruvilla, F. G., A. F. Shamji, S. M. Sternson and P. J. Hergenrother. 2002. Dissecting glucose signaling with diversity-oriented synthesis and small-molecule microarrays. *Nature* 416:653–657.

MacBeath, G., and S. L. Schreiber. 2000. Printing proteins as microarrays for high-throughput function determination. *Science* 289:1760–1763.

Schwikowski, B., P. Uetz, and S. Fields. 2000. A network of protein–protein interactions in yeast. *Nat. Biotechnol.* 18:1257–1261.

Uetz, P. et al. 2000. A comprehensive analysis of protein–protein interactions in *Saccharomyces cerevisiae. Nature* 403:623–627.

Von Mering, C., R. Krause, B. Snel and M. Cornell. 2002. Comparative assessment of large-scale data sets for protein–protein interactions. *Nature* 417:399–403.

Zhu, H. et al. 2001. Global analysis of protein activities using proteome chips. *Science* 293:2101–2105.

11

The Functional Genomics of Model Organisms: Addressing Old Questions from a New Perspective

Dorothea K. Thompson and Jizhong Zhou

11.1 INTRODUCTION

Developing a comprehensive and integrative description of the inner workings of living cells is a formidable task. Not surprisingly, the most rapid progress in delineating gene function and interaction will initially be achieved using genome sequence data accumulated for simpler prokaryotic and eukaryotic model organisms. The prokaryotes *Escherichia coli* and *Bacillus subtilis*, and the unicellular eukaryote *Saccharomyces cerevisiae* (yeast) have traditionally served as model organisms because of their reduced structural and functional complexity and their intrinsic advantages as experimental systems. The nonpathogenic laboratory strain (K-12) of *E. coli* was one of the earliest candidates proposed for whole-genome sequencing because of the bacterium's preferred status as a model for prokaryotic genetics, molecular biology, and biotechnology, especially recombinant DNA technology (Blattner et al., 1997). *B. subtilis* was the first gram-positive bacterium to have its genome fully sequenced (Kunst et al., 1997) and has served as a paradigm for investigating the biochemistry, physiology, and genetics of this phylogenetic group of bacteria. Because *S. cerevisiae* performs all the basic functions of eukaryotic cells, and as many as 30% of positionally cloned genes implicated in human disease have yeast homologs (Bassett et al., 1997), determining the actual biological roles

Microbial Functional Genomics, Edited by Jizhong Zhou, Dorothea K. Thompson, Ying Xu, and James M. Tiedje.
ISBN 0-471-07190-0   © 2004 John Wiley & Sons, Inc.

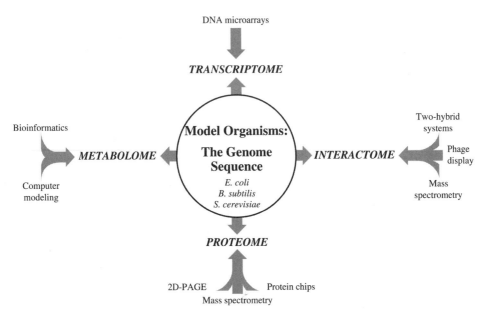

Figure 11.1 *Elucidation of the cellular domains of model organisms using a functional genomics approach and comprehensive technologies.* The recent abundance of genome sequence data for model organisms is driving the systematic analysis of global mRNA levels (the transcriptome) in response to diverse environmental or growth conditions, the encoded protein content (the proteome), the protein interaction networks (the interactome) that dictate cellular function, and metabolic networks (the metabolome).

of *S. cerevisiae* gene products will be an extremely important step in advancing our understanding of more complex and less genetically tractable metazoans (Lashkari et al., 1997; Winzeler et al., 1999).

Despite over 40 years of intensive investigation, genomic sequence annotation has revealed that more than 30% of the open reading frames (ORFs) comprising the chromosomes of *E. coli* (Blattner et al., 1997) and *B. subtilis* (Kunst et al., 1997) have no attributed function. This theme is reiterated in the sequenced genome of *S. cerevisiae*, where one-third of its approximately 6,000 predicted genes remain classified as ORFs of unknown cellular function (Goffeau et al., 1996; Uetz et al., 2000). Thus, it is evident that systematic studies other than genomic structural analyses are necessary to produce information that can be used to place the enormous amount of sequence data within a biologically meaningful context. Functional genomics and related high-throughput comprehensive technologies and methodologies (e.g., DNA microarrays, whole-genome mutagenesis, two-dimensional gel electrophoresis, two-hybrid systems, protein micro-arrays) attempt to define novel genes within the cellular domains of the transcriptome, proteome, metabolome, and interactome (Fig. 11.1). Some of the best examples of the impact of functional genomics come from research on *E. coli*, *B. subtilis*, and *S. cerevisiae*; therefore, these organisms will be the focus of this chapter. We will conclude this chapter with a brief discussion of additional model organisms.

11.2 *Escherichia coli:* A MODEL EUBACTERIUM

The status of *E. coli* as a model experimental organism in prokaryotic biology is unsurpassed. It is without a doubt the best-characterized free-living, single-celled organism and has served as a biological model for such cellular processes as DNA replication and repair, transcription, metabolic pathways, adaptive stress responses, signal transduction, and genetic regulation. Elucidation of the *E. coli* K-12 genome presents an enormous challenge, because many uncharacterized genes bear no significant homology to known sequences in databases. Despite decades of genetic research, 38% of the 4,288 protein-coding genes predicted in the *E. coli* K-12 genome still have no attributed biological role and 1,853 of those genes were previously described (Blattner et al., 1997). This demonstrates that significant gaps in our knowledge exist, even for well-studied model organisms. Understanding functionally unclassified genes within a biological context will require going beyond the level of sequence annotation to integrative whole-genome functional analysis of the transcriptome, proteome, metabolome, and interactome. The utility of these functional genomics approaches to providing informative hints for functional prediction will be discussed in the sections that follow.

11.2.1 *E. coli* Genome

What new information have we learned from a structural analysis of the *E. coli* K-12 genome? Besides enabling methods for genome-wide parallel functional analysis, the *E. coli* sequence data have revealed a number of newly proposed genes. Six previously undiscovered transfer RNA (tRNA)-encoding genes, for example, were identified in the process of annotating the *E. coli* genome (Blattner et al., 1997). Four of these new tRNA genes—namely, *valZ*, *lysY*, *lysZ*, and *lysQ*—are organized in the *lysT* operon, whereas the other two (*asnW* and *ileY*) form single-gene transcriptional units. Bioinformatic analysis of the genome sequence has also predicted the structural and regulatory components missing from our knowledge of various biochemical pathways or cellular machineries. On the basis of sequence similarity searches, gaps in a pathway for the degradation of aromatic compounds (e.g., phenylpropionate) were filled in with four other putative *mhp* genes known to exist but not identified prior to the determination of the genomic sequence. The presence of conserved sequence motifs identified one of these genes as a possible transcriptional regulator for the operon encoding the degradation pathway. Sequence scrutiny revealed a second, previously unrecognized, operon for the degradation of aromatic compounds composed of genes that resemble *Pseudomonas* genes for the decomposition of toluene, benzene, and biphenyl (Blattner et al., 1997; Tan et al., 1993). The proposed operon consists of three genes comprising the enzyme 1,2-dioxygenase, which opens the aromatic ring and oxidizes carbons 1 and 2, an ORF resembling the enzyme dihydro-1,2-dioxygenase, and the gene encoding ferredoxin reductase, the last component of the dioxygenase. A divergently transcribed ORF preceding the proposed operon has been postulated to function in the regulation of the genes. In addition, the remaining 12 flagellar synthesis genes were discovered as a result of structural analysis and were shown to be nearly identical to those of *Salmonella*. The value of this information lies in the biological questions and hypotheses that are framed based on sequence data and the emergence of new directions in experimentation.

11.2.2 *E. coli* Transcriptomics

As discussed in detail in Chapter 6, knowledge of the nucleotide sequence of complete genomes enables researchers to apply the method of cDNA- or oligonucleotide-based microarray analysis to assess in parallel genome-wide transcriptional patterns in response to a specific stimulus, genetic mutation, or physiological perturbation. After sequencing and bioinformatic analysis of an entire genome, gene expression profiling constitutes the next step in the process toward understanding an organism's physiological potential. With sequence information, a complete set of primers can be designed to amplify all the annotated ORFs in a genome, and the polymerase chain reaction (PCR) products can then be arrayed for monitoring changes in the transcript (mRNA) abundance levels for each gene. A powerful approach to the functional interpretation of fully sequenced genomes is to examine alterations in mRNA levels in response to various environmental stimuli and physiological perturbations or mutations in specific regulatory genes. This section uses a number of examples to illustrate how functional genomics is adding new details to *E. coli* cellular processes that have been extensively studied in the past with traditional approaches.

The Heat Shock Response   The heat shock response, sometimes referred to generally as a stress response, is a homeostatic mechanism exhibited by living cells when exposed to suboptimal elevated temperatures. The hallmark of this evolutionarily conserved molecular response to heat is the induced synthesis of a limited class of proteins called heat shock (or stress) proteins. In addition to thermal stress, other suboptimal physiological conditions, such as exposure to ethanol and transition heavy metals, can elicit the enhanced production of heat shock proteins (Morimoto et al., 1992). Besides protecting cellular proteins from denaturation and preventing the formation of aggregates during excessive heat, stress proteins also perform functions necessary for normal physiological growth, such as assisting in the correct assembly of multimeric structures and the folding of newly translated polypeptides to their native tertiary conformation (for a review, see Hendrick and Hartl, 1993; Georgopoulos and Welch, 1993; Lund, 2001); consequently, heat shock proteins are more generally called molecular chaperones.

The cellular response to heat (i.e., the heat shock response) is well studied and is evolutionarily conserved among many diverse organisms (Ang et al., 1991). In bacteria, the cellular response to heat stress was first discovered in *E. coli* and has been studied intensively. Because of its conserved nature, the heat shock response has been used as a model system for investigating regulated gene expression in other prokaryotes such as the Archaea (Kuo et al., 1997). The preferential and transient overproduction of heat shock proteins in *E. coli* and other bacteria is controlled at the transcriptional level by modulating the amount and activity of the alternative sigma (σ) factors *rpoH* (σ32) (Grossman et al., 1984) and *rpoE* (σE) (Erickson and Gross, 1989; Wang and Kaguni, 1989). In *E. coli*, promoters that dictate the expression of heat-inducible genes are recognized by RNA polymerase holoenzymes carrying σ32 instead of the σ70 vegetative subunit (reviewed in Mager and de Kruijff, 1995). When complexed with the core RNA polymerase, the *E. coli* σ32 transcription factor permits the transcription machinery to initiate transcription specifically from heat shock-regulated promoters (Grossman et al., 1984; Cowing et al., 1985) for both steady-state and stress-activated levels of heat shock gene expression (Cowing et al., 1985; Zhou et al., 1988). However, it is important to note that, although much of the research on the prokaryotic heat shock response has been conducted using *E. coli* as the experimental model, other novel heat shock regulatory mechanisms, such as negative *cis*-acting inverted repeat elements, have been shown to operate in some gram-positive bacteria (Hecker et al., 1996).

Microarrays containing PCR-amplified full-length coding sequences representing 97% of the total protein-coding capacity of the *E. coli* K-12 genome were employed to revisit the heat shock stimulon of this bacterium (Richmond et al., 1999). Global gene expression profiles during optimal growth at 37°C were compared with those produced under growth at 50°C (heat shock temperature) using cyanine (Cy3)-labeled control cDNA and cyanine 5 (Cy5)-labeled heat shock samples. The expression of the majority of *E. coli* genes was unaffected by heat stress, while a specific subset of genes displayed altered transcript levels following heat shock treatment at 50°C. The most significant discovery was the observation that 35 ORFs of unknown function (i.e., genes assigned as encoding hypothetical proteins) based on homology searches were transcriptionally affected by heat stress, providing, for the first time, molecular evidence of their expression and potential biological role (Richmond et al., 1999). In addition to genes previously identified as members of the *E. coli* heat shock stimulon, ORFs up-regulated in response to thermal stress also included those assigned putative functions based on sequence similarities but not described as heat shock inducible prior to this microarray study. Some of these genes were previously reported to be transcriptionally induced in response to other stress conditions, such as low pH or high lysine concentrations (*cadAB* operon encoding lysine decarboxylase and a transporter of lysine/cadaverine, respectively; Watson et al., 1992) and nutritional stress (*cspD* encoding a cold shocklike protein; Yamanaka and Inouye, 1997). Newly identified members of the *E. coli* heat shock stimulon also included *rseA*, a negative regulator of σE (Missiakas et al., 1997; De Las Penas et al., 1997); *prlC*, a trypsinlike proteinase (Conlin et al., 1992; Jiang et al., 1998); and *clpA*, the ATPase component of the ClpAP protease (Katayama et al., 1988; Gottesman et al., 1990).

The study by Richmond and coworkers (1999) validated the use of DNA microarray technology to accurately detect alterations in bacterial transcript abundance and illustrated how functional genomics allows well-characterized cellular processes to be examined from a new and global perspective. However, the study also points out the limitation of investigating a complex cellular response based solely on genome-wide measurements of transcript abundance. Although microarrays provided a list of potential participants in *E. coli*'s response to increased temperatures, other methods combined with gene expression profiling are needed to give a complete description of how the cell integrates and regulates these functions to produce a cohesive and rapid adaptive response to environmental stress (Richmond et al., 1999).

Transcriptome Analysis of Cellular Metabolism and Growth

Although the essential operons or genes concerned with a particular biosynthetic or metabolic process (e.g., tryptophan amino acid biosynthesis) may be known, other genes not required for this purpose, yet influencing or influenced by certain metabolic events, may not have been identified prior to the development of more integrated, global experimental methods. For example, DNA microarrays have been used to analyze physiological and genetic alterations that affect tryptophan metabolism in *E. coli* (Khodursky et al., 2000). The *trp* (tryptophan) operon of *E. coli* is one of the most extensively analyzed bacterial biosynthetic operons. The five genes of the *trp* operon (*trpE*, *trpD*, *trpC*, *trpB*, and *trpA* in that order) encode the enzymes required for the conversion of chorismate, a branch-point intermediate in the aromatic amino acid pathway, to tryptophan (for a review, see Pittard, 1996). Transcription of the *trp* operon is governed by repression control via the repressor protein TrpR and an entirely different mode of regulation termed transcription attenuation (Pittard, 1996). By viewing tryptophan metabolism from a genome-wide perspective using cDNA microarrays, Khodursky and his colleagues (2000) were able to determine the expression profile that is predominantly dictated by the TrpR repressor and to identify genes whose transcription is influenced by changes in tryptophan metabolism. Previous work by Pittard (1996) demonstrated that all five principal operons connected with tryptophan biosynthesis, transport, and regulation (i.e., *trp*, *aroH*, *mtr*, *trpR*, and *aroL*) were regulated at the transcriptional level by the tryptophan-activated *trp* repressor. Using microarrays to monitor global changes in mRNA abundance, Khodursky and coworkers (2000) found that only the genes of the *trp* (tryptophan biosynthesis), *mtr* (tryptophan-specific permease), and *aroH* (specifies one of three nearly identical enzymes that catalyze the initial reaction in the common pathway of aromatic amino acid biosynthesis) operons constitute the core, highly responsive *trp* repressor regulon. This was based on the fact that only these genes met the criteria of being down-regulated by excess tryptophan, up-regulated by tryptophan starvation, and up-regulated on inactivation of the *trp* repressor. Although no new TrpR repressor-specific transcriptional targets were identified in this microarray-based analysis, the expression for a number of genes was indirectly affected by changes in tryptophan metabolism, the most prominent being the sensitivity of arginine biosynthesis genes to tryptophan starvation.

Functional genomics, as embodied predominantly in microarray-mediated transcription profiling, allows one to visualize not only a focused aspect of microbial physiology, such as tryptophan metabolism, and how a specific gene or regulon interacts with all other aspects of gene expression, but also provides a window into the genomic expression underlying such broad physiological capabilities as cell growth on glucose. The value of functional genomics was clearly illustrated in a study analyzing genomic expression during late logarithmic growth of *E. coli* on minimal medium and on rich (Luria) broth medium, both containing 0.2% glucose (Tao et al., 1999). PCR-amplified, ORF-specific DNA fragments corresponding to the 4,290 annotated genes of the *E. coli* K-12 genome were arrayed on nylon membranes and probed with radioactively labeled cDNA prepared from cells grown on the two different media. As expected, *E. coli* cells grew more than twice as fast in rich medium containing glucose (generation time [G] = 25 minutes) than in minimal glucose media (G = 57 minutes) (Tao et al., 1999). Differences in growth rate were reflected in the cellular macromolecule composition (e.g., rRNA and tRNA), which was, in turn, realized at the level of gene expression. Overall, DNA arrays revealed that differences in the transcription of certain functionally grouped genes paralleled the cellular physiology of the two growth conditions, thus providing insight into growth rate-dependent gene expression and the global regulation of biosynthetic regulons (Tao et al., 1999). Growth on minimal glucose medium resulted in significantly higher expression levels (ratio ≥ 2.5-fold) for 225 genes (5.2% of the total annotated gene content) compared to 119 genes (2.8%) for growth in rich medium. Growth rate-responsive ORFs were grouped into the following functional categories: (1) translation apparatus; (2) nitrogen metabolism; (3) amino acid biosynthesis; (4) biosynthesis of vitamins, cofactors, prosthetic groups, and carriers; (5) nucleotide biosynthesis; (6) fatty acid biosynthesis and degradation; (7) carbon and energy metabolism; and (8) cellular processes and global regulators.

The hallmark features of *E. coli* growth in rich medium with glucose as the carbon and energy source were rapid growth rates, the shutdown of biosynthetic pathways, and the increased expression of genes involved in macromolecule synthesis, most notably protein synthesis. All these aspects of the physiology were revealed at the level of genomic expression by exploiting the whole-genome sequence data and employing the DNA array technique. It has been known for years that faster-growing cells show a corresponding increase in protein synthesis and ribosome abundance (Grunberg-Manago, 1996; Keener and Nomura, 1996). Of the 128 genes encoding the components of the *E. coli* translation apparatus, 53 (41.4%) were transcribed at significantly higher levels in rich medium-grown cells in contrast to cells grown in minimal medium. The majority of these genes (42 out of 53) encoded ribosomal proteins, while other genes coded for factors involved in translation and ribosomal modification (e.g., translation elongation factor genes *tsf, tufB, tufA, efp,* and *fusA*). This result was consistent with the coupled synthesis of translation factors and ribosome components (Grunberg-Manago, 1996; Tao et al., 1999). In contrast to the expression profiles of cells grown in rich medium, the transcription patterns indicated that genes involved in amino acid biosynthesis were generally induced for growth in minimal medium, a finding that parallels the need for cells to generate amino acids de novo from a sole carbon source (in this case, glucose). These biosynthetic genes included the first gene of the *ilvGMEDA* operon (for isoleucine and valine synthesis), the entire *leuABCD* operon (for leucine synthesis), and four of the five genes of the *trpEDCBA* operon (for tryptophan synthesis). The high expression ratios of the leucine and valine biosynthetic genes indicate a relatively high abundance of these amino acids in the cell, which is consistent with *E. coli* physiology (Neidhardt and Umbarger, 1996; Tao et al., 1999). Eight of the 22 highly expressed genes encoding enzymes for amino acid biosynthesis corresponded to the initial step in the pathway, thus suggesting that this is a common regulatory scheme employed by cells grown on minimal medium in order to control the flow of precursor metabolites into biosynthetic pathways (Tao et al., 1999).

Microarray analysis also revealed that the number of carbon and energy metabolism genes that were expressed in cells grown in minimal glucose media was four times greater than in cells grown in rich media. Of these, the most notable were genes involved in D-lactate utilization (*dld*), acetate formation (*poxB*), the regulation of *poxB* expression (*rpoS*, which encodes the stationary phase sigma factor), acetate utilization (*aceA, aceB, gltA, icd,* and *mdh*), and the coupling of glucose and acetate cometabolism (*uspA*, a gene encoding a universal stress protein). The induction of these genes implicates acetate metabolism as an important feature of *E. coli* growth on glucose as the sole carbon and energy source (Tao et al., 1999). Finally, growth on both kinds of media resulted in the increased expression of different regulatory genes, some of which control cellular responses indicative of the physiological state. For example, *rpoS* expression was substantially elevated in cells grown on minimal medium (Tao et al., 1999) and is known to be regulated by the stringent-response signal molecule, ppGpp, which is overproduced in cells challenged by amino acid limitation (Hengge-Aronis, 1996). The functions of RpoS-dependent genes suggested the possible role of RpoS regulation in these cells. As mentioned, a prominent feature of growth on glucose minimal medium is the production of overflow metabolites, namely acetate. The microarray results suggest that RpoS might control the expression of genes involved in protecting the cell from self-imposed acid stress (Tao et al., 1999). The RpoS-dependent expression of two unknown genes, designated *hdeA* and *hdeB*, implicated these ORFs in acid tolerance.

In the microarray profiling analysis conducted by Tao and coworkers (1999), 43 of the 225 genes showing higher expression ratios on minimal medium and 26 of the 119 genes showing increased transcription on rich medium were classified as ORFs of unknown biological function. Similarly, in another study applying microarray technology to monitoring gene expression as a function of growth stage and medium, 25% of the genes exhibiting altered expression levels in response to growth conditions lacked a functional assignment (Wei et al., 2001). With comprehensive transcript profiling, we can begin to describe a putative function for these unclassified genes based on their coregulation with similar and related genes. In addition, testable hypotheses that emerge from microarray experiments may lead to evidence confirming the biological functions of unknown genes. Hence, microarray-based transcription profiling provides further impetus for the continued study of such well-characterized model organisms as *E. coli* (Wei et al., 2001).

The NtrC regulon   Among the many powerful advantages offered by microarray-based genomics technology is the capability to detect all the genes and operons that are under the transcriptional control of a specific regulatory protein. DNA microarrays are facilitating the elucidation of multigene networks, like the Ntr (nitrogen-regulated) system, which perform a cellular function in response to information processed about the physiological state of the cell. *E. coli* can acquire nitrogen for amino acid biosynthesis by assimilating environmentally available ammonia. The assimilation of nitrogen requires the synthesis of two central intermediates, glutamate and glutamine (Ikeda et al., 1996). The cell senses restricted supplies of external ammonia by a reduction in the intracellular concentration of glutamine (Ikeda et al., 1996). The molecular response to external nitrogen limitation is the activation of transcription of genes under control of nitrogen regulatory protein C (NtrC). The NtrC protein activates transcription of sigma 54-dependent genes, whereas the nitrogen assimilation control (Nac) protein serves as an adapter between NtrC and sigma 70-dependent operons by activating transcription of sigma 70-dependent genes in response to nitrogen limitation (Pomposiello et al., 1998; Zimmer et al., 2000). In this way, the cell integrates the expression of sets of genes and operons to achieve a global adaptive response. This multigene regulation results in the expression of gene products that allow the cell to utilize any remaining traces of ammonia in the environment and then turn to other sources of nitrogen, thus minimizing the slowing of growth under nitrogen-limiting conditions.

To reveal the complete NtrC/Nac regulon of *E. coli*, Zimmer and colleagues (2000) used DNA microarrays to compare the global transcript levels in a mutant strain overexpressing NtrC-activated genes to those in a strain possessing an *ntrC* null allele. Despite the fact that the nitrogen network has been extensively studied in *E. coli*, comprehensive gene expression profiling identified a number of new NtrC regulon members. Some of these new operons specified ATP-binding cassette transporters for putrescine (*potFGHI*), oligopeptides (*oppABCDF*), and dipeptides (*dppABCDF*), and secondary ion-coupled transporters for nucleosides (*nupC*) and D-alanine/D-serine/glycine (*cycA*) (Zimmer et al., 2000). Other newly identified NtrC/Nac-controlled genes included several operons encoding hypothetical proteins (*ycdGHIJKLM*, *yeaGH*, and *yedL*), thus again demonstrating the power of microarray expression profiling for implicating unknown genes in the performance of certain cellular functions. Examination of the NtrC/Nac regulon from a whole genomic perspective indicated that NtrC controls approximately 2% of *E. coli* genes ($\sim$75), most of which were operons involved in substrate transport (Zimmer et al., 2000). The putative functions of these genes underscore the capacity of *E. coli* to scavenge its environment for nitrogen-containing compounds as a first line of defense against nitrogen starvation.

11.2.3   *E. coli* Proteomics

In addition to genomic structural analysis and transcriptome characterization using array technologies, proteomic analyses constitute an important component of functional studies, because they enable the most basic level of gene expression to be visualized. The central questions associated with proteomic analyses concern the authenticity of the ORFs predicted from sequence annotation and whether the physical properties (e.g., isoelectric point [pI] and molecular mass) of the proteins are consistent with those predicted by the ORFs. Furthermore, studying the proteome of a sequenced organism reveals important features that cannot be deduced from the theoretical proteome derived from genomic sequence alone. This other information includes in vivo protein abundance, posttranslational modifications, and proteolysis (Link et al., 1997b). The strategy for proteome analysis commonly consists of two-dimensional gel electrophoresis (2-DE) followed by spot identification using *N*-terminal sequencing or mass spectrometry (MS) [e.g., matrix-assisted laser desorption ionization-time of flight (MALDI-TOF MS); see Chapter 9 for a detailed discussion of MS]. The availability of the sequence of whole genomes is crucial for the rapid identification of protein species extracted from two-dimensional gels.

The proteome of *E. coli* K-12 was surveyed using 2-DE, followed by the identification and quantitation of 364 2-DE spots using amino (*N*)-terminal Edman sequence analysis (Link et al., 1997b; for a review of Edman sequencing, see Walker, 1994). The *N*-terminal protein sequence tags generated from Edman sequencing were used to query the completed *E. coli* genomic sequence as well as the databases. The *N*-terminal sequencing of a protein expressed under in vivo conditions serves to verify its corresponding predicted ORF in the genome. Amino-terminal processing events such as initiator methionine and signal peptide cleavage can also be determined by comparing the observed protein sequences with those predicted from the conceptual translation of the genomic sequence. A number of findings resulting from the study by Link and coworkers (1997b) could not have been predicted from the genomic sequence alone and represent new information about an "old" model organism. For example, a number of highly abundant proteins were identified: YjbJ, YjbP, YggX, HdeA, and AhpC. Although YjbJ is one of the most abundant proteins observed in *E. coli* during early stationary phase, it was not previously characterized (Link et al., 1997b). The most abundant proteins analyzed in this study were detected in a narrow pI range of 4 to 7 and a molecular mass range of 10 to 100 kDa, which points to the limitations of 2-DE analysis as a global proteomic tool (see Chapter 9). Other properties of the *E. coli* proteome included the observations that 60% of the 223 uniquely identified loci encoded proteins that were proteolytically processed, a cellular strategy for regulating protein activity, and that 18% of 2-DE spots identified by Edman sequencing represented isoforms. These isoforms constituted protein products of the same gene that possessed different observed isoelectric points and molecular masses, suggesting the involvement of posttranslational processing (Link et al., 1997b).

Genomic sequence alone does not provide an entirely accurate description of the biochemical properties and functional capabilities of the translated product that is eventually produced in the cell. For the most part, the experimentally determined values for isoelectric point and molecular mass showed reasonable concordance with the values predicted from the genomic sequence, with the exception of several discrepancies. Significant deviations between observed and expected values could be the result of highly processed proteins or misinterpretations of the genomic sequence (Link et al., 1997b).

As with transcriptome analyses, major differences in protein expression and abundance can be surveyed at different cellular states. In the study reported by Link and colleagues (1997b), the authors examined the dynamic nature of the *E. coli* proteome under exponential phase growth in glucose-minimal media and stationary phase growth in rich media. Although a comprehensive view of the proteome of the cell is not afforded by 2-DE analysis, several highly abundant proteins identified under early stationary phase growth in rich media were not observed in cells growing in glucose-minimal media. These proteins were (1) tryptophanase (TnaA), which catalyzes both the degradation and synthesis of tryptophan (Morino and Snell, 1967); (2) galactose-binding protein (MglB), which is involved in the transport of galactose into the cell; and (3) a starvation-inducible protein (Dps), which forms stable complexes with DNA and thus protects DNA from oxidative damage (Almiron et al., 1992). This study illustrates the biological principle that cells change the contents of their proteome to adapt to changing environmental conditions (Link et al., 1997b). In addition, it shows how more global methods of analysis, such as recently developed genomic and proteomic tools, are altering the way we view living cells, even well-described model organisms such as *E. coli.*

Proteomics can be employed to enhance and support predictive information supplied by genomic sequence analysis. Elucidation of the *E. coli* K-12 genome offered the opportunity to define the protein composition of the *E. coli* outer membrane (OM) using two-dimensional electrophoresis for protein arraying and MS for protein identification (see Chapter 9 for a detailed discussion of proteomic techniques). Using a new method for the isolation of bacterial outer membrane proteins (OMPs) based on carbonate incubation, Molloy and coworkers (2000) were able to identify transmembrane cell surface proteins as the most abundant protein species in the *E. coli* OM (Fig. 11.2A); in total, 78% of the predicted *E. coli* integral OMPs known to be transcriptionally active were verified using a combination of 2-DE and MS (Molloy et al., 2000). In addition, a number of these proteins (e.g., iron receptors FhuF, FepA, and CirA) were previously missing from *E. coli* two-dimensional gel maps, and two OMPs were known only from their genomic sequences prior to this proteomic study. Some new and profound findings for *E. coli* were obtained as well. The monitoring of comparative OMP expression under conditions of iron limitation led to the observation of markedly increased expression for a hypothetical protein designated YbiL, thus suggesting a putative functional role for YbiL in iron transport (Fig. 11.2B; Molloy et al., 2000). YbiL may represent a previously unidentified iron receptor in *E. coli.* In an additional experiment, Molloy and his colleagues (2000) demonstrated that the expression of another OMP (Ag43) was repressed at low growth temperatures.

11.2.4 Modeling *E. coli* Metabolism: *In silico* Metabolomics

An important aim of recent genomic science is to relate annotated nucleotide sequence information to physiological functions in the cell (Edwards et al., 2001). As massive data from genome sequencing and functional studies continue to accumulate, the need to formulate and develop *in silico* representations or mathematical models of complex, integrated cellular systems has emerged. With the inception of the genomics era, the traditional *reductionist* approach to biology is being supplanted by *integrated* approaches that address multicomponent genetic circuits in their entirety (Edwards and Palsson, 2000a; McAdams and Shapiro, 1995). A goal of mathematical modeling in microbial biology is the complete simulation of cellular metabolism based on fundamental physicochemical principles that integrates conventionally derived biochemical kinetic data (Edwards and Palsson, 2000a). The challenge is how to integrate and synthesize the extensive "molecular parts catalogue" and the connectedness of these parts in order to generate whole-cell models with interpretative and predictive capability (Edwards and Palsson, 2000b). The metabolic flux distributions within a multigenic metabolic network

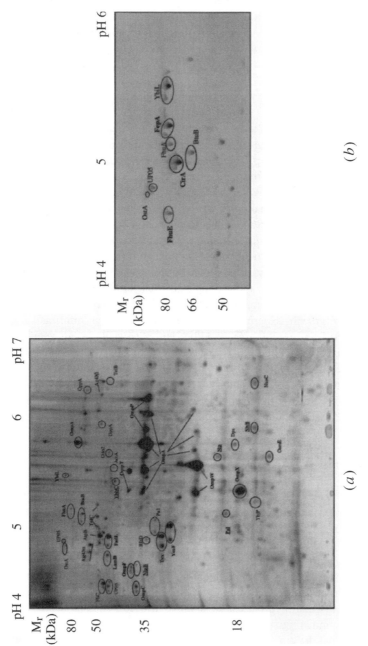

(a)

(b)

Figure 11.2 *Proteomic analysis of the outer membrane of Escherichia coli (Molloy et al., 2000). (a)* Proteins in carbonate-treated membranes of *E. coli* were separated by two-dimensional gel electrophoresis and then identified using mass spectrometry. Proteins identified as integral outer membrane proteins are presented in bold in the two-dimensional gel image. Outer membrane lipoproteins are indicated by underlined bold text. *(b) E. coli* cells grown under conditions of iron limitation were subjected to carbonate treatment and then separated by two-dimensional gel electrophoresis. [Reprinted with permission from Molloy et al., *Eur. J. Biochem.*, vol. 267. Copyright (2000) Blackwell Publishing Ltd.]

can be used to define "metabolic phenotypes"; flux balance analysis can be used to analyze the capabilities of a reconstructed metabolic network based on mass-balance and reaction constraints (i.e., physicochemical constraints) (Edwards et al., 2001; Edwards and Palsson, 2000a,b; Varma and Palsson, 1994). Interpreting and predicting metabolic flux distributions require mathematical modeling and computer simulation (Edwards et al., 2001). In cellular modeling, a physicochemical constraints-based approach attempts to describe the "best" the cell can do and what it cannot do, instead of predicting how the cell will actually behave under a particular set of conditions (Fig. 11.3; Edwards and Palsson, 2000b).

E. coli has been selected as a model for exploring the possibility and potential utility of developing an in silico representation of metabolic capabilities (Edwards and Palsson, 2000b). Edwards and Palsson (2000b) used biochemical data obtained from conventional experimentation, annotated genome sequence information, and strain-specific information to reconstruct the E. coli metabolic map. The constraints-based computational model of E. coli metabolism was used to assess the biological effects of gene deletions in central metabolic pathways on cell growth. When compared with experimental observation, in silico analysis of E. coli metabolic capabilities was able to qualitatively predict the growth potential of various mutant strains in 86% of the gene-deletion cases examined (Edwards and Palsson, 2000b). Similarly, other studies demonstrated that in silico metabolic models could be used to interpret mutant behavior (Edwards and Palsson, 2000a) and, in combination with experimental biology, could provide useful and insightful information on the genotype–phenotype relationships for metabolism in bacterial cells (Edwards et al., 2001). These studies suggested that computational analysis of metabolic behavior could facilitate the design of growth experiments and identify the most informative gene knockouts (Edwards and Palsson, 2000b). In addition, in silico analysis might aid in the identification of missing (i.e., hypothetical) or incorrect functional assignments derived from genomic sequence annotation (Edwards and Palsson, 2000a,b). However, this recent field of study is very much in its infancy, and such computer modeling and simulation will need to be used iteratively with experimental research to continually improve in silico models (Edwards and Palsson, 2000a).

11.3 *Bacillus subtilis*: A PARADIGM FOR GRAM-POSITIVE BACTERIA

B. subtilis is the best-characterized representative of the gram-positive bacterial lineage. It is a facultatively aerobic, endospore-forming, rod-shaped bacterium that is commonly found in soil and water. B. subtilis and its close relatives are of commercial interest because of their metabolic diversity and, in particular, their ability to produce extracellular hydrolytic enzymes (e.g., amylases and proteases) that degrade polysaccharides, nucleic acids, and lipids, which then serve as carbon sources for the organism. In addition to the production of macromolecular hydrolases, B. subtilis also responds to conditions of nutritional starvation by initiating such secondary metabolic processes as antibiotic (e.g., surfactin, fengycin, and difficidin) production. The failure to reestablish growth under conditions of nutritional starvation culminates in the bacterium's most distinctive characteristic, the formation of chemically, irradiation- and desiccation-resistant endospores (see Levin and Grossman, 1998, for a review of B. subtilis sporulation). These features of its physiology and its amenability to genetic manipulation have resulted

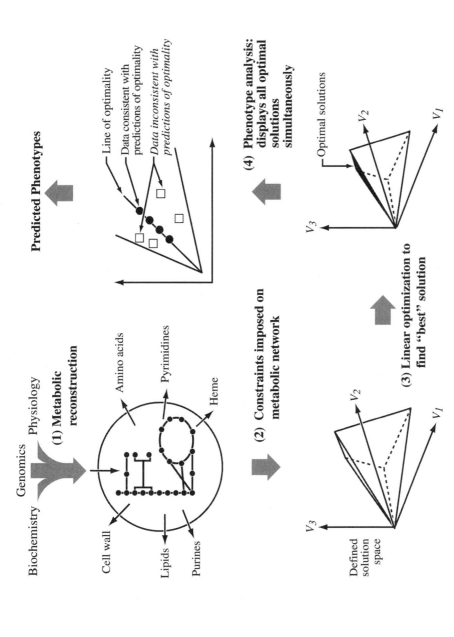

Figure 11.3 *Reconstruction of microbial metabolic networks from annotated genome sequence, biochemical data, and physiology data.* By using annotated genome sequence as a framework, along with biochemical and physiological experimental data, it is possible to construct an *in silico* representation of metabolic pathways in bacteria. Step 1 involves reconstructing a metabolic map from all relevant genomic, biochemical, and physiological data currently available for an organism. Constraints are then imposed on metabolic functions in step 2. Constraints placed on the cell are "tightened" to further narrow the range of phenotypes it can display (steps 3 and 4), leading to the prediction of metabolic phenotypes (e.g., the prediction of cellular and biochemical consequences of gene deletions).

in *B. subtilis*'s use as a model experimental organism for studying gram-positive bacteria. The entire genome sequence was published in 1997, representing the first complete genome sequence of a gram-positive eubacterium (Kunst et al., 1997). This section will discuss how genomic sequencing, functional analyses, and proteomics have advanced *B. subtilis* biology.

11.3.1 *B. subtilis* Genome

A recurrent theme among sequenced complete genomes, even for well-studied model organisms, is that a substantial percentage, as much as 30 to 60%, of the total predicted protein-coding genes cannot be functionally classified based on sequence homology. *B. subtilis* is no exception, with 42% of its gene products having no known function (Kunst et al., 1997). Despite decades of intensive research, only ~1,200 gene functions (~30%) in *B. subtilis* have been experimentally verified (Kunst et al., 1997), suggesting that much more work on the discovery of new gene functions and the confirmation of annotated ones remains before we gain a comprehensive understanding of the physiology of *B. subtilis*.

Soil bacteria, such as *B. subtilis*, have evolved complex regulatory systems in order to respond rapidly to the adverse environmental conditions and stresses they encounter. Many regulatory proteins found in both prokaryotes and eukaryotes belong to the helix-turn-helix (HTH) family of transcription factors. Sequence analysis of the *B. subtilis* genome revealed that the specific cellular functions of a number of putative HTH transcription regulatory genes remain to be ascertained, including 18 of 20 GntR family regulators, 15 of 19 LysR family regulators, 5 of 12 LacI family regulators, and 10 of 11 AraC family regulators to name a few (Kunst et al., 1997). The regulation of adaptive cellular responses to environmental stress is, to a large extent, effected by two-component signal transduction pathways, which are widespread among prokaryotes. Two-component regulatory systems consist of a sensor protein kinase and its cognate response regulator. In *B. subtilis*, 37 sensor kinases and 34 genes encoding response regulators have been identified based on sequence similarity to known proteins in the databases (Kunst et al., 1997). As will be illustrated in the subsection below, complete genomic sequence information provides the opportunity to design experimental strategies for the functional characterization of these two-component regulatory systems.

Bacteria are often challenged by a variety of stress conditions (e.g., changes in temperature, humidity, or nutrient source availability). Not surprisingly, the genome sequence of *B. subtilis* revealed 43 temperature shock and general stress proteins that play an important role in the adaptation of *B. subtilis* to changes in environmental conditions (Kunst et al., 1997). These proteins displayed a strong similarity to their *E. coli* counterparts, suggesting that the cellular response to stress is evolutionarily conserved among gram-negative and gram-positive bacteria. ABC transporters constituted the most frequent functional class of proteins identified in *B. subtilis* (Kunst et al., 1997) and may reflect differences in the architecture of the outer envelope for the two eubacterial groups. The single-membrane-containing envelope of gram-positive bacteria is undoubtedly a weaker protective barrier compared to the multilayered outer envelope of gram-negative bacteria. In addition, a greater number of ATP-binding transport proteins are likely to enhance the ability of *B. subtilis* to resist the toxic action of various compounds. A total of 77 ABC transporters were predicted in the *B. subtilis* genome, which had been greatly expanded by gene duplication events (Kunst et al., 1997).

The determination of complete genome sequences for both *E. coli* and *B. subtilis* allows their relative genomic diversity to be studied from a comprehensive perspective. Such comparative analyses may offer insight into the evolutionary divergence of eubacteria into the gram-negative and gram-positive lineages, which probably diverged more than 1 billion years ago. Although the *B. subtilis* genome (4.2 Mb) resembles the *E. coli* genome (4.6 Mb) in terms of size and about 1,000 (25%) of the predicted *B. subtilis* genes have clear orthologous counterparts in *E. coli*, some striking distinctions in encoded functional capabilities and operon structure exist between the two genomes. In contrast to *E. coli*, for example, many *B. subtilis* genes are involved in the synthesis of secondary metabolites, such as antibiotics. Almost 4% of the *B. subtilis* genome codes for large multifunctional enzymes with sequence similarity to those proteins involved in antibiotic synthesis in *Streptomyces* species. Sequence analysis has also indicated that bacteriophage infection has played an important role in the horizontal transfer of genetic information, as evident from the presence of at least 10 prophages or vestiges of prophages (Kunst et al., 1997; Nicolas et al., 2002). As discussed in Chapter 4, horizontal gene transfer is considered a driving force in bacterial evolution (Ochman et al., 2000). Finally, differences in gene organization for the two genomes were observed for amino acid and purine biosynthetic genes, to name a few. *E. coli* genes for arginine and purine biosynthesis are dispersed throughout the chromosome, whereas the *B. subtilis* counterparts are organized into operons. Functional characterization of the numerous unknown genes contained within the *E. coli* and *B. subtilis* genomes will likely provide even greater insight into the evolutionary, physiological, and functional divergence of these model organisms.

11.3.2 *B. subtilis* Transcriptomics

After the entire *B. subtilis* genome was sequenced in 1997, a number of systematic functional analyses were conducted to delineate gene function and regulatory networks in response to different growth conditions, environmental perturbations or stress shock, and genetic mutations. In this section, we will discuss several microarray-based functional analyses on the heat shock response in *B. subtilis*, two-component regulatory systems, and *B. subtilis* growth under different conditions. These studies illustrate how functional genomics has contributed to the rich repository of experimental evidence on the physiology of *B. subtilis*.

***Global Characterization of Heat Shock in* B. subtilis**   As mentioned previously in our discussion of the *E. coli* heat shock stimulon, the transcription of many genes in bacteria is activated as part of an adaptive response to heat stress. *B. subtilis* responds to diverse growth-limiting stresses such as heat shock, osmotic stress, and energy stress by activating genes in the large general stress regulon controlled by the alternative sigma transcription factor, σ^B (Hecker et al., 1996; Price, 2000), while the expression of other heat-inducible genes is governed by the HrcA (Schultz and Schumann, 1996) or CtsR (Derre et al., 1999; Kruger and Hecker, 1998) transcriptional regulators. Transcription sigma factor σ^B is the general stress sigma factor of gram-positive bacteria. The critical event in the induction of the stress regulon is the activation of σ^B activity following metabolic or environmental stress or starvation. Transcription of many general stress genes occurs at a basal level from vegetative σ^A-dependent promoters, but is increased dramatically in a σ^B-dependent manner in response to stress or starvation. The discovery and functional characterization of new σ^B-regulated genes should improve our understanding of the physiological role that the general stress regulon plays in stress adaptation by *B. subtilis* (Petersohn et al., 1999). Global approaches such as DNA array technology permit a comprehensive analysis of the transcriptional response to stress.

DNA microarrays containing PCR-amplified DNA fragments corresponding to approximately 90% of the ~4,100 annotated *B. subtilis* ORFs were used to monitor the global transcriptional response to heat shock (Helmann et al., 2001). In this study, growing cultures of *B. subtilis* cells were shifted from 37°C (optimal growth temperature) to 48°C (heat shock) to elicit the heat shock response. Total RNA was isolated from cells grown under the two different temperatures, labeled by reverse transcription in the presence of two fluorophores, and applied to *B. subtilis* arrays for differential display. The full magnitude of the heat shock response in *B. subtilis* was revealed in a single microarray experiment with genome-wide analysis: Over 10% of the *B. subtilis* genome exhibited detectable increases in gene expression in response to heat stress, while more than 5% of transcriptionally active genes showed at least a 3-fold induction in expression (Helmann et al., 2001). The microarray experiments confirmed the transcriptional induction of the 70 known or previously proposed members of the σ^B-dependent general stress regulon and more comprehensively defined the magnitude of the σ^B regulon by identifying another 72 new candidate members, including 24 genes encoding hypothetically conserved proteins or proteins with no described homolog. The classification of these functionally undefined genes as members of the σ^B regulon suggests that their protein products perform functions that aid the cell in resisting the adverse effects of environmental or metabolic stress. The presence of putative σ^B-dependent promoter consensus sequences proximal to the newly identified heat-induced genes supports the transcriptional profiling data (Helmann et al., 2001). A limitation of microarray-based transcription profiling is that the method only suggests the involvement of genes in a particular cellular process; additional detailed functional analyses will be required to define the precise functions of these proteins in stress adaptation. The most surprising result obtained from the microarray experiments was the very strong transcriptional induction (≥ 50-fold) observed for three operons involved in arginine biosynthesis (*argCJBDcarABargF* and *argGHytzD*) and transport (*yqiXYZ*). The regulatory factor(s) mediating their heat-induced expression, however, are currently not known.

  B. subtilis gene arrays have also been used to investigate the global transcriptional response to ethanol stress (Price et al., 2001) and to identify σ^B-independent stress phenomena, most notably the induction of the extracytoplasmic function sigma factor σ^W and its entire regulon in response to salt shock (Petersohn et al., 2001).

Two-component Regulatory Systems An important class of adaptive response systems in prokaryotes, lower eukaryotes, and plants consists of two signal transduction proteins: a sensor histidine kinase that "senses" extracellular stimuli and a cognate response regulator that mediates the adaptive cellular response at the level of transcriptional regulation (reviewed in Stock et al., 2000). Two-component regulatory systems serve as a basic stimulus-response coupling mechanism that allows organisms to detect and rapidly respond to environmental changes. Two-component regulatory systems govern the expression of target genes through controlled changes in protein phosphorylation. Signal reception alters the ability of a membrane-bound histidine kinase protein to transfer phosphate from ATP to a highly conserved histidine residue, typically located in the phosphoryl transfer–dimerization domain of the kinase protein.

The transfer of phosphate from the histidine to an aspartate residue on the cognate response regulator increases the affinity of the latter protein for its target promoters.

The genomic sequencing of *B. subtilis* has indicated the presence of numerous putative two-component regulatory systems: 37 sensor kinases and 34 response regulators were identified, and among these, 30 kinase-response regulator partners reside adjacently in the *B. subtilis* chromosome (Kunst et al., 1997). The environmental signals inducing many of these two-component regulatory systems are not known, thus making the identification of their target genes by stimulating the cells with the appropriate stimulus difficult, if not impossible (Ogura et al., 2001). To circumvent this problem, Ogura and coworkers (2001) placed the response regulator gene under the transcriptional control of an isopropyl-β-D-thiogalactopyranoside-inducible promoter on a multicopy plasmid (pDG148), and overexpressed the response regulator genes in *B. subtilis* mutant strains harboring disruptions in their cognate sensor kinase genes. The expectation was that overproduction of the response regulator might result in altered expression of the target genes in the absence of the specific environmental signal and cognate sensor kinase responsible for its phosphorylation. The genome-wide transcriptional effect on gene expression caused by the overproduction of the response regulator was determined using DNA microarray analysis.

As a proof of concept, this experimental strategy for defining the regulons of two-component regulators was evaluated using the well-characterized DegU, ComA, and PhoP as model response regulators, because parts of their gene targets were already known (Ogura et al., 2001). The *B. subtilis* DegS/DegU two-component system is known to control such cellular processes as exoprotease production, competence development, and motility (Msadek et al., 1995). Cell density signals activate the ComP/ComA two-component system (Lazazzera et al., 1999), and induction of the Pho regulon in response to phosphate starvation is regulated by the PhoP/PhoR system (Hulett, 1996). Microarray-based global analysis of the regulons of the two-component regulators detected the majority of the known target genes. In some cases, microarray analysis failed to detect previously identified target genes, such as *rapC* and *rapE* for the ComP/ComA regulatory system and *tagAB* and *tagDEF* for the PhoP/PhoR system (Ogura et al., 2001). However, with this comprehensive approach to the analysis of *B. subtilis* two-component regulatory systems, a number of genes were identified whose response regulator-dependent expression had not been previously observed. DNA microarray analysis, for example, revealed that the expression of 116 target gene candidates was affected by DegU overproduction, including known target genes (e.g., *aprE*, *nprE*, and *ispA*) and a number of newly identified members of the DegU regulon (e.g., *bpr*, *yukL*, *ycdA*, and *murD*). Transcription profiling of the ComA and PhoR regulons indicated 33 and 23 target candidate genes, respectively, including the newly identified genes *rapF* (ComA regulon) and *yycP* and *yjdB* (PhoR regulon).

In a similar study, Kobayashi and colleagues (2001) employed the strategy of overproduction of nonphosphorylated response regulators and DNA microarray analysis to comprehensively identify target gene candidates for 24 functionally undefined two-component regulatory systems found in the *B. subtilis* genome sequence. Genes encoding the 24 different response regulators were cloned into the multicopy plasmid pDG148 and then transferred to *B. subtilis* mutants deficient in the cognate sensor kinases for analysis. For some of the unknown two-component systems analyzed, a probable cellular function was implicated based on the list of affected genes (Kobayashi et al., 2001). Altered expression levels for the *mcpA*, *mcpB*, *flgK*, and *flgM* genes in the presence of overproduced YdbF suggested that the YdbG/YdbF two-component system might be involved in chemotaxis. Furthermore, it is likely that the YufL/YufM system plays a role in regulating competence because of the observation that many of its target gene candidates are also members of the regulon controlled by ComK, a transcription factor regulating the development of genetic competence in *B. subtilis* (Hamoen et al., 1998). Finally, a putative regulatory role was unveiled for the YvrG/YvrH system, which appears to be associated with cell membrane and cell wall function based on its target gene candidates coding for membrane proteins, transporters, and wall-related proteins. Of course, further molecular functional analyses are needed to definitively ascertain the cellular function of these two-component systems. An interesting outcome of the study by Kobayashi and coworkers (2001) was the observed interaction that appeared to occur between several two-component regulatory systems as revealed by DNA microarray analysis. The target gene candidates identified for YxjM/YxjL and YvqE/YvqC, for instance, showed extensive overlap, with 17 out of the 19 target genes of YxjM/YxjL also regulated by the YvqE/YvqC system. Two additional two-component systems exhibiting overlap among their target genes were DesK/DesR and YvfT/YvfU. These studies demonstrate how microarray-based transcription profiling, in combination with other more conventional genetic techniques, can be used to detect a majority of the target gene candidates of functionally uncharacterized two-component regulatory systems, thus providing clues as to the roles these systems play in the signal-transduction network of a cell. Although such experimental approaches are largely heuristic in nature, the results obtained from global transcription profiling can lead to more definitive avenues of functional analysis.

Global Gene Expression During Growth and Sporulation of B. subtilis

Anaerobic Respiration   Bacteria frequently encounter fluctuating levels of external oxygen. In contrast to strict aerobes or anaerobes, facultative bacteria cope with changes in environmental oxygen levels by sensing oxygen tension and making the appropriate adjustments in their cellular metabolism (Gunsalus and Park, 1994). The adaptation of organisms to changing redox environments is typically achieved by altering the underlying gene expression or by modulating protein activity. It was long thought that *B. subtilis* was a strict aerobe, requiring carbon (in the form of simple carbohydrates or organic acids) as an energy source for growth and the biosynthesis of cellular components. However, recent studies have demonstrated that *B. subtilis* can respire anaerobically by using nitrate or nitrite as a terminal electron acceptor, or by fermentation in the absence of external electron acceptors (Nakano et al., 1997; Nakano and Zuber, 1998). The process of dissimilatory reduction of nitrate to ammonia requires two enzymes, the membrane-bound nitrate reductase (encoded by the *narGHJI* operon) and the NADH-dependent nitrite reductase (encoded by the *nasDEF* operon) (Hoffmann et al., 1995). Fumarate/nitrate reduction regulator (FNR) is a transcriptional activator of anaerobically induced genes that regulates expression of the *narGHJI* operon (Kiley and Beinert, 1998). The two-component signal transduction system, ResD/ResE, performs an essential role in the regulatory pathway governing anaerobic respiration by activating *fnr* transcription upon oxygen limitation (Nakano and Zuber, 1998). In the absence of nitrate and nitrite, the efficiency of *B. subtilis* growth on glucose under anaerobic conditions is quite poor but can be substantially improved by the addition of pyruvate to the media (Nakano et al., 1997).

    The global changes in gene expression associated with anaerobic growth are presently not known. To explore the metabolic and genetic control of anaerobic gene expression on a genomic scale, *B. subtilis* whole-genome microarrays containing 4,020 (out of 4,100)

ORFs were used to examine differential gene expression patterns of aerobic and anaerobic cultures during exponential growth (Ye et al., 2000). The conversion from aerobic to anaerobic growth resulted in the induction or repression of approximately several hundred genes involved in diverse cellular functions, including carbon metabolism, electron transport, iron uptake and transport, antibiotic production, and adaptive stress responses. Most notably, the expression of the following genes was induced: (1) *ydjL*, which encodes a gene product similar to the 2,3-butanediol dehydrogenase of *Pseudomonas putida* (Huang et al., 1994); (2) *ytkA*, a gene located upstream from the transcriptionally affected stress-response gene *dps*; and (3) the unknown region *yolIJK*, which encodes gene products similar to an aspartyl protease (YolJ) and disulfide-bond oxidoreductases (YolK and YolI). The mRNA levels of some other unknown genes were elevated significantly and preferentially under certain anaerobic conditions (Ye et al., 2000). High transcriptional levels for *ykzH* and *ykjA* genes, for example, were detected when *B. subtilis* was grown on nitrate or nitrite, whereas transcript levels for *ywcJ* and *yumD* genes were increased by 30- or 50-fold under fermentative conditions in the absence of pyruvate. Additional unknown gene clusters, *yjlCD* and *yxxG-yxiM*, were observed to be down-regulated during dissimilatory reduction of nitrate and nitrite and during fermentative growth without pyruvate. Although microarray expression profiling is largely exploratory for gene function, the results reported by Ye and colleagues (2000) at least suggest a putative function for genes previously annotated as unclassified.

The signature gene expression profile of nitrate and nitrite respiration in *B. subtilis* is the dramatic induction of *narGHJI* (nitrate reductase), *narK* (nitrite extrusion protein), *fnr* (global anaerobic regulator), and *hmp* (putative flavohemoglobin of unknown physiological function), followed by *nasDEF* (nitrite reductase), *cydABCD* (cytochrome oxidase and ABC membrane transporter), *sbo-alb* (subtilosin A-unknown protein), *ywiD* (unknown physiological function), and *ywiC* (unknown physiological function). The hallmark of inefficient fermentative growth on glucose is a reduction in *pdhAB* (pyruvate dehydrogenase) expression and a marked increase in *lctPE* (L-lactate permease and L-lactate dehydrogenase) expression (Ye et al., 2000). Global transcriptional profiling indicated that the regulatory circuitry controlling anaerobic metabolism in *B. subtilis* is complex and dynamic, with certain subsets of genes specifically regulated during anaerobic respiration with nitrite and different genes primarily affected during fermentative growth (Ye et al., 2000). Similarly, other studies used microarrays to systematically analyze glucose-repressive genes of *B. subtilis* that are expressed following the shift from glycolytic to gluconeogenic growth in nutrient medium (Yoshida et al., 2001). Such research is beginning to define the transcriptomes in cells at different physiological states, thus providing valuable information for the further investigation of important cellular processes.

Sporulation   Historically, scientific interest in *B. subtilis* as a model experimental organism began, in large part, because of the bacterium's ability to undergo sporulation, a highly specialized adaptive response to starvation (see reviews by Kroos et al., 1999; Stragier and Losick, 1996). In bacteria, sporulation represents a relatively simple, experimentally tractable developmental process that results in prolonged transformation of cell structure. *B. subtilis* responds to starvation by undergoing asymmetric division to produce two different cell types, a small forespore ("prespore") and much larger mother cell. Briefly, the forespore is engulfed by the mother cell, which provides a nutritional and protective environment for the forespore as it continues its process of differentiation.

Ultimately, the mother cell lyses to release a mature, dormant, and environmentally resistant spore. Under more favorable growth conditions, the spore germinates to produce a vegetative cell again.

Because sporulation is a sophisticated differentiation process, it requires a complex and intricate program of transcriptional gene regulation to coordinate gene expression with dramatic morphological change. The process requires the activity of a number of sigma transcription factors that are themselves carefully controlled by multiple mechanisms (Kroos et al., 1999). Several decades of research in numerous laboratories have led to the isolation and characterization of the major genes involved in the regulation of sporulation (primarily called *spo* genes), including the transcription activator and repressor protein Spo0A, which governs the entry of *B. subtilis* cells into sporulation (Stragier and Losick, 1996). Major information inputs derived from external environmental and cellular physiological signals are integrated by a phosphorelay system that controls the activity of Spo0A, a member of the response regulator family of proteins, through phosphorylation (Burbulys et al., 1991).

The availability of the complete sequence of the *B. subtilis* chromosome permits an analysis of global changes in gene expression during spore formation. Such a comprehensive molecular description of *B. subtilis* sporulation was not possible prior to the functional genomics era. To rectify this deficiency in the research, Fawcett and coworkers (2000) used nylon-substrate DNA arrays (macroarrays) representing approximately 96% of the potential protein-coding capacity in the *B. subtilis* genome to compare the gene expression profiles of wild-type cells with those of mutant cells for the sporulation transcription factors Spo0A or σ^F, a regulatory protein specific for directing transcription in the forespore (Stragier and Losick, 1996). Genes displaying significant differences in transcript abundance during sporulation were grouped into three distinct categories (Fig. 11.4): (1) genes whose expression depended on Spo0A but not σ^F; (2) genes whose expression was inhibited by Spo0A; and (3) genes whose expression depended on both Spo0A and σ^F (Fawcett et al., 2000). Out of the 586 genes (10% of the total predicted ORFs) identified as being under sporulation control, the largest group of genes (283) was those exhibiting Spo0A-dependent expression. This category included sporulation genes whose transcription was previously known to be directly regulated by Spo0A (e.g., *spo0F*, *spoIIA*, and *spoIIG*) or indirectly dependent on Spo0A (e.g., *kinA* and *spoVG*). However, a large majority of genes under the direct control of Spo0A were newly identified and of unknown function. These genes included *yjcP*, *yneE*, and two apparent operons (the gene cluster that includes *yxbB*, *yxbA*, and *yxnB* and the cluster that includes *ybcO*, *ybcP*, *ybcQ*, *ybcS*, *ybcT*, *ybdA*, and *ybdB*). Transcriptional profiling also revealed genes whose expression was directly or indirectly repressed by Spo0A and, as a result, were overexpressed in the spo0A mutant. Most notably, the expression of various chemotaxis genes (*mcpA*, *fliJ*, *cheB*, *cheA*, *cheW*, and *cheC*), motility genes (*flgC*, *flgE*, *flhO*, *fliD*, *fliF*, *fliG*, *fliK*, and *fliY*), and autolysin genes (*lytD* and *lytE*) was inhibited by Spo0A. Finally, in the third category (Spo0A- and σ^F-dependent expression), some of the newly identified sporulation-controlled genes not previously recognized as being expressed under sporulation encoded proteins with putative functions in the metabolism of long-chain fatty acids (*yngJ*, *I*, *H*, *G*, *F*, *E*), in peptidoglycan biosynthesis (*murA*), and a gene (*prkA*) coding for a serine protein kinase of unknown function (Fawcett et al., 2000).

In addition to the identification of many previously uncharacterized genes whose expression is under sporulation control, two adjacently positioned genes, *yabP* and *yabQ*, which appear to constitute an operon, were detected by transcriptional profiling and later

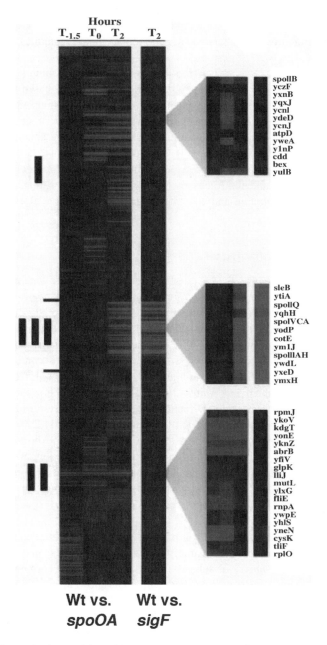

Figure 11.4 *See color insert. Hierarchical cluster analysis (see Chapter 7 for an explanation) of microarray-derived transcription profiles of 586* Bacillus subtilis *genes whose expression levels depended on SpoOA (Fawcett et al., 2000).* Genes with similar expression patterns were grouped together. As shown, the genes fell into three categories: (I) genes whose expression was dependent on SpoOA but not on σ^F; (II) genes whose expression was inhibited by SpoOA; and (III) genes whose expression was under the control of σ^F or some downstream transcription factor in sporulation. For each of the categories, representative genes are shown on the right. Red and green colors indicate higher and lower mRNA expression levels, respectively. [Reprinted with permission from P. Fawcett, P. Eichenberger, R. Losick, and P. Youngman. The transcriptional profile of early to middle sporulation in *Bacillus subtilis*, PNAS vol. 97, 8063–8068. Copyright (2000) National Academy of Sciences, U.S.A.]

shown to be essential for spore formation in *B. subtilis* (Fawcett et al., 2000). Null mutations of *yabP* and *yabQ*, generated by gene-replacement techniques, led to a severe sporulation defect.

Furthermore, the putative *yabPQ* operon resides on the chromosome near and between the known sporulation genes *spoVT* and *spoIIE*. Overall, this array-based investigation by Fawcett and his colleagues (2000) is an excellent example of how transcriptional profiling can enhance the molecular description of a well-studied cellular process like bacterial sporulation by leading to the identification of many functionally annotated as well as undefined genes whose expression was previously unrecognized as being under sporulation control. Microarray expression profiling, unlike traditional single-gene experiments, establishes and reinforces the view that complex cellular processes, such as morphological differentiation and adaptational responses, involve global alterations in gene expression.

11.3.3 *B. subtilis* Proteomics

A number of research groups have captured the specific pattern of gene expression of *B. subtilis* in a defined physiological state by generating proteomic maps of two-dimensional gels (Antelmann et al., 2001, 2002; Buttner et al., 2001; Coppee et al., 2001; Eymann et al., 2002; Hirose et al., 2000; Hoffmann et al., 2002; Movahedi and Waites, 2000; Ohlmeier et al., 2000; Yoshida et al., 2001). Some of these investigations, for example, revealed the dynamic nature of the *B. subtilis* proteome under conditions of sulfur limitation (Coppee et al., 2001), low and high salinity (Hoffmann et al., 2002), heat stress during sporulation (Movahedi and Waites, 2000), glucose (catabolite) repression (Yoshida et al., 2001), and nutrient starvation (Eymann et al., 2002). Other studies have focused on elucidating the extracellular complement (Antelmann et al., 2001, 2002; Hirose et al., 2000), cytosolic complement (Buttner et al., 2001), and alkaline protein composition (Ohlmeier et al., 2000) of the *B. subtilis* proteome. In this section, we will focus on analyses of the extracellular proteome and the induction of stress response proteins in sporulating *B. subtilis* cells as illustrations of how proteomics is improving our understanding of overall cell physiology.

The Extracellular Proteome   As soil microorganisms, *B. subtilis* and related species secrete large quantities of proteins, largely degradative enzymes, that enable them to acquire nutrients from a variety of substrates and to survive in a complex, continuously changing environment (for a review, see Simonen and Palva, 1993). In addition, eubacterial secretory proteins perform other vital functions such as cell-to-cell communication, detoxification of environmental compounds, or the elimination of potential competitors. A two-dimensional map of cytosolic proteins of *B. subtilis* indicated that the most abundant proteins of exponentially growing cells were those that performed mainly housekeeping functions in glycolysis, tricarboxylic acid (TCA) cycle, amino acid biosynthesis, translation, and protein quality control (Buttner et al., 2001). In *B. subtilis*, protein secretion is primarily a cellular event typical of the postexponential (i.e., stationary) stage of cell growth. Two proteins essential for normal cell growth and protein translocation are SecA and Ffh. Thus far, the Ffh protein, which is homologous to the 54-kDa subunit of the eukaryotic signal recognition particle (Honda et al., 1993), is the only molecular chaperone definitely implicated in the process of protein secretion in *B. subtilis*. Molecular chaperones are required early in the secretion process to retain the targeted precursor protein in an unfolded, translocation-competent state and to target the precursor protein to the secretory machinery. SecA, the translocation ATPase, interacts with the precursor–chaperone complex and subsequently directs the precursor into the translocation channel composed of SecY, SecE, SecDF, and a SecG homolog for protein export (Simonen and Palva, 1993). During or shortly following translocation of the preprotein across the cytoplasmic membrane, its amino-terminal signal peptide is removed by one of the five type-I signal peptidases present in *B. subtilis* (Tjalsma et al., 1997; van Dijl et al., 1992).

Sequence analysis of the *B. subtilis* genome predicted 180 secretory and 114 lipoprotein signal peptides (Tjalsma et al., 1997, 1998, 1999). The most significant structural difference between the signal peptides of lipoproteins and secretory proteins is the presence in lipoprotein precursors of a conserved "lipobox," which contains an invariable cysteine residue possessing a lipid modification introduced by the lipoprotein diacylglyceryl transferase prior to peptidase cleavage (Antelmann et al., 2001). In one two-dimensional protein electrophoretic study, approximately 100 to 110 spots were visualized in a gel of *B. subtilis* 168 extracellular proteins (Hirose et al., 2000). When two-dimensional extracellular protein profiles from the parental strain *B. subtilis* 168 were compared to those from a *secA* temperature-sensitive mutant and an *ffh* conditional mutant, over 90% and 80% of the exported proteins detected in the *B. subtilis* 168 gel disappeared in the absence of SecA and Ffh, respectively. Thus, the appearance of most of these extracellular proteins is dependent on SecA and Ffh, requiring cooperation between the signal-recognition particle and Sec protein-secretion pathways. In contrast, the appearance of a protein spot identified by *N*-terminal amino acid sequencing as Hag, an exported flagellin protein, was demonstrated to be SecA- and Ffh-independent (Hirose et al., 2000). The identity of 23 proteins was determined by *N*-terminal sequencing. Of these, 17 were confirmed extracellular proteins containing signal peptides in their preprotein form, whereas two were membrane proteins that appeared to be released into the culture medium after processing (Hirose et al., 2000).

In a proteomic approach to genome-based signal peptide predictions, Antelmann and coworkers (2001) visualized approximately 200 extracellular proteins by two-dimensional gel electrophoresis and identified 82 of those proteins by MS. Of the 82 identified extracellular proteins, 50 proteins had previously assigned functions, whereas the remaining 32 proteins were of unknown function. This work built on the proteome analysis reported by Hirose and colleagues (2000) (described above) when defining an additional 62 extracellular proteins of *B. subtilis*. Besides providing proteomic verification of genome-wide predictions of protein export signals of *B. subtilis*, the study by Antelmann and coworkers (2001) revealed a variety of unexpected results that could not have been predicted solely based on genomic structural analysis. Fifty extracellular proteins identified in the study possessed a typical signal peptide motif with a type-I signal peptidase cleavage site. Strikingly, proteomic analysis disclosed the export of 41 proteins previously predicted to be cell-associated because of the apparent lack of a signal peptide (in the case of cytoplasmic proteins) or the presence of specific cell-retention signals in addition to an export signal (in the case of membrane-bound lipoproteins) (Antelmann et al., 2001). It should be noted, however, that cytoplasmic proteins (e.g., aldolase, enolase, elongation factor G, GroEL, and various dehydrogenases) have been found in the extracellular proteomes of other eubacteria (Jungblut et al., 1999; Lei et al., 2000; Rosenkrands et al., 2000). A number of predicted cell-associated lipoproteins, mostly ABC transporters, were unexpectedly identified in the extracellular proteome of a *B.*

subtilis mutant defective in prelipoprotein modification and processing. The data as a whole indicate that the identified portion of the *B. subtilis* extracellular proteome includes enzymes related to the metabolism of carbohydrates, proteases or peptidases, enzymes involved in amino acid metabolism, enzymes involved in the decay of nucleic acids (DNA or RNA), lipases, alkaline phosphatases, phosphodiesterases, proteins involved in cell-wall biogenesis, lipoproteins (including substrate-binding components of different transport systems), detoxification proteins, flagella-related proteins, a putative transcriptional regulator, proteins involved in protein synthesis and folding (including the chaperonin GroEL), prophage-related proteins, a sporulation-specific protein, and 13 proteins of unknown function (Antelmann et al., 2001). The relative abundance of most of the identified extracellular proteins increased substantially during the stationary phase, when the availability of nutrients becomes limiting. The study demonstrates the value of proteomic analyses in casting new light on the composition of extracellular proteomes, the significance of signal peptide-independent protein export pathways, and the function of extracellular proteins.

Heat Stress Proteome during Sporulation   Studies attempting to define the heat stress proteome of *B. subtilis* have centered almost exclusively on vegetative cells (Bernhardt et al., 1997), where a large group of sigma factor σ^B-controlled stress proteins were shown to be induced nonspecifically by such physical stresses as heat, salt, ethanol, or acid and by glucose, oxygen, or phosphate starvation (Hecker and Volker, 1990). Recently, Movahedi and Waites (2000) described the first detailed two-dimensional electrophoresis study of the stress response of sporulating *B. subtilis* cells. The induction of stress proteins during sporulation correlated with the heat resistance of spores formed subsequently. In their analysis, Movahedi and Waites (2000) showed that *B. subtilis* cells subjected to sublethal heat shock early in the sporulation process produce spores with increased heat resistance. High-resolution two-dimensional gel electrophoresis revealed that 60 stress proteins were synthesized de novo and/or overexpressed in *B. subtilis* during sporulation concurrent to acquired thermotolerance in spores. Eleven of these detected proteins were ascertained to be heat shock-specific proteins, because they were synthesized or induced in sporulating cells specifically in response to heat stress but not cold shock or glucose starvation (Movahedi and Waites, 2000). In contrast to the heat stress proteome of vegetative cells, none of the stress-specific proteins identified in this proteomic analysis exhibited σ^B-dependent expression. Time-course induction studies indicated that stress proteins disappeared later in sporulation, suggesting that the functional role of these proteins is to increase heat resistance by affecting spore structure rather than by repairing heat-damaged proteins.

11.4 *Saccharomyces cerevisiae*: A MODEL FOR HIGHER EUKARYOTES

The budding yeast *Saccharomyces cerevisiae* is a simple unicellular eukaryote that has been promoted as a model species for higher eukaryotes. Research on this organism is valuable for studying the basic mechanisms of cell life, in particular the molecular basis of human genetic diseases. In contrast to humans, genes in *S. cerevisiae* are experimentally very manipulable, capable of being easily deleted, mutated, and reintroduced into yeast cells, overexpressed, tagged, and comprehensively analyzed. Consequently, it is of great interest that the yeast proteome contains structural homologs for 46% of the human proteins that have been identified (International Human Genome Sequencing Consortium, 2001; Foury and Kucej, 2001). The proteins conserved between human and yeast are involved in the fundamental processes of cellular life, such as DNA replication, recombination and repair, RNA transcription, translation, intracellular trafficking, and general metabolism.

Of medical significance is the finding that yeast genes share substantial sequence identity with 30 to 40% of human disease-associated genes (Basset et al., 1996; Foury, 1997). Although it is difficult to assess functional conservation between structurally homologous yeast and human genes, *S. cerevisiae* represents, nonetheless, an invaluable experimental system for approaching the function of uncharacterized human disease-associated genes (Foury, 1997). An example of the potential value of yeast as an experimental model for more complex eukaryotes is given by the human *ALD* gene associated with adrenoleukodystrophy (Mosser et al., 1993), which is a neurodegenerative disease characterized by defective β-oxidation of saturated long-chain fatty acids in peroxisomes (a major type of microbody or membrane-bound eukaryotic cytoplasmic organelle containing a specific enzymatic content). The human *ALD* gene partially encodes an ABC transporter protein localized in the peroxisomal membrane that displays structural homology to two yeast ORFs, Pal1p and Ykl188c (Bossier et al., 1994; Shani et al., 1995). It was reported by Hettema and colleagues (1996) that the yeast Pal1p and Ykl188c transporters are involved in importing activated long-chain fatty acids into peroxisomes, thus suggesting that human ALD may have an analogous function (Foury, 1997). However, because gene sequence and gene function are not always faithfully conserved between species, it is necessary to be cautious about extrapolating the relevance of homologous genes from model organisms such as yeast to human disease.

This section is devoted to a description of the functional genomics of *S. cerevisiae*, starting with the specific contribution of the systematic sequencing of the yeast genome. We will then discuss how functional genomics approaches such as genome-wide transcription profiling and proteomic descriptions are helping to define cellular functions for the approximately one-third uncharacterized novel yeast genes discovered by genomic sequencing and to elucidate regulatory schemes. Finally, we will conclude this section by describing an emerging avenue of investigation focusing on the yeast protein *interactome* (protein–protein interaction maps), that is, how the $\sim 6,000$ gene products interact to create a eukaryotic organism by performing crucial roles in the execution of various biological functions. Knowledge of direct and indirect interactions among members of the yeast proteome is necessary for a complete understanding of the molecular basis of cellular functions.

11.4.1 Yeast Genome

The complete genomic sequencing of the budding yeast *S. cerevisiae* was accomplished by an international collaboration among scientists from Belgium, the United Kingdom, Canada, the United States, France, Germany, Japan, and Switzerland. With its publication in 1996, the yeast genome represented the first completely elucidated genome sequence of a eukaryotic organism (Goffeau et al., 1996). Sequence analysis of its 12 million bases, organized among 16 discrete chromosomes, led to the prediction that 6,274 potential ORFs are likely to encode protein products in the yeast cell (Mewes et al., 1997). The total number of known and predicted proteins has since been revised to 6,145 (Costanzo et al.,

2001). Several years after the completion of the *S. cerevisiae* genome sequence, nearly all yeast genes have been extensively annotated with sequence and genetic information. The online Yeast Genome Database (MYGD), available through the Munich Information Center for Protein Sequences (MIPS), displays annotated functional properties, homologies, and structures of ORFs, RNA genes, and other genetic elements as well as classical genetic, biochemical, and cell biological knowledge mined from the literature (Mewes et al., 2000). The genome database of sequence homologies has been constructed based on the FASTA algorithm for sequence database searches (Mewes et al., 1997). Another accessible online database is the Yeast Proteome Database (YPD^{TM}), which serves as a comprehensive resource for organized information about *S. cerevisiae* proteins (Costanzo et al., 2000). The YPD contains protein-specific information derived from an exhaustive, in-depth curation of the available scientific literature. Despite the rich history of systematic research on yeast, approximately 32% of the proteins predicted from the genome sequence were functionally undefined at the time sequencing was complete. Here, we discuss new information obtained from a comprehensive survey of the complete *S. cerevisiae* genome sequence.

Genomic sequence determination has been used to identify probable novel genes that escaped functional characterization by classical genetic approaches and to provide information about the higher-order organization of the 16 chromosomes present in yeast and the distribution of genes and other sequence elements (e.g., transposons, repetitive motifs) among the different chromosomes (Goffeau et al., 1996). Sequencing uncovered considerable genetic redundancy, especially at the ends of yeast chromosomes. The two terminal domains of chromosome III, for instance, exhibit substantial nucleotide sequence homology both to one another and to the terminal domains of chromosomes V and XI. The sequences that comprise the two ends of yeast nuclear chromosome I are characterized by (1) a much lower gene density and largely untranscribed DNA, (2) the presence of several apparent pseudogenes and a 15-kbp redundant sequence, and (3) the absence of genes essential for vegetative growth (Bussey et al., 1995). The low gene density, presence of pseudogenes, and transcriptional inactivity are consistent with the view that these terminal chromosomal regions represent the yeast equivalent of heterochromatin (Bussey et al., 1995), highly compacted DNA that is generally devoid of coding sequence or informational content and therefore is likely to play a role in chromosome structure and stability. Bussey and coworkers (1995) suggested that these terminal domains provide the chromosome with the critical length required for proper function.

Informatic analysis of the theoretical yeast proteome deduced from the genomic sequence permitted the functional classification of about 50% of the proteins based on standard homology criteria. Since the public disclosure of the yeast genome sequence and the emergence of functional genomics techniques, significant progress has been made in empirically defining the functions of genes encoding novel proteins. Information derived from computational approaches provides valuable guides to experimentation, but functional analyses in the form of DNA microarray experiments, gene deletion studies, and biochemical studies are needed to verify protein function, as will be discussed in the sections that follow.

Conservative *in silico* analysis revealed that the yeast cell dedicates 11% of its proteome to metabolism (which includes both fermentative and oxidative metabolism); 7% to transcription; 6% to translation; 3% to energy production and storage; and 3% to DNA replication, repair, and recombination (Goffeau et al., 1996). Another result of genomic sequencing is the unveiling of gene products whose existence was previously in doubt

(Goffeau et al., 1996). Early views of yeast chromosomes perpetuated the notion that *S. cerevisiae* lacks the histone protein H1. Histones are basic proteins around which helical DNA is wound to form a chromatin fiber structure called a nucleosome. Contrary to what was originally thought, sequence analysis indicated that *S. cerevisiae* contains the full repertoire of eukaryotic histones, including H1, which resides on chromosome XVI (Ushinsky et al., 1997). In another example, the discovery and chromosomal localization of the gamma tubulin gene had eluded yeast geneticists for years despite extensive research efforts until genomic sequencing identified this gene on chromosome XII (Johnston et al., 1997).

11.4.2 Yeast Transcriptomics

Since publication of the complete yeast genome sequence in 1996, considerable progress has been achieved in deciphering *S. cerevisiae* gene function, with approximately 1,900 ORFs requiring functional assignments (Goffeau, 2000). High-throughput microarray analysis has been used in numerous investigations to assess the global genomic expression patterns in yeast cells in response to a given stimulus or environmental perturbation or genetic alteration. Assessing the mRNA expression levels of novel genes under various physiological or developmental conditions is indeed a powerful approach in the process of elucidating gene function. Not only have cDNA and oligonucleotide arrays provided new details on the basic cellular processes of chromosome replication (Raghuraman et al., 2001), chromatin remodeling (Fazzio et al., 2001; Sudarsanam et al., 2000), meiosis (Primig et al., 2000), sporulation (Chu et al., 1998), and cell-cycle regulation (Iyer et al., 2001; Spellman et al., 1998), but they have shed light on the impact of uncharacterized perturbations on the dynamic composition of the yeast transcriptome (e.g., Alexandre et al., 2001; De Sanctis et al., 2001; Gasch et al., 2001; Gross et al., 2000; Mercier et al., 2001). Functional genomics tools have been particularly useful in identifying yeast genes that are directly regulated in response to changes in environmental concentrations of metals such as iron, copper, and zinc (Eide, 2001). In this section, we will focus on the contribution of genome-wide expression profiling to our understanding of metal homeostasis, the regulation of growth, and the in vivo role of chromatin remodeling complexes in eukaryotic gene regulation.

Genetic Basis of Metal Homeostasis All living organisms require metal ions such as iron, copper, and zinc for various biochemical processes. Iron and copper, for example, are important cofactors in electron transport and in many redox-active metalloenzymes; zinc is a catalytic component of many enzymes and plays a critical role in such structural motifs as zinc figures. However, overaccumulation of any of these metal ions can be toxic to the cell. To control intracellular metal-ion levels, organisms like *S. cerevisiae* have evolved homeostatic regulatory systems that govern the uptake, distribution, storage, and detoxification of metal ions (Eide, 2001). In addition to yeast, significant studies of metal metabolism using functional genomics have also been conducted on the model prokaryotes *E. coli* (Brocklehurst and Morby, 2000) and *B. subtilis* (Ye et al., 2000).

To better understand the regulation of metal metabolism in *S. cerevisiae*, Lyons and colleagues (2000) examined the yeast genes regulated by the Zap1p transcription factor using DNA microarrays and *in silico* promoter motif analysis. The transcription factor Zap1p senses the status of cellular zinc levels and stimulates expression of its target genes in response to zinc limitation (Zhao and Eide, 1997). Zap1p has been previously shown to control the expression of the *ZRT1*, *ZRT2*, and *ZRT3* zinc transporter genes by binding to a conserved, zinc-responsive element (ZRE) in the promoters of these target genes (MacDiarmid et al., 2000; Zhao et al., 1998). Initially, a consensus ZRE sequence was derived based on elements in the promoters of *ZRT1*, *ZRT2*, and *ZRT3*: 5′-ACCYYNAAGGT-3′ (where Y = a pyrimidine and N = any nucleotide) (Zhao et al., 1998). The effects of zinc and ZAP1 on the yeast transcriptome were determined by subjecting wild-type and *zap1* mutant strains grown under different zinc concentrations (deficient, replete, and excess) to microarray analysis. The motif analysis program multiple expectation-maximization for motif elicitation (MEME; Bailey and Elkan, 1994) was used to analyze the promoter regions of the 111 genes displaying Zap1p-dependent expression based on microarray hybridization. MEME can identify potential regulatory elements shared among the promoters of similarly regulated genes and, in some cases, can redefine the consensus recognition-site motif for a transcription regulator. As presented in Fig. 11.5, this study strongly implicated 46 genes, 18 of which encoded proteins of unknown function, as potential target genes for Zap1p-dependent regulation (Lyons et al., 2000). Furthermore, MEME analysis indicated that the most probable form of the ZRE consensus motif is 5′-ACCTTNAAGGT-3′ (Lyons et al., 2000). The findings in this study extend the definition of the regulon for a specific eukaryotic transcription factor, thus enhancing our understanding of zinc metabolism at the molecular level.

The uptake and utilization of iron and copper are tightly regulated in yeast by precise homeostatic regulatory systems. Studies employing DNA array technology have identified new target gene candidates in the regulons of the yeast transcription factors Aft1p and Mac1p (Gross et al., 2000; Protchenko et al., 2001; Yun et al., 2000). In *S. cerevisiae*,

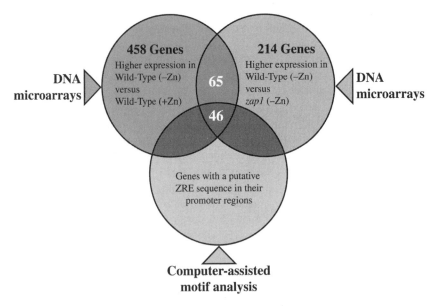

Figure 11.5 *Schematic representing a functional genomics approach to identifying target genes in the Zap1p zinc-responsive regulon in* Saccharomyces cerevisiae *(yeast).* This strategy identified 46 (shaded area) candidate Zap1p target genes. [Adapted from T. J. Lyons, A. P. Gasch, L. A. Garther, D. Batstein, P. O. Brown, and D. J. Eide. Genome-wide characterization of the Zap1p zinc-responsive regulon in yeast, *PNAS* vol. 97, 7957–7962. Copyright (2000) National Academy of Sciences, U.S.A.]

Aft1p is the major iron-dependent transcription factor that mediates transcriptional regulation of the iron regulon (Yamaguchi-Iwai et al., 1995), which consists of many genes involved in iron uptake or acquisition (e.g., *FET3, FTR1*, and *FRE1, 2*), siderophore uptake (e.g., *ARN1-4* and *FIT1-3*), iron transport across the vacuole membrane or compartmentalization (e.g., *FTH1*), and iron-sulfur cluster formation (*ISU1, 2*). Aft1p controls target gene expression by binding to a conserved promoter sequence in an iron-responsive manner to activate transcription under conditions of iron deprivation (Yamaguchi-Iwai et al., 1996). Transcript levels in a wild-type strain and an *aft1Δ* mutant (containing a deleted AFT1 gene) cultivated under varying iron concentrations were compared using cDNA microarray analysis to interrogate the yeast genome for new genes involved in iron metabolism (Yun et al., 2000). Although many members of the Aft1p regulon are already known, this study identified four highly homologous genes (designated *ARN1, 2, 3,* and *4*) exhibiting Aft1p-regulated expression. These four newly identified target genes, *ARN1–ARN4*, are predicted to encode highly homologous members of a subfamily of the major facilitator superfamily of transporters (Nelissen et al., 1997; Pao et al., 1998). Inspection of upstream sequences flanking the 5'-end of each ARN gene revealed an Aft1p consensus-binding site (Yun et al., 2000). Furthermore, deletion of the genes *ARN3* and *FET3*, a component of the high-affinity ferrous iron transport system, resulted in yeast cells that were deficient in growth on ferrioxamine as an iron source and in the uptake of ferrioxamine-complexed iron (Yun et al., 2000). Ferrioxamine B is a hydroxamate-type siderophore (i.e., a low-molecular-weight compound that has a high affinity for iron) that is synthesized and secreted in its iron-free form, desferrioxamine B. Together, the findings described by Yun and coworkers (2000) suggest that *S. cerevisiae* utilizes two discrete pathways for desferrioxamine-mediated iron uptake, one that is *FET3*-dependent and one that requires *ARN3*. In a complementary Aft1p regulon study, Protchenko and colleagues (2001) used cDNA microarrays representing virtually the entire protein-coding capacity of the *S. cerevisiae* genome to identify three additional new genes that are transcriptionally regulated by Aft1p. These genes—called *FIT1, FIT2*, and *FIT3* (for facilitator of iron transport)—function to facilitate the uptake of iron but apparently are not critical components of the iron uptake machinery since yeast cells can compensate for their absence by augmenting the expression of other iron-acquisition components (Protchenko et al., 2001).

Rutherford and coworkers (2001) have extended our understanding of iron homeostasis by examining the functional role of *AFT2*, a paralog of the iron-dependent transcription factor-encoding *AFT1* gene, in iron metabolism. The mapping of duplicated chromosomal segments in the *S. cerevisiae* genome revealed the presence of paralogs *AFT1* and *AFT2* on chromosomes VII and XVI, respectively (Seoighe and Wolfe, 1999). The deduced protein product of *AFT2* exhibited 39% amino acid sequence identity to Aft1p in a region predicted to represent the Aft1p DNA-binding domain and a region containing the Cys-X-Cys motif, which confers iron sensitivity. To ascertain whether *AFT2* plays a role in iron homeostasis, phenotypes were determined for strains harboring either a single deletion of *AFT1* or *AFT2*, or a double *aft1Δaft2Δ* deletion under growth conditions of iron deprivation. While no iron-dependent phenotype was observed for the single *aft2Δ* strain of *S. cerevisiae*, growth of the *aft1Δaft2Δ* double mutant strain was significantly more sensitive to iron-deficient conditions and oxidative stress than the single *aft1Δ* mutant, thus suggesting that Aft2p is a functional transcription factor that responds to iron (Rutherford et al., 2001). Experiments with DNA microarrays indicated *AFT2*-controlled expression for a number of genes, including two genes of unknown function (*YOL083w* and

YDL124w) and four transporter-encoding genes (*ZRT1*, *FTR1*, *FTH1*, and *SMF3*). The exaggerated iron-deficient growth defect of the *aft1Δaft2Δ* yeast strain and the sequence homology between Aft1p and Aft2p are consistent with proteins possessing partial overlapping cellular functions (Rutherford et al., 2001). The microarray results supported this idea, because a subset of the *AFT2*-regulated genes is controlled by Aft1p (*FIT1*, *FIT3*, *FTR1*, *FTH1*, *FRE1*, and *TIS11*). In addition, two other genes in the Aft2p regulon are regulated by the zinc-responsive transcription factor Zap1p (*ZRT1* and *YOL154w*). As determined by global transcription profiling, the expression of yeast AFT2 itself is elevated in response to such stress conditions as stationary phase growth, nitrogen starvation, and treatment with alkylating agents (Rutherford et al., 2001). Rutherford and his colleagues, therefore, surmised that the regulatory activity of Aft1p and Aft2p might predominate under different environmental conditions. This study indicates a second iron regulatory system in *S. cerevisiae*, one dependent on Aft2p, and underscores the importance of microarrays as valuable tools for revealing fundamental information on cellular processes.

In *S. cerevisiae*, copper ions are required as functional cofactors for at least three (and probably more) key enzymes: (1) an active cytochrome oxidase complex, which enables yeast cells to grow on nonfermentable carbon sources; (2) the copper-metalloenzyme superoxide dismutase, which protects the cell against the detrimental effects of reactive superoxide anions; and (3) the copper-metalloenzyme Fet3, a ferro-oxidase that is critical for Fe(II) uptake. Copper ion homeostasis in yeast is maintained through copper-regulated gene expression mediated by two functionally distinct transcriptional activators, Mac1 and Ace1. Mac1 activates the expression of genes, such as high-affinity copper uptake genes, under growth conditions of copper deficiency (Jungmann et al., 1993), whereas Ace1 mediates the induction of gene expression in response to copper ion stress or excess copper (Buchman et al., 1989; Thiele, 1988). A study of copper-regulated gene expression using whole-genome microarray analysis was performed using *S. cerevisiae* as a model for copper metalloregulation (Gross et al., 2000). DNA microarray hybridization revealed the differential expression of a limited set of genes under growth conditions of copper deficiency or excess. In addition to previously established targets of Mac1 and Ace1, a number of new target genes were identified as playing a role in copper homeostasis. For example, two genes (*FRE7* and YJL217w) identified as new Mac1 target genes encode proteins of unknown function, while a third newly identified target, YFR055w, is predicted to encode one of several cystathionine γ-lyase isozymes in *S. cerevisiae* that produce cysteine from cystathionine (Gross et al., 2000). The observed Mac1-dependent induction of YFR055w suggests that part of the cellular response to copper depletion is an expansion of intracellular cysteine pools (Gross et al., 2000). An unexpected result of genomic screening was the Ace1-mediated activation of *FET3* and *FTR1*, genes functioning in high-affinity iron uptake and regulated by the Aft1 transcriptional activator in iron-deficient cells (Askwith et al., 1996). This finding underscores the interconnection between copper and iron metabolism. However, it is not clear how applicable yeast is as a model system for investigating copper metalloregulation in animal cells.

Functional Genomics of Metabolic Reprogramming The important position of *S. cerevisiae* as a model eukaryote can be attributed to the intrinsic advantages it offers as an experimental organism. Unlike many eukaryotes of greater morphological and genomic complexity, yeast can be cultivated in experimenter-controlled chemical and physical environments using defined media. In addition, its life cycle and highly efficient homologous recombination properties make *S. cerevisiae* ideally suited for classical genetic analysis, including gene deletion or replacement. This experimental tractability was recently exploited in a functional study of the yeast genome using gene deletion and microarray analysis (Winzeler et al., 1999). By using the sequence of the yeast genome, precise deletions were generated for 2,026 ORFs (more than one-third of the ORFs) in the *S. cerevisiae* genome. Homologous portions of the targeted gene were placed at each end of a selectable marker gene (an antibiotic resistance cassette) using PCR amplification. When introduced into yeast cells, the high rate of homologous recombination permitted replacement of the targeted gene with the selectable marker gene. Phenotype analysis of the yeast deletion strains indicated that 17% of the deleted ORFs were essential for cell viability in rich medium, while 40% of the deletion strains exhibited measurable growth defects in either rich or minimal medium (Winzeler et al., 1999). With a few exceptions, mutants that grew poorly in rich medium also generally grew poorly in minimal medium.

In a seminal expression profiling study, DeRisi and coworkers (1997) used whole-genome microarrays for *S. cerevisiae* to comprehensively examine the temporal and global patterns of gene expression underlying the metabolic transition from fermentative to respiratory growth. In yeast, the utilization of glucose by fermentation results in the production of ethanol. The depletion of glucose leads to a temporary arrest of growth. During this diauxic transition phase, cells switch from fermentative to oxidative metabolism. DNA microarray analysis of batch yeast cultures during exponential growth on glucose indicated that genomic profiles of gene expression remained relatively stable. However, a dramatic change was observed in gene expression profiles as glucose was progressively depleted from the growth media. Moreover, the widespread changes in gene expression correlated with the metabolic reprogramming that occurs during the diauxic transition in yeast. For example, large increases in mRNA abundance were observed for genes encoding the enzymes aldehyde dehydrogenase (*ALD2*) and acetyl-coenzyme A (CoA) synthase (*ACS1*) (DeRisi et al., 1997). These enzymes function in combination to convert the products of alcohol dehydrogenase to acetyl-CoA, which is then utilized by the cell to fuel the TCA cycle and the glyoxylate cycle. Classes of genes that were coordinately induced upon glucose depletion included cytochrome c-related genes and those involved in the TCA/glyoxylate cycle and carbohydrate storage. Along with the induced transcription of approximately 710 genes, DeRisi and colleagues (1997) detected a concomitant decline in the mRNA levels for about 1,030 genes. Genes coding for ribosomal proteins, tRNA synthetases, and translation, elongation, and initiation factors showed a coordinated decrease in expression. More than 400 of the differentially expressed *S. cerevisiae* ORFs displayed no apparent sequence similarity to genes encoding protein products of known function. Thus, this early microarray analysis provided the first small indication of their probable function in yeast cells (DeRisi et al., 1997).

In another study, genome-wide transcription patterns were analyzed under aerobic and anaerobic growth conditions for steady-state chemostat cultures of *S. cerevisiae* (ter Linde et al., 1999). In contrast to batch cultures, the experimental advantage of chemostat cultivation is that it permits the molecular effect of individual physiological parameters of growth to be examined. ter Linde and coworkers (1999) observed similar expression patterns for the majority of the yeast genes under aerobic and anaerobic conditions. In response to aerobiosis, only 219 genes displayed a greater than 3-fold higher transcription level, while 140 genes showed a greater than 3-fold increase in transcript levels in response to anaerobiosis (ter Linde et al., 1999). These findings appeared to contradict those reported by DeRisi and coworkers (1997). However, an important difference between the two studies was that DeRisi and his colleagues used batch cultures of *S. cerevisiae* growing on glucose to analyze global gene expression profiles in response to the physiological shift from fermentative metabolism to respiratory metabolism. In the experiments described by ter Linde and coworkers (1999), genomic expression patterns under aerobic and anaerobic growth were monitored for cells in glucose-limited chemostat cultures in which glucose repression was alleviated by low residual glucose concentrations. It would appear that under these growth conditions the metabolic flux through the TCA cycle and respiration is controlled primarily at the level of posttranscription (ter Linde et al., 1999). As was the case for the study by DeRisi and colleagues (1997), the physiological roles for a number of the aerobically and anaerobically induced genes remain unclear and will require further functional characterization. A microarray-based comparison of transcript profiles generated in wild-type *S. cerevisiae* under different growth conditions does not by itself enable a precise description of the molecular mechanisms of transcriptional regulation dictating metabolic reprogramming. Genetic mutations affecting the activity of key transcriptional regulators, in conjunction with other functional genomics tools, are needed to dissect and characterize regulatory pathways and networks controlling cell growth.

Loss-of-function studies combined with array-based gene expression monitoring is a powerful approach to understanding the biological role of a specific gene product in a cellular process and assigning putative cellular functions to new, nonannotated gene products. In a study by Haurie and coworkers (2001), for example, high-density DNA filters (macroarrays) were used to identify genes whose expression at the diauxic transition is affected by a *cat8* null mutation in *S. cerevisiae*. Cat8p is a zinc cluster-containing transcriptional activator that is essential for the growth of *S. cerevisiae* on nonfermentable carbon sources (Hedges et al., 1995). Glucose regulates expression of the *CAT8* gene itself as well as Cat8p-dependent transcriptional activation. To better understand the contribution of Cat8p to the reprogramming of carbon metabolism during the diauxic shift, DNA arrays containing 6,144 ORFs of the yeast genome were used to compare gene expression profiles in the wild-type *S. cerevisiae* strain and an isogenic *cat8* deletion strain. Although over 6,000 ORFs were represented on the array, only about 3,000 ORFs of the yeast genome were expressed at high enough transcript levels to be measurable by the macroarray filter approach. Transcriptome analysis revealed 34 genes whose expression at the diauxic shift is dependent on a functional Cat8p. For 25 of these genes, including eight genes of unknown function, this study represented the first demonstration of Cat8p-dependent expression. Although Cat8p-regulated expression appears to concern only a small fraction of the yeast genome, Cat8p plays an important role in the reprogramming of yeast metabolism during the diauxic shift by controlling the expression of genes whose products are necessary for the initial steps of ethanol utilization, the glyoxylate cycle, and gluconeogenesis (Haurie et al., 2001).

Nucleosome Remodeling Complexes in Gene Regulation

Regulation of gene expression is fundamental to the functioning of biological systems, and much of the cell's regulatory activity is exerted at the level of transcription (Holstege and Young, 1999). It has become increasingly apparent that modification of the compact chromatin structure is an important regulatory mechanism in eukaryotic organisms, including yeast. The repeating structural subunit of chromatin is the nucleosome, characterized by 200-base-pair-long segments of DNA negatively supercoiled around histone cores (aggregates of small, positively charged, arginine- and lysine-rich proteins). Genes residing in these DNA segments are effectively packaged within nucleosomes. Studies have demonstrated both in vivo and in vitro that nucleosomes inhibit transcription initiation by occluding access of the eukaryotic RNA polymerase II holoenzyme and transcriptional factors to DNA (reviewed in Workman and Kingston, 1998). Recently, it has become evident that classes of multiprotein complexes, such as the conserved Swi/Snf and RSC complexes, function to alleviate nucleosome-mediated transcriptional repression by modulating the topology of the chromatin structure in an ATP-dependent manner (Kingston and Narlikar, 1999). The nucleosome remodeling activity of these chromatin architectural factors facilitates transcription initiation on chromatin templates by perturbing nucleosomal structure and allowing transcription factors to bind to their specific recognition sites in the DNA sequence (Fig. 11.6; Workman and Kingston, 1998). Some basic aspects concerning the function of the chromatin-remodeling complex Swi/Snf remain unresolved. However, the development of high-density parallel tools for measuring the mRNA levels of an entire genome provides an excellent opportunity to elucidate Swi/Snf function.

In a recent study, Sudarsanam and colleagues (2000) employed a combination of functional genomics and more traditional methods (Northern blot hybridization and genetic analyses) to obtain a more complete picture of the in vivo function of the *S.*

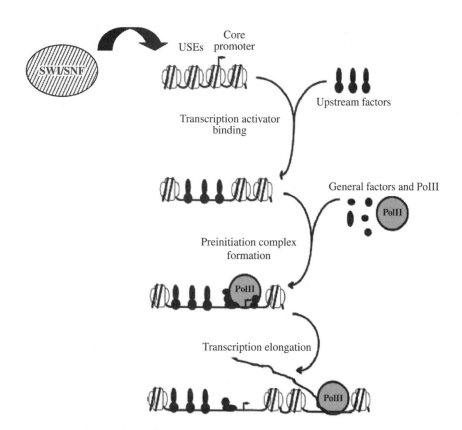

Figure 11.6 *Nucleosomal inhibition of gene transcription can occur at the stages of transcription factor binding, preinitiation complex formation, or transcription elongation (adapted from Workman and Kingston, 1998).* Multiprotein complexes, like Swi/Snf, facilitate the binding of activator proteins to upstream elements (USEs) and promote transcription elongation through nucleosomes by RNA polymerase II (pol II).

cerevisiae Swi/Snf, an ATP-dependent chromatin-remodeling complex. Their research unveiled some important features of the regulatory role of Swi/Snf in yeast. For example, the entire subset of yeast genes whose expression depends on both subunits of the complex (i.e., the conserved Snf2 and the unconserved Swi1 components) was determined using whole-genome DNA microarrays. To accomplish this, mRNA expression levels in wild-type and *swi/snf* mutant strains were compared for each *S. cerevisiae* gene by differential display on microarrays. Sudarsanam and coworkers found that the loss of Snf2 or Swi1 produced similar changes in genome-wide transcript levels ($\sim$1% of the 6,014 genes examined), thus leading to the conclusion that most Swi/Snf-controlled genes appear to require both subunits of the complex. Furthermore, Swi/Snf dependence is influenced by different nutrient conditions. Decreased transcript levels for genes encoding the hexose transporters *HXT1*, *HXT3*, *HXT6*, and *HXT7*, for example, were observed only in minimal medium, whereas transcription of the acid phosphatase genes (*PHO5*, *PHO11*, *PHO12*) is Swi/Snf-dependent in rich medium but unaffected in minimal medium (Sudarsanam et al., 2000). Three members of the *MATα* (mating type alpha)-specific gene family (*STE3*, *MFα2*, and *SAG1*) exhibited transcriptional changes in *swi/snf* mutants in both rich and minimal media. An unexpected consequence of microarray analysis was the observation that the yeast Swi/Snf remodeling complex, which is not essential for growth, also directly controls the transcription of the essential gene *MCM1*, the yeast homolog of human serum response factor. Inspection of their chromosomal positions revealed that the genes regulated by Swi/Snf-dependent nucleosomal remodeling are dispersed throughout the genome. This suggests that Swi/Snf-mediated perturbations in nucleosomal structure are highly localized, with transcriptional control exerted at the level of individual and specific genes rather than over large chromosomal domains (Sudarsanam et al., 2000).

Finally, whole-genome expression analysis of *swi/snf* mutants revealed that the mRNA abundance for a number of genes was elevated in *swi/snf* mutants (Sudarsanam et al., 2000). This rather surprising finding suggested that the *S. cerevisiae* Swi/Snf complex also functions in a negative capacity to control transcription, although the transcriptional effect may be indirect (e.g., the Swi/Snf complex controls the expression of a repressor). Other expression studies indicate that proteins involved in chromatin-mediated gene regulation possess a dual regulatory role, both repressing as well as potentiating the transcription of genes functioning in various cellular processes (Moreira and Holmberg, 2000; Murphy et al., 1999; Trouche et al., 1997). Functional genomics, in particular global expression profiling using DNA microarray technology, has contributed significantly to the emerging view of nucleosome remodeling complexes as facilitators of a dynamic and reversible transcriptional state of chromatin.

The Swi/Snf complex is the prototype chromatin-remodeling complex in yeast (Cote et al., 1994). However, a survey of the yeast genome indicates approximately 17 other ORFs with significant homology to the Swi/Snf helicaselike ATPase subunit (Snf2p) (Chervitz et al., 1998; Muchardt and Yaniv, 1999), and global analysis of yeast genes using microarrays has also contributed to the identification of additional chromatin-remodeling complexes, along with delineating the full range of genes repressed or activated by these factors (Jonsson et al., 2001). The deciphered yeast genome contains two highly conserved proteins, Rvb1p and Rvb2p, that are related to the helicase subset of the AAA + class chaperonelike ATPases, which contain sequence motifs representing active sites for ATP- (or dNTP-) binding and hydrolysis (Jonsson et al., 2001; Qiu et al., 1998). Conserved orthologs of the yeast *RVB1* and *RVB2* exist in human, fly, and worm. Such a high degree of sequence conservation immediately suggested that Rvb1p and Rvb2p perform important functions in vivo. To shed light on the cellular function of the Rvb proteins, Jonsson and coworkers (2001) generated yeast mutant strains deficient in either the *RVB1* or *RVB2* alleles and used whole-genome high-density oligonucleotide arrays to analyze the *S. cerevisiae* genes affected by the removal of Rvb1p or Rvb2p. The requirement of Rvb1p and Rvb2p in yeast transcription appears to be relatively widespread, with the expression of over 5% of active yeast genes affected in the absence of Rvb1p or Rvb2p (Fig. 11.7; Jonsson et al., 2001). In addition, microarray analysis revealed that the two Rvb proteins had similar effects on transcription in terms of the number of genes being repressed and activated, thus suggesting the possibility that Rvb1p and Rvb2p associate with each other in vivo. Immunoprecipitation experiments followed by MS-based protein identification demonstrated that Rvb1p and Rvb2p associate with each other in a high-molecular-weight complex, and this complex exhibited ATP-dependent chromatin-remodeling activity in vitro (Jonsson et al., 2001). This study illustrates how microarray-based genomics tools can open experimental avenues that eventually lead to the functional characterization of genes. In addition, genome-wide microarray analysis of the regulatory activities of the Rvb1p/Rvb2p-containing complex permitted comparison with those displayed by the

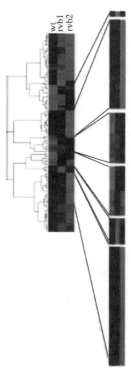

Figure 11.7 *See color insert. Hierarchical cluster analysis (see Chapter 7) of whole-genome microarray data from a study assessing the genome-wide effects of Rvb inactivation (Jonsson et al., 2001).* Yeast genes whose expression levels were affected by the removal of Rvb1p or Rvb2p relative to the wild type are shown to the right of the cluster image. Green and red indicates a decrease or an increase in mRNA abundance, respectively. [Reprinted with permission from Z. O. Jónsson et al., Rvb1p nd Rvb2p are essential components of a chromatin remodeling complex that regulates transcription of over 5% of yeast genes, *The Journal of Biological Chemistry*, vol. 276, 16279–16288. Copyright (2001) The American Society for Biochemistry and Molecular Biology, Inc.]

prototypic Swi/Snf complex in similar studies. Microarray results indicate that the Swi/Snf complex and the Rvb1p/Rvb2p-containing complex largely regulate independent subsets of yeast promoters, but for some genes, transcription appears to be controlled by both chromatin-remodeling complexes (Jonsson et al., 2001; Sudarsanam et al., 2000). Examples of these genes include HO, which encodes a homothallic switching endonuclease, and GAL1, which codes for a galactokinase needed in galactose utilization. The implication of this comparison of microarray data is that multiple chromatin-remodeling factors can act at a given promoter to facilitate transcription of the downstream gene (Jonsson et al., 2001).

11.4.3 Yeast Proteomics

Uncharacterized yeast genes discovered by genomic sequencing can be examined by evaluating biochemical activities, protein–protein interactions, and subcellular localization of the expressed gene products constituting the proteome. Indeed, systematic proteomics is needed to decipher the encoded protein networks and interactions that determine cellular function and, in a sense, bring the "genome to life" (Iyer et al., 2001; Pawson and Nash, 2000). Generally, the proteome of an organism is determined by two-dimensional gel electrophoresis followed by MS-based protein identification given a completely sequenced genome. Although still early in its applications, rapid and accurate MS methods enable one to monitor compositional and abundance changes in the proteome under different physiological conditions and to identify the covalent modification of proteins. An early proteomic analysis of yeast, for example, established the value and utility of desorption ionization MS in connecting information at the protein level to information at the genome level (Shevchenko et al., 1996b). Mass spectrometry positively identified 80% of the protein spots (150 proteins in total) on two-dimensional gels, of which 32 proteins were novel and matched to previously uncharacterized ORFs in the yeast genome. Additionally, the proteins identified in this study exhibited a wide range of molecular weights, isoelectric points, and abundances (Shevchenko et al., 1996b).

Subcellular Localization of Yeast Proteins Techniques other than two-dimensional gel electrophoresis and MS have been utilized to catalog the subcellular distribution of proteins within the *S. cerevisiae* proteome. The subcellular localization of a protein is considered to be an indicator of its general molecular function in the cell and can often be a clue to its mechanism of action (Kumar et al., 2002). However, the subcellular protein distribution in yeast is a fundamental data set that until recently has remained virtually unexplored. In 2002, Michael Snyder and his colleagues described the first proteome-scale analysis of protein localization within a eukaryote (Kumar et al., 2002). The subcellular localization of 2,744 yeast proteins (60% of the theoretical proteome of *S. cerevisiae*) was determined by high-throughput immunolocalization of epitope-tagged gene products. Yeast proteins were tagged either by directed cloning of PCR-amplified ORFs into a yeast V5 epitope-tagging/expression vector or by transposon mutagenesis-mediated random tagging. Cloned genes were inserted downstream of the galactose-inducible *GAL1* promoter, so that galactose induction could be imposed to drive expression of the target gene as a fusion protein with a *C*-terminal V5 epitope. Transposon-tagged proteins carried the HA epitope. Proteins were localized by indirect immunofluorescence using monoclonal antibodies specific to the V5 or HA epitopes and a Cy3-conjugated secondary antibody.

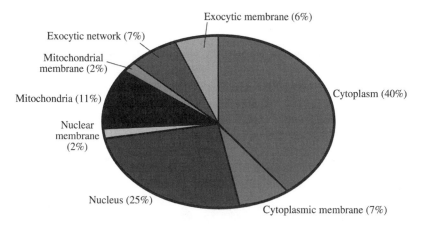

Figure 11.8 *Subcellular compartmentalization of the* Saccharomyces cerevisiae *proteome.* The percentage of the yeast proteome contained within the different membrane and soluble fractions of each compartment is indicated. [Data from Kumar et al., 2002.]

High-throughput immunolocalization revealed that the tagged proteins in yeast were distributed to a wide variety of intracellular structures and organelles, including the cytoplasm, nucleus, mitochondria, endoplasmic reticulum, plasma membrane, and vacuole (Kumar et al., 2002). The majority (47%) of the yeast proteins examined were found to be localized in the cytoplasm, while 13% of yeast proteins were in the mitochondria, 13% in the endoplasmic reticulum and secretory vesicles, and 27% in the nucleus (Fig. 11.8). A significant number of proteins showed mixed compartmentalization (i.e., localized predominantly to a single organelle but also displayed measurable localization to another subcellular structure). For instance, many proteins with assigned functions in the cellular processes of transcription or cytoskeletal organization colocalized to the cytoplasm and nucleus. Under physiologically normal conditions, the transcriptional activator Pho4p, which is involved in the cellular stress response to phosphate starvation, localized predominantly to the cytoplasm, with some appreciable amounts detected in the nucleus. This observation was consistent with a previous investigation that reported Pho4p to be concentrated in the nucleus only under conditions of phosphate deficiency (O'Neill et al., 1996). Furthermore, protein localization correlated strongly with the molecular function of a protein (Kumar et al., 2002). This is of particular significance given the fact that a substantial fraction of the transcriptionally active yeast genome encodes putative proteins of unknown function. In the study by Kumar and coworkers (2002), experimental localization data were derived for 955 proteins whose function was previously undefined.

Proteome Microarrays New technologies, such as proteome microarrays or *protein chips*, have emerged in the postgenomics era and are currently being explored for their application in proteomics. Recently, a novel protein microarray technology was tested by overexpressing all 119 of the yeast protein kinases (both known and predicted kinases), covalently attaching the purified proteins to a solid substrate, and then using the protein microarray to perform large-scale kinase specificity assays (Zhu et al., 2000b). As illustrated in Fig. 11.9A, the protein microarray design consisted of an array of micro-wells (1.4 mm in diameter and 300 μm deep) in a disposable silicone elastomer,

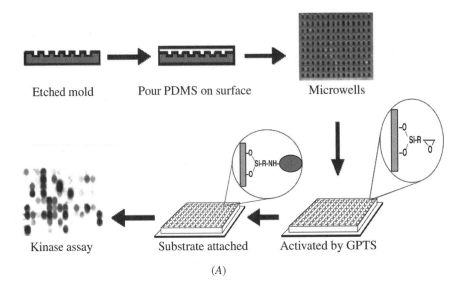

Etched mold   Pour PDMS on surface   Microwells

Kinase assay   Substrate attached   Activated by GPTS

(A)

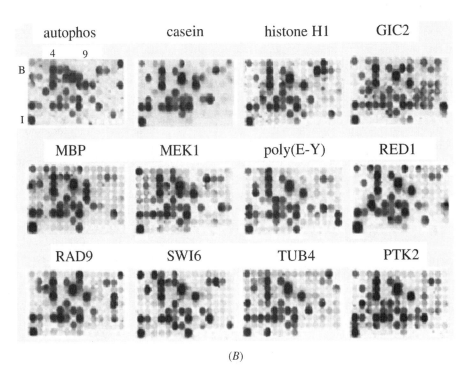

autophos   casein   histone H1   GIC2

MBP   MEK1   poly(E-Y)   RED1

RAD9   SWI6   TUB4   PTK2

(B)

Figure 11.9 *Protein chip fabrication and analysis of yeast protein kinases (Zhu et al., 2000b).* (A) Protein chips were constructed by pouring PDMS over an acrylic mold, curing, and mounting the wells on a glass slide. The surface of the wells was then modified, followed by protein attachment. (B) Kinase activities were detected using protein chips. Images of phosphorylation signals in the presence of 12 substrates are shown. [Reprinted with permission from H. Zhu et al., Analysis of yeast protein kinases using protein chips, *Nature Genetics*, vol. 26, 283–289. Copyright (2000) Nature Publishing Group http://www.nature.com/.]

poly(dimethylsiloxane) or PDMS, placed on top of a standard microscope slide. The arrays covered slightly more than one-third of the slide surface, and generally two arrays per slide were used. Proteins were covalently attached to the microwells using the cross-linker 3-glycidoxypropyltrimethoxysilane (GPTS) and then assessed for in vitro kinase activity in 17 different assays using radiolabelled ATP (γ-^{33}P-ATP) and 17 substrates [e.g., bovine histone H1, bovine casein, myelin basic protein, and the tyrosine substrate poly(Tyr-Glu)] (Fig. 11.9B; Zhu et al., 2000b). This initial analysis of yeast proteins using microarray technology identified a number of novel kinase activities. In addition, Zhu and colleagues (2000b) found that particular proteins are preferred substrates for particular protein kinases and that a large number of protein kinases in yeast are capable of phosphorylating tyrosine.

A eukaryotic proteome microarray was recently fabricated for *S. cerevisiae* to screen for diverse biochemical activities for the ultimate purpose of understanding gene function (Zhu et al., 2001). This analysis was performed on a global scale: A yeast proteome array was constructed by overexpressing 5,800 open reading frames (approximately 80% of the predicted total proteins in yeast) as amino-terminal glutathione S-transferase-polyhistidine (GST-HisX6) fusion proteins and printing the purified corresponding proteins onto nickel-coated slides, in which the fusion proteins attached through their HisX6 tags. The proteins represented on the proteome microarray were screened for the ability to interact with calmodulin, a highly conserved calcium-binding protein involved in numerous calcium-regulated cellular processes (Hook and Means, 2001), and phospholipids. Both the calmodulin and liposome probes were biotinylated; bound probes were detected using Cy3-labeled streptavidin, which binds strongly to biotin.

Analysis of calmodulin-binding proteins revealed 33 potential novel partners in addition to known calmodulin-interacting partners (Zhu et al., 2001). A total of 150 different protein targets were identified for the various phosphoinositide (PI) lipids tested, and of these, 52 (35%) of the lipid-binding proteins corresponded to functionally nonannotated proteins (Zhu et al., 2001). The remaining 98 PI lipid-binding proteins of known function included many membrane-associated proteins (e.g., protein kinases, the Atp1p subunit of the F1-ATP synthase of the mitochondrial membrane, and the prospore membrane-associated protein Sps2p) and a number of proteins involved in glucose metabolism (the phosphoglycerate mutase, enolase, and pyruvate kinase enzymes of glycolysis; hexokinase; and two protein kinases). The power of the protein microarray-based approach to proteomic analysis lies in its ability to screen a comprehensive set of individual proteins for a wide variety of in vitro activities, including enzymatic and biochemical activities, protein–lipid interactions, and protein–protein interactions (Zhu et al., 2001). However, it is important to note that the comprehensive in vitro approach enabled by protein microarrays removes the interacting proteins from a physiological context that is likely to be fundamental to a complete understanding of function. The next subsection focuses on protein–protein interaction mapping (the elucidation of 'interactomes') as a viable approach to identifying a putative function for undefined proteins in yeast.

11.4.4 Yeast Interactome: Mapping Protein–Protein Interactions

The biochemical circuitry of a cell involves numerous intimate protein–protein and protein–DNA interactions. In fact, proteins rarely act alone but, instead, function in combination with other biomolecules through physical interactions to perform specific and critical cellular tasks, thus enabling the entire cell to ultimately function as a cohesive unit that can respond in an organized fashion to its environment (Pawson and Nash, 2000). A complete description of protein–protein interactions is therefore indispensable to a thorough understanding of the cell at the molecular level. An important functional genomics strategy to defining proteins newly discovered by genome sequencing projects is to classify novel proteins based on their physical association with functionally characterized proteins (Ito et al., 2001; Schwikowski et al., 2000). Interactions of proteins with other biomolecules can provide testable hypotheses and highly informative hints for functional prediction. This is of particular significance given that approximately 40% of *S. cerevisiae* proteins have been conserved throughout eukaryotic evolution (Chervitz et al., 1998), thus suggesting that elucidation of the yeast protein interactome (i.e., the global protein interaction map) will provide a partial framework for understanding more complex eukaryotic proteomes (Ho et al., 2002). Traditionally, the yeast two-hybrid system (Fields and Song, 1989; see Chapter 10 for a detailed explanation) has been employed to generate large-scale protein–protein interaction maps. Recent high-throughput and ultrasensitive mass spectrometric methods are beginning to be used to identify multiprotein complexes in yeast (Gavin et al., 2002; Ho et al., 2002). The application of these methods and technologies to a comprehensive exploration of the *S. cerevisiae* protein interactome is discussed here.

Genomic Two-hybrid Screens   The yeast two-hybrid system takes advantage of the *S. cerevisiae* GAL4 protein, a transcriptional activator that controls expression of genes involved in galactose utilization (Fields and Song, 1989). The GAL4 protein contains two separable and functionally distinct domains that are both essential for activation of target gene expression: the *N*-terminal domain is responsible for specific DNA-binding activity of the protein, and the *C*-terminal domain contains acidic regions that are required for activation of transcription. In this system to detect binary interactions, two hybrid proteins are generated, one with a protein fused to the GAL4 DNA-binding domain and a second different protein fused to the GAL4 activating region. Physical interaction between the two proteins is detected through the activation of reporter gene expression.

A number of recent investigations have used the two-hybrid system to analyze protein–protein interactions in yeast on a proteomic scale (e.g., Ito et al., 2000, 2001; Schwikowski et al., 2000; Uetz et al., 2000; for a review, see Legrain et al., 2001). Toward the goal of a protein–protein interaction map of *S. cerevisiae*, Ito and coworkers (2000) established a comprehensive two-hybrid screening system in which all the yeast ORFs were cloned individually as both a DNA-binding domain fusion ("bait") and an activation domain fusion ("prey"). Mutually interacting proteins were screened by systematically mating bait and prey clone pools with each other. Transformants (diploid cells bearing a pair of bait and prey) generated by the mating were plated onto media lacking adenine, histidine, and uracil to select for clones activating the three reporter genes (*ADE2*, *HIS3*, and *URA3*), which were transcriptionally driven by unique Gal4-responsive promoters (Ito et al., 2000). In this study, approximately 4×10^6 different combinations between yeast proteins ($\sim$10% of all possible ones) were initially examined, revealing 183 independent two-hybrid interactions. Most of these independent two-hybrid associations (163 in number) constituted previously unreported interactions. Those interactions that appeared to be biologically relevant or "highly likely" included the physical association of Srp14 and Srp21, which are verified subunits of the signal recognition particle (SRP) although no direct evidence for their interaction has been substantiated previously (Ito et al., 2000).

Other probable two-hybrid interactions were those detected between components of various multiprotein complexes, such as the spindle pole body, ribosome, vacuolar H^+-ATPase complex, TRAPP (transport particle protein) complex, spliceosome, and small nuclear ribonucleoproteins. The unknown category of binary interactions also included 105 novel ones for which no strong evidence currently exists for their biological relevance (Ito et al., 2000). Based on its association with Bet3, a protein component of the TRAPP complex (Sacher et al., 1998), Ito and colleagues (2000) proposed a putative cellular role for the unclassified Ybr254c in TRAPP complex function. The TRAPP multiprotein complex plays a key role in the targeting and/or fusion of endoplasmic reticulum-to-Golgi transport vesicles (Sacher et al., 1998). Ybr254c was shown by another research group to be identical to the 20K component of the TRAPP complex (Sacher et al., 1998).

The initial experiments for the comprehensive examination of bait–prey interactions suggest that it is a useful, informative approach to genome-wide hunting of protein–protein interactions. However, the authors emphasized several potential problems or limitations associated with two-hybrid analysis that should be considered when performing such studies. First, two-hybrid assays are prone to false positives and false negatives (i.e., interactions that escape detection in the screening process sometimes because of masking effects). Hence, it is a good idea to integrate two-hybrid data with those of genetic interactions, subcellular localizations, and microarray-based expression profiles to eliminate biologically meaningless observations (Ito et al., 2000). Second, some proteins must be in an "activated" state in order to interact with their partners in the cell. Small GTPases are examples of proteins that exhibit a substantially stronger interaction with their effector proteins in activated or GTP-bound forms. Third, when conducting two-hybrid screens, it is important to be cognizant of the fact that two-hybrid interactions are not always a reflection of direct binding between the bait and prey (Ito et al., 2000). The interaction between two yeast proteins is sometimes mediated by a third protein, which acts to physically link the bait to the prey (Ito et al., 2000).

In completing their systematic two-hybrid analysis, Ito and his colleagues provided a description of pairwise protein interactions that significantly extends our knowledge of the yeast interactome. A total of 4,549 two-hybrid interactions among 3,278 proteins were identified using the screening strategy briefly described above (Ito et al., 2000, 2001). The two-hybrid data set revealed various intriguing and unexpected interactions between yeast proteins, resulting in the proposal of biologically relevant subnetworks and the prediction of gene function for novel interacting proteins. Based on the two-hybrid data, for example, a network of protein interactions for spindle pole body function was proposed that contains the components of spindle pole body and the proteins affecting its function (Ito et al., 2001). Spindle pole bodies are microtubule-organizing centers that are integrated constituents of the nuclear membrane and participate in yeast cell division, or the separation of daughter chromosomes. Three hypothetical proteins, designated Ydr016c, Ykr083c, and Ylr423c, are implicated in spindle pole body function based on their interactions with proteins of known function that are involved in the same biological process (see Fig. 11.10). The potential involvement of Ydr016c in spindle pole body function is consistent with its observed localization to intranuclear spindles and spindle pole bodies (Ito et al., 2001). The participation of the product of the essential, unknown gene YKR083C in the interaction network for spindle pole body function suggests the testable hypothesis that mutants for this novel gene will exhibit defects of the spindle pole body. Careful analysis of the two-hybrid data also led to the construction of a protein interaction model for vesicular transport, for which nine hypothetical proteins of unknown

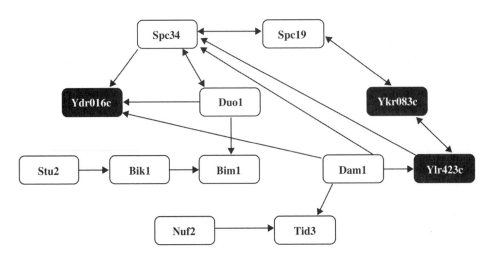

Figure 11.10 *Model of a biological subnetwork in* Saccharomyces cerevisiae *consisting of proteins involved in spindle pole body formation.* The protein–protein interactions depicted in this network within the yeast interactome are based on two-hybrid analyses. Directionality of protein interaction is indicated by arrows. Hypothetical proteins of unknown functions are depicted by black ovals. [Reproduced with permission from T. Ito, T. Chiba, R. Ozawa, M. Yoshida, M. Hattori, and Y. Sakaki, A comprehensive two-hybrid analysis to explore the yeast protein interactome, *PNAS*, vol. 98, 4569–4574. Copyright (2001) National Academy of Sciences, U.S.A.]

function were implicated (Ito et al., 2001). This rather large network is comprised of at least 25 proteins, many of which are known to function in the membrane fusion step of the vesicular transport process. A part of this network was revealed previously in the pilot phase of the study (Ito et al., 2000) and was expanded substantially upon completion of the comprehensive two-hybrid analysis (Ito et al., 2001). Such hypothetical networks or protein complexes derived from large-scale two-hybrid analysis would serve as the most appropriate targets for additional proteomics-based experimentation.

The limitations of the comprehensive two-hybrid approach are illustrated by comparing data sets obtained from two independent but similar investigations of the yeast interactome. For the most part, the core data from the study by Ito and coworkers (2001) failed to overlap with the data generated by Uetz and coworkers (2000). The two independently derived two-hybrid data sets shared only 141 interactions, representing 16.8% of the total data described by Ito and colleagues (2001) and 20.4% of the data reported by Uetz and colleagues (2000). The reasons for this significant disparity in the protein interaction data are unclear; however, some important differences existed in the strategy and stringency of selection used in the two projects. For example, Ito and coworkers (2001) used three reporter genes and multicopy two-hybrid plasmids, whereas Uetz and his colleagues used a single reporter gene (*HIS3*) and low-copy vectors. The small overlap between the two-hybrid data sets emphasizes the necessity of exploring the protein interactome of an organism from the perspectives of multiple independent analyses (Ito et al., 2001).

***Visualizing Protein–Protein Interaction Networks* in silico** The complexity and connectivity of the protein–protein interaction networks in yeast were recently emphasized in a study employing a computer-based graphics approach to visualizing direct protein associations. Schwikowski and colleagues (2000) used the AGD software library to create graphic representations of 2,709 published interactions involving 2,039 proteins of *S. cerevisiae.* The data for this computational *in silico* mapping came from public databases (Costanzo et al., 2000; Mewes et al., 2000) and from the two recent genomic two-hybrid studies previously discussed (Ito et al., 2000; Uetz et al., 2000). The MYGD (Mewes et al., 2000, 2002) and the YPD (Costanzo et al., 2000) serve as comprehensive database resources for the organization and comparison of genomic and protein information. Comprehensive interaction maps for yeast were diagrammed from the information in these databases and provided insight into the functional relationships among characterized and uncharacterized proteins. Unexpectedly, a single large network of yeast protein interactions was generated that contained 1,548 proteins linked by a total of 2,358 protein interactions (Schwikowski et al., 2000). The next largest protein interaction network was comprised of only 19 proteins. Proteins in the large interaction map (containing 1,548 proteins) had functional roles, as defined by the YPD (Costanzo et al., 2000), in membrane fusion, chromatin structure, cell structure, lipid metabolism, and cytokinesis.

Interestingly, numerous cross-connections or instances of cross-talk may be observed between and within the different functional groups and subcellular compartments. Within the large network, proteins of known function and cellular location tend to be linked by protein interactions, with 63% of the assembled connections occurring between proteins assigned a common functional role and 76% occurring between proteins residing in the same subcellular compartment (Schwikowski et al., 2000). In addition to interactions between proteins sharing a similar functional assignment, biologically meaningful connections occurred between the 21 proteins involved in membrane fusion and the 141 proteins involved in vesicular transport. Proteins controlling cell cycle events in yeast displayed the most interactions with proteins assigned to other functional classes. Other examples of cross-talk were surprising. RNA processing proteins, for instance, showed the expected connections to RNA splicing, RNA turnover, and RNA polymerase II transcription but also exhibited interactions with proteins assigned functional roles in mitosis, chromatin, and protein synthesis (Schwikowski et al., 2000). A common form of cross-talk between subcellular compartments occurred, as expected, between nuclear and cytoplasmic proteins, although somewhat inexplicable connections between the nuclear and mitochondrial compartments also were visualized within the diagram of protein interactions. Finally, network modeling and visualization is a useful approach to predicting functions for uncharacterized proteins (Schwikowski et al., 2000). The validity of the methodology was demonstrated by the correct prediction of functional categories for 72% of the 1,393 characterized proteins showing interaction with at least one partner of known biological function. This approach, therefore, has been employed to predict functional roles for 364 formerly undefined proteins. However, as with any purely computational method or genome-wide interaction strategy, uncertainties in the biological relevance of data remain that will require additional experimentation.

Analysis of Yeast Multiprotein Complexes by Mass Spectrometry The emerging technology of ultrasensitive, high-throughput mass spectrometric analysis of multiprotein complexes, in addition to proteome arrays (discussed briefly above), is beginning to significantly expand our knowledge of the yeast protein interactome. A limitation of the yeast two-hybrid strategy is that it detects binary interactions between proteins, leaving an entire higher level of functional organization virtually unexplored. The eukaryotic proteome can be visualized as a network of multiprotein complexes that operate at an organizational level which extends beyond binary interactions. Mass spectrometry is currently being applied to the direct identification of protein complexes on a proteome-wide scale (Gavin et al., 2002; Ho et al., 2002). The contributions of this technology to dissecting the complex yeast interactome are discussed here.

High-throughput mass spectrometric protein complex identification (HMS-PCI) has been used with 725 bait proteins (10% of predicted *S. cerevisiae* proteins) to detect 3,617 associated proteins, representing 25% of the yeast proteome (Ho et al., 2002). A one-step immuno-affinity purification based on the Flag epitope tag was used to capture the set of bait yeast proteins, which included 100 protein kinases, 36 phosphatases and regulatory subunits, and 86 proteins functionally implicated in the cellular response to DNA damage. Multiprotein complexes were immunoprecipitated, and the individual proteins were resolved by SDS-polyacrylamide gel electrophoresis, visualized by staining, and then removed from the gel. Tryptic digests of the excised proteins were then analyzed using MS, and the proteins subsequently identified by database search algorithms. Numerous protein assemblies from a variety of subcellular compartments (e.g., the cytoplasm, cytoskeleton, nucleus, nucleolus, plasma membrane, mitochondrion, and vacuole) were identified using the HMS-PCI method. In addition, a number of proteins (531) identified as complex components corresponded to hypothetical, functionally undefined proteins predicted from the informatic analysis of the yeast genome sequence.

The molecular machinery enabling cellular signaling events is characterized by a complex connectivity of protein interactions between kinases and phosphatases, often in combination with regulatory factors. For example, the primary cyclin-dependent kinase for cell division control, Cdc28 (reviewed in Morgan, 1997), is at the center of an extensive network of interactions that include associations with its known cyclin partners (Cln1, Cln2, Clb2, Clb3, and Clb5) and the cyclin-dependent kinase-binding subunit Cks1 (Fig. 11.11; Ho et al., 2002). Cyclin-dependent protein kinases play a well-established role in the regulation of eukaryotic cell division cycle by processing and integrating extracellular as well as intracellular signals to ensure coordination of cell cycle events under changing environmental conditions (Morgan, 1997). Intermediary partners often served to bridge the interactions detected for Cdc28. Similarly, the HMS-PCI method uncovered numerous known protein–protein interactions and new ones of likely biological importance within the global network for the DNA damage response, which includes DNA repair processes and checkpoint pathways governing cell cycle progression, transcription, protein degradation as well as DNA repair. Besides recovering most of the known nucleotide excision repair factors in their subcomplexes, HMS-PCI revealed associations involving new proteins, such as protein phosphatase 2C and the uncharacterized gene product Ydr071c. The interaction of Ydr071c with Rad53 (a component of the transcription factor TFIIH complex required for RNA polymerase II-dependent transcription and nucleotide excision repair) and PP2C-type phosphatases implicates this functionally unknown protein in the DNA damage response-specific regulation of PP2C-type phosphatases (Ho et al., 2002). In addition, the Dun1 protein kinase was found to interact with two proteins (Ymr226c and Ygr086c) of unknown biological functions that are induced in response to general cell stress, thus suggesting a role for Dun1 in processes other than DNA damage.

In another MS-based characterization of yeast multiprotein complexes, Gavin and coworkers (2002) used a tandem affinity purification (TAP) tag methodology (Rigaut et al., 1999) to purify protein assemblies from different cellular compartments, including

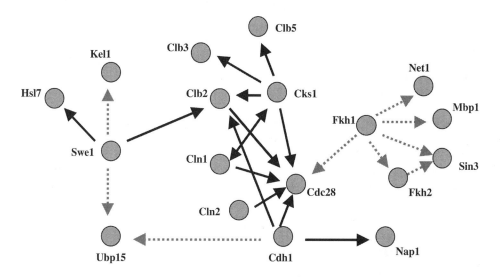

Figure 11.11 *Protein–protein interaction map for Cdc28 and Fkh1/2 complexes involved in signaling pathways.* Arrows point from the bait protein to the interaction partner. Known interactions are denoted with solid black arrows. Gray dotted arrows indicate new interactions determined by high-throughput mass spectrometric protein complex identification (HMS-PCI). [Modified with permission from Y. Ho et al., Systematic identification of protein complexes in *Saccharomyces cerevisiae* by mass spectrometry, *Nature*, vol. 415, 180–183. Copyright (2002) Nature Publishing Group http://www.nature.com/.]

complexes associated with membranes. Briefly, tagged proteins were generated by inserting gene-specific cassettes containing the TAP tag at the 3′-end of genes through homologous recombination. Assemblies in yeast cells expressing a tagged protein were purified from total cellular lysates by tandem affinity purification. Gel-separated proteins were digested with trypsin, and the resulting peptides were analyzed by matrix-assisted laser desorption/ionization–time-of-flight MS (MALDI–TOF MS). Using the TAP/MS-based functional proteomics approach, 1,739 genes were processed, including 1,143 genes representing human orthologs (i.e., genes evolved by vertical descent from a common ancestor and encoding products that perform a similar cellular function). This large-scale analysis of yeast protein complexes led to confirmation of the expression of 1,440 ORFs annotated in the yeast genome ($\sim$25% of the total protein-coding capacity) (Gavin et al., 2002). Applying the "guilt by association" concept in which proteins of similar function tend to cluster together, new cellular roles were proposed for proteins having no previous YPD functional annotation and for proteins already having a known function. Purified protein assemblies were organized into 134 new complexes. The functional classification of individual components listed in YPD (Costanzo et al., 2000) and the research literature were used to assign cellular roles to multiprotein complexes, which showed a wide functional distribution over the following categories: cell cycle (6%), cell polarity and structure (3%), intermediate and energy metabolism (19%), membrane biogenesis/turnover (9%), protein synthesis/turnover (14%), protein/RNA transport (5%), RNA metabolism (12%), signaling (9%), and transcription/DNA maintenance/chromatin structure (24%) (Gavin et al., 2002).

Two major themes underlying the higher-order organization map of the yeast proteome are (1) that complex composition can be dynamic and (2) most complexes are linked not only by physical interaction but also share a common regulation, localization, turnover, or architecture (Gavin et al., 2002). For example, cellular signaling complexes formed around the protein phosphatase 2A (PP2A) illustrate quite well the dynamics of complex composition. Tagging different known PP2A components revealed trimeric complexes containing PP2A in association with different sets of proteins. In addition, orthologous gene products appear to interact preferentially with complexes largely composed of other orthologs (Gavin et al., 2002). This same tendency was observed for protein products of essential genes, thus raising the possibility that complexes of orthologs and essential gene products largely constitute the eukaryotic "core proteome" responsible for basic cellular functions (Gavin et al., 2002).

In summary, mass spectrometric approaches seem to be especially suited for the dissection of multiprotein complexes, whereas comprehensive two-hybrid methods do not. A comparison of data sets indicated that the HMS-PCI methodology demonstrated an average 3-fold higher success rate in the detection of known multiprotein complexes than compared with large-scale two-hybrid studies (Ho et al., 2002). Similarly, the TAP/MS data set covered 56% of the known YPD protein complexes in contrast to the 10% of two-hybrid approaches (Gavin et al., 2002). Another advantage of MS is the identification of low-abundance proteins that normally would elude detection by methods involving expression proteomics. Finally, the generation of higher-order interaction maps using MS technology is likely to be more reflective of the intricate complexity of the yeast interactome.

11.5   COMPARATIVE GENOMICS OF MODEL EUKARYOTIC ORGANISMS

Determination of the complete genome sequence of the free-living nematode *Caenorhabditis elegans* represented the first from a multicellular eukaryote (The *C. elegans* Sequencing Consortium, 1998). The 97-megabase genomic sequence of *C. elegans* contains 19,717 predicted protein-coding genes, of which only 1,877 genes have been examined using either classical genetics or biochemistry (Kim, 2001; The *C. elegans* Sequencing Consortium, 1998). Although the worm genome is 30 times smaller than the human genome (3,000 Mb and 31,000 predicted genes), the number of genes in the *C. elegans* genome is only 1.6-fold fewer (Lander et al., 2001; Venter et al., 2001). A theme emerging from the rich tradition of *C. elegans* research is that many aspects of the biological processes in *C. elegans* are conserved between invertebrates and vertebrates. Because of this evolutionary conservation and the experimental tractability of *C. elegans*, the worm is an excellent model for functional genomics (Kim, 2001). Approximately 40% of *C. elegans* genes share DNA sequence homology with characterized genes in other organisms, thus pointing to the potential relevance of *C. elegans* research to our understanding of the biology of other metazoans (multicellular organisms).

Since the sequencing of *C. elegans*, complete genomic sequences for the fly *Drosophila melanogaster* (Adams et al., 2000), the flowering plant *Arabidopsis thaliana* (The *Arabidopsis* Genome Initiative, 2000), and humans (Lander et al., 2001; Venter et al., 2001) have been elucidated. The full genome sequence of the small mustard weed *A. thaliana* provides a unique opportunity to understand, at a comprehensive molecular level, the organizational and physiological differences between plants and eukaryotic organisms of the animal kingdom (The *Arabidopsis* Genome Initiative, 2000). It also provides a means for identifying plant-specific gene functions and a foundation for characterizing the function of plant genes. Furthermore, the deciphered genomes of three model organisms, *D. melanogaster*, *C. elegans*, and *S. cerevisiae*, permit comparisons between cellular and developmental processes in diverse phyla to gain insight into the evolution of eukaryotes (Rubin et al., 2000). This section focuses on the comparative genomics of *S. cerevisiae* and higher eukaryotic organisms.

***Orthology and Divergence between* C. elegans *and* S. cerevisiae** With the publication of the complete genomic sequence of *C. elegans* 2 years following that of *S. cerevisiae*, it was possible, for the first time in history, to compare the entire complement of predicted protein sequences of two highly diverged eukaryotic species, one representing a unicellular microorganism and the other a multicellular animal (Chervitz et al., 1998). Chervitz and colleagues (1998) conducted a computational comparative analysis of the complete protein sets of the nematode worm and yeast by identifying orthologous proteins (i.e., proteins that evolved from a common ancestor by vertical descent and presumed to perform the same cellular function) as well as shared and novel protein domains. Rather surprising was the finding that unequivocal, one-to-one orthologous relationships were identified for a significant fraction of yeast and worm genes: Worm homologs for 2,497 yeast ORFs (40% of total yeast ORFs) and yeast homologs for 3,653 worm ORFs (19% of total worm ORFs) were found (Chervitz et al., 1998). Furthermore, various core biological processes of the two organisms are carried out by orthologous (closely related) proteins. The core biological functions conserved in both yeast and worm included intermediary metabolism, DNA and RNA metabolism, protein folding and degradation, and transport and secretion (see Fig. 11.12). Some of these orthologous pairs, such as the yeast *CDC28* and the worm *ncc-1* orthologs in the cyclin-dependent kinase family, have already been demonstrated experimentally to be functionally interchangeable in vivo and, thus, functionally conserved (Mori et al., 1994). In addition, worm and yeast genes were found in clusters of DNA-dependent RNA polymerases and the large cluster of HSP70 heat shock proteins. Comparative analysis of the two model genomes also revealed highly conserved *C. elegans* homologs for 108 mitochondrial yeast proteins. These orthologous pairs had assigned functions in such diverse mitochondrial processes as the TCA cycle, electron transport, lipid metabolism, amino acid biosynthesis, intermediary metabolism, membrane transport, protein processing, RNA metabolism, and protein synthesis. A significant conclusion from this comparative genomics study was that organisms as disparate as yeast and worm possess a shared core biology, and proteins conserved in yeast and worm are likely to have orthologs throughout the eukaryotic domain.

Of primary interest is a comprehensive understanding of the functions that define multicellularity. What sets of protein sequences or domains are involved in specialized cellular processes that are unique to multicellular organisms and may have contributed to the advent of multicellular life? In their computational comparison of the predicted proteins of yeast and worm, Chervitz and coworkers (1998) discovered that most of the genes known or expected to be involved in the specialized processes of signal transduction and regulatory control have no orthologs in yeast but sometimes contain protein domain sequences shared by both the worm and yeast. Biological processes characteristic of multicellularity, such as the regulation of gene expression and signal transduction, are performed by novel protein species that vary significantly from proteins dedicated to core

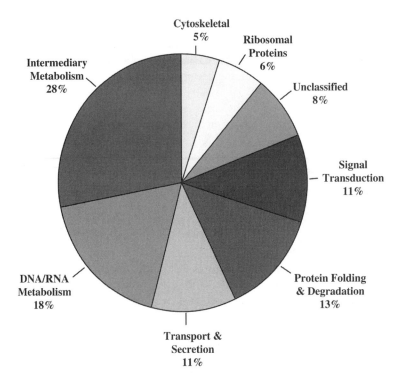

Figure 11.12 *Distribution of core biological functions conserved in both yeast and worm.* [Data from Chervitz et al., 1998.]

processes. Comparison of worm and yeast proteins revealed instances of invention of domains de novo, acquisition of domains (domain shuffling) by proteins enabling novel forms of signal transduction, and duplications with subsequent divergence (Chervitz et al., 1998). Furthermore, the level of complexity associated with multicellularity is linked to relatively small but important sets of regulatory and signal transduction domains found in *C. elegans* but not in *S. cerevisiae*. The most prominent examples of these domains are those of extracellular signaling and adhesion molecules (e.g., epidermal growth factor domains) as well as components of programmed cell death machinery. Similarly, a small set of regulatory domains is found only in yeast and includes the zinc-binding cluster C6 finger, a DNA-binding domain. Finally, there are regulatory domains that are conserved in both *C. elegans* and *S. cerevisiae* but have functional roles in very different cellular processes. In yeast, these conserved domains act in DNA binding or intracellular protein–protein interactions, while in the worm they function in signal transduction pathways not found in yeast, acting as extracellular adhesion and signaling modules.

Comparative Genomics of Fly, Worm, and Yeast   With the availability of complete genomic sequences for a number of model prokaryotes and eukaryotes, much of our knowledge concerning the molecular foundation distinguishing unicellular organisms from more complex multicellular organisms will derive from comparative analyses of protein families and domains and intracellular protein networks underlying developmental and cellular processes. Such studies are beginning to appear in the literature (Ledent and

Vervoort, 2001; Maglich et al., 2001; Mar Alba et al., 2001; Rubin et al., 2000). One study compared predicted proteins encoded in the genomes of *D. melanogaster*, *C. elegans*, and *S. cerevisiae* within the context of cellular, developmental, and evolutionary processes (Rubin et al., 2000). The core proteome, defined as the number of distinct protein families in an organism, was determined for fly, worm, and yeast using computational informatic approaches. Remarkably, the core proteome of *D. melanogaster* (8,065 proteins), a complex metazoan, is only twice as large as the core proteome of yeast (4,383 proteins), a simple unicellular eukaryote (Rubin et al., 2000). In addition, the striking differences exhibited by the fly and worm in terms of development and morphology do not appear to be attributed to differences in the size of their respective core proteomes (8,065 and 9,453 proteins, respectively). Indeed, this seems to be a major theme emerging from genomic comparison investigations: The complexity apparent in diverse metazoans is not achieved by possessing proportionately larger numbers of genes or molecular components (Rubin et al., 2000).

Similar to the study by Chervitz and coworkers (1998) on orthologous proteins, Rubin and his colleagues found that the fly, worm, and yeast share a set of proteins that most likely perform basic functions common to all eukaryotic cells. Nearly 20% of *D. melanogaster* proteins, for example, have a putative ortholog or counterpart in both *C. elegans* and *S. cerevisiae*. In addition, 744 protein families or domains were common to all three organisms. Multidomain proteins in the fly and worm, however, are considerably more complex compared to those in yeast (Rubin et al., 2000). The architecture of proteins is typically mosaic, containing two or more distinct identifiable structural domains, and different proteins often consist of different combinations of domains. *D. melanogaster* and *C. elegans* actually contain a similar number of recognizable multidomain proteins (2,130 and 2,261 proteins, respectively), whereas yeast possesses substantially fewer multidomain proteins (672). A portion of this difference is due to the presence of extracellular domains involved in signaling pathways (i.e., cell–cell and cell–substrate interactions), which are characteristic of multicellularity. Expansion or contraction of certain classes of proteins may also reflect differences in growth and development in *D. melanogaster*, *C. elegans*, and *S. cerevisiae*. For instance, genomic comparisons revealed that a single class of trypsinlike (S1) peptidases is greatly expanded in the fly (199 S1 peptidases) in contrast to the worm (7 S1 peptidases) and yeast (1 S1 peptidase). In humans, trypsin-related peptidases carry out diverse functions in digestion as well as in a number of signaling pathways, and their cellular role in the fly may be similar. In conclusion, *in silico* comparative genomic analyses will contribute in significant ways to improving our understanding of the molecular basis of multicellular life. Such computational studies also provide a foundation for designing biochemical and genetic experiments to ascertain the biological relevance of differences observed at the genomic level.

11.6 SUMMARY

A recurrent theme among sequenced genomes, including those of extensively investigated model organisms, is that a substantial portion (e.g., 30–50%) of the potential protein-coding content is functionally unknown. Functional genomics and associated genome-based technologies have begun to accelerate progress in placing uncharacterized genes predicted from annotated genomes within a biologically meaningful context. In this chapter, we have focused on the impact of structural and functional genomics on elucidating the dynamics of the transcriptome, proteome, metabolome, and interactome of the well-studied model organisms *E. coli*, *B. subtilis*, and *S. cerevisiae* (budding yeast). Functional and comparative genomics is likely to reveal the true utility and contribution of research on *E. coli*, *B. subtilis*, and *S. cerevisiae* to our understanding of gene functions and fundamental cellular processes in more complex, less experimentally tractable organisms.

Despite decades of research on *E. coli* and *B. subtilis*, over 30% of the protein-coding genes predicted in the genomes of these model organisms had no attributed biological role at the completion of their sequencing. Functional interpretation of fully sequenced genomes can be investigated by examining changes in mRNA abundance in response to various environmental stimuli and physiological perturbations or mutations in specific regulatory genes. Using whole-genome microarray technology, for example, researchers have investigated the transcriptomic dynamics of the *E. coli* response to heat stress and nitrogen limitation. A number of new heat shock stimulon and NtrC regulon members were identified by comprehensive gene expression profiling, many of which corresponded to genes encoding proteins of unknown function. Similarly, transcriptomic analyses have examined global gene expression in response to heat shock and during growth and sporulation of *B. subtilis*. In addition, the regulons for response regulators in *B. subtilis* two-component systems have been investigated using DNA microarrays. Such studies demonstrated how functional genomic technologies, in combination with traditional genetic techniques, can be utilized to detect a majority of the target gene candidates of functionally uncharacterized two-component regulatory systems, thus providing clues to the roles these phosphorelay systems play in the signal-transduction network of the cell. The major biological theme emerging from proteomic studies is that bacterial cells alter the contents of their proteome to adapt to changing environmental conditions, and proteomics can be employed to enhance and support predictive information supplied by genomic sequence analysis. Examination of the extracellular proteome of *B. subtilis*, for example, has been particularly informative in light of the fact that this model soil microorganism secretes large quantities of proteins, mainly degradative enzymes, that enable the bacterium to acquire nutrients from a variety of substrates and to survive in a continuously changing environment. Furthermore, proteomic analyses, like microarray-based transcrip-tion profiling, can implicate hypothetical proteins in the performance of cellular processes. Both genomic and proteomic information can be integrated into a framework that is used to develop computer-derived models of complex metabolic pathways, an area referred to as *in silico* metabolomics.

S. cerevisiae has long been considered a model for higher eukaryotes, and it is expected that research on this unicellular simple eukaryotic organism will be of value in understanding basic cellular processes in more complex multicellular organisms, as well as the molecular basis of human genetic diseases. The yeast proteome, for instance, contains structural homologs for 46% of the human proteins that have been identified. Genome-wide transcription profiling and proteomic descriptions are expediting the definition of biological functions for the approximately one-third uncharacterized novel yeast genes uncovered by genomic sequencing and the delineation of regulatory networks. In addition to transcriptomic and proteomic information, a complete description of protein–protein interactions is indispensable to a complete understanding of the cell at the molecular level. Traditionally, the yeast two-hybrid system has been employed to generate large-scale protein–protein interaction maps. Recent high-throughput, ultrasensitive MS methods are beginning to be used to generate higher-order protein interaction maps that are more reflective of the exquisite complexity of the yeast interactome.

FURTHER READING

DeRisi, J. L., V. R. Iyer, and P. O. Brown. 1997. Exploring the metabolic and genetic control of gene expression on a genomic scale. *Science* 278:680–686.

Fawcett, P., P. Eichenberger, R. Losick, and P. Youngman. 2000. The transcriptional profile of early to middle sporulation in *Bacillus subtilis. Proc. Natl. Acad. Sci. USA* 97:8063–8068.

Ho, Y., A. Gruhler, and A. Heilbut, et al. 2002. Systematic identification of protein complexes in *Saccharomyces cerevisiae* by mass spectrometry. *Nature* 415:180–183.

Ito, T., T. Chiba, and R. Ozawa, et al. 2001. A comprehensive two-hybrid analysis to explore the yeast protein interactome. *Proc. Natl. Acad. Sci. USA* 98:4569–4574.

Khodursky, A. B., B. J. Peter, N. R. Cozzarelli, and D. Botstein, et al. 2000. DNA microarray analysis of gene expression in response to physiological and genetic changes that affect tryptophan metabolism in *Escherichia coli. Proc. Natl. Acad. Sci. USA* 97:12170–12175.

Kumar, A., S. Agarwal, and J. A. Heyman, et al. 2002. Subcellular localization of the yeast proteome. *Genes Dev.* 16:707–719.

Ye, R. W., W. Tao, L. Bedzyk, and T. Young, et al. 2000. Global gene expression profiles of *Bacillus subtilis* grown under anaerobic conditions. *J. Bacteriol.* 182:4458–4465.

Zimmer, D. P., E. Soupene, H. L. Lee, and V. F. Wendisch, et al. 2000. Nitrogen regulatory protein C-controlled genes of *Escherichia coli*: scavenging as a defense against nitrogen limitation. *Proc. Natl. Acad. Sci. USA* 97:14674–14679.

12

Functional Genomic Analysis of Bacterial Pathogens and Environmentally Significant Microorganisms

Dorothea K. Thompson and Jizhong Zhou

12.1 INTRODUCTION

The molecular basis and regulation of bacterial pathogenesis, in particular the interaction between animal hosts and their pathogens, are complex and involve the intricate and temporal control of multiple virulence determinants. The broad spectrum of physiological and virulence properties is a reflection of the tightly coordinated expression of diverse suites of gene functions. Generally, the infectious cycle of a pathogen begins with adherence to and colonization of the host, followed (sometimes) by bacterial invasion of host tissues or cells, multiplication and persistence within the host cell, and finally, the exit of the pathogen and transmission to new hosts (for a review, see Finlay and Falkow, 1997). Microbial infection is mediated by complex regulatory networks that involve species-specific cell-to-cell communication. The advent of the genomics era, initiated by the publication of the complete genome sequence of *Haemophilus influenzae* (Fleischmann et al., 1995) in the mid-1990s, has transformed studies of bacterial pathogenesis by enabling a global resolution of the molecular features responsible for virulence traits. The

Microbial Functional Genomics, Edited by Jizhong Zhou, Dorothea K. Thompson, Ying Xu, and James M. Tiedje.
ISBN 0-471-07190-0 © 2004 John Wiley & Sons, Inc.

complete genomes of at least 25 distantly related pathogenic bacteria have been deciphered, and whole-genome sequence analysis of many other microbial pathogens is currently in progress (see Chapter 2). Complete genomic sequence information, for example, is available for infectious species such as *Helicobacter pylori* (Tomb et al., 1997; Alm et al., 1999), *Mycobacterium tuberculosis* (Cole et al., 1998), *Mycoplasma genitalium* (Fraser et al., 1995), *Pseudomonas aeruginosa* (Stover et al., 2000), *Streptococcus pneumoniae* (Hoskins et al., 2001), *Vibrio cholerae* (Heidelberg et al., 2000), *Listeria monocytogenes* (Glaser et al., 2001), and most recently, *Bacillus anthracis* (Read et al., 2003). The genome sequences from a number of environmentally important eubacteria and archaea have also been determined (see Chapter 2), including the metal ion-reducing bacterium *Shewanella oneidensis* (Heidelberg et al., 2002), the extreme radiation-resistant bacterium *Deinococcus radiodurans* (White et al., 1999b), and the hyperthermophilic archaeon *Pyrococcus furiosus* (Robb et al., 2001).

Bioinformatic analysis of an organism's total gene content permits a comprehensive search for candidate genes associated with infection and the identification of potentially new targets for antimicrobial drug design. Comparative analysis of genomes from multiple phylogenetically divergent bacteria and closely related strains within a pathogenic species has uncovered variability in gene content within natural populations and revealed basic mechanisms involved in species diversification and the evolution of bacterial pathogens, as well as the strain-specific basis for differences in severity of pathology. As more members of the Archaea domain are sequenced, comparative genomics will reveal and reinforce the major differences between archaea (particularly between the crenarchaeal and euryarchaeal branches), and bacteria and eukarya. In addition, functional genomics tools, such as DNA microarrays, are having an increasingly important impact on our understanding of the modulation of host gene expression in response to microbial infection and the progression of molecular events that eventually result in the clinical manifestation of disease. A very limited number of microarray-based functional studies of environmentally significant microorganisms are beginning to define gene function and to reveal the regulatory networks and unique enzyme systems underlying phenotypes important for cell survival in harsh environments. Finally, proteomics is beginning to extend our knowledge of gene functionality by determining if and when the encoded products of predicted virulence and unannotated genes are translated and to provide insight into the roles of posttranscriptional regulation and posttranslational modification in bacterial pathogenesis.

In this chapter, we will discuss the contribution of genome sequence and *in silico* bioinformatic analyses to virulence gene identification, the impact of comparative genomics on revealing genetic diversity and evolutionary trends among pathogenic bacteria, genetic and microarray-based approaches to elucidating bacterial gene function and host–pathogen interactions, and the proteomics of bacterial pathogenesis. Because of the number of bacterial pathogens that have been sequenced and the pace at which additional microbial genomes are currently being elucidated, it is beyond the scope of this chapter to present a comprehensive description of each sequenced microbial pathogen. Instead, we will discuss how functional and comparative genomics have contributed to our understanding of bacterial pathogenesis by focusing on selected studies from the literature as general illustrations. Finally, we will describe how genomic data facilitate a basic understanding of bacterial phenotypes of environmental relevance (e.g., the bioremediation potential of metal-reducing bacteria) and provide insight into the molecular mechanisms or characteristics of cells living in extreme habitats (e.g., high temperature, acidic pH, high salinity).

12.2 ADVANCING KNOWLEDGE OF BACTERIAL PATHOGENESIS THROUGH GENOME SEQUENCE AND FUNCTION ANNOTATION

Bacterial pathogens have evolved the ability to invade both animal and plant host cells, to elude host immune defenses and counteract antimicrobial agents, and to survive and multiply within very different intracellular and extracellular environments. For such diverse pathogenic strategies, bacteria require a repertoire of tightly regulated virulence factors that vary in terms of their number and nature. Although many virulence determinants appear to be host-specific, a limited number of mechanisms are employed by phylogenetically diverse bacterial species to express a broad range of virulence phenotypes (for a detailed review, see Finlay and Falkow, 1997). These commonalities suggest that at least some of the underlying virulence mechanisms of pathogens possess ancient evolutionary origins that have been preserved across bacterial taxa (Rahme et al., 2000).

Traditionally, bacterial pathogenesis has been investigated by focusing on a single gene or protein in a time-consuming, linear fashion. Genome sequencing and high-throughput tools such as DNA microarrays (see Chapter 6) now enable many genes and proteins relevant to bacterial infection to be simultaneously identified and functionally analyzed in parallel. Complete genome sequence data for bacterial pathogens can be exploited in many ways, from identifying novel virulence-associated genes and pathogenicity islands to designing new antimicrobial agents (discussed in Chapter 13). As discussed in this section, searches for multiple candidate virulence genes, repeated DNA elements, and horizontally acquired DNA sequences, such as pathogenicity islands, are now being performed on a genome-wide scale using bioinformatic approaches.

12.2.1 Predicting Virulence Genes from Sequence Homology

Knowledge of an organism's entire gene content permits an exhaustive search for predicted virulence genes based on sequence homology to known virulence determinants in public databases. An important caveat of sequence interpretation, however, is that sequence similarity does not equal function in all cases. Along with the identification of putative virulence genes, it is necessary to empirically determine whether the candidate gene is an actual functional homolog of a gene with confirmed virulence-associated activities. A limitation of using genomic sequence data exclusively to identify putative genes contributing to bacterial pathogenicity is that the *in silico* bioinformatic approach is really only useful in recognizing candidates for known genes whose functions have been elucidated. Comparative analysis of sequences from bacterial genomes reveals that a large portion (typically, $\sim 30-40\%$) of predicted genes have unknown or hypothetical functions. Virulence genes with novel functions or unrecognized motifs cannot be easily predicted based on sequence information alone (Weinstock, 2000). To identify genes of possible relevance to pathogenicity, it may be necessary to query databases for more general characteristics of deduced protein sequences, such as those sequence motifs associated with transmembrane and/or secreted proteins. The rationale for such an approach is that virulence factors involved in host–pathogen interactions are likely to be localized to the cell surface or destined for extracellular export (Weinstock, 2000). In one study, for example, the entire genome sequence of a virulent serogroup B strain (MC58) of *Neisseria meningitidis* was explored for the purpose of identifying potentially novel virulence factors that could serve as vaccine candidates (Pizza et al., 2000). Open reading frames (ORFs) of the *N. meningitidis* genome were analyzed for sequence or structural motifs and other signatures that typically characterize surface-associated or exported proteins: transmembrane domains, leader peptides, homologies to known surface proteins, lipoprotein signature, outer membrane anchoring motives, and host-cell binding domains. Finally, genome sequence-based microarray analysis of global gene expression can be used to find new virulence factors by identifying genes whose expression is coregulated with known virulence factors.

The complete genome sequence of an organism has been utilized to access novel biological information in a rapid, comprehensive manner (Hood et al., 1996a,b; Pizza et al., 2000). The availability of the 1.8-Mb sequence of the *H. influenzae* strain Rd genome, for example, has greatly facilitated the identification of putative virulence genes and structural components involved in the biosynthesis of lipopolysaccharide (LPS) (Hood et al., 1996b). Besides being a major virulence determinant, LPS is also crucial in maintaining the integrity and function of the outer cell membrane of gram-negative bacteria (Strauss and Falkow, 1997). Twenty-five candidate LPS genes were identified by searching the *H. influenzae* genomic database for sequences exhibiting DNA and amino acid similarity to known LPS biosynthetic genes from other organisms. Sequence information allowed the design and construction of clones necessary for generating targeted disruptions in the 25 genes identified. Immunochemical techniques, polyacrylamide gel electrophoresis (PAGE) fractionation, and electrospray mass spectrometry (MS) confirmed a potential role in LPS biosynthesis for the majority of the candidate genes and permitted the estimation of the minimal LPS structure required for intravascular dissemination in the infant rat (Hood et al., 1996b). Studies like this one illustrate how sophisticated computational techniques can be used to mine genomic sequence for genes encoding putative virulence factors and how the availability of such data can substantially expedite the task of testing experimental hypotheses.

12.2.2 Repeated DNA Elements Indicate Potential Virulence Factors

In addition to sequence homology, repetitive DNA sequences can implicate genes in virulence. Repetitive DNA is associated with virulence factor phenotype or phase variation in a number of pathogenic bacteria, including *H. influenzae*, *Neisseria* species, and *Staphylococcus aureus* (Table 12.1). A feature of many genes encoding cell surface-accessible virulence determinants is the presence of repeated nucleotides in mononucleotide (homopolymeric) tracts or tandemly iterated dinucleotides, tetranucleotides, and other repeats, typically located within the 5′-end of the translated reading frame. Changes in the number of repeated nucleotides, presumably through a slipped-strand mispairing (Levinson and Gutman, 1987) or genetic recombination mechanism (van Belkum et al., 1998), mediate phenotypic variation of surface-exposed proteins by altering the reading frame of genes. In slipped-strand mispairing, the unusual tertiary structure of highly repetitive DNA causes mismatching of neighboring repeats. As a result, DNA repeats can be inserted or deleted, depending on the orientation of the strand, during DNA polymerase-mediated DNA synthesis, thus producing a frameshift in the open reading frame (Coggins and O'Prey, 1989; van Belkum et al., 1998). Phenotypic variation of virulence-associated factors is a strategy employed by pathogenic bacteria to evade the host immune response and adapt to different microenvironments within the host (for a review, see Finlay and Falkow, 1997). As an example, the phase variation of *H. influenzae* fimbriae, an adhesin protein required for bacterial attachment to respiratory epithelial cells, is associated with

TABLE 12.1 Examples of Short-sequence DNA Repeats and Their Associated Gene Functions in Bacterial Pathogens

Species	Repeat Motif	Gene	Gene Function	Reference(s)
H. influenzae	CAAT	*lic1–lic3*	LPS biosynthesis	Hood et al. (1996a)
	GCAA	*yadA*	Adhesin	Hood et al. (1996a)
	GACA	*lgtC*	Glycosyltransferase	Hood et al. (1996a)
	TTGG	ND[a]	Iron-binding proteins	Hood et al. (1996a)
	AGTC	ND	Restriction modification, methyltransferase	Hood et al. (1996a)
	TTTA	ND	Unknown *Bacillus* homolog	Hood et al. (1996a)
	TA	*hifA/B*	Synthesis of fimbriae	Van Ham et al. (1993, 1994)
N. meningitidis	G	*lsi2*	LPS biosynthesis	Burch et al. (1997)
	CTCTT	*opa*	Opacity surface proteins	Meyer et al. (1990)
	A	*opa*	Opacity surface proteins	Meyer et al. (1990)
	G	*porA*	Outer membrane protein	Van der Ende et al. (1995)
S. aureus	93 bp	*fnb*	Fibronectin binding protein	Patti et al. (1994)
	561 bp	*cna*	Collagen adhesin	Patti et al. (1992)
	81 bp	*coa*	Coagulase	Goh et al. (1992)
	GAAGAX₄AAXAAXCCTXGXAAA	*spa*	Protein A	Clewell (1993)
	GAXTCXGAXTCXGAXAGX	*clf*	Clumping factor, fibrinogen receptor	McDevitt and Foster (1995)

[a]ND, not defined.

381

dinucleotide TA repeats (van Alphen et al., 1991). Similarly, the 5'-termini of the translated reading frames of lipopolysaccharide biosynthetic genes contain multiple tandem DNA repeats of CAAT or GCAA that are involved in translational switching (Jarosik and Hansen, 1994; Weiser et al., 1989).

Given the importance of repetitive DNA in bacterial pathogenicity, one approach to identifying potential novel virulence factors would be to scan genomic sequences for tandem oligonucleotide repeated sequences. Hood and coworkers (1996a) performed such an inspection using the whole-genome sequence of *H. influenzae* strain Rd. In their analysis of the Rd genome sequence, nine novel loci containing multiple (6–36) tandem tetranucleotide repeats within the putative ORFs were identified (Hood et al., 1996a). These genes encoded homologs of hemoglobin receptor proteins and a glycosyltransferase (*lgtC* gene product) from *Neisseria*, and an adhesin protein (*yadA* gene product) from *Yersinia*. In addition, three previously characterized *Haemophilus* genes involved in LPS biosynthesis contained multiple repeats of CAAT, thus validating the utility of complete microbial genome sequences for identifying virulence factors. Further characterization of one of the novel loci, *lgtC*, revealed that this gene is involved in phenotypic switching of a lipopolysaccharide epitope and is required for the full virulence of *H. influenzae* in an infant rat model. This study clearly demonstrates how information derived from a complete genome sequence can yield experimental data of biological relevance and may be used to investigate the biology of pathogenic bacteria (Hood et al., 1996a).

12.2.3 Evolution of Bacterial Pathogens: Gene Acquisition and Loss

The availability of complete genomic sequences and their comparison provide unparalleled opportunities to assess the impact of lateral gene transfer events on the evolution of microbial pathogens (Ochman and Moran, 2001). Lateral (or horizontal) gene transfer, in effect, remodels a genome, altering the genome repertoire, by introducing or deleting substantial amounts of DNA from the chromosome. The mechanisms underlying gene acquisition and loss (namely, transformation, transduction, and conjugation) are described in detail in Chapter 4. The accumulation of point mutations over time has contributed to the diversification of microorganisms through modulating virulence phenotype or altering the expression of existing genes; however, stepwise mutational changes rarely confer novel functions or enable organisms to explore or exploit new environments (Lawrence, 1997; Lawrence and Ochman, 1998). Instead, there is mounting evidence that lateral gene transfer—acquisition or loss of genetic information as a result of bacterial spread of mobile genetic elements—plays an integral role in genome evolution and bacterial speciation, particularly in the emergence of new pathogens (Ochman et al., 2000; Ochman and Moran, 2001).

The occurrence of lateral or horizontal genetic exchange can be investigated using genetic analysis of DNA nucleotide sequences. Genes within a particular bacterial species' genome are relatively homogeneous with respect to their base composition [e.g., guanine plus cytosine (G + C) content], patterns of codon usage bias, and frequencies of di- and tri-nucleotides (Muto and Osawa, 1987; Karlin et al., 1998). New sequences, such as pathogenicity islands (discussed below), originating from a foreign source through lateral gene transfer, retain the sequence features of the donor genome and thus can be distinguished from ancestral or vertically transmitted DNA by variations in G + C content and patterns of codon usage (Ochman et al., 2000). Determination of the scope of lateral gene transfer has only recently been possible through comparisons of completely sequenced genomes. The cumulative amount (percentage) of laterally acquired foreign DNA varies among sequenced bacterial genomes: For example, 12.8% of the *Escherichia coli K12* genome, 3.3% of the *M. tuberculosis* genome, 4.5% of the *H. influenzae* genome, and 6.2% of the *H. pylori 26695* genome constitute foreign DNA (Ochman et al., 2000). The contribution of horizontal gene transfer to genome plasticity (fluidity), genetic diversity, and the evolution of bacterial pathogens is discussed below.

Pathogenicity Islands   Virulence genes specify a disease condition by coding for toxins, adhesins (proteins that mediate adhesion to host cell surfaces), secretion system components, proteins conferring serum resistance, as well as other factors. Such disease-determining genes can occur on transmissible ("mobile") genetic elements (e.g., transposons, plasmids, or bacteriophages) or reside on discrete segments of the bacterial chromosome, termed pathogenicity islands (PAIs) (reviewed in Hacker at al., 1997; Hacker and Kaper, 2000; Groisman and Ochman, 1996). Sequence analysis of PAIs indicated that these virulence cassettes most likely originated by horizontal gene transfer and, therefore, may play a central role in the creation of new pathogenic variants or pathotypes (Hacker and Kaper, 2000). The capacity of lateral gene transfer to alter the nature of a bacterial species, however, does not appear to be indiscriminate. In other words, certain organisms are preadapted to become pathogens upon acquiring pathogenicity islands, because they already possess capabilities that facilitate survival under conditions encountered in mammalian hosts, such as mechanisms to counteract host defenses and to metabolically compensate for nutrient deficiencies (Ochman and Moran, 2001). The sequence features distinguishing PAIs from other genomic regions are discussed in this section, as well as specific examples of microbial PAIs and the regulation of PAIs.

Discerning PAIs from Whole-Genome Sequences   A common theme emerging from genomic analyses is that microbial genomes are substantially homogeneous in G + C content and codon usage (Hacker and Kaper, 2000; Karlin, 2001). A gene is categorized as a '*putative alien* (pA)' if its codon usage deviates significantly from the rest of the genome (Karlin, 2001), suggesting that it was acquired by lateral gene transfer (Finlay and Falkow, 1997; Hacker and Kaper, 2000; Ochman et al., 2000). Putative alien or anomalous gene clusters (sometimes referred to as genomic islands) in bacteria are of particular relevance in identifying pathogenicity islands. The presence of anomalous gene regions in sequenced genomes can be determined by applying five criteria to the analysis of DNA sequences (Karlin, 2001).

(i)   Differences in (G + C) Content. The standard DNA computational method for discerning anomalous gene regions, such as pathogenicity islands, is to compare the G + C frequency within a sliding window W (W equals 10, 20, or up to 50 kb of sequence in length) to the average genomic G + C frequency. Windows possessing a significantly different G + C content compared to the rest of the genome indicate potential genomic islands.

(ii)   *Genome Signature Contrasts*. Each genome has a characteristic "signature" that distinguishes it from the genomic sequence of other organisms (Campbell et al., 1999; Karlin, 1998; Karlin et al., 1997). A *genome signature* is a set of dinucleotide relative abundance values, obtained by measuring the ratios between observed dinucleotide frequencies and the frequencies expected if neighboring sequences were randomly selected (Campbell et al., 1999). Deviations from the average genomic signature or dinucleotide bias suggest DNA of a foreign origin.

(iii) *Extremes of Codon Bias.* The codon bias for all the genes is calculated and compared to the average gene in a genome. An outlying gene or cluster of genes deviating from the typical codon usage may represent a PAI.

(iv) *Divergence in Amino Acid Usage.* The amino acid biases of deduced proteins are compared to the average amino acid frequencies determined for the proteome. Genomic regions where the amino acid usage in translated reading frames diverges from the average protein may constitute a genomic island or PAI. Codon bias and amino acid usage are both complementary measures of gene composition.

(v) *Putative Alien Gene Clusters.* Genes are labeled as putative alien genes if codon usage differences from the average gene and from ribosomal protein genes, translation and transcription processing factor genes, and chaperone gene classes are high. Such putative alien genes may represent virulence-associated genes contained within a pathogenicity island.

PAIs have been identified in both uropathogenic and enteropathogenic strains of *E. coli, Salmonella typhimurium, V. cholerae,* and *H. pylori* and in certain gram-positive bacteria (see Table 12.2; Hacker et al., 1997). In addition, multiple pathogenicity islands can occur in a single bacterial strain (Mecsas and Strauss, 1996). As PAIs continue to be discovered among diverse groups of pathogenic bacteria, especially as a result of microbial genome sequencing, shared features have emerged, which permit PAIs to be defined according to a number of criteria (Hacker and Kaper, 2000; Hacker et al., 1997). (1) PAIs harbor virulence-associated genes, such as adherence factors, toxins, and type-III secretion systems. (2) PAIs, which are widely distributed among the genomes of pathogenic bacteria, are absent from the genomes of nonvirulent variants of the same species or closely related commensal species. (3) PAIs occupy relatively large regions [$\geq 10-200$ kilobases (kb) in length] of the host bacterial chromosome that are distinguished from the core genome by significant differences in G + C content and codon usage. (4) PAIs are typically flanked by short [e.g., 9–135 base pairs (bp)] directly repeated sequences or insertion sequence (IS) elements. The presence of such specific boundary sequences and integrase determinants as well as other mobility loci argue for the introduction of PAIs into the chromosome by a recombination event (Hacker and Kaper, 2000; Mecsas and Strauss, 1996). Direct repeats and IS elements may serve as targets for recombinases, leading to the genetic instability of some PAIs (Hacker et al., 1997). (5) PAIs are often associated with transfer RNA (tRNA) genes, which may function as "genomic landmarks" for the integration of foreign DNA (Hacker et al., 1997).

PAIs Contribute to Virulence Phenotypes   This section illustrates the diversity of PAI-associated virulence determinants by focusing on selected gram-negative bacteria of pathogenic importance.

Members of the family *Enterobacteriaceae* are etiologic agents for intestinal and urinary tract infections. The chromosomal PAIs of uropathogenic and enteropathogenic *E. coli* strains were the first virulence-determining regions to be described and intensively investigated (Hacker et al., 1983, 1997; Blum et al., 1994; Swenson et al., 1996). The uropathogenic *E. coli* (UPEC) strain 536, for example, contains two pathogenicity islands,

TABLE 12.2 Selected Examples of Pathogenicity Islands

Bacterium	PAI Designation	Size (kb)	G + C content[a]	Associated tRNA	Function
Uropathogenic E. coli (536)	PAI I	70	51/41	selC	Hemolysin production and P-related fimbriae
	PAI II	190	51/41	leuX	Hemolysin production and P-related fimbriae
Uropathogenic E. coli (J96)	PAI IV	170	51/41	pheV	Hemolysin production
	PAI V	110	51/41	pheR	Hemolysin production
Enteropathogenic E. coli (E2348/69)	LEE[b]	35	51/39	selC	Attachment and effacement, invasion (encodes type-III secretion system)
Salmonella typhimurium	SPI-1	40	52/40–47	—	Invasion into nonphagocytic cells (encodes type-III secretion system)
	SPI-2	40	52/40–47	valV	Bacterial survival in host (encodes type-III secretion system)
	SPI-3	17	—	selC	Invasion, survival in monocytes
	SPI-4	25	—	putative tRNA	Invasion, survival in monocytes
Helicobacter pylori	Cag PAI	40	38–45/35	glr[c]	Encodes CagA[d] and VacA regulator, needed for full virulence
Vibrio cholerae	VPI	39.5	47–49/35	ssrA	Contains TCP-ACF[e] element and toxT[f] gene, regulation of the cholera toxin
Yersinia pestis	HPI (pgm locus)	102	46–50/46–50	asnT	Iron acquisition, hemin uptake, Yersiniabactin synthesis
Clostridium difficile		19.6			Toxins A and B
Listeria monocytogenes	prf vir Gene cluster	10			Virulence regulation, escape from vacuole, cell-to-cell spread

[a]G + C (%) content of the bacterial host organism vs. G + C (%) content of the PAI.
[b]LEE = locus of enterocyte effacing.
[c]glr = glutamate racemase.
[d]CagA = cytotoxin-associated antigen.
[e]TCP = toxin-corregulated pili; ACF = accessory colonization factor.
[f]toxT = transcriptional activator gene.

PAI I and PAI II, which are 70 and 190 kb in size, respectively, and are flanked by direct repeats (Blum et al., 1994; Ritter et al., 1995). PAI I encodes α-hemolysin (*hly*), a pore-forming toxin capable of lysing erythrocytes and other eukaryotic cells via insertion into cell membranes. PAI II carries the *hly* virulence gene linked to the *prf* determinant, which codes for P-related fimbriae, an important adherence factor of uropathogenic *E. coli* strains (Hacker and Kaper, 2000). PAI II-encoded P fimbriae enable uropathogenic *E. coli* to adhere to uroepithelial cells via galactose-α 1-4-galactose-specific receptor molecules. Both UPEC-specific PAIs are associated with tRNA genes, with PAI I located in the selenocysteine (*selC*) tRNA gene and PAI II integrated into *leuX*, a minor leucyl tRNA (Blum et al., 1994). Similarly, enteropathogenic strains of *E. coli* (EPEC) harbor a 35-kb virulence cassette at the chromosomal *selC* locus (McDaniel et al., 1995), but, in contrast to the UPEC-specific PAI, the EPEC-specific pathogenicity island encodes a highly specialized type-III secretion system that exports proteins essential for the attachment and effacing lesions of intestinal cells (Jarvis et al., 1995), thereby resulting in a different disease condition. Evolutionarily related to the flagellar apparatus, type-III secretion systems in gram-negative bacteria are both highly conserved and uniquely adapted as a virulence mechanism (Hueck, 1998). The core structural components of type-III systems, for example, are substantially conserved, but the effector proteins that are delivered into the host cell cytosol to modulate host cellular functions are unique for each bacterial species. Proteins comprising type-III secretion machineries are also encoded in the PAIs of *Yersinia*, *Shigella*, *Salmonella*, and a number of plant pathogens (e.g., *Erwinia* spp., *Xanthomonas campestris*, and *Ralstonia solanacearum*) (reviewed by Hacker and Kaper, 2000).

Another pathogenicity island, which fits the definition for PAIs described earlier, was discovered recently in *V. cholerae*, the causative bacterial agent of the diarrheal disease known as cholera. While both harmless aquatic strains and virulent strains of *V. cholerae* exist, chromosomal pathogenicity islands have only been identified in epidemic and pandemic strains (Karaolis et al., 1998). The *V. cholerae* pathogenicity island (VPI), which has markedly low %G + C content (35%) relative to the average genomic %G + C content (47–49%), is genetically stable and occupies a 39.5-kb region in the genome of the bacterial host. Sequence analysis revealed that VPI contains the TCP-ACF cluster and *toxT* gene, which is involved in the regulation of the cholera toxin (Karaolis et al., 1998). TCP or toxin-coregulated pilus, encoded by the *tcpA-tcpF* genes, is a type-IV pilus that acts as an essential adherence factor in the colonization of the intestinal epithelium (Cotter and DiRita, 2000; Herrington et al., 1988) and is coordinately regulated with the filamentous bacteriophage (CTXΦ)-encoded cholera toxin by the ToxR regulatory system (DiRita, 1992; DiRita et al., 1991; Taylor et al., 1987). Also encoded by the *V. cholerae* pathogenicity island is a second gene cluster, designated ACF for accessory colonization factor. Tn*phoA* insertions in *acf* genes generated *V. cholerae* mutants that showed reduced colonization ability in a mouse model (Peterson and Mekalanos, 1988). Besides containing genes with direct roles in cholera pathogenesis, the VPI carries putative integrase and transposase genes, which are important in the transfer and mobility of the VPI (Karaolis et al., 1998). Pathogenicity islands, therefore, can contribute substantially to the emergence of new epidemic and pandemic strains of *V. cholerae*.

H. pylori, a microaerophilic gram-negative bacterium, is the causative agent of acute and chronic gastritis and peptic ulcer disease in humans. Strains of *H. pylori* have been broadly classified in terms of their association with severe disease pathology (type I) or attenuated virulence (type II) based on the presence or absence of *cagA* (cytotoxin-associated gene A) (Censini et al., 1996; Xiang et al., 1995). In contrast to type-II strains, only *H. pylori* strains associated with severe forms of gastroduodenal disease (type-I strains) express the immunodominant antigen encoded by *cagA* (Xiang et al., 1995). The interaction of type-I *H. pylori* with human gastric epithelial cells is a complex one, resulting in the effacement of microvilli at the site of attachment, cytoskeletal rearrangements, and production of the cytokine interleukin 8 (IL-8) (Segal et al., 1996, 1997). The enhanced virulence of type-I *H. pylori* strains is due largely to the presence of a 40-kb foreign DNA region that contains *cagA* as well as other genes coding for disease-associated virulence factors. Analysis of the genetic locus containing *cagA* revealed that it is a pathogenicity island flanked by 31-bp direct repeats and inserted into the chromosomal glutamate racemase (*glr*) gene instead of a tRNA gene (Censini et al., 1996; reviewed by Hacker et al., 1997). Applying the methods for detecting PAIs described previously, the *cag* PAI showed the highest codon bias and was lower in its %G + C content (35%) compared to the rest of the genome (38–45%) (Karlin, 2001). The *cag* PAI encodes proteins showing similarities to components of specialized secretory pathways, including the type-IV secretion system that is involved in the export of virulence factors (Censini et al., 1996). The *cag* region of *H. pylori* illustrates how type-specific virulence factors can be acquired through chromosomal insertion of pathogenicity islands, leading to the differentiation of a more virulent type of bacterium within a genus (Censini et al., 1996).

Acquisition of Antibiotic Resistance The development of new vaccines and intervention therapies for infectious disease is a critical need now more than ever because of the increasing incidence of antibiotic-resistant strains of a number of important human pathogens, including pneumococci, enterococci, staphylococci, *Plasmodium falciparum*, and *M. tuberculosis* (Fauci, 2001; Fraser at al., 2000). A commonly recognized case of lateral gene transfer is the conferment of antibiotic resistance to a previously sensitive bacterium, thus expanding the microorganism's ecological niche and providing it with a survival advantage. Because of the selective advantages of antimicrobial resistance, genes conferring such traits tend to be associated with highly mobile genetic elements, usually plasmids, which are readily transferred between taxa and are often maintained extrachromosomally without elimination by segregation. The transfer of antibiotic resistance genes between bacterial genomes can also be mediated by transposition (i.e., transposable elements such as insertion sequences flanking the resistant determinant). Occasionally, antibiotic resistance genes are propagated by integrons, which are gene expression elements that drive the transcription of incorporated promoterless genes. Integrons are defined by three sequence elements: (1) an attachment site that serves as the site for integration of horizontally acquired sequences; (2) a gene encoding a site-specific recombinase (or integrase); and (3) a promoter that controls the expression of the incorporated sequences (Hall, 1997; Rowe-Magnus and Mazel, 1999). The mobilization of integrons requires insertion sequences, transposons, or conjugative plasmids.

Recently, microarray genomic technology (discussed in Chapter 6) provided insights into the pathogenicity and evolution of methicillin-resistant strains of *S. aureus*, which causes septicemia, endocarditis, and toxic shock syndrome in humans (Fitzgerald et al., 2001). A DNA microarray representing >90% of the genome of *S. aureus* strain COL was fabricated to investigate the genomic diversity, evolutionary genomics, and virulence gene distribution among 36 *S. aureus* clinical strains, including 11 methicillin-resistant strains, isolated from different human disease types and other mammalian infections. The results of this study indicated extensive genetic variation among a limited number of clonal lineages responsible for a large proportion of *S. aureus* infections (Fitzgerald et al., 2001). Approximately 22% of *S. aureus* genes were different among the strains compared and thus constituted strain-specific sequences, some of which encoded colonization factors for specialized hosts and other factors ensuring survival in certain environments. DNA microarray analysis also revealed the fundamental importance of lateral gene transfer in the evolution of pathogenic *S. aureus* (Fitzgerald et al., 2001). For example, in the case of methicillin resistance, the *mec* gene has been horizontally transmitted into distinct *S. aureus* genomes at least five times, indicating that methicillin resistance has evolved mutiple, independent times, instead of along a single ancestral line.

Loss of Genetic Information in Pathogen Evolution   It is evident from the comparison of genome sequences from pathogens and related nonpathogenic strains or species that bacterial virulence largely results from the acquisition of laterally transmitted virulence determinants that are absent from avirulent forms (Ochman and Moran, 2001). For example, benign *E. coli* strains have been converted into virulent forms by acquiring a single pathogenicity island (Groisman and Ochman, 1996; Hacker et al., 1997). However, as comparative sequence analysis of genomes suggests, the pathogenic nature of bacterial species is determined not only by the acquisition of virulence genes but also, in some cases, by the loss of certain genes that diminish pathogenic potential. For example, the gene encoding OmpT, a surface protease, acts as a virulence suppressor and is absent from *Shigella* (the causative agent of dysentery) but is present in closely related nonpathogenic *E. coli*. When the *ompT* gene is introduced into *Shigella* cells, the protein product attenuates virulence by disrupting the intercellular spreading ability of the organism (Nakata et al., 1993). Similarly, comparison of the *Shigella* flexneri and *E. coli K-12* genomes revealed a large deletion ("black hole") in the region of *cadA* in *Shigella* (Maurelli et al., 1998). The presence of a functional *cadA* gene, which encodes lysine decarboxylase, was found to be detrimental to the pathogenicity of *Shigella* spp. by inhibiting enterotoxin activity. The creation of genome deletions complements the horizontal spread of virulence genes, enabling commensal bacteria to evolve into pathogens (Maurelli et al., 1998).

12.3 COMPARATIVE GENOMICS: CLUES TO BACTERIAL PATHOGENICITY

Comparative analyses of sequenced microbial genomes have provided an unprecedented opportunity to explore the considerable amount of genetic variation and genomic plasticity underlying differences in the physiology, biochemistry, pathogenesis, and evolution of bacterial species. As discussed in the previous section, the elucidation of the total genomic content of organisms, ranging from bacteria to archaea to eukaryotes, has permitted the assessment of the overall impact of horizontal gene transfer on bacterial speciation. Completion of the *H. influenzae Rd* sequence in 1995 represented the first elucidated genome from a free-living organism (Fleischmann et al., 1995) and, in essence, launched the era of microbial genomics. Scientific focus shifted from single genes to the genome and how multiple, individual genes interact to produce complex phenotypic traits. The road for comparative genomics was paved with the reporting of the complete sequence of the *M. genitalium* genome (Fraser et al., 1995; Razin et al., 1998), followed shortly thereafter by the publication of the *H. influenzae* genome. Sequencing of the second

H. pylori genome (Alm et al., 1999) represented a landmark in comparative genomics, because it signified the beginning of a trend in sequencing the genomes of different strains of a single pathogenic species (Field et al., 1999). As more genomic data from phylogenetically diverse bacteria and closely related species are added to the existing collection of sequences, the impact of comparative genomics on studies of gene content, genomic organization, and gene acquisition (bacterial species evolution) is expected to be substantial, revealing not only basic principles but also unique features of individual bacteria. For example, comparison of whole-genome sequences from different organisms is likely to reveal informative patterns that provide insight into the fundamental distinctions between pathogens and commensals (Strauss and Falkow, 1997). Within-species comparisons of multiple strains are likely to reveal the genetic basis for differences in the ability to cause disease and disease severity (Whittam and Bumbaugh, 2002). In this section, we will discuss the contributions of comparative genomics to unraveling the molecular basis of microbial pathogenesis, including the identification of strain-specific genes that influence infection and the acquisition of such genes.

12.3.1 The Genomics of *Mycobacterium tuberculosis*: Virulence Gene Identification and Genome Plasticity

The tubercle bacillus is the etiological agent for tuberculosis, a chronic infectious disease that continues to pose a serious human health problem worldwide. Despite the intensive research efforts that have been devoted to the gram-positive bacterium *M. tuberculosis*, little is known about the molecular basis of its pathogenicity except that virulence is not associated with toxin production. Excessive cell-mediated immune and inflammatory responses, produced by the host in response to mycobacterial antigens, largely account for the pathology resulting from infection (Brosch et al., 2000). In recent years, the widespread emergence of drug-resistant strains of *M. tuberculosis* has created a critical need for new prophylactic and therapeutic strategies. Sequencing of the complete genome of *M. tuberculosis* (Cole et al., 1998) has allowed comparative and functional genomic studies of this significant human pathogen, and these studies, in turn, have accelerated progress in understanding the molecular basis of the pathogenesis, evolution, and phenotypic differences of mycobacteria. Further genomic studies using microarray technology and investigations of the mycobacterial proteome (discussed later in this chapter) will potentially enable the development of new and more effective therapies for preventing and treating this contagious airborne disease.

The complete genome sequence of the well-characterized *M. tuberculosis* strain H37Rv was determined using systematic sequence analysis of both selected large-insert clones [cosmids and bacterial artificial chromosomes (BACs)] and random small-insert clones from a whole-genome library prepared by the shotgun method (Cole et al., 1998). The 4.4-megabase-pair (Mb) circular chromosome of *M. tuberculosis* H37Rv contains approximately 4,000 predicted protein-encoding genes and exhibits a high $G + C$ content (65.6%). Interestingly, the high $G + C$ content remains relatively constant throughout the entire H37Rv genome, suggesting the absence of horizontally acquired pathogenicity islands of divergent base composition (Cole et al., 1998). Annotation of the genomic sequence indicated that precise functions could be attributed to 40% of the predicted proteins, while some functional similarity was indicated for 44% of the protein-coding genes. Functional descriptions could not be assigned for the remaining 16%, and these genes likely encode proteins of mycobacterial-specific functions. Sequence analysis also revealed that the *M. tuberculosis* H37Rv genome is rich in repetitive DNA, particularly insertion sequence (IS) elements, and duplicated housekeeping genes. IS elements are small DNA segments (< 2.5 kb) that can insert at multiple sites within a genome via a transposition reaction catalyzed by the IS-encoded transposase enzyme (Mahillon and Chandler, 1998). The H37Rv genome contains 56 loci with homology to IS elements belonging to at least nine different families, including a cluster of six IS elements that form a new family, designated the IS*1535* family (Gordon et al., 1999). Most of these insertion sequences appear to have inserted in intergenic or noncoding regions of the genome, with a high frequency of transposition occurring near tRNA genes. IS elements play a role in the horizontal transfer of virulence or drug-resistance genes, and IS-catalyzed chromosomal deletions may constitute important events that contribute to mycobacterial genome plasticity (Brosch et al., 1999; Gordon et al., 1999).

One remarkable feature that distinguishes *M. tuberculosis* from other bacteria is the presence of two new families of glycine-rich proteins with repetitive structures (Cole et al., 1998). Analysis of the genome sequence of *M. tuberculosis* H37Rv led to the identification of two large unrelated families of acidic, glycine-rich proteins, whose genes are clustered in the genome and occupy about 7% of the total coding sequence (Cole et al., 1998; Cole, 1999). These multigene families have been designated PE and PPE, signifying the presence of Pro-Glu (PE) and Pro-Pro-Glu (PPE) motifs found in the highly conserved *N*-terminal domains of the proteins. Both the PE and PPE proteins contain *C*-terminal segments that vary in size, sequence, and repeat copy number. The largest subfamily of the PE proteins, referred to as the polymorphic repetitive sequence (PGRS) class, is characterized by *C*-terminal multiple tandem repetitions of the motifs Gly-Gly-Ala or Gly-Gly-Asn. Members of the major polymorphic tandem repeat (MPTR) class of PPE proteins contain *C*-terminal segments rich in repeats with the signature Asn-X-Gly-X-Gly-Asn-X-Gly. The existence of these protein families was not known prior to the elucidation of the *M. tuberculosis* genome. Consequently, there is little information on the biological functions of PE and PPE proteins. The repetitive nature and variability exhibited by these proteins, however, suggest that they may constitute a major source of antigenic variation in the mycobacterial cell and serve as immunologically relevant antigens (Cole et al., 1998; Cole, 1999).

Prior to the completion of the genome sequence, only a small number of virulence factors in *M. tuberculosis* had been determined experimentally. These virulence determinants included genes encoding a macrophage-colonizing factor (*mce*), a sigma transcription factor (*sigA*), and a catalase-peroxidase (Arruda et al., 1993; Collins, 1996; Collins et al., 1995). However, the identity of the virulence factors that enable the tubercle bacillus to survive in macrophages and to induce the clinical manifestations of tuberculosis remains largely unknown. Genomics should expedite the identification of infection-related factors through predictive computational (bioinformatic) approaches or by genetic methods such as signature-tagged mutagenesis (Brosch et al., 2000), which will be discussed later in this chapter. Database homology searches have already uncovered a homolog of the *S. typhimurium* gene *smpB*, which has been implicated in intracellular survival (Baumler et al., 1994), homologs of the p60 secreted virulence factor of *L. monocytogenes*, and other virulence factor candidates in the form of phospholipases C, lipases, and esterases, which might degrade cellular or vacuolar membranes (Brosch et al., 2000; Cole et al., 1998). The genome sequence of *M. tuberculosis* also revealed two hemoglobinlike proteins that may be involved in oxidative stress protection and a flavohemoglobin (*hmp*) that may confer resistance to oxidative or nitrosative stress (Hu et al., 1999).

Genomes within the species *M. tuberculosis* have been compared using microarray technology for the purpose of investigating genetic variability within natural bacterial populations. An Affymetrix oligonucleotide GeneChip, representing all 3924 ORFs and 738 intergenic regions of *M. tuberculosis* H37Rv, was used to detect genomic deletions among 19 epidemiologically well-characterized clinical isolates of *M. tuberculosis* (Kato-Maeda et al., 2001). Compared to other bacterial pathogens, relatively little genetic variability has been detected within the species *M. tuberculosis*, and sequence-based analysis of the highly conserved *M. tuberculosis* complex suggests that single-nucleotide polymorphisms and horizontal gene exchange rarely contribute to genome plasticity. In the study by Kato-Maeda and colleagues (2001), none of the 16 *M. tuberculosis* clones examined by an array-based comparative genomics approach differed from the sequenced reference strain H37Rv by more than 38 ORFs. In comparison with H37Rv, clones were missing, on average, approximately 0.3% (13,248 bp) of the sequenced genome, and 25 different deleted sequences (76,839 bp in total) were detected (Kato-Maeda et al., 2001), suggesting that deletion events are a principal source of genetic variation in the evolution of mycobacterial pathogens. In addition, the work of Kato-Maeda and coworkers (2001) suggested that the pathogenicity of *M. tuberculosis* species tends to diminish as the amount of genomic deletions (or mutations in general) increases. Similarly, genomic analysis revealed variable loci between an attenuated tubercle bacillus and the fully virulent *M. tuberculosis* H37Rv that were the apparent result of IS-mediated genomic deletions from the H37Rv chromosome (Brosch et al., 1999). Such comparative genomic analyses offer unique perspectives on mycobacterial evolution and pathogenesis and, coupled with microarrays, serve as a suitable genotyping methodology for epidemiologic studies of *M. tuberculosis* infection (Brosch et al., 2001; Kato-Maeda et al., 2001).

12.3.2 Microarray-based Comparative Genomics of *Helicobacter pylori*

H. pylori is a gram-negative, flagellated bacterium that is implicated in a wide spectrum of pathologies ranging from chronic active gastritis to more severe gastroduodenal diseases, such as peptic and gastric ulcers, gastric cancer, and mucosa-associated lymphoid tissue (MALT) lymphoma (Cover and Blaser, 1996). *H. pylori* is distinguished among bacterial pathogens by its unusual ability to successfully colonize the gastric mucosa, an environment characterized by high acidity. It has been proposed that the wide variation in patient symptomology associated with *H. pylori* infection is at least partly attributable to the extensive strain-specific genetic diversity observed among *H. pylori* isolates (Atherton et al., 1997; Blaser, 1997). The entire genomic sequence of two independent *H. pylori* strains, 26695 and J99, has been elucidated using a whole-genome random (*shotgun*) sequencing method (Alm et al., 1999; Tomb et al., 1997). The genomes of 26695 and J99 consist of a single, relatively small circular chromosome of 1.67 and 1.64 Mb, respectively, with an average G + C content of 39%. Both strains also contain a 40-kb pathogenicity island, which encodes a bacterial type-IV secretory system responsible for the secretion and translocation of the CagA protein into host cells. In addition, genomic sequence analysis indicated a higher occurrence of the basic amino acids arginine and lysine in the *H. pylori* proteome (set of proteins encoded by the genome) relative to the proteomes of *H. influenzae* and *E. coli* (Tomb et al., 1997). Tomb and colleagues (1997) proposed that the greater frequency of positively charged amino acids might be a reflection of the adaptation of *H. pylori* to gastric acidity.

The sequencing of two *H. pylori* genomes from distinct isolates enabled genomic-sequence comparisons using bioinformatic and microarray-based methods. A computational comparative genomics approach revealed the existence of intraspecies variation, even though the overall genomic organization, gene order, and predicted proteins of strains 26695 and J99 were quite similar (Alm et al., 1999). Comparative sequence analysis also indicated that approximately 6 to 7% of the total gene content was specific to each strain (Alm et al., 1999). DNA restriction–modification system genes accounted for 15 to 20% of the J99- and 26695-specific genes, while genes with predicted functions in cell envelope synthesis, DNA transfer systems and competence, DNA replication, energy metabolism, and phospholipid metabolism constituted minor sources of strain variation (Alm et al., 1999). Genome sequence comparison of *H. pylori* strains 26695 and J99 indicated that they possess an unusually high number of proteins with predicted roles in type-II restriction–modification systems. In strain 26695, eleven restriction–modification systems were identified based on gene order and sequence similarity to known endonucleases, methyltransferases, and specificity subunits (Tomb et al., 1997). Recently, 22 type-II restriction endonucleases with 18 distinct functional specificities, of which three represented completely novel endonucleases, were identified among six different *H. pylori* strains (Xu et al., 2000). Type-II restriction–modification systems are defined by two separate enzymes: The restriction endonuclease recognizes specific DNA sequences and cleaves unmodified foreign DNA at a particular site, whereas the cognate methyltransferase modifies DNA within the same endonuclease recognition sequence, thus protecting endogenous DNA from endonucleolytic digestion (Wilson and Murray, 1991). Different strains of *H. pylori* possess extremely diversified restriction–modification systems, a characteristic that makes *H. pylori* unique (Lin et al., 2001; Xu et al., 2000). Although the set of endonucleases varies in different strains, biochemical analysis has demonstrated that all these enzymes specifically cleave DNA at four- or five-base recognition sequences (Xu et al., 2000). The biological significance of such an unusually high complement of restriction–modification systems is not entirely clear. However, the natural competence of *H. pylori* (Suerbaum et al., 1998) suggests that the large number of restriction–modification systems prevents the integration of exogenous (foreign) DNA into the host genome (Xu et al., 2000).

Genome-wide surveys of genetic diversity among different bacterial strains are facilitated by DNA microarray technology. In another comparative genomics study, high-density DNA microarrays were used to examine the genome composition of 15 *H. pylori* clinical isolates of varying virulence (Salama et al., 2000). In cases where both sequenced *H. pylori* strains 26695 and J99 contained the same ORF, strain 26695 was selected as the reference strain for microarray fabrication. Ninety-one additional ORFs present only in strain J99 were also represented on the array. Of the 1,643 genes analyzed, a minimal functional core of 1,281 genes was derived (Salama et al., 2000). These genes were common to all the strains evaluated and encoded primarily metabolic, biosynthetic, and regulatory functions. Most noteworthy was the finding that at least 12 to 18% of each strain's genome was composed of strain-specific genes, of which those of unknown function constituted the most abundant class (Fig. 12.1). The two largest classes of strain-specific genes with known biological function were restriction–modification system components and transposase genes, which play important roles in regulating the exchange of DNA between bacteria and likely promote the genetic diversification of *H. pylori* strains

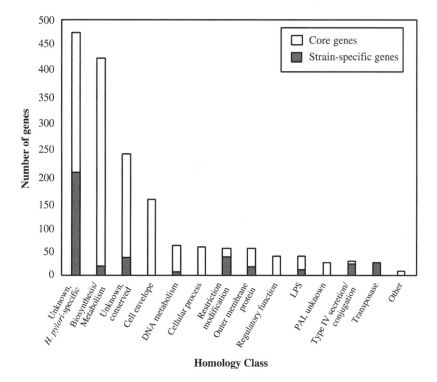

Figure 12.1 *Distribution of* Helicobacter pylori *core and strain-specific genes among the different functional classes using microarray-based comparative genomics of 15* H. pylori *clinical isolates (Salama et al., 2000).* Each bar represents the number of strain-specific (gray) and core genes (white) found for the indicated homology class. [Reproduced with permission from N. Salama, K. Guillemin, T.K. McDaniel, G. Sherlack, L. Tompkins, and S. Falkow, A whole-genome microarray reveals genetic diversity among *Helicobacter pylori* strains, *PNAS*, vol. 97, 14668–14673. Copyright (2000) National Academy of Sciences, U.S.A.]

(Salama et al., 2000). Other strain-specific genes encoded various outer membrane proteins, lipopolysaccharide synthesis proteins, as well as three *virB4* homologs and a *traG* homolog, both ATPases involved in the assembly of type-IV secretion structures (Salama et al., 2000). Hierarchical clustering analysis (see Chapter 7 for a general discussion) of the strain-specific genes identified a group of new potential virulence factors that showed coinheritance with the *H. pylori* pathogenicity island and may modulate PAI function or act synergistically with the PAI to cause persistence or disease in the host (Salama et al., 2000). The fact that the strain-specific DNA restriction–modification genes and other strain-specific genes have a lower G + C content than the remainder of the 26695 or J99 genome (35% vs. 39%, respectively) suggests that these genes may have been acquired more recently from other bacterial species through horizontal gene transfer (Alm et al., 1999). A comparative genomics study of the restriction–modification systems in *H. pylori* showed that all the strain-specific restriction–modification genes are active, whereas those genes that are conserved among strains are functionally inactive, which supports the notion that the strain-specific functional restriction–modification genes have been acquired more recently through horizontal gene transfer (Lin et al., 2001). Studies on the comparative genomics of *H. pylori* will continue to identify the strain-specific gene content of different

H. pylori isolates, and differences in strain-specific genes are likely to explain differences in strain evolution, adaptation to genetically diverse hosts, and disease outcomes.

The availability of complete genomic sequence information for *H. pylori* 26695 and J99 permitted a sequence-based search for shared families of genes predicted to encode outer membrane proteins (Alm et al., 2000). Such surface-accessible proteins are likely to be involved in the adaptation of *H. pylori* to the unique gastric environment and often represent the most significant antigenic determinants of a particular bacterial species. Based on comparative sequence analysis, five paralogous gene families, comprising approximately 4% of each strain's protein-coding potential, were identified in the genomes of J99 and 26695. Paralogs are genes related by duplication within a genome and encode proteins expected to possess similar biochemical functions but distinct biological roles. These families include the Hop outer membrane proteins, whose members have been implicated as adhesins (Odenbreit et al., 1999; Peck et al., 1999); the newly identified Hom outer membrane proteins, which have conserved *N*- and *C*-termini flanking a variable central domain (Alm et al., 2000); iron-regulated outer membrane proteins (e.g., FecA-like proteins); and efflux pump outer membrane proteins. Two of the families (Hop and Hom) contained members that were specific for either *H. pylori* J99 or 26695. Genomic sequence analysis suggests that the expression of a number of genes may be regulated by slipped-strand repair (see Section 12.2.2) at either homopolymeric tracts or dinucleotide repeats, resulting in antigenic variation and adaptive evolution (Alm et al., 1999; Tomb et al., 1997). For example, the same members of the large *hop* paralogous gene family in both strains contain CT dinucleotide repeats in their signal sequence; however, the number of repeats differs without affecting the predicted expression status (Alm et al., 1999).

12.3.3   Comparative Analysis of the *Borrelia Burgdorferi* and *Treponema pallidum* Genomes

The two pathogenic spirochetes *Borrelia burgdorferi* and *Treponema* pallidum are moderately related bacteria that have a common ancient ancestry, similar morphologies, small genomes (1.5 and 1.1 Mb, respectively), and approximately equivalent protein-coding capacities but have diverged over time, driven possibly by adaptation to different biological niches (Porcella and Schwan, 2001; Subramanian et al., 2000). A result of this divergence is the clinical manifestation of distinct chronic diseases in the human host. *B. burgdorferi* is the causative agent of Lyme disease, whereas *T. pallidum* causes venereal syphilis. However, much remains to be discovered concerning the specific mechanisms contributing to disease and persistent, long-term infections caused by both pathogenic spirochetes (Porcella and Schwan, 2001).

The sequencing and subsequent comparative bioinformatic analysis of the *B. burgdorferi* genome (Fraser et al., 1997) and *T. pallidum* genome (Fraser et al., 1998) have already uncovered striking disparities in chromosomal structure, lipoprotein content, the core functions of DNA repair, and more specialized functions, such as signal transduction, metabolism, and host and environment response. These differences at the genomic level underscore the physiological and adaptation strategies unique to each spirochete. In contrast to *T. pallidum*, the genomic structure of *B. burgdorferi* is complex, containing a linear chromosome of 910,725 bp (with an average $G + C$ content of 28.6%) and 21 extrachromosomal elements, including 12 linear and 9 circular plasmids (Casjens et al., 2000; Fraser et al., 1997). Interestingly, the majority ($> 90\%$) of the predicted open reading frames residing on the *B. burgdorferi* plasmids show no similarity to any known bacterial sequences (Casjens et al., 2000). These novel genes likely contribute to the organism's unusual ability to survive a life cycle that alternates between existence in warm-blooded animals and cold-blooded ticks (Porcella and Schwan, 2001). The more simplistic genomic structure of *T. pallidum* consists of a single circular chromosome of 1,138,006 bp (with an average G + C content of 52.8%) and no extrachromosomal elements (Fraser et al., 1998). Comparative sequence analysis of the two genomes revealed a proportionately greater coding capacity for lipoproteins in *B. burgdorferi* compared to *T. pallidum* and other bacteria (Porcella and Schwan, 2001). Lipoprotein-encoding genes comprise 5% of the total chromosomal ORFs predicted in *B. burgdorferi*, whereas 2.1% of the total coding capacity in *T. pallidum* and 1.3% of that in *H. pylori* represent lipoprotein genes. In addition, the cell surface-exposed proteins in *B. burgdorferi* are largely lipoproteins in contrast to *T. pallidum*, for which no lipoproteins have been definitively localized to the outer surface. The substantial disparity in the number and cellular location of lipoproteins produced by the two pathogens is consistent with their different modes of parasite–host interaction and may reflect the diverse host adaptation of *B. burgdorferi* (Porcella and Schwan, 2001; Subramanian et al., 2000).

Local sequence similarity searches, profile searches, and analysis of individual domains and protein families were used in another bioinformatic study of the *B. burgdorferi* and *T. pallidum* genomes. The findings of this comparative analysis provided insights into the evolutionary trends and adaptive strategies of the two spirochetes (Subramanian et al., 2000). For example, the absence of a major DNA repair protein, the proofreading 3′-5′ exonuclease (the ε subunit) of DNA polymerase III, in the repair protein repertoire of *B. burgdorferi* suggests that the organism may require error-prone replication to generate antigenic variation, which is commonly used by bacterial pathogens to elude host immune defenses. Other conspicuous differences between *B. burgdorferi* and *T. pallidum* were found among genes involved in transcription regulation (Subramanian et al., 2000). Three genes encoding transcription sigma factors 24 (RpoE), 28, and 43 (SigA) are present in *T. pallidum* but not in *B. burgdorferi*. The implication of this finding is that the additional sigma factors are used to regulate the expression of *T. pallidum*-specific gene sets. Noteworthy differences were also identified in the metabolic processes inferred from the genomic sequences. In contrast to *T. pallidum*, genes required for the utilization of glycerol (e.g., glycerol uptake genes and the gene encoding FAD-dependent glycerol-3-phosphate dehydrogenase) and chitin (genes encoding glucosamine deacetylase and glucosamine deaminase) were found in *B. burgdorferi*. Such metabolic capabilities may be indicative of *B. burgdorferi*'s adaptation to survival in an arthropod host, in which chitin would be available as a substrate for cell wall biosynthesis or as an energy source (Subramanian et al., 2000).

Protein profile searches disclosed the presence of previously undetected components of the signal transduction machinery and novel proteins containing domains that had been detected only in eukaryotes (Subramanian et al., 2000). Three von Wille-brand A factor (vWA) domain-containing proteins (Ponting et al., 1999) were identified in each of the spirochetes. The vWA domain, which is a Mg^{2+}-binding protein module, participates in adhesion and protein–protein interactions in a variety of eukaryotic proteins. By analogy, the secreted or membrane-associated vWA-domain proteins may be involved in the adhesion of *B. burgdorferi* and *T. pallidum* to the extracellular matrix or to cells of the host connective tissues (Subramanian et al., 2000). Another hitherto undetected protein of *B. burgdorferi* was identified by genomic sequence analysis. This secreted or periplasmic protein contains a PR1 domain, which has been primarily found in plant pathogenesis-related proteins and in proteins expressed in the animal immune system (Szyperski et al., 1998). Conversely, an unusual secreted (periplasmic) protein containing an oligonucleotide/oligosaccharide binding (OS)-fold domain (Murzin, 1993) was identified only in *T. pallidum* based on predicted protein sequences encoded in the genome. These species-specific proteins may participate in extracellular protein–protein interactions (PR1-domain proteins) or mediate adhesion of the spirochete to host cells and function as virulence factors (OB-fold-containing proteins).

12.3.4 Sequence Comparison of Pathogenic and Nonpathogenic Species of *Listeria*

Listeriosis is a severe opportunistic food-borne disease caused by the gram-positive, facultative intracellular bacterium *L. monocytogenes*. The pathogen normally enters the host by ingestion of contaminated food, after which the bacteria cross the gastrointestinal barrier and disseminate via the lymph and the blood to distant tissues (Cossart and Lecuit, 1998). *L. monocytogenes* infects host cells by macrophage-mediated phagocytosis or by inducing its own phagocytosis (invasion) into cells that are normally benign. Residence within the membrane-bound phagocytic vacuoles occurs for approximately 30 minutes before the membrane is lysed, releasing the bacterium into the cell cytoplasm where it replicates and spreads cell to cell using actin-based motility. Listerial infections of the central nervous system, which appear clinically as meningitis or meningoencephalitis, are of critical importance because of the high mortality rate. One approach to investigating the genetic basis of virulence is to take advantage of naturally occurring differences in bacterial pathogenicity between variants of the same species or between closely related species (Tinsley and Nassif, 1996). To better understand the organism's pathogenicity and ability to colonize and grow in diverse ecosystems, the genomes of *L. monocytogenes* and the noninvasive, nonpathogenic species *Listeria innocua* were both sequenced by the whole-genome random sequencing method and compared (Glaser et al., 2001). The aim of this comparative genomics study was to identify genomic regions in *L. monocytogenes* that do not have a counterpart in *L. innocua* and that therefore may determine factors responsible for the pathogenesis of listeriosis.

As illustrated in Fig. 12.2, *L. monocytogenes* and *L. innocua* are similar in terms of genome size (2.9 and 3.0 Mbp, respectively) and G + C content (39 and 37%, respectively). The distribution of the 270 *L. monocytogenes*-specific and 149 *L. innocua*-specific genes within the various functional categories is presented graphically in Fig. 12.3. Genomic sequence comparison revealed that 23 of the 86 genes encoding secreted proteins in the pathogenic *L. monocytogenes* are absent from the genome of the nonpathogenic *L. innocua*. Three of the secreted proteins missing in *L. innocua* are soluble internalins, invasion proteins of which the surface-exposed internalins (InlA and InlB) have been shown to be necessary for the entry of *L. monocytogenes* into mammalian cells (Cossart and Lecuit, 1998). Genomic sequence analysis confirmed that the 10-kb virulence gene cluster of *L. monocytogenes* is absent from the nonpathogenic species. Present in this locus are genes encoding virulence-related determinants that promote escape from the phagocytic vacuole (LLO and PlcA) and are required for intracellular actin-based movement and cell-to-cell spread (ActA and PlcB). Interestingly, the *Listeria* virulence gene cluster differs from other well-characterized pathogenicity islands, because it is relatively small in terms of sequence length and has a G + C content similar to the rest of the chromosome (Cossart and Lecuit, 1998). Comparison of the region containing the

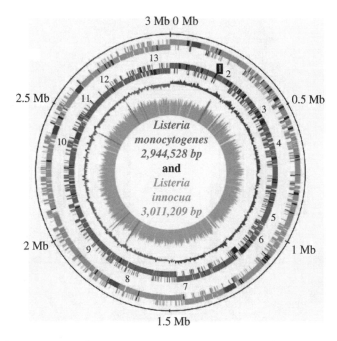

Figure 12.2 *See color insert. Circular genome maps of* listeria monocytogenes *and* listeria innocua *(Glaser et al., 2001).* Circles 1 and 2 (from the outside) represent the genomes of *L. innocua* and *L. monocytogenes*, respectively. *L. innocua* genes are shown in green; *L. monocytogenes* genes are shown in red. Black indicates genes that are specific for *L. monocytogenes* or *L. innocua*. The number 1 indicates the position of the virulence locus, prfA-plcA-hly-mpl-actA-plcB. [Reprinted with permission from P. Glaser et al., Comparactive genomics of *Listeria* species, *Science*, vol. 294, 849–852. Copyright (2001) American Association for the Advancement of Science.]

virulence gene cluster of *L. monocytogenes* and the homologous regions of the *L. innocua* and *B. subtilis* genomes suggested that the virulence gene cluster was likely acquired by a common ancestor of *Listeria* (Glaser et al., 2001).

Another important difference in gene content revealed by genomic sequence comparison was the absence of *prfA* in *L. innocua*. PrfA is a transcriptional regulatory protein in *L. monocytogenes* that activates virulence genes by binding specifically to recognition sequences in promoter regions (Vazquez-Boland et al., 2001). Sequence analysis identified palindromic PrfA recognition sequences preceding a number of genes in both genomes, suggesting that at some point in the evolution of *L. innocua*, this critical virulence determinant was lost. Recently, microarray-based transcriptome analysis comparing wild-type *L. monocytogenes* and a *prfA* deletion mutant indicated that PrfA positively regulates a core set of 12 genes, including two previously unknown genes, and negatively controls a second group of genes, thus suggesting that PrfA can act as an activator or a repressor (Milohanic et al., 2003).

Also included in the 270 *L. monocytogenes*-specific genes were three genes putatively involved in the degradation of bile salts, which possibly equip the bacterium with the capacity to survive in the mammalian gut (Glaser et al., 2001). One of these genes contained a PrfA-box in its upstream region. This kind of bioinformatic analysis of sequenced genomes can be used to identify genes of potential significance to pathogenesis.

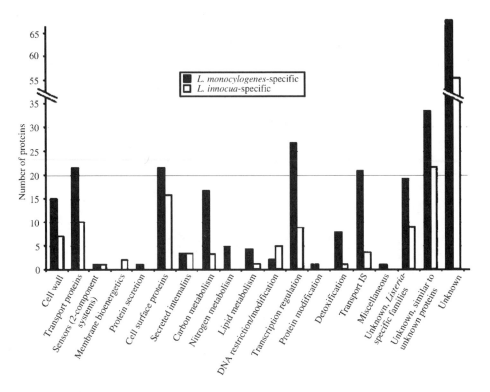

Figure 12.3 *Distribution of the 270* Listeria monocytogenes-*specific genes and the 149* L. innocua-*specific genes within the different functional categories.* [Reprinted with permission from P. Glaser et al., Comparative genomics of *Listeria* species, *Science*, vol. 294, 849–852. Copyright (2001) American Association for the Advancement of Science.]

In addition, sequence analysis of the two *Listeria* genomes revealed that a large proportion of the proteins with virulence-associated functions reflected species-specific properties of *L. monocytogenes* in contrast to *L. innocua*. This study is an excellent example of the application of a *species-filter* approach (Field et al., 1999) to sequence data mining in order to define shared and unshared genes. By subtracting all gene homologs of a pathogenic species from a nonpathogenic species, the contents of a species-specific gene pool are revealed, providing insight into what makes a certain bacterial species or strain pathogenic.

12.3.5 Comparative Genomics of *Chlamydia pneumoniae* and *Chlamydia trachomatis*: Two Closely Related Obligate Intracellular Pathogens

In this section, the comparative analysis of genomes from the closely related species *Chlamydia pneumoniae* (Kalman et al., 1999) and *Chlamydia trachomatis* (Stephens et al., 1998) is used to illustrate the contribution of comparative genomics to extending our knowledge of bacterial pathogenesis. A similar comparative study has been conducted for sequenced genomes of *Mycoplasma pneumoniae* and *M. genitalium* (Himmelreich et al., 1997).

C. pneumoniae and *C. trachomatis* are phylogenetically related obligate intracellular pathogens that differ significantly in their tissue tropism and disease outcomes. *C. pneumoniae* causes pneumonia and bronchitis in humans, while *C. trachomatis* infection results in trachoma, an ocular infection that leads to blindness (Kuo et al., 1995; Schachter, 1990). Comparison of the 1.2-Mb *C. pneumoniae* genome with the 1.0-Mb *C. trachomatis* genome suggested common biological processes required for infection and survival in mammalian cells and revealed genomic differences, which likely are responsible for the unique properties that differentiate the two species in terms of their pathogenesis (Kalman et al., 1999). Computational comparison of the chlamydial genomes revealed a high level of functional conservation between the two bacteria, with 80% of the predicted ORFs for *C. pneumoniae* identified as orthologs to *C. trachomatis* genes (Kalman et al., 1999). The central metabolic pathways and fundamental mechanisms of transcription regulation inferred from the genomic sequences are the same in both *C. pneumoniae* and *C. trachomatis*. In addition, genomes of both of these obligate intracellular pathogens contained a type-III secretion virulence system, which is required for invasion by other pathogenic bacteria. Genomic analysis also indicated that 214 protein-coding sequences in *C. pneumoniae* were not found in *C. trachomatis*, and 186 of the 214 genes did not have any detectable homologs in other bacteria (Kalman et al., 1999). Similarly, 60 of the 70 *C. trachomatis* genes that lacked an identifiable counterpart in *C. pneumoniae* showed no sequence similarity to any other known protein. Genes encoding a new family of chlamydial polymorphic membrane proteins (Pmp) represented 22% of this increased coding capacity of *C. pneumoniae*. The biological function of the Pmp protein family, however, is not known. The unique genes were predicted to be essential for the specific traits that differentiate the biology, tropism, and pathogenesis of *C. trachomatis* and *C. pneumoniae* (Kalman et al., 1999). Results obtained from studies like this one will generate new avenues of experimental research, identifying new potential infection-related genes for functional characterization.

12.4 DISCOVERY OF NOVEL INFECTION-RELATED GENES USING SIGNATURE-TAGGED MUTAGENESIS

As illustrated in the preceding section, comparative *in silico* bioinformatic analysis of genome sequences provides valuable insights into the pathogenesis, evolution, adaptive strategies, and genetic variability of bacterial pathogens. In computational approaches to genome analysis, gene function is inferred directly from nucleotide sequences or conserved motifs indicative of particular cellular functions. However, sequence annotation of microbial genomes typically ascribes an unknown or hypothetical function to a large fraction (in excess of 30%) of the total predicted ORFs. Novel genes important for bacterial pathogenicity, therefore, can be missed based on sequence similarity alone. As a result, a number of functional genomic techniques based on transposon mutagenesis have emerged to complement the genome sequencing of bacterial pathogens (Lehoux et al., 2001). One method, termed signature-tagged mutagenesis (STM), is used to globally screen bacterial genomes for genes essential to or critical for virulence in different bacterial hosts (Shea and Holden, 2000). The STM strategy, in particular, has been lauded as a method for investigating virulence genes (Pelicic et al., 1998). The technique of STM is discussed in detail in Chapter 8. Briefly, in STM each transposon mutant of a bacterial pathogen is tagged with a unique DNA signature sequence [polymerase chain reaction

(PCR)-amplifiable oligonucleotide region], which permits identification of large numbers of bacteria by hybridization following negative selection in an animal model (Hensel et al., 1995). A mutant expressing attenuated virulence in a living host is selected by its failure to be recovered from the host. The application of STM to *S. typhimurium* (Hensel et al., 1995), for instance, led to the discovery of the *Salmonella* pathogenicity island 2 (SPI2), a chromosomal region encoding a second type-III secretion system that is critical for virulence (Shea et al., 1996). It is interesting that STM is essentially a "pregenome" method for the global investigation of gene functionality that is being used in a postgenomic era (Saunders and Moxon, 1998). Although genome sequence information can aid and accelerate STM, performance of the method does not require such data, and, in fact, the technique was pioneered before the availability of complete genomic sequences.

The encoded functions of virulence-associated genes have been classified into three main categories: (1) bacterial factors required for host and tissue tropism; (2) factors required for survival and replication of the pathogen within the host, including those critical for the evasion of host immune defenses and nutrient synthesis/acquisition; and (3) factors involved in host toxicity (Shea and Holden, 2000). Although the first two classes of virulence factors have been identified using STM, the third category of factors (e.g., toxins) is more difficult to identify by this method, because such functions can be potentially transcomplemented by the presence of other strains (Shea and Holden, 2000). Genome-wide STM screens for virulence genes have been performed in a number of bacterial pathogens, including *S. aureus*, *S. pneumoniae*, *V. cholerae*, *Yersinia enterocolitica*, *Legionella pneumophila*, *L. monocytogenes*, *P. aeruginosa*, and others (Hamer et al., 2001a; Shea and Holden, 2000). A better understanding of the genes essential for bacterial pathogenicity could potentially lead to the development of novel therapeutic and preventive strategies for infection. In this section, we discuss how the STM method has been successfully used to identify virulence factors in four bacterial pathogens: *V. cholerae*, *S. aureus*, *E. coli K1*, and *S. pneumoniae*.

12.4.1 *Vibrio cholerae* Genes Critical for Colonization

Colonization of the host upper intestinal milieu by *V. cholerae* initiates the pathogenesis of cholera and requires a fimbriallike factor called toxin coregulated pilus (TCP) (Attridge et al., 1993). Clinical manifestations of the disease (i.e., profuse secretory diarrhoea) are caused by the cholera toxin, which is produced by the pathogen during infection. Previously, the search for additional colonization factors was severely hampered by the impracticality of screening large numbers of individual random mutants to determine whether a specific genetic change resulted in deficient colonization. Transposon-based approaches such as STM, however, have greatly facilitated the search for bacterial virulence determinants.

STM was used to screen for random insertion mutants of *V. cholerae* that affected colonization in the suckling mouse model for cholera (Chiang and Mekalanos, 1998). A number of genetic loci critical for colonization were identified. Of the approximately 1,100 mutants screened, five colonization-defective mutants were isolated with transposon insertions in TCP biogenesis genes, thus confirming the ability of STM to identify mutants of attenuated virulence (Chiang and Mekalanos, 1998). More important, colonization-defective mutants were isolated that contained transposon insertions in homologs of genes that had not been previously implicated in cholera pathogenesis and in genes with little or no similarity to any known genes. STM identified lipopolysaccharide, biotin, and purine biosynthetic genes as contributors to successful *V. cholerae* colonization. Other loci implicated in colonization included two genes involved in phosphate transfer (homologs of *pta* and *ptfA*), a homolog of *E. coli yabN*, *mgtE*, and a gene of unknown function based on sequence comparisons. Chiang and Mekalanos hypothesized that phosphate transfer may modulate expression of the ToxR regulon, which is involved in regulating the *tcp* (toxin coregulated pilus) operon at the transcriptional level. Based on sequence homology to proteins of known function, they also postulated that YabN might act to protect *V. cholerae* against bacteriocidal substances in the gut, whereas MgtE, as a Mg^{2+} transporter, might regulate virulence determinants in *V. cholerae* (Chiang and Mekalanos, 1998). With STM, both novel genes and genes with a previously unrecognized functional role in the regulation of known virulence determinants were shown to be necessary for successful colonization by *V. cholerae*.

12.4.2 Virulence Genes of *Staphylococcus aureus* Infection

S. aureus is a gram-positive bacterium that causes a wide range of disease states (e.g., endocarditis, arthritis, cutaneous infections, and pneumonia) and infects a variety of different host tissues (Sherris and Plorde, 1990). *S. aureus* infections are typically established via a bacteremic stage in which the pathogen must survive and grow in the blood for infection to progress. To understand the functions of genes crucial for survival and replication in the blood, STM was applied to *S. aureus*, using a murine model of bacteremia to select for mutants attenuated in virulence (Mei et al., 1997). Of the 50 virulence-deficient mutants isolated, approximately half possessed transposon insertions in genes of unknown function, while the remainder had mutations in genes affecting nutrient biosynthesis and cell surface metabolism. The disruption of genes affecting tryptophan and purine biosynthesis had a deleterious effect on virulence (Mei et al., 1997). This could be explained by limitations in the nutritional environment of the host, which requires the pathogen to synthesize de novo various amino acids, cofactors, and nucleotides (Groisman and Ochman, 1994; Mei et al., 1997). In another study using the STM approach, a gene encoding a proline permease important for proline uptake was found to contribute to the in vivo survival of *S. aureus* in animal models (Schwan et al., 1998). Other genes critical for survival and the proliferation of *S. aureus* in blood are those that show sequence similarity to genes involved in diaminopimelic acid (a component of peptidoglycan) synthesis, surface adhesin integrity, membrane transport, and lipoprotein modification (Mei et al., 1997). An unexpected finding was the effect of mutations in tricarboxylic acid (TCA) cycle enzymes on virulence. Such mutations may have an indirect effect on virulence by affecting capsule production, which occurs during certain stages of *S. aureus* growth and requires respiratory activity (Dassy and Fournier, 1996; Mei et al., 1997). Finally, STM implicated a number of functionally unknown genes in *S. aureus* infection. Unlike purely bioinformatic approaches, STM studies like this one can provide the basis for the further physiological and genetic characterization of genes of unknown or hypothetical function in order to define their specific roles in bacterial pathogenesis.

12.4.3 *Escherichia coli* K1: Identification of Invasion Genes

E. coli K1 continues to be a leading cause of gram-negative neonatal meningitis in humans, primarily because of our limited understanding of the molecular basis of *E. coli* K1 pathogenesis. To identify genes that contribute to the invasion of human brain microvascular endothelial cells in the blood–brain barrier, STM was applied to *E. coli* K1 (Badger et al., 2000). Seven STM mutants that exhibited loss or decreased invasion of endothelial cells were isolated, including five mutants with disruptions in previously uncharacterized loci. The gene products encoded by *traJ*, responsible for the positive regulation of DNA conjugation genes, and *cnf1*, a gene previously demonstrated to play a role in bacterial invasion, also contributed to the *E. coli* K1 invasion of human brain microvascular endothelial cells. The protein product of *cnf1* is a cytotoxic necrotizing factor, a toxin that has been shown to induce bacterial phagocytosis in epithelial cells and is produced by several pathogenic *E. coli* strains (Falzano et al., 1993). This STM study provided the first in vivo evidence of the involvement of *traJ* in the pathogenesis of *E. coli* K1 meningitis. In addition to identifying genes necessary for tissue culture cell invasion, the STM technique has been used to discover novel virulence factors required for intestinal colonization (Martindale et al., 2000) and systemic infection (Gonzalez et al., 2001) by *E. coli* K1.

12.4.4 Diverse Genes Implicated in *Streptococcus pneumoniae* Virulence

S. pneumoniae is the primary causative agent for bacterial pneumonia and is responsible for such infections as otitis media and meningitis in children. Despite the fact that many years of research have been devoted to *S. pneumoniae*, the virulence factors of this pathogen are not completely understood (DeVelasco et al., 1995). Therefore, with the recent development of STM and other technical advances in genomics, the focus of studies on *S. pneumoniae* pathogenesis has shifted to the identification of virulence genes on a genome-wide scale. The STM technique was adapted to *S. pneumoniae* for the large-scale identification of virulence genes using selection of virulence-attenuated mutants in a mouse pneumonia model and mouse septicemia model (Polissi et al., 1998). The 126 putative virulence genes identified by this methodology could be grouped into six categories: (1) genes corresponding to previously described pneumococcal virulence factors; (2) genes involved in metabolic pathways; (3) genes encoding proteases; (4) genes coding for ATP-binding cassette transporters; (5) genes involved in DNA recombination/ repair; and (6) genes of unknown function, as listed in Table 12.3 (Polissi et al., 1998). Some of these genes exhibited sequence similarity to virulence genes found in other bacteria. Negative in vivo screening, for example, identified a virulence-defective mutant with an insertion in the gene coding for an IgA1 protease, which was previously shown to be important for the survival of *Neisseria gonorrhoeae* within epithelial cells due to its ability to cleave lysosomal integral membrane proteins (Lin et al., 1997; Polissi et al., 1998).

A major class of mutants displaying attenuated virulence included those with disruptions in genes involved in metabolic pathways and nutrient uptake, indicating that host adaptation is a key component of bacterial pathogenicity (Polissi et al., 1998). Nucleotide-limiting conditions in the host impose a requirement for de novo synthesis of purines by pathogens. This aspect of pathogenesis is reflected in the isolation of mutants with insertions in genes comprising the purine biosynthetic pathway (*purE, purK, purC,* and *purL*). Other mutations were isolated in genes involved in glutamine metabolism, suggesting that reduced glutamine synthesis or loss of glutamine transport results in virulence attenuation in *S. pneumoniae*. The implication of proteases and DNA recombination/repair proteins in virulence emphasizes the multifactorial nature of *S. pneumoniae* pathogenesis and points to the importance of genes that confer to the

TABLE 12.3   Predicted Genes of Unknown Function Implicated in *S. pneumoniae* Virulence by STM (Data from Polissi et al., 1998)

Strain	Gene[a]	Similar to (source and/or product)	Septicemia[b]
SPN-1031		*yejD (Escherichia coli)*	High
SPN-1101		*ykrA (Bacillus subtilis)*	Medium
SPN-1119		*PHBQ002 (Pyrococcus horikoshii)*	High
SPN-113		*TPTC (Streptococcus crista, ABC protein)*	High
SPN-1145		*MTCY21C12.02 (Mycobacterium tuberculosis, putative cation-transporting ATPase)*	High
SPN-1200		*Dictyostelium discoideum* α-L-fucosidase precursor	High
SPN-1338		No homology	Medium
SPN-1471		*iga (Streptococcus sanguis)*	High
SPN-1583		*yqeT (B. subtilis, probable methyltransferase)*	Medium
SPN-162		No homology	Low
SPN-1631		*ydeG (Schizosaccharomyces pombe)*	High
SPN-1808		*BB0831 (Borrelia burgdorferi,* xylose operon regulatory protein)	High
SPN-1818		*ydhP (B. subtilis)*	ND
SPN-224		*yidA (E. coli)*	High
SPN-627		*ydiL (B. subtilis)*	High
SPN-631		*ypsA (B. subtilis)*	High
SPN-633	*ypoA*		High
SPN-636		*radA (B. subtilis,* possible DNA repair protein)	ND
SPN-655		*ORF8 (Enterococcus faecalis)*	Medium
SPN-233		*ybhL (E. coli,* hypothetical membrane protein)	High
SPN-634		No homology	High
SPN-641		*yitS (B. subtilis)*	Medium
SPN-962		*N5,N10*-Methylenetetrahydromethanopterin reductase gene *(Staphylococcus carnosus)*	High
SPN-1297		*pstB (Methanococcus jannaschii,* hypothetical ABC transporter)	High

[a]Gene already identified in *S. pneumoniae.*
[b]To further evaluate the attenuated virulence of individual mutants, clones were individually analyzed in a mouse septicemia model. Those mutants showing the same virulence as the wild-type *S. pneumoniae* strain G54 were classified as high, whereas those showing the same virulence attenuation as the nonencapsulated PR212 strain were classified as low. Mutants falling in between the two classes were defined as medium.

pathogen the ability to survive and grow in a specific host environment (Polissi et al., 1998).

12.5   APPLICATION OF MICROARRAYS TO DELINEATING GENE FUNCTION AND INTERACTION

We have already described the impact of genomic sequence information and comparative genomics on the field of bacterial pathogenesis in terms of identifying new potential infection-related genes, assessing the contribution of horizontal gene transfer to bacterial evolution and diversification, and elucidating intra- and interspecies genomic variation. However, sequence alone cannot provide information about whether putative virulence factors are actively expressed and under what circumstances or how they are regulated in the cell, nor can it significantly advance our knowledge of how genes interact to form complex traits such as virulence. The value of microarrays in discovering genes involved in complex human diseases, such as hereditary diseases and cancer, is already beginning to grow in importance (Meltzer, 2001). In addition, microarrays are a promising high-throughput, diagnostic tool for the detection and discrimination of human pathogens based on hybridization patterns to genes encoding antigenic determinants and known virulence factors (Chizhikov et al., 2001). Microarray-based research has already demonstrated that the power of this high-throughput, parallel technology lies in generating new biological hypotheses, directing scientists to unexpected or unpredicted relationships between diverse genes, and implicating genes of unknown function in cellular processes (Barry and Schroeder, 2000). DNA microarrays, in conjunction with other genetic tools such as signature-tagged mutagenesis (see Section 12.4), will help to unravel the biology of pathogenic bacteria by revealing differential gene expression, the dynamics of the host–pathogen interaction, and new potential drug targets. This section discusses some of the important ways in which DNA microarray technology is bridging the gap between genome and biology.

12.5.1 Exploring the Transcriptome of Bacterial Pathogens

Most bacterial virulence factors are tightly regulated, usually in a coordinate fashion, and their biological effect on the host cell often depends on the controlled expression of other gene products. Different stages in the infectious cycle demand the induction of certain genes and the repression of others. The expression of virulence factors is often linked to various biochemical and physical (environmental) parameters, including temperature, osmolarity, nutrient availability, ion concentrations, iron levels, pH, growth phase, and oxygen tension (Gross, 1993; Mekalanos, 1992). These cues are used by the bacterium to sense the conditions of the microenvironment that they occupy and subsequently to elicit the appropriate cellular response to achieve adaptation. Whole-genome microarray analyses have been applied to studying the global transcriptional effect of acid stress (Ang et al., 2001) and growth phase (Thompson et al., 2003) in *H. pylori*, hypoxic (reduced oxygen tension) conditions in *M. tuberculosis* (Sherman et al., 2001), and iron limitation in *Pasteurella multicoda* (Paustian et al., 2001). Some functional genomics studies profiling temperature-induced changes in gene expression and exploring microbial transcriptomes in response to biofilm formation and targeted genetic mutations are highlighted below.

Temperature-responsive Genes   DNA microarray analysis has been used to investigate the influence of various biochemical and environmental parameters on global gene expression in bacterial pathogens. One such study employed array technology to identify genome-wide alterations in Group A *Streptococcus* (GAS) gene expression in response to physiologically relevant temperature changes (Smoot et al., 2001). As a human-specific pathogen, GAS causes various infections ranging from superficial skin infections to severe invasive illnesses such as streptococcal toxic shock syndrome. As a result, GAS is able to survive and multiply in diverse anatomic sites (e.g., skin, throat, female urogenital tract, lower gastrointestinal tract, and blood), characterized by extremes in temperature varying from about 25 to 40°C. GAS differentially regulates expression of virulence and other genes at these different sites of infection (Smoot et al., 2001). To explore the spectrum of temperature-responsive gene expression in GAS, microarrays containing 1,605 open reading frames in the pathogen's genome were fabricated, and RNA was isolated from GAS cells grown at 29, 37, and 40°C for array hybridization. Microarray analysis showed that 9% of the arrayed genes exhibited differential transcription in cells cultivated at 29°C compared with 37°C. A large proportion of these differentially expressed genes encode extracellular proteins, suggesting that the GAS extracellular proteome is substantially influenced by temperature (Smoot et al., 2001). Twenty-six of these genes encoded proteins with predicted secretion signal sequences and included proteases, cell wall-associated proteins, transporters, and previously uncharacterized hypothetical proteins. Microarray hybridization also revealed that the transcription of genes coding for several transcriptional regulator homologs was up-regulated at 29°C relative to 37°C, while GAS hemolysin (virulence factor) genes were up-regulated at 40°C relative to 37°C.

In another study, temperature-induced gene expression changes in *B. burgdorferi*, the etiologic agent of Lyme disease, were profiled using macroarrays containing 1,662 putative *B. burgdorferi* ORFs prepared on nylon membranes (Ojaimi et al., 2003). Because this spirochete cycles between a tick vector and a mammalian host during its life cycle, temperature is an important environmental factor that affects *B. burgdorferi* gene expression. Whole-genome arrays were therefore constructed to identify specific temperature-responsive genes. A total of 215 ORFs were differentially expressed at 23 and 35°C, with 133 genes being expressed at significantly greater levels at the higher temperature (Ojaimi et al., 2003). Furthermore, transcriptome analysis revealed that the majority (134) of these 215 ORFs are annotated as genes of unknown function, and 63% of the differentially expressed genes reside on plasmids. These findings implicate plasmid-encoded genes as playing an important role in the adjustment of *B. burgdorferi* to growth under different temperatures.

Biofilm Formation and Quorum Sensing The gram-negative bacterium *P. aeruginosa* is an ecologically versatile, highly adaptable organism found in various environmental niches (e.g., soil and marine habitats, plants, animals, and humans). It constitutes a significant source of opportunistic human infections, such as bacteremia, urinary tract infections, and hospital-acquired pneumonia (Bodey et al., 1983). *P. aeruginosa* can exist as free-living (planktonic) cells or in biofilms, structured communities of bacterial cells that are attached to surfaces and enmeshed in a self-produced hydrated polymeric matrix (Costerton et al., 1999). The complex architecture of mature biofilms is characterized by mushroom and pillarlike structures with intervening channels in which water and nutrients can circulate. Bacteria existing in biofilms exhibit remarkable phenotypic resistance to microbicidal agents such as antibiotics and to the immune defenses of hosts (Costerton et al., 1995, 1999). The ability of *P. aeruginosa* to adhere to surfaces and differentiate into biofilm is a major virulence trait in chronic or persistent infections (Costerton et al., 1999).

Whole-genome microarrays were used to gain insight into the potential differences in gene expression between planktonic *P. aeruginosa* cells and those in biofilms (Whiteley et al., 2001). Based on the complete genome sequence for *P. aeruginosa* strain PAO1 (Stover et al., 2000), an array consisting of 5,500 of the 5,570 predicted *P. aeruginosa* genes was constructed to compare global gene expression in planktonic and biofilm cells. Microarray analysis indicated that approximately 1% of the genes (34 activated genes and 39 repressed genes) showed differential transcription in biofilm populations, suggesting that gene expression, generally, is similar in the two growth modes (Whiteley et al., 2001). Interestingly, 34% of the 73 biofilm-regulated genes encode hypothetical proteins of unknown function, emphasizing the limitations of computational approaches based on genome sequence. Further experimental studies will be needed to define the specific function of these unknown genes in biofilm growth.

Quorum sensing, or the ability of bacteria to sense information from other cells in the population when cells reach a critical concentration, has been shown to be important in regulating the production of *P. aeruginosa* virulence factors and to be involved in biofilm formation and development (Davies et al., 1998; for a review, see de Kievit and Iglewski, 2000). *P. aeruginosa* possesses two interrelated acyl-homoserine lactone quorum-sensing-signaling systems, designated the LasR-LasI and RhlR-RhlI systems, which consist of a transcriptional regulatory protein (LasR and RhlR) and the autoinducer synthase (LasI and RhlI). High-density oligonucleotide microarrays have been used to define quorum-sensing-controlled genes and to investigate global gene expression patterns modulated by quorum-sensing regulons for the *P. aeruginosa* PAO1 genome (Schuster et al., 2003; Wagner et al., 2003). Microarray analysis identified numerous novel quorum-sensing-regulated *P. aeruginosa* genes, including 222 genes that were repressed by quorum sensing, and these quorum-sensing-controlled genes were expressed differently depending on the cell growth phase (Wagner et al., 2003).

Transcription Profiling of Mutants   Another approach to the array-based analysis of bacterial pathogen transcriptomes is to compare gene expression in wild-type cells vs. mutant cells. For example, an oligonucleotide microarray representing greater than 86% of the *S. aureus* genome was used to identify genes regulated by Agr and/or SarA, two well-characterized regulators of the pathogen's virulence response (Dunman et al., 2001). Genome-wide expression profiles in wild-type cells were compared to those in mutant cells that were defective in *agr*, *sarA*, and *agr* + *sarA*. Microarray hybridization demonstrated that several putative virulence factors were regulated by Agr and/or SarA as well as certain genes involved in several biological processes not known to be associated with pathogenesis. Similarly, array-based transcriptome analysis of a *prfA* mutant strain of *L. monocytogenes* identified three groups of genes regulated differentially by PrfA, a member of the Crp/Fnr family of transcription regulators (Milohanic et al., 2003; see Section 12.3.4 in this chapter). Such studies allow the researcher to gain a deeper understanding of the regulon controlled by the transcription factor of interest.

12.5.2 Elucidating the Molecular Intricacies of Host–Pathogen Interactions

Infection of a host by pathogenic bacteria is characterized by drastic changes in the physiology of both organisms. These cellular changes are largely the reflection of tightly regulated alterations in host and microbe gene expression (Diehn and Relman, 2001). Hence, the genome-wide transcription profiling capability of array technology seems particularly well suited for analyzing the complex gene expression responses of host cells to various infectious stimuli. However, it is important to note that posttranscription regulatory events cannot be detected using microarray hybridization, and many host cell events (e.g., cytoskeletal rearrangements) occur posttranscriptionally (Cummings and Relman, 2000). Some aspects of the host molecular program initiated in response to bacterial infection, therefore, may not be readily defined by gene expression profiling. Nevertheless, microarrays and other functional genomic technologies hold great promise for accelerating our understanding of the molecular events that follow infection. In the next section, we highlight some of the studies that have used host-specific gene arrays to explore the transcriptional response of the host to bacterial infection (Belcher et al., 2000;

Boldrick et al., 2002; Coombes and Mahoney, 2001; Detweiler et al., 2001; Eckmann et al., 2000; Huang et al., 2001; Ichikawa et al., 2000; Maeda et al., 2001; Ragno et al., 2001).

Molecular Response of Epithelial Cells to Pseudomonas aeruginosa ***and Other Pathogen Infections*** In the study by Ichikawa and coworkers (2000), high-density microarrays, containing 1,506 human cDNA clones, were used to delineate differentially expressed genes in the A549 lung pneumocyte cell line in response to *P. aeruginosa* infection. The approximately 1,500 human cDNA clones on the array represented known and uncharacterized genes. Typically, in such host–pathogen interaction experiments, total RNA is isolated from uninfected host cells (control or reference sample) and infected host cells (experimental sample), differentially labeled using reverse transcription and different fluorochromes (typically, Cy3- and Cy5-dCTP or -dUTP), and hybridized to the microarray. In the microarray experiment performed by Ichikawa and colleagues (2000), gene expression profiles from host epithelial cells infected with *P. aeruginosa* PAK were compared to those from host cells inoculated with nonpiliated derivative PAK-NP. Adherence of *P. aeruginosa* in the in vitro model depends on the expression of type-IV pili (Cervin et al., 1994; Chi et al., 1991). Expression microarray analysis indicated that a number of host cell genes are up-regulated in response to adherence by *P. aeruginosa*, including several transcription factors, an epithelial-specific transcription factor, an interferon regulatory factor (IRF-1), several inflammatory response genes, and several genes implicated in signal transduction pathways (Ichikawa et al., 2000). Activation of IRF-1 transcription as a result of *P. aeruginosa* attachment suggested the importance of bacterial adherence in activating a part of a host pathway previously shown to play a role in the host-cell response to viral infection (Ichikawa et al., 2000).

Further evidence of the utility of DNA microarrays in understanding host–pathogen interactions was demonstrated in the analysis of host transcriptional profiles during the interaction of *Bordetella pertussis* (the causative agent of whooping cough) with a human bronchial epithelial cell line (Belcher et al., 2000). The early transcriptional response revealed, for the first time, a dominant proinflammatory state in *B. pertussis*-exposed bronchial epithelial cells. In addition, microarray-based gene expression profiling helped to uncover some of the mechanisms by which *B. pertussis* attempts to subvert host immune defenses (Belcher et al., 2000). In another study, Guillemin and coworkers (2002) took a novel approach to investigating host–pathogen interactions. They used human cDNA microarrays to examine the temporal transcriptional profiles of gastric epithelial cells infected with wild-type *H. pylori* (the reference strain) and a panel of isogenic mutants in order to dissect the contributions of individual virulence determinants encoded in the *cag* pathogenicity island (PAI) (discussed in Section 12.2.3). Infection with the wild-type strain induced expression of genes involved in signal transduction, cell shape regulation, and host cell cytoskeletal response, while a mutant lacking the *cagA* gene induced the expression of fewer cytoskeletal genes in gastric epithelial cells (Guillemin et al., 2002). The host-cell response to CagE, CagN, VacA, and the entire *cag* PAI was examined in a similar fashion to determine their individual contributions to virulence. Other cDNA array analyses have started to illuminate the transcriptional response of endothelial cells to infection with *C. pneumoniae* (Coombes and Mahoney, 2001), the stimulation of intracellular signaling pathways in human gastric cancer cells by *H. pylori* (Maeda et al., 2001), and changes in gene expression in a human macrophage cell line following infection with *M. tuberculosis* (Ragno et al., 2001).

Specific Gene Expression Programs in Host Cell Responses to Diverse Bacteria The two studies described in this section illustrate how microarray analysis can be used to reveal the modulation of host gene expression in response to different pathogens. Huang and colleagues (2001) used oligonucleotide arrays to measure the gene expression profiles of human monocyte-derived dendritic cells in response to phylogenetically diverse pathogens. The primary question addressed was whether dendritic cells, which are involved in the initiation of both innate and adaptive immunity, modulate their gene expression patterns as a means of discriminating among different pathogens. To determine whether dendritic cells elicit tailored pathogen-specific immune responses, dendritic cells were exposed to *E. coli* (a gram-negative bacterium), *Candida albicans* (a fungus), and influenza virus (Huang et al., 2001). RNA isolated from the treated dendritic cells was then labeled and hybridized to an oligonucleotide microarray containing probes representing ~ 6,800 genes. Microarray analysis revealed that dendritic cells elicit a shared core response and pathogen-specific programs of gene expression in response to the different pathogens. A common set of 166 highly regulated genes was observed for all three pathogen responses. Shared up-regulated genes included immune cytokines, chemokines, cytoskeletal genes, signaling genes, transcription factors, and several antigen processing and presentation genes. In addition, pathogen-regulated gene expression was observed in human monocyte-derived dendritic cells, suggesting a certain plasticity of dendritic cell responses to pathogens (Huang et al., 2001). While a subset (242 genes) of the *E. coli* and *C. albicans* responses was common to both pathogens, a smaller number of differential response genes were identified as *E. coli*-specific or *C. albicans*-specific (Fig. 12.4). A similar result was found when comparing microarray-derived gene profiles for the interaction of dendritic cells with *E. coli* and influenza. Furthermore, this study showed that gene expression in dendritic cells was most rapidly induced by *E. coli* and most slowly by influenza virus.

In a similar study, David Relman and his colleagues (Boldrick et al., 2002) used human cDNA microarrays to develop a more complete molecular picture of the ability of human peripheral blood mononuclear cells (PBMCs) to initiate distinct gene expression programs in response to different infectious microorganisms. To explore common and unique host transcriptional responses to diverse bacteria, Boldrick and coworkers (2002) compared expression profiles of PBMCs treated with heat-killed gram-negative bacteria, *B. pertussis* and *E. coli*, or the gram-positive bacterium *S. aureus*. Each bacterial challenge elicited the induction of approximately 206 common genes, although gram-negative bacteria were substantially more potent inducers of the transcriptional response than gram-positive bacteria.

Host Response to Bacterial Virulence Determinants Microarrays have also been used to specifically address the host molecular response to individual bacterial virulence determinants (Rappuoli, 2000). In a study of *Salmonella enterica* subspecies *typhimurium* infection, Detweiler and coworkers (2001) used a microarray, consisting of 22,571 human cDNAs, to identify the host molecular pathways that are affected by the *Salmonella phoP* gene, a response transcription regulator of a two-component system that is required for virulence. *S. typhimurium* mutants with a transposon insertion in the *phoP* gene are avirulent (Miller et al., 1989) and such mutants fail to cause typhoid fever in humans (Hohmann et al., 1996). The transcriptional response of a human monocytic cell line to infection with wild-type (virulent) *S. typhimurium* and an isogenic (avirulent) *phoP::Tn10* mutant of *S. typhimurium* was compared by hybridization on cDNA

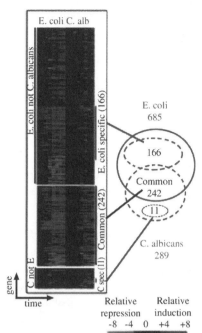

Figure 12.4 *See color insert. Gene expression levels in human dendritic cells in response to* Escherichia coli *and* Candida albicans *exposure (Huang et al., 2001).* Dendritic host cells were treated with *E. coli* and *C. albicans* for 0, 1, 2, 4, 8, 12, and 24 hours, and mRNA expression levels in the host were measured using microarray hybridization. Individual genes are represented by a single row and each exposure time point is represented by a single column. The number of genes showing both common and pathogen-specific alterations in expression patterns is shown. [Reprinted with permission from Q. Huang et al., The plasticity of dendritic cell responses to pathogens and their components, *Science*, vol. 294, 870–875. Copyright (2001) American Association for the Advancement of Science.]

microarrays (Detweiler et al., 2001). Host microarray analysis revealed that differentially expressed genes included those that affected human macrophage cell death, suggesting the role of *phoP* in the regulation of bacterial genes that influence macrophage signal. In all, host genomic transcriptional profiling, in combination with functional assays, is a powerful approach to unraveling the complexities of the interaction between host and pathogen.

12.5.3 Identification of Antimicrobial Drug Targets

As discussed earlier, gene expression profiling of bacterial pathogens has provided valuable insights into the molecular basis of the infection process and testable hypotheses (Kato-Maeda et al., 2001). First, microarray analysis has suggested biological functions for uncharacterized genes derived from genome sequence annotation. Second, such an approach has identified virulence genes that promote different aspects of infection (e.g., colonization and tissue damage). Third, microarrays have facilitated the investigation of regulatory mechanisms controlling gene expression in response to different environmental parameters. Such information is crucial for the selection of antibacterial drug targets and the development of new therapeutics. It is imperative to accelerate the drug discovery process in order to address the widespread emergence of antibiotic resistant strains of pathogenic bacteria. Whole-genome sequencing of bacterial pathogens and microarray expression profiling have made it possible, for the first time in history, to perform a comprehensive search for new potential drug and vaccine targets.

As an example, Wilson and colleagues (1999) examined changes in the gene expression profile of *M. tuberculosis* in response to the antituberculosis drug isoniazid, a drug that blocks the mycolic acid biosynthesis pathway, by using a DNA microarray containing 97% of the total ORFs predicted from the *M. tuberculosis* genome (Cole et al., 1998). *M. tuberculosis* is distinguished from other bacteria by a cell wall that contains unusual lipid moieties (Barry, 2001). A large portion ($\sim 9\%$) of the protein-coding capacity of *M. tuberculosis* H37Rv is devoted to the synthesis of enzymes involved in novel biosynthetic pathways that generate a diverse array of cell-wall components, such as mycolic acids, mycocerosic acid, phenolthiocerol, lipoarabinomannan, and arabinogalactan (Cole et al., 1998). The low permeability of the mycobacterial cell wall protects the organism from chemical damage, dehydration, and certain antibiotics; however, it is unclear whether the cell wall can be accurately labeled as a "virulence factor" (Barry, 2001). Isoniazid was selected for study, because it is a drug to which resistance emerges most frequently (Pablos-Mendez et al., 1998). By using microarray hybridization to explore isoniazid-induced alterations in *M. tuberculosis* gene expression, Wilson and coworkers (1999) were able to identify genes potentially involved in mycolic acid metabolism, whose role was previously unrecognized in this process. The study identified other proteins functioning in the isoniazid-affected enzymatic pathway that might serve as appropriate targets for new drug development. In addition, it may be possible to use a microarray-based approach to predict the mode of action of a novel compound based on the organism's gene expression response to that compound.

12.6   THE PROTEOMICS OF BACTERIAL PATHOGENESIS

The availability of genome sequence data permits the prediction of the total protein complement encoded by a genome. Analysis of the expressed protein complement of a genome, termed the proteome (Wasinger et al., 1995) of an organism, has developed into a new discipline, proteomics. Proteome analysis typically combines the techniques of two-dimensional polyacrylamide gel electrophoresis (2-D PAGE) and mass spectrometry for large-scale protein separation and sensitive protein identification, respectively (for a review, see Aebersold and Goodlett, 2001). The various protein characterization tools employed in the field of proteomics are reviewed in Chapter 9. Proteome analysis by 2-D PAGE and mass spectrometry, in combination with protein chemical methods, is a powerful approach to postgenomic characterization of important bacterial pathogens and complements RNA transcriptional analysis by DNA microarray hybridization. Although methodologies in the field are continuing to be developed and improved, proteomics, in particular, two-dimensional gel electrophoresis, is currently limited in its ability to detect proteins in low abundance, cell membrane-spanning proteins, and proteins of high molecular charge or very low molecular mass (Ragno et al., 2001). Nevertheless, proteomics will enable the comparison of virulent strains vs. attenuated strains by differential protein display to identify infection-related factors that may have been lost in attenuated or nonpathogenic strains. Proteomics will also allow the identification of proteins that are differentially expressed in response to growth conditions of pathogenic relevance and the different stress challenges (e.g., low pH, heat shock, low nutrient supply, low oxygen pressure) the pathogen faces in intracellular environments. The impact of proteomics on bacterial pathogenesis is discussed here.

12.6.1 Comparative Proteomics

Sequence determination of the genome of the intracellular pathogen *M. tuberculosis* strain H37Rv (Cole et al., 1998), the causative agent of human tuberculosis, has facilitated proteome analysis of this important pathogen. Annotation of the complete genome sequence predicted 3,924 ORFs. Jungblut and colleagues (1999) conducted a comparative proteome analysis of two nonvirulent vaccine strains of *M. bovis* Bacille Calmette-Guerin or BCG (Chicago and Copenhagen) and two virulent strains of *M. tuberculosis* (H37Rv and Erdman). A combination of two-dimensional electrophoresis and matrix-assisted laser desorption/ionization mass spectrometry (MALDI-MS) was utilized to analyze the mycobacterial proteomes. One purpose of this study was to identify *M. tuberculosis*-specific gene products, those absent in the vaccine strain *M. bovis* BCG, that could serve as potential vaccine targets in order to accelerate the design of new strategies for the prevention and therapy of tuberculosis (Jungblut et al., 1999). Using two-dimensional PAGE methodology, the composition of the mycobacterial cellular proteome was resolved into 1,800 distinct protein species, whereas 2-D PAGE patterns for the extracellular proteome (culture supernatants) contained 800 proteins. A total of 263 proteins were identified by MS.

The 2-D PAGE patterns reflected the highly conserved nature of the *M. tuberculosis* complex, which included the genomes of both the *M. bovis* strains and *M. tuberculosis* strains (Jungblut et al., 1999; Sreevatsan et al., 1997). However, some clear differences were observed among the strains in terms of spot intensity (reflective of protein abundance), presence or absence, and position of the spots (reflective of amino acid exchanges or posttranslational modifications such as phosphorylation, glycosylation, or acylation) (Jungblut et al., 1999). Thirty-one protein variants were detected between subsets of the proteomes of BCG Chicago and H37Rv. The following six proteins identified in strain H37Rv did not have a detectable counterpart in BCG: L-alanine dehydrogenase (40 kDa antigen), isopropyl malate synthase, nicotinate-nucleotide pyrophosphatase, MPT64, and two conserved hypothetical proteins. MPT64 is a culture filtrate protein that has previously been shown to induce a delayed-type hypersensitivity response in guinea pigs (Kaufmann and Andersen, 1998). Alanine dehydrogenase, which has been implicated as a component of the cell wall synthesis machinery, may be of value in vaccine development or diagnostics (Jungblut et al., 1999). Proteomic comparison between *M. tuberculosis* Erdman and *M. bovis* BCG Chicago indicated four mobility variants, including the transcriptional regulator MoxR, which may be involved in the formation of an active methanol dehydrogenase (Jungblut et al., 1999). Compared to *M. tuberculosis* H37Rv, the proteome of *M. tuberculosis* Erdman contained six proteins of increased abundance (i.e., increased intensity), two additional proteins not represented in H37Rv, six absent proteins, and two mobility variants. This study also demonstrated, for the first time, the expression of three lipoproteins out of the 65 lipoprotein-encoded genes predicted in the mycobacterial genome. These predicted proteins are putatively associated with the mycobacterial cell envelope (Cole et al., 1998).

In a similar study, the proteomes of *M. tuberculosis* strain H37Rv and a recent clinical isolate of *M. tuberculosis*, CDC 1551, were compared using 2-D PAGE and MS to detect any protein expression differences that might explain the differing phenotypes of these two strains (Betts et al., 2000). Although the sequenced genomes of H37Rv and CDC 1551 show > 99% identity at the nucleotide level (Delcher et al., 1999), CDC 1551 differs phenotypically from H37Rv in that the former induces a more rapid and vigorous immune response in the host (Manca et al., 1999). Approximately 1,750 proteins, accounting for 44% of the 3924 predicted genes in *M. tuberculosis* H37Rv, were resolved using two-dimensional gel electrophoresis. Comparative proteomic analysis indicated that the protein profiles of the two strains were highly similar, with only a total of 17 protein spot differences observed between H37Rv and CDC 1551. Of these proteins, seven are unique to the isolate CDC 1551 (including a probable alcohol dehydrogenase and two transcriptional regulator MoxR homologs), and three are exclusive to H37Rv. Other proteins exhibited differences in spot intensity or gel mobility. However, Betts and coworkers (2000) were not able to deduce an obvious association of the proteomic differences with phenotype. This may be attributed to the limitations of 2-D PAGE techniques mentioned previously, such as the under-representation of membrane (hydrophobic) proteins due to solubility problems (Betts et al., 2000). Despite these technical drawbacks, two-dimensional gel-based proteomics is a useful (but not a comprehensive) approach to uncovering differential protein expression between bacterial strains displaying different virulence phenotypes. In contrast to genomic sequence comparisons, differences at the proteome level are likely to mirror differences in translational control and posttranslational modification between variable strains (Betts et al., 2000).

12.6.2 Defining the Proteome of Individual Bacterial Pathogens

With the availability of complete sequences for a number of pathogenic bacterial genomes, the next step in the process of characterizing gene functionality is to determine protein composition and expression. To elucidate the molecular basis of bacterial pathogenesis, it is necessary to ascertain whether changes in the steady-state mRNA levels of genes (derived from DNA microarray studies) correlate with changes in protein expression or abundance levels (derived from proteomic studies). Only by incorporating both a functional genomics approach and a proteomic approach can we obtain a global, integrated perspective of disease processes. A number of investigations have focused on defining the proteomes of various microbial pathogens and relating protein expression data to information obtained from complete genome sequences. In one such study, *M. tuberculosis* extracellular proteins from the culture medium and proteins in cellular extracts were examined by 2-D PAGE followed by MS and antibody-based immunodetection of selected protein species (Rosenkrands et al., 2000). Of the 49 culture filtrate proteins and 118 lysate proteins identified by MS, 83 of the identified *M. tuberculosis* proteins were novel. A fraction of these novel proteins were conserved hypotheticals or were encoded by genes of unknown function. Demonstrating the expression of genes of unknown function is relevant to investigating their potential involvement in specific mycobacterial functions (Rosenkrands et al., 2000).

A similar proteomics study was applied to the mollicute *M. genitalium*, the smallest autonomously replicating organism that has been completely sequenced. The 580-kb genome of *M. genitalium* can potentially express 480 gene products (Fraser et al., 1995). In an analysis of the *M. genitalium* proteome, 427 distinct proteins were resolved in association with the exponential growth of the bacterial pathogen (Wasinger et al., 2000), making the description of this proteome one of the most complete ever reported. Proof of expression for 201 proteins involved in adherence, environmental stress response, energy metabolism, replication, transcription, translation, regulatory functions, cofactor biosynthesis, fatty acid metabolism, and unknown functions was obtained using MS. As a human surface parasite, *M. genitalium* colonizes host cells largely through adherence mechanisms (Razin et al., 1998). The expression of cytadherence proteins may enhance close contact with host cells, and three proteins associated with cytadherence were identified in the *M. genitalium* proteome (Wasinger et al., 2000). In addition, proteome analysis confirmed the expression of 17 hypothetical proteins and provided a basis for further investigations into the functionality of these proteins. As with the analysis of the *M. tuberculosis* proteome, cataloguing the expressed gene products of *M. genitalium* extended the information derived from the genome sequence by indicating those genes that can serve a useful biological purpose (Wasinger et al., 2000).

Other proteomic analyses have detected the most abundant components of the *H. influenzae* type-strain NCTC 8143 proteome (Link et al., 1997c), characterized the structurally and functionally unknown proteins in the small genome of *M. genitalium* (Balasubramanian et al., 2000), or examined the modulation of bacterial protein synthesis in response to specific environmental stress conditions of physiologic relevance or inhibitors of cellular processes (Evers et al., 2001; Garbe et al., 1996; Lee and Horwitz, 1995; Wong et al., 1999). As an illustration, one study investigated differential protein expression in *M. tuberculosis* during intracellular residence in human macrophages and extracellular growth in response to stresses such as heat shock, low pH, and hydrogen peroxide (Lee and Horwitz, 1995). Proteome analysis revealed that the pattern of induced and repressed proteins was unique to each stress condition. Other proteomic investigations have advanced our knowledge of the biology of *M. tuberculosis* by demonstrating the existence of open reading frames in *M. tuberculosis* H37Rv that were not predicted by genomics (Jungblut et al., 2001).

12.6.3 Proteomic Approach to Host–Pathogen Interactions

The outcome of infection with a bacterial pathogen depends on a complex set of interactions involving changes in the expression of genes and proteins in both the host cell and bacterium. In a rare study integrating both genomic and proteomic approaches, host-genome microarrays and proteomic tools (i.e., 2-D PAGE and MS) were employed in parallel to investigate altered gene expression in the human macrophage cell line in response to infection with *M. tuberculosis* (Ragno et al., 2001). The transcriptome and proteome of uninfected macrophages were compared to those of infected macrophages during the initial period of interaction between the host and pathogen. Microarray analysis indicated that the gene encoding interleukin IL-1β was the most highly induced gene following *M. tuberculosis* infection, and the protein IL-1β was detected as being the most highly up-regulated species in the proteome subset. In addition to the induction of IL-1β expression, proteomic analysis also revealed that macrophage-derived superoxide dismutase, an enzyme involved in cellular protection during oxidative stress, was significantly up-regulated in response to *M. tuberculosis* infection. An unexpected observation was the presence of the mycobacterial protein ATP synthase β chain, a protein thought to be involved in the regulation of cytoplasmic pH, in infected macrophage extracts (Ragno et al., 2001). An important limitation of proteomic methodologies, as demonstrated in this study as well as many others, is that typically only a small fraction of the total proteome can be investigated and regulatory proteins, which are generally not major components of the cell compared to structural proteins, are often missed (Ragno et al., 2001). Technical advances over time in the field of proteomics should lead to improved methodologies for resolving a greater proportion of whole-cell lysates.

12.7 GENOME SEQUENCE AND FUNCTIONAL ANALYSIS OF ENVIRONMENTALLY IMPORTANT MICROORGANISMS

Many microorganisms exhibit properties that have potentially useful industrial applications or important implications with regard to the bioremediation of metal and radioactive contaminants in the environment. Knowledge of the complete gene complement provides a valuable foundation for research on the genetic and biochemical capacities of environmentally significant microorganisms. Genome sequence information for a number of bacteria and archaea of potential environmental or biotechnological relevance is beginning to accumulate and includes representatives of dissimilatory metal-reducing bacteria [*S. oneidensis* (Heidelberg et al., 2002)], extreme radiation-resistant bacteria [*D. radiodurans* (White et al., 1999b)], photosynthetic cyanobacteria [*Anabaena* sp. strain PCC 7120 (Kaneko et al., 2001), *Synechocystis* sp. strain PCC6803 (Kaneko et al., 1996)], thermophilic and hyperthermophilic archaea *Pyrococcus horikoshii* (Kawarabayasi et al., 1998), *Aeropyrum pernix* (Kawarabayasi et al., 1999), *Thermotoga maritima* (Nelson et al., 1999), *Thermoplasma volcanium* (Kawashima et al., 2000), *P. furiosus* (Robb et al., 2001), *Pyrobaculum aerophilum* (Fitz-Gibbon et al., 2002)], thermoacidophilic archaea [*Sulfolobus tokodaii* (Kawarabayasi et al., 2001), *Sulfolobus solfataricus* (She et al., 2001)], methanogens [*Methanococcus jannaschii* (Bult et al., 1996), *Methanobacterium thermoautotrophicum* (Smith et al., 1997a), *Methanopyrus kandleri* (Slesarev et al., 2002), *Methanosarcina acetivorans* (Galagan et al., 2002)], sulfate-reducing archaea [*Archaeoglobus fulgidus* (Klenk et al., 1997)], and halophilic archaea [*Halobacterium* sp. strain NRC-1 (Ng et al., 2000)]. Despite the large number of environmentally significant microorganisms that have been sequenced thus far, very few investigations have been reported in the literature that describe the application of DNA microarray technology to genomic expression analysis of these organisms. Array-based transcriptome analyses have focused predominantly on microbial pathogens and such model organisms as *E. coli*, *Bacillus subtilis*, and *Saccharomyces cerevisiae* (yeast). DNA microarrays have been used to monitor changes in *Synechocystis* gene expression in response to acclimation to high light intensity (Hihara et al., 2001) and to the redox state of the photosynthetic electron transport chain (Hihara et al., 2003), but we will limit our discussion in this section to microarray profiling of gene expression in three organisms of environmental significance, namely, *S. oneidensis*, *D. radiodurans*, and *P. furiosus*.

12.7.1 Dissimilatory Metal Ion-reducing Bacterium *Shewanella oneidensis*

S. oneidensis MR-1 [formerly *Shewanella putrefaciens* strain MR-1 (Venkateswaran et al., 1999)] is a facultatively anaerobic γ-proteobacterium that is noted for its remarkably diverse respiratory capacities. In addition to utilizing oxygen as a terminal electron acceptor during aerobic respiration, *S. oneidensis* can anaerobically respire various organic and inorganic substrates, including oxidized metals [e.g., Mn(III) and (IV), Fe(III), Cr(VI), U(VI)], fumarate, nitrate, nitrite, thiosulfate, elemental sulfur, trimethylamine *N*-oxide (TMAO), dimethyl sulfoxide (DMSO), and anthraquinone-2,6-disulphonate (AQDS)

(Lovley, 1991; Moser and Nealson, 1996; Nealson and Saffarini, 1994). This unusual versatility in the use of alternative electron acceptors for anaerobic respiration is conferred in part by complex electron transport networks, composed of cytochromes, reductases, iron-sulfur proteins, and quinones (Richardson, 2000). The metal ion-reducing capabilities of this bacterium have important implications with regard to the potential for in situ bioremediation of metal contaminants in the environment. In theory, microbial-mediated remediation of metal pollution can occur as a result of the immobilization of metals in a nonbioavailable form (e.g., from a soluble to insoluble form) or their transformation into less toxic forms (Valls and de Lorenzo, 2002). However, application of organisms of bioremediation potential to contaminated sites is often complicated by unpredictable interactions with other indigenous microorganisms, environmental stresses, and the inability to effectively predict or assess bioremediation performance or activity due to insufficient knowledge concerning the gene networks and regulatory mechanisms enabling microbial metal reduction.

To expedite our understanding of metal reduction by *S. oneidensis* MR-1, The Institute for Genomic Research (TIGR) under the support of the U.S. Department of Energy (DOE) recently sequenced its $\sim$5-Mb genome. (Heidelberg et al., 2002), making it feasible to apply microarray technology to the study of energy metabolism in this bacterium. The *S. oneidensis* MR-1 genome includes a circular chromosome (4,969,803 bp in size), which contains a total of 4,758 predicted protein-encoding ORFs (CDSs), and a megaplasmid (161,613 bp) with 173 CDSs (Heidelberg et al., 2002). Genome analysis revealed that *S. oneidensis* possesses 39 c-type cytochromes (including eight genes encoding decaheme cytochrome c proteins), which is more than any other organism sequenced thus far (Heidelberg et al., 2002). This is consistent with the bacterium's considerable anaerobic respiratory capacity.

Several studies have examined the transcriptional response of *S. oneidensis* to different respiratory growth conditions (Beliaev et al., 2002b) and to the disruption of genes encoding putative transcriptional regulators (Beliaev et al., 2002a; Thompson et al., 2002) using DNA microarrays containing 691 arrayed genes. The partial genome microarrays consisted of PCR-amplified MR-1 ORFs putatively involved in energy metabolism, transcriptional regulation, adaptive responses to environmental stress, iron acquisition, and transport systems. These arrays were constructed prior to the closure of the *S. oneidensis* genome sequence. To identify genes specifically involved in anaerobic respiration, differential mRNA expression profiles of *S. oneidensis* were monitored under aerobic and fumarate-, Fe(III)-, or nitrate-reducing conditions using partial genome microarrays (Beliaev et al., 2002b). Gene expression profiling indicated that 121 of the 691 arrayed ORFs showed at least a 2-fold difference in transcript abundance in response to changes in growth conditions (Beliaev et al., 2002b). Not surprisingly, the transition from aerobic to anaerobic respiration resulted in the repression of a number of genes required for aerobic growth, including genes encoding cytochrome c and d oxidases and TCA cycle enzymes.

Genes induced in a general response to anaerobic respiration, irrespective of the terminal electron acceptor, belonged to several cellular functional categories: cofactor biosynthesis and assembly, substrate transport, and anaerobic energy metabolism. Most noteworthy was the observation that certain genes preferentially displayed increased transcript levels in response to specific electron acceptors. The expression of genes encoding a periplasmic nitrate reductase (*napBHGA* operon), cytochrome c_{552}, and prismane, for example, was elevated 8- to 56-fold specifically in response to the presence nitrate, while genes encoding a tetra-heme cytochrome c (*cymA*), a flavocytochrome c (*ifcA*), and a fumarate reductase (*frdA*) were preferentially induced 3- to 8-fold under conditions of fumarate reduction. In addition, the mRNA abundance levels for two oxidoreductase-related genes of unknown function and several cell envelope genes involved in multidrug resistance increased specifically under Fe(III)-reducing conditions. To complement the gene expression study, 2-D PAGE and protein identification using MS (specifically, microliquid chromatography-electrospray ionization tandem MS) were performed to analyze a subset of the *S. oneidensis* proteome under aerobic and anaerobic respiratory conditions (Beliaev et al., 2002b). Although some of the protein expression data correlated with the microarray data, there were several inconsistencies noted between mRNA and protein abundance levels for certain ORFs that the authors postulated could have resulted from different turnover rates or posttranscriptional regulation (Beliaev et al., 2002b). Nonetheless, this work represented the first attempt to characterize a complex system in *S. oneidensis* on a genome scale. Other microarray-based transcriptomic studies have focused on defining the function of putative *S. oneidensis* regulatory genes encoding a ferric uptake regulator (*fur*; Thompson et al., 2002a) and an electron transport regulator (*etrA*; Beliaev et al., 2002a).

12.7.2 Extreme Radiation-resistant Bacterium *Deinococcus radiodurans*

D. radiodurans strain R1 is the most characterized member of the DNA damage-resistant bacterial family *Deinococcaceae*, which is comprised of at least seven different species that form a distinct eubacterial phylogenetic lineage most closely related to the *Thermus* genus (Makarova et al., 2001). *D. radiodurans* is a gram-positive, nonsporulating bacterium that was originally isolated in 1956 from canned meat that had spoiled following exposure to X-rays (Anderson et al., 1956). Species in the genus *Deinococcus*, particularly *D. radiodurans*, are extremely resistant to a number of physicochemical agents and environmental conditions that damage DNA, including ionizing and ultraviolet (UV) radiation, desiccation, heavy metals, and oxidative stress (reviewed in Battista, 1997; Battista et al., 1999; Minton, 1996). Studies have demonstrated that *D. radiodurans* not only can survive acute exposures to gamma radiation that exceed 15,000 Gy without lethality or induced mutation (Daly et al., 1994; Daly and Minton, 1995) but can also flourish in the presence of high-level chronic irradiation (60 Gy per hour) (Lange et al., 1998; Venkateswaran et al., 2000). *D. radiodurans* also expresses an intrinsic ability to reduce metals and radionuclides (Fredrickson et al., 2000) and thus has potential applications for the bioremediation of metal- and radionuclide-contaminated sites where the presence of radioactivity prohibitively restricts the activity of more sensitive dissimilatory metal-reducing bacteria such as *Shewanella*.

To further understanding of the biology of Deinococci and the molecular basis of extreme DNA damage resistance, the complete genome of *D. radiodurans* R1 was sequenced by TIGR (White et al., 1999b) under DOE support. The *D. radiodurans* genome consists of two chromosomes (2.65 and 412 kbp), one megaplasmid (177 kbp), and one plasmid (46 kbp) encoding a total of 3,187 predicted ORFs (Makarova et al., 2001; White et al., 1999b). Of these ORFs, more than 30% are of unknown function based on sequence homology searches. Sequence analysis indicates that essentially the entire repertoire of recombinational DNA repair genes identified in *D. radiodurans* has functional homologs in other prokaryotes (Makarova et al., 2001; White et al., 1999b), suggesting that the extreme radioresistance of R1 may be attributable to novel genes, repair pathways, and mechanisms yet to be described. Thus far, homology-based sequence comparisons between the *D. radiodurans* genome and genomes from other phylogenetically diverse organisms have failed to establish a unified molecular basis for extreme radiation resistance in *D. radiodurans*. Detailed computational genomic analyses alone, therefore, are unlikely to uncover the fundamental answers underlying the remarkable ability of *D. radiodurans* to withstand DNA-damaging conditions.

To understand the molecular response of *D. radiodurans* to ionizing radiation and to gain some insight into the regulatory mechanism(s) underlying its radiation-resistant phenotype, DNA microarrays covering $\sim 94\%$ of the organism's predicted protein-encoding genes were used to analyze the transcriptome dynamics in response to cellular recovery from acute ionizing radiation (Liu et al., 2003). In this study, *D. radiodurans* cells exposed to acute irradiation (15 kGy) were allowed to recover at 37°C for 0 to 24 hours. *Deinococcus* transcriptome dynamics were monitored in cells representing early (0–3 hours), middle (3–9 hours), and late (9–24 hours) phases of recovery from ionizing radiation and compared to nonirradiated control cells. Microarray analysis of genomic expression patterns revealed a wide repertoire of *D. radiodurans* genes responding to acute irradiation: 832 genes (28% of the genome) were induced and 451 genes (15% of the genome) were repressed 2-fold or greater during *D. radiodurans* recovery (Liu et al., 2003). Genes exhibiting increased transcription in the early phase of cell recovery belonged to various functional groups, including DNA replication, DNA repair, recombination, cell wall metabolism, cellular transport, and uncharacterized proteins. Hierarchical clustering of genes showing changes in mRNA abundance levels revealed similar expression patterns and clusters of presumably coregulated genes (Fig. 12.5). Genes responding to recovery from irradiation clustered into three distinct groups: (1) *recA*-like activation pattern (based on the expression profile of *recA*, which is critical for *D. radiodurans* recovery and is substantially up-regulated during early phase recovery and down-regulated before the onset of late phase), (2) growth-related activation pattern, and (3) repressed patterns. An unexpected, potentially important finding to emerge from this research was the observation that genes encoding TCA cycle components were repressed in the early and middle phases of recovery, whereas genes encoding the glyoxylate shunt pathway were induced during this interval (Liu et al., 2003). In addition, a number of poorly characterized genes showed high induction folds in expression during at least one phase of recovery, thus implicating their encoded proteins in a functional role, that of cell recovery, for the first time. The response of metabolic gene systems is not immediately clear and will require further experimentation.

The study by Liu and colleagues (2003) represents the first published description of the application of DNA microarrays to the functional analysis of *D. radiodurans*. The microarray data suggest that the recovery process for *D. radiodurans* cells involves the complicated coordination of DNA repair and metabolic functions as well as other cellular functions. This work also implicated a distinct ATP-dependent DNA ligase as the dominant ligase during postirradiation repair. Microarray analysis showed that the ATP-dependent DNA ligase is up-regulated early in cell recovery, while the typical bacterial NAD-dependent DNA ligase is down-regulated. The data generated in this study are also valuable from the perspective of the questions they pose, requiring new lines of research.

12.7.3 Hyperthermophilic Archaeon *Pyrococcus furiosus*

P. furiosus is a member of a phylogenetically distinct group of prokaryotes called the Archaea, which constitutes a primary, separate domain in the universal tree of life (Woese

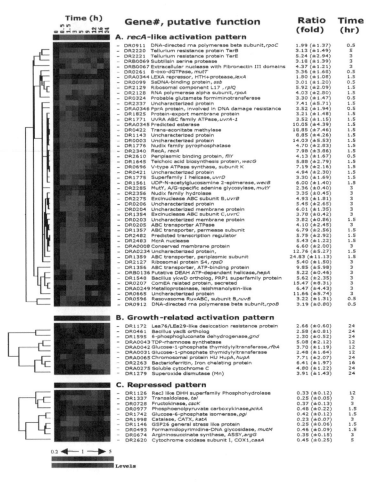

Figure 12.5 *See color insert. Hierarchical clustering analysis of gene expression profiles in* Deinococcus radiodurans *(Liu et al., 2003).* Genes showing changes in expression in response to recovery from ionizing radiation are clustered into three distinct groups using hierarchical clustering: (A) recA-like activation pattern, (B) growth-related activation pattern, and (C) repressed patterns. [Reprinted with permission from Y. Liu et al., Transcriplame dynamics of *Deinococcus radiodurans* recovering from ionizing radiation, *PNAS*, vol. 100, 4191–4196. Copyright (2003) National Academy of Sciences, U.S.A.]

et al., 1990a; Olsen and Woese, 1997). The Archaea domain is composed of organisms with diverse phenotypes, such as methane-producing methanogens, extreme halophiles, and extremely thermophilic sulfur-metabolizing species (Woese, 1987), and is divided into two taxonomic kingdoms: the *Euryarchaeota* (consisting of the methanogens, extreme halophiles, sulfate-reducing species, and two types of thermophiles) and the *Crenarchaeota* (containing the extreme thermophiles and thermoacidophiles) (Woese et al., 1990a). Typically, archaeal genes involved in energy production, cell division, cell wall biosynthesis, and metabolism have homologs in bacteria, whereas genes encoding proteins that function in the informational processes of DNA replication, transcription, and translation are more similar to their eucaryal counterparts (Bult et al., 1996). For example,

Archaea possess an eucaryallike transcription initiation machinery, consisting of the TATA-binding protein (TBP), a TFIIB homolog (TFB), and a structurally complex RNA polymerase (for a review, see Reeve et al., 1997). Archaea also share certain RNA processing components with Eucarya, such as fibrillarin (a pre-rRNA processing protein) (Bult et al., 1996; Belfort and Weiner, 1997) and tRNA splicing endonucleases (Belfort and Weiner, 1997; Kleman-Leyer et al., 1997). The mosaic nature of archaea makes this group of organisms extremely interesting from an evolutionary perspective. The sequencing and analysis of archaeal genomes should provide valuable insights into the origin or evolution of eukaryotes, as well as the molecular mechanisms enabling their adaptation to extreme environments.

The hyperthermophilic archaeon *P. furiosus* is able to grow optimally at a temperature of 100°C (Fiala and Stetter, 1986) and can metabolize peptides and carbohydrates. A maltose utilization operon is uniquely present in *P. furiosus* compared to other *Pyrococcus* species and is identical to the *mal* operon in *Thermococcus litoralis*, apparently due to a recent lateral gene transfer between *Thermococcus* spp. and *P. furiosus* (DiRuggiero et al., 2000). Studies support a highly regulated fermentation-based metabolism in *P. furiosus* (Adams et al., 2001), which can utilize the disaccharide maltose in the presence or absence of elemental sulfur ($S°$). In addition, *P. furiosus* can couple the reduction of $S°$ to the oxidation of catabolism-generated, reduced ferredoxin, but the molecular mechanism for this metabolic coupling is not presently known (Schut et al., 2001).

The availability of the complete genome sequence of *P. furiosus* (Robb et al., 2001) permits the global analysis of gene function and expression using high-density DNA microarray technology. To investigate the molecular basis of $S°$ metabolism, Schut and coworkers (2001) used DNA microarrays containing 271 ORFs (of the $\sim 2,200$ total ORFs predicted) from the *P. furiosus* genome (1.9 Mb) to measure differential gene expression in cells grown at 95°C on maltose in the presence or absence of $S°$. The arrayed PCR products represented ORFs with proposed functions in sugar and peptide catabolism, metal utilization, and the biosynthesis of cofactors, amino acids, and nucleotides. This study by Schut and colleagues (2001) represents the first and, to date, only account of the application of DNA microarray analysis to a member of the Archaea. As yet published genomic analyses of archaea have been almost exclusively limited to the sequencing, annotation, and *in silico* comparative analysis of archaeal genomes.

DNA microarray analysis revealed a number of ORFs whose expression was dramatically down-regulated ($>$5-fold decrease) by $S°$, including 18 genes encoding various subunits associated with three different hydrogenase systems (Schut et al., 2001). Other genes displaying decreased transcription when *P. furiosus* cells were grown with $S°$ encoded a hypothetical protein and two homologs (ornithine carbamoyltransferase and HypF) involved in hydrogenase biosynthesis. In the presence of $S°$, the expression of two previously uncharacterized ORFs (encoding products designated SipA and SipB for "sulfur-induced proteins") increased by a strikingly $>$ 25-fold amount. The encoded proteins of these ORFs were proposed by Schut et al. (2001) to be part of a novel $S°$-reducing, membrane-associated, iron-sulfur cluster-containing complex in *P. furiosus*. The research reported by Schut and coworkers (2001) clearly illustrates the power of DNA microarray analysis in generating new lines of experimentation and in implicating previously uncharacterized ORFs identified by genome sequencing in biological processes. No doubt, the continuing determination of archaeal genomes will spawn more microarray-based functional studies of extremophiles.

12.8 SUMMARY

Whole-genome sequencing, microarray expression profiling, and advances in computer-based homology search methodologies and functional predictions are transforming the study of pathogenic bacteria and environmentally significant prokaryotes. The growing list of completed bacterial genome sequence projects includes the majority of important human pathogenic species, as well as a number of different strains within the same species. Similarly, bacteria exhibiting phenotypes of environmental relevance and a number of extremophiles from the phylogenetic domain of Archaea can be counted among the sequenced genomes. The development of data-mining tools for bioinformatic analysis of genome sequences is improving the starting points for experimental work and providing insights into specialized functions involved in virulence, metal reduction, and radiation resistance.

Complete knowledge of a pathogenic bacterium's gene content permits a genome-scale search for predicted virulence genes based on sequence homology and the identification of candidates for vaccines and antimicrobial targets and diagnostics. For example, potential infection-related genes can be identified by scanning the genome for the presence of repeated nucleotides in mononucleotide (homopolymeric) tracts or tandemly iterated dinucleotides, tetranucleotides, and other repeats, typically located within the 5'-end of the translated reading frame. Pathogenicity islands (PAIs) can be readily identified from genomic sequence information by differences in $G + C$ content and codon bias, divergences in amino acid usage, and the presence of flanking directly repeated sequences or insertion sequence elements. PAIs are relatively large discrete segments ($\geq 10-200$ kb) of the bacterial chromosome that harbor virulence-associated genes, such as adherence factors for invasion and colonization, toxins, and secretion systems. In addition, the availability of complete genomic sequences and their comparison afford the unprecedented opportunity to assess the impact of lateral gene transfer on the evolution of microbial pathogens. For example, a good correlation exists between regions of the genome that are responsible for pathogenicity and regions that have undergone lateral gene transfer.

Microarray expression profiling is a powerful method for identifying genes of bacterial pathogens that are differentially regulated in response to host-specific signals. The application of microarray technology provides whole-genome snapshots of the transcriptomes of the pathogen and host in a time-, tissue-, and cell-specific manner. Besides the use of in vitro microarray-based expression profiling in virulence gene identification, genetic strategies like signature-tagged mutagenesis (STM) are being used to identify genes in pathogens that are essential for growth or colonization in the host animal.

The input of multiple genomic sequences also allows a comparative genomics approach to the study of bacterial pathogenicity. Recently, such comparisons have been performed with the genomes of two strains of *H. pylori*, the spirochetes *B. burgdorferi* and *T. pallidum*, the food-borne pathogen *L. monocytogenes* and the commensal *L. innocua*, *C. trachomatis* and *C. pneumoniae*, and *M. genitalium* and *M. pneumoniae*. Microarray-based comparative genomics of closely related bacterial strains and species provides a means of assessing the genetic variability within natural populations of clinical isolates and reveals crucial differences between a pathogen and commensal. Furthermore, the comparison of strains within the same genus that differ significantly in their tissue tropism and disease outcomes, like *C. trachomatis* and *C. pneumoniae*, can lead to the discovery of tissue-specific virulence factors. Such information is important in the selection of molecular targets for vaccine or antimicrobial therapy (see Chapter 13). Ultimately, the integration of proteomic approaches with computational, comparative, and gene expression-based approaches is needed to obtain a comprehensive description of disease processes. Proteomic strategies will enable the comparison of virulent strains and attenuated strains by differential protein display to identify infection-related factors that may have been lost in attenuated or nonpathogenic strains. In addition, proteomics will allow the identification of proteins that are differentially expressed in response to growth conditions of pathogenic relevance and different stress conditions that the pathogen faces in host intracellular environments.

Finally, although the genomes of a large number of environmentally significant microorganisms (e.g., archaea and metal-reducing bacteria) have been determined, functional studies incorporating DNA microarray profiling have focused largely on bacterial pathogens and model organisms. At the time of this writing, microarray analyses have been reported for only the metal-reducing bacterium *S. oneidensis* MR-1, the extreme radiation-resistant bacterium *D. radiodurans*, and the hyperthermophilic archaeon *P. furiosus*. Because of the potential bioremediation and biotechnological applications of such organisms, it is expected that the number of functional genomic analyses for environmentally important prokaryotes will increase dramatically in the future.

FURTHER READING

Alm, R. A., L.-S. L. Ling, and D. T. Moir, et al. 1999. Genomic-sequence comparison of two unrelated isolates of the human gastric pathogen *Helicobacter pylori*. *Nature* 397:176–180.

Finlay, B. B. and S. Falkow. 1997. Common themes in microbial pathogenicity revisited. *Microbiol. Mol. Biol. Rev.* 61:136–169.

Fleischmann, R. D., M. D. Adams, and O. White, et al. 1995. Whole-genome random sequencing and assembly of *Haemophilus influenzae* Rd. *Science* 269:496–512.

Glaser, P., L. Frangeul, and C. Buchrieser, et al. 2001. Comparative genomics of *Listeria* species. *Science* 294:849–852.

Hacker, J. and J. B. Kaper. 2000. Pathogenicity islands and the evolution of microbes. *Annu. Rev. Microbiol.* 54:641–679.

Huang, Q., D. Liu, P. Majewski, and L. C. Schulte, et al. 2001. The plasticity of dendritic cell responses to pathogens and their components. *Science* 294:870–875.

Ichikawa, J. K., A. Norris, M. G. Bangera, and G. K. Geiss, et al. 2000. Interaction of *Pseudomonas aeruginosa* with epithelial cells: identification of differentially regulated genes by expression microarray analysis of human cDNAs. *Proc. Natl. Acad. Sci. USA* 97:9659–9664.

Jungblut, P. R., U. E. Schaible, H.-J. Mollenkopf, and U. Zimny-Arndt, et al. 1999. Comparative proteome analysis of *Mycobacterium tuberculosis* and *Mycobacterium bovis* BCG strains: towards functional genomics of microbial pathogens. *Mol. Microbiol.* 33:1103–1117.

Ochman, H. and N. A. Moran. 2001. Genes lost and genes found: evolution of bacterial pathogenesis and symbiosis. *Science* 292:1096–1098.

Porcella, S. F. and T. G. Schwan. 2001. *Borrelia burgdorferi* and *Treponema pallidum*: a comparison of functional genomics, environmental adaptations, and pathogenic mechanisms. *J. Clin. Invest.* 107:651–656.

Salama, N., K. Guillemin, and T. K. McDaniel, et al. 2000. A whole-genome microarray reveals genetic diversity among *Helicobacter pylori* strains. *Proc. Natl. Acad. Sci. USA* 97:14668–14673.

Weinstock, G. M. 2000. Genomics and bacterial pathogenesis. *Emerging Infect. Dis.* 6:496–504.

13

The Impact of Genomics on Antimicrobial Drug Discovery and Toxicology

Dorothea K. Thompson and Jizhong Zhou

13.1 INTRODUCTION

Presently, there is an urgent need for novel drugs of desirable clinical utility due to the growing incidence of antibiotic resistance among microbial pathogens (Allsop, 1998). As microorganisms continue to evolve and adapt at a remarkable rate, many have developed resistance mechanisms to every class of antibiotic currently in clinical use, and an escalating number of bacteria have acquired multiple resistance systems for avoiding or reducing the action of an antibiotic (Gold and Moellering, 1996; Sefton, 2002). Hence, the immediate challenge of drug discovery is to develop novel antimicrobial agents possessing new mechanisms of action instead of continuing to focus on the development of new analogs of known drug classes. Genome-scale DNA sequence information offers an unprecedented variety of potential molecular targets for drug development, and newly developed genomic technologies promise to impact all facets of the drug discovery and development process, from the initial stage of molecular target selection to lead ("drug candidate") identification, optimization, and validation (Cockett et al., 2000; see Fig. 13.1). A significant challenge faced by the pharmaceutical industry is the efficient screening of new compounds to assess drug toxicity at an early stage in the development process. In addition to enhancing the drug discovery process, genomics is expected to impact the assessment of drug safety by reducing the high failure rate of new drug candidates due to toxicity (Lakkis et al., 2002). The applications of microarray technologies in drug discovery are expanding rapidly and presently include basic research

Microbial Functional Genomics, Edited by Jizhong Zhou, Dorothea K. Thompson, Ying Xu, and James M. Tiedje.
ISBN 0-471-07190-0 © 2004 John Wiley & Sons, Inc.

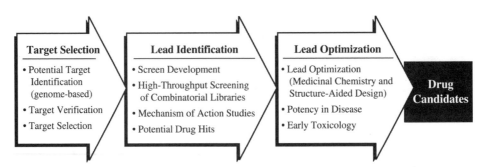

Figure 13.1 *Major stages in the drug discovery process.* [Modified with permission from J. Rosamond and A. Allsop, Harnessing the power of the genome in the search for new antibiotics, *Science*, Vol. 287, 1973–1976. Copyright (2000) American Association for the Advancement of Science.]

and drug discovery, biomarker determination, pharmacology, toxicogenomics, target selectivity, and the development of prognostic tests (reviewed in Butte, 2002).

In this chapter, we discuss the impact of bacterial genome sequencing, bioinformatics, and genome-based technologies on the discovery of new drugs to treat microbial infections and on studies of toxicity. The integration of computational genome analysis (e.g., comparative genomics, motif analysis) and experimental functional genomics (e.g., microarrays, proteomics) has already begun to affect the process of antibacterial drug discovery by facilitating the paradigm shift from direct antimicrobial screening programs to rational target-based strategies. Here we first present a brief historical overview of antibacterial drug discovery, followed by discussions on the challenges of new drug discovery, the impact of microbial genomics on target identification, and the application of genomic-scale experimental technologies to target validation and drug screening. We will conclude with a discussion of the related topic of toxicogenomics, which is concerned with identifying the molecular mechanisms of action that underlie the potential toxicity of chemicals, environmental pollutants, and drug candidates through the use of genomics resources. As we will illustrate, microarray technology offers an ideal platform for analyzing toxicological processes at the transcriptome level for the prediction of potential drug toxicity (Nuwaysir et al., 1999).

13.2 ANTIBACTERIAL DRUG DISCOVERY: A HISTORICAL PERSPECTIVE

The discovery of antibacterial agents like penicillin, cephalosporin, streptomycin, and the bacitracins during the 1930s and 1940s ushered in the "antibiotic age" in the pharmaceutical industry. Following the introduction of penicillin into medical practice, most of the existing classes of antibiotics were discovered by systematic screening of large libraries of natural products (many of which were generated by soil microorganisms) for their ability to kill or inhibit bacteria. The assay systems employed for such random screening of natural-product libraries were phenomenological rather than target-based (Vidal and Endoh, 1999).

The production of antibiotics as a result of microbial secondary metabolism, indeed, occupies a prominent position in the history of drug discovery. As much as 40% of the drugs in clinical use today were derived from such natural sources as microorganisms and plants, and of the medically important antibiotics, the majority were produced by the species of three main bacterial and fungal genera, *Streptomyces*, *Aspergilli*, and *Penicilli*

(Hutchinson, 1998). Furthermore, current widely used antibiotics discovered by using the classical approach were found to inhibit a small number of vital cellular functions: (1) cell wall peptidoglycan synthesis (e.g., beta-lactams, glycopeptides); (2) DNA or RNA synthesis (e.g., quinolones, novobiocin, rifampin, metronidazole); and (3) protein synthesis (e.g., tetracyclines, chloramphenicol, fusidic acid) (Moir et al., 1999). For example, tetracyclines target the 30S ribosomal subunit required for the synthesis of cellular protein, and rifampin inhibits RNA synthesis by targeting the beta subunit of DNA-dependent RNA polymerase (for a review, see Chopra, 1998). A number of other critical cellular functions or pathways, however, such as protein secretion, signal transduction, cell division, and many metabolic activities (e.g., energy production), remain untargeted and, therefore, fertile areas for drug target exploration. For the past 30 years, the pharmaceutical industry has responded to the absence of new antibiotic classes by synthetically modifying antibiotics of existing chemical and structural classes (Hancock and Knowles, 1998).

In an alternative approach, drug discovery and development in pharmaceutical companies have relied historically on targeting proteins involved in well-studied biochemical pathways implicated in pathophysiological processes (Gmuender, 2002). Targets selected for therapeutic intervention were typically enzymes or receptors. The enzyme, preferably one that catalyzes a rate-limiting step in an essential pathway, was characterized, purified, and then screened against collections of structurally diverse small molecules to identify lead molecules possessing inhibitory activity against the purified enzyme (Debouck and Goodfellow, 1999; Gmuender, 2002). Occasionally, a screened chemical library could be limited to more specific classes of small molecules if sufficient information concerning the mechanism of action and structure of the enzyme were available. Toward the goal of optimizing lead compounds identified by biochemical in vitro screening, medicinal chemists worked at endowing potential therapeutic compounds with desirable drug-associated properties, such as bioavailability, high specificity for the target enzyme and antimicrobial activity, the ability to penetrate into bacterial cells, and no adverse drug effects (Gmuender, 2002). This stage of the drug discovery and development process, referred to as lead optimization (or chemical synthesis with structural design support), was then followed by in vivo validation using animal models. In some cases, however, no lead compounds were found. In other cases, lead compounds that scored well in terms of selectivity and sensitivity in vitro turned out to be ineffective as therapeutic agents, because these same compounds exhibited poor bioavailability or stability in animal models (Gmuender, 2002). In addition, the molecular mechanisms of action and toxicities of lead compounds were often incompletely defined. Nonetheless, the traditional biochemistry-based approach to drug discovery provided a number of effective antibiotics and therapeutic agents for a variety of diseases.

With the advent of genomics, the evolution of the drug-discovery field has entered a new phase. Historically, target identification was a major hurdle in the drug discovery process, because the number of potential antiinfective targets was limited by the number of cloned genes (Gmuender, 2002). The prodigious wealth of available genomic data, however, has substantially increased the number of potential therapeutic targets for drug design and discovery. For example, it is estimated that the approximately 1,000 therapeutic molecular targets currently used by the pharmaceutical industry will increase dramatically to as many as 10,000 as a result of the explosion of information from whole-genome sequencing (Dean et al., 2001). The integration of genomic data, bioinformatic analysis, novel genetic approaches, and genomics-based technologies (in particular, microarrays) has altered the traditional strategy of drug discovery from a random screening exercise to a process of rational target-based drug design (Allsop, 1998). The genomics-based approach to drug discovery is schematically depicted in Fig. 13.2. Also of relevance to the drug discovery and development process is the advent of preclinical chemical genomics, toxicogenomics (discussed later in this chapter), and pharmocogenomics within clinical research. The chemical genomics approach has emerged as a result of the combined utilization of chemical libraries, screening, and genomics technologies in an attempt to validate drug targets (Cockett et al., 2000; Ward, 2001). In this synergistic approach, large numbers of potential target proteins are used in standardized high-throughput, drug-screening assays to investigate protein–protein interactions. The merging of genomics and pharmacology has created the subdiscipline, pharmacogenomics, which seeks to identify genetic markers (e.g., genetic polymorphisms) that predict the response of individuals to drugs (i.e., interindividual differences in the efficacy and toxicity of drugs) (Evans and Relling, 1999; Gmuender, 2002). The primary goal of pharmacogenomics is to associate human sequence polymorphisms with drug metabolism, adverse effects, and therapeutic efficacy (Harris, 2000). It is important to point out, however, that genomics-based drug discovery and development are still in their infancy, and the full extent to which the

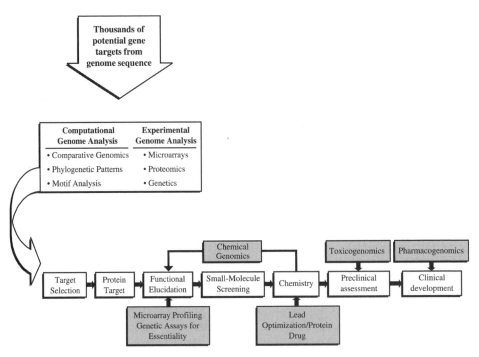

Figure 13.2 *Target-based drug discovery (adapted from Cockett et al., 2000).* The availability of genomic sequence information from numerous bacterial pathogens has dramatically increased the choice of potential molecular targets. Mining microbial genomes for target and drug discovery involves both computational and experimental genome analyses, which can be used in an integrated manner to select gene targets and prioritize drug candidates. Target validation (functional elucidation) can potentially be expedited using genomic-scale technologies, such as microarray-based gene expression profiling. Chemical genomics, the combination of chemistry and screening, may help to streamline the drug discovery process. Toxicogenomics and pharmacogenomics are expected to impact the stages of preclinical assessment and clinical trials, respectively.

perceived value of genomics will translate into innovation and enhanced productivity in pharmaceutical research is not yet known (Ward, 2001). In the sections that follow, we will focus on how microbial genomics has transformed the pharmaceutical industry's capability for developing new antibacterial drugs. An in-depth discussion of chemical genomics and pharmacogenomics, however, is beyond the scope of this chapter.

13.3 CHALLENGES OF NEW DRUG DISCOVERY

Almost all antimicrobial agents currently in medical use are the products of semirational optimization programs based largely on natural compounds demonstrated to possess antimicrobial activity in whole-cell screening assays (Rosamond and Allsop, 2000). These limited classes of antibiotics have failed to deliver a sufficient range of molecular diversity in the clinical arena, resulting in the erosion of their efficacy by the emergence of resistant bacteria. Presently, a major challenge for the pharmaceutical industry is the discovery and development of new classes of drugs with novel mechanisms of action that will provide effective therapy, at least for a period of time. Sequenced genomes offer a plethora of potential molecular targets for pharmaceutical drug research and development. Along with this advantage, however, is the challenge of identifying the most promising novel targets from the numerous possibilities that will reduce the occurrence of preclinical and clinical failures. In this section, we present a brief review of the problem of antibiotic resistance and then discuss the criteria that each new antibiotic must meet to be an effective therapeutic agent.

13.3.1 Resistance to Antimicrobial Agents and the Need for New Antibiotic Discovery

The greatest threat to the treatment of infectious diseases is the acquisition of resistance mechanisms in many clinically important microorganisms. The traditional reliance of antimicrobial drug discovery on random screening or semirational modification of known structural classes has failed to generate the chemical variability needed to prevent a serious expansion in clinical resistance (Rosamond and Allsop, 2000). Bacteria have evolved a variety of effective resistance mechanisms to survive exposure to every class of antibiotic used therapeutically (Allsop, 1998). Indeed, it is an illustration of the remarkable adaptive power of microorganisms that every antibiotic placed into clinical use has engendered resistance, regardless of its chemical class or molecular target (Davies, 1997). This microbial adaptation has been rapid and widespread, leading to the emergence of antibiotic-resistant clinical profiles among phylogenetically diverse gram-negative and gram-positive infectious bacteria. Compounding this important public health problem, microorganisms also harbor mechanisms for antibiotic resistance gene capture and dissemination between different bacterial strains and species (Davies, 1994; Rowe-Magnus et al., 2002). Indeed, evidence suggests that the evolution of heritable resistance to any new antimicrobial agent is probably inevitable, given time and selection pressure (Silver and Bostian, 1993).

Resistance to antimicrobial drugs is not only a serious clinical problem in the hospital environment but in the community as well (Cohen, 2000). More than 40% of the *Staphylococcus aureus* strains acquired in the hospital are resistant to methicillin, and an increasing number of methicillin-resistant strains are exhibiting resistance to vancomycin, often the therapeutic agent of last resort. Other nosocomial (hospital-acquired) infections are caused by vancomycin-resistant enterococci and azole-resistant *Candida*. In communities, the emergence of multidrug-resistant pneumococci is particularly worrisome. The incidence of antibiotic-resistant *Salmonella*, *Shigella*, *Neisseria gonorrheae*, and *Mycobacterium tuberculosis* is also rising in the community. Obviously, there is a critical need to discover structurally novel classes of drugs that are not susceptible to existing antimicrobial resistance mechanisms and thus can forestall the emergence of acquired resistance. This is a major challenge for the pharmaceutical industry in the twenty-first century. One approach to enhancing bacterial susceptibility to antimicrobials is to target the components of the molecular mechanisms that enable drug-resistance phenotypes.

Integrons are the primary vehicles enabling the capture and spread of multiple antibiotic resistance determinants among diverse gram-negative bacteria (Rowe-Magnus et al., 2001, 2002); however, plasmids and transposons can effect the mobilization of antibiotic resistance genes as well. Integrons are genetic elements that contain the components of a site-specific recombination system, which is capable of capturing and mobilizing gene cassettes (Hall and Collis, 1995). An integron can harbor a single antibiotic resistance gene or various combinations of antibiotic resistance gene cassettes for the dissemination of multidrug resistance. A resident promoter drives expression of the proteins encoded by the cassette genes, thus making integrons natural cloning and expression systems (Hall and Collis, 1995). The architecture of the integron includes an integrase (*intI*) that mediates site-specific recombination between an adjacent site (*attI*) and a second site (*attC*), which is usually found to be associated with a gene (most commonly an antibiotic resistance gene). The *attC*-ORF structure is referred to as a gene cassette. A single site-specific recombination event involving the integron-associated *attI* site and the *attC* site results in the insertion of the gene cassette into the recipient integron at the *attI* site, which is located downstream of a promoter responsible for expression of the cassette-encoded gene.

In order for an antibiotic to inhibit an intracellular molecular target and subsequently kill the bacterium, it has to accumulate within the bacterial cytoplasm. A widespread mechanism in bacteria that contributes to antibiotic resistance is drug exclusion by various efflux systems. Both gram-negative and gram-positive bacteria express numerous membrane transporters that promote the efflux of antibiotics and other drugs from the cell to the surrounding milieu (see reviews by Markham and Neyfakh, 2001; Poole, 2001), and microbial genomics has confirmed the wide distribution of these efflux systems among bacteria. The restricted permeability of the outer membrane of gram-negative bacteria works in synergy with active efflux mechanisms or enzymes that disable antibiotics (e.g., β-lactamases) to make gram-negative microorganisms generally more resistant to antimicrobials than gram-positive microorganisms (Hancock, 1997; Poole, 2001). Efflux systems can be drug-specific or can display broad substrate specificity, as in the case of multidrug transporters.

Five classes of efflux systems capable of transporting multiple antimicrobial compounds have been identified in gram-negative bacteria: (1) the major facilitator superfamily (MFS), (2) the ATP-binding cassette (ABC) family, (3) the resistance-nodulation-division (RND) family, (4) the small multidrug resistance (SMR) family, and (5) the multidrug and toxic compound extrusion (MATE) family. Members of the RND family figure prominently in multidrug resistance to clinically relevant antimicrobials

(Poole, 2001; Zgurskaya and Nikaido, 2000). RND transporters act in conjunction with a periplasmic membrane fusion protein (MFP) and an outer membrane factor (OMF) (Zgurskaya and Nikaido, 2000). RND/MFP/OMF-type multidrug efflux systems, which are typically chromosomally encoded, have been described for a number of organisms, including *Escherichia coli*, *Salmonella typhimurium*, *Haemophilus influenzae*, *Neisseria* spp., *Pseudomonas aeruginosa*, *Pseudomonas putida*, and *Burkholderia* spp. (Poole, 2001). RND family multidrug transporters are predicted to function as drug–proton antiporters, using the transmembrane electrochemical gradient or proton motive force to exchange protons in the outer membrane for drug molecules. This molecular mechanism, however, has been demonstrated only in the case of the *E. coli* AcrAB-TolC efflux system (Zgurskaya and Nikaido, 1999).

Drug efflux transporters of gram-positive bacteria belong to the MF, SMR, and ABC families of membrane proteins. Active efflux mechanisms contribute significantly to the resistance of gram-positive pathogens (e.g., *Staphylococcus epidermidis*) to macrolide, lincosamide, and streptogramin (MLS) antibiotics, which are functionally related but distinct chemical classes that inhibit protein synthesis by attacking the 50S ribosomal subunit of the protein synthesis machinery (Markham and Neyfakh, 2001). Active efflux of macrolides accounts for as much as a 64-fold increase in resistance levels in *Streptococcus pneumoniae* (Sutcliffe et al., 1996). The efflux-mediated decreased accumulation of macrolides requires the chromosomally located *mefA* or *mefE* genes, which encode transporters belonging to the major facilitator (MF) superfamily (Tait-Kamradt et al., 1997). The Mef transporter is specific for macrolides and does not confer resistance to the antibiotics lincosamides and stretogramins. Furthermore, the macrolide efflux gene *mef* of *S. pneumoniae* resides within a transposable genetic element (Santagati et al., 2000), and recently, the transfer of *mef* genes has been further substantiated by their identification in a number of other gram-positive bacteria (Luna et al., 1999), as well as gram-negative species (Luna et al., 2000). In addition, Mef transporters are prevalent in macrolide-resistant strains: The Mef transporter accounts for the majority (55–70%) of all macrolide-resistant clinical isolates of *S. pneumoniae* reported in the United States and Canada (Johnston et al., 1998). In the gram-positive bacterium *S. epidermidis*, resistance to macrolides and streptogramins is provided by the plasmid-borne genes, *msrA* and *msrB*, which code for an efflux-mediated resistance mechanism. Msr belongs to the ABC (ATP-binding cassette) family of transporters. However, the Msr efflux system appears to be of limited importance in *S. aureus*, for which only 13.6% of isolates resistant to at least one MLS antibiotic actually contain the *msrA* gene (Lina et al., 1999).

Bacterial resistance to tetracyclines, an antibiotic class that inhibits bacterial protein synthesis by targeting the 30S ribosomal subunit, has compromised the treatment of a variety of bacterial diseases of the respiratory, urinary, and digestive tracts. More than 20 different tetracycline (Tet) resistance determinants (designated by letters) have been identified (for a review, see Roberts, 1996). Tet resistance determinants mediate resistance either by active efflux or ribosomal protection. Genes of tetracycline efflux transporters, mainly from the MF superfamily, have been identified in both gram-negative bacteria (classes A–E, G, and H) and gram-positive bacteria (K, L, P, V, Z, and OtrB). Tet resistance classes M, O, and S among *Streptococcus* spp., *Staphylococcus* spp., and *Listeria* spp. and class Q among *Bacteroides* species consist of cytoplasmic proteins that impart resistance by protecting ribosomes from the inhibitory action of tetracyclines (Burdett, 1996). To combat tetracycline resistance in bacteria, two therapeutic approaches are currently being adopted. The first approach is based on the synthesis of tetracycline derivatives, for example, glycyclines, that possess a higher activity for the ribosomal binding site and that are not susceptible to exclusion by efflux mechanisms (Someya et al., 1995). The second approach uses tetracycline analogs to block efflux transporters, so that intracellular concentrations of tetracycline can increase to levels that effectively nullify Tet resistance (Nelson and Levy, 1999).

13.3.2 Desirable Properties of Antimicrobial Targets

The availability of complete genome sequences and high-throughput genomic technologies promise to accelerate molecular target discovery and facilitate the validation of new molecular targets. With the birth of microbial genomics, the challenge confronting pharmaceutical research has changed from one of searching for potential drug targets to selecting the best target candidates from a complete list of gene products (Galperin and Koonin, 1999). The preselection of appropriate drug targets from whole-genome data is the crucial first step in drug discovery and development, and the application of bioinformatics/sequence annotation plays an important role in this initial stage. Potential

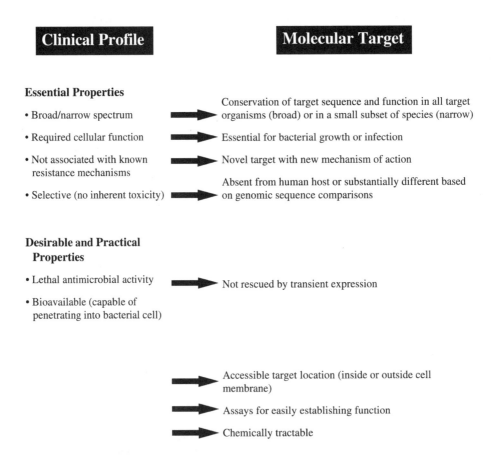

Figure 13.3 *Translation of essential, desirable, and practical clinical properties to molecular target properties (information source: Allsop, 1998).*

molecular targets in a genome sequence must be filtered through a set of key criteria that evaluate the likelihood a target candidate will offer effective therapeutic intervention. In other words, as depicted in Fig. 13.3, molecular targets can be defined according to the required clinical properties for a given antibacterial treatment (Loferer et al., 2000).

In general, antimicrobial targets should meet the criteria of spectrum, selectivity, functionality, and essentiality (Moir et al., 1999). First, a molecular target should provide adequate spectrum and selectivity, yielding an antimicrobial agent that is active against a desired spectrum of bacterial pathogens but is selective in that it is not harmful or toxic to the human host. For broad-spectrum applications, bioinformatic screening of targets via *in silico* comparative genomic analysis is performed to identify genes that have orthologs in many evolutionarily distant microorganisms but exhibit little or no conservation in humans (Loferer et al., 2000; Read et al., 2001). Similarly, gene candidates can be chosen for their presence in a small subset of sequenced bacterial genomes for the development of narrow-spectrum antibacterial compounds (Loferer et al., 2000). Second, a target should be essential for growth or viability of the pathogen, so that inhibiting or disrupting a specific target will either kill the bacterial cell or affect its viability sufficiently to cure an infection. This category could also include factors essential for infectivity or survival of the pathogen in the host cell. The essentiality of a selected target is then commonly validated by gene knockout strategies. Recently, molecular targets involved in resistance against antibacterial agents are also being considered in the development of therapeutics (Schmid, 1998). Third, some knowledge on the function of the target is necessary in order to design assays and high-throughput screens for identifying candidate lead molecules, which are eventually developed into new antimicrobial drugs.

13.4 MICROBIAL GENOMICS AND DRUG TARGET SELECTION

Comprehensive genome sequencing and bioinformatics are driving the discovery and development of novel classes of antimicrobial agents. In the "pregenomic sequencing era," the selection of potential targets for therapeutic purposes was limited to the small number of cloned and functionally characterized genes. With the publication of the complete *Saccharomyces cerevisiae* (yeast) genome sequence (Goffeau et al., 1996) and more than 70 complete microbial genomes, including those of numerous bacterial pathogens, all the genes residing in these organisms are now available as potential molecular targets for therapeutic intervention. The goal of the computational analysis of microbial genomes is to simplify and prioritize target selection for antibacterial screening by extracting the maximum amount of information *in silico* to facilitate this process (Loferer, 2000). This section describes the application of bioinformatics and comparative genomics to the search for potential molecular targets. Transcript profiling using DNA microarrays is discussed from the perspective of target identification (establishing the functionality of annotated and unknown genes).

13.4.1 Mining Genomes for Antimicrobial Drug Targets

Knowledge of gene function is paramount in rational target-based drug discovery. In contrast, the identification of genes, whether they be functionally known or unknown, that encode surface-exposed antigens is most important in vaccine discovery (Zagursky and Russell, 2001). Understanding the functionality associated with bacterial genes is necessary for facilitating the development of suitable target-based screening assays. Once open reading frames (ORFs) contained within a genomic sequence are predicted, the next step is to perform homology-based analyses to determine whether the gene encodes a protein similar to one of known function (see Chapter 5 for a detailed description of the bioinformatic tools used to define gene function). In computational genome analysis, the two primary algorithms used for homology searching are basic local alignment search tool (BLAST; Altschul et al., 1990) and FASTA (Pearson and Lipman, 1988; see Chapter 5). In addition, the concept of orthologous families in establishing the clusters of orthologous groups (COGs) provides a framework for functional and evolutionary genome analysis (see Chapter 4 for a detailed discussion). From a drug discovery viewpoint, genes that have orthologs in a wide variety of evolutionarily distant species are most useful as potential molecular targets for broad-spectrum antimicrobial drugs (Loferer, 2000).

The size and completeness of genome sequences can present challenges in the search for drug targets. The development of improved informatics is one approach to selecting better candidates from the impracticably large number of potential candidates generated as a consequence of genome sequencing. The computational method of phylogenetic profiling (Pellegrini et al., 1999), for example, enhances conventional sequence-alignment techniques by predicting putative functions for uncharacterized genes. Phylogenetic profiling is based on the assumption that proteins participating in a common metabolic pathway, structural complex, biological process, or closely related physiological function (i.e., proteins that are functionally linked) evolve in a correlated fashion. Functionally linked proteins, in general, do not possess any amino acid sequence similarity with each other. As illustrated in Fig. 13.4, each protein is characterized by its phylogenetic profile, a string that encodes the presence or absence of homologs across a set of organisms whose genomes have been sequenced (Pellegrini et al., 1999). A functional linkage between two proteins is inferred if the proteins have the same phylogenetic profile (i.e., the same pattern of presence and absence) in all the genomes surveyed. In this way, the function of an uncharacterized protein can be equated with that of its neighbors in phylogenetic-profile space (Pellegrini et al., 1999). Additional members of a relevant pathway, from a drug discovery perspective, can be identified by searching the genome for proteins with phylogenetic profiles similar to those of known pathway components.

Another computational approach to mining microbial genomes for antibacterial targets is motif analysis. The combination of extensive DNA and protein sequence archives and substantial amounts of experimental data has enabled the identification of signature sequence motifs indicative of certain biochemical activities (e.g., proteases). Among the genes of unknown function identified by Arigoni and colleagues (1998) as potential molecular targets for antimicrobial therapy, three targets were predicted to encode nucleotide-binding proteins and a metalloprotease using motif analysis. In prioritizing the target candidates, the putative metalloprotease would be favored for drug design and development since inhibitors of metalloproteases already exist (Loferer, 2000). Recently, the need for new antibiotics with innovative modes of action has led to the proposal of targeting signal transductory histidine kinases in order to combat the growing problem of hospital-acquired multidrug-resistant bacteria (Matsushita and Janda, 2002). As integral members of two-component regulatory systems, histidine kinases play an important role in essential signal transduction for the adaptation of bacterial cells to changes in their environment. Two-component systems are central in the ability of bacterial pathogens to mount and establish a successful infection within the host; consequently, components of such systems are viewed as targets for new antimicrobial agents.

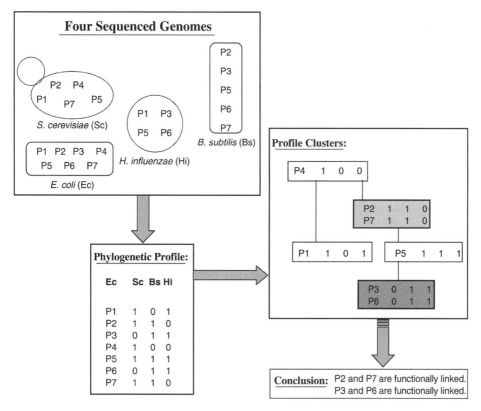

Figure 13.4 *Representation of phylogenetic profiling method for predicting protein function (Pellegrini et al., 1999).* This method is schematically illustrated using a hypothetical case focusing on seven proteins (P1-P7) from four sequenced organisms (*Escherichia coli, Haemophilus influenzae, Bacillus subtilis,* and *Saccharomyces cerevisiae*). Phylogenetic profiles are first constructed for *H. influenzae, B. subtilis,* and *S. cerevisiae* in order to show which organisms contain homologs of the seven *E. coli* proteins. Clustering of phylogenetic profiles then allows one to determine which proteins share the same profiles. Functionally linked proteins are indicated by identical (or similar) profiles. [Reprinted with permission from M. Pellegrini, E. M. Marcotte, M. J. Thompson, D. Eisenberg, and T. O. Yeates, *PNAS,* Vol. 96, 4285–4288. Copyright (1999) National Academy of Sciences, U.S.A.]

Recently, two new computational methods have been reported that do not rely on direct sequence homology for the prediction of protein function (Enright et al., 1999; Marcotte and coworkers 1999b). In the method described by Marcotte et al. (1999b), proteins are grouped by phylogeny (correlated evolution), correlated messenger RNA expression patterns, and patterns of domain fusion to identify functional linkages between proteins. The phylogenetic profiling method (Pellegrini et al., 1999) identifies functionally associated proteins based on the assumption that proteins which are always inherited together operate together. In the fused domain method (Marcotte et al., 1999a), protein interactions are inferred from genome sequences based on the observation that some pairs of functionally related proteins (called *component* proteins) have homologs in different species that are fused into the same polypeptide chain (called a *composite* protein). By combining theoretical prediction (phylogeny and domain fusion) with experimental data (gene expression profiles), the function of uncharacterized proteins can be predicted by links with characterized proteins (Marcotte et al., 1999b). The utility of such bioinformatic approaches to protein function prediction has been demonstrated by the assignment of a general function to approximately half of the 2,557 previously uncharacterized yeast proteins (Marcotte et al., 1999b). A potentially even more powerful approach to assigning previously unknown molecular function to a protein is direct comparison of three-dimensional (or 3D) protein structures, because protein function is more directly a consequence of protein conformation than amino acid sequence (Grigoriev and Kim, 1999; Rosamond and Allsop, 2000). Proteins with weak sequence similarity have been found to assume similar three-dimensional folds, and this is consistent with the relatively limited number (∼ 700) of protein folds observed from three-dimensional structures. As more three-dimensional protein structures are resolved, the field of structural genomics is likely to have a significant impact on the drug discovery and development process by providing computational modeling strategies that offer improved functional predictions compared to alignments of amino acid sequences (Grigoriev and Kim, 1999).

13.4.2   Comparative Genomics: Assessing Target Spectrum and Selectivity

From the perspective of drug discovery, a complementary approach to searching individual microbial genomes for potential molecular targets is provided by comparative genomics (for an in-depth treatment of this field, see Chapter 4). Comparative genomics is a useful approach to the assessment of target spectrum and selectivity, two criteria used by the pharmaceutical industry to evaluate potential therapeutic utility. As determined by comparison of microbial genome sequences, pools of gene families that are highly conserved among different bacteria but absent in eukaryotes offer the opportunity to develop broad-spectrum antibiotics (Tatusov et al., 1997). In this direction, Arigoni and colleagues (1998) used comparative genomics integrated with targeted gene disruptions to identify essential bacterial genes. Twenty-six genes with no predictable function in the model organism *E. coli* were broadly conserved in the genomes of *Bacillus subtilis*, *Mycoplasma genitalium*, *H. influenzae*, *Helicobacter pylori*, *S. pneumoniae*, and *Borrelia burgdorferi*. Arigoni and coworkers (1998) reasoned that this set of conserved, functionally unknown genes represented novel molecular targets for broad-spectrum antibiotic development. With a gene knockout strategy, null mutations in the 26 genes were screened for lethality, and six of these genes were subsequently verified to be essential in *E. coli*. Furthermore, orthologs of the six essential *E. coli* genes were shown to be essential in *B. subtilis* (Arigoni et al., 1998). Experimental demonstration of the essentiality of a potential target gene, however, is only the first step in the process of developing an antimicrobial drug. Considerable followup research is needed to establish a function for the protein product of an uncharacterized essential gene and to validate it as an effective therapeutic target. Comparative genomic analysis also indicated that 15 of the 26 *E. coli* proteins displayed significant sequence similarity to proteins in the unicellular eukaryote *S. cerevisiae*, suggesting that antimicrobial drugs directed toward these targets might possess inherent toxicity in humans (Arigoni et al., 1998). Another criterion used by the pharmaceutical industry to judge the potential utility of drug candidates is whether they offer selectivity over humans. However, some sequence conservation with mammalian proteins among marketed antimicrobial agents is not without precedence (Moir et al., 1999).

In addition to revealing highly conserved targets for broad-spectrum applications, comparative analysis of microbial genomes has also indicated that a significant proportion (approximately 30–40%) of sequenced genomes encode proteins with no known function based on sequence homology. Some of these uncharacterized genes are likely to be specific to the individual organism and may serve as potential molecular targets for the development of narrow-spectrum antibiotics or drugs with a high degree of specificity for a single organism (Rosamond and Allsop, 2000). This approach is useful for identifying target candidates that are specific to the spectrum of bacterial species which cause a given disease, such as isolates of the human gastric pathogen *H. pylori* in the case of duodenal ulcers (Alm et al., 1999; Loferer, 2000). While such narrow specificity potentially reduces occurrences of cross-resistance, the development of narrow-spectrum antibiotics is hindered by the fact that most species-specific genes are unannotated and require functional elucidation, making it difficult to conduct target-based screens for potential drug candidates (Rosamond and Allsop, 2000). Similarly, genes of general importance to pathogenicity would appear to be the most obvious choice for a drug target. *Differential genome display* suggests that genes present in the genome of a pathogenic bacterium, but absent in a closely related genome of a noninfectious bacterium, are therefore likely to be of particular relevance to pathogenicity. Comprehensive comparison of *H. influenzae* and *E. coli* gene products, for example, revealed 40 *H. influenzae* genes that exclusively reside in the genomes of pathogenic bacteria and thus hold great promise as drug targets (Huynen et al., 1997). In addition, comparative analysis of complete genomes has revealed that pathogens possess diminished biosynthetic capabilities and consequently depend on well-conserved membrane-associated transport systems to obtain essential nutrients from their hosts (Clayton et al., 1997). Therefore, bacterial transport proteins constitute attractive drug targets as well.

13.4.3 Genetic Strategies: Verifying the Essentiality or Expression of Gene Targets

Drug discovery based on approaches that combine molecular biology and conventional drug screening has the potential of expediting the identification of antimicrobials with unique modes of action. Depending on the genetic amenability of the microbial species targeted for antimicrobial therapy and the availability of genetic tools, a variety of molecular genetic methods can be used to validate essential genes or discover genes important for establishing pathogenic growth. Traditionally, individual genes can be assessed for essentiality by gene knockout strategies using transposon systems or directed allele exchange, or by mutating genes to a conditional-lethal form. A more recent strategy utilizes an antisense-based functional genomics approach to identify new antifungal drug targets. De Backer and colleagues (2001) combined antisense RNA inhibition and promoter interference (see Chapter 8 for a description) to identify genes critical for the growth of *Candida albicans*, the major pathogen causing human fungal infections. The genes identified in this study were subsequently used as targets to identify new antifungals in a drug screen (De Backer et al., 2001).

The clinical need to identify genes associated with pathogenic processes has been the impetus behind the development of at least three different genetic approaches. Signature-tagged mutagenesis (STM), a transposon-based gene tagging method, has been used to identify genes that are essential for the establishment and maintenance of a bacterial infection. This technique is described in detail in Chapters 8 and 12. Briefly, a pool of sequence-tagged insertion mutants is used to infect an animal host. Genes essential for virulence are revealed when mutations represented in the initial inoculum are not recovered from the host (Hensel et al., 1995). For example, Chiang and Mekalanos (1998) used STM to identify genes that are critical for the colonization of *Vibrio cholerae* during an infection.

Another approach to identifying genes associated with pathogenic processes is a genetic promoter trap method termed in vivo expression technology (IVET). IVET is designed to identify genes specifically induced during an infection by selecting bacterial promoters driving the expression of a gene required for growth within the host (Heithoff et al., 1997; Mahan et al., 1993, 1995). A novel aspect of this genetic system is that infected mammalian tissues are used directly to induce the expression of candidate virulence genes rather than attempting to reproduce the in vivo milieu in the laboratory. The IVET genetic system, for example, was used to identify more than 100 *S. typhimurium* genes that are specifically expressed during infection of BALB/c mice and murine-cultured macrophages (Mahan et al., 1993, 1995). In vivo induction profiles indicate a collection of in vivo induced (*ivi*) genes exhibiting a broad array of regulatory, metabolic, and virulence functions that contribute to enhanced growth and persistence in host tissues (Heithoff et al., 1997). Although pioneered in the gram-negative bacterium *S. typhimurium*, IVET has been applied to the study of virulence in *S. aureus* (Lowe et al., 1998) and *P. aeruginosa* (Handfield et al., 2000). The IVET system developed for *S. aureus* uses genetic recombination as a reporter of gene activation in vivo. With this approach, Lowe and coworkers (1998) identified 45 staphylococcal genes that are induced preferentially during infection in a murine renal abscess model.

Differential fluorescence induction (DFI) is a third selection strategy for identifying genes specifically expressed when a bacterium associates with its host cell. DFI is a promoter trap method that takes advantage of high-throughput, semiautomated fluorescence activated cell sorting to measure intracellular fluorescence in individual bacterial cells (Valdivia and Falkow, 1997). Two features distinguish this strategy: (1) the gene used for selection encodes a modified green fluorescent protein, and (2) the selection is accomplished with a fluorescence-activated cell sorter (Moir et al., 1999). Genetic methods such as STM, IVET, and DFI are important for establishing the essentiality or significance of potentially new antimicrobial target genes associated with virulence.

13.4.4 Microarray Analysis: Establishing Functionality for Novel Drug Targets

Genome-wide changes in gene expression are a powerful indicator of the effects of therapeutic agents on cells, tissues, or whole animals. Tissue-specific gene expression and the preferential transcription of genes during disease or infection states can implicate the potential relevance of genes to antimicrobial drug discovery (Cockett et al., 2000). Hence, the information provided by microarray (DNA chip)-based expression profiling is expected to be of great importance in the drug development process (Braxton and Bedilion, 1998). DNA microarrays may be used to generate clues to gene function that can facilitate the identification of novel molecular targets for therapeutic intervention and to monitor changes in global gene expression underlying cellular responses to drug treatments. Thus, microarrays have the potential to define the mechanistic basis of many drugs. The use of DNA microarrays in drug target validation (i.e., confirmation that a compound inhibits the intended target) will be discussed later in this chapter. At this point, we will focus on the application of microarray-based transcription profiling to deciphering the function of potential molecular targets for drug therapy and as a comparative genomics tool for screening clinical isolates of disease relevance for conserved genes.

In modern drug discovery, genomics is used to assign a cellular function to new gene targets in a high-throughput fashion (Cockett et al., 2000). DNA microarray technology, which permits the assessment of gene expression in a massively parallel manner, is primarily driving this endeavor. However, the integration of DNA technology within pharmaceutical research and development is a relatively recent event; as a result, the published information in this area is not extensive (Rosamond and Allsop, 2000). One particular study illustrates how hierarchical clustering techniques (see Chapter 7 for details) can be used to organize microarray expression data in a way that allows function to be associated with gene sequence. Clustering analysis of gene expression is based on the assumption that genes of similar function are transcriptionally regulated in a similar fashion (Eisen et al., 1998). Hughes and colleagues (2000a) predicted functional roles for eight previously uncharacterized yeast genes by applying clustering analysis to a large compendium of gene expression profiles. These eight yeast genes encode proteins required for sterol metabolism, cell wall function, mitochondrial respiration, or protein synthesis based on their expression profiles in response to 300 diverse mutations and chemical treatments in *S. cerevisiae* (Hughes et al., 2000a). In another study, differential RNA display was employed to investigate the induction of gene expression in uropathogenic *E. coli* following pilus-mediated adherence to its host cell receptor (Zhang and Normark, 1996). A gene encoding an uncharacterized, sensor-regulator protein essential for the bacterial response to iron starvation was included in the set of genes induced following adhesion. An insertion mutation of the sensor-regulator gene abolished the ability of uropathogenic *E. coli* to grow in urine, indicating that this gene of the sensor-regulatory process is involved in uropathogenic *E. coli* urinary tract infection (Zhang and Normark, 1996) and constitutes a likely target candidate for drug therapy.

Microarray analysis also enables the identification of new regulatory pathways or networks on which the design of novel drugs can be based. Using oligonucleotide microarray technology for whole-genome expression profiling, de Saizieu and coworkers (2000) identified a novel *S. pneumoniae* regulon controlled by the Blp (bacteriocinlike peptide) two-component system, which is closely related to quorum-sensing systems regulating cell density-dependent phenotypes. Generally, bacteriocins are defined as compounds produced by bacteria that selectively inhibit or kill closely related species. The Blp two-component system is a peptide-sensing system. Microarray expression profiling revealed that a synthetic oligopeptide corresponding to the processed form of BlpC induces a distinct set of 16 genes, including genes for the regulation, synthesis, export, and processing of Blps (de Saizieu et al., 2000). The *blp* genes are transcriptionally induced in a cell density-dependent manner, suggesting that the *blp* regulon is a functional quorum-regulated system (de Saizieu et al., 2000). Quorum-sensing systems regulate numerous cellular functions, including virulence (Ji et al., 1995), the development of genetic competence (i.e., the natural ability to take up exogenous DNA) (Pestova et al., 1996), and the production of antimicrobial peptides (Brurberg et al., 1997; Diep et al., 1996). In this way, bacterial expression profiling using microarray technology can lead to new insights that allow the development of novel ways to prevent or treat bacterial infections.

For an antimicrobial drug to be truly effective, it is vital that the targets are conserved in the spectrum of disease-relevant, naturally occurring isolates. DNA microarrays can be used as a comparative genomics tool to screen populations of clinical isolates or closely related strains for conserved genes as well as genetic variation. In a recent study, a high-density oligonucleotide array representing 1,968 genes of a *S. pneumoniae* type-4 strain was used to investigate genomic variation in 20 *S. pneumoniae* isolates (Hakenbeck et al., 2001). These isolates represented major antibiotic-resistant clones. Microarray hybridization showed that 75% of the protein-coding genes of the reference type-4 strain were conserved in all 20 isolates and thus represent genetic information common to pneumococci (Hakenbeck et al., 2001). Variable loci detected between individual isolates and the reference strain included mosaic genes encoding antibiotic resistance determinants and gene clusters encoding capabilities for bacteriocin production. In addition, DNA microarrays were used to compare the genome of *S. pneumoniae* with the genomes of commensal *Streptococcus mitis* and *Streptococcus oralis* strains. Most of the pneumococcal specific virulence gene loci were not detected in the oral streptococci (Hakenbeck et al., 2001), and commensal strains that had acquired pneumococcal virulence factors could easily be identified using the microarray-based hybridization technique. This study illustrates the value of microarrays for investigating the pathogenicity potential of species and for identifying genes of potential relevance to pneumococcal pathogenicity that can be further explored for their therapeutic utility.

13.5   DETERMINING THERAPEUTIC UTILITY: DRUG TARGET SCREENING AND VALIDATION

As discussed in the previous section, the drug discovery and development process begins with the selection of a molecular target, whether essential to bacterial cell survival or critical for the establishment of pathogenic growth. Once a target for pharmacological intervention has been selected, the next stage in the process is lead, or drug candidate, identification (Fig. 13.1), which involves identifying compounds with antimicrobial activity (traditionally through whole-cell screening) or biochemical inhibitors that specifically inhibit the selected molecular target (through target-based screening). Although whole-cell screening for novel antibacterial compounds has historically been used successfully and is highly reproducible, target-based screening is generally more sensitive (i.e., can detect weak or poorly penetrating compounds or inhibitors suitable for chemical optimization), can be employed to target new areas of biology, and facilitates rational drug design (Rosamond and Allsop, 2000).

The standard approach to identifying a lead molecule specific for a protein target is high-throughput screening (HTS) of compound banks (Dean et al., 2001), followed by the analysis of hits and mechanism of action studies. Numerous chemical agents that are archived in extensive databases with the annotation of associated biological activity can be screened in high-throughput assays for potential therapeutic utility. An emerging, possibly more efficient, approach than HTS is to directly identify potential lead molecules *in silico* by exploiting current knowledge of three-dimensional structures of proteins and using advancements in computer-aided drug design (Dean et al., 2001). *In silico* drug design will depend heavily on structural genomics and bioinformatics. Lead optimization, the final stage of the drug discovery process, is used to methodically select for improvements in drug-associated properties, such as broad-spectrum antibacterial activity, bioavailability, metabolic stability, and antimicrobial potency, until potential drug candidates are developed and ready for efficacy testing in clinical trials. Lead optimization includes early toxicological studies, which will be the subject of the section on toxicogenomics.

In this section, we will discuss the impact of genomics and proteomics on the processes of drug target screening and validation in the pharmaceutical industry. Traditionally, drug screening strategies directly searched for compounds that killed microorganisms; however, such whole-cell screening methods provide no rational basis for compound optimization, because the molecular target is unknown. The application of genomics-based technologies and proteomics led to changes in the drug discovery paradigm by shifting the focus to rational design guided by structural information. Drug compounds are now being screened using strain-based, structure-based, and surrogate-ligand-based approaches. In addition, once a lead compound is identified, it is imperative that the relationship between inhibition of the biochemical target and antimicrobial action is established (Rosamond and Allsop, 2000). As we will illustrate, genomics technologies like DNA microarrays are contributing significantly to the elucidation of mechanisms of drug action.

13.5.1 Target-based Drug Screening

Chemistry and drug screening are among the core competencies of the pharmaceutical industry. The development of organic chemistry permitted the construction of complex libraries of artificially synthesized drugs to be screened, whereas molecular biology allowed the use of defined protein targets to be used in drug screening. It is expected that genome-based technologies will be used in combination with chemical libraries and screening assays to identify novel antimicrobial compounds and validate their molecular targets (Cockett et al., 2000). The mounting threat of drug- and antibiotic-resistant bacteria has spawned new screening strategies that will potentially advance the drug discovery process. Below we describe several new screening methods for discovering lead compounds that inhibit specific molecular targets.

Strain Array for Antibacterial Discovery One recently developed screening methodology for antimicrobial drug discovery incorporates target specificity and antibacterial activity into a high-throughput, primary whole-cell screen. DeVito and colleagues (2002) describe the construction of an array of *E. coli* strains engineered to be more sensitive (hypersusceptible) for specific enzyme inhibitors than for antibacterial activity alone. This was accomplished by generating bacterial strains that exhibited low-level expression of an essential gene for a specific target (e.g., helicase [*dnaβ*], which functions in DNA replication). To achieve regulated expression of essential target genes in *E. coli*, the chromosomal copy of the selected gene is deleted, and a complementary plasmid-borne copy of the essential gene is expressed under the control of the arabinose regulon promoter (P_{BAD}) (DeVito et al., 2002). For each engineered bacterial cell, the intracellular level of one essential target protein is modulated by adjusting a specific inducer for the P_{BAD} promoter. Each strain is then screened against a chemical library using growth inhibition as an end point (DeVito et al., 2002). This strategy identifies lead compounds that are active against whole cells and are also specific for a molecular target. The advantages of using a strain array for antimicrobial activity are (1) the assay can be used for any target, (2) the molecular target of an inhibitor of a hypersusceptible strain can be confirmed with a functional assay of the biochemical target, (3) the strategy can be used to identify compounds with good inhibitory activity against more than one essential target (i.e., multimodal inhibitors), and (4) engineered strains constitute permanent tools that can be screened at any time and at low cost (DeVito et al., 2002).

In silico *Screening and Structure-based Drug Design*   Analysis of the protein sequences and structures encoded in a genome provides fundamentally important information for drug discovery, because proteins are the physical targets of most drugs. Furthermore, three-dimensional molecular details of protein-binding domains and/or catalytic sites are critical for understanding the specificity and mechanistic properties of newly emerging therapeutic targets (Burley et al., 1999). A recent approach used to find and develop agents of pharmaceutical potential is structure-based screening and design. This approach couples information on the three-dimensional molecular structure of the target molecule with specialized computer programs in order to propose novel enzyme inhibitors and other therapeutic agents (Kuntz, 1992). The three-dimensional molecular structure of the target molecule is determined through X-ray crystallography, nuclear magnetic resonance (NMR) spectroscopy in solution, or inferred by homology modeling (Dean et al., 2001). However, X-ray crystallography is currently biased toward those proteins that are easy to purify and crystallize (Dean and Zanders, 2002). *In silico* solutions offer a way to evaluate the pharmaceutical relevance of small molecules before a commitment is made to invest in expensive chemical analyses. The process of *in silico* drug design is particularly useful with small binding sites, such as enzyme catalytic sites or allosteric regulatory sites and ligand-binding sites on receptors (Dean and Zanders, 2002). Determination of the structure of any form of receptor, for example, provides a starting point for direct modeling activities using the computer program DOCK (Kuntz, 1992; Kuntz et al., 1982). A schematic representing the general approach to the structure-based

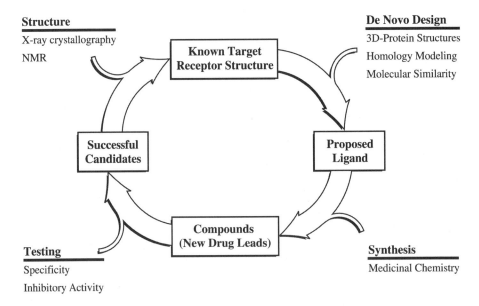

Figure 13.5 *Flow of structure-based design from protein target to synthesis of small molecules.* The general approach to structure-based design of biological inhibitors begins with the determination of the structure of the target receptor or molecule. A series of putative ligands are then proposed using computer modeling based on theoretical principles and the three-dimensional molecular structure of the target molecule. Compounds generated *in silico* are synthesized and tested for specificity and inhibitory activity.

design of biological inhibitors is presented in Fig. 13.5. Using computer programs such as DOCK, a series of putative ligands are proposed based on whether they have the correct geometric and electronic features to fit into appropriate sites on the receptor (Dean et al., 2001; Kuntz, 1992). The docking algorithm searches three-dimensional databases of small molecules and ranks each ligand candidate on the basis of the quality of fit, that is, the best orientations that can be found for a particular molecular conformation (Kuntz, 1992). Compounds based on the putative ligands identified by *in silico* virtual screening are then chemically synthesized and tested for pharmacological activity. Thus, computational programs such as DOCK can serve as valuable screening tools for generating leads in drug development. Virtual screening effectively bridges proteomics and drug discovery by providing a foundation for small molecule generation *in silico* (Dean and Zanders, 2002).

Surrogate-ligand-based Screening The accumulation of genomic sequence information has resulted in a rapidly expanding list of potential molecular targets for pharmacological intervention. Screening methods that are independent of functional assays are needed to discover new inhibitors of protein function. Traditionally, protein–protein interactions are not widely utilized or explored as targets for therapeutic drugs, despite the biological relevance of such interactions to human disease (Vidal and Endoh, 1999). Screening strategies have been developed recently that are based on protein–protein interactions. One in vitro approach utilizes "surrogate" ligands in combinatorial peptide libraries to detect small-molecule enzyme inhibitors with a wide range of potencies (Hyde-DeRuyscher et al., 2000). Surrogate ligands are defined as short peptides that bind specifically with high affinity to functional sites of target proteins, thereby inhibiting their function (Loferer et al., 2000). Surrogate ligands that bind to regions of biological interaction on target proteins are isolated from combinatorial peptide libraries by phage display (see Chapter 10 for a detailed discussion; for a review of phage display in pharmaceutical biotechnology, see Sidhu, 2000).

In a study by Hyde-DeRuyscher and coworkers (2000), the enzymes targeted for phage display fell into four diverse enzyme classes: (1) ligases [tyrosyl tRNA synthetase (*H. influenzae*), prolyl-tRNA synthetase (*E. coli*)], (2) oxidoreductases [alcohol dehydrogenase (*S. cerevisiae*)], (3) hydrolases [carboxypeptidase B [pig], β-glucosidase (*Agrobacterium faecaelis*)], and (4) transferases [hexokinase (*S. cerevisiae*), glycogen phosphorylase a (rabbit)]. Peptide ligands were detected for each target protein and found to bind one or two sites per target. Despite similar amino acid sequences, the peptides were highly specific for their cognate target protein, showing no cross-reaction with other target proteins (Hyde-DeRuyscher et al., 2000). When these peptides were tested for their ability to inhibit the biochemical activities of the various enzymes, 13 of 17 peptides were identified as specific inhibitors of enzyme function. In addition, Hyde-DeRuyscher and colleagues (2000) tested the ability of two peptides (Tyr-1 and Tyr-4) specific for *H. influenzae* tyrosyl-tRNA synthetase (TyrRS) to detect known TyrRS inhibitors. Competitive binding assays demonstrated that the binding of phage displaying peptide Tyr-1 to TyrRS could be blocked by a competitive inhibitor. This analysis suggested that peptidic surrogate ligands could be utilized in high-throughput binding assays to detect small-molecule inhibitors of target protein function (Hyde-DeRuyscher et al., 2000). Furthermore, the findings of this study provide evidence that peptide interaction with biochemically diverse enzymes is specific rather than random and occurs at functional sites on the target protein. An important aspect of the surrogate-ligand approach is the fact that knowledge of target protein activity is not required to identify peptide ligands that bind specifically, thus suggesting this method can be used to identify functional sites on proteins of unknown function (Hyde-DeRuyscher et al., 2000).

As mentioned, the discovery of antimicrobial drugs typically involves the massive screening of extensive compound libraries to identify leads exhibiting a moderate affinity (i.e., binding capacity) for the target. Consequently, weaker binding compounds, capable of being optimized by using medicinal chemistry and structure-aided design, are often not detected by conventional drug screening methods. As an alternative to traditional screening assays, Erlanson and coworkers (2000) developed a strategy, called *tethering*, to rapidly and reliably identify small soluble drug fragments (approximately 250 Da in molecular weight) that bind with low affinity to a specifically targeted site on a protein or macromolecule. Site-directed ligand discovery depends on the formation of a disulfide bond between the ligand and a cysteine residue in the protein target. A cysteine-containing target protein is allowed to react reversibly with a library of disulfide-containing molecules (~1,200 compounds) in the presence of a reducing agent (e.g., 2-mercaptoethanol) at concentrations normally used in drug screening (10–200 μM). The majority of the library members will possess little or no inherent affinity for the target protein, so that the associated disulfide bond to the protein will be easily reduced. However, the presence of a molecule with even weak inherent affinity for the target protein will shift the equilibrium toward the modified, or tethered, protein (Erlanson et al., 2000). Mass spectrometry is then used to identify the tethered compounds. Utilizing this strategy, Erlanson and colleagues (2000) identified a potentially potent inhibitor for thymidylate synthase, an essential enzyme in pyrimidine metabolism with therapeutic applications in cancer and infectious diseases.

In another approach to drug screening, the yeast two-hybrid genetic system (see Chapter 10) is used to validate that a peptide has a significant effect on the target protein in vivo. An excellent example of the combined utilization of combinatorial libraries, screening, and two-hybrid technology for the selection of peptides that break specific protein–protein interactions is provided in the study by Cohen and coworkers (1998). Through two-hybrid selection, a peptide aptamer, designated pep8, was isolated from a combinatorial library that binds and competitively inhibits cyclin-dependent kinase 2 (Cdk2). Peptide aptamers constitute a new class of 20-residue molecules that were designed to mimic the recognition function of the complementarity-determining regions of immunoglobulins and, therefore, were designed to interfere with protein interactions inside cells (Colas et al., 1996). The pep8 peptide aptamer inhibits the kinase activity of Cdk2 by binding at or near the enzyme's active site (Cohen et al., 1998). Cdk2 promotes the transition between the G1 and S phase of the cell cycle in mammalian cells. Cohen and colleagues (1998) found that expression of the anti-Cdk2 peptide aptamer, pep8, in human cells inhibits cell-cycle progression by interfering specifically with the interaction between Cdk2 and one of its substrates.

13.5.2 Microarrays and Drug Target Validation

Among the major challenges in developing new drugs is drug target validation (confirmation that a compound inhibits the intended target) and toxicity studies (identification of undesirable or toxic secondary effects). The development of DNA microarray technology has significantly contributed to facilitating the processes of drug target validation and the identification of secondary drug target effects. Gene expression profiling can be used to describe and predict adverse drug effects. For rational drug design to advance, the molecular mechanisms of drug actions must be clearly understood. Such knowledge can help identify other, possibly more effective, targets for pharmacological intervention. Recent studies illustrate that microarrays can be used to determine the primary genetic response to a drug application and thus provide important insights into the mode of drug action. The parallel measurement of global gene expression provides a comprehensive framework to determine how a compound affects cellular metabolism and the regulation of gene expression on a genomic scale (Marton et al., 1998). Ideally, the inhibitory activity of an antimicrobial drug should be so potent and specific that it would appear as if the target gene product did not even exist (Marton et al., 1998). Treatment of cells with a drug should engender alterations in gene expression similar to those resulting from deletion of the selected target gene. The gene expression profile generated in response to drug exposure can serve as a molecular signature of the drug used and shed light on the modes of action of uncharacterized inhibitors, thus incriminating the affected pathway and perhaps even the specific targeted protein (Wilson et al., 1999). Here we discuss genomic approaches toward elucidating mechanisms of drug action by focusing on the use of microarray hybridization in exploring drug-induced alterations in gene expression.

Microarray Analysis of Genetic and Pharmacologic Inhibition of Gene Function

Marton and colleagues (1998) evaluated the utility of microarray profiling of genome-wide gene expression patterns as a method for drug target validation and identification of secondary drug target effects at the transcriptional level. The method was demonstrated by examining the effects of gene expression associated with pharmacological (drug inhibitor-mediated) or genetic (deletion mutation-mediated) inhibition of calcineurin function in yeast cells. Calcineurin, a highly conserved calcium- and calmodulin-activated serine/threonine protein phosphatase, is involved in various cellular processes dependent on calcium signaling, such as intracellular ion homeostasis and the regulation of the onset of mitosis (Cardenas et al., 1994; Klee et al., 1998; Mizunuma et al., 1998; Tanida et al., 1995). Calcineurin activity is inhibited by the immunosuppressant drugs FK506 and cyclosporin A (CsA).

Whole-genome microarrays for *S. cerevisiae* were used to compare the transcriptional profiles of cells grown in the presence of FK506 or CsA and cells grown in the absence of the drug but possessing deletions in the genes encoding the catalytic subunits of calcineurin (*CNA1* and *CNA2*) (Marton et al., 1998). Pharmacologic inhibition of calcineurin produced a "signature" pattern of altered gene expression that was very similar to the pattern obtained with the calcineurin mutant strain, indicating that null mutants of drug targets "phenocopy" drug-targeted cells on a genomic scale (Marton et al., 1998). This approach, therefore, can be used to establish whether a putative target is required to generate the drug signature. For a drug having a single biochemical target, microarray analysis of pharmacologic and genetic inhibition of the target may be particularly useful for target validation; however, a compound often may affect unintended components of multiple pathways and thus elicit a complex gene expression pattern or signature (Marton et al., 1998). When null mutants were treated with FK506, microarray analysis revealed additional affected pathways distinct from the drug's primary target. In those cases, gene expression profiling can unmask secondary drug effects on genomic expression patterns, providing a means for evaluating a compound's specificity.

Effect of Isoniazid Treatment on Genomic Expression in **Mycobacterium tuberculosis** A serious threat to combating the chronic infectious disease, tuberculosis, is the emerging epidemic of multiply drug-resistant isolates of *M. tuberculosis.* The recently completed and annotated genomic sequence of *M. tuberculosis* (Cole et al., 1998) has enabled, for the first time, a comprehensive genomic approach to the discovery and development of novel antimycobacterial agents. Using a DNA microarray containing 97% of the predicted ORFs for *M. tuberculosis*, Wilson and coworkers (1999) monitored alterations in this bacterium's global gene expression in response to the antituberculosis drug isoniazid, which selectively blocks the mycolic acid biosynthetic pathway. Mycolic acids constitute the major component of the waxy, outer lipid envelope of mycobacteria. Although the molecular target of isoniazid inhibition is known, the precise mechanism of isoniazid-mediated killing remains unresolved, even though it is the most frequently used drug for treating tuberculosis (Wilson et al., 1999). In their microarray-based study, Wilson and his colleagues used isoniazid as a model system for investigating drug-induced transcriptional profiles of *M. tuberculosis.* Microarray analysis of the transcriptional effects of isoniazid exposure indicated increased expression levels for several genes that encode proteins of physiological relevance to the drug's mode of action. For example, isoniazid induced genes belonging to an operonlike cluster of five genes (*fabD, acpM, kasA, kasB,* and *accD6*) that encode polypeptide components of the type-II fatty acid synthase (FAS-II) complex (Wilson et al., 1999). A combination of genetic and biochemical evidence has shown that isoniazid blocks the FAS-II complex in the mycolic acid biosynthetic pathway (Banerjee et al., 1994; Mdluli et al., 1998; Rozwarski et al., 1998). This result demonstrates that microarray hybridization can be used to highlight genes directly involved in the cellular processes inhibited by drugs. In addition, isoniazid induced *fbpC*, which encodes a protein (trehalose dimycolyl transferase) involved in mycolate maturation. Other isoniazid-responsive genes encoding two fatty-acyl-CoA dehydrogenases (*fadE24* and *fadE23*), an alkyl hydroperoxide reductase subunit (*ahpC*), and an efflux protein (*efpA*) were apparently not directly associated with affected biosynthetic pathways and probably mediate processes linked to toxic consequences of the drug (Wilson et al., 1999). Isoniazid-mediated induction of *efpA*, which encodes a predicted proton-energized transporter, suggests that the gene product may be involved in transporting molecules relevant to mycolic acid production (Wilson et al., 1999). The *efpA* gene product may serve as an appropriate novel drug target if additional experimentation can confirm that EfpA mediates an essential mycolate biosynthetic function. Hence, microarray analysis can provide insights that may help to define new drug targets that heretofore were not known to be acting within drug-inhibited pathways. By analyzing changes in genomic expression in response to a characterized antituberculosis drug, Wilson and colleagues (1999) demon-strated that microarray expression profiling is a viable approach to predicting a drug's mode of action based on a physiologically derived interpretation of the transcriptional response elicited by drug application.

Gene Expression Response of **Haemophilus influenzae** *to Novobiocin or* *Ciprofloxacin Exposure* Gmuender and coworkers (2001) used microarrays to determine changes in global gene expression triggered by exposure of *H. influenzae* to the antibiotics novobiocin and ciprofloxacin. Novobiocin (a coumarin) and ciprofloxacin (a quinolone) are well-characterized DNA gyrase inhibitors representing two different functional classes, that is, these antibiotics inhibit the same target enzyme but through different molecular mechanisms (Gmuender et al., 1997; Kampranis et al., 1999). DNA gyrase, which consists of two subunits (A and B), is a prokaryotic topoisomerase II enzyme that is essential for cell viability and has no direct mammalian equivalent (for a review, see Sharma and Mondragon, 1995). The enzyme introduces negative supercoils into DNA by using energy derived from ATP hydrolysis and thus changes the topology of DNA. Novobiocin, a nonbactericidal antibiotic, inhibits the ATPase activity of DNA gyrase by binding to the ATP-binding site on subunit B and, as a result, indirectly changes the degree of DNA supercoiling. Promoter activity has been shown to be influenced by the degree of supercoiling (Wang and Lynch, 1993). Ciprofloxacin obstructs DNA supercoiling by inhibiting DNA gyrase-mediated double-stranded cleavage and resealing of DNA. As a bactericidal DNA gyrase inhibitor, ciprofloxacin induces the RecA (SOS) DNA repair system (Piddock and Wise, 1987).

A goal of the study by Gmuender and colleagues (2001) was to determine whether novobiocin and ciprofloxacin induce different mechanism-related expression patterns. For this purpose, high-density oligonucleotide microarrays were used to investigate the expression levels of more than 80% of the genes in *H. influenzae* in response to the two different inhibitors of DNA gyrase. Because regulation at the transcriptional level is not the only means by which cells can respond to changing growth conditions, simultaneous analysis of translational patterns was also performed using two-dimensional polyacrylamide gel electrophoresis (2D-PAGE). Gmuender and coworkers (2001) found that novobiocin and ciprofloxacin induced different responses at the level of transcription and/ or translation, even though these antibiotics inhibit the same target enzyme. The expression levels of numerous genes were altered when *H. influenzae* cells were treated with the ATPase inhibitor novobiocin. Genes exhibiting altered expression levels included not only ∼ 50 ORFs encoding hypothetical proteins but also genes encoding DNA gyrase subunit B, ribosome releasing factor, and topoisomerase I. The transcriptional response induced by novobiocin reflected the fact that DNA supercoiling influences the initiation of transcription for many genes (Gmuender et al., 2001).

Microarray hybridization showed that ciprofloxacin treatment primarily stimulated the expression of DNA repair systems. The induction of DNA repair systems is a response to the DNA damage caused by the stable ternary complexes that ciprofloxacin forms with DNA gyrase and DNA (Gmuender et al., 2001). Genes involved in the SOS repair mechanism (e.g., *recA*, *uvrA*, *lexA*) and other DNA repair systems (e.g., *ruvB*, *recO*, *recN*, *impA*, *recF*), for example, showed increased expression levels in response to ciprofloxacin. In addition to clear differences in signature gene expression patterns compared to novobiocin, the onset of the transcriptional response was delayed for ciprofloxacin, whereas the response was immediate for novobiocin (Gmuender et al., 2001). Microarray-based expression profiling also indicated some common transcriptional effects of novobiocin and ciprofloxacin. For example, genes encoding amino acid biosynthesis enzymes, amino acid transporters, ribosomal proteins, and tRNA synthetases were affected by antibiotic exposure. For both novobiocin and ciprofloxacin, changes at the mRNA level were, qualitatively, in agreement with changes at the proteome level. As expected, however, the sensitivity and reproducibility of the expression analysis using microarray technology were better than those of expression analysis using 2D-PAGE followed by computerized image analysis (Gmuender et al., 2001).

Microarray analysis of novobiocin- and ciprofloxacin-responsive genes illustrates how profiling the transcriptional response to an antibiotic can yield important information concerning its mode of action. Microarray-based genome technology will be particularly useful in providing indications for the classification of an unknown inhibitor, because its specific expression profile or molecular "signature" can be compared with those signatures characteristic of known inhibitors (Gmuender et al., 2001). Thus, it is expected that high-throughput, parallel expression analysis will accelerate the characterization of novel pharmaceuticals needed to address the problem of multiple drug resistance among infectious bacteria.

13.6   GENOMICS AND TOXICOLOGY: THE EMERGENCE OF TOXICOGENOMICS

As we have explored in previous sections, microbial genomics has had and will continue to have a major impact on the identification and validation of molecular drug targets and on the development of high-throughput screening assays. It is expected that genome sequencing and genomic technologies will also be important for other aspects of drug development, such as toxicology and clinical studies, although the potential value of genomics to such subsequent stages of drug development has yet to be fully realized.

The biological revolution initiated by genomic sequencing has led to the development of numerous new scientific subdisciplines, one of which represents the convergence of the fields of toxicology and genomics. The subdiscipline of toxicogenomics uses genomics resources to identify potential toxicants harmful to humans and the environment and to understand the molecular mechanisms underlying toxic processes (Nuwaysir et al., 1999). In terms of drug discovery and development, toxicogenomics is specifically concerned with the prediction and mechanistic analysis of toxicological problems associated with drug compounds at the transcriptome level of a target organ or cell (Storck et al., 2002). Gene-array technology is gaining prominence as a commercial application for gene expression analysis in the field of toxicogenomics. A hallmark of direct or indirect exposure to a toxicant is altered gene expression. Microarray technology offers an ideal platform for assessing gene expression profiles or signatures that can be used as highly sensitive and informative markers for toxicity (Nuwaysir et al., 1999). Toxicogenomics will be discussed here primarily in regard to the applications of microarray technology in mechanistic or investigative toxicological research and in predictive toxicology.

13.6.1   Microarrays in Mechanistic Toxicology

Assessment of potential toxicity is classically performed using numerous in vivo model systems, such as the rat, mouse, and rabbit, or using appropriate primary or established cell lines. A long-term rodent cancer bioassay (Chhabra et al., 1990), for instance, is normally utilized to assess nongenotoxic carcinogenesis. Examples of in vitro techniques that have been developed to measure toxicity (typically by assaying for toxicant-induced DNA damage) include the Ames test, the Syrian hamster embryo cell transformation assay, micronucleus assays, and measurements of sister chromatid exchange (Nuwaysir et al., 1999). Toxicologists have used these traditional bioassays to identify and assess the safety of chemicals and drug candidates for human use. Furthermore, established genetic approaches, such as antisense oligonucleotides to inhibit endogenous proteins and reporter genes to measure gene promoter activity, allow mechanistic questions concerning toxicity effects to be addressed (Burchiel et al., 2001). The biological role of selected gene products in the in vivo response to toxicants can be studied using transgenic and knockout mice.

Recently, some toxicologists have been promoting genomic expression technologies as a superior alternative to traditional animal bioassays. Microarray-based gene expression profiling holds promise as an approach to identifying the mechanisms of action that underlie the potential toxicity of chemicals and drug candidates. The modulation of gene expression in a biological system in response to chemical exposure can provide clues to that chemical compound's mechanism of action. Nuwaysir and colleagues (1999), therefore, have argued that a method based on measurements of genome-wide steady-state mRNA levels in an organism following toxicant exposure is fundamentally informative and complements established methods used in toxicology testing. They developed a method for the identification of toxicants and the determination of their putative modes of action by using toxicant-induced gene expression profiles (Nuwaysir et al., 1999). In this approach, dose and time-course parameters are established for a series of toxicants within a given mechanistic class [e.g., polycyclic aromatic hydrocarbons (PAHs)] using one or more defined model systems. Target cells are then exposed to these agents at a fixed toxicity level, as measured by cell survival. Total cellular RNA is isolated and used to measure changes in genomic expression by hybridization to a cDNA microarray. For such studies, Nuwaysir and coworkers (1999) have developed the custom cDNA microarray ToxChip v1.0, which is comprised of 2,090 human genes selected for their well-established involvement in basic cellular processes (e.g., DNA replication and repair genes) and in response to different types of toxic exposure (e.g., genes responsive to PAHs, dioxinlike compounds, and oxidant stress). Subsequent analysis of alterations in global gene expression induced by the application of the test agent reveals the common set of changes unique to that particular class of toxicants. The modulation of gene expression represents a "signature" of the cellular response to the test agent, and the "toxicant signature" is different for each prototypic toxicant class (Nuwaysir et al., 1999). Collections of determined signatures can then be compared to gene expression profiles induced by unknown agents (see Fig. 13.6 for a schematic of the method). A match between a new signature and an established signature indicates a putative mechanism of action for the unknown test compound.

Pennie (2000) describes the development of a series of custom cDNA microarrays, called ToxBlot arrays, for the specific purpose of investigating toxicity processes. Approximately 600 marker genes of either human or murine origin were selected for array construction and consisted of genes of particular interest to mechanistic or investigative toxicology research (e.g., basic transcription factors, cell adhesion, cell surface receptors, drug metabolism, heat shock proteins, oxidative stress, steroid receptors). These DNA sequences were identified as potential diagnostic markers for various toxicity purposes. For reproducibility, each gene to be profiled is represented on the ToxBlot array by four individual spots. In addition, a database of background literature, including information on biochemical/enzymatic function, tissue distribution, and known allelic variation, is available for each arrayed gene in order to assist in the interpretation of differential gene expression patterns (Pennie, 2000). An immediate goal of toxicogenomics is to perform microarray analysis on cells treated with different classes of compounds in order to identify groups of genes that are tightly correlated with known classes of toxicants. A condensed set of "very informative" genes could then be derived and used to fabricate a next-generation array with a small number of genes (Lovett, 2000).

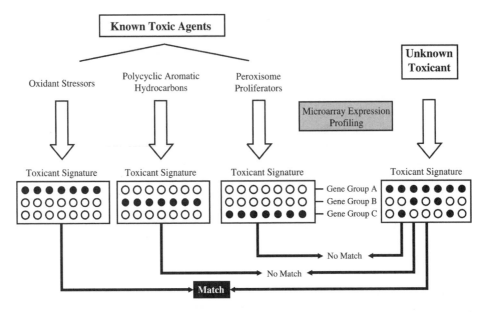

Figure 13.6 *Using microarray-based gene expression profiling to identify an unknown toxicant's mechanism of action (Nuwaysir et al., 1999).* In this method, microarrays are used to analyze alterations in global gene expression patterns of model systems in response to exposure to known toxicants. These gene expression signatures, which are characteristic of a certain toxicant, are compared to the signature derived for an unknown, suspected toxicant. As shown in its expression profile (signature), the unknown toxicant elicited consistent changes in the set of group A genes, which matches the profile of oxidant stressors. In this way, a putative mechanism of action can be assigned to an unknown toxicant. [Adapted with permission from E. F. Nuwaysir, M. Bittner, J. Trent, J. C. Barrett, and C. A. Afstiari, Microarrays and toxicology: the Advent of Toxicogenomics, *Molecular Carcinogenesis*, vol. 24, 153–159. Copyright (1999) Wiley-Liss, Inc.]

A study by Hamadeh and colleagues (2002) substantiated the notion that analysis of gene expression patterns should allow classification of toxicants and provide important mechanistic insights. Microarray analyses were performed using liver RNA derived from rats that had been treated with structurally unrelated agents from a common chemical class of compounds, peroxisome proliferators, and with a well-studied enzyme inducer, phenobarbital. Expression profiling revealed similarity in the gene expression patterns induced by the three different peroxisome proliferators (clofibrate, Wyeth 14,643, and gemfibrozil), but a very distinct transcription profile was produced using the enzyme inducer phenobarbital (Hamadeh et al., 2002). Distinguishable patterns in gene expression profiles could be discerned within a chemical class of compounds, indicating that microarray analysis reveals chemical-specific profiles.

As discussed above, there are high expectations that RNA expression profiling technologies will revolutionize the way toxicologists examine the molecular basis of adverse effects of chemicals and drugs. The major advantage of using gene arrays is that they provide a global approach to understanding the complex molecular mechanisms involved in toxicology (Burchiel et al., 2001). However, it is important to acknowledge the challenges and limitations associated with gene expression profiling in mechanistic and predictive toxicology. As Fielden and Zacharewski (2001) point out, the complexity of mechanisms of action may not permit delineation using microarray expression profiling alone, which measures a single end point (i.e., RNA levels). Furthermore, toxicants and drugs that initiate toxicity can affect enzyme activity, DNA integrity, membrane integrity, and other processes that are not decipherable at the level of gene expression (Fielden and Zacharewski, 2001). In other words, the initiation of toxicity is not always directly dependent on the induction of gene expression, and changes in gene expression do not necessarily imply toxicity. Hence, it is important to use gene expression profiling in combination with other studies that measure multiple end points at the molecular, cellular, tissue, and physiological levels within a whole-organism context (Fielden and Zacharewski, 2001). Gene expression data will need to be incorporated into larger studies designed to assess effects at higher levels of biological organization. Even in cases where altered gene expression precedes or coincides with toxic events, our understanding of the mechanisms of drug action will be limited to our knowledge of the function of affected genes, their regulation, and their involvement in cellular pathways (Fielden and Zacharewski, 2001). In addition, there are practical issues that must be addressed before the full potential of microarray-based expression profiling in mechanism-based risk assessment can be realized (Nuwaysir et al., 1999). For example, in which biological system, at what dose, and at what time is toxicant-induced gene expression monitored? Other issues, like standardizing experimental conditions, will need to be addressed to reduce variability in microarray data between laboratories.

13.6.2 Microarrays in Predictive Toxicology

The validity and utility of predictive toxicology and toxicogenomics depend on the ability of different groups or classes of chemicals (grouped by toxic end point, mechanism, structure, target organ, etc.) to generate distinct and diagnostic gene expression profiles or signature patterns under a specific set of conditions (Hamadeh et al., 2002; Nuwaysir et al., 1999). Proof-of-principle experiments in rat have demonstrated that compound classification based on gene expression profiles is feasible (Hamadeh et al., 2002). The aim of predictive toxicology is to use gene expression analysis to describe and predict adverse drug or chemical effects and will naturally require a large database, or compendium, of expression profiles associated with compounds of known toxic or biological endpoints. Unknown chemicals and drugs will be assigned a classification based on the similarity of their induced gene expression pattern by comparison with expression profiles induced by chemicals or drugs with known mechanisms of action (Fielden and Zacharewski, 2001). From a practical perspective, predictive toxicology will help to reduce the substantial amount of time and expense invested in pharmaceutical research and discovery experiments (Gmuender, 2002). The likelihood that a compound will show adverse drug effects will be indicated at a much earlier stage in the drug development process by matching a gene expression profile of a new drug candidate with profiles of known compounds. Decisions can then be made concerning whether it is worth pursuing the development of that particular compound, thus potentially avoiding costly investments in clinical failures.

13.7 SUMMARY

In this chapter we have discussed how the tremendous influx of genomic sequence data generated from recent microbial sequencing projects can be used to enhance the quest for molecular targets for antimicrobial drugs. The surge in genomic sequence information comes at a critical time as the growing number of drug-resistant microbes erodes the efficacy of current therapeutic agents. Historically, antimicrobial drug discovery in the pharmaceutical industry has relied on random screening campaigns or the chemical modification of compounds belonging to known structural and mechanistic classes. This approach, however, has failed to deliver a sufficient range of chemical diversity to counteract the acquisition and emergence of drug resistance within the clinical arena. Through a combination of computational and experimental approaches, microbial genome sequencing has uncovered thousands of new potential targets for therapeutic intervention, thereby transforming drug discovery from a process involving direct antimicrobial screening programs to one of rational target-based strategies. The development of genome-based technologies such as DNA microarrays and advances in bioinformatics have already begun to impact the identification, validation, and screening of molecular targets and candidate compound series for drug discovery. In addition, as we have discussed, gene expression profiling using microarrays holds promise as a valuable approach for research and diagnostic studies in mechanistic and predictive toxicology. Both classical (e.g., the generation of null mutants) and recent (e.g., STM and IVET) genetic strategies will continue to play important roles in assessing the relevance of bacterial genes as potential antibiotic targets. The investigation of the proteome as it relates to drug discovery is still in a formative stage. Because proteins constitute the actual cellular targets for small-molecule drugs, technologies that improve our understanding of three-dimensional protein structure and associated biological function, posttranslational modifications, and protein–protein interaction networks should greatly increase the development of novel antimicrobial drug targets, as well as the number of successful outcomes of drug discovery. Similarly, the wide utility of genomics is beginning to be realized in the new discipline of toxicogenomics, which seeks to identify potential human and environmental toxicants and to characterize the molecular mechanisms leading to toxicity. Microarray-based gene expression profiling, in particular, is rapidly becoming a standard analysis in mechanistic toxicology studies and is providing important insights into the mechanisms of action that underlie the potential toxicity of chemicals and drug candidates. Toxicogenomics is expected to impact the safety assessment of new drugs and to reduce the number of drug failures that occur in the later stages of development.

FURTHER READING

Cockett, M., N. Dracopoli, and E. Sigal. 2000. Applied genomics: integration of the technology within pharmaceutical research and development. *Curr. Opin. Biotechnol.* 11:602–609.

Debouck, C. and P. N. Goodfellow. 1999. DNA microarrays in drug discovery and development. *Nat. Genet.* 21:48–50.

Gmuender, H., K. Kuratli, and K. Di Padova, et al. 2001. Gene expression changes triggered by exposure of *Haemophilus influenzae* to novobiocin or ciprofloxacin: combined transcription and translation analysis. *Genome Res.* 11:28–42.

Marton, M. J., J. L. DeRisi, and H. A. Bennett, et al. 1998. Drug target validation and identification of secondary drug target effects using DNA microarrays. *Nat. Med.* 4:1293–1301.

Nuwaysir, E. F., M. Bittner, and J. Trent, et al. 1999. Microarrays and toxicology: the advent of toxicogenomics. *Mol. Carcinog.* 24:153–159.

Wilson, M., J. DeRisi, and H.-H. Kristensen, et al. 1999. Exploring drug-induced alterations in gene expression in *Mycobacterium tuberculosis* by microarray hybridization. *Proc. Natl. Acad. Sci. USA* 96:12833–12838.

14

Application of Microarray-Based Genomic Technology to Mutation Analysis and Microbial Detection

Jizhong Zhou and Dorothea K. Thompson

14.1 INTRODUCTION

DNA or oligonucleotide microarrays provide a powerful tool for the multiplexed detection of nucleic acids. Two major current applications of microarrays include gene expression profiling (e.g., DeRisi et al., 1997; Wodicka et al., 1997) and genetic mutation analysis (Hacia, 1999). Gene expression profiling with microarrays has yielded genome-wide data that were impossible to achieve a decade ago. Genetic mutation studies with microarrays, however, are still in the development stage and thus have not been widely described in the literature. Among the variety of genetic mutations, single nucleotide polymorphisms (SNPs) are the most suitable targets for microarrays although challenges still exist (Broude et al., 2001). Multiple mutations, insertions, deletions, and rearrangements present difficulties when analyzed using microarray technology.

Recently, the application of microarray-based genomic technologies has been extended to the detection of microorganisms in natural environments (for a review, see Zhou and Thompson, 2002; Zhou, 2003). Although DNA microarray technology has been used successfully to analyze global gene expression in pure cultures, it has not been rigorously tested and evaluated within the context of complex environmental samples. In theory, microarray-based genomic technology provides the advantages necessary for the comprehensive and quantitative characterization of complex microbial communities.

Microbial Functional Genomics, Edited by Jizhong Zhou, Dorothea K. Thompson, Ying Xu, and James M. Tiedje.
ISBN 0-471-07190-0   © 2004 John Wiley & Sons, Inc.

Adapting microarray hybridization for use in environmental studies, however, has a number of challenges associated with specificity, sensitivity, and quantitation (Zhou and Thompson, 2002; Zhou, 2003).

This chapter reviews the basic principles of microarray technology and recent advances in applying this technology to the analysis of genetic mutations and detection of microorganisms in natural environments. The various methods for the analysis of mutations using microarrays are presented, as well as descriptions of various types of microarrays specifically developed for analyzing microbial community structure within the context of environmental samples.

14.2 OLIGONUCLEOTIDE MICROARRAYS FOR MUTATION ANALYSIS

SNPs are the most frequent type of variation in the human genome and the genomes of other natural and experimental organisms. It is estimated that there is one nucleotide difference in every 1,000 between any two copies of a chromosome (Landegren et al., 1998). SNPs are important markers in genetic analysis, because they are typically located near or within any locus of interest, and many SNPs can be expected to directly affect protein structure or gene expression levels. In addition, the inheritance of SNPs is very stable. Genotyping large numbers of SNPs in appropriate samples should provide insights into the basis of heritable variations in disease susceptibility and resistance, complex genetic trait differences, and human evolution (Hacia, 1999). Locating, identifying, and cataloguing sequence differences due to SNPs are the initial steps in relating genetic variation to phenotypic variation in both normal and diseased states. However, such studies will require rapid and cost-effective large-scale sequence analysis of hundreds or thousands of SNPs present in thousands of samples.

A variety of traditional approaches, such as minisequencing, molecular beacons, oligonucleotide ligation, and 5′ exonuclease assays, have been developed and used to genotype SNPs (Landegren et al., 1998; Hirschhorn et al., 2000). Although these methods have been used successfully to genotype small numbers of SNPs, they cannot meet the high-throughput demands of large-scale sequence comparisons and mutational analyses. To perform large-scale genetic studies efficiently, high-throughput parallel genotyping methods are needed. The following microarray-based experimental strategies have been developed and tested that can genotype large numbers of SNPs: differential hybridization with allele-specific oligonucleotide probes and arrayed primer extension assay (Hacia, 1999). This section briefly describes the principles and applications of each experimental strategy.

14.2.1 Microarray-based Hybridization Assay with Allele-specific Oligonucleotides

A key requirement for a scoring method used for detecting genomic SNPs is that it should unequivocally distinguish between homozygous and heterozygous allelic variants in diploid genomes. Differential hybridization with allele-specific oligonucleotide (ASO) probes is most commonly used in the microarray format (Yershov et al., 1996; Wang et al., 1998; Hacia, 1999). Such hybridization assays rely on the differences in hybridization stability of the short oligonucleotides to perfectly matched and mismatched target sequence variants. However, the specificity of genotyping by ASO strongly depends on probe characteristics and detection conditions. Probe design is critical to achieving specific detection.

Probe and Array Design   Discrimination in hybridization between perfectly matched and single-base mismatched DNA duplexes relies on differences in stability, which depends on probe characteristics and hybridization conditions. For large-scale analysis, it is ideal to identify a single set of hybridization conditions for which unequivocal discrimination between matches and mismatches can be accomplished for all SNPs of interest. This can be accomplished by selecting ASO probes having comparable melting temperatures, which depends on probe length, base composition, and the mismatch position within the ASO sequences.

Probe length is a key factor affecting duplex stability. In general, to obtain greater discrimination, a shorter probe sequence is desirable but has an overall lower duplex stability. While longer probe sequences form more stable duplexes, they offer less discrimination, because the percentage of mismatched sequences decreases. In addition, secondary structures from single-stranded DNA of the target samples affect the choice of probe length. Under high salt conditions, single-stranded DNA can form internal secondary structures. If the thermodynamic stability of such structures is greater than the stability of the duplex being formed between the ASO probe and target DNA, the hybridizing regions of the single-stranded target DNA may not be accessible to the surface-bound probes on the microarrays. Such a problem can be partially alleviated by selecting longer probe sequences, which enables hybridization to be performed at higher temperatures. Hybridization at higher temperatures can melt any internal secondary structures of the single-stranded target DNA. With all these factors taken into consideration, ASO probes are designed to have a length that generally ranges from 15 to 25 bp (Guo et al., 1994; Hacia and Collins, 1999).

The effects of probe length on specificity were determined using ASO probes of 12, 15, and 20 bp (Guo et al., 1994). While all the probes gave approximately equal hybridization signals, the 15-mer ASO probe yielded the best single-base discrimination. The 12-mer ASO probe was difficult to use due to low melting temperature, and no reproducible discrimination was obtained with the 20-mer ASO probe. Although G + C content has a significant effect on duplex stability, there is less choice in probe design due to probe sequence restriction. Studies have shown that probes with less than 50% G + C content yielded excellent single-base discrimination. Good single-nucleotide discrimination can be achieved with a probe containing 65% G + C. The location of the mismatched base also has a significant effect on duplex stability. The greatest discrimination between matches and mismatches was obtained with mismatched bases located near the center of the probe sequence (Pease et al., 1994; Hacia, 1999). Thus, mismatches should be placed near the center of ASO probe sequences to maximize discriminatory power. Many previous studies have shown that single-nucleotide discrimination can be obtained using the microarray-based hybridization approach.

To detect all possible single-nucleotide substitutions, the microarray for analyzing genotypes is designed to interrogate each nucleotide position of a target sequence of interest with four probes. One probe (perfect match or PM) is designed to be perfectly complementary to a short section of the target sequence, while the other three probes (mismatch probes or MM) are identical to the PM except at the interrogation position where one of the other three bases is substituted (Fig. 14.1). For example, while the PM probe has a T at the central position, the MM probes are designed such that they have the

```
Reference sequence      5'-TGCATAGGAGATAATCATAGGAATCC-3'
              PM        5'-ACGTATCCTCTATTAGTATCCT-3'
              MM        5'-ACGTATCCTCAATTAGTATCCT-3'
              MM        5'-ACGTATCCTCCATTAGTATCCT-3'
              MM        5'-ACGTATCCTCGATTAGTATCCT-3'
              PM         5'-CGTATCCTCTATTAGTATCCTT-3'
              MM         5'-CGTATCCTCTTTTAGTATCCTT-3'
              MM         5'-CGTATCCTCTCTTAGTATCCTT-3'
              MM         5'-CGTATCCTCTGTTAGTATCCTT-3'
              PM          5'-GTATCCTCTATTAGTATCCTTA-3'
              MM          5'-GTATCCTCTAATAGTATCCTTA-3'
              MM          5'-GTATCCTCTACTAGTATCCTTA-3'
              MM          5'-GTATCCTCTAGTAGTATCCTTA-3'
```

Figure 14.1 *Standard probe tilling design.* Each nucleotide of the reference sequence is interrogated by a set of four probes in which one probe is a perfect match (PM) to the reference sequence and the other probes are completely complementary to the reference sequence except that the base at the central position is substituted as shown in bold (mismatch probe or MM). Three probe sets are shown to query the adjacent bases in the reference sequence.

same flanking sequences but contain an A, C, or G at the central position. Generally, for any given nucleotide position, two sets of probes are designed to be complementary to both sense and antisense strands of the target sequence. Therefore, for detecting all the substitutions in a target sequence with N base pairs, 8N probes are needed (Hacia, 1999). This approach is usually referred to as the standard tiling design. To interrogate all possible deletions and insertions in both strands, many more probes are needed. As a result, the current array technology is not feasible for detecting large numbers of deletions and insertions (Hacia, 1999; Lipshutz et al., 1999). The advantage of an array design with redundant probes is that higher specificity and sensitivity can be achieved. The use of multiple probes minimizes random sources of errors associated with hybridization signal fluctuations.

Gain-of-signal Approach   The gain-of-signal approach compares the hybridization signals obtained with probes perfectly matching mutant (test) and wild-type (reference) sequences (Fig. 14.2). When a heterozygous mutant sample is labeled with a fluorescent dye, for example, cyanine 5 (Cy5), and hybridized with the genotyping microarray, hybridization signals will be observed for the oligonucleotide probes representing a perfect match to the mutant sequence. Thus, relative to the wild-type counterparts, hybridization signals derived from the mutant-specific probes for both strands will be gained. By scoring the hybridization signal gaining patterns, the sequence variations of the test heterozygous mutant samples can be identified. However, only mutations with corresponding complementary probes represented on the microarrays can be detected with this approach. In addition, the gain-of-signal approach is not sensitive to the presence of larger deletions and single-base insertions because of the increased cross-hybridization of the wild-type sequence to the mutant probes. Although hybridization to the mutation-specific probe can be used to identify the nature of a sequence change, unambiguous assignment is sometimes difficult to achieve, and hence sequence verification by other independent methods is generally needed (Hacia and Collins, 1999).

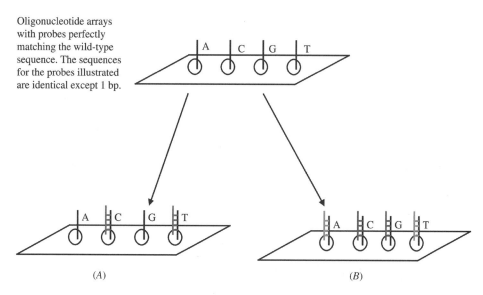

Oligonucleotide arrays with probes perfectly matching the wild-type sequence. The sequences for the probes illustrated are identical except 1 bp.

Figure 14.2 *See color insert. Gain- or loss-of-signal approach for detecting single nucleotide polymorphisms.* (A) Gain-of-signal analysis. The heterozygote has C and T residues at a specific nucleotide position. After hybridization of red-labeled test sequence with microarrays, the probes perfectly match the heterozygous sequence and will show hybridization, that is, a signal is gained, C or T in this case. (B) Loss-of-signal analysis with two colors. For a heterozygous variation (e.g., changes from C:G to C:T or A:G), a 50% loss signal intensity relative to the wild-type target will be observed; thus, light green color (50% green plus 25% red) of the spots will be observed, C or G in this case. If no mutations occur, the green-labeled wild-type reference and red-labeled mutant sequences will hybridize equally well with the wild-type probes, thus, yellow color spots (50% green plus 50% red) will be observed, A and T in this case.

Loss-of-signal Approach In the loss-of-signal approach, sequence variations between mutant and wild-type samples can be detected by quantifying the relative losses of the hybridization signals of the wild-type PM probes in test samples compared to those in the reference sample (Fig. 14.2). In an ideal situation, 50% of the signal intensity of the probes that perfectly match the wild-type sequence will be lost for a heterozygous sequence change, whereas a complete signal loss will be observed for a homozygous change.

A two-color assay with internal reference standards is used to assess the relative signal loss of the wild-type PM probes (Hacia et al., 1996, 1998). In this scheme, the known reference sequence and unknown test sequences are first labeled with two dyes, for example, fluorescein (green) and biotin (red), and cohybridized with the genotyping microarrays. Subsequently, the signal intensity of the two dyes is normalized, and the ratio of the wild-type PM probe signal intensities from reference samples (green) and test (red) samples is calculated (green/red). Finally, to display the presence of sequence variation, these ratios are plotted against the wild-type reference nucleotide position (Fig. 14.3). While regions of identical sequence should express a ratio close to 1.0, in regions with sequence changes, a peak centered near the point of mutation will be observed (Hacia et al., 1996). Under ideal conditions, a heterozygous mutation will result in a peak of 2.0, because there are two wild-type alleles in the reference sample and only a single wild-type allele present in the heterozygous mutant samples. Due to cross-hybridization of the

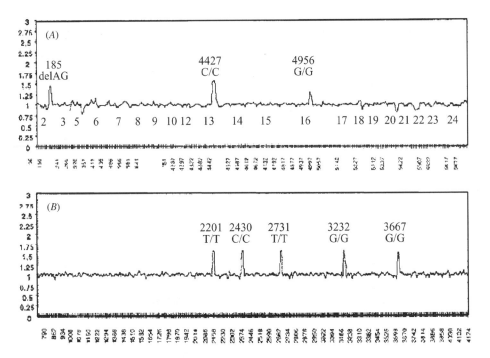

Figure 14.3 *Detection of sequence variations through loss-of-signal analysis.* The target and reference samples of the entire *BRCA1* coding sequences (~5.4 kb) were labeled with biotin (red) and fluorescein (green), respectively, and cohybridized to oligonucleotide arrays interrogating *BRCA1* sense and antisense strand sequences. Then the signal intensity of the two dyes is normalized, and the ratios are plotted against the wild-type reference nucleotide position. (A) Results from all coding exons except 11. (B) Results from the 3.43-kb *BRCA1* exon 11. While the labeled peak in (A) is at the site of the heterozygous 185delAG mutation found in the test sample, the other labeled peaks in (A) and (B) correspond to seven common homozygous polymorphisms found in this sample. [Reprinted with permission from J. G. Hacia, Resequencing and mutational analysis using oligonucleatide microarrays, *Nature Genetics*, vol. 21, 42–47. Copyright (1999) Nature Publishing Group http://www.nature.com/.]

mutant allele to the probe perfectly matched to the wild-type reference sequences, this ratio will be less than 2.0. In practice, a cutoff value of 1.2 represents a good threshold for evaluating sequence variation (Hacia et al., 1996). For homozygous changes, the peak height will be theoretically infinite, because the signal from a wild-type allele is absent. However, in practice, the peak is usually within a value of 10 due to cross-hybridization (Hacia et al., 1996). A drawback of the loss-of-signal assay is that the mutation cannot be directly discerned. The identified sequence change must be verified by other methods. The gain-of-signal assay, however, can be used to verify questionable losses of hybridization signal signatures.

Since both gain- and loss-of-signal approaches are complementary, it is advantageous to use a combination of both for mutation detection. For example, Hacia and colleagues (1996) described a two-tiered strategy for mutation analysis using oligonucleotide microarrays. The first level of analysis involves identifying candidate regions where sequence changes may exist through loss-of-signal analysis. The wild-type reference and test samples can be hybridized with the array containing only wild-type probes from both strands as well as probes for common polymorphisms for gain-of-signal analysis. If a putative mutation is identified at a known polymorphism site via loss-of-signal analysis, it can be checked for signal gain at this site for verification. However, it is possible that many candidate mutations could be explained. The second level of analysis involves confirming the unexplained candidate mutations via a gain-of-signal approach with a second, more complex, microarray containing base substitutions, deletions, and insertions. This two-level approach may prove to be more efficient and cost-effective in the future.

Application Both the gain- and loss-of-signal approaches have been used independently or together to analyze sequence variations (Kozal et al., 1996; Chee et al., 1996; Gingeras et al., 1998; Hacia et al., 1996, 1998, 1999; Wang et al., 1998). One of the most representative studies demonstrating the feasibility and utility of DNA microarrays for genotying focused on the 3.43-kb exon 11 of the hereditary breast and ovarian cancer gene BRCA1 (Hacia et al., 1996, 1998). An oligonucleotide array consisting of more than 96,000 probes was designed to interrogate all single-nucleotide substitutions, single-nucleotide insertions, and 1- to 5-bp deletions. Fourteen of the 15 heterozygous mutations were correctly identified by microarray hybridization. The results from this study indicated that single-nucleotide substitutions generally produced more robust gain and loss of hybridization signal signatures than small insertions and deletions. It was also important to analyze data from both strands, because some sequence changes were more readily detected on one strand than the other. In addition, the loss-of-signal assay was more sensitive and specific than the gain-of-signal assay. Because of the potential cross-hybridization of wild-type sequence to the mutant-specific probes, the gain-of-signal assay was insensitive to larger deletions and single-base insertions. Finally, this study indicated that a target sequence containing a single-nucleotide deletion could cross-hybridize to a single-nucleotide substitution probe of similar sequence, and thus great attention should be paid when mutation-specific probes are used to identify specific sequence changes.

With oligonucleotide microarrays consisting of 12,224 probes, the gain-of-signal approach has been used to determine all the single-nucleotide changes in the HIV-1 protease gene (297 bp) of 167 viral isolates from 102 patients (Kozal et al., 1996). Sequence identities were evaluated based on the hybridization signals and verified by DNA sequencing. The results indicated that an accuracy level of 98.3% in base calling was achieved.

Another successful application of microarray technology to genotying large-scale SNPs in the human genome was demonstrated with 2.3 Mb of human sequence (Wang et al., 1998). A total of 3,241 putative SNPs were identified with 149 arrays, each containing 150,000 to 300,000 probes, through both gain- and loss-of-signal approaches. A separate prototype array was developed to allow simultaneous identification of 500 SNPs. This study demonstrated the feasibility of using microarrays for large-scale genotying SNPs.

Most recently, photolithography-based oligonucleotide microarrays were used to detect mutations in the ataxia telangiectasia mutated (ATM) gene (9.45-kb coding region with 62 exons) found in some cases of lymphoid neoplasia (Fang et al., 2003). This microarray contained >250,000 probes of 25 bp to interrogate both ATM strands for all possible sequence variations in all 62 coding exons and their splice junctions. Some of the probes were designed to complement all possible single-nucleotide substitutions, single-base insertions, single- and two-base deletions. A two-color, loss-of-signal assay was used to detect all possible heterozygous sequence changes in samples from 120 patients. Microarray analysis indicated that the highest deleterious and nonsense mutations existed within the mantle cell subtype, which is in agreement with previous studies. This study demonstrates that the oligonucleotide-based microarray hybridization assay could be very useful in examining tumor samples for mutations in the presence of wild-type alleles. However, detecting all possible mutations in the ATM gene, especially for the less abundant alleles, in the complex mixtures of tumor and normal cells is difficult.

One of the main hurdles for high parallel screening of SNPs is the need for polymerase chain reaction (PCR) amplification of the DNA fragments containing SNPs to achieve sufficient detection sensitivity and specificity. It is generally difficult to achieve reproducible multiplex PCR amplification with tens or hundreds of different primers. Thus, a method without PCR amplification has been examined (Dong et al., 2001a). In this approach, the genotypic DNA was digested with certain restriction enzymes and separated by gel electrophoresis. The DNA fraction containing the expected SNPs of interest was recovered from the gel, ligated to a common adaptor, and amplified with one adaptor-specific primer in a single PCR reaction. The amplified PCR products were hybridized with the arrays containing allele-specific primers. A total accuracy of 98.7% was obtained for screening SNPs (Dong et al., 2001a).

Technical Challenges   Although it is possible to use oligonucleotide microarrays for detecting sequence changes on a large scale and analyzing many samples simultaneously, there are several technical challenges (Hacia, 1999). First, accuracy should be improved by reducing false negative and false positive errors. The current array-based methodologies are more suitable for detecting sequence changes when 5 to 10% false negative error is allowed (Hacia and Collins, 1999). Decreasing the false negative rate is the biggest challenge for array hybridization-based mutational analysis.

The nucleotide sequence composition of the probes and the DNA fragment being analyzed is critical in defining the specificity and sensitivity of array-based assays. Short repeated sequences and duplications present a great challenge to any hybridization-based assay. Although using buffer systems containing tetramethylammonium chloride can alleviate the effects of nucleotide sequence composition on hybridization, it is still difficult to simultaneously achieve specific and sensitive hybridization for all arrayed probes with different G + C content. Suboptimal conditions are needed to allow sufficient hybridization for probes with high AT content and to permit single mismatch discrimination for probes with high G + C content.

Secondary structures present in the target and/or probes can make the array hybridization less predictable. The sequence change in a test sequence could disrupt the secondary structure present in the original wild-type sequence and thus may show better hybridization with the wild-type probe compared to the wild-type reference sequence. It is also difficult to fully interrogate all target sequences due to poor hybridization signals caused by secondary structures.

14.2.2   Microarray-based Single-base Extension for Genotyping

Although microarray hybridization-based genotyping methods are useful in analyzing SNPs, the inherent problem of cross-hybridization between target and probe sequences is a limiting factor for these methods. It is impossible to design a single set of hybridization conditions to achieve optimal signal intensities and maximum discrimination for a large number of sequence mutations simultaneously. Therefore, accuracy is a great challenge for array hybridization-based assays, because the signal-to-noise ratios for many probes are generally quite low. To improve the accuracy of detection, other enzyme-assisted microarray assays, such as single-base extension (SBE), also referred to as minisequencing, have been proposed and evaluated. In SBE, a detection primer anneals to the target nucleotide acid sequence immediately adjacent to a variable nucleotide position. The 3' end of the primer is extended by DNA polymerase with a labeled nucleotide analog that is complementary to the nucleotide at the SNP site. One of the advantages of the SBE approach, compared to the hybridization approach described previously, is that all SNPs can be discriminated with optimal discrimination using the same reaction conditions. In addition, since discrimination between genotypes is based on the DNA polymerase reaction with high sequence specificity, arrays with high probe redundancy are not required. Two types of microarray-based SBE approaches are described below.

Microarray-based Allele-specific Primer Extension   In this approach, two allele-specific oligonucleotide probes from both strands are designed to terminate at the base 5' to a SNP (Pastinen et al., 1997, 2000). The probes are typically tethered to the array surface via a 5'-linkage, leaving an exposed 3'-OH. The test target sequences are then hybridized with the probes on the microarrays. The hybridized target sequences and oligonucleotide probes serve as templates and primers for single-base extension, respectively. Generally, a mixture of all four dideoxyribonucleotide triphosphates, each labeled with different fluorescent dyes, was used for single-base extension. The identity of the added dideoxyribonucleotide is determined with a fluorescence microscope and used to determine the base composition of the target nucleotide adjacent to the 3'-end of each probe. Heterozygous single-nucleotide substitutions will result in two signals corresponding to the identity of the two alleles.

This method was validated with genomic fragments containing nine human disease mutations (Pastinen et al., 1997). The results indicated a 10-fold improvement in discriminating genotypes using DNA polymerase-assisted SBE compared to array-based hybridization. This approach was also applied by Raitio and coworkers (2002) to analyze 25 SNPs in the Y chromosome. The results showed that the microarray-based allele-specific primer extension provided more than 5-fold discrimination between the Y chromosome genotypes. Finally, Pastinen and colleagues (2000) examined array-based allele-specific extension using RNA as templates and reverse transcriptase for extension. This can be effective, because single-stranded RNA templates produce higher signal-to-noise ratios. Arrays containing about 80 primers for 40 SPNs were used to screen more than 8,000 genotypes and all the known genotypes were correctly identified. These results demonstrate that the array-based SBE method is a powerful tool for detecting SNPs.

Microarray-based Tagged Single-base Extension   The microarry-based tagged single-base extension approach [SBE-TAGS (Hirschhorn et al., 2000) or TAG-SBE (Fan et al., 2000)] combines microarray hybridization with single-base extension. In contrast to the array hybridization-based SBE analysis described previously, this approach uses unique sequence tags attached to locus-specific primers. SNPs are detected by single-base extension using biofunctional primers containing a unique sequence tag in addition to a locus-specific sequence. Since each locus has a distinct tag, the sequence reaction can be used for genotying a large number of SNPs in parallel fashion.

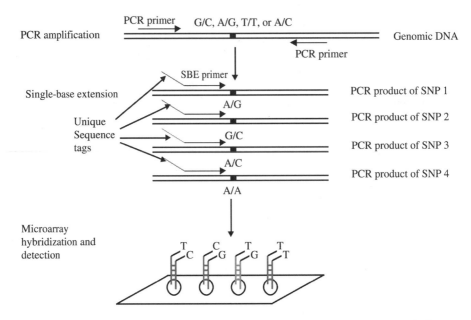

Figure 14.4 *See color insert. Schematic of microarray-based single base extension (SBE).* Locus-specific primers are designed to amplify individual single nucleotide polymorphisms (SNPs) from genomic DNA. The double-stranded PCR products are used as templates for the SBE reaction. Each SBE primer contains a 5′-end unique sequence complementary to a sequence tag on the array and is followed by a 3′-end locus-specific sequence complementary to the genomic sequence and terminating one base before a SNP. Each SBE primer is therefore uniquely associated with a specific sequence tag on the microarrays. In the SBE reaction, the SBE primers corresponding to different locus are mixed together and extended with fluorescent dideoxynucleotides. The labeled SBE reaction products are then hybridized to the tag microarrays.

The TAGS-SBE approach includes the following steps (Fig. 14.4). First, unique sequence tags are designed, and arrays containing sequence tags unique to each locus are constructed. Second, the genomic regions containing SNPs are amplified with locus-specific primers, and the PCR products are pooled together for SBE reactions. Third, the SBE reaction is carried out with pooled PCR products in the presence of labeled dideoxyribonucleotide triphosphates using multiple chimeric SBE primers. Each SBE primer contains a unique sequence tag at the 5′-end and the locus-specific sequence, which terminates one base before a polymorphic site. Thus, each SBE primer is uniquely associated with a specific tag on the array. Fourth, the labeled multiplex SBE reaction products are pooled together and hybridized with the array containing unique sequence tags. Finally, genotypes are inferred based on fluorescence intensity ratios of different dyes (Fan et al., 2000; Hirschhorn et al., 2000). This method is highly accurate. Over 100 SNPs were genotyped using this method and 99% accuracy was obtained (Hirschhorn et al., 2000). The hybridization results appear to be quantitative, and thus, this method can also be used to estimate allele frequency in pooled DNA samples.

A critical step in this approach is the selection of unique sequence tags. The sequence tags should have a similar melting temperature and similar G + C content. Also, the primer sequence tags should not form any internal secondary structure and should not cross-hybridize with each other, as well as with locus-specific primers, under standard hybridization conditions. Hence, careful pairing of the tags and locus-specific primers is needed (Hirschhorn et al., 2000). In addition, to enable background and cross-hybridization subtraction, each tag generally has a PM probe and a MM probe at the central position (Fan et al., 2000).

One of the main advantages of this method is its flexibility. Using a standard array containing generic tags eliminates the need for designing and manufacturing custom arrays for specific sets of SNPs (Fan et al., 2000), making this approach rapid and inexpensive (Hirschhorn et al., 2000).

14.2.3 Microarray-based Ligation Detection Reaction for Genotyping

An array-based method using allele-specific ligation has been developed for genotyping sequence variations (Gerry et al., 1999; Favis et al., 2000). This approach is similar to the TAG-SBE approach except that ligase is used for discrimination instead of DNA polymerase for single-base extension. The principle of this method is based on the fact that single-base mismatch prevents ligation. It has been shown that a G/T mismatch at the 3′-end to be ligated inhibits the reaction by up to 1,000-fold under standardized reaction conditions.

This approach includes the following steps: First, the regions containing sequence variation are amplified by PCR. Second, the ligation detection reaction is performed with the PCR products as a template using two adjacent locus-specific primers. The ligation site between the two primers is located at the polymorphic site of interest. Similar to TAG-SBE, the first locus-specific primer contains a universal sequence tag, called a zip code, at its 5′-end. In contrast to the TAG-SBE primer, which terminates one base before a polymorphic site, the first primer for this approach contains the discriminating base at its 3′-end. Several primers with different sequence tags at the 5′ end and different possible sequence variations at the polymorphic site are usually designed for the same locus to detect various wild and mutant types. The second primer is phosphorylated at the 5′-end and fluorescently labeled at the 3′-end. Ligation occurs between the primers only when there is a perfect complementary base at the junction. Third, the ligation reaction products are hybridized with the array containing the zip code sequence tags. Based on fluorescent hybridization signal patterns, the sequence variations in the target samples can be detected.

This approach has been used successfully to identify mutations in the K-*ras* gene, and insertions and deletions in the *BRCA1* and *BRCA2* genes (Gerry et al., 1999; Favis et al., 2000). The potential advantages of this approach are increased sensitivity and accuracy in detecting sequence variations, particularly small insertions and deletions. However, the throughput of this method remains uncertain, and the requirement for locus-specific primers increases the cost of this approach.

14.3 MICROARRAYS FOR MICROBIAL DETECTION IN NATURAL ENVIRONMENTS

14.3.1 Limitations of Conventional Molecular Methods for Microbial Detection

Microorganisms play an integral and unique role in ecosystem function and sustainability. Understanding the structure and composition of microbial communities and their responses and adaptations to environmental perturbations, such as toxic contaminants, climate change, and agricultural and industrial practices, is critical in maintaining or restoring desirable ecosystem functions. However, the detection, characterization, and quantification of microbial population diversity are formidable tasks for microbial ecologists. Traditional culture enrichment techniques for studying microbial communities have proven difficult and, ultimately, they provide an extremely limited view of microbial diversity, because the majority of naturally occurring species are not culturable (Amann et al., 1995). The development and application of nucleic acid-based techniques largely eliminated the reliance on cultivation-dependent methods and, consequently, greatly advanced the detection and characterization of microorganisms in natural habitats (Amann et al., 1995). However, the limitations of conventional nucleic acid-based detection methods prevent them from being readily adapted as high-throughput, cost-effective assessment tools for monitoring microbial communities.

To assess microbial community dynamics and activities in natural environments, microbial detection tools need to be (1) simple, rapid, and hence real-time and field-applicable; (2) specific and sensitive; (3) quantitative; (4) capable of high throughput; and (5) cost-effective. Although conventional nucleic acid detection approaches [e.g., SSU rRNA gene-based cloning methods, denatured gradient gel electrophoresis (DGGE), terminal restriction fragment length polymorphism (T-RFLP), quantitative PCR, and in situ hybridization and PCR amplification] are vital to studies of microbial communities, they meet these requirements with difficulty. Microarray-based technology has the potential to overcome the limitations of traditional molecular methods for studying microbial community structure.

14.3.2 Advantages and Challenges of Microbial Detection in Natural Environments

In addition to the advantages mentioned in Chapter 6 (e.g., high density, parallel analysis, two color detection and low background), microarray-based technology is well suited for detecting microorganisms in natural environments. Many target functional genes involved in biogeochemical cycling in environments are highly diverse, and it is difficult, sometimes even impossible, to identify conserved regions for designing PCR primers or oligonucleotides. The microarray-based approach does not require such sequence conservation, because all the diverse gene sequences from different populations of the same functional group can be fabricated on arrays and used as probes to monitor their corresponding populations. In contrast to studies using pure cultures, microarray-mediated analysis of environmental nucleic acids presents a number of technical challenges that must be addressed. In environmental samples, target and probe sequences can be very diverse and the performance of microarrays, including how sequence divergence is reflected in hybridization signal intensity, may not be similar to that with pure culture samples.

Environmental samples are generally contaminated with substances such as humic matter, organic contaminants, and metals, which may interfere with DNA hybridization on microarrays. In contrast to pure cultures, the retrievable biomass in environmental samples is generally low; consequently, microarray hybridization may not be sensitive enough to detect microorganisms in all types of environmental samples. Finally, microarray-based detection may not be quantitative. Environmental and ecological studies require experimental tools that not only detect the presence or absence of particular groups of microorganisms but also provide quantitative data on their in situ biological activities.

In the following sections, we will discuss different types of microarrays used in environmental studies, namely, functional gene arrays, phylogenetic oligonucleotide arrays, and community genome arrays.

14.3.3 Functional Gene Arrays

The genes encoding functional enzymes involved in various biogeochemical cycling processes (e.g., carbon, nitrogen, sulfate, and metals) are very useful as signatures for monitoring the physiological status and functional activities of microbial populations and communities in natural environments. Microarrays containing functional gene sequence information are referred to as functional gene arrays (FGAs), because they are primarily used for the functional analysis of microbial community activities in environments (Wu et al., 2001). Similar to the microarrays used for monitoring gene expression, both oligonucleotides and PCR-amplified DNA fragments derived from functional genes can be used for fabricating FGAs.

Selection of Gene Probes FGAs are designed for studying functional gene diversity in natural environments. To construct FGAs, the gene probes should be carefully defined and selected based on the specific research questions to be addressed. For example, microarrays can consist of gene probes that are involved in various biogeochemical processes, including nitrification (ammonia monooxygenase, *amoA*), denitrification (nitrite reductases, *nirS* and *nirK*), nitrogen fixation (nitrogenases, *nifH*), sulfite reduction (sulfite reductase, *dsvA/B*), methanogenesis (methyl coenzyme M reductase genes, *mcrA*), methane oxidation (methane mono-oxygenases, *mmo*), and plant and fungal polymer degradation (cellulases, xylanases, lignin peroxidases, chitinases).

There are three general approaches to obtaining FGA probes. The first strategy is to amplify the desired gene fragment from genomic DNA extracted from pure bacterial cultures with specific primers or from cloned plasmids containing the desired gene insert with vector-specific primers. However, the availability of pure cultures and plasmid clones can be limited. The second approach is to recover the desired gene fragments from natural environments using PCR-based cloning methods (Zhou et al., 1997). The sequences that show > 85% identity can be used as specific probes for FGAs. These two approaches were utilized to construct FGAs containing nitrite reductase genes and ammonia monooxygenase genes for monitoring bacteria involved in nitrification and denitrification (Wu et al., 2001). The third approach is to use oligonucleotide probes. Longer oligonucleotides, such as 50 to 70 mers, can be designed based on the functional sequences available in databases and synthesized for microarray fabrication (Zhou, 2003).

Specificity An important parameter of any detection method is hybridization specificity. It is influenced by many factors, such as $G + C$ content, degree of sequence divergence, length, secondary structure of the probe, temperature, and salt concentrations. To determine the specificity of DNA microarray hybridization, we have constructed and used FGAs consisting of heme- and copper-containing nitrite reductase genes, ammonia monooxygenase (*amoA*), and methane monooxygenase genes [*pmoA*] (Wu et al., 2001). SSU rRNA genes and yeast genes were used as positive and negative controls. Cross-hybridization among different gene groups was not observed at either low (45°C) or high

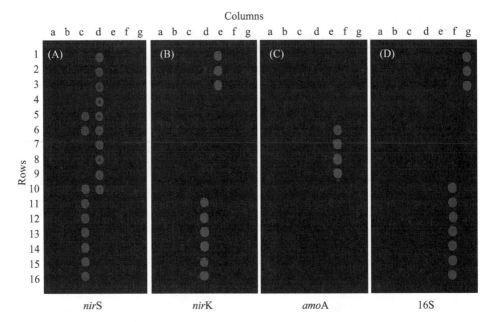

Figure 14.5 *See color insert. Fluorescence images showing the specificity of* nirS, nirK, amoA, *and* 16S rRNA target genes in DNA microarray hybridization. Target DNA was labeled with either cyanine 3 (Cy3; *nirS* and 16S rRNA genes from pure cultures) or cyanine 5 (Cy5; *nirK* and *amoA* genes from pure cultures) using the method of polymerase chain reaction amplification and hybridized separately at high stringency (65°C) to functional gene arrays containing *nirS, nirK,* and *amoA* gene probes from both pure bacterial cultures and environmental clones. 16S rRNA and yeast genes served as positive and negative controls, respectively. (A) *nirS,* (B) *nirK,* (C) *amoA,* and (D) 16S rDNA. [Reprinted with permission from Wu, L. Y., D. K. Thompson, G.-S. Li, R. Hurt, H. Hung, J. M. Tiedje, and J.-Z. Zhou, Development and evaluation of functional gene arrays for detection of selected genes in the environment, *Appl. Environ. Microbiol.,* vol. 67, 5780–5790. Copyright (2001) American Society for Microbiology.]

(65°C) stringency. Furthermore, no hybridization was observed with any of the five yeast genes, which served as negative controls for hybridization on the microarray (Fig. 14.5). These results indicate that specific hybridization can be achieved using the glass slide-based microarray format with bulk community DNA extracted from environmental samples. Based on the sequence similarities, it was estimated that microarray hybridization can differentiate between sequences exhibiting a dissimilarity of approximately 15% at 65°C and 10% at 75°C (Wu et al., 2001). In addition, at low stringency, most *nirS, nirK,* or *amoA* genes hybridized well with their respective homologous target DNA, suggesting that a broad range of detection can be achieved by adjusting the conditions for microarray hybridization.

To determine the potential performance of the oligonucleotide-based microarrays for environmental studies, a 50-mer FGA was constructed and evaluated using 1,033 genes involved in nitrogen cycling (*nirS, nirK, nifH, amoA,* and *pmoA*) and sulfite reduction (*dsrA/B*) from public databases and our own sequence collections (Tiquia et al., unpublished). Under the hybridization conditions of 50°C and 50% formamide, genes having <86 to 90% sequence identity were clearly differentiated. As expected, the hybridization specificity of the 50-mer FGAs is higher than that of the PCR product-based

FGAs. Based on the comparisons of the probe sequences from the pure cultures involved in nitrification, denitrification, nitrogen fixation, methane oxidation, and sulfate reduction, the average similarity of these functional genes at the species level ranged from 74 to 84%. Such results suggest that the 50-mer FGAs could provide species-level resolution for analyzing microorganisms involved in these biogeochemical processes.

Compared to the PCR product-based FGA, the 50-mer oligonucleotide arrays offer the following main advantages. First, higher hybridization specificity can be achieved with the 50-mer FGAs than with the PCR product-based FGA, because the probe sizes in oligonucleotide microarrays are much smaller than those used in the PCR product-based FGA. This type of FGA could therefore provide a higher level of resolution than PCR product-based FGA in differentiating microbial populations. In addition, the construction of oligonucleotide microarrays is much easier than that of the PCR product-based FGAs, because the probes can be directly designed and synthesized based on sequence information from public databases. To construct the microarrays containing large DNA fragments, the probes used for microarray fabrication are generally amplified by PCR from environmental clones or from pure genomic DNA. However, obtaining all the diverse environmental clones and bacterial strains from various sources as templates for amplification can be very challenging. As a result, constructing comprehensive microarrays to represent all functional genes of interest is very difficult. With oligonucleotide microarrays, a greater number of genes can easily be arrayed in order to conduct a comprehensive survey of the populations and activities of diverse microbial communities in the environment. In addition, since no PCR amplification is involved in oligonucleotide microarray fabrication, potential cross-contamination due to PCR amplification is minimized.

Sensitivity   Sensitivity is another critical parameter that impacts the effectiveness of microarray-based detection of microorganisms. The detection sensitivity of hybridization with a prototype PCR product-based FGA was determined using genomic DNA from both pure cultures and soil community samples (Fig. 14.6). At high stringency, strong hybridization signals were observed with 5 ng of DNA for both *nirS* and SSU rRNA genes, whereas hybridization signals were weaker but detectable with 1 ng of DNA (Fig. 14.6A). The hybridization signal at low DNA concentrations was stronger for SSU rRNA genes than for *nirS* genes. Hybridization signals were measurable with 0.5 ng of genomic DNA, but the fluorescence intensity was poor (Fig. 14.6A). Therefore, the detection limit with randomly labeled pure genomic DNA under these hybridization conditions was estimated to be approximately 1 ng.

The detection sensitivity of microarray hybridization was also evaluated with community genomic DNA isolated from surface soil that contained a high level of chromium and organic matter. All the arrayed genes, with the exception of the five yeast genes, showed hybridization with 50 and 25 ng of labeled community DNA (Fig. 14.6B). Only the SSU rRNA genes could be detected when as little as 10 ng of soil community DNA were used in the hybridization reaction (Fig. 14.6B). Thus, with this microarray hybridization system, the detection sensitivity of *nirS* and SSU rRNA genes in this soil sample was considered to be approximately 25 and 10 ng of the total environmental DNA, respectively. These approximate levels of detection sensitivity should be sufficient for many studies in microbial ecology. These results suggest that microarray hybridization can potentially be used as a sensitive tool for analyzing microbial community structure in environmental samples.

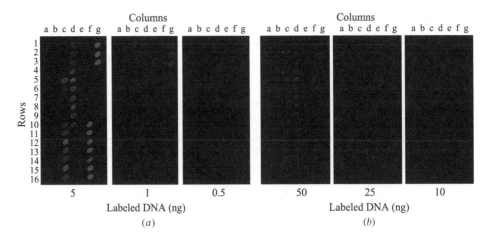

Figure 14.6 *See color insert. Array hybridization images showing the detection sensitivity with labeled pure genomic DNA and bulk community DNA from soil.* (*a*) Genomic DNA from a pure culture of *nirS*-containing *Pseudomonas stutzeri* E4-2 was labeled with cyanine 5 (Cy5) using the random primer labeling method. The target DNA was hybridized to the microarrays at total concentrations of 0.5, 1, and 5 ng. (*b*) Genomic DNA from surface soil (10, 25, and 50 ng) was labeled with Cy5 as described in (*a*) and hybridized with the microarrays. [Reprinted with permission from Wu, L. Y., D. K. Thompson, G.-S. Li, R. Hurt, H. Hung, J. M. Tiedje, and J.-Z. Zhou, Development and evaluation of functional gene arrays for detection of selected genes in the environment, *Appl. Environ. Microbiol.*, vol. 67, 5780–5790. Copyright (2001) American Society for Microbiology.]

The detection limit with the 50-mer FGAs was approximately 8 ng of pure genomic DNA in the absence of heterogeneous nontarget DNAs. As expected, the sensitivity of the 50-mer FGAs is 10 times lower than that of the PCR product-based FGAs and 100 times lower than that of community genome arrays (CGAs; Tiquia et al., unpublished; Zhou, 2003). Recently, we found that the sensitivity of 50-mer oligonucleotide microarrays is further decreased and was about 50- to 100-fold lower than that of the PCR product-based FGAs in the presence of heterogeneous nontarget DNA. One of the main reasons for the lower sensitivity of the 50-mer FGAs was that the probes were much shorter (50 mer) than the probes used in DNA-based FGA and CGA, which have more binding sites available for capturing the labeled target DNAs. In addition, microarray hybridization signals have been observed for some probes when 2 μg of bulk community DNA from marine sediments were labeled. These results implied that the 50-mer microarray hybridization can be used to detect dominant populations but that it is still not sensitive enough for detecting microorganisms of low abundance in environmental samples. It should be noted that the detection sensitivity is dependent on reagents, especially with fluorescent dyes.

Quantitation   Many environmental and ecological studies require quantitative data on the in situ abundance and activities of microbial communities. Because of the inherently high variation associated with array fabrication, probe labeling, hybridization, and image processing, the accuracy of microarray-based quantitative assessment is still uncertain. Comparison of microarray hybridization results with previously known results suggested that microarray hybridization is quantitative enough for detecting differences in gene expression patterns under various conditions (DeRisi et al., 1997; Lockhart et al., 1996; Taniguchi et al., 2001; Liu et al. 2003). DNA microarrays have also been used to measure differences in DNA copy number in breast tumors (Pinkel et al., 1998; Pollack et al., 1999). Single-copy deletions or additions can be detected (Pollack et al., 1999), suggesting that microarray-based detection is potentially quantitative. A recent study in which lambda (λ) DNA was cospotted with DNA from reference bacterial strains also indicated that microarrays can accurately quantify genes in DNA samples (Cho and Tiedje, 2001).

To evaluate whether microarray hybridization can be used as a quantitative tool with environmental samples, the relationship between target DNA concentration and hybridization signal was examined with the PCR product-based FGAs (Wu et al., 2001). A strong linear relationship ($r^2 = 0.96$) was observed between signal intensity and target DNA concentration with DNA from a pure bacterial culture within a range of 1 to 100 ng (Fig. 14.7A). Similar to the PCR product-based FGAs, a strong linear relationship was observed with the 50-mer FGAs between signal intensity and target DNA concentrations from 8 to 1,000 ng for all six different functional gene groups ($r^2 = 0.96-0.98$) (Tiquia et al., unpublished). These results suggest that microarray hybridization is quantitative for pure bacterial cultures within a limited range of DNA concentrations. With our optimized protocol, experimental variation between array slides can be reduced to below 15% with environmental samples (Wu et al., 2001). This is consistent with the findings of microarray studies on gene expression (Bartosiewicz et al., 2000).

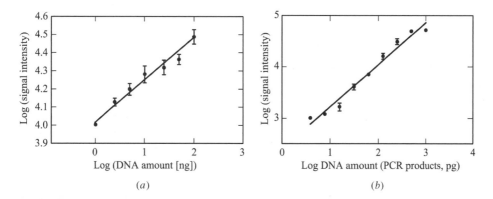

Figure 14.7 *Quantitative analysis of functional gene arrays.* (*a*) Relationship of hybridization signal intensity to DNA target concentration from a single pure culture. Genomic DNA from *nirS*-containing *P. stutzeri* E4-2 was labeled with cyanine 5 (Cy5) and hybridized to the microarrays at target concentrations of 1, 2.5, 5, 10, 25, 50, and 100 ng. The plot shows the log-transformed average hybridization intensity vs. the log-transformed target DNA concentration. (*b*) Relationship of hybridization signal intensity to DNA target concentration using a mixture of target DNAs. The polymerase chain reaction products from nine strains were mixed together in different quantities (pg): E4-2 (*nirS*), 1,000; G179 (nirK), 500; wc301-37 (*amoA*), 250; ps-47 (*amoA*), 125; pB49 (*nirS*), 62.5; Y32K (*nirK*), 31.3; wA15 (*nirS*), 15.6; ps-80 (*amoA*), 7.8; wB54 (*nirK*), 3.9. All these genes are less than 80% identical. The mixed templates were labeled with Cy5. The plot shows the log-transformed average hybridization intensity vs. the log-transformed target DNA concentration for each strain. The target DNA was prepared by labeling MR-1 genomic DNA with Cy5 using a Klenow fragment with random hexamer primers. For both (*a*) and (*b*), the data points are mean values derived from three independent microarray slides, with three replicates on each slide (a total of nine data points). Error bars represent the standard deviation.

Because environmental samples contain a mixture of target and nontarget templates, the presence of other nontarget templates could affect microarray-based quantification. To determine whether microarray hybridization is quantitative for targeted templates within the context of environmental samples, 11 different genes, exhibiting less than 80% sequence identity, were labeled and hybridized with the PCR product-based FGAs. For this mixed DNA population, a linear relationship ($r^2 = 0.94$) was observed between signal intensity and target DNA concentration (Fig. 14.7B), further suggesting that microarray hybridization holds promise as a quantitative tool for studies in environmental microbiology.

The target genes within functional groups present in environmental samples may have different degrees of sequence divergence. Such sequence differences will affect microarray hybridization signal intensities and hence its quantitative power. Although it was shown that microarray hybridization could be used to quantify mixed DNA templates, the difficult challenge in quantifying the abundance of microbial populations in natural environments, based on hybridization signal intensity, is how to distinguish differences in hybridization intensity due to population abundance from those due to sequence divergence. One possible solution is to carry out microarray hybridization under conditions of varying stringency. Based on the relationships among signal intensity, sequence divergence, hybridization temperature, and washing conditions, it should be possible to distinguish, to some extent, the contributions of population abundance and sequence divergence to hybridization intensity (Wu et al., 2001). For instance, Wu and colleagues (2001) showed that at about 55 to 60°C, sequence divergence had little or no effect on signal intensity for *amoA* genes with greater than 80% identity to the labeled target DNA. This suggests that under such hybridization conditions, the effect of sequence divergence on signal intensity is negligible for genes with $> 80\%$ sequence identity; therefore, any significant differences in signal intensity are most likely due to differences in population abundance. Another possible solution to this problem is to use microarrays containing probes that are extremely specific to the target population of interest, such as those used in oligonucleotide microarrays.

Applications   Since FGAs for microbial detection are presently in the development stage, their applications are still being explored. To demonstrate the applicability of DNA microarrays for microbial community analysis, Wu and coworkers (2001) used functional gene arrays to analyze the distribution of denitrifying and nitrifying microbial populations in marine sediment and soil samples. The prototype functional gene array revealed differences in the apparent distribution of *nirS*, *nirK*, and *amoA/pmoA* gene families in sediment and soil samples. Recently, a 70-mer oligonucleotide microarray containing 64 *nirS* genes (14 from cultured microorganisms and 50 from environmental clones) was evaluated in studying functional gene diversity in the Choptank River–Chesapeake Bay system (Taroncher-Oldenburg et al., 2003). Significant differences in the hybridization patterns were observed between the sediment samples from two stations in the Choptank River. The changes in the *nirS*-containing denitrifier population could have been caused by differences in salinity, inorganic nitrogen, and dissolved organic carbon between these two stations.

Thus far, very limited studies have been carried out to evaluate specificity, sensitivity, sequence divergence, and quantitation of DNA microarrays for environmental applications. Although this tool is potentially valuable for environmental studies, more development is needed, especially for improved sensitivity, quantitation, and to determine the biological meaning of a detectable specificity before DNA microarrays can be used broadly and interpreted meaningfully within the context of microbial ecology.

14.3.4 Phylogenetic Oligonucleotide Arrays

Ribosomal RNA genes are powerful molecules for studying phylogenetic relationships among different organisms and for analyzing microbial community structure in natural environments. These genes exist in all organisms and contain both highly conserved and highly variable regions, which are useful for differentiating microorganisms at different taxonomic levels (e.g., kingdom, phyla, family, genus, species, and strain). Ribosomal RNA genes have the largest representative database, making them ideal molecules for developing microarray-based detection tools. In addition, cells generally have multiple copies of rRNA genes, and the majority (> 95%) of total RNA isolated from samples is rRNA. Consequently, the detection sensitivity will be higher for rRNA genes than for functional genes. Therefore, rRNA genes are very useful targets for developing microarray-based detection approaches.

Oligonucleotide microarrays containing information from rRNA genes are referred to as phylogenetic oligonucleotide arrays (POAs), because such microarrays are used primarily for the phylogenetic analysis of microbial communities. POAs can be constructed for different phylogenetic taxa and used in community analysis studies. The oligonucleotide probes can be designed in a phylogenetic framework to survey different levels of sequence conservation, from highly conserved sequences giving broad taxonomic groupings to hypervariable sequences giving genus- (and potentially species-) level groupings. Because highly conserved universal primers for amplifying rRNA genes are available, POA-based hybridization can be easily coupled with PCR amplification. Therefore, highly sensitive assays may be implemented.

Challenges of Phylogenetic Oligonucleotide Arrays   Although non-rRNA gene-based oligonucleotide microarrays have been used successfully for monitoring genome-wide gene expression (e.g., de Saizieu et al., 1998, Lockhart et al., 1996) and detecting genetic polymorphisms (e.g., Wang et al., 1998), rRNA gene-based oligonucleotide arrays present some unique technical challenges (Zhou and Thompson, 2002; Zhou, 2003).

Specificity   Since the rRNA gene is highly conserved and present in all microorganisms, specific detection with rRNA-targeted oligonucleotide microarrays can be difficult. First, the probe length and G + C content can significantly impact microarray hybridization (Guschin et al., 1997a). Second, probe selection is limited by the sequence differences among the target genes, and cross-hybridization can be a problem for oligonucleotide arrays. Oligonucleotide microarrays typically contain many probes. Ideally, all the oligonucleotides should have similar or identical melting kinetics, so that all the probes on an array element can be subjected to the same hybridization conditions at once. This may be difficult to achieve, because the melting temperature depends on the length and composition of the oligonucleotide probe as well as the target 16S rRNA molecules in the samples.

Secondary Structure   The hybridization of oligonucleotide probes to target nucleic acids possessing stable secondary structure can be particularly challenging, since low stringency conditions (i.e., hybridization temperatures between 0–30°C) are required for stable association of a long target nucleic acid with a short immobilized oligonucleotide probe (Guschin et al., 1997a,b; Drobyshev et al., 1997; Southern et al., 1999). Any stable secondary structure of the target DNA or RNA must be overcome in order to make complementary sequence regions available for duplex formation. The stable secondary structure of SSU rRNA will have serious effects on hybridization specificity and detection sensitivity.

Specificity and Sensitivity In a study by Guschin and coworkers (1997a), gel-pad oligonucleotide microarrays were constructed using oligonucleotides complementary to SSU rRNA sequences from key genera of nitrifying bacteria. The results showed that specific detection could be achieved with this type of microarray. However, the probe specificity depends on various factors, such as probe length. Guschin and colleagues (1997a) showed that, as the length of the oligonucleotide probe increases, mismatch discrimination is lost; conversely, as the length of the probe decreases, hybridization signal intensity (i.e., sensitivity) is sacrificed. A recent study showed that gel-pad-based oligonucleotide microarrays also can be used to distinguish *Bacillus* species, namely, *B. thuringiensis* and *B. subtilis* (Bavykin et al., 2001). Using glass-based two-dimensional microarrays, Small and colleagues (2001) detected metal-reducing bacteria, such as *Geobacter chapellei* and *Desulfovibrio desulfuricans.*

The potential advantage of oligonucleotide probes is that target sequences containing single-base mismatches can be differentiated by microarray hybridization. However, this has not been fully demonstrated with SSU rRNA gene-based probes. To systematically determine whether single mismatch discrimination can be achieved for SSU rRNA genes using microarray hybridization, we constructed a model oligonucleotide microarray consisting of probes derived from three different regions of the SSU rRNA molecule corresponding to different bacterial taxa (X. Zhou and J. Zhou, unpublished data). The probes had one to five mismatches in different combinations along the length of the oligonucleotide probe with at least one mismatch at the central position. Hybridization signal intensity with a single-base mismatch was decreased by 10 to 30% depending on the type of mismatched nucleotide base. The signal intensity of probes with two-base mismatches was 5 to 25% of that of the perfect match probes. Probes with three or four base-pair mismatches yielded signal intensities that were 5% of that of the perfect match probes. Maximum discrimination and signal intensity were achieved with 19-base probes. These results indicated that single-base discrimination for SSU rRNA genes can be achieved with glass slide-based array hybridization, but complete discrimination appears to be problematic with SSU rRNA genes (Bavykin et al., 2001; Small et al., 2001; Urakawa et al., 2002). Urakawa and coworkers (2002) demonstrated that the single-base-pair near-terminal and terminal mismatches have a significant effect on hybridization signal intensity. With SSU rRNA gene-based oligonucleotide microarrays, the level of detection sensitivity obtained using the *G. chapellei* 16S rRNA gene is about 0.5 μg of total RNA extracted from soils (Small et al., 2001).

Applications SSU rRNA gene-based oligonucleotide microarrays are still in the early stages of development, and, therefore, only a few studies have applied POAs to the analysis of microbial structure within the context of environmental samples. Using photolithography-based Affymetrix technology, Wilson and colleagues (2002) designed a microarray containing 31,179 20-mer oligonucleotide probes specific for SSU rRNA genes. All the probes were derived from a small SSU rRNA gene region (i.e., *E. coli* positions 1409 to 1491), which is bounded on both ends by universally conserved segments. The microarray also contained control sequences, which were paired with the probe sequences. A control sequence was identical to the paired probe sequence except that there was a mismatch nucleotide at the 11th position. Thus, the microarray contained a total of 62,358 features. The number of probes for individual sequences contained in the Ribosomal Database Project (RDP version 5.0, with about 3,200 sequences) ranges from zero to 70. A total of 17 pure bacterial cultures were used to assess the performance of this microarray, and 15 bacterial species were identified correctly. However, it failed to resolve the individual sequences comprising complex mixed samples (Wilson et al., 2002).

Rudi and colleagues (2000) constructed a small microarray containing 10 SSU rRNA probes derived from cyanobacteria, and used it to analyze their presence and abundance in lakes with both low and high biomass. The probes were specific to the cultures analyzed, and reproducible abundance profiles were obtained with these lake samples. Relatively good qualitative correlations were observed between the community diversity and standard hydrochemical data, but the levels of correlation were lower for the quantitative data.

Loy and coworkers (2002) constructed a glass-based microarray containing 132 SSU rRNA-targeted oligonucleotide probes, which represented all recognized groups of sulfate-reducing prokaryotes. Microarray hybridizations with 41 reference strains showed that, under the hybridization conditions used, clear discrimination between PM and MM probes was obtained for most, but not all, of the 132 probes. They used this microarray to determine the diversity of sulfate-reducing prokaryotes in periodontal tooth pockets and a hypersaline cyanobacterial mat. The microarray hybridization results were consistent with those obtained using well-established conventional molecular methods. These results suggest that microarray hybridization could be a potentially powerful tool in analyzing community structure, but great caution needs to be used in data interpretation because of the potential for cross-hybridization.

14.3.5 Community Genome Arrays

Many microorganisms have been isolated from a variety of natural habitats. However, little or nothing is known about the genomic sequences for the majority of these microorganisms. This large collection of pure cultures would be very useful for monitoring microbial community composition, structure, and dynamics in natural environments if microarrays could be developed that did not require prior knowledge of gene sequences. Thus, a novel prototype microarray containing whole genomic DNA, termed community genome array (CGA), was developed and evaluated in our laboratory.

The CGA is conceptually analogous to membrane-based reverse sample genome probing (RSGP; Voordouw et al., 1991), but CGA hybridization is distinctly different from RSGP in terms of the arraying substrate and signal detection strategies. In contrast to RSGP, CGA uses a nonporous surface for fabrication and fluorescence-based detection. The capability of accurate and precise miniaturization with robots on nonporous substrates is one of the two key advances of microarray-based genomic technologies. The miniaturized microarray format coupled with fluorescent detection represents a fundamental revolution in biological analysis. Like RSGP, the main disadvantage of the CGA is that only the cultured components of a community can be monitored, because the construction requires the availability of individual pure isolates, although CGA-based hybridization itself does not require culturing (Voordouw, 1998). With the recent advances in environmental genomics, high-molecular-weight DNA from uncultivated microorganisms could be accessed through a bacterial artificial chromosomes (BAC)-based cloning approach. BAC clones could also be used to fabricate CGAs, thus allowing the investigation of uncultivated components of a complex microbial community. In the following sections, we will briefly describe the performance of CGA-based hybridization in terms of specificity, sensitivity, and quantitation.

Specificity   To examine hybridization specificity under varying experimental conditions and to determine threshold levels of genomic differentiation, a prototype microarray was fabricated that contained genomic DNA isolated from 67 different representative environmental microorganisms classified as α-, β-, and γ-proteobacteria and gram-positive bacteria. Many of the selected species are closely related to each other based on small subunit (SSU) rRNA and *gyrB* gene phylogenies and belong primarily to three major bacterial genera (*Pseudomonas*, *Shewanella*, and *Azoarcus*). The G + C content of the genomes varies from 37 to 69.3%. By adjusting the hybridization temperature and the concentration of additives such as formamide (which increases hybridization stringency), different threshold levels of phylogenetic differentiation could be achieved using the CGAs. For instance, under hybridization conditions of 55°C and 50% formamide, strong signals were obtained for genomic DNAs of species corresponding to the labeled target. Little or no cross-hybridization ($\sim 0-4\%$) was observed for nontarget species as well as for negative controls (yeast genes), thus indicating that species-specific differentiation can be achieved with CGAs under the hybridization conditions used. However, different strains of *P. stutzeri*, *A. tolulyticus*, *Bacillus methanolicus*, and *Shewanella algae* could not be clearly distinguished under these conditions (Wu et al., unpublished). By further increasing hybridization temperature (65 and 75°C), strain-level differentiation was obtained for closely related *Azoarcus* strains (Wu et al., unpublished).

Due to the complicated nature of microarray hybridization, it is less likely that such assays will completely eliminate some degree of hybridization to nontarget strains. The central question is how to distinguish true hybridization signals from nonspecific background noise. One common approach is to determine signal-to-noise ratios (SNRs) and to discard values below a certain threshold value. Our studies showed that the average SNR for hybridizations with different species within a genus is about 3.35 $\pm$ 0.32, which is substantially lower than hybridizations with different strains from the same species. This value is very close to the commonly used threshold value (SNR = 3.0).

CGAs could be used to determine the genetic distance between different bacteria at the taxonomic levels of species and strain. Significant linear relationships were observed between CGA hybridization ratios and sequence similarity values derived from SSU rRNA and *gyrB* genes, DNA–DNA reassociation, or REP- and BOX-PCR fingerprinting profiles ($r^2 = 0.80-0.95$) (Wu et al., unpublished), suggesting that CGAs could provide meaningful insights into relationships between closely related strains. Because of its high capacity, one can construct CGAs containing bacterial type strains plus appropriately related strains. By hybridizing genomic DNA from unknown strains with this kind of microarray, one should be able to quickly and reliably identify unknown strains, provided a suitably related probe is on the array. When using CGAs for strain identification, less stringent hybridization conditions (e.g., 45°C and 50% formamide) should be used first to ensure that good hybridization signals can be obtained for distantly related target species. If multiple probes have significant hybridization with the unknown target strains, highly stringent hybridization conditions should then be used.

Compared to the traditional DNA–DNA reassociation approach, CGAs have several advantages for determining species relatedness. Since many bacterial genomes can be deposited on microarray slides, the tedious and laborious pairwise hybridizations associated with the traditional DNA–DNA reassociation approach among different species are not needed with CGAs. In contrast to the traditional DNA–DNA reassociation approach, which generally requires about 100 μg DNA, CGA-based hybridization requires only about 2 μg of genomic DNA. This is important for determining the relationships between bacterial species that are recalcitrant to cultivation or grow very slowly.

Sensitivity and Quantitative Potential To determine CGA sensitivity, genomic DNA from a pure bacterial culture was fluorescently labeled and hybridized with the community genome array at different concentrations. Under stringent hybridization conditions (i.e., 65°C), the detection limit with randomly labeled pure genomic DNA was estimated to be approximately 0.2 ng in the absence of heterogeneous non-target DNA, whereas genomic DNA concentrations of 0.1 ng were barely detectable above background levels. The detection limit with randomly labeled mixed genomic DNAs from 16 species was estimated to be approximately 2.5 ng (Wu et al., unpublished). The level of CGA detection sensitivity should be sufficient for many studies in microbial ecology. The detection sensitivity was approximately 10-fold higher·than that of DNA-based FGAs and about 100 times higher than that of the 50-mer FGAs. These results were expected, because the CGA probes represent entire genomes rather than a single gene.

The capacity of CGA hybridization to serve as a quantitative tool was explored by examining the relationship between the concentration of labeled target DNA and hybridization signal intensity. Quantitative potential was determined using labeled genomic DNA from a single pure culture and from 16 targeted bacteria representing different genera and species. In both cases, strong linear relationships between fluorescence intensity and DNA concentration were observed within a certain range of concentrations ($r^2 = 0.92-0.95$) (Wu et al., unpublished). The results indicated that CGAs could be used for the quantitative analysis of microorganisms in environmental samples. The quantitative feature of CGA is similar to those of the PCR product- and oligonucleotide-based FGAs (Wu et al., 2001).

14.3.6 Whole-genome Open Reading Frame Arrays for Revealing Genome Differences and Relatedness

Whole-genome sequence information for microorganisms that are closely related based on SSU rRNA gene sequences can be used to understand the genetic basis for observed phenotypic differences. Sequence similarity and variability will provide insights into the conservation of gene functions, physiological plasticity, and evolutionary processes. However, sequencing the entire genomes of all closely related species is expensive and time-consuming. DNA microarrays containing individual ORFs of a sequenced microorganism to view genome diversity and relatedness of other closely related microorganisms may eliminate the need for sequencing multiple genomes, because substantial portions of the genomic sequence will be common among closely related species.

Genome diversity and relatedness among closely related organisms have been examined using the whole-genome ORF array-based hybridization approach in several studies. The genome diversity and relatedness of several related metal-reducing bacteria within the

Shewanella genus were evaluated using partial ORF microarrays for the sequenced metal-reducing bacterium, *S. oneidensis* MR-1 (Murray et al., 2001). Among the nine species tested, both conserved and poorly conserved genes were identified. The hybridization results were most informative for the closely related organisms with SSU rRNA sequence similarities greater than 93% and *gyrB* sequence similarities greater than 80%. Above this level of homology, the similarities of microarray hybridization profiles were strongly correlated with *gyrB* sequence divergence. It appears that this approach can be used to identify genes or operons that were horizontally transferred, because most genes present in operons had high levels of DNA relatedness (Murray et al., 2001).

Dong and colleagues (2001) identified the genes in *Klebsiella pneumoniae* 342, a common endophyte of maize that is closely related to *E. coli*. Using ORF arrays for *E. coli* K-12, they determined that about 24% of the *E. coli* genes were absent in strain 342 and about 3,000 (70%) of *E. coli* genes were present with greater than 55% identity. The genes with low sequence identity were involved in carbon compound metabolism, membrane proteins, structural proteins, central intermediary metabolism, and proteins involved in adaptation and protection, whereas the genes with high sequence identity were those involved in cell division, DNA replication, transcription, translation, transport, regulatory proteins, energy, amino acid and fatty acid metabolism, and cofactor synthesis. Genes that were not identified in strain 342 included mobility proteins, putative enzymes, putative regulatory proteins, putative chaperones, surface structure proteins, and hypothetical proteins. These results on genomic diversity are consistent with the physiological properties of these two strains, suggesting that the microarray-based whole-genome comparison is a powerful approach to revealing the relatedness and genomic diversity of closely related organisms.

The whole-genome ORF microarray approach was also used successfully to detect the deletions existing in other strains of *Mycobacterium tuberculosis* and *M. bovis* (Behr et al., 1999) and to identify genome differences among 15 *Helicobacter pylori* strains with more or less virulence (Salama et al., 2000). These studies suggest that whole-genome ORF microarrays will be useful for revealing relatedness and genome difference. Whole-genome arrays for *S. oneidensis* MR-1, *D. radiodurans* R1, *Rhodopseudomonas palustris*, *Nitrosomonas europaea*, *Desulfovibrio vulgaris*, and *Geobacter metallireducens* are available at Oak Ridge National Laboratory (ORNL) and many other whole-genome ORF arrays are available for additional microorganisms. At ORNL, we are also currently using these whole-genome ORF arrays to understand the genome diversity and relatedness of some important environmental isolates.

14.3.7   Other Types of Microarrays for Microbial Detection and Characterization

When genome sequence information is not available, DNA microarrays containing random genomic fragments can be used to determine species relatedness. Cho and Tiedje (2001) randomly selected 60 to 96 genomic fragments of about 1 kb from four fluorescent *Pseudomonas* species as reference genomes for microarray fabrication. Cluster analysis of hybridization profiles from 12 well-characterized fluorescent *Pseudomonas* species indicated that such types of microarray hybridization could provide species to strain level resolution. Compared to the CGA, this approach may have higher resolution, because extensive component information is obtained rather than an average for the whole genome. However, this approach is more costly and time-consuming to develop than CGAs and would be more limited in scope since many array positions would be used for reference microorganisms (L. Wu, personal communication).

Kingsley and coworkers (2002) have recently developed a random nonamer oligonucleotide microarray. They used the micorarry to obtain fingerprinting profiles among closely related strains instead of using a gel electrophoresis-based method. A prototype array containing 47 randomly selected nonamer oligonucleotides was constructed and used to differentiate 14 closely related *Xanthomonas* strains. The REP-PCR was first carried out to obtain the fingerprints from different strains, then the amplified REP-PCR products were hybridized with the nonamer array, and fingerprinting profiles for each strain were obtained based on microarray hybridization. The microarray hybridization-based approach appeared to provide higher resolution in strain differentiation than conventional gel electrophoresis, because the microarray-based fingerprinting methods provided clear resolution among all strains examined, including two strains (*X. oryzae* 43836 and 49072) that could not be resolved using traditional gel electrophoresis of REP-PCR amplification products. The microarray hybridization-based approach is attractive, because a universal nonamer array could be developed to generate fingerprints from any microorganisms.

14.4 SUMMARY

Single nucleotide polymorphisms (SNPs) are the most frequent type of variation in the human genome and experimental organisms. Genotying large numbers of SNPs in appropriate samples should lead to insights into the basis of heritable variations in disease susceptibility and resistance, complex genetic trait differences, and human evolution. Although many conventional molecular methods have been used successfully to genotype small numbers of SNPs, they cannot meet the high-throughput demands of large-scale sequence comparisons and mutational analyses. Several microarray-based experimental strategies have been developed and tested that can genotype large numbers of SNPs.

Differential hybridization with allele-specific oligonucleotide (ASO) probes is the most commonly used microarray format. Such hybridization assays rely on the differences in hybridization stability of the short oligonucleotides to perfectly matched and mismatched target sequence variants. However, the specificity of genotyping by ASO strongly depends on probe characteristics and detection conditions. Probe design in terms of probe length, G + C content, and melting temperature, and mismatch location is important to achieving specific detection. Although the gain- and loss-of-signal approaches can be used independently to analyze sequence variations, it is advantageous to use a combination of both for mutation detection. It is possible to use oligonucleotide microarrays for detecting sequence changes on a large scale and analyzing many samples simultaneously, but there are several technical challenges in terms of detection accuracy and hybridization specificity.

To improve the accuracy of detection, other enzyme-assisted microarray assays such as single-base extension (SBE), also referred to as minisequencing, have been developed. In this approach, a detection primer anneals to the target nucleotide acid sequence immediately adjacent to a variable nucleotide position. The 3'-end of the primer is extended by DNA polymerase with a labeled nucleotide analog that is complementary to the nucleotide at the SNP site. Two SBE approaches, microarray-based allele-specific primer extension and tagged single-base extension, have been developed and evaluated for genotyping. A microarray-based ligation detection reaction has also been developed for genotyping. The experimental results demonstrated that these array-based methods are powerful tools for detecting single nucleotide polymorphisms.

Although DNA microarray technology has been used successfully to analyze global gene expression in pure cultures, it has not been rigorously tested and evaluated within the context of complex environmental samples. Adapting microarray hybridization for use in environmental studies has a number of challenges associated with specificity, sensitivity, and quantitation. Recently, several types of microarrays, such as functional gene arrays (FGAs) containing functional gene sequence information, phylogenetic oligonucleotide arrays (POAs) containing probe information from rRNA genes, and community genome arrays (CGAs) constructed with whole genomic DNA from many individual pure cultures, have been developed and evaluated within the context of environmental applications. The results suggest that specific, sensitive, and quantitative detections can be obtained with DNA- and oligonucleotide-based FGAs and CGAs. Specific detection at the single-base level is problematic with POAs due to the high conservation of rRNA genes and their secondary structure. SSU rRNA gene-based oligonucleotide arrays are still in the very early stages of development. The whole-genome microarrays containing individual ORFs of a sequenced microorganism are also useful in viewing the genome diversity and relatedness of other closely related microorganisms.

FURTHER READING

Fan, J. B., X. Chen, M. K. Halushka, and A. Berno, et al. 2000. Parallel genotyping of human SNPs using generic high-density oligonucleotide tag arrays. *Genome Res.* 10:853–860.

Hacia, J. G. 1999. Resequencing and mutational analysis using oligonucleotide microarrays. *Nat Genet.* 21:42–47.

Hacia, J. G. and F. S. Collins. 1999. Mutational analysis using oligonucleotide microarrays. *J. Med Genet.* 36:730–736.

Hirschhorn, J. N., P. Sklar, K. Lindblad-Toh, and Y. M. Lim, et al. 2000. SBE-TAGS: an array-based method for efficient single-nucleotide polymorphism genotyping. *Proc. Natl. Acad. Sci. USA*: 97:12164–12169.

Lipshutz, R. J., S. P. Fodor, T. R. Gingeras, and D. J. Lockhart. 1999. High density synthetic oligonucleotide arrays. *Nat. Genet.* 21:20–24.

Pastinen, T., M. Raitio, K. Lindroos, and P. Tainola, et al. 2000. A system for specific, high-throughput genotyping by allele-specific primer extension on microarrays. *Genome Res.* 10: 1031–1042.

Voordouw, G. 1998. Reverse sample genome probing of microbial community dynamics. *ASM News* 64:627–633.

Wu, L., D. K. Thompson, G. Li, and R. A. Hurt, et al. 2001. Development and evaluation of functional gene arrays for detection of selected genes in the environment. *Appl Environ Microbiol.* 67:5780–5790.

Zhou, J. and D. K. Thompson. 2002. Challenges in applying microarrays to environmental studies. *Curr. Opin. Biotechnol.* 13:204–207.

Zhou, J.-Z. 2003. Microarrays for bacterial detection and microbial community analysis. *Curr. Opin. Microbiol.* 6:288–294.

15

Future Perspectives: Genomics Beyond Single Cells

James M. Tiedje and Jizhong Zhou

15.1 INTRODUCTION

Genomics is a rapidly accelerating field, and in such circumstances, it is both useful and risky to attempt to provide a vision for the future: useful because it can help direct the field in diverse and needed directions; risky because new developments can carry the field in unanticipated directions. Furthermore, some of the longer-range visions may be naïve, especially those in which the development proves more complex and problematic than anticipated. Given these caveats, we present our view of a path forward from today's vantage point by focusing on genomic studies beyond single cells.

The foundation for today's new biology has been the genome sequence. Now scientists in diverse fields are ready to use this new resource for a more in-depth and comprehensive understanding of biology. The microbiologists share in this general vision, but with some important distinctions from those of the macrobiologists. The prokaryotic cell is much simpler; no organelles, fewer genes and proteins, no or only simple cell differentiation, and no complex life cycle. Instead, the microbial world has much more biochemical diversity with likely more than 10 million species (Curtis et al., 2002), and with considerable divergence of gene sequence within and among species. This sequence divergence can lead to important functional differences within gene families, but this divergence also makes study of the microbial world by PCR amplification and hybridization incomplete. However, because of the importance of this diversity to microbial processes on this planet, a large diversity of microorganisms will need to be studied to gain sufficient understanding of the functional roles of populations and communities. Hence, one general difference

Microbial Functional Genomics, Edited by Jizhong Zhou, Dorothea K. Thompson, Ying Xu, and James M. Tiedje.
ISBN 0-471-07190-0   © 2004 John Wiley & Sons, Inc.

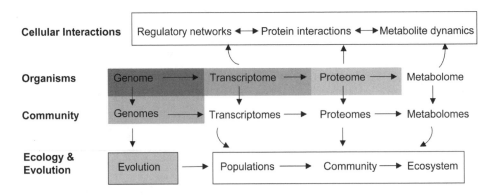

Figure 15.1 *Directions for the future development of the field of microbial genomics. The degree of shading indicates the level of current effort, with the darkest gray shading indicating the greatest effort. Expansion into metabolomes, cellular interactions, and ecology await future developments.*

in the general direction of genomic studies between the macrobiologists and the microbiologists is that the latter will be able to, and in fact should, give substantially more effort to the study of organisms in their communities.

Our vision for the development of microbial genomics is illustrated in Figure 15.1. The shading indicates current effort and shows the expansion from the initial focus on genome sequence to the transcriptome and to the proteome. The microbial genomics effort so far has been accomplished almost entirely with single organisms, although some groundbreaking work has recently been done with genomes of communities (c.f. Beja et al., 2000; Rondon et al., 2000). As illustrated by the figure, the opportunity for development of microbial genomics is broad, since research is needed in all directions. Expansion to the right will lead to a comprehensive understanding of cell functions, and the metabolome will be the test of how well the genome, transcriptome, and proteome can be interpreted with regard to function. Expansion downward, from single organisms to combinations of microbes or simple communities, will allow studies of microbial interactions and can be carried out in a manner parallel to that done with single organisms. The genomic field is also likely to bifurcate into the two major directions, which are illustrated in the upper and lower areas of the figure, and reflect the interests and expertise of the investigators. One direction (upward) is reductionistic, or subcellular, in which the goal is to understand how the cell works: all of its regulatory control, the functions of its protein complexes, and their integration into metabolism and growth. The other direction (downward) is more holistic, and that is to study organisms in the context of their environment, i.e., the populations, communities, abiotic effectors, and finally, their role in ecosystems. This direction is generally described as the ecological approach.

15.2   THE INFORMATIONAL BASE OF MICROBIAL BIOLOGY: GENOME SEQUENCES

15.2.1   Determination of the Genetic Content of Both Cultured and Uncultured Microorganisms

Microorganisms inhabit almost every imaginable environment on earth. In contrast to plant and animal diversity, however, the extent of microbial diversity is largely unknown.

Although microbial genome sequencing projects reveal enormous amounts of information about a particular microorganism, these projects only scratch the surface of microbial diversity (Fraser and Dujon, 2000). Clearly, insight into microbial diversity will require the sequencing of the genomes of many microbial species from various environments.

While progress in sequencing microbial genomes is moving forward at a rapid pace, continued determination of such genomes will be important, because genome sequences will remain the essential base resource for studying the biology of any important microbe. This area should not be deemphasized prematurely. Along with continued sequencing, improvements are needed for rapid annotation and functional analysis of the discovered genes, both current bottlenecks to realizing the benefits of the sequence information. Organisms to be sequenced should include both phylogenetically divergent and close relatives, and those known to have important functions. Information from distantly related species is needed to explore the diversity of the microbial world and to understand early evolution. Information from close relatives will be useful in understanding adaptations to ecological niches, the basis of phenotype, and speciation. With the availability of perhaps as many as 1,000 microbial genomes within the next few years, genomic sequence mining should begin to reveal patterns informative of ecologically important traits, evolutionary history, and complex traits such as competitiveness.

It is well known that most microbes in nature are uncultured, and many of these belong to divisions with no cultured members (e.g., Huggenholtz et al., 1998a,b; Zhou et al., 2003). Hence, their energy conversion pathways, physiology, and biochemical capabilities are completely unknown. Obtaining whole genome sequences from these microorganisms is the most straightforward approach to gaining insights into their metabolic capacities and ecological roles. Some important and interesting scientific questions can be addressed based on sequence information: (1) How diverse genetically are the uncultured microorganisms? (2) How are the uncultured microorganisms related to known cultured microorganisms at the whole-genome level? (3) Do the uncultured microorganisms use similar genes, pathways, regulatory networks, and protein machines compared to those found in cultured microorganisms for survival, growth, replication, and environmental adaptation? (4) What is the genetic basis for our failure to culture the uncultured microorganisms?

Recent advances in whole-genome sequencing and associated technologies should make it possible to obtain whole-genome sequence information from uncultivated micro-organisms. One approach to accessing the genetic content of uncultured micro-organisms is to separate cell populations of interest from other cells. The DNA can be isolated from single cells, amplified, and used as templates for DNA sequencing. Although fluorescence labeling and flow cytometry-based approaches have been successfully demonstrated for isolating some populations, the feasibility and application of this approach for analyzing diverse microorganisms from environmental samples is still unknown. Improving the cytometry-based technology, magnetic-linked antibody, micromanipulation, and/or developing other new technologies for recovering targeted and uncultured cells are needed. Another technical challenge is how to isolate DNA from single cells and then perform DNA amplification with such tiny DNA templates to generate enough PCR product for DNA sequencing. For instance, once a desired cell population is isolated from an environmental sample, considerable amounts of genomic DNA will be needed to determine genome sequences, and thus, the DNA from single cells must be amplified without bias. Current PCR-based amplification methods suffer from the problems of amplification bias and poor quantification. For single cell biological experimentation to be successful, experiments must be performed in volumes similar to that of a single cell ($\approx 1\,\mathrm{pL}$ or less).

Microfluidic devices or lab-on-chip (e.g., Dolnik et al., 2000; Auroux et al., 2002) might provide an excellent approach to isolating and manipulating DNA from single cells in a high-throughput fashion. Microfluidic devices (microchips) are fabricated of glass or plastic for performing simple chemical and biochemical assays. They integrate sample handling and sample processing operations with analyte detection on a single, monolithic substrate. Such integration allows for the efficient automation of chemical analyses. They have the potential to revolutionize the traditional analytical chemistry.

Another emerging technology for accessing uncultured microorganisms is single-molecule-based DNA sequencing (e.g., Braslavsky et al., 2003). Genomic sequence information can be obtained from single DNA molecules. For example, the single-molecule nanotechnology from Solexa (http://www.solexa.co.uk) could allow simultaneous analysis of hundreds of millions of individual molecules on a chip. The single-molecule-based sequencing technology developed at Nanofluidics (http://www.nanofluidics.com) has great potential to radically improve DNA sequencing 10,000 times faster than current sequencing technology and has the potential to sequence a human genome for less than $1,000. With the single-molecule sequencing strategy, the difficult steps in the separation of single cells, followed by DNA isolation and amplification from single cells, appear to be unnecessary. Once such technologies mature, they could provide a great opportunity for assessing genetic content of unculturable microbes.

15.2.2 Community Genomics or Metagenomics

One of the exciting areas of microbial genomics is the exploration of microbial communities through genomics, their informational base. Because most of environmental microorganisms are uncultured, DNA cannot be obtained for sequencing by pure culture-based methods. The approach most commonly used to assess the genetic content of uncultured microorganisms is to directly isolate DNA from environmental samples and then clone high-molecular-weight DNA fragments into artificial bacterial chromosomes (BACs) or fosmid vectors for sequencing. This approach, known as community genomics or metagenomics, provides information on both the culturable and the unculturable members of a community. Sequence-based community genomics can be oriented to several different goals.

Determination of the Extent of Microbial Diversity   The 16S rRNA gene-based phylogenetic studies with a variety of environmental samples indicate that microbial communities are extremely diverse. However, the extent of this diversity in environments is still unknown. Communities are comprised of largely cooperating populations with the capacity to synergistically perform all functions necessary for a viable and robust community. Sequencing of entire community genomes would reveal, for the first time, community genetic properties such as the extent of genetic diversity and metabolic capabilities necessary for the vital functions of a community and the patterns of diversity within community guilds. Some interesting questions can be addressed based on sequence information: (1) What are the extent and patterns of phylogenetic and metabolic diversity in microbial communities from different environments? (2) How are microbial communities adapted to different environments at the genetic level? (3) Is there conservation of metabolic function among communities in spite of extensive phylogenetic diversity of individuals?

Discovery of Novel Pathways By cloning large DNA segments, entire pathways may be discovered and recovered for expression in a heterologous host. Pathways potentially important for pharmaceutical products have been and are examples of this use of community genomics (Schloss and Handelsman, 2003). This approach has value beyond the obvious commercial goals.

Exploration of the Diversity of Targeted Genes The long evolutionary history of the microbial world has yielded a large diversity of genes for particular processes. These genes can vary in features important to their function such as kinetic parameters (Km, Vmax), substrate specificity, adaption to high or low temperatures, and tolerance to environmental extremes such as high and low pH, etc. Community genomics can be used to recover large families of complete genes, which then can be screened and sorted for their desired properties. Genes confirming important ecological functions such as biogeochemical cycle processes, pollutant degradation, and pathogenesis are particularly important areas to target with this approach. Since diversity of function is such an important area in microbial ecology and ecosystem science, this is a particularly important theme to pursue.

Identification of Traits of Currently Unculturable Microbes Cloning of large fragments of environmental DNA provides the possibility of identifying traits in dominant unculturable microbes as long as phylogenetic markers reside on the same DNA fragment. If the cloning is comprehensive, a reasonable sampling of the microbial community and its gene content can be obtained. This is currently a vigorous area of exploratory research, and one that should continue, but in balance with the other approaches.

Patterns of Community Versus Population Diversity Communities are made of up different species, and most likely there is also considerable subspecies diversity. Understanding how the environment controls the distribution and dynamics of species versus subspecies diversity is needed to inform microbial ecology on the expected outcomes of succession and eventually evolution. The degree of subspecies variation will be important to community genome assembly.

One of the main challenges of studying natural microbial communities using metagenomic approaches is that, in most of the cases, communities are very diverse with many complex interactions between community members. This complexity limits our ability to resolve the cause-and-effect relationships needed to develop a mechanistic understanding of the communities. Initial studies on simpler communities should be more tractable. Communities that have experienced a high degree of ecological selection through growth, physical separation, or due to extreme conditions offer this potential.

The second technical challenge of the metagenomic approach is to recover high-molecular-weight DNA fragments (e.g. >100kb) of sufficient purity and without bias. Recovery of high-molecular-weight DNA is essential for genome sequence assembly. To minimize recovery bias, generally, rigorous cell lysis approaches such as grinding are needed, but the cell lysis process might shear DNA fragments. Recovering high-molecular-weight DNA without bias is still very difficult, and new methods are needed.

The third challenge of the metagenomic approach is the unequal abundance of community members. While random shotgun sequencing is a powerful technique for rapidly obtaining genome sequence information from pure cultures, it is inefficient when applied to the sequencing of whole communities, because the coverage will be excessive for the most dominant members and too little or absent for the more infrequent members. Also, some of the microbial populations within the different communities are identical or closely related. Thus, new strategies for prioritizing BAC or fosmid clones for sequencing are needed. Screening by microarrays, some form of subtractive hybridization, or other ways to normalize the library are needed.

Another technical challenge is the assembly of genomes in mixed microbial communities, especially if they are complex, i.e., > 20 to 100 species. New assembly tools, along with other mapping and quantitative population tools, are needed.

15.3   GENE FUNCTIONS AND REGULATORY NETWORKS

Large-scale community genome sequencing efforts will identify a massive number of putative genes whose functions are completely unknown as well as known genes but likely with unique characteristics. The latter genes may carry improved kinetic properties or tolerance to stresses, which may be key to efficient and stable functioning in nature. It is critical to understand the cellular and biochemical functions of these unknown genes and their regulatory networks within the context of communities. Some possible questions are: (1) What are the cellular and biochemical functions of unknown genes discovered in uncultured microorganisms? (2) Are the protein complexes discovered in the uncultured microorganisms different from those in the known, cultured microorganisms? (3) Are there unique structural and kinetic characteristics of the known genes found in uncultured microorganisms? Can these unique characteristics be used for protein engineering? (4) How are the genes regulated at the community level and what are the signal molecules for different populations communicating and interacting with each other?

Since many of the unknown genes could be from uncultured microbes, determining their cellular and biochemical functions and their regulation within the context of communities will be extremely difficult. A basic strategy for understanding their functions is to express these genes in a heterologous host, and subsequently, examine their catalytic function if possible. Characterization of protein structure with X-rays, neutron scattering, NMR, and mass spectrometry will also help to understand their functions. Also, high-throughput biochemical screening will be necessary for establishing protein function. In addition, bioinformatic and computational tools for predicting protein structures should be useful in providing information for guiding experiments. Finally, insights into biochemical function may be gained from gene expression analyses using nucleic acid or protein microarray technologies.

To understand the functions and regulation of the genes from uncultured microbes, developing appropriate experimental systems to monitoring their changes under different environmental conditions is key. Laboratory-based bioreactor systems that simulate the natural environmental conditions and allow the growth of the uncultured microbes in a mixed simplified community is one approach. Then advanced microarray-based and mass spectrometry-based genomic technologies can be used to analyze gene expression in such bioreactor samples in a rapid, high-throughput fashion.

Although DNA microarrays and mass spectrometry are very powerful in analyzing gene expression at both transcriptome and proteome levels in pure culture (DeRisi et al., 1997; Tyers and Mann, 2003), the current technology is not adequate to meet the requirement of analyses at the community scale. The number of genes needed for monitoring on a whole-community scale could be far beyond the capacity of the current microarray and mass spectrometry technologies. Thus, more advanced microarray- and mass spectrometry-based technologies with higher capacity and sensitivity are needed. In addition, many effective genomic approaches developed for defining gene functions and regulatory networks in pure cultures, such as growth manipulations and genetic mutagenesis, will not be feasible. Novel strategies and approaches are needed to facilitate the functional characterization of novel genes from uncultured microorganisms.

15.4 ECOLOGY AND EVOLUTION

Once the whole-genome sequences from cultured and uncultured microorganisms are obtained, the next step is to understand their ecological functions and evolution in the environment. Some examples of ecological and evolutionary questions are: (1) Are the novel uncultured microorganisms ecologically important? (2) Are these organisms active in particular environments? (3) How do the uncultured microorganisms interact with other microbial populations? (4) How do they respond and adapt to environmental changes? (5) Can their activities be enhanced for desired functions and what are the possible evolutionary consequences that result from such manipulations?

Because of the uncultured status of most of the microorganisms present in natural environments, a comprehensive understanding of their functions and metabolic capacities in environments is a great challenge. A basic strategy to understand the physiological and ecological roles and dynamics of uncultured organisms is to extensively evaluate their abundance, distribution, gene expression, and biochemical functions in response to environmental changes in both laboratory and field studies using a variety of technologies, such as microarrays and isotope analysis.

For understanding the ecological functions of microbes in natural environments, microarrays will be a central technology. However, such types of studies consider a whole microbial community as a working unit. The scale and scope presented are much larger than any pure culture-based genomic study. High-throughput automatic genomic facilities are extremely critical. Automation of the processes involved in microarray technologies, such as sample processing, oligonucleotide synthesis, PCR amplification, gel electrophoresis, nucleic acid purification, microarray construction, microarray hybridization and scanning, data processing and analysis will be needed to allow rapid high-throughput analysis. Advanced computational and bioinformatic tools for probe designing, data handling, and analysis are also important for the elucidation of the ecological functions and evolution of uncultured microbes in natural environments.

15.5 SYSTEM-LEVEL UNDERSTANDING OF MICROBIAL COMMUNITY DYNAMICS

Functional stability and adaptation are two important properties of biological systems. The relationship between diversity and stability of biological communities is a long-standing controversy in macro-community ecology. Although it seems "intuitive" that more diverse systems would exhibit greater stability, a number of theoretical and experimental studies show the opposite. Several studies with microbial communities also indicate that functional stability of microbial communities does not correlate with their phylogenetic diversity. It is believed that the functional stability and adaptation of microbial communities will be determined by the genetic and metabolic diversity of the individual microbial populations, which includes their ability to respond to environmental changes. However, there is no evidence to support this hypothesis. Understanding the genetic basis and factors controlling microbial community stability and adaptation is of great importance in managing microbial communities for desired functions, such as bioremediation of contaminated sites, sequestration of carbon from the atmosphere, biological control of disease, enhancement of plant productivity, and efficient cycling of nitrogen.

With the availability of whole-community sequences, it is possible to address some fundamental ecological questions at the ecosystem level such as: (1) What is the genetic basis for functional stability and adaptation of microbial communities? (2) How is the functional stability of a microbial community related to its genetic and metabolic diversity as well as environmental disturbance? (3) Can the functional stability and future status of a microbial community be predicted based on the metabolic functional conservation and differentiation of individual microbial populations? (4) Can a microbial community be manipulated to achieve a desired stable function by manipulating the metabolic traits of a microbial community?

The availability of whole community sequences and associated high-throughput genomic technologies will also allow us to address major ecological questions about population interactions in natural communities. For example, how does the natural physical chemical environment of a cell aggregate or an intestinal tract affect gene expression among community members? Do these changes trigger succession? How do invading species and the invaded respond? What defenses are used against grazers? How does the cell use its coded information to drive ecological responses?

Conceptual and predictive system-level understanding of the genetic basis of microbial community dynamics at the ecosystem level is a great challenge. One strategy could be to compare the commonality and differences of microbial community diversity, metabolic capacities, and functional activities under different stress conditions in a similar habitat using high-throughput genomic technologies. While this approach provides information on the distribution and abundance of microbes, it provides neither mechanistic knowledge of the factors that determine the observed distributions nor knowledge of their metabolic activities. Another complementary approach is to establish well-controlled laboratory systems such as bioreactors with simplified communities to study the responses of microbial communities to environmental stressors. Such laboratory systems are important to establish cause-and-effect relationships. Laboratory systems have great advantages in terms of system controls, monitoring, and data collection. Determination of cause-and-effect relationships is easier with simple, engineered, laboratory-based systems than with complex, natural communities, because input and output parameters can be controlled, along with environmental conditions. Although a bioreactor community is not a natural community, it offers the best opportunity to acquire mechanistic understanding of the fundamental principles of interactions between microbes, and how selection affects communities.

A central theme of genome-based biology is to develop the necessary experimental and computational capabilities to enable a predictive understanding of the dynamic behavior of microbes and microbial communities. As we briefly indicated in Chapter 1, such a biological systems-level approach faces several grand computational challenges. First, modeling global cellular behavior is very difficult due to the complexity of metabolic pathways and our lack of understanding of the dynamic behavior and regulatory mechanisms. Second, genomic data from analyses of transcriptomes, proteomes, and metablomes, as well as physiological data, are heterogeneous. Synthesizing various types of data together to make biological sense is also difficult. In addition, because the dynamic behaviors of biological systems at various levels (cells, individuals, populations, communities, and ecosystems) are measured on different temporal and spatial scales, linking cellular-level genomic information to ecosystem-level functional information for predicting ecosystem dynamics is even more challenging. Thus, the development of novel mathematical framework and computational tools is also critical to systems-level understanding of microbial community dynamics.

15.6 SUMMARY

Large-scale sequencing of entire genomes represents a new age in biology, but the greatest challenges are to define genetic structure, gene functions, and regulatory networks in the vast majority of uncultured microorganisms; resolve the extreme complexity of microbial communities; link cellular responses of individual cells to the functional stability and adaptation of microbial communities; and simulate and predict the dynamics of biological systems at cellular, population, community, and ecosystem levels. Although many representative microorganisms have been sequenced, more effort is needed to sequence microbial genomes from a variety of environments to obtain comprehensive insight into the microbial diversity present in nature. Since the majority of microorganisms cannot be cultured, the technologies of isolating and manipulating single cells or single DNA molecules and cloning and sequencing large DNA fragments directly from environmental samples will be vital to access the genetic content of uncultured microorganisms. The genome sequence information from uncultured microorganisms will greatly advance our understanding of uncultured microbes in terms of genetic diversity, metabolic capacity, and microbial evolution.

Large-scale community genome sequencing efforts will identify an abundance of putative genes whose functions are completely unknown, and most of them are from uncultured microorganisms. Since these microorganisms are unculturable, determining their cellular, biochemical, physiological, and ecological functions and their regulatory networks within the context of communities will be a formidable task. Moreover, linking cellular behaviors to ecosystem dynamics will be more challenging, because a whole microbial community is considered as a working unit, and hence the scale and scope presented are much larger than any pure culture-based genomic study. Novel experimental and theoretical frameworks and approaches are needed to attack the challenges resulting from such complexity of biological systems at various levels across different temporal and spatial scales.

FURTHER READING

Beja, O., L. Aravind, E. V. Koonin, and M. T. Suzuki, et al. 2000. Bacterial rhodopsin: Evidence for a new type of phototrophy in the sea. *Science* 289:1902–1906.

Braslavsky, I., B. Hebert, E. Kartalov, and S. R. Quake. 2003. Sequence information can be obtained from single DNA molecules. *Proc. Natl. Acad. Sci. USA* 100:3960–3964.

Fraser, C. M. and B. Dujon. 2000. The genomics of microbial diversity. *Curr. Opin. Microbiol.* 3:443–444.

Hugenholtz, P., B. M. Goebel, and N. R. Pace. 1998a. Impact of culture-independent studies on the emerging phylogenetic view of bacterial diversity. *J. Bacteriol.* 180:4765–4774.

Schloss, P. D. and J. Handelsman. 2003. Biotechnological prospects from metagenomics. *Curr. Opin. Biotechnol.* 14:303–310.

Tyers, M. and M. Mann. 2003. From genomics to proteomics. *Nature* 422:193–197.

GLOSSARY

***Ab initio* gene finding** Computational methods for gene identification, based on statistical features common in known gene sequences. These features may include dicodon bias, periodicity, G + C composition, etc.

Additive and multiplicative errors An error to a data point is called an additive error if its quantity is independent of the data value itself. A multiplicative error's quantity is proportional to the data value.

Allele One of the variant forms of a gene at a particular locus, or location, on a chromosome. Different alleles produce variation in inherited characteristics that are reflected in the phenotype.

Allele exchange Genetic exchange through recombination between two DNA strands carrying homologous sequences. Also called homologous recombination.

Antisense mRNA/antisense silencing A piece of RNA, which has a sequence exactly opposite to that of an mRNA molecule made by the organism. mRNA molecules serve as templates for the synthesis of protein. Since the "antisense" mRNA molecule binds tightly to its mirror image, it can prevent a particular protein from being made. The inhibition of gene expression using antisense mRNA is utilized in antisense silencing.

β-galactosidase An enzyme widely used as a reporter of gene expression because of its ability to cleave the chromogenic substrate 5-bromo-4-chloro-3-indoyl β-D-galactopyranoside (X-Gal) to yield a blue product.

Biochemical genomics The large-scale identification of genes encoding biochemical activities, such as enzyme catalysis.

Bioinformatics The development and application of computational tools to assist with data analysis from analytical technologies as well as *in silico* predictions of biological structures and interactions.

Biological pathway inference To build biological pathway models, using mathematical modeling methods, which are consistent with available experimental data (i.e., microarray gene expression data).

Bottom-up proteomics The characterization of a proteome by proteolytic digestion of the complex protein mixture, followed by identification of the resulting peptides. The proteins present are deduced from the peptides measured and identified.

Cluster identification To identify regions in a data set that have significantly different characteristics, such as data density, from its local background.

Coding potential It measures how probable a sequence may be coding for protein peptides.

Coexpressed genes Genes that exhibit similar expression patterns over a time series and/or under a common set of experimental conditions.

Community genome arrays Microarrays constructed with whole genomic DNA from many individual pure cultures.

Community genomics Study that aims to understand the diversity, structure, functions, and evolution of a biological community at the genome level.

Comparative approach to gene finding Computational methods for gene finding, based on direct sequence comparisons with known genes, cDNA sequence, or EST sequences.

Comparative genomics Comparative studies of genome information (e.g., genome sequences, mRNA and protein expression, protein interactions, metabolite dynamics) from a variety of organisms to reach a genome-wide understanding of biological processes and phenomena using both computational and experimental high-throughput approaches.

Computational genomics Genomic studies using integral computational analysis to model *in silico* whole-genome information for the prediction of metabolic pathways, regulatory networks, phylogenetic relationships, and protein structure and interactions.

Concerted evolution The members of a multigene family in a species coevolve together to maintain their sequence homogeneity.

Convergent evolution A phenomenon in which the same character state (e.g., gene function, catalytic enzyme mechanisms, structural motifs, and sequences) appears and evolves independently in separate evolutionary lineages.

Counterselectable marker Gene encoding sensitivity to a certain chemical and under appropriate conditions promoting the death of the microorganism harboring it.

Cross-talk The phenomenon in which an emission signal from one channel in a microarray detection system is detected in another channel.

Data-constrained clustering To computationally cluster a data set into groups that are consistent with known information, say, two particular genes known to be transcriptionally coregulated.

Data transformation To apply mathematical transformation to a data set (e.g., applying $\log_2$ to each data point).

Differentially expressed genes Genes that exhibit expression pattern changes under designed experimental stimuli.

Dihydrofolate reductase (DHFR) Enzyme involved in one-carbon metabolism that is required for the survival of prokaryotic and eukaryotic cells. The enzyme catalyzes the reduction of dihydrofolate to tetrahydrofolate, which is required for the biosynthesis of serine, methionine, purines, and thymidylate.

Direct labeling Fluorescent tags are directly incorporated into the nucleic acid target mixture before hybridization through enzymatic synthesis or chemical methods.

Disordered vs. ordered regions Disorder represents a conformational state in which a protein or a protein segment does not have a well-defined (ordered) tertiary structural conformation, that is, it forms a dynamic molten globule or random coil, and hence is partially or completely unfolded. Recent studies have indicated that disordered regions are often associated with molecular recognition domains, protein folding inhibitors, flexible linkers, etc.

Diversity-oriented synthesis A synthetic chemistry strategy to efficiently synthesize a large collection of diverse small molecules.

DNA microarrays Ordered arrays containing DNA fragments typically generated with the polymerase chain reaction (PCR).

Drug target validation Drug target validation is a stage in drug discovery and development concerned with verifying that a compound inhibits the intended target. Genomics tools like microarray-based gene expression profiling have facilitated the validation of drug targets.

EC class A classification scheme of all known enzymes. At the top level, all enzymes are classified into six classes: *oxidoreductases*, *transferases*, *hydrolases*, *lyases*, *isomerases*, and *ligases*, which are further divided into over 200 subclasses of enzymatic functions organized as a functional hierarchy.

Ecological genomics   Studies that aim to reach a genome-wide understanding of the molecular basis and mechanisms for determining the composition, structure, functions, and dynamics of biological systems within the context of ecology.

Ecosystem genomics   Studies using genomic sequence information and genomic technologies to understand the molecular mechanisms controlling the fluxes of materials and energy among various trophic levels within an ecosystem, and the genetic basis and factors controlling community stability, adaptation, and responses to environmental changes.

Efflux systems   Efflux systems are utilized by a wide distribution of microorganisms as a mechanism enabling antimicrobial drug resistance. Both gram-negative and gram-positive bacteria express numerous membrane transporters that promote the efflux of antibiotics and other drugs from the cell to the surrounding milieu. Efflux systems can be specific to a certain drug or can display broad substrate specificity, as in the case of multidrug transporters.

Electrospray ionization (ESI)   The process of transferring preformed biological ions from a solution directly into the gas phase for detection by mass spectrometry. This is accomplished by flowing the sample solution through a metal needle that is biased at a high voltage relative to the entrance aperture of the mass spectrometer.

Enzymatic chain termination method   Also called the dideoxynucleotide method, it is currently the method of choice for determining the nucleotide sequence of a DNA fragment. The hallmark feature of the enzymatic sequencing method, which was developed by Frederick Sanger, is the termination of primer extension during DNA synthesis by the incorporation of a dideoxyribonucleotide in the growing chain. Because it lacks a $3'$-hydroxyl group, the dideoxynucleotide (a synthetic molecule) cannot form a phosphodiester bond with the next incoming nucleotide. After enzymatic DNA synthesis with DNA polymerase, each reaction tube will contain a unique set of oligonucleotides that can be separated by polyacrylamide gel electrophoresis. Traditionally, the sequence of the DNA segment was "read" off an autoradiograph of the DNA sequencing gel. Today, manual sequencing largely has been replaced by automated sequencing.

Enzyme classification   Enzymes are among the most studied class of proteins. The ENZYME database is a classification database of enzymes in SWISS-PROT. It classifies all enzymes into six classes: oxidoreductases, transferases, hydrolases, lyases, isomerases, and ligases, which are further divided into over 200 subclasses of enzymatic functions organized as a functional hierarchy.

Equifunctionalization   Both copies of the duplicated genes may persist in the genome with perfect or near-perfect sequence identity and cause a higher level of the gene product.

Evolutionary genomics   Studies using genome information to decipher the evolutionary relationships, history, and processes among extant organisms.

Fluorenylmethoxycarbonyl (Fmoc)   Protection group used to block the amino group of incoming amino acids during solid phase peptide synthesis to limit addition to one amino acid per cycle.

Functional gene arrays   Microarrays containing functional gene sequence information for the functional analysis of microbial community activities in environments.

Functional genomics   Studies seeking a system-level understanding of gene function and regulatory networks using global genome-wide approaches based on the information provided by structural genomics.

Functional motifs   Regions of DNA sequences that facilitate or are directly involved in biological functional activities of a gene or protein. Typically, these regions are conserved, on the sequence level, across members of the same protein family or superfamily.

Gap closure   This is the final phase of a genome-sequencing project using the random shotgun-sequencing approach. Closure entails ordering the assembled contigs and linking them together in a single contiguous sequence with specific PCR products or cloned inserts that span gaps in the genomic sequences.

Gel electrophoresis (GE)   A method to separate proteins by their differential electrophoretic mobility through a gel as a function of the voltage difference applied to the gel.

Gene conversion   The nonreciprocal recombination process resulting in a sequence becoming identical by replacing one copy of a gene with all or part of the sequence of a second copy of the gene.

Gene expression data clustering   To group genes into *clusters* so that genes of the same cluster exhibit similar expression patterns and genes of different clusters exhibit different expression patterns, under designed experimental stimuli.

Genome   The complete set of genes and chromosomes of an organism.

Genomics   Study that aims to reach a genome-level understanding of the molecular basis of the structure, functions, and evolution of biological systems using whole-genome sequence information and high-throughput genomic technologies.

Genome annotation   A process of identifying and labeling functional units and special signals existing in a genomic sequence. These may include protein-coding genes, RNA-coding genes, promoters, regulatory binding sites, terminators, operons, etc.

Genome rearrangement   Study of how locations of corresponding genes are different in related genomes. This can provide one measure of evolutionary relationships among different genomes.

Glutathione-S-transferase (GST)   Enzyme commonly fused to open reading frames to serve as an affinity purification tag. Can be used to purify proteins with glutathione-conjugated beads.

Guilt by association   The assumption that genes expressed together may function together. Thus, genes function can be predicted based on gene expression profiles.

Hierarchical clustering   A data clustering scheme that imposes a hierarchical view on a data set by representing the data set as a tree structure. The tree root represents the cluster for the whole data set and each tree leaf represents a data cluster consisting of a single data point. Each tree node in the middle represents a cluster that merges the clusters represented by the node's offspring nodes in the tree.

Homologous genes   Genes that have a common evolutionary ancestor.

Horizontal gene transfer   The transfer of genetic information from one species to another.

Indel   A position in a sequence alignment where an insertion or deletion has occurred.

Indirect labeling Fluorescence is introduced into the microarray procedure after hybridization.

Informational genes The genes involved in information processing such as those functions in transcription, translation, replication, homologs of ATPases, GTPases, and most tRNA synthetases.

Ink-jet ejection Noncontact microarray printing method by ejecting a droplet of sample to the surface of substrates.

Interactome The global protein–protein interaction networks encoded in a genome. The global protein interaction map, or interactome, dictates cellular function.

K-means clustering A data clustering scheme that partitions a data set into K groups with the data points of each group centering around the mean of the group.

Lead molecule The term used by the pharmaceutical industry to indicate a potential drug candidate that is selected for further development and validation.

Linear and nonlinear errors An error to a data point whose quantity can be represented as a linear or nonlinear function of the data point.

Liquid chromatography (LC) An analytical technique for separating molecules based on their selective partitioning between a flowing mobile phase and suitable stationary material. A common method employed for proteins and peptides that utilizes an alaphatic stationary phase and an aqueous/organic solution phase, in which the biomolecules are separated on the basis of their hydrophobicities.

Markov chain model A statistical model for a sequence of events; a key assumption of the model is the probability that each event depends only on a small number of events, which have happened prior to this event.

Mass spectrometry (MS) An analytical technique for measuring and interrogating ions corresponding to the components of a sample.

Matrix-assisted laser desorption/ionization (MALDI) The process of transferring biological species from a sample directly into the gas phase for detection by mass spectrometry. This is accomplished by mixing a biological solution with a molar excess of laser-absorbing matrix compound (such as sinapinic acid) and then drying this material onto a target plate. Gas phase ions are generated by laser desorption/ionization of the resulting sample.

Mechanical microspotting Microarray printing method accomplished by direct contact with the surface using a computer-controlled robot with various pins.

Mechanistic toxicology Studies of mechanistic toxicology focus on identifying the mechanisms of action that underlie the potential toxicity of chemicals and drug candidates. Traditionally, assessment of potential toxicity is performed using in vivo animal or cell model systems. Most recently, gene expression profiling using DNA microarrays is being explored as a method for determining modes of action in cases where alterations in gene expression precede or coincide with toxicity.

Metabolome The complete set of metabolites in an organism.

Metabolomics The large-scale investigation of the dynamics and interactions of the metabolome.

Metagenome Refers to the collective genomes in an environmental community (e.g., soil sample) and contains substantially more genetic information than is available in the culturable subset. Studies of the metagenome typically involve cloning large fragments of DNA isolated directly from microbes in natural environments into a bacterial artificial chromosome (BAC) vector for sequence and functional analyses.

Microarray An orderly arrangement of thousands of known biomolecules immobilized on solid substrates.

Microarray-based allele-specific hybridization The hybridization of an oligonucleotide microarray to determine which allele is contained in a DNA molecule.

Microarray-based allele-specific primer extension Two allele-specific oligonucleotide probes from both strands are tethered to the array surface via a 5′-linkage, leaving an exposed 3′-OH, which can be extended for an additional nucleotide in the presence of a mixture of all four dideoxyribonucleotide triphosphates, each labeled with different fluorescent dyes. The identity of the added dideoxyribonucleotide is determined with a fluorescence microscope. Heterozygous single-nucleotide substitutions will result in two signals corresponding to the identity of the two alleles.

Microarray-based subtractive hybridization This approach involved the parallel comparison of genomes by hybridization to DNA or oligonucleotide microarrays representing the complete genome of a sequenced related microorganism. It is a strategy for accelerating progress in comparative microbial genomics without the need for de novo sequencing of closely related species. The procedure involves combining differentially labeled genomic DNA from the test (unsequenced) species and the reference (sequenced) species and hybridizing the mixture to a microarray representing the complete genome of the reference species. Unique or highly divergent sequences in the reference organism are identified by image display of the microarray hybridization.

Microarray-based tagged single-base extension Unique sequence tags are attached to locus-specific primers. SNPs are detected by single-base extension using biofunctional primers containing a unique sequence tag in addition to a locus-specific sequence.

Microarray data normalization Mathematical procedures that can be used to put data points collected under different conditions on an equal footing so their values can be compared with each other.

Minimal genome The smallest possible group of genes necessary to support a functioning cellular life form under the most favorable conditions with the availability of all essential nutrients and the absence of environmental stress.

Molecular barcoding Modification of the gene replacement mutagenesis where the gene of interest is replaced by a unique sequence tag (also see **Signature-tagged mutagenesis**).

Molecular clock The evolution rate of amino acid or nucleotide substitution in a given gene is roughly constant in different evolutionary lineages, and thus the molecule ticks like a clock, which enables divergence times to be assigned to the branch points in a gene tree.

Monophyly A phenomenon in which two or more organisms or sequences share a same ancestor.

Multistage LC-MS The use of two or more stages of liquid chromatography in conjunction with mass spectrometry. This provides extensive separation of biological components prior to mass spectrometric detection.

Mutation An alteration of genetic material, such as the number, arrangement, or molecular sequence of a gene.

Nanospray   The technique of conducting electrospray ionization at very low flow rates (nL per minute) with small capillaries. This technique affords enhanced detection sensitivity over conventional electrospray ionization.

Neo-functionalization   Random mutations cause one copy of the duplicate genes to diverge functionally and the diverged copy may acquire a new beneficial function, while the other copy retains the original function.

Nonfunctionalization   One copy of the duplicated genes is suppressed, either by physical deletion or by accumulating point mutations until it becomes a pseudogene.

Nonhomologous approaches   While homology-based approaches (like sequence-based and structure-based methods) are powerful tools for functional inference of genes/proteins, they have their limitations. Methods that do not rely on homology information, like phylogenetic profiling, gene fusion, have proven to be complementary to homology-based methods in functional inference of genes/proteins.

Nonorthologous gene replacement   The phenomenon that a gene coding for a protein responsible for a particular function is replaced with a nonorthologous (distantly related or unrelated) but functionally analogous gene.

Nonsynonymous nucleotide substitutions   A nucleotide substitution that changes a codon to that specifying another amino acid.

Objective function   Mathematical functions that define how the result of an optimization procedure should be measured.

Oligonucleotide microarrays   Ordered arrays of oligonucleotides (typically, < 75 mer) on a solid surface.

Online LC-MS   The process of connecting a liquid chromatograph directly to a mass spectrometer. This minimizes sample losses and provides for automated analyses.

Operational genes   Genes that function in cell operation, for example, amino acid synthesis, biosynthesis of cofactors, cell envelope proteins, energy metabolism, intermediary metabolism, fatty acid and phospholipid biosynthesis, nucleotide synthesis and regulation.

Operon   A genetic unit or cluster with one or more genes that are transcribed into the same mRNA and adjacent *cis*-acting regulatory elements.

Optical mapping   Optical mapping is a system for the construction of whole-genome maps from genomic DNA molecules. Ordered restriction maps of an entire genome are created by binding individual DNA molecules to derivatized glass surfaces, cleaving the DNA with restriction enzymes, fluorescently staining the cleavage products, and generating an image by digital fluorescence microscopy. Optical mapping facilitates the genome sequencing process by providing a scaffold for guiding contig closure and validating the finished sequence.

Ordered-clone approach   A cloning and sequencing strategy used in genome projects that utilizes a large-insert library to construct a physical map of overlapping clones that covers the entire genome. Given a physical map, a minimal tiling set of the inserts covering the genome can be selected for shotgun sequencing.

Orthologs   Homologous genes that are vertically descended from a common ancestor as a consequence of a speciation event and encode proteins with the same functions in different species.

Paralogs   Homologous genes that evolved through duplications and encode for proteins with similar but not identical functions.

Pharmacogenomics   Study that aims to identify genetic markers to enable the prediction of drug responses in clinical diagnosis.

Phenotype   Any observable or measurable feature of an organism that is the result of one or more genes, as well as nongenetic factors such as temperature, nutrients, exposure to stress, etc. Phenotypic traits are not necessarily genetic.

Photolithography   Method for in situ synthesis of oligonucleotide probes on solid surface through a combination of modified phosphoramidite, ultraviolet (UV) light, and photomasks.

Phylogenetic oligonucleotide arrays   Oligonucleotide microarrays containing information from rRNA genes for phylogenetic analysis of microbial communities.

Phylogeny   A classification scheme that illustrates the evolutionary history of a group of taxa or genes and their ancestors.

Physiological genomics   Studies that aim to define the genetic pathways and protein interactions which mediate physiological responses of an organism.

Pixels   Numerical picture elements that are used to store data in a graphic computer file.

Polyphyly   A phenomenon in which two or more organisms or sequences descended from different ancestors.

Population genomics   Study that aims to understand genome-wide genetic variations within a population.

Predictive toxicology   It may be possible to potentially predict toxicity based on an understanding of gene expression changes that direct a specific cellular outcome, or mode of action. Predictive toxicology evaluates the potential for a chemical or drug to be toxic by comparing the similarity of its induced gene expression profile with the expression profiles induced by chemicals or drugs with known mechanisms of action.

Preference model   A statistical model for a sequence of events; it assumes that the probability of each event is independent of other events.

Primer walking   This directed sequencing approach is often used for closing the remaining gaps after the finished shotgun phase. In primer walking, sequence data from one sequence reaction are used to design a new primer, which is complementary to a sequence region near the end of sequenced DNA, for the next sequencing reaction. The procedure of sequencing and primer synthesis is repeated until the complete nucleotide sequence is determined.

Principle component analysis (PCA)   A multivariate technique for examining relationships among a group of data points in Euclidean space. It has been widely used in the analysis of gene expression data for various purposes, including identification of outliers in a data set, reduction of dimensionality, etc.

Probe   The nucleic acids deposited on the surface of microarray substrate that react with the complementary target molecule labeled with fluorescent dyes in a solution.

Promoter   A promoter of a gene represents the genomic sequence region that encodes the information about the regulation of the gene. A promoter of a gene typically appears in the upstream region of the gene, and it generally is composed of multiple transcription factor binding sites.

Protein domains   Some percentage of genes in a (microbial) genome might have multiple domains, where each domain could have its own biological function and exist independently of other domains of the gene. Some statistics suggest that the average size of a protein domain is about 150 amino acids. Hence for genes significantly longer than this, one may want to consider the possibility of multiple-domain genes. It is often the case that one domain of a gene could appear as a domain in another gene.

Protein family, superfamily, and fold family   A gene family is generally used to represent a group of genes that are homologous. Operationally, gene families are generally defined in terms of sequence similarities or the similarities of their protein structures. A group of gene families form a superfamily if their similar structural features and similar functions suggest that these proteins evolved from the same origin. The determination of a superfamily relies on structural and functional comparisons. Proteins from different superfamilies form a fold family if they have the same secondary structures in similar three-dimensional (or 3D) topological arrangements. Proteins of the same fold family only indicate that they have similar three-dimensional structures but not necessarily similar functions.

Protein family classification   Family classification is often done based on gene/protein sequence or protein structure similarities, using sequence comparison tools like BLAST or sequence-structure comparison tools like PROSPECT.

Protein fragment complementation   Assays based on an enzyme reassembly strategy whereby a protein–protein interaction promotes the efficient refolding and complementation of enzyme fragments to restore an active enzyme.

Protein threading   Protein threading makes a structural fold prediction from the protein sequence by recognizing a nativelike fold of the query protein (if there is any) in a protein structure database, like PDB. It aligns (threads) each amino acid of a query protein sequence onto consecutive positions in a protein structure in such a way as to optimize some potential energy function. Typically, a threading energy function is statistics-based, that is, derived from known protein structures.

Proteome   The complete set of proteins of an organism.

Proteomics   Large-scale study of expression dynamics and interactions of the proteome.

Quantitation   The measurement of the concentration levels of the components of a sample.

Random shotgun-sequencing approach   Currently, random shotgun sequencing is the most widely used strategy for sequencing complete microbial genomes. In this approach, a large number of clones containing DNA inserts that are collectively representative of the whole genome are maintained in libraries and randomly selected for shotgun sequencing. The sequence data are then combined and assembled into contigs, or contiguous stretches of sequence, using complex computer algorithms.

Representational difference analysis   This analysis uses subtractive hybridization and polymerase chain reaction (PCR) amplification enrichment to selectively purify restriction endonuclease fragments that are present in one population of DNA fragments but not in another. Representational difference analysis was originally developed as a tool for identifying genetic polymorphisms in eukaryotic organisms but has since been used to detect and clone genomic differences between closely related bacterial species.

Self-organizing map   A class of neural networks often used for data clustering. On the conceptual level, an SOM approach has a similar objective to that of a *K*-means approach, that is, it tries to identify a good representative (e.g., centroid) for a group of nearby (similar) data points and to group data points around these representatives.

Sequence tag A short section of sequential amino acids from a given protein sequence. This information is often obtained by tandem mass spectrometry investigations of peptides and enables protein identification by database querying.

Shotgun proteomics The technique of digesting an entire proteome with a suitable protease and then directly examining the mixture by online liquid chromatography-mass spectrometry (i.e., no gel separations are employed).

Signal-to-noise ratio (SNR) Ratio of the difference between signal and background divided by standard deviation of background intensity.

Signature-tagged mutagenesis Modification of the random transposon mutagenesis approach, where the inactivated genes are tagged with unique 20-bp sequences. The pool of tagged mutants allows one to study en masse the relative abundance and survivability of all mutants in any given environment.

Single-nucleotide polymorphism A single nucleotide difference in a gene that is carried by some individuals of a population.

Single-base extension (or minisequencing) A detection primer anneals to the target nucleotide acid sequence immediately adjacent to a variable nucleotide position. The 3'-end of the primer is extended by DNA polymerase with a labeled nucleotide analog that is complementary to the nucleotide at the SNP site.

SPOT synthesis Positionally addressable parallel synthesis on continuous membranes. It is a method for performing solid phase peptide synthesis on cellulose membranes to generate peptide arrays.

Structural genomics The genome-wide structural study of genes and proteins using genome mapping, sequencing, and organizational as well as genome-scale protein structure characterization.

Structure-based drug screening and design This *in silico* approach to discovering and developing agents of pharmaceutical potential couples information on the three-dimensional molecular structure of a target molecule with specialized computer programs to propose novel enzyme inhibitors and other therapeutic agents.

Subfunctionalization Random mutations may cause both copies to diverge functionally to the point at which their total functional capacity is reduced to the level of the single copy of ancestral genes.

Suicide vector DNA delivery vehicle that requires the presence of certain factors or conditions for replication.

Surface plasmon resonance (SPR) An optical phenomenon used to measure the change in refractive index of the solvent near the surface that occurs during complex formation or dissociation. It is used to measure interactions between biomolecules.

Surrogate-ligand-based screening This in vitro screening strategy is based on protein–protein interactions. Surrogate-ligand-based screening utilizes "surrogate" ligands in combinatorial peptide libraries to detect small-molecule enzyme inhibitors with a wide range of potencies. Surrogate ligands are short peptides that bind specifically with high affinity to functional sites of target proteins, thereby inhibiting their function. Phage display is used to isolate surrogate ligands that bind to regions of biological interaction on target proteins.

Symbiosis A phenomenon in which two or more organisms coexist in a mutually beneficial manner.

Synonymous nucleotide substitution A nucleotide substitution that changes a codon into a second codon specifying the same amino acid.

Tandem mass spectrometry (MS/MS) The process of isolating a selected molecular ion in a mass spectrum, followed by induced dissociation inside the mass spectrometer into characteristic fragment ions. For proteins and peptides, this often provides the sequence information necessary to identify the molecule.

Target The nucleic acids labeled with fluorescent dyes in a solution that react with the complementary probe molecule fabricated on the surface of microarray substrates.

Target spectrum The spectrum of an antimicrobial agent refers to the set of target organisms in which the target sequence and function are conserved. An essential property of an antimicrobial agent is that it is active against a desired spectrum or population of bacterial pathogens, but is selective in that it is not toxic to the human host.

Terminator A terminator marks the end of an operon structure and provides a signal for the RNA polymerase to stop transcription.

Three levels of biological functions A protein's function can be described at three levels: molecular, cellular, and phenotypic, representing, respectively, (1) its activity with respect to interactions with other molecules, (2) its role in a cellular system, and (3) its influence on the properties of an organism as a whole.

Top-down proteomics The characterization of a proteome by the direct measurement of the intact proteins in a complex mixture. This provides information about the molecular forms of the proteins present, including details about protein posttranslational processing.

Toxicogenomics This new subdiscipline of toxicology is concerned with identifying and characterizing the molecular mechanisms of action that underlie the potential toxicity of chemicals, environmental pollutants, and drug candidates through the application of genomics resources. Toxicogenomics is expected to revolutionize the process of drug toxicity assessment in the pharmaceutical industry.

Transcriptome The entire assembly of mRNAs in a cell.

Transcriptomics Large-scale study of the expression dynamics and regulation of the transcriptome.

Transformation A process by which an organism acquires plasmids. This term most commonly refers to a bench procedure performed by the investigator that introduces experimental plasmids into bacteria.

Transposon mutagenesis Insertion of a transposable element into a genome to generate a pool of random mutants.

t-test A classical statistical method for accepting or rejecting a hypothesis with a rigorously defined confidence level. It is well known from statistical theory that the average value over a large number of random variables approximately follows a normal distribution, and the difference between two such average values for two different sets of variables follows a distribution called the *Student distribution*. Assessing a hypothesis that the two normal distributions are statistically different can be done by using the Student distribution table, which provides a confidence value for accepting such a hypothesis.

Ubiquitin A 76-amino-acid protein that is present in cells either free or covalently linked to other proteins. Ubiquitin fusions with other proteins are rapidly cleaved by ubiquitin-specific proteases, which recognize the folded conformation of ubiquitin.

Vector DNA vehicle for cloning, transforming, and expressing experimental DNA. The vector provides all sequences essential for replicating the test DNA. Typical vectors include plasmids, cosmids, and phages.

Xenologous gene replacement The ancestral orthologous gene in the recipient strain is replaced by a distantly related orthologous gene acquired by horizontal gene transfer and then the ancestral gene is eliminated from the recipient strain.

Yeast two-hybrid system An in vivo protein–protein interaction detection method that relies on the reconstitution of a functional transcriptional activator through the recruitment of a transcriptional activation domain to a DNA-binding domain via interaction of the proteins being tested for interaction.

REFERENCES

Abdeddaim, S. and B. Morgenstern. 2001. Speeding up the DIALIGN multiple alignment program by using the 'Greedy Alignment of BIOlogical Sequences LIBrary' (GABIOS-LIB). *Lect. Notes Comp. Sci.* 2066:1–11.

Adams, M. D., S. E. Celniker, R. A. Holt, and C. A. Evans, et al. 2000. The genome sequence of *Drosophila melanogaster. Science* 287:2185–2195.

Adams, M. W. W., J. F. Holden, A. L. Menon, and G. J. Schut, et al. 2001. Key role for sulfur in peptide metabolism and in regulation of three hydrogenases in the hyperthermophilic archaeon *Pyrococcus furiosus. J. Bacteriol.* 183:716–724.

Adler, K., et al. 2000. MICROMAX^{TM}: A highly sensitive system for differential gene expression on microarrays. In M. Schena (ed.), *Microarray Biochip Technology*, pp. 221–230, Eaton Publishing, Natick, Mass.

Aebersold, R. and D. R. Goodlett. 2001. Mass spectrometry in proteomics. *Chem. Rev.* 101:269–295.

Afanassiev, V., V. Hanemann, and S. Wolfl. 2000. Preparation of DNA and protein micro arrays on glass slides coated with an agarose film. *Nucl. Acids Res.* 28:E66.

Agrawal, S., Z. Jiang, Q. Zhao, and D. Shaw, et al. 1997. Mixed-backbone oligonucleotides as second generation antisense oligonucleotides: in vitro and in vivo studies. *Proc. Natl. Acad. Sci. USA* 94:2620–2625.

Aho, A. V., J. E. Hopcroft, and J. D. Ullman. 1974. *The Design and Analysis of Computer Algorithms*. Addison-Wesley, Reading, Mass.

Akerley, B. J., E. J. Rubin, A. Camilli, and D. J. Lampe, et al. 1998. Systematic identification of essential genes by in vitro mariner mutagenesis. *Proc. Natl. Acad. Sci. USA* 95:8927–8932.

Akerley, B. J., E. J. Rubin, V. L. Novick, and K. Amaya, et al. 2002. A genome-scale analysis for identification of genes required for growth or survival of *Haemophilus influenzae. Proc. Natl. Acad. Sci. USA* 99:966–971.

Akutsu, T., S. Miyano, and S. Kuhara. 1999. Identification of genetic networks from a small number of gene expression patterns under the Boolean network model. *Proc. Pacific Symp. Biocomp.* 4:17–28.

Akutsu, T., S. Miyano, and S. Kuhara. 2000a. Infering qualitative relations in genetic networks and metabolic pathways. *Bioinformatics* 16:727–734.

Akutsu, T., S. Miyano, and S. Kuhara. 2000b. Algorithms for identifying Boolean networks and related biological networks based on matrix multiplication and fingerprint function. *J. Comput. Biol.* 7:331–344.

Akutsu, T., S. Miyano, and S. Kuhara. 2000c. Algorithms for inferring qualitative models of biological networks. *Proc. Pacific Symp. Biocomput.* 293–304.

Alberts, M. M., D. Botstein, S. Brenner, and C. R. Cantor, et al. 1988. *Report of the Committee on Mapping and Sequencing the Human Genome, Report of the Committee on Mapping and Sequencing the Human Genome*. National Research Council, Washington, D.C.

Alexandre, H., V. Ansanay-Galeote, S. Dequin, and B. Blondin. 2001. Global gene expression during short-term ethanol stress in *Saccharomyces cerevisiae. FEBS Lett.* 498:98–103.

Alexandrov, N. N., R. Nussinov, and R. M. Zimmer. 1996. Fast protein fold recognition via sequence to structure alignment and contact capacity potentials. In L. Hunter and T. Klein (eds.), *Biocomputing: Proceedings of the 1996 Pacific Symposium*, pp. 53–72. World Scientific Publishing Co., Singapore.

Alexeyev, M. F. 1999. The pKNOCK series of broad-host-range mobilizable suicide vectors for gene knockout and targeted DNA insertion into the chromosome of gram-negative bacteria. *Biotechniques* 26:824–828.

Alizadeh, A. A., M. B. Eisen, R. E. Davis, and C. Ma, et al. 2000. Distinct types of diffuse large B-cell lymphoma identified by gene expression profiling. *Nature* 403:503–511.

Allsop, A. E. 1998. Bacterial genome sequencing and drug discovery. *Curr. Opin. Biotechnol.* 9:637–642.

Alm, R. A., J. Bina, B. M. Andrews, and P. Doig, et al. 2000. Comparative genomics of *Helicobacter pylori*: analysis of the outer membrane protein families. *Infect. Immun.* 68:4155–4168.

Alm, R. A., L. S. Ling, D. T. Moir, and B. L. King, et al. 1999. Genomic-sequence comparison of two unrelated isolates of the human gastric pathogen *Helicobacter pylori*. *Nature* 397:176–180.

Almiron, M., A. J. Link, D. Furlong, and R. Kolter. 1992. A novel DNA-binding protein with regulatory and protective roles in starved *Escherichia coli*. *Genes Dev.* 6:2646–2654.

Altschul, S. F., T. L. Madden, A. A. Schäffer, and J. Zhang, et al. 1997. Gapped BLAST and PSI-BLAST: a new generation of protein database search programs. *Nucl. Acids Res.* 25:3389–3402.

Altschul, S. F., W. Gish, W. Miller, and E. W. Myers, et al. 1990. Basic local alignment search tool. *J. Mol. Biol.* 215(3):403–410.

Altuvia, S., D. Weinstein-Fischer, A. Zhang, and L. Postow, et al. 1997. A small, stable RNA induced by oxidative stress: role as a pleiotropic regulator and antimutator. *Cell* 90:43–53.

Amann, R. I., W. Ludwig, and K. H. Schleifer. 1995. Phylogenetic identification and in situ detection of individual microbial cells without cultivation. *Microbiol. Rev.* 59:143–169.

Anderson, A., H. Nordan, R. Cain, and G. Parrish, et al. 1956. Studies on a radioresistant *micrococcus*. I. Isolation, morphology, cultural characteristics, and resistance to gamma radiation. *Food Technol.*. 10:575–578.

Andersson, S. G. and C. G. Kurland. 1998. Reductive evolution of resident genomes. *Trends in Microbiology.* 6:263–268.

Andersson, S. G. and C. G. Kurland. 1999. Origins of mitochondria and hydrogenosomes. *Curr. Opin. Microbiol.* 2:535–541.

Ang, D., K. Liberek, D. Skowyra, and M. Zylicz, et al. 1991. Biological role and regulation of the universally conserved heat shock proteins. *J. Biol. Chem.* 266:24233–24236.

Ang, S., C. Z. Lee, K. Peck, and M. Sindici, et al. 2001. Acid-induced gene expression in *Helicobacter pylori*: study in genomic scale by microarray. *Infect. Immun.* 69:1679–1686.

Antelmann, H., H. Tjalsma, B. Voigt, and S. Ohlmeier, et al. 2001. A proteomic view on genome-based signal peptide predictions. *Genome Res.* 11:1484–1502.

Antelmann, H., H. Yamamoto, J. Sekiguchi, and M. Hecker. 2002. Stabilization of cell wall proteins in *Bacillus subtilis*: a proteomic approach. *Proteomics* 2:591–602.

Arantes, O. and D. Lereclus. 1991. Construction of cloning vectors for *Bacillus thuringiensis*. *Gene* 108:115–119.

Arap, W., M. G. Kolonin, M. Trepel, and J. Lahdenranta, et al. 2002. Steps toward mapping the human vasculature by phage display. *Nat. Med.* 8:121–127.

Aravind, L., H. Watanabe, D. J. Lipman, and E. V. Koonin. 2000. Lineage-specific loss and divergence of functionally linked genes in eukaryotes. *Proc. Natl. Acad. Sci. USA* 97:11319–11324.

Aravind, L., R. L. Tatusov, Y. I. Wolf, and D. R. Walker, et al. 1998. Evidence for massive gene exchange between archaeal and bacterial hyperthermophiles. *Trends Genet.* 14:442–444.

Arigoni, F., F. Talabot, and M. Peitsch, et al. 1998. A genome-based approach for the identification of essential bacterial genes. *Nat. Biotechnol.* 16:851–856.

Arruda, S., G. Bomfim, R. Knights, and T. Huima-Byron, et al. 1993. Cloning of an M. tuberculosis DNA fragment associated with entry and survival inside cells. *Science* 261:1454–1457.

Asai, K., S. Hayamizu, and K. Handa. 1993. Prediction of protein secondary structure by the hidden Markov model. *Comp. Appl. Biosci.* (CABIOS) 9(2):141–146.

Asai, T., D. Zaporojets, C. Squires, and C. L. Squires. 1999. An *Escherichia coli* strain with all chromosomal rRNA operons inactivated: complete exchange of rRNA genes between bacteria. *Proc. Natl. Acad. Sci. USA* 96:1971–1976.

Askwith, C. C., D. de Silva, and J. Kaplan. 1996. Molecular biology of iron acquisition in *Saccharomyces cerevisiae. Mol. Microbiol.* 20:27–34.

Atherton, J. C. Jr., R. M. Peek, K. T. Tham, and T. L. Cover, et al. 1997. Clinical and pathological importance of heterogeneity in vacA, the vacuolating cytotoxin gene of *Helicobacter pylori. Gastroenterology* 12:92–99.

Attridge, S. R., E. Voss, and P. A. Manning. 1993. The role of toxin-coregulated pili in the pathogenesis of *Vibrio cholerae* O1 El Tor. *Microb. Pathog.* 15:421–431.

Attwood, T. K., M. Blythe, D. R. Flower, and A. Gaulton, et al. 2002. PRINTS and PRINTS-S shed light on protein ancestry. *Nucl. Acids Res.* 30(1):239–241.

Augenlicht, L., J. Taylor, L. Anderson, and M. Lipkin. 1991. Patterns of gene expression that characterize the colonic mucosa in patients at genetic risk for colonic cancer. *Proc. Natl. Acad. Sci. USA* 88:3286–3289.

Augenlicht, L., M. Wahrman, H. Halsey, and L. Anderson, et al. 1987. Expression of cloned sequences in biopsies of human colonic tissue and in colonic-carcinoma cells induced to differentiate in vitro. *Canc. Res.* 47:6017–6021.

Auroux, P.-A., D. Iossifidis, D. R. Reyes, and A. Manz. 2002. Micro total analysis systems 2. Analytical standard operations and applications. *Anal. Chem.* 74:2637–2652.

Avery, O. T., C. M. MacLeod, and M. McCarty. 1944. Studies on the chemical nature of the substance inducing transformations of pneumococcal types. *J. Exp. Med.* 79:137.

Ayala, F. J. 1997. Vagaries of the molecular clock. *Proc. Natl. Acad. Sci. USA* 94:7776–7783.

Badger, J. L., C. A. Wass, S. J. Weissman, and K. S. Kim. 2000. Application of signature-tagged mutagenesis for identification of *Escherichia coli* K1 genes that contribute to invasion of human brain microvascular endothelial cells. *Infect. Immun.* 68:5056–5061.

Bafna, V. and D. H. Huson. 2000. The conserved exon method for gene finding. In *Proceedings of International Conference on Intelligent Systems for Molecular Biology*, pp. 3–12. AAAI Press.

Bai, C. and S. J. Elledge. 1997. Gene identification using the yeast two-hybrid system. *Meth. Enzymol.* 283:141–156.

Bailey, T. L. and C. Elkan. 1994. Fitting a mixture model by expectation maximization to discover motifs in biopolymers. *Proc. Int. Conf. Intell. Syst. Mol. Biol.* 2:28–36.

Bains, W. and G. Smith. 1988. A novel method for nucleic acid sequence determination. *J. Theoret. Biol.* 135:303–307.

Bairoch, A. 1993. The ENZYME data bank. *Nucl. Acids Res.* 21:3155–3156.

Bairoch, A. and R. Apweiler. 2000. The SWISS-PROT protein sequence database and its supplement TrEMBL in 2000. *Nucl. Acids Res.* 28:45–48.

Balasubramanian, S., T. Schneider, M. Gerstein, and L. Regan. 2000. Proteomics of *Mycoplasma genitalium*: identification and characterization of unannotated and atypical proteins in a small model genome. *Nucl. Acids Res.* 28:3075–3082.

Baldi, P., and A. D. Long 2001. A Bayesian framework for the analysis of microarray expression data-regularized t-test and statistical inferences of gene changes. *Bioinformatics* 17:509–519.

Banerjee, A., E. Dubnau, A. Quemard, and V. Balasubramanian, et al. 1994. inhA, a gene encoding a target for isoniazid and ethionamide in *Mycobacterium tuberculosis. Science* 263:227–230.

Bannai, H., Y. Tamada, O. Maruyama, and K. Nakai, et al. 2002. Extensive feature detection of N-terminal protein sorting signals. *Bioinformatics* 18(2):298–305.

Barabasi, A. L. and R. Albert. 1999. Emergence of scaling in random networks. *Science* 285:509–512.

Barker, W. C., J. S. Garavelli, P. B. McGarvey, and C. R. Marzec, et al. 1999. The PIR-international protein sequence database. *Nucl. Acids Res.* 27:39–42.

Barlev, N. A. and S. N. Borchsenius. 1991. Continuous distribution of *Mycoplasma* genome sizes. *Biomed. Sci.* 2:641–645.

Barns, S. M., C. F. Delwiche, J. D. Palmer, and N. R. Pace. 1996. Perspectives on archaeal diversity, thermophily and monophyly from environmental rRNA sequences. *Proc. Natl. Acad. Sci. USA* 93:9188–9193.

Baron, U. and H. Bujard. 2000. Tet repressor-based system for regulated gene expression in eukaryotic cells: principles and advances. *Methods Enzymol.* 327:401–421.

Barry, C. E. III and B. G. Schroeder. 2000. DNA microarrays: translational tools for understanding the biology of *Mycobacterium tuberculosis. Trends Microbiol.* 8:209–210.

Barry, C. E. III. 2001. Interpreting cell wall 'virulence factors' of *Mycobacterium tuberculosis. Trends Microbiol.* 9:237–241.

Bartel, P. L., J. A. Roecklein, D. SenGupta, and S. Fields. 1996. A protein linkage map of *Escherichia coli* bacteriophage T7. *Nat. Genet.* 12:72–77.

Bartosiewicz, M., M. Trounstine, D, Barker, R. Johnston, and A. Buckpitt. 2000. Development of a toxicological gene array and quantitative assessment of this technology. *Arch. Biochem. Biophys.* 376:66–73.

Basarsky, T., D. Verdnik, J. Y. Zhai, and D. Wellis. 2000. Overview of a microarray scanner: design essentials for an integrated acquisition and analysis platform. In M. Schena (ed.), *Microarray Biochip Technology,* pp. 265–284. Eaton Publishing, Natick, Mass.

Basset, D. E, M. S. Boguski, and P. Hieter. 1996. Yeast genes and human disease. *Nature* 379:589–590.

Bassett, Jr., D., M. Eisen, and M. Boguski. 1999. Gene expression informatics—it's all in your mine. *Nat. Genet. Suppl.* 21:51–55.

Bassett, Jr., D., E. M. S. Boguski, F. Spencer, and R. Reeves, et al. 1997. Genome cross-referencing and XREFdb: implications for the identification and analysis of genes mutated in human disease. *Nat. Genet.* 15:339–344.

Bassingthwaighte, J. B. 2000. Strategies for the physiome project. *Ann. Biomed. Eng.* 28:1043–1058.

Bateman, A., E. Birney, L. Cerruti, and R. Durbin, et al. 2002. The Pfam protein families database. *Nucl. Acids Res.* 30(1):276–280.

Bateman, A., E. Birney, R. Durbin, S. R. Eddy, R. D. Finn, and E. L. Sonnhammer. 1999. Pfam 3.1-1313 multiple alignments and profile HMMs match the majority of proteins. *Nucl. Acids Res.* 27:260–262.

Battaglia, C., G. Salani, C. Consolandi, and L. R. Bernardi, et al. 2000. Analysis of DNA microarrays by non-destructive fluorescent staining using SYBR green II. *Biotechniques* 29:78–81.

Battista, J. R. 1997. Against all odds: the survival strategies of *Deinococcus radiodurans. Annu. Rev. Microbiol.* 51:203–224.

Battista, J. R., A. M. Earl, and M.-J. Park. 1999. Why is *Deinococcus radiodurans* so resistant to ionizing radiation? *Trends Microbiol.* 7:362–365.

Batzoglou, S., L. Pachter, J. P. Mesirov, and B. Berger, et al. 2000. Human and mouse gene structure: comparative analysis and application to exon prediction. *Genome Res.* 7:950–958.

Baudin, A., O. Ozier-Kalogeropoulos, A. Denouel, and F. Lacroute, et al. 1993. A simple and efficient method for direct gene deletion in *Saccharomyces cerevisiae. Nuc. Acids Res.* 21:3329–3330.

Baumler, A. J., J. G. Kusters, I. Stojiljkovic, and F. Heffron. 1994. *Salmonella typhimurium loci* involved in survival within macrophages. *Infect. Immun.* 62:1623–1630.

Bavykin, S. G., J. P. Akowski, V. M. Zakhariev, and V. E. Barsky, et al. 2001. Portable system for microbial sample preparation and oligonucleotide microarray analysis. *Appl. Environ. Microbiol.* 67:922–928.

Beattie, K. L., M. D. Eggers, J. M. Shumaker, and M. E. Hogan, et al. 1992. Genosensor technology. *Clin. Chem.* 39:719–722.

Beattie, W. G., L. Meng, S. Turner, and R. S. Varma, et al. 1995. Hybridization of DNA targets to glass-tethered oligonucleotide probes. *Mol. Biotechnol.* 4:213–225.

Behr, M. A., M. A. Wilson, W. P. Gill, and H. Salamon, et al. 1999. Comparative genomics of BCG vaccines by whole-genome DNA microarray. *Science* 284:1520–1523.

Beier, M. and J. Hoheisel. 1999. Versatile derivatisation of solid support media for covalent bonding on DNA-microchips. *Nucl. Acids Res.* 27:1970–1977.

Beja, O., L. Aravind, E. V. Koonin, and M. T. Suzuki, et al. 2000a. Bacterial rhodopsin: evidence for a new type of phototrophy in the sea. *Science* 289:1902–1906.

Beja, O., M. T. Suzuki, E. V. Koonin, and L. Aravind, et al. 2000b. Construction and analysis of bacterial artificial chromosome libraries from a natural microbial assemblage, *Environ. Microbiol.* 2:516–529.

Beja, O., E. N. Spudich, J. L. Spudich, and M. Leclerc, et al. 2001. Proteorhodopsin phototrophy in the ocean. *Nature* 411:786–789.

Beja, O., M. T. Suzuki, J. F. Heideberg, and W. C. Nelson, et al. 2002. Unsuspected diversity among marine aerobic anoxygenic phototrophs. *Nature* 415:630–633.

Belas, R., A. Mileham, M. Simon, and M. Silverman. 1984. Transposon mutagenesis of marine *Vibrio* spp. *J. Bacteriol.* 158:890–896.

Belcher, C. E., J. Drenkow, B. Kehoe, and T. R. Gingeras, et al. 2000. The transcriptional responses of respiratory epithelial cells to *Bordetella pertussis* reveal host defensive and pathogen counter-defensive strategies. *Proc. Natl. Acad. Sci. USA* 97:13847–13852.

Belfort, M. and A. Weiner. 1997. Another bridge between kingdoms: tRNA splicing in Archaea and Eukaryotes. *Cell* 89:1003–1006.

Beliaev, A. S., D. A. Saffarini, J. L. McLaughlin, and D. Hunnicutt. 2001. MtrC, an outer membrane decaheme c cytochrome required for metal reduction in *Shewanella putrefaciens* MR-1. *Mol. Microbiol.* 39:722–730.

Beliaev, A. S., D. K. Thompson, M. W. Fields, and L. Wu, et al. 2002a. Microarray transcription profiling of a *Shewanella oneidensis* etrA mutant. *J. Bacteriol.* 184:4612–4616.

Beliaev, A. S., D. K. Thompson, T. Khare, and H. Lim, et al. 2002b. Gene and protein expression profiles of *Shewanella oneidensis* during anaerobic growth with different electron acceptors. *OMICS* 6:39–60.

Bennett, K.L., T. Matthiesen, and P. Roepstorff. 2000. Probing protein surface topology by chemical surface labeling, crosslinking, and mass spectrometry. *Methods Mol. Biol.* 146:113–131.

Benson, D. A., I. Karsch-Mizrachi, D. J. Lipman, and J. Ostell, et al. 2002. Genbank. *Nucl. Acids Res.* 30(1):17–20.

Benson, G. 1999. Tandem repeats finder: a program to analyze DNA sequences. *Nucl. Acids Res.* 27:573–580.

Bentley, S. D., K. F. Chater, A. M. Cerdeno-Tarraga, and G. L. Challis, et al. 2002. Complete genome sequence of the model actinomycete *Streptomyces coelicolor* A3(2). *Nature* 9:141–147.

Berg, C. M. and D. E. Berg. 1996. Transposable element tools for microbial genetics. In F. C. Neidhardt, R. Curtiss, III, C. Gross, and J. Ingraham, et al. (eds.), *Escherichia coli* and *Salmonella typhimurium*: Cellular and Molecular Biology, 2nd ed., pp. 2588–2612. American Society for Microbiology, Washington, D.C.

Berg, C. M. and K. J. Shaw. 1981. Organization and regulation of the ilvGEDA operon in *Salmonella typhimurium* LT2. *J. Bacteriol.* 145:468–470.

Berg, C. M., K. J. Shaw, J. Vender, and M. Borucka-Mankiewicz. 1979. Physiological characterization of polar Tn5-induced isoleucine-valine auxotrophs in *Escherichia coli* K12: evidence for an internal promoter in the ilvGEDA operon. *Genetics* 93:309–319.

Berg, D. E. 1989. Transposon Tn5. In D. E. Berg and M. M. Howe (eds.), *Mobile DNA*, pp. 185–210. American Society for Microbiology, Washington, D.C.

Berg, O. G. and C. G. Kurland. 2000. Why mitochondrial genes are most often found in nuclei. *Mol. Biol. Evol.* 17:951–961.

Berger, B., D. B. Wilson, E. Wolf, and T. Tonchev, et al. 1995. Predicting coiled coils by use of pairwise residue correlations. *Proc. Natl. Acad. Sci. USA* 92:8259–8263.

Bergthorsson, U. and H. Ochman. 1998. Distribution of chromosome length variation in natural isolates of *Escherichia coli. Mol. Biol. Evol.* 15:6–16.

Bernhardt, J., U. Volker, A. Volker, and H. Antelmann, et al. 1997. Specific and general stress proteins in *Bacillus subtilis*—a two-dimensional electrophoresis study. *Microbiology* 143:999–1017.

Betts, J. C., P. Dodson, S. Quan, and A. P. Lewis, et al. 2000. Comparison of the proteome of *Mycobacterium tuberculosis* strain H37Rv with clinical isolate CDC 1551. *Microbiology* 146:3205–3216.

Biek, D. P. and S. N. Cohen. 1986. Identification and characterization of recD, a gene affecting plasmid maintenance and recombination in *Escherichia coli. J. Bacteriol.* 167:594–603.

Biemann, K. 1988. Contributions of mass spectrometry to peptide and protein structure. *Biomed. Environ. Mass Spectrom.* 16:99–111.

Bintrim, S. B., T. J. Donohue, J. Handelsman, and G. P. Roberts, et al. 1997. Molecular phylogeny of Archaea from soil. *Proc. Natl. Acad. Sci. USA* 94:277–282.

Bird, A. P. 1987. CPG Islands as gene markers in the vertebrate nucleus. *Trends Gen.* 3(12):342–347.

Bishop, C. M. 1995. *Neural Networks for Pattern Recognition.* Oxford University Press, Oxford, U.K.

Biswas, I., A. Gruss, S. D. Ehrlich, and E. Maguin. 1993. High-efficiency gene inactivation and replacement system for Gram-positive bacteria. *J. Bacteriol.* 175:3628–3635.

Bjellqvist, B., K. Ek, P. G. Righetti, and E. Gianazza, et al. 1982. Isoelectric focusing in immobilized pH gradients: principle, methodology and some applications. *J. Biochem. Biophys. Methods* 6:317–339.

Black, W. C. T., C. F. Baer, M. F. Antolin, and N. M. DuTeau. 2001. Population genomics: genome-wide sampling of insect populations. *Annu. Rev. Entomol.* 46:441–469.

Blackwell, H. E., L. Perez, R. A. Stavenger, and J. A. Tallarico, et al. 2001. A one-bead, one-stock solution approach to chemical genetics: part 1. *Chem. Biol.* 8:1167–1182.

Blank, P. S., C. M. Sjomeling, P. S. Backlund, and A. L. Yergey. 2002. Use of cumulative distribution functions of characterize mass spectra of intact proteins. *J. Am. Soc. Mass Spectrom.* 13:40–46.

Blaser, M. J. 1997. Not all *Helicobacter pylori* strains are created equal: should all be eliminated? *Lancet* 349:1020–1022.

Blattner, F. R., G. Plunkett III, and C. A. Bloch, et al. 1997. The complete genome sequence of *Escherichia coli* K-12. *Science* 277:1453–1462.

Blonder, J., M. B. Goshe, R. J. Moore, and L. Pasa-Tolic, et al. 2002. Enrichment of integral membrane proteins for proteomic analysis using liquid chromatography-tandem mass spectometry. *J. Proteome Res.* 1:351–360.

Blum, G., M. Ott, A. Lischewski, and A. Ritter, et al. 1994. Excision of large DNA regions termed pathogenicity islands from tRNA-specific loci in the chromosome of an *Escherichia coli* wild-type pathogen. *Infect. Immun.* 62:606–614.

Blum, P., D. Holzschu, H. S. Kwan, and D. Riggs, et al. 1989. Gene replacement and retrieval with recombinant M13mp bacteriophages. *J. Bacteriol.* 171:538–546.

Bodey, G. P., R. Bolivar, V. Fainstein, and L. Jadeja. 1983. Infections caused by *Pseudomonas aeruginosa. Rev. Infect. Dis.* 5:279–313.

Boguski, M. S., T. M. Lowe, and C. M. Tolstoshev. 1993. dbEST—database for "expressed sequence tags." *Nat. Genet.* 4(4):332–333.

Boldrick, J. C., A. A. Alizadeh, M. Diehn, and S. Dudoit, et al. 2002. Stereotyped and specific gene expression programs in human innate immune responses to bacteria. *Proc. Natl. Acad. Sci. USA* 99:972–977.

Borisy, G. G. and E. W. Taylor. 1967. The mechanism of action of colchicine. Binding of colchicine-3H to cellular protein. *J. Cell Biol.* 34:525–533.

Bork, P., T. Dandekar, Y. Diaz-Lazcoz, and F. Eisenhaber, et al. 1998. Predicting functions: from gene to genomes and back. *J. Mol Biol.* 283:707–725.

Borodovsky, M. and J. McIninch. 1993. GENMARK: parallel gene recognition for both DNA strands. *Comp. Chem.* 17:123–133.

Borodovsky, M., Y. U. Sprizhitskii, E. Golovanov, and N. Alexandrov. 1986. Statistical patterns in the primary structures of functional regions of the genome in *E. coli*. II. Nonuniform Markov models. *Molek. Biol.* 20:1114–1123.

Bossier, P., L. Pernandes, C. Vilela, and C. Rodrigues-Pousada. 1994. The yeast YKL74i gene situated on the left arm of chromosome XI codes for a homologue of the human ALD protein. *Yeast* 10:681–686.

Bouchez, D. and H. Hofte. 1998. Functional genomics in plants. *Plant Physiol.* 118:725–732.

Bousquet, J., S. H. Strauss, A. H. Doerksen, and R. A. Price. 1992. Extensive variation in evolutionary rate of rbcL gene sequences among seed plants. *Proc. Natl. Acad. Sci. USA* 89:7844–7848.

Bowie, J. U., R. Luthy, and D. Eisenberg. 1991. A method to identify protein sequences that fold into a known three-dimensional structure. *Science* 253:164–170.

Bowtell, D. and J. Sambrook. 2002. *A Molecular Cloning Manual: DNA Microarrays*. Cold Spring Harbor Laboratory Press, Plainview, N.Y.

Brantl, S. 2002. Antisense-RNA regulation and RNA interference. *Biochim. Biophys. Acta* 1575:15–25.

Braun, E. L., A. L. Halpern, M. A. Nelson, and D. O. Natvig. 2000. Large-scale comparison of fungal sequence information: mechanisms of innovation in *Neurospora crassa* and gene loss in *Saccharomyces cerevisiae. Genome Res.* 10:416–430.

Braxton, S. and T. Bedilion. 1998. The integration of microarray information in the drug development process. *Curr. Opin. Biotechnol.* 9:643–649.

Brazma, A. and J. Vilo. 2000. Gene expression data analysis. *FEBS Lett.* 480:17–24.

Brendel, V. and E. N. Trifonov. 1986. A computer algorithm for testing potential prokaryotic terminators. *Nucl. Acids Res.* 12:4411–4427.

Brenner, D. J., J. T. Staley, and N. R. Krieg. 2000. Classification of prokaryotic organisms and the concept of speciation. D. Boone, R. Castenholz, G. Garrity (eds), In *Bergey's Manual of Systematic Bacteriology*, 2nd ed. Springer-Verlag, New York.

Brenner, S. E. 1999. Errors in genome annotation. *Trends Genet.* 15:132–133.

Brent, R. 1999. Functional genomics: learning to think about gene expression data. *Curr. Biol.* 9:R338–R341.

Brent, R. and R. L. Finley, Jr. 1997. Understanding gene and allele function with two-hybrid methods. *Annu. Rev. Genet.* 31:663–704.

Bringel, F., L. Frey, and J. C. Hubert. 1989. Characterization, cloning, curing, and distribution in lactic acid bacteria of pLP1, a plasmid from *Lactobacillus plantarum* CCM 1904 and its use in shuttle vector construction. *Plasmid* 22:193–202.

Britten, R. J. 1986. Rates of DNA sequence evolution differ between taxonomic groups. *Science* 231:1393–1398.

Brochier, C., H. Philippe, and D. Moreira. 2000. The evolutionary history of ribosomal protein RpS14: horizontal gene transfer at the heart of the ribosome. *Trends Genet.* 16:529–533.

Brocklehurst, K. R. and A. P. Morby. 2000. Metal-ion tolerance in *Escherichia coli*: analysis of transcriptional profiles by gene array technology. *Microbiology* 146:2277–2282.

Brosch, R., A. S. Pym, S. V. Gordon, and S. T. Cole. 2001. The evolution of mycobacterial pathogenicity: clues from comparative genomics. *Trends Microbiol.* 9:452–458.

Brosch, R., S. V. Gordon, K. Eiglmeier, and T. Garnier, et al. 2000. Comparative genomics of the leprosy and tubercle bacilli. *Res. Microbiol.* 151:135–142.

Brosch, R., W. Philipp, E. Stavropolous, and M. J. Colston, et al. 1999. Genomic analysis reveals variation between *Mycobacterium tuberculosis* H37Rv and the attenuated M. tuberculosis H37Ra. *Infect. Immun.* 67:5768–5774.

Broude, N. E., K. Woodward, R. Cavallo, and C. R. Cantor, et al. 2001. DNA microarrays with stem-loop DNA probes: preparation and applications. *Nucl. Acids Res.* 29:E92.

Brown, J. K. 1994. Bootstrap hypothesis tests for evolutionary trees and other dendrograms. *Proc. Natl. Acad. Sci. USA* 91:12293–12297.

Brown, J. R. and W. F. Doolittle. 1995. Root of the universal tree of life based on ancient aminoacyl-tRNA synthetase gene duplications. *Proc. Natl. Acad. Sci. USA* 92:2441–2445.

Brown, J. R. and W. F. Doolittle. 1997. Archaea and the prokaryote-to-eukaryote transition. *Microbiol. Mol. Biol. Rev.* 61:456–502.

Brown, J. R., C. J. Douady, M. J. Italia, and W. E. Marshall, et al. 2001. Universal trees based on large combined protein sequence data sets. *Nat. Genet.* 28:281–285.

Brown, P. O. and D. Botstein. 1999. Exploring the new world of the genome with DNA microarrays. *Nat. Genet.* 21:33–37.

Brown, T. A. 1999. *Genomes*. John Wiley & Sons, New York.

Brownstein, M. J. and A. B. Khodursky. 2003. *Functional Genomics: Methods and Protocols.* Humana Press, Totowa, N.J.

Brurberg, M. B., I. F. Nes, and V. G. Eijsink. 1997. Pheromone-induced production of antimicrobial peptides in *Lactobacillus. Mol. Microbiol.* 26:347–360.

Brush, M. 1998. Dye hard—protein gel staining products. *Scientist* 12:16.

Bucher, P. and A. Bairoch. 1994. A generalized profile syntax for biomolecular sequences motifs and its function in automatic sequence interpretation. In R. Altman, D. Brutlag, P. Karp, and R. Lathrop (eds.), *ISMB-94, Proceedings 2nd International Conference on Intelligent Systems for Molecular Biology,* 53–61.

Buchman, C., P. Skroch, J. Welch, and S. Fogel, et al. 1989. The CUP2 gene product, regulator of yeast metallothionein expression, is a copper-activated DNA-binding protein. *Mol. Cell. Biol.* 9:4091–4095.

Bult, C. J., O. White, G. J. Olsen, and L. Zhou, et al. 1996. Complete genome sequence of the methanogenic archaeon, *Methanococcus jannaschii. Science* 273:1058–1073.

Bulyk, M. L., X. Huang, Y. Choo, and G. M. Church. 2001. Exploring the DNA-binding specificities of zinc fingers with DNA microarrays. *Proc. Natl. Acad. Sci. USA* 98:7158–7163.

Bumann, D., T. F. Meyer, and P. R. Jungblut. 2001. Proteome analysis of the common human pathogen *Helicobacter pylori. Proteomics* 1:473–479.

Burbulys, D., K. A. Trach, and J. A. Hoch. 1991. Initiation of sporulation in *B. subtilis* is controlled by a multicomponent phosphorelay. *Cell* 64:545–552.

Burch, C. L., R. J. Danaher, and D. C. Stein. 1997. Antigenic variation in *Neisseria gonorrhoeae*: production of multiple lipooligosaccharides. *J. Bacteriol.* 179:982–986.

Burchiel, S. W., C. M. Knall, and J. W. Davis II, et al. 2001. Analysis of genetic and epigenetic mechanisms of toxicity: potential roles of toxicogenomics and proteomics in toxicology. *Toxicol. Sci.* 59:193–195.

Burdett, V. 1996. Tet(M)-promoted release of tetracycline from ribosomes is GTP dependent. *J. Bacteriol.* 178:3246–3251.

Burge, C., A. M. Campbell, and S. Karlin. 1992. Over- and under-representation of short oligonucleotides in DNA sequences. *Proc. Natl. Acad. Sci. USA* 89:1358–1362.

Burge, C. and S. Karlin. 1997. Prediction of complete gene structures in human genomic DNA. *J. Mol. Biol.* 268(1):78–94.

Burger, G., B. F. Lang, M. Reith, and M. W. Gray. 1996. Genes encoding the same three subunits of respiratory complex II are present in the mitochondrial DNA of two phylogenetically distant eukaryotes. *Proc. Natl. Acad. Sci. USA* 93:2328–2332.

Burley, S. K., S. C. Almo, and J. B. Bonanno, et al. 1999. Structural genomics: beyond the human genome project. *Nat. Genet.* 23:151–157.

Burris, J., R. Cook-Deegan, and B. Alberts. 1998. The Human Genome Project after a decade: policy issues. *Nat. Genet.* 20:333–335.

Bury, K. V. 1975. *Statistical Methods in Applied Science.* John Wiley & Sons, New York.

Bussey, H., D. B. Kaback, W. Zhong, and D. T. Vo, et al. 1995. The nucleotide sequence from chromosome I from *Saccharomyces cerevisiae. Proc. Natl. Acad. Sci. USA* 92:3809–3813.

Butte, A. 2002. The use and analysis of microarray data. *Nat. Rev. Drug Disc.* 1:951–960.

Buttner, K., J. Bernhardt, C. Scharf, and R. Schmid, et al. 2001. A comprehensive two-dimensional map of cytosolic proteins of *Bacillus subtilis. Electrophoresis* 22:2908–2935.

Campbell, A., J. Mrazek, and S. Karlin. 1999. Genome signature comparisons among prokaryote, plasmid and mitochondrial DNA. *Proc. Natl. Acad. Sci. USA* 96:9184–9189.

Cardenas, M. E., M. Lorenz, C. Hemenway, and J. Heitman. 1994. Yeast as model T cells. *Perspect. Drug Disc. Design* 2:103–126.

Carter, R. J., I. Dubchak, and S. R Holbrook. 2001. A computational approach to identify genes for functional RNAs in genomic sequences. *Nucl. Acids Res.* 29(19):3928–3938.

Casjens, S. 1998. The diverse and dynamic structure of bacterial genomes. *Annu. Rev. Genet.* 32:339–377.

Casjens, S., M. Delange, H. L. Ley, and P. Rosa, et al. 1995. Linear chromosomes of Lyme disease agent spirochetes: genetic diversity and conservation of gene order. *J. Bacteriol.* 177:2769–2780.

Casjens, S., N. Palmer, R. van Vugt, and W. M. Huang, et al. 2000. A bacterial genome in flux: the twelve linear and nine circular extrachromosomal DNAs in an infectious isolate of the Lyme disease spirochete *Borrelia burgdorferi. Mol. Microbiol.* 35:490–516.

CASP3, CASP3 Special Issue. 1999. *Prot.: Struct., Func. Gen.*, Suppl. 3:1–237.

CASP4, CASP4 Special Issue. 2001. *Prot.: Struct., Func. Gen.,* Suppl. 4:1–399.

Causton, H., J. Quackenbush, and A. Brazma. 2003. *Microarray Gene Expression Data Analysis: A Beginner's Guide.* Blackwell Publishing, Malden, Mass.

Cavalier-Smith, T. 1992. Bacteria and eukaryotes. *Nature* 356:570.

Cavalier-Smith, T. 1993. Kingdom protozoa and its 18 phyla. *Microbiol. Rev.* 57:953–994.

Cavanaugh, C. M., S. L. Gardiner, M. L. Jones, and H. W. Jannasch, et al. 1981. Prokaryotic cells in the hydrothermal vent tube worm *Riftia pachyptila*: Possible chemoautotrophic symbionts. *Science* 213:340–342.

Cech, N. B. and C. G. Enki. 2001. Practical implications of some recent studies in electrospray ionization fundamentals. *Mass Spec. Rev.* 20:362–387.

Celia, H., E. Wilson-Kubalek, R. A. Milligan, and L. Teyton. 1999. Structure and function of a membrane-bound murine MHC class I molecule. *Proc. Natl. Acad. Sci. USA* 96:5634–5639.

Censini, S., C. Lange, Z. Xiang, and J. E. Crabtree, et al. 1996. cag, a pathogenicity island of *Helicobacter pylori*, encodes type I-specific and disease-associated virulence factors. *Proc. Natl. Acad. Sci. USA* 93:14648–14653.

Cervin, M. A., D. A. Simpson, A. L. Smith, and S. Lory. 1994. Differences in eucaryotic cell binding of *Pseudomonas*. *Microb. Pathog.* 17:291–299.

Cestra, G., L. Castagnoli, L. Dente, and O. Minenkova, et al. 1999. The SH3 domains of endophilin and amphiphysin bind to the proline-rich region of synaptojananin 1 at distinct sites that display an unconventional binding specificity. *J. Biol. Chem.* 274:32001–32007.

Chee, M., R. Yang, E. Hubbell, and A. Berno, et al. 1996. Accessing genetic information with high-density DNA arrays. *Science* 274:610–614.

Cheetham, B. F. and M. E. Katz. 1995. A role for bacteriophages in the evolution and transfer of bacterial virulence determinants. *Mol. Microbiol.* 18:201–208.

Chen, H. I., A. Einbond, S.-J. Kwak, and H. Linn, et al. 1997a. Characterization of the WW domain of human Yes-associated protein and its polyproline-containing ligands. *J. Biol. Chem.* 272:17070–17077.

Chen, T., H. L. He., and G. M. Church. 1999. Modeling gene expression with differential equations. *Proc. Pacific Symp. Biocomput.* 29–40.

Chen, Y., E. R. Dougherty and M. L. Bittner. 1997b. Ratio-based decisions and the quantitative analysis of cDNA microarray images. *J. Biomed. Optics* 2:364–374.

Chernushevich, I. V., A. V. Loboda, and B. A. Thomson. 2001. An introduction to quadrupole-time-of-flight mass spectrometry. *J. Mass Spec.* 36:849–865.

Chervitz, S. A., L. Aravind, G. Sherlock, and C. A. Ball, et al. 1998. Comparison of the complete protein sets of worm and yeast: orthology and divergence. *Science* 282:2022–2028.

Chhabra, R. S., J. E. Huff, B. S. Schwetz, and J. Selkirk. 1990. An overview of prechronic and chronic toxicity/carcinogenicity experimental study designs and criteria used by the National Toxicology Program. *Environ. Health Perspect.* 86:313–321.

Chi, E., T. Mehl, D. Nunn, and S. Lory. 1991. Interaction of *Pseudomonas aeruginosa* with A549 pneumocyte cells. *Infect. Immun.* 59:822–828.

Chiang, S. L. and J. J. Mekalanos. 1998. Use of signature-tagged transposon mutagenesis to identify *Vibrio cholerae* genes critical for colonization. *Mol. Microbiol.* 27:797–805.

Chiang, S. L. and J. J. Mekalanos. 1999. rfb mutations in *Vibrio cholerae* do not affect surface production of toxin-coregulated pili but still inhibit intestinal colonization. *Infect. Immun.* 67:976–980.

Chiang, S. L., J. J. Mekalanos, and D. W. Holden. 1999. In vivo genetic analysis of bacterial virulence. *Annu. Rev. Microbiol.* 53:129–154.

Chizhikov, V., A. Rasooly, K. Chumakov, and D. D. Levy. 2001. Microarray analysis of microbial virulence factors. *Appl. Environ. Microbiol.* 67:3258–3263.

Cho, J. C. and J. M. Tiedje. 2000. Biogeography and degree of endemicity of fluorescent *Pseudomonas* strains in soil. *Appl. Environ. Microbiol.* 66:5448–5456.

Cho, J. C. and J. M. Tiedje. 2001. Bacterial species determination from DNA-DNA hybridization by using genome fragments and DNA microarrays. *Appl. Environ. Microbiol.* 67:3677–3682.

Chopra, I. 1998. Research and development of antibacterial agents. *Curr. Opin. Microbiol.* 1:495–501.

Chow, W. Y. and D. E. Berg. 1988. Tn5tac1, a derivative of transposon Tn5 that generates conditional mutations. *Proc. Natl. Acad. Sci. USA* 85:6468–6472.

Chrisey, L. A., G. U. Lee, and C. E. O'Ferrall. 1996. Covalent attachment of synthetic DNA to self-assembled monolayer films. *Nucl. Acids Res.* 24:3031–3039.

Chu, S., J. DeRisi, M. Eisen, and J. Mulholland, et al. 1998. The transcriptional program of sporulation in budding yeast. *Science* 282:699–705.

Claus, H., M. Frosch, and U. Vogel. 1998. Identification of a hotspot for transformation of *Neisseria meningitidis* by shuttle mutagenesis using signature-tagged transposons. *Mol. Gen. Genet.* 259:363–371.

Claverie, J. and L. Bougueleret. 1986. Heuristic informational analysis of sequences. *Nucl. Acids Res.* 10:179–196.

Claverie, J., I. Sauvaget, and L. Bougueleret. 1990. K-tuple frequency analysis: from intron/exon discrimination to T-cell epitope mapping. *Methods Enzymol.* 183:237–252.

Claverie, M. 1999. Computational methods for the identification of differential and coordinated gene expression. *Hum. Molec. Gen.* 8(10):1821–1832.

Clayton, R. A., O. White, K. A. Ketchum, and J. C. Venter. 1997. The first genome from the third domain of life. *Nature* 387:459–462.

Clewell, D. R. 1993. Sex pheromones and the plasmid encoded mating response in *Enterococcus faecalis*. In D. B. Clewell (ed.), *Bacterial Conjugation*, pp. 349–367. Plenum Press, New York.

Cockett, M., N. Dracopoli, and E. Sigal. 2000. Applied genomics: integration of the technology within pharmaceutical research and development. *Curr. Opin. Biotechnol.* 11:602–609.

Coggins, L. W. and M. O'Prey. 1989. DNA tertiary structures formed in vitro by misaligned hybridization of multiple tandem repeat sequences. *Nucl. Acids Res.* 17:7417–7426.

Cohen, B. A., P. Colas, and R. Brent. 1998. An artificial cell-cycle inhibitor isolated from a combinatorial library. *Proc. Natl. Acad. Sci. USA* 95:14272–14277.

Cohen, C. and D. A. Parry. 1994. Alpha-helical coiled coils: more facts and better predictions. *Science* 264:1068.

Cohen, M. L. 2000. Changing patterns of infectious disease. *Nature* 406:762–767.

Coissac, E., E. Maillier, and P. Netter. 1997. A comparative study of duplications in bacteria and eukaryotes: the importance of telomeres. *Mol. Biol. Evol.* 14:1062–1074.

Colas, P., B. Cohen, T. Jessen, and I. Grishina, et al. 1996. Genetic selection of peptide aptamers that recognize and inhibit cyclin-dependent kinase 2. *Nature* 380:548–550.

Cole, J. R., B. Chai, T. L. Marsh, and R. J. Farris, et al. 2003. The Ribosomal Database Project (RDP-II): previewing a new autoaligner that allows regular updates and the new prokaryotic taxonomy. *Nucleic Acids Res.* 31:442–443.

Cole, S. T. 1999. Learning from the genome sequence of *Mycobacterium tuberculosis* H37Rv. *FEBS Lett.* 452:7–10.

Cole, S. T., K. Eiglmeier, J. Parkhill, and K. D. James, et al. 2001. Massive gene decay in the leprosy bacillus. *Nature* 409:1007–1011.

Cole, S. T., R. Brosch, J. Parkhill, T. Garnier, et al. 1998. Deciphering the biology of *Mycobacterium tuberculosis* from the complete genome sequence. *Nature* 393:537–544.

Colegio, O. R., T. J. Griffin, N. D. Grindley, and J. E. Galan. 2001. In vitro transposition system for efficient generation of random mutants of *Campylobacter jejuni*. *J. Bacteriol.* 183:2384–2388.

Coleman, J., P. J. Green, and M. Inouye. 1984. The use of RNAs complementary to specific mRNAs to regulate the expression of individual bacterial genes. *Cell* 37:429–436.

Collins, D. M. 1996. In search of tuberculosis virulence genes. *Trends Microbiol.* 4:426–430.

Collins, D. M., R. P. Kawakami, G. W. de Lisle, and L. Pascopella, et al. 1995. Mutation of the principal sigma factor causes loss of virulence in a strain of the *Mycobacterium tuberculosis* complex. *Proc. Natl. Acad. Sci. USA* 92:8036–8040.

Colovos, C., D. Cascio, and T. O Yeates. 1998. The 1.8A crystal structure of the ycaC gene product from *E. coli* reveals an octameric hydrolase of unknown specificity. *Structure* 6:1329–1337.

Conlin, C. A., N. J. Trun, T. J. Silhavy, and C. G. Miller. 1992. *Escherichia coli* prlC encodes an endopeptidase and is homologous to the *Salmonella typhimurium* opdA gene. *J. Bacteriol.* 174:5881–5887.

Coombes, B. K. and J. B. Mahony. 2001. cDNA array analysis of altered gene expression in human endothelial cells in response to *Chlamydia pneumoniae* infection. *Infect. Immun.* 69:1420–1427.

Coppee, J. Y., S. Auger, E. Turlin, and A. Sekowska, et al. 2001. Sulfur-limitation-regulated proteins in *Bacillus subtilis*: a two-dimensional gel electrophoresis study. *Microbiology* 147:1631–1640.

Corpet, F., F. Servant, J. Gouzy, and D. Kahn. 2000. ProDom and ProDom-CG: tools for protein domain analysis and whole genome comparisons. *Nucl. Acids Res.* 28:267–269.

Corsaro, D., D. Venditti, M. Padula, and M. Valassina. 1999. Intracellular life. *Crit. Rev. Microbiol.* 25:39–79.

Cossart, P. and M. Lecuit. 1998. Interactions of *Listeria* monocytogenes with mammalian cells during entry and actin-based movement: bacterial factors, cellular ligands and signaling. *EMBO J.* 17:3797–3806.

Costanzo, M. C., J. D. Hogan, M. E. Cusick, and B. P. Davis, et al. 2000. The yeast proteome database (YPD) and *Caenorhabditis elegans* proteome database (WormPD): comprehensive resources for the organization and comparison of model organism protein information. *Nucl. Acids Res.* 28:73–76.

Costanzo, M. C., M. E. Crawford, J. E. Hirschman, and J. E. Kranz, et al. 2001. YPD, PombePD and WormPD: model organism volumes of the BioKnowledge library, an integrated resource for protein information. *Nucl. Acids Res.* 29:75–79.

Costerton, J. W., P. S. Stewart, and E. P. Greenberg. 1999. Bacterial biofilms: a common cause of persistent infections. *Science* 284:1318–1322.

Costerton, J. W., Z. Lewandowski, D. E. Caldwell, and D. R. Korber, et al. 1995. Microbial biofilms. *Annu. Rev. Microbiol.* 49:711–745.

Cote, J., J. Quinn, J. L. Workman, and C. L. Peterson. 1994. Stimulation of GAL4 derivative binding to nucleosomal DNA by the yeast SWI/SNF complex. *Science* 265:53–60.

Cotter, P. A. and V. J. DiRita. 2000. Bacterial virulence gene regulation: an evolutionary perspective. *Annu. Rev. Microbiol.* 54:519–565.

Cottrell, M. T., J. A. Moore, and D. L. Kirchman. 1999. Chitinases from uncultured marine microorganisms. *Appl. Environ. Microbiol.* 65:2553–2557.

Cover, T. L. and M. J. Blaser. 1996. *Helicobacter pylori* infection, a paradigm for chronic mucosal inflammation: pathogenesis and implications for eradication and prevention. *Adv. Intern. Med.* 41:85–117.

Cowing, D. W., J. C. A. Bardwell, E. A. Craig, and C. Woolford, et al. 1985. Consensus sequence for *Escherichia coli* heat shock gene promoters. *Proc. Natl. Acad. Sci. USA* 82:2679–2683.

Craig, N. L. 1996. Transposition. In F. C. Neidhardt, R. Curtiss III, J. L. Ingraham, and E. C. C. Lin, et al. (eds.), *Escherichia coli* and *Salmonella: Molecular and Cellular Biology*, 2nd ed., pp. 2339–2362. ASM Press, Washington, D.C.

Craig, N. L. 1997. Target site selection in transposition. *Annu. Rev. Biochem.* 66:437–474.

Craven, M., D. Page, J. Shavlik, and J. Bockhorst, et al. 2000. A probabilistic learning approach to whole-genome operon prediction. *Proc. Int. Conf. Intell. Syst. Mol. Biol.* 8:116–127.

Cristianini, N. and J. Shawe-Taylor. 2000. *An Introduction to Support Vector Machines and Other Kernel-based Learning Methods*. Cambridge University Press, Cambridge, U.K.

Crooke, S. T. 2000. Potential roles of antisense technology in cancer chemotherapy. *Oncogene* 19:6651–6659.

Cummings, C. A. and D. A. Relman. 2000. Using DNA microarrays to study host-microbe interactions. *Emerging Infect. Dis.* 6:513–525.

Curtis, T. P., W. T. Sloan, and J. W. Scannell. 2002. Estimating prokaryotic diversity and its limits. *Proc. Natl. Acad. Sci. USA.* 99:10494–10499.

Daly, M. J. and K. W. Minton. 1995. Interchromosomal recombination in the extremely radioresistant bacterium *Deinococcus radiodurans. J. Bacteriol.* 177:5495–5505.

Daly, M. J., O. Ling, and K. W. Minton. 1994. Interplasmidic recombination following irradiation of the radioresistant bacterium *Deinococcus radiodurans. J. Bacteriol.* 176:7506–7515.

Dandekar, T. and M.W. Hentze. 1995. Finding the hairpin in the haystack: searching for RNA motifs. *Trends Genet.* 11:45–50.

Darji, A., K. Niebuhr, M. Hense, and J. Wehland, et al. 1996. Neutralizing monoclonal antibodies against listeriolysin: mapping of epitopes involved in pore formation. *Infect. Immun.* 64:2356–2358.

Darnell, J. E., Jr. 1978. Implications of RNA-RNA splicing in evolution of eukaryotic cells. *Science* 202:1257–1260.

Dass, C. 2001. *Principles and Practice of Biological Mass Spectrometry.* Wiley-Interscience, New York.

Dassy, B. and J. M. Fournier. 1996. Respiratory activity is essential for post-exponential-phase production of type 5 capsular polysaccharide by *Staphylococcus aureus. Infect. Immun.* 64:2408–2414.

Datsenko, K. A. and B. L. Wanner. 2000. One-step inactivation of chromosomal genes in *Escherichia coli* K-12 using PCR products. *Proc. Natl. Acad. Sci. USA* 97:6640–6645.

d'Aubenton Carafa, Y., E. Brody, and C. Thermes. 1990. Prediction of rho-independent *Escherichia coli* transcription terminators. A statistical analysis of their RNA stem-loop structures. *J. Mol. Biol.* 216:835–858.

Davies, D. G., M. R. Parsek, J. P. Pearson, and B. H. Iglewski, et al. 1998. The involvement of cell-to-cell signals in the development of a bacterial biofilm. *Science* 280:295–298.

Davies, H., L. Lomas, and B. Austen. 1999. Profiling of amyloid b peptide variants using SELDI ProteinChip arrays. *Biotechniques* 27:1258–1261.

Davies, J. 1994. Inactivation of antibiotics and the dissemination of resistance genes. *Science* 264:375–382.

Davies, J. E. 1997. Origins, acquisition and dissemination of antibiotic resistance determinants. *Ciba Found. Symp.* 207:15–27.

Davis, M. T. and T. D. Lee. 1998. Rapid protein identification using a microscale electrospray LC/ MS system on an ion trap mass spectrometer. *J. Am. Soc. Mass Spectrom.* 9:194–201.

Davis, M. T., C. S. Spahr, M. D. McGinley, and J. H. Robinson, et al. 2001. Towards defining the urinary proteome using liquid chromatography-tandem mass spectrometry. II. Limitations of complex mixture analyses. *Proteomics* 1:108–117.

De Backer, M. D., B. Nelissen, M. Logghe, and J. Viaene, et al. 2001. An antisense-based functional genomics approach for identification of genes critical for growth of *Candida albicans. Nat. Biotechnol.* 19:235–241.

Delong, E. F., G. Wickham, and N. R. Pace. 1989. Phylogenetic stains: Ribosomal RNA-based probes for identification of single microbial cells. *Science* 243:1360–1363.

Delong, E. F., L. L. King, R. Massana, and H. Cittone, et al. 1998. Dibiphytanyl ether lipids in nonthermophilic crenarchaeotes. *Appl. Environ. Microbiol.* 64:1133–1138.

DeLong, E. F., L. T. Taylor, T. L. Marsh, and C. M. Preston. 1999. Visualization and enumeration of marine planktonic archaea and bacteria by using polyribonucleotide probes and fluorescent in situ hybridization. *Appl. Environ. Microbiol.* 65:5554–5563.

DeLong, E. F. and N. R. Pace. 2001. Environmental diversity of bacteria and archaea. *Syst. Biol.* 50:470–478.

D'Hondt, S., S. Rutherford, and A. J. Spivack. 2002. Metabolic activity of subsurface life in deep-sea sediments. *Science.* 295:2067–2070.

de Kievit, T. R. and B. H. Iglewski. 2000. Bacterial quorum sensing in pathogenic relationships. *Infect. Immun.* 68:4839–4849.

De Las Penas, A., L. Connolly, and C. A. Gross. 1997. The sigmaE-mediated response to extracytoplasmic stress in *Escherichia coli* is transduced by RseA and RseB, two negative regulators of sigmaE. *Mol. Microbiol.* 24:373–385.

de Saizieu, A., C. Gardes, and N. Flint, et al. 2000. Microarray-based identification of a novel *Streptococcus pneumoniae* regulon controlled by an autoinduced peptide. *J. Bacteriol.* 182:4696–4703.

de Saizieu, A., U. Certa, J. Warrington, and C. Gray, et al. 1998. Bacterial transcript imaging by hybridization of total RNA to oligonucleotide arrays. *Nat. Biotechnol.* 16:45–48.

Dean, D. 1981. A plasmid cloning vector for the direct selection of strains carrying recombinant plasmids. *Gene* 15:99–102.

Dean, N. M. 2001. Functional genomics and target validation approaches using antisense oligonucleotide technology. *Curr. Opin. Biotechnol.* 12:622–625.

Dean, P. M., E. D. Zanders, and D. S. Bailey. 2001. Industrial-scale, genomics-based drug design and discovery. *Trends Biotechnol.* 19:288–292.

Debouck, C. and P. N. Goodfellow. 1999. DNA microarrays in drug discovery and development. *Nat. Genet.* 21:48–50.

Del Rosario, M., J. C. Stephans, J. Zakel, and J. Escobedo, et al. 1996. Positive selection system to screen for inhibitors of human immunodeficiency virus-1 transcription. *Nat. Biotechnol.* 14:1592–1596.

del Solar, G., P. Acebo, and M. Espinosa. 1995. Replication control of plasmid pLS1: efficient regulation of plasmid copy number is exerted by the combined action of two plasmid components, CopG and RNA II. *Mol. Microbiol.* 18:913–924.

Delcher, A. L., S. Kasif, R. D. Fleischmann, and J. Peterson, et al. 1999. Alignment of whole genomes. *Nucl. Acids Res.* 27:2369–2376.

Delihas, N. and S. Forst. 2001. MicF: an antisense RNA gene involved in response of *Escherichia coli* to global stress factors. *J. Mol. Biol.* 313:1–12.

Delneri, D., F. L. Brancia, and S. G. Oliver. 2001. Towards a truly integrative biology through the functional genomics of yeast. *Curr. Opin. Biotechnol.* 12:87–91.

Demirev, P. A., J. Ramirez, and C. Fenselau. 2001. Tandem mass spectrometry of intact proteins for characterization of biomarkers from *Bacillus cereus* T spores. *Anal. Chem.* 73:5725–5731.

Dempster, A. P., N. M. Laird, and D. B. Rubin. 1977. Maximum likelihood from incomplete data via EM algorithm. *J. Roy. Statist. Soc. Series B—Methodological* 39(1):1–38.

DeRisi, J. L., V. R. Iyer, and P. O. Brown. 1997. Exploring the metabolic and genetic control of gene expression on a genomic scale. *Science* 278:680–686.

Derre, I., G. Rapoport, and T. Msadek. 1999. CtsR, a novel regulator of stress and heat shock response, controls clp and molecular chaperone gene expression in gram-positive bacteria. *Mol. Microbiol.* 31:117–131.

Desai, R. P. and E. T. Papoutsakis. 1999. Antisense RNA strategies for metabolic engineering of *Clostridium acetobutylicum. Appl. Environ. Microbiol.* 65:936–945.

De Sanctis, V., C. Bertozzi, G. Costanzo, and E. Di Mauro, et al. 2001. Cell cycle arrest determines the intensity of the global transcriptional response of *Saccharomyces cerevisiae* to ionizing radiation. *Rad. Res.* 156:379–387.

Detweiler, C. S., D. B. Cunanan, and S. Falkow. 2001. Host microarray analysis reveals a role for the *Salmonella* response regulator phoP in human macrophage cell death. *Proc. Natl. Acad. Sci. USA* 98:5850–5855.

DeVelasco, E. A., A. F. M. Verheul, J. Verhoef, and H. Snippe. 1995. *Streptococcus pneumoniae*: virulence factors, pathogenesis, and vaccines. *Microbiol. Rev.* 59:591–603.

DeVito, J. A., J. A. Mills, and V. G. Liu, et al. 2002. An array of target-specific screening strains for antibacterial discovery. *Nat. Biotechnol.* 20:478–483.

Devreese, B., F. Vanrobaeys, and J. Van Beeumen. 2001. Automated nanoflow liquid chromatography/tandem mass spectrometric identification of proteins from *Shewanella putrefaciens* separated by two-dimensional polyacrylamide gel electrophoresis. *Rapid Commun. Mass Spectrom.* 15:50–56.

DeVries, G. E., C. K. Raymond, and R. A. Ludwig. 1984. Extension of bacteriophage λ host range: selection, cloning, and characterization of a constitutive λ receptor gene. *Proc. Natl. Acad. Sci. USA* 81:6080–6084.

Dharmasiri, K. and D. L. Smith. 1996. Mass spectrometric determination of isotopic exchange rates of amide hydrogens located on the surfaces of proteins. *Anal. Chem.* 68:2340–2344.

Di Gennaro, J. A., N. Siew, B. T. Hoffman, and L. Zhang, et al. 2001. Enhanced functional annotation of protein sequences via the use of structural descriptors. *J. Struct. Biol.* 134(2–3):232–245.

Diehl, F., S. Grahlmann, M. Beier, and J. D. Hoheisel. 2001. Manufacturing DNA microarrays of high spot homogeneity and reduced background signal. *Nucl. Acids Res.* 29:E38.

Diehn, M. and D. A. Relman. 2001. Comparing functional genomic datasets: lessons from DNA microarray analyses of host-pathogen interactions. *Curr. Opin. Microbiol.* 4:95–101.

Diep, D. B., L. S. Havarstein, and I. F. Nes. 1996. Characterization of the locus responsible for the bacteriocin production in *Lactobacillus plantarum* C11. *J. Bacteriol.* 178:4472–4483.

DiRita, V. J. 1992. Co-ordinate expression of virulence genes by ToxR in *Vibrio cholerae. Mol. Microbiol.* 6:451–458.

DiRita, V. J., C. Parsot, G. Jander, and J. J. Mekalanos. 1991. Regulatory cascade controls virulence in *Vibrio cholerae. Proc. Natl. Acad. Sci. USA* 88:5403–5407.

DiRuggiero, J., J. D. Dunn, D. L. Maeder, and R. Holley-Shanks, et al. 2000. Evidence for recent lateral gene transfer among hyperthermophilic Archaea. *Mol. Microbiol.* 38:684–693.

Dolnik, V., S. Liu, and S. Jovanovich. 2000. Capillary electrophoresis on a microchip. *Electrophoresis* 21:41–54.

Dong, S., E. Wang, L. Hsie, and Y. Cao, et al. 2001a. Flexible use of high-density oligonucleotide arrays for single-nucleotide polymorphism discovery and validation. *Genome Res.* 11:1418–1424.

Dong, Y., J. D. Glasner, F. R. Blattner, and E. W. Triplett. 2001b. Genomic interspecies microarray hybridization: rapid discovery of three thousand genes in the maize endophyte, *Klebsiella pneumoniae* 342, by microarray hybridization with *Escherichia coli* K-12 open reading frames. *Appl. Environ. Microbiol.* 67:1911–1921.

Doolittle, R. F. 1994. Convergent evolution: the need to be explicit. *Trends Biochem. Sci.* 19:15–18.

Doolittle, R. F. 2002. Biodiversity: microbial genomes multiply. *Nature* 416:697–700.

Doolittle, R. F., D. F. Feng, S. Tsang, and G. Cho, et al. 1996. Determining divergence times of the major kingdoms of living organisms with a protein clock. *Science* 271:470–477.

Doolittle, W. F. 1998. A paradigm gets shifty. *Nature* 392:15–16.

Doolittle, W. F. 1999a. Phylogenetic classification and the universal tree. *Science* 284:2124–2129.

Doolittle, W. F. 1999b. Lateral genomics. *Trends Cell Biol.* 9:M5–M8.

Doolittle, W. F. 2000. Uprooting the tree of life. *Sci. Amer.* 282:90–95.

Drmanac, R., I. Labat, I. Brukner, and R. Crkvenjakov. 1989. Sequencing of megabase-plus DNA by hybridization: theory of the method. *Genomics* 4:114–128.

Drobyshev, A., N. Mologina, V. Shik, and D. Pobedimskaya, et al. 1997. Sequence analysis by hybridization with oligonucleotide microchip: identification of beta-thalassemia mutations. *Gene* 188:45–52.

Drobyshev, A. L., A. S. Zasedatelev, G. M. Yershov, and A. D. Mirzabekov. 1999. Massive parallel analysis of DNA-Hoechst 33258 binding specificity with a generic oligodeoxyribonucleotide microchip. *Nucl. Acids Res.* 27:4100–4105.

Duan, Y. and P. A. Kollman. 1998. Pathways to a protein-folding intermediate observed in a 1-microsecond simulation in aqueous solution. *Science* 282:740–744.

Ducklow, H. W. and C. A. Carlson. 1992. Oceanic bacterial production. *Adv. Microb. Ecol.* 12:113–181.

Duda, R. O. and P. E. Hart. 1973. *Pattern Classification and Scene Analysis*. Wiley-Interscience, New York.

Dudoit, S., Y. H Yang, T. P. Speed, and M. J. Callow, Statistical methods for identifying differentially expressed genes in replicated cDNA microarray experiments. Statistica Sinica. *UC Berkeley Statistics tech. Report*, 2001.

Duggan, D. J., M. Bittner, Y. Chen, P. Mattzer, and J. M. Trent. 1999. Expression profiling using cDNA microarrays. *Nat. Genet.* 21(Suppl.):10–14.

Dunker, A. K. and Z. Obradovic. 2001. The protein trinity—linking function and disorder. *Nat. Biotechnol.* 19(9):805–806.

Dunker, A. K., C. J. Brown, J. D. Lawson, and L. M. Iakoucheva, et al. 2002. Intrinsic disorder and protein function. *Biochemistry* 41(21):6573–6582.

Dunman, P. M., E. Murphy, S. Haney, and D. Palacios, et al. 2001. Transcription profiling-based identification of *Staphylococcus aureus* genes regulated by the agr and/or sarA loci. *J. Bacteriol.* 183:7341–7353.

Dykhuizen, D. E. 1998. Santa Rosalia revisited: why are there so many species of bacteria? *Antonie Van Leeuwenhoek* 73:25–33.

Dzau, V. J. and S. Glueck. 2001. Physiological genomics: who we are and where we're going. *Physiol. Genomics* 7:65–67.

Easteal, S., C. Collet, and D. Betty. 1995. *The Mammalian Molecular Clock*. Landes, Austin, TX.

Eckerskorn, C., K. Strupat, D. Schleuder, and D. Hochstrasser, et al. 1997. Analysis of proteins by direct-scanning infrared-MALDI mass spectrometry after 2D-PAGE separation and electroblotting. *Anal. Chem.* 69:2888–2892.

Eckmann, L., J. R. Smith, M. P. Housley, and M. B. Dwinell, et al. 2000. Analysis of high density cDNA arrays of altered gene expression in human intestinal epithelial cells in response to infection with the invasive enteric bacteria *Salmonella*. *J. Biol. Chem.* 275:14084–14094.

Edgell, D. R. and W. F. Doolittle. 1997. Archaea and the origin(s) of DNA replication proteins. *Cell* 89:995–998.

Edwards, J. S. and B. O. Palsson. 2000a. Metabolic flux balance analysis and the *in silico* analysis of *Escherichia coli* K-12 gene deletions. *BMC Bioinform.* 1:1.

Edwards, J. S. and B. O. Palsson. 2000b. The *Escherichia coli* MG1655 *in silico* metabolic genotype: its definition, characteristics, and capabilities. *Proc. Natl. Acad. Sci. USA* 97:5528–5533.

Edwards, J. S., R. U. Ibarra, and B. O. Palsson. 2001. *In silico* predictions of *Escherichia coli* metabolic capabilities are consistent with experimental data. *Nature Biotechnol.* 19:125–130.

Efron, B., E. Halloran, and S. Holmes. 1996. Bootstrap confidence levels for phylogenetic trees. *Proc. Natl. Acad. Sci. USA* 93:7085–7090.

Eggers, M. D., M. E. Hogan, and R. K. Reich et al. 1994. A microchip for quantitative detection of molecular utilizing luminescent and radioisotope reporter groups. *Biotechniques* 14:516–525.

Eide, D. J. 2001. Functional genomics and metal metabolism. *Genome Biol.* 2: reviews 1028.1–1028.3.

Eisen, J. A. 2000. Horizontal gene transfer among microbial genomes: new insights from complete genome analysis. *Curr. Opin. Genet. Dev.* 10:606–611.

Eisen, M. and P. Brown 1999. DNA microarrays for analysis of gene expression. *Method. Enzymol.* 303:179–205.

Eisen, M. B., P. T. Spellman, P. O. Brown, and D. Botstein. 1998. Cluster analysis and display of genome-wide expression patterns. *Proc. Natl. Acad. Sci. USA* 95:14863–14868.

Eisenberg, D., E. M. Marcotte, I. Xenarios, and T. O. Yeates. 2000. Protein function in the post-genomic era. *Nature* 405:823–826.

Elder, J. F., Jr. and B. J. Turner. 1995. Concerted evolution of repetitive DNA sequences in eukaryotes. *Q. Rev. Biol.* 70:297–320.

Ellis, H. M., D. Yu, T. DiTizio, and D. L. Court. 2002. High efficiency mutagenesis, repair, and engineering of chromosomal DNA using single-stranded oligonucleotides. *Proc. Natl. Acad. Sci. USA* 98:6742–6746.

Ellison, M. J., R. J. Kelleher, and A. Rich. 1985. Thermal regulation of beta-galactosidase synthesis using anti-sense RNA directed against the coding portion of the mRNA. *J. Biol. Chem.* 260:9085–9087.

Embley, T. M. and B. J. Finlay. 1994. The use of small subunit rRNA sequences to unravel the relationships between anaerobic ciliates and their methanogen endosymbionts. *Microbiology* 140:225–235.

Eng, J. K., A. L. McCormack, and J. R.Yates, 3rd. 1994. An approach to correlate tandem mass spectral data of peptides with amino acid sequences in a protein database. *J. Am. Mass Spectrom.* 5:976–989.

Engdahl, H. M., T. A. Hjah, and E. G. Wagner. 1997. A two unit antisense RNA cassette test system for silencing of target genes. *Nucleic Acids Res.* 25:3218–3227.

Englert, D. 2000. Production of microarrays on porous substrates using noncontact piezoelectric dispensing. In M. Schena (ed.), *Microarray Biochip Technology*, pp. 231–246. Eaton Publishing, Natick, Mass.

Engst, S. and S. M. Miller. 1998. Alternative routes for entry of HgX2 into theactive site mercuric ion reductase depend on the nature of the X ligands. *Biochemistry* 37:11496–11507.

Ennifar, E., A. Nikulin, S. Tishchenko, and A. Serganov, et al. 2000. The crystal structure of UUCG tetraloop. *J. Mol. Biol.* 304:35–42.

Enright, A. J., I. Iliopoulos, N. C. Kyrpides, and C. A. Ouzounis. 1999. Protein interaction maps for complete genomes based on gene fusion events. *Nature* 402:86–90.

Erickson, J. W. and C. A. Gross. 1989. Identification of the sigma E subunit of *Escherichia coli* RNA polymerase: a second alternate sigma factor involved in high-temperature gene expression. *Genes Dev.* 3:1462–1471.

Erlanson, D. A., A. C. Braisted, and D. R. Raphael, et al. 2000. Site-directed ligand discovery. *Proc. Natl. Acad. Sci. USA* 97:9367–9372.

Ermolaeva, M. D., H. G. Khalak, O. White, and H. O. Smith, et al. 2000. Prediction of transcription terminators in bacterial genomes. *J. Mol. Biol.* 301:27–33.

Ermolaeva, M. D., O. White, and S. L. Salzberg. 2001. Prediction of operons in microbial genomes. *Nucl. Acids Res.* 29:1216–1221.

Evans, S. V. and C. R. MacKenzie. 1999. Characterization of protein-glycolipid recognition at the membrane biolayer. *J. Mol. Recog.* 12.

Evans, W. E. and M. V. Relling. 1999. Pharmacogenomics: translating functional genomics into rational therapeutics. *Science* 286:487–491.

Evers, S., K. Di Padova, M. Meyer, and H. Langen, et al. 2001. Mechanism-related changes in the gene transcription and protein synthesis patterns of *Haemophilus influenzae* after treatment with transcriptional and translational inhibitors. *Proteomics* 4:522–544.

Evertsz, E., P. Starink, R. Gupta, and D. Watson. 2000. Technology and applications of gene expression microarrays. In M. Schena (ed.), *Microarray Biochip Technology*, pp. 149–166. Eaton Publishing, Natick, Mass.

Ewing, B. and P. Green. 1998. Base-calling of automated sequencer traces using phred. II. Error probabilities. *Genome Res.* 8:186–194.

Ewing, B., L. Hillier, M. C. Wendl, and P. Green. 1998. Base-calling of automated sequencer traces using phred. I. Accuracy assessment. *Genome Res.* 8:175–185.

Eymann, C., G. Homuth, C. Scharf, and M. Hecker. 2002. *Bacillus subtilis* functional genomics: global characterization of the stringent response by proteome and transcriptome analysis. *J. Bacteriol.* 184:2500–2520.

Falquet, L., M. Pagni, P. Bucher, and N. Hulo, et al. 2002. The PROSITE database, its status in 2002. *Nucl. Acids Res.* 30:235–238.

Falzano, L., C. Fiorentini, G. Donelli, and E. Michel, et al. 1993. Induction of phagocytic behaviour in human epithelial cells by *Escherichia coli* cytotoxic necrotizing factor type 1. *Mol. Microbiol.* 9:1247–1254.

Fan, J. B., X. Chen, M. K. Halushka, and A. Berno, et al. 2000. Parallel genotyping of human SNPs using generic high-density oligonucleotide tag arrays. *Genome Res.* 10:853–860.

Fang, N. Y., T. C. Greiner, D. D. Weisenburger, and W. C. Chan, et al. 2003. Oligonucleotide microarrays demonstrate the highest frequency of ATM mutations in the mantle cell subtype of lymphoma. *Proc. Natl. Acad. Sci. USA* 100(9):5372–5377.

Farrelly, V., F. A. Rainey, and E. Stackebrandt. 1995. Effect of genome size and rrn gene copy number on PCR amplification of 16S rRNA genes from a mixture of bacterial species. *Appl. Environ. Microbiol.* 61:2798–2801.

Fauci, A. S. 2001. Infectious diseases: considerations for the 21st century. *Clini. Infec. Dis.* 32:675–685.

Favis, R., J. P. Day, N. P. Gerry, and C. Phelan, et al. 2000. Universal DNA array detection of small insertions and deletions in BRCA1 and BRCA2. *Nat. Biotechnol.* 18:561–564.

Fawcett, P., P. Eichenberger, R. Losick, and P. Youngman. 2000. The transcriptional profile of early to middle sporulation in *Bacillus subtilis*. *Proc. Natl. Acad. Sci. USA* 97:8063–8068.

Fayet, O., T. Ziegelhoffer, and C. Georgopoulos. 1989. The groES and groEL heat shock gene products of *Escherichia coli* are essential for bacterial growth at all temperatures. *J. Bacteriol.* 171:1379–1385.

Fazzio, T. G., C. Kooperberg, J. P. Goldmark, and C. Neal, et al. 2001. Widespread collaboration of Isw2 and Sin3-Rpd3 chromatin remodeling complexes in transcriptional repression. *Mol. Cell. Biol.* 21:6450–6460.

Fedorova, N. D. and S. K. Highlander. 1997a. Plasmids for heterologous expression in *Pasteurella haemolytica*. *Gene* 186:207–211.

Fedorova, N. D. and S. K. Highlander. 1997b. Generation of targeted nonpolar gene insertions and operon fusions in *Pasteurella haemolytica* and creation of a strain that produces and secretes inactive leukotoxin. *Infect. Immun.* 65:2593–2598.

Felici, F., A. Luzzago, A. Folgori, and R. Cortese. 1993. Mimicking of discontinuous epitopes by phage-displayed peptides, II. Selection of clones recognized by a protective monoclonal antibody against the *Bordetella pertussis* toxin from phage peptide libraries. *Gene* 128:21–27.

Felske, A., A. Wolterink, R. van Lis, and W. M. de Vos, et al. 2000. Response of a soil bacterial community to grassland succession as monitored by 16S rRNA levels of the predominant ribotypes. *Appl. Environ Microbiol.* 66:3998–4003.

Feng, D. F., G. Cho, and R. F. Doolittle. 1997. Determining divergence times with a protein clock: update and reevaluation. *Proc. Natl. Acad. Sci. USA* 94:13028–13033.

Fenn, J. B., M. Mann, C. K. Meng, and S. F. Wong, et al. 1989. Electrospray ionization for mass spectrometry of large biomolecules. *Science* 246:64–71.

Ferdows, M., P. Serwer, G. Griess, and S. Norris, et al. 1996. Conversion of a linear to a circular plasmid in the relapsing fever agent *Borrelia hermsii. J. Bacteriol.* 178:793–800.

Ferea, T. L., D. Botstein, P. O. Brown, and R. F. Rosenzweig. 1999. Systematic changes in gene expression patterns following adaptive evolution in yeast. *Proc. Natl. Acad. Sci. USA* 96:9721–9726.

Fiala, G. and K. O. Stetter. 1986. *Pyrococcus furiosus* sp. nov. represents a novel genus of marine heterotrophic archaebacteria growing optimally at 100°C. *Arch. Microbiol.* 145:56–61.

Fichant, G. A. and C. J. Burks. 1991. Identifying potential tRNA genes in genomic DNA sequences. *J. Mol Biol.* 220(3):659–671.

Fickett, J. W. and C. S. Tung. 1992. Assessment of protein coding measures. *Nucl. Acids Res.* 20(24):6441–6450.

Fiehn, O. 2002. Metabolomics—the link between genotypes and phenotypes. *Plant Mol. Biol.* 48:155–171.

Field, D., D. Hood, and R. Moxon. 1999. Contribution of genomics to bacterial pathogenesis. *Curr. Opin. Genet. Dev.* 9:700–703.

Fielden, M. R. and T. R. Zacharewski. 2001. Challenges and limitations of gene expression profiling in mechanistic and predictive toxicology. *Toxicolog. Sci.* 60:6–10.

Fields, S. and O. Song. 1989. A novel genetic system to detect protein-protein interactions. *Nature* 340:245–246.

Fields, S., Y. Kohara, and D. J. Lockhart. 1999. Functional genomics. *Proc. Natl. Acad. Sci. USA* 96:8825–8826.

Finkelstein, D. B., J. Gollub, R. Ewing, and F. Sterky, et al. 2001. Iterative linear regression by sector. In *Methods of Microarray Data Analysis. Papers from CAMDA 2000.* S. M. Lin and K. F. Johnson (eds.). Kluwer Academic, pp. 57–68.

Finlay, B. B. and S. Falkow. 1997. Common themes in microbial pathogenicity revisited. *Microbiol. Mol. Biol. Rev.* 61:136–169.

Fisher, R. A. and F. Yates. 1963. *Statistical Tables for Biological, Agricultural and Medical Research,* 6th ed. Hafner Press (Macmillan), New York, NY.

Fitzgerald, J. R., D. E. Sturdevant, S. M. Mackie, and S. R. Gill, et al. 2001. Evolutionary genomics of *Staphylococcus aureus*: insights into the origin of methicillin-resistant strains and the toxic shock syndrome epidemic. *Proc. Natl. Acad. Sci. USA* 98:8821–8826.

Fitz-Gibbon, S. T. and C. H. House. 1999. Whole genome-based phylogenetic analysis of free-living microorganisms. *Nucl. Acids Res.* 27:4218–4222.

Fitz-Gibbon, S. T., H. Ladner, U.-J. Kim, and K. O. Stetter, et al. 2002. Genome sequence of the hyperthermophilic crenarchaeon *Pyrobaculum aerophilum. Proc. Natl. Acad. Sci. USA* 99:984–989.

Fleischmann, R. D., M. D. Adams, O. White, and R. A. Clayton, et al. 1995. Whole-genome random sequencing and assembly of *Haemophilus influenzae* RD. *Science* 269:496–512.

Fleischmann, R. D., D. Alland, J. A. Eisen, and L. Carpenter, et al. 2002. Whole-genome comparison of *Mycobacterium tuberculosis* clinical and laboratory strains. *J. Bacteriol.* 184:5479–5490.

Fodor, S., R. Rava, X. Huang, and A. Pease, et al. 1993. Multiplexed biochemical arrays with biological chips. *Nature* 364:555–556.

Fodor. S. P. A., J. Read, M. Pirrung, and L. Stryer, et al. 1991. Light-directed, spatially addressable parallel chemical synthesis. *Science* 251:767–773.

Folgori, A., R. Tafi, A. Meola, and F. Felici, et al. 1994. A general strategy to identify mimotopes of pathological antigens using only random peptide libraries and human sera. *EMBO J.* 13:2236–2243.

Forbes, A. J., M. T. Mazur, H. M. Patel, and C. T. Walsh, et al. 2001. Toward efficient analysis of >70 kDa proteins with 100% sequence coverage. *Proteomics* 1:927–933.

Forsyth, R. A., R. J. Haselbeck, K. L. Ohlsen, and R. T. Yamamoto, et al. 2002. A genome-wide strategy for the identification of essential genes in *Staphylococcus aureus*. *Mol. Microbiol.* 43:1387–1400.

Forterre, P. 1997. Protein versus rRNA: rooting the universal tree of life? *ASM News* 63:89–92.

Forterre, P. and H. Philippe. 1999. Where is the root of the universal tree of life? *Bioessays* 21:871–879.

Fotin, A., A. Drobyshev, D. Proudnikov, and A. Perov, et al. 1998. Parallel thermodynamic analysis of duplexes on oligodeoxyribonucleotide microchips. *Nucl. Acids Res.* 26:1515–1521.

Foury, F. 1997. Human genetic disease: a cross-talk between man and yeast. *Gene* 195:1–10.

Foury, F. and M. Kucej. 2001. Yeast mitochondrial biogenesis: a model system for humans? *Curr. Opin. Chem. Biol.* 6:106–111.

Fox, B. and C. T. Walsh. 1983. Mercuric reductease: homology to glutathione reductase and lipoamide dehydrogenase: iodoacetamide alkylation and sequence of the active site peptide. *Biochemistry* 22:4082–4088.

Fox, G. E., E. Stackebrandt, R. B. Hespell, and J. Gibson, et al. 1980. The phylogeny of prokaryotes. *Science* 209:457–463.

Frank, R. 1992. Spot synthesis: an easy technique for the positionally addressable, parallel chemical synthesis on a membrane support. *Tetrahedron* 48:9217–9232.

Frank, R. and H. Overwin. 1996. SPOT synthesis. Epitope analysis with arrays of synthetic peptides prepared on cellulose membranes. *Meth. Mol. Biol.* 66:149–169.

Fraser, C. M. and B. Dujon. 2000. The genomics of microbial diversity. Editorial overview. *Curr. Opin. Microbiol.* 3:443–444.

Fraser, C. M., J. Eisen, R. D. Fleischmann, and K. A. Ketchum, et al. 2000. Comparative genomics and understanding of microbial ecology. *Genomics* 6:505–512.

Fraser, C. M., J. A. Eisen, and S. L. Salzberg. 2000. Microbial genome sequencing. *Nature* 406:799–803.

Fraser, C. M., J. D. Gocayne, O. White, and M. D. Adams, et al. 1995. The minimal gene complement of *Mycoplasma genitalium*. *Science* 270:397–403.

Fraser, C. M., S. Casjens, and W. M. Huang, et al. 1997. Genomic sequence of a Lyme disease spirochete, *Borrelia burgdorferi*. *Nature* 390:364–370.

Fraser, C. M., S. J. Norris, G. M. Weinstock, and O. White, et al. 1998. Complete genome sequence of *Treponema pallidum*, the syphilis spirochete. *Science* 281:375–388.

Fredrickson, J. K., H. M. Kostandarithes, S. W. Li, and A. E. Plymale, et al. 2000. Reduction of Fe(III), Cr(VI), U(VI), and Tc(VII) by *Deinococcus radiodurans* R1. *Appl. Environ. Microbiol.* 66:2006–2011.

Friedman, N., M. Linial, I. Nachman, and D. Pe'er. 2000. Using Bayesian networks to analyze expression data. *J. Comput. Biol.* 7:601–620.

Frutos, R., M. Pages, M. Bellis, and G. Roizes, et al. 1989. Pulsed-field gel electrophoresis determination of the genome size of obligate intracellular bacteria belonging to the genera *Chlamydia, Rickettsiella,* and *Porochlamydia. J. Bacteriol.* 171:4511–4513.

Fuhrman, J. A., K. McCallum, and A. A. Davis. 1992. Novel major archaebacterial group from marine plankton. *Nature (Lond.)* 356:148–149.

Fukuda, Y., T. Washio, and M. Tomita. 1999. Comparative study of overlapping genes in the genomes of *Mycoplasma genitalium* and *Mycoplasma pneumoniae. Nucl. Acids Res.* 27(8):1847–1853.

Fulda, S., F. Huang, F. Nilsson, and M. Hagemann, et al. 2000. Proteomics of *Synechocystis* sp. strain PCC 6803. Identification of periplasmic proteins in cells grown at low and high salt concentrations. *Eur. J. Biochem.* 267:5900–5907.

Fung, E. T., V. Thulasiraman, S. R. Weinberger, and E. A. Dalmasso. 2001. Protein biochips for differential profiling. *Curr. Opin. Biotechnol.* 12:65–69.

Fu, R. and G. Voordouw. 1997. Targeted gene-replacement mutagenesis of dcrA, encoding an oxygen sensor of the sulfate-reducing bacterium *Desulfovibrio vulgaris* Hildenborough. *Microbiology* 143:1815–1826.

Furka, A., F. Sebestyen, M. Asgedom, and G. Dibo. 1991. General method for rapid synthesis of multicomponent peptide mixtures. *Int. J. Pept. Prot. Res.* 37:487–493.

Furste, J. P., W. Pansegrau, G. Ziegelin, and M. Kroger, et al. 1989. Conjugative transfer of promiscuous IncP plasmids: interaction of plasmid-encoded products with the transfer origin. *Proc. Natl. Acad. Sci. USA* 86:1771–1775.

Futcher, B. 2000. Microarrays and cell cycle transcription in yeast. *Curr. Opin. Cell Biol.* 12:710–715.

Gaasterland, T. 1998. Structural genomics: bioinformatics in the driver's seat. *Nat. Biotechnol.* 16:625–627.

Galagan, J. E., C. Nusbaum, A. Roy, and M. G. Endrizzi, et al. 2002. The genome of *M. acetivorans* reveals extensive metabolic and physiological diversity. *Genome Res.* 12:532–542.

Galas, D. J. and M. Chandler. 1989. Bacterial insertion sequences. In D. E. Berg and M. M. Howe (eds.), *Mobile DNA*, pp. 109–162. American Society for Microbiology, Washington, D.C.

Galperin, M. Y. and E. V. Koonin. 1999. Searching for drug targets in microbial genomes. *Curr. Opin. Biotechnol.* 10:571–578.

Galperin, M. Y., D. R. Walker, and E. V. Koonin. 1998. Analogous enzymes: independent inventions in enzyme evolution. *Genome Res.* 8:779–790.

Garbe, T. R., N. S. Hibler, and V. Deretic. 1996. Response of *Mycobacterium tuberculosis* to reactive oxygen and nitrogen intermediates. *Mol. Med.* 2:134–142.

Gary, P. A., M. C. Biery, R. J. Bainton, and N. L. Craig. 1996. Multiple DNA processing reactions underlie Tn7 transposition. *J. Mol. Biol.* 257:301–316.

Gasch, A. P., M. Huang, S. Metzner, and D. Botstein, et al. 2001. Genomic expression responses to DNA-damaging agents and the regulatory role of the yeast ATR homolog Mec1p. *Mol. Biol. Cell* 12:2987–3003.

Gasch, A. P., P. T. Spellman, C. M. Kao, and O. Carmel-Harel, et al. 2000. Genomic expression programs in the response of yeast cells to environmental changes. *Mol. Biol. Cell* 11:4241–4257.

Gatlin, C. L., G. R. Kleemann, L. G. Hays, and A. J. Link, et al. 1998. Protein identification at the low femtomole level from silver-stained gels using a new fritless electrospray interface for liquid chromatography-microspray and nanospray mass spectrometry. *Anal. Biochem.* 263:93–101.

Gausepohl, H., C. Boulin M. Kraft, and R. W. Frank. 1992. Automated multiple peptide synthesis. *Pept. Res.* 5:315–320.

Gavin, A. C., M. Bosche, R. Krause, and P. Grandi, et al. 2002. Functional organization of the yeast proteome by systematic analysis of protein complexes. *Nature* 415:141–147.

Gay, P., D. Le Coq, M. Steinmetz, and E. Ferrari, et al. 1983. Cloning structural gene sacB, which codes for exoenzyme levansucrase of *Bacillus subtilis*: expression of the gene in *Escherichia coli*. *J. Bacteriol.* 153:1424–1431.

Gay, P., D. Le Coq, M. Steinmetz, and T. Berkelman, et al. 1985. Positive selection procedure for entrapment of insertion sequence elements in gram-negative bacteria. *J. Bacteriol.* 164:918–921.

Ge, Y., B. G. Lawhorn, M. ElNaggar, and E. Strauss, et al. 2002. Top down characterization of larger proteins (45 kDa) by electron capture dissociation mass spectrometry. *J. Am. Chem. Soc.* 124:672–678.

Gegg, C. V., K. E. Bowers, and C. R. Matthews. 1997. Probing minimal independent folding units in dihydrofolate reductase by molecular dissection. *Prot. Sci.* 6:1885–1892.

Gehring, A. M., J. R. Nodwell, S. M. Beverley, and R. Losick. 2000. Genome-wide insertional mutagenesis in *Streptomyces coelicolor* reveals additional genes involved in morphological differentiation. *Proc. Natl. Acad. Sci. USA* 97:9642–9647.

Genetics Computer Group. 1994. *GCG Program Manual for the Wisconsin Package*, Madison, Wisconsin.

Georgopoulos, C. and W. J. Welch. 1993. Role of the major heat shock proteins as molecular chaperones. *Annu. Rev. Cell. Biol.* 9:601–634.

Gerry, N. P., N. E. Witowski, J. Day, and R. P. Hammer, et al. 1999. Universal DNA microarray method for multiplex detection of low abundance point mutations. *J. Mol. Biol.* 292:251–262.

Geysen, H. M., R. H. Meloen, and S. J. Barteling. 1984. Use of peptide synthesis to probe viral antigens for epitopes to a resolution of a single amino acid. *Proc. Natl. Acad. Sci. USA* 81:3998–4002.

Geysen, H. M., S. J. Rodda, and T. J. Mason. 1986. A priori delineation of a peptide which mimics a discontinuous antigenic determinant. *Mol. Immunol.* 23:709–715.

Gibbs, P. E., W. F. Witke, and A. Dugaiczyk. 1998. The molecular clock runs at different rates among closely related members of a gene family. *J. Mol. Evol.* 46:552–561.

Gil, R., B. Sabater-Munoz, A. Latorre, and F. J. Silva, et al. 2002. Extreme genome reduction in *Buchnera* spp.: toward the minimal genome needed for symbiotic life. *Proc. Natl. Acad. Sci. USA* 99:4454–4458.

Gillespie, J. H. 1991. *The Causes of Molecular Evolution*. Oxford University Press, New York.

Gingeras, T. R., D. Sciaky, and R. E. Gelinas, et al. 1982. Nucleotide sequences from the adenoviruse-2 genome. *J. Biol. Chem.* 257:13475–13491.

Gingeras, T. R., G. Ghandour, E. Wang, and A. Berno, et al. 1998. Simultaneous genotyping and species identification using hybridization pattern recognition analysis of generic *Mycobacterium* DNA arrays. *Genome Res.* 8:435–448.

Glaser, P., L. Frangeul, C. Buchrieser, and C. Rusniok, et al. 2001. Comparative genomics of *Listeria* species. *Science* 294:849–852.

Gmuender, H. 2002. Perspectives and challenges for DNA microarrays in drug discovery and development. *Biotechniques* 32:152–158.

Gmuender, H., K. Kuratli, and W. Keck. 1997. In the presence of subunit A inhibitors DNA gyrase cleaves DNA fragments as short as 20 bp at specific sites. *Nucl. Acids Res.* 25:604–610.

Gmuender, H., K. Kuratli, and K. Di Padova, et al. 2001. Gene expression changes triggered by exposure of *Haemophilus influenzae* to novobiocin or ciprofloxacin: combined transcription and translation analysis. *Genome Res.* 11:28–42.

Godzik, A., J. Skolnick, and A. Kolinski. 1992. A topology fingerprint approach to the inverse folding problem. *J. Mol. Biol.* 227:227–238.

Goesmann, A., M. Haubrock, F. Meyer, and J. Kalinowski, et al. 2002. PathFinder: reconstruction and dynamic visualization of metabolic pathways. *Bioinformatics* 18:124–129.

Goff, S. A., D. Ricke, T. H. Lan, and G. Presting, et al. 2002. A draft sequence of the rice genome (*Oryza sativa L.* ssp. *japonica*). *Science* 296:92–100.

Goffeau, A. 2000. Four years of post-genomic life with 6,000 yeast genes. *FEBS Lett.* 480:37–41.

Goffeau, A., B. G. Barrell, H. Bussey, and R. W. Davis, et al. 1996. Life with 6000 genes. *Science* 274:546–567.

Gogarten, J. P. 1994. Which is the most conserved group of proteins? Homology-orthology, paralogy, xenology, and the fusion of independent lineages. *J. Mol. Evol.* 39:541–543.

Gogarten, J. P., H. Kibak, P. Dittrich, and L. Taiz, et al. 1989. Evolution of the vacuolar H^+-ATPase: implications for the origin of eukaryotes. *Proc. Natl. Acad. Sci. USA* 86:6661–6665.

Goh, S. H., S. K. Byrne, and A. W. Chow. 1992. Molecular typing of *Staphylococcus aureus* on the basis of coagulase gene polymorphisms. *J. Clin. Microbiol.* 30:1642–1645.

Gold, H. S. and R. C. Moellering. 1996. Antimicrobial-drug resistance. *N. Engl. J. Med.* 335:1445–1453.

Gold, T. The deep, hot biosphere. 1992. *Proc. Natl. Acad. Sci. USA* 89:6045–6049.

Goldberg, I. and J. J. Mekalanos. 1986. Cloning of the *Vibrio cholerae* recA gene and construction of a *Vibrio cholerae* recA mutant. *J. Bacteriol.* 165:715–722.

Golub, T. R., D. K. Slonim, P. Tamayo, and C. Huard, et al. 1999. Molecular classification of cancer: class discovery and class prediction by gene expression monitoring. *Science* 286: 531–537.

Gomez, S. M., J. N. Nishio, K. F. Faull, and J. P. Whitelegge. 2002. The chloroplast grana proteome defined by intact mass measurements from liquid chromatography mass spectrometry. *Mol. Cell Proteomics* 1:46–59.

Gonnet, G. H., M. A. Cohen, and S. A. Benner. 1992. Exhaustive matching of the entire protein sequence database. *Science* 257:1609–1610.

Gonzalez, M. D., C. A. Lichtensteiger, and E. R. Vimr. 2001. Adaptation of signature-tagged mutagenesis to *Escherichia coli* K1 and the infant-rat model of invasive disease. *FEMS Microbiol. Lett.* 198:125–128.

Gonzalez, R. C. and P. Wintz. 1987. *Digital Image Processing*, 2nd ed. Addison-Wesley, Reading, Mass.

Goodfellow, M. and A. G. O'Donnell. 1993. Roots of bacterial systematics. In Goodfellow, M., O'Donnell, A.G. (eds.), *Handbook of New Bacterial Systematics*. Academic Press Inc., San Diego. pp. 3–56.

Goodner, B., G. Hinkle, S. Gattung, and N. Miller, et al. 2001. Genome sequence of the plant pathogen and biotechnology agent *Agrobacterium tumefaciens* C58. *Science* 294:2323–2328.

Gordon, S. V., B. Heym, J. Parkhill, and B. Barrell, et al. 1999. New insertion sequences and a novel repeated sequence in the genome of *Mycobacterium tuberculosis* H37Rv. *Microbiology* 145:881–892.

Gorg, A., C. Obermaier, G. Boguth, and A. Harder, et al. 2000. The current state of two-dimensional electrophoresis with immobilized pH gradients. *Electrophoresis* 21:1037–1053.

Goryshin, I. Y. and W. S. Reznikoff. 1998. Tn5 in vitro transposition. *J. Biol. Chem.* 273:7367–7374.

Goryshin, I. Y., J. Jendrisak, L. M. Hoffman, and R. Meis, et al. 2000. Insertional transposon mutagenesis by electroporation of released Tn5 transposition complexes. *Nat. Biotechnol.* 18:97–100.

Gossen, M. and H. Bujard. 1992. Tight control of gene expression in mammalian cells by tetracycline-responsive promoters. *Proc. Natl. Acad. Sci. USA* 89:5547–5551.

Gottesman, S., W. P. Clark, and M. R. Maurizi. 1990. The ATP-dependent Clp protease of *Escherichia coli.* Sequence of clpA and identification of a Clp-specific substrate. *J. Biol. Chem.* 265:7886–7893.

Graham, D. E., R. Overbeek, G. J. Olsen, and C. R. Woese. 2000. An archaeal genomic signature. *Proc. Natl. Acad. Sci. USA* 97:3304–3308.

Grant, P. A., D. Schieltz, M. G. Pray-Grant, and D. J. Steger, et al. 1998. A subset of TAF(II)s are integral components of the SAGA complex required for nucleosome acetylation and transcriptional stimulation. *Cell* 94:45–53.

Gray, M. W. 1992. The endosymbiont hypothesis revisited. *Int. Rev. Cytol.* 141:233–357.

Gray, M. W. 1996. The third form of life. *Nature* 383:299–300.

Gray, M. W., B. F. Lang, R. Cedergren, and G. B. Golding, et al. 1998a. Genome structure and gene content in protist mitochondrial DNAs. *Nucl. Acids Res.* 26:865–878.

Gray, M. W., G. Burger, and B. F. Lang. 1999. Mitochondrial evolution. *Science* 283:1476–1481.

Gray, N. D. and I. M. Head. 2001. Linking genetic identity and function in communities of uncultured bacteria. *Environ. Microbiol.* 3:481–492.

Gray, N. S., L. Wodicka, A. M. Thunnissen, and T. C. Norman, et al. 1998b. Exploiting chemical libraries, structure, and genomics in the search for kinase inhibitors. *Science* 281:533–538.

Gregory, S., C. Soderlund, and A. Coulson. 1997. Contig assembly by fingerprinting. In P. Dear. (ed.) *Genome Mapping: A Practical Approach*, pp. 227–254. Oxford University Press, Oxford, U.K.

Grey, T. and S. Williams. 1971. Microbial productivity in soil. *Symposia of the Society for General Microbiology* 21:255–286.

Gribaldo, S. and P. Cammarano. 1998. The root of the universal tree of life inferred from anciently duplicated genes encoding components of the protein-targeting machinery. *J. Mol. Evol.* 47:508–516.

Griffin, T. J., L. Parsons, A. E. Leschziner, and J. DeVost, et al. 1999. In vitro transposition of Tn552: a tool for DNA sequencing and mutagenesis. *Nucl. Acids Res.* 27:3859–3865.

Griffin, T. J., S. P. Gygi, T. Ideker, and B. Rist, et al. 2002. Complementary profiling of gene expression at the transcriptome and proteome levels in *Saccharomyces cerevisiae*. *Mol. Cell Proteomics* 1:323–333.

Grigorenko, E. V. 2002. *DNA Arrays: Technologies and Experimental Strategies*. CRC Press, Boca Raton, FL.

Grigoriev, I. V. and S.-H. Kim. 1999. Detection of protein fold similarity based on correlation of amino acid properties. *Proc. Natl. Acad. Sci. USA* 96:14318–14323.

Grishin, N. V., Y. I. Wolf, and E. V. Koonin. 2000. From complete genomes to measures of substitution rate variability within and between proteins. *Genome Res.* 10:991–1000.

Groisman, E. A. and H. Ochman. 1994. How to become a pathogen. *Trends Microbiol.* 2:289–293.

Groisman, E. A. and H. Ochman. 1996. Pathogenicity islands: bacterial evolution in quantum leaps. *Cell* 87:791–794.

Gross, C., M. Kelleher, V. R. Iyer, and P. O. Brown, et al. 2000. Identification of the copper regulon in *Saccharomyces cerevisiae* by DNA microarrays. *J. Biol. Chem.* 275:32310–32316.

Gross, R. 1993. Signal transduction and virulence regulation in human and animal pathogens. *FEMS Microbiol. Rev.* 10:301–326.

Grossman, A. D., J. W. Erickson, and C. A. Gross. 1984. The htpR gene product of *E. coli* is a sigma factor for heat-shock promoters. *Cell* 38:383–390.

Grunberg-Manago, M. 1996. Regulation of the expression of aminoacyl-tRNA synthetases and translation factors. In F. C. Neidhardt, R. Curtiss III, J. L. Ingraham, and E. C. C. Lin, et al. *Escherichia coli and Salmonella: Cellular and Molecular Biology*, 2nd ed., pp. 1432–1457. ASM Press, Washington, D.C.

Grunenfelder, B., G. Rummel, J. Vohradsky, and D. Roder, et al. 2001. Proteomic analysis of the bacterial cell cycle. *Proc. Natl. Acad. Sci. USA* 98:4681–4686.

Guigo, R. 1999. DNA comparison, codon usage and exon prediction. In M. J. Bishop (ed.), Genetics Database, pp. 54–80, Academic Press, London.

Guiney, D. G. and E. Yakobson. 1983. Location and nucleotide sequence of the transfer origin of the broad host range plasmid RK2. *Proc. Natl. Acad. Sci. USA* 80:3595–3598.

Gunsalus, R. P. and S.-J. Park. 1994. Aerobic-anaerobic gene regulation in *Escherichia coli*: control by the ArcAB and Fnr regulons. *Res. Microbiol.* 145:437–450.

Guo, B. P. and J. J. Mekalanos. 2001. *Helicobacter pylori* mutagenesis by mariner in vitro transposition. *FEMS Immunol. Med. Microbiol.* 30:87–93.

Guo, Z., R. A. Guilfoyle, A. J. Thiel, and R. Wang, et al. 1994. Direct fluorescence analysis of genetic polymorphisms by hybridization with oligonucleotide arrays on glass supports. *Nucl. Acids Res.* 22:5456–5465.

Guillemin, K., N. R. Salama, L. S. Tompkins, and S. Falkow. 2002. Cag pathogenicity island-specific responses of gastric epithelial cells to *Helicobacter pylori* infection. *Proc. Natl. Acad. Sci. USA* 99:15136–15141.

Gupta, R. S. 1998. Protein phylogenies and signature sequences: a reappraisal of evolutionary relationships among archaebacteria, eubacteria, and eukaryotes. *Microbiol. Mol. Biol. Rev.* 62:1435–1491.

Gupta, R. S. and B. Singh. 1992. Cloning of the HSP70 gene from *Halobacterium marismortui*: relatedness of archaebacterial HSP70 to its eubacterial homologs and a model for the evolution of the HSP70 gene. *J. Bacteriol.* 174:4594–4605.

Guschin, D. Y., B. K. Mobarry, D. Proudnikov, and D. A. Stahl, et al. 1997a. Oligonucleotide microchips as genosensors for determinative and environmental studies in microbiology. *Appl. Environ. Microbiol.* 63:2397–2402.

Guschin, D., G. Yershov, A. Zaslavsky, and A. Gemmell, et al. 1997b. Manual manufacturing of oligonucleotide, DNA, and protein microchips. *Anal. Biochem.* 250:203–211.

Gygi, S. P., B. Rist, S. A. Gerber, and F. Turecek, et al. 1999. Quantitative analysis of complex protein mixtures using isotope-coded affinity tags. *Nat. Biotechnol.* 17:994–999.

Hacia, J. 1999. Resequencing and mutational analysis using oligonucleotide microarrays. *Nat. Gen. Suppl.* 21:42–47.

Hacia, J. G. and F. S. Collins. 1999. Mutational analysis using oligonucleotide microarrays. *J. Med. Genet.* 36:730–736.

Hacia, J. G., B. Sun, N. Hunt, and K. Edgemon, et al. 1998. Strategies for mutational analysis of the large multiexon ATM gene using high-density oligonucleotide arrays. *Genome Res.* 8:1245–1258.

Hacia, J. G., J. B. Fan, O. Ryder, and L. Jin, et al. 1999. Determination of ancestral alleles for human single-nucleotide polymorphisms using high-density oligonucleotide arrays. *Nat. Genet.* 22:164–167.

Hacia, J. G., L. C. Brody, M. S. Chee, and S. P. Fodor, et al. 1996. Detection of heterozygous mutations in BRCA1 using high density oligonucleotide arrays and two-colour fluorescence analysis. *Nat. Genet.* 14:441–447.

Hacker, J. and J. B. Kaper. 2000. Pathogenicity islands and the evolution of microbes. *Annu. Rev. Microbiol.* 54:641–679.

Hacker, J., G. Blum-Oehler, I. Muhldorfer, and H. Tschape. 1997. Pathogenicity islands of virulent bacteria: structure, function and impact on microbial evolution. *Mol. Microbiol.* 23:1089–1097.

Hacker, J., S. Knapp, and W. Goebel. 1983. Spontaneous deletions and flanking regions of the chromosomal inherited hemolysin determinant of an *Escherichia coli* O6 strain. *J. Bacteriol.* 154:1145–1152.

Hakenbeck, R., N. Balmelle, and B. Weber, and C. Gardes, et al. 2001. Mosaic genes and mosaic chromosomes: intra- and interspecies genomic variation of *Streptococcus pneumoniae*. *Infect. Immun.* 69:2477–2486.

Hall, R. M. 1997. Mobile gene cassettes and integrons: moving antibiotic resistance genes in Gram-negative bacteria. *CIBA Found. Symp.* 207:192–205.

Hall, R. M. and C. M. Collis. 1995. Mobile gene cassettes and integrons: capture and spread of genes by site-specific recombination. *Mol. Microbiol.* 15:593–600.

Hallet, B., R. Rezsohazy, J. Mahillon, and J. Delcour. 1994. IS231A insertion specificity: consensus sequence and DNA bending at the target site. *Mol. Microbiol.* 14:131–139.

Hamadeh, H. K., P. R. Bushel, and S. Jayadev, et al. 2002. Gene expression analysis reveals chemical-specific profiles. *Toxicolog. Sci.* 67:219–231.

Hamer, L., K. Adachi, M. V. Montenegro-Chamorro, and M. M. Tanzer, et al. 2001b. Gene discovery and gene function assignment in filamentous fungi. *Proc. Natl. Acad. Sci. USA* 98:5110–5115.

Hamer, L., T. M. DeZwaan, M. V. Montenegro-Chamorro, and S. A. Frank, et al. 2001. Recent advances in large-scale transposon mutagenesis. *Curr. Opin. Chem. Biol.* 5:67–73.

Hamoen, L. W., A. F. Van Werkhoven, J. J. Bijlsma, and D. Dubnau, et al. 1998. The competence transcription factor of *Bacillus subtilis* recognizes short A/T-rich sequences arranged in a unique, flexible pattern along the DNA helix. *Genes Dev.* 12:1539–1550.

Han, D. K., J. Eng, H. Zhou, and R. Aebersold. 2001. Quantitative profiling of differentiation-induced microsomal proteins using isotope-coded affinity tags and mass spectrometry. *Nat. Biotechnol.* 19:946–951.

Hancock, R. E. 1997. The bacterial outer membrane as a drug barrier. *Trends Microbiol.* 5:37–42.

Hancock, R. E. W. and D. Knowles. 1998. Are we approaching the end of the antibiotic era? *Curr. Opin. Microbiol.* 1:493–494.

Handfield, M., D. E. Lehoux, and F. Sanschagrin, et al. 2000. In vivo-induced genes in *Pseudomonas aeruginosa*. *Infect. Immun.* 68:2359–2362.

Haniford, D. B. and G. Chaconas. 1992. Mechanistic aspects of DNA transposition. *Curr. Opin. Genet. Dev.* 2:698–704.

Hao, C. Y. and R. E. March. 2001. A survey of recent research activity in quadrupole ion trap mass spectrometry. *Int. J. Mass Spec.* 212:337–357.

Harkki, A. and E. T. Palva. 1984. Application of phage lambda technology to *Salmonella typhimuruim*: construction of a lambda-sensitive *Salmonella* strain. *Mol. Gen. Genet.* 195:256–259.

Harris, T. 2000. Genetics, genomics, and drug discovery. *Med. Res. Rev.* 20:203–211.

Harte, J., A. Kinzig, and J. Green. 1999. Self-similarity in the distribution and abundance of species. *Science* 284:334–336.

Hartley, J. L., G. F. Temple, and M. A. Brasch. 2000. DNA cloning using in vitro site-specific recombination. *Genome Res.* 10:1788–1795.

Hashimoto, T. and M. Sekiguchi. 1976. Isolation of temperature-sensitive mutants of R plasmid by in vitro mutagenesis with hydroxylamine. *J. Bacteriol.* 127:1561–1563.

Hasunuma, K. and M. Sekiguchi. 1977. Replication of plasmid pSC101 in *Escherichia coli* K12: requirement for dnaA function. *Mol. Gen. Genet.* 154:225–230.

Haurie, V., M. Perrot, T. Mini, and P. Jeno, et al. 2001. The transcriptional activator Cat8p provides a major contribution to the reprogramming of carbon metabolism during the diauxic shift in *Saccharomyces cerevisiae*. *J. Biol. Chem.* 276:76–85.

Haussler, D., A. Krogh, I. S. Mian, and K. Sjolander. 1993. *Protein Modeling Using Hidden Markov Models: Analysis of Globins, Proceedings of Hawaii International Conference on Systems Science*. IEEE Computer Society Press, Los Alamitos, CA, N.J.

Hayashi, T., K. Makino, M. Ohnishi, and K. Kurokawa, et al. 2001. Complete genome sequence of enterohemorrhagic *Escherichia coli* O157:H7 and genomic comparison with a laboratory strain K-12. *DNA Res.* 8:11–22.

Hazen, T. C., L. Jimenez, G. Lopez de Victoria, and C. B. Fliermans. 1991. Comparison of Bacteria from deep subsurface sediment and adjacent groundwater. *Microb. Ecol.* 22:293–304.

Head-Gordon, T. and C. L. Brooks 3rd. 1991. Virtual rigid body dynamics. *Biopolymers* 31:77–100.

Hecker, M. and U. Volker. 1990. General stress proteins in *Bacillus subtilis. FEMS Microbiol. Ecol.* 74:197–213.

Hecker, M., W. Schumann, and U. Volker. 1996. Heat-shock and general stress response in *Bacillus subtilis. Mol. Microbiol.* 19:417–428.

Hedges, D., M. Proft, and K. D. Entian. 1995. CAT8, a new zinc cluster-encoding gene necessary for derepression of gluconeogenic enzymes in the yeast *Saccharomyces cerevisiae. Mol. Cell. Biol.* 15:1915–1922.

Hedges, S. B. 2002. The origin and evolution of model organisms. *Nat. Rev. Genet.* 3:838–849.

Hegde, P., R. Qi, K. Abernathy, and C. Gay, et al. 2000. A concise guide to cDNA microarray analysis. *Biotechniques* 29:548–560.

Heidelberg, J. F., I. T. Paulsen, K. E. Nelson, and E. J. Gaidos, et al. 2002. Genome sequence of the dissimilatory metal ion-reducing bacterium *Shewanella oneidensis. Nat. Biotechnol.* 20:1118–1123.

Heidelberg, J. F., J. A. Eisen, W. C. Nelson, and R. A. Clayton, et al. 2000. DNA sequence of both chromosomes of the cholera pathogen *Vibrio cholerae. Nature* 406:477–484.

Heithoff, D. M., C. P. Conner, and P. C. Hanna, et al. 1997. Bacterial infection as assessed by in vivo gene expression. *Proc. Natl. Acad. Sci. USA* 94:934–939.

Heller, H. M., M. Schena, A. Chai, and D. Shalon, et al. 1997. Discovery and analysis of inflammatory disease-related genes using cDNA microarrays. *Proc. Natl. Acad. Sci. USA* 94:2150–2155.

Helmann, J. D., M. F. Wu, P. A. Kobel, and F. J. Gamo, et al. 2001. Global transcriptional response of *Bacillus subtilis* to heat shock. *J. Bacteriol.* 183:7318–7328.

Hendrick, J. P. and F.-U. Hartl. 1993. Molecular chaperone functions of heat-shock proteins. *Annu. Rev. Biochem.* 62:349–384.

Hendrickson, C. L. and M. R. Emmett. 1999. Electrospray ionization Fourier transform ion cyclotron resonance mass spectrometry. *Annu. Rev. Phys. Chem.* 50:517–536.

Hengge-Aronis, R. 1996. Regulation of gene expression during entry into stationary phase. In F. C. Neidhardt, R. Curtiss III, J. L. Ingraham, and E. C. C. Lin, et al. (eds.), Escherichia coli *and* Salmonella*: Cellular and Molecular Biology*, 2nd ed., pp. 1497–1512. ASM Press, Washington, D.C.

Henikoff, J. G., E. A. Greene, S. Pietrokovski, and S. Henikoff. 2000. Increased coverage of protein families with the blocks database servers. *Nucl. Acids Res.* 28:228–230.

Henikoff, S. and J. G. Henikoff. 1992. Amino acid substitution matrices from protein blocks. *Proc. Natl. Acad. Sci. USA* 89(22):10915–10919.

Henikoff, S., J. G. Henikoff, and S. Pietrokovski. 1999. Blocks + : A non-redundant database of protein alignment blocks derived from multiple compilations. *Bioinformatics* 15(6):471–479.

Henne, A., R. A. Schmitz, M. Bomeke, and G. Gottschalk, et al. 2000. Screening of environmental DNA libraries for the presence of genes conferring lipolytic activity on *Escherichia coli. Appl. Environ. Microbiol.* 66:3113–3116.

Hensel, M. and D. W. Holden. 1996. Molecular genetic approaches for the study of virulence in both pathogenic bacteria and fungi. *Microbiology* 142:1049–1058.

Hensel, M., J. E. Shea, C. Gleeson, and M. D. Jones, et al. 1995. Simultaneous identification of bacterial virulence genes by negative selection. *Science* 269:400–403.

Hentschel, U., M. Steinert, and J. Hacker. 2000. Common molecular mechanisms of symbiosis and pathogenesis. *Trends Microbiol.* 8:226–231.

Hergenrother, P. J., K. M. Depew, and S. L. Schreiber. 2000. Small-molecule microarrays: covalent attachment and screening of alcohol-containing small molecules on glass slides. *J. Am. Chem. Soc.* 122:7849–7850.

Hermann, T., W. Pfefferle, C. Baumann, and E. Busker, et al. 2001. Proteome analysis of *Corynebacterium glutamicum. Electrophoresis* 22:1712–1723.

Hernandez, J. and C. V. Robinson. 2001. Dynamic protein complexes: insights from mass spectrometry. *J. Biol. Chem.* 276:46685–46688.

Hernychova, L., J. Stulik, P. Halada, and A. Macela, et al. 2001. Construction of a *Francisells tularensis* two-dimensional electrophoresis protein database. *Proteomics* 1:508–515.

Herre, E. A., N. Knowlton, U. G. Mueller, and S. A. Rehner. 1999. The evolution of mutualisms: exploring the paths between conflict and cooperation. *Trends Ecol. Evol.* 14:49–53.

Herrington, D. A., R. H. Hall, G. Losonsky, and J. J. Mekalanos, et al. 1988. Toxin, toxin-coregulated pili, and the toxR regulon are essential for *Vibrio cholerae* pathogenesis in humans. *J. Exp. Med.* 168:1487–1492.

Herwig, R., A. J. Poustka, C. Müller, and C. Bull, et al. 1999. Large-scale clustering of cDNA-fingerprinting data. *Genome Res.* 9:1093–1105.

Hess, D., T. C. Covey, R. Winz, and R. W. Brownsey, et al. 1993. Analytical and micropreparative peptide mapping by high performance liquid chromatography/electrospray mass spectrometry of proteins purified by gel electrophoresis. *Prot. Sci.* 2:1342–1351.

Hettema, E. H., C. W. van Roermund, B. Distel, and M. van den Berg, et al. 1996. The ABC transporter proteins Pat1 and Pat2 are required for import of long-chain fatty acids into peroxisomes of *Saccharomyces cerevisiae. EMBO J.* 15:3813–3822.

Heyer, L. J., S. Kruglyak, and S. Yooseph. 1999. Exploring expression data: identification and analysis of coexpressed genes. *Genome Res.* 9:1106–1115.

Heyman, J. A., J. Cornthwaite, L. Foncerrada, and J. R. Gilmore, et al. 1999. Genome-scale cloning and expression of individual open reading frames using topoisomerase I-mediated ligation. *Genome Res.* 9:383–392.

Hieter, P. and M. Boguski. 1997. Functional genomics: it's all how you read it. *Science* 278:601–602.

Higgins, D., J. Thompson, T. Gibson, and J. D. Thompson, et al. 1994. CLUSTAL W: improving the sensitivity of progressivemultiple sequence alignment through sequence weighting, position-specific gap penalties and weight matrix choice. *Nucl. Acids Res.* 22:4673–4680.

Hihara, Y., A. Kamei, M. Kanehisa, and A. Kaplan, et al. 2001. DNA microarray analysis of cyanobacterial gene expression during acclimation to high light. *Plant Cell* 13:793–806.

Hihara, Y., K. Sonoike, M. Kanehisa, and M. Ikeuchi. 2003. DNA microarray analysis of redox-responsive genes in the genome of the cyanobacterium *Synechocystis* sp. strain PCC 6803. *J. Bacteriol.* 185:1719–1725.

Hilario, E. and J. P. Gogarten. 1993. Horizontal transfer of ATPase genes—the tree of life becomes a net of life. *Biosystems* 31:111–119.

Hillenkamp, F.M. Karas, R. C. Beavis, and B. T. Chait. 1991. Matrix-assisted laser desorption/ionization mass spectrometry of biopolymers. *Anal. Chem.* 63:1193A-1203A.

Himmelreich, R., H. Hilbert, H. Plagens, and E. Pirkl, et al. 1996. Complete sequence analysis of the genome of the bacterium *Mycoplasma pneumoniae. Nucl. Acids Res.* 24:4420–4449.

Himmelreich, R., H. Plagens, H. Hilbert, and B. Reiner, et al. 1997. Comparative analysis of the genomes of the bacteria *Mycoplasma pneumoniae* and *Mycoplasma genitalium. Nucl. Acids Res.* 25:701–712.

Hirose, I., K. Sano, I. Shioda, and M. Kumano, et al. 2000. Proteome analysis of *Bacillus subtilis* extracellular proteins: a two-dimensional protein electrophoretic study. *Microbiology* 146:65–75.

Hirschhorn, J. N., P. Sklar, K. Lindblad-Toh, and Y. M. Lim, et al. 2000. SBE-TAGS: an array-based method for efficient single-nucleotide polymorphism genotyping. *Proc. Natl. Acad. Sci. USA* 97:12164–12169.

Hirt, R. P., J. M. Logsdon, Jr., B. Healy, and M. W. Dorey, et al. 1999. Microsporidia are related to fungi: evidence from the largest subunit of RNA polymerase II and other proteins. *Proc. Natl. Acad. Sci. USA* 96:580–585.

Hjalt, T.A. and E. G. Wagner. 1995. Bulged-out nucleotides protect an antisense RNA from RNase III cleavage. *Nucl. Acids Res.* 23:571–579.

Ho, Y., A. Gruhler, A. Heilbut, and G. D. Bader, et al. 2002. Systematic identification of protein complexes in *Saccharomyces cerevisiae* by mass spectrometry. *Nature* 415:180–183.

Hoch, J. A. 1991. Genetic analysis in *Bacillus subtilis. Methods Enzymol.* 204:305–320.

Hoffmann, T., A. Schutz, M. Brosius, and A. Volker, et al. 2002. High-salinity-induced iron limitation in *Bacillus subtilis. J. Bacteriol.* 184:718–727.

Hoffmann, T., B. Troup, A. Szabo, and C. Hungerer, et al. 1995. The anaerobic life of *Bacillus subtilis*: cloning of the genes encoding the respiratory nitrate reductase system. *FEMS Microbiol. Lett.* 131:219–225.

Hoheisel, J. 1997. Oligomer-chip technology. *Trends Biotech.* 15:465–469.

Hohmann, E. L., C. A. Oletta, K. P. Killeen, and S. I. Miller. 1996. phoP/phoQ-deleted *Salmonella* typhi (Ty800) is a safe and immunogenic single-dose typhoid fever vaccine in volunteers. *J. Infect. Dis.* 173:1408–1414.

Holm, L. and C. Sander. 1996. Mapping the protein universe. *Science* 273:595–602.

Holstege, F. C. P. and R. A. Young. 1999. Transcriptional regulation: contending with complexity. *Proc. Natl. Acad. Sci. USA* 96:2–4.

Honda, K., K. Nakamura, M. Nishiguchi, and K. Yamane. 1993. Cloning and characterization of a *Bacillus subtilis* gene encoding a homolog of the 54-kilodalton subunit of mammalian signal recognition particle and *Escherichia coli* Ffh. *J. Bacteriol.* 175:4885–4894.

Hood, D. W., M. E. Deadman, M. P. Jennings, and M. Bisercic, et al. 1996a. DNA repeats identify novel virulence genes in *Haemophilus influenzae. Proc. Natl. Acad. Sci. USA* 93:11121–11125.

Hood, D. W., M. E. Deadman, T. Allen, and H. Masoud, et al. 1996b. Use of the complete genome sequence information of *Haemophilus influenzae* strain Rd to investigate lipopolysaccharide biosynthesis. *Mol. Microbiol.* 22:951–965.

Hook, S. S. and A. R. Means. 2001. Ca2 + /CaM-dependent kinases: from activation to function. *Annu. Rev. Pharmacol. Toxicol.* 41:471–505.

Horn, D.M., R. A. Zubarev, and F. W. McLafferty. 2000b. Automated de novo sequencing of proteins by tandem high-resolution mass spectrometry. *Proc. Natl. Acad. Sci. USA* 97:10313–10317.

Horn, D.M., R. A. Zubarev, and F. W. McLafferty. 2000c. Automated reduction and interpretation of high resolution electrospray mass spectra of large molecules. *J. Am. Soc. Mass Spectrom.* 11:320–332.

Horn, D.M., Y. Ge, and F. W. McLafferty. 2000a. Activated ion electron capture dissociation for mass spectral sequencing of larger (42 kDa) proteins. *Anal. Chem.* 72:4778–4784.

Hoskins, J., W. E. Alborn, Jr., J. Arnold, and L. C. Blaszczak, et al. 2001. Genome of the bacterium *Streptococcus pneuomoniae* strain R6. *J. Bacteriol.* 183:5709–5717.

Houry, W. A., D. Frishman, C. Eckerskorn, and F. Lottspeich, et al. 1999. Identification of in vivo substrates of the chaperonin GroEL. *Nature* 402:147–154.

Hu, J. C. 2000. A guided tour in protein interaction space: coiled coils from the yeast proteome. *Proc. Natl. Acad. Sci. USA* 97:12935–12936.

Hu, Y., P. D. Butcher, J. A. Mangan, and M.-A. Rajandream, et al. 1999. Regulation of hmp gene transcription in *Mycobacterium tuberculosis*: effects of oxygen limitation and nitrosative and oxidative stress. *J. Bacteriol.* 181:3486–3493.

Huang, M., F. B. Oppermann, and A. Steinbuchel. 1994. Molecular characterization of the *Pseudomonas putida* 2,3-butanediol catabolic pathway. *FEMS Microbiol. Lett.* 124:141–150.

Huang, Q., D. Liu, P. Majewski, and L. C. Schulte, et al. 2001. The plasticity of dendritic cell responses to pathogens and their components. *Science* 294:870–875.

Huber, H. M., J. Hohn, R. Rachel, and T. Fuchs, et al. 2002. A new phylum of Archaea represented by a nanosized hyperthermophilic symbiont. *Nature* 417:63–67.

Hudson, J. R., E. P. Dawson, K. L. Rushing, and C. H. Jackson, et al. 1997. The complete set of predicted genes from *Saccharomyces cerevisiase* in a readily usable form. *Genome Res.* 7:1169–1173.

Hueck, C. J. 1998. Type III protein secretion systems in bacterial pathogens of animals and plants. *Microbiol. Mol. Biol. Rev.* 62:379–433.

Hugenholtz, P., C. Pitulle, K. L. Hershberger, and N. R. Pace. 1998b. Novel division level bacterial in a Yellowstone Hot Spring. *J. Bacteriol.* 180:366–376.

Hugenholtz, P., B. M. Goebel, and N. R. Pace. (1998) Impact of culture-independent studies on the emerging phylogenetic view of bacterial diversity. *J Bacteriol* 180:4765–4774.

Hughes, J. B., J. J. Hellmann, T. H. Ricketts, and B. J. Bohannan. 2001. Counting the uncountable: statistical approaches to estimating microbial diversity. *Appl. Environ. Microbiol.* 67:439–406.

Hughes, T. R., M. J. Marton, A. R. Jones, and C. J. Roberts, et al. 2000a. Functional discovery via a compendium of expression profiles. *Cell* 102:109–126.

Hughes, T. R., C. J. Roberts, H. Dai, and A. R. Jones, et al. 2000b. Widespread aneuploidy revealed by DNA microarray expression profiling. *Nat. Genet.* 25:333–337.

Hulett, F. M. 1996. The signal-transduction network for Pho regulation in *Bacillus subtilis*. *Mol. Microbiol.* 19:933–939.

Hunt, D.F., A. M. Buko, J. M. Ballard, and J. Shabanowitz, et al. 1981. Sequence analysis of polypeptides by collision activated dissociation on a triple quadrupole mass spectrometer. *Biomed. Mass Spectrom.* 8:397–408.

Hunt, D.F., J. R. Yates, 3rd, J. Shabanowitz, and S. Winston, et al. 1986. Protein sequencing by tandem mass spectrometry. *Proc. Natl. Acad. Sci. USA* 83:6233–6237.

Huston, M. A. 1997. Hidden treatments in ecological experiments: re-evaluating the ecosystem function of biodiversity. *Oecologia* 110:449–460.

Huston, M. A., L. W. Aarssen, M. P. Austin, and B. S. Cade, et al. 2000. No consistent effect of plant diversity on productivity. *Science* 289:1255.

Hutchinson, C. R. 1998. Combinatorial biosynthesis for new drug discovery. *Curr. Opin. Microbiol.* 1:319–329.

Hutchinson, G. B. and M. R Hayden. 1992. The prediction of exons through an analysis of spliceable open reading frames. *Nucl. Acids Res.* 20:3453–3462.

Hutchison, C. A., S. N. Peterson, S. R. Gill, and R. T. Cline, et al. 1999. Global transposon mutagenesis and a minimal *Mycoplasma* genome. *Science* 286:2165–2169.

Huynen, M. A. and P. Bork. 1998. Measuring genome evolution. *Proc. Natl. Acad. Sci. USA* 95:5849–5856.

Huynen, M. A., Y. Diaz-Lazcoz, and P. Bork. 1997. Differential display of genomes. *Trends Genet.* 13:389–390.

Hyde-DeRuyscher, R., L. A. Paige, and D. J. Christensen, et al. 2000. Detection of small-molecule enzyme inhibitors with peptides isolated from phage-displayed combinatorial peptide libraries. *Chem. Biol.* 7:17–25.

Iborra, F. and J. M. Buhler. 1976. Protein subunit mapping. a sensitive high resolution method. *Anal. Biochem.* 74:503–511.

Ichikawa, J. K., A. Norris, M. G. Bangera, and G. K. Geiss, et al. 2000. Interaction of *Pseudomonas aeruginosa* with epithelial cells: identification of differentially regulated genes by expression microarray analysis of human cDNAs. *Proc. Natl. Acad. Sci. USA* 97:9659–9664.

Ideker, T., V. Thorsson, J. A. Ranish, and R. Christmas, et al. 2001. Integrated genomic and proteomic analyses of a systematically perturbed metabolic network. *Science* 292:929–934.

Ikeda, T. P., A. E. Shauger, and S. Kustu. 1996. *Salmonella typhimurium* apparently perceives external nitrogen limitation as internal glutamine limitation. *J. Mol. Biol.* 259:589–607.

International Human Genome Sequencing Consortium. 2001. Initial sequencing and analysis of the human genome. *Nature* 409:860–921.

Israel, D. A., N. Salama, and C. N. Arnold, et al. 2001a. *Helicobacter pylori* strain-specific differences in genetic content, identified by microarray, influence host inflammatory responses. *J. Clin. Invest.* 107:611–620.

Israel, D. A., N. Salama, and U. Krishna, et al. 2001b. *Helicobacter pylori* genetic diversity within the gastric niche of a single human host. *Proc. Natl. Acad. Sci. USA* 98:14625–14630.

Itaya, M. and T. Tanaka. 1997. Experimental surgery to create subgenomes of *Bacillus subtilis* 168. *Proc. Natl. Acad. Sci. USA* 94:5378–5382.

Ito, T., K. Tashiro, S. Muta, and R. Ozawa, et al. 2000. Toward a protein-protein interaction map of the budding yeast: a comprehensive system to examine two-hybrid interactions in all possible combinations between the yeast proteins. *Proc. Natl. Acad. Sci. USA* 97:1143–1147.

Ito, T., T. Chiba, R. Ozawa, and M. Yoshida, et al. 2001. A comprehensive two-hybrid analysis to explore the yeast protein interactome. *Proc. Natl. Acad. Sci. USA* 98:4569–4574.

Iwabe, N., K. Kuma, M. Hasegawa, and S. Osawa, et al. 1989. Evolutionary relationship of archaebacteria, eubacteria, and eukaryotes inferred from phylogenetic trees of duplicated genes. *Proc. Natl. Acad. Sci. USA* 86:9355–9359.

Iyer, V. R., M. B. Eisen, D. T. Ross, and G. Schuler, et al. 1999. The transcriptional program in the response of human fibroblasts to serum. *Science* 283:83–87.

Iyer, V. R., C. E. Horak, C. S. Scafe, and D. Botstein, et al. 2001. Genomic binding sites of the yeast cell-cycle transcription factors SBF and MBF. *Nature* 409:533–538.

Jacobsson, K. and L. Frykberg. 1995. Cloning of ligand-binding domains of bacterial receptors by phage display. *Biotechniques* 18:878–885.

Jacobsson, K. and L. Frykberg. 1996. Phage display shot gun cloning of ligand-binding domains of prokaryotic receptors approaches 100% correct clones. *Biotechniques* 20:1070–1080.

Jacobsson, K., H. Jonsson, H. Lindmark, and B. Guss, et al. 1997. Shot-gun phage display mapping of two streptococcal cell-surface proteins. *Microbiol. Res.* 152:121–128.

Jain, R., M. C. Rivera, and J. A. Lake. 1999. Horizontal gene transfer among genomes: the complexity hypothesis. *Proc. Natl. Acad. Sci. USA* 96:3801–3806.

James, P., J. Halladay, and E. A. Craig. 1996. Genomic libraries and a host strain designed for highly efficient two-hybrid selection in yeast. *Genetics* 144:1425–1436.

Jamshidi, N., J. S. Edwards, T. Fahland, and G. M. Church, et al. 2001. Dynamic simulation of the human red blood cell metabolic network. *Bioinformatics* 17:286–287.

Jansen, R. C. and J. P. Nap. 2001. Genetical genomics: the added value from segregation. *Trends Genet.* 17:388–391.

Jarosik, G. P. and E. J. Hansen. 1994. Identification of a new locus involved in expression of *Haemophilus influenzae* type b lipooligosaccharide. *Infect. Immun.* 62:4861–4867.

Jarvis, K. G., J. A. Giron, A. E. Jerse, and T. K. McDaniel, et al. 1995. Enteropathogenic *Escherichia coli* contains a putative type III secretion system necessary for the export of proteins involved in attaching and effacing lesion formation. *Proc. Natl. Acad. Sci. USA* 92:7996–8000.

Jaworski, D. D. and D. B. Clewell. 1995. A functional origin of transfer (oriT) on the conjugative transposon Tn916. *J. Bacteriol.* 177:6644–6651.

Jeffords, J. M. and T. Daschle. 2001. Policy issues. Political issues in the genome era. *Science* 291:1249–1251.

Jensen, O. N., M. R. Larsen, and P. Roepstorff. 1998. Mass spectrometric identification and microcharacterization of proteins from electrophoretic gets: strategies and applications. *Proteins* 2:74–89.

Jeong, H., B. Tombor, R. Albert, and Z. N. Oltvai, et al. 2000. The large-scale organization of metabolic networks. *Nature* 407:651–654.

Jeong, H., S. P. Mason, A.-L. Barabasi, and Z. N. Oltvai. 2001. Lethality and centrality in protein networks. *Nature* 411:41–42.

Jerome, L. J., T. van Biesen, and L. S. Frost. 1999. Degradation of FinP antisense RNA from F-like plasmids: the RNA-binding protein, FinO, protects FinP from ribonuclease E. *J. Mol. Biol.* 285:1457–1473.

Ji, G., R. C. Beavis, and R. P. Novick. 1995. Cell density control of staphylococcal virulence mediated by an octapeptide pheromone. *Proc. Natl. Acad. Sci. USA* 92:12055–12059.

Ji, Y., A. Marra, M. Rosenberg, and G. Woodnutt. 1999. Regulated antisense RNA eliminates alpha-toxin virulence in *staphylococus aureus* infection. *J. Bacteriol.* 181:6585–6590.

Ji, Y., B. Zhang, S. F. Van Horn, and P. Warren, et al. 2001. Identification of critical staphylococcal genes using conditional phenotypes generated by antisense RNA. *Science* 293:2266–2269.

Jiang, T., Y. Xu, and M. Zhang (eds.). 2002. *Current Topics in Computational Molecular Biology.* MIT Press, Cambridge, Mass.

Jiang, X., M. Zhang, Y. Ding, and J. Yao, et al. 1998. *Escherichia coli* prlC gene encodes a trypsin-like proteinase regulating the cell cycle. *J. Biochem.* (Tokyo) 124:980–985.

Jing, J., Z. Lai, C. Aston, and J. Lin, et al. 1999. Optical mapping of *Plasmodium falciparum* chromosome 2. *Genome Res.* 9:175–181.

Joachims, T. 1998. Making large-scale support vector machine learning practical. In B. Scholkopf, C. J. C. Burges, and S. Mika (eds.). *Advances in Kernel Methods-Support Vector Learning.* pp 169–184, MIT Press, Cambridge, Mass.

Johnsson, N. and A. Varshavsky. 1994. Split ubiquitin as a sensor of protein interactions in vivo. *Proc. Natl. Acad. Sci. USA* 91:10340–10344.

Johnston, M. 1998. Gene chips: array of hope for understanding gene regulation. *Curr. Biol.* 8:R171–R174.

Johnston, M., L. Hillier, L. Riles, and K. Albermann, et al. 1997. The nucleotide sequence of *Saccharomyces cerevisiae* chromosome XII. *Nature* 387:87–90.

Johnston, N. J., J. C. de Azavedo, J. D. Kellner, and D. E. Low. 1998. Prevalence and characterization of the mechanisms of macrolide, lincosamide and streptogramin resistance in isolates of *Streptococcus pneumoniae. Antimicrob. Agents Chemother.* 42:2425–2426.

Jones, D. T., W. R. Taylor, and J. M. Thornton. 1992. A new approach to protein fold recognition. *Nature* 358:86–89.

Jonsson, U., L. Fagerstam, B. Ivarsson, and B. Johnsson, et al. 1991. Real-time biospecific interaction analysis using surface plasmon resonance and a sensor chip technology. *Biotechniques* 11:620–627.

Jonsson, Z. O., S. K. Dhar, G. J. Narlikar, and R. Auty, et al. 2001. Rvb1p and Rvb2p are essential components of a chromatin remodeling complex that regulates transcription of over 5% of yeast genes. *J. Biol. Chem.* 276:16279–16288.

Jordan, I. K., K. S. Makarova, J. L. Spouge, and Y. I. Wolf, et al. 2001. Lineage-specific gene expansions in bacterial and archaeal genomes. *Genome Res.* 11:555–565.

Jorde, L. B., W. S. Watkins, and M. J. Bamshad. 2001. Population genomics: a bridge from evolutionary history to genetic medicine. *Hum. Mol. Genet.* 10:2199–2207.

Jucker, F. M., H. A. Heus, P. F. Yip, E. H. Moors, and A. Pardi. 1996. A network of heterogeneous hydrogen bonds in GNRA tetraloops. *J. Mol. Biol.* 264:968–980.

Judson, N. and J. J. Mekalanos. 2000a. TnAraOut, a transposon-based approach to identify and characterize essential bacterial genes. *Nat. Biotechnol.* 18:740–745.

Judson, N. and J. J. Mekalanos. 2000b. Transposon-based approaches to identify essential bacterial genes. *Trends Microbiol.* 8:521–526.

Jungblut, P. R., E. C. Muller, J. Mattow, and S. H. Kaufmann. 2001. Proteomics reveals open reading frames in *Mycobacterium tuberculosis* H37Rv not predicted by genomics. *Infect. Immun.* 69:5905–5907.

Jungblut, P. R., U. E. Schaible, H.-J. Mollenkopf, and U. Zimny-Arndt et al. 1999. Comparative proteome analysis of *Mycobacterium* tuberculosis and *Mycobacterium bovis* BCG strains: towards functional genomics of microbial pathogens. *Mol. Microbiol.* 33:1103–1117.

Jungblut, P. and B. Thiede. 1997. Protein identification from 2-DE gels by MALDI mass spectrometry. *Mass Spectrom. Rev.* 16:145–162.

Jungmann, J., H. A. Reins, J. Lee, and A. Romeo, et al. 1993. MAC1, a nuclear regulatory protein related to Cu-dependent transcription factors is involved in Cu/Fe utilization and stress resistance in yeast. *EMBO J.* 12:5051–5056.

Jurka, J., P. Klonowski, V. Dagman, and P. Pelton. 1996. CENSOR—a program for identification and elimination of repetitive elements from DNA sequences. *Comp. Chem.* 20(1):119–122.

Kaback, D. B., P. W. Oeller, H. Yde Steensma, and J. Hirschman, et al. 1984. Temperature-sensitive lethal mutations on yeast chromosome I appear to define only a small number of genes. *Genetics* 108:67–90.

Kalman, S., W. Mitchell, R. Marathe, and C. Lammel, et al. 1999. Comparative genomes of *Chlamydia pneumoniae* and *C. trachomatis. Nat. Gene.* 21:385–389.

Kalocsai, P. and S. Shams. 2001. Use of Bioinformatics in Arrays. In J. B. Rampal (ed.), DNA Arrays, Methods and Protocols, Methods in Molecular Biology Volume 170, Humana Press.

Kalocsai, P. and S. Shams. 2001. Use of bioinformaties in arrays. *Methods Mol. Biol.* 170:223–236.

Kalogeraki, V. S. and S. C. Winans. 1997. Suicide plasmids containing promoterless reporter genes can simultaneously disrupt and create fusions to target genes of diverse bacteria. *Gene* 188:69–75.

Kaltashov, I. A. and S. J. Eyles. 2002. Studies of biomolecular conformations and conformational dynamics by mass spectrometry. *Mass Spec. Rev.* 21:37–71.

Kampranis, S. C., N. A. Gormley, R. Tranter, and G. Orphanides, et al. 1999. Probing the binding of coumarins and cyclothialidines to DNA gyrase. *Biochemistry* 38:1967–1976.

Kane, M. D., T. A. Jatkoe, C. R. Stumpf, and J. Lu, et al. 2000. Assessment of the sensitivity and specificity of oligonucleotide (50 mer) microarrays. *Nucl. Acids Res.* 28:4552–4557.

Kanehisa, M., S. Goto, S. Kawashima, and A. Nakaya. 2002. The KEGG databases at GenomeNet. *Nucl. Acids Res.* 30:42–46.

Kaneko, T., S. Sato, H. Kotani, and A. Tanaka, et al. 1996. Sequence analysis of the genome of the unicellular cyanobacterium *Synechocystis* sp. strain PCC6803. II. Sequence determination of the entire genome and assignment of potential protein-coding regions. *DNA Res.* 3:109–136.

Kaneko, T., Y. Nakamura, C. P. Wolk, and T. Kuritz, et al. 2001. Complete genomic sequence of the filamentous nitrogen-fixing cyanobacterium *Anabaena* sp. strain PCC 7120. *DNA Res.* 8:205–213.

Kaniga, K., I. Delor, and G. R. Cornelis. 1991. A wide-host-range suicide vector for improving reverse genetics in gram-negative bacteria: inactivation of the blaA gene of *Yersinia enterocolitica. Gene* 109:137–141.

Kaper, J. B., H. Lockman, M. M. Baldini, and M. M. Levine. 1984. Recombinant nontoxinogenic *Vibrio cholerae* strains as attenuated cholera vaccine candidates. *Nature* 308:655–668.

Karaolis, D. K. R., J. A. Johnson, C. C. Bailey, and E. C. Boedeker, et al. 1998. A *Vibrio cholerae* pathogenicity island associated with epidemic and pandemic strains. *Proc. Natl. Acad. Sci. USA* 95:3134–3139.

Karaolis, D. K., S. Somara, D. R. Maneval, Jr., and J. A. Johnson, et al. 1999. A bacteriophage encoding a pathogenicity island, a type-IV pilus and a phage receptor in cholera bacteria. *Nature* 399:375–379.

Karlin, S. 1998. Global dinucleotide signatures and analysis of genomic heterogeneity. *Curr. Opin. Microbiol.* 1:598–610.

Karlin, S. 2001. Detecting anomalous gene clusters and pathogenicity islands in diverse bacterial genomes. *Trends Microbiol.* 9:335–343.

Karlin, S. and L. Brocchieri. 2000. Heat shock protein 60 sequence comparisons: duplications, lateral transfer, and mitochondrial evolution. *Proc. Natl. Acad. Sci. USA* 97:11348–11353.

Karlin, S. and S. F. Altschul. 1990. Methods for assessing the statistical significance of molecular sequence features by using general scoring schemes. *Proc. Natl. Acad. Sci. USA* 87:2264–2268.

Karlin, S., A. M. Campbell, and J. Mrazek. 1998. Comparative DNA analysis across diverse genomes. *Annu. Rev. Genet.* 32:185–225.

Karlin, S., J. Mrazek, and A. M. Campbell. 1997. Compositional biases of bacterial genomes and evolutionary implications. *J. Bacteriol.* 179:3899–3913.

Karlyshev, A. V., P. C. Oyston, K. Williams, and G. C. Clark, et al. 2001. Application of high-density array-based signature-tagged mutagenesis to discover novel *Yersinia* virulence-associated genes. *Infect. Immun.* 69:7810–7819.

Karplus, K., C. Barrett, and R. Hughey. 1998. Hidden Markov models for detecting remote protein homologies. *Bioinformatics* 14(10):846–856.

Katayama, Y., S. Gottesman, J. Pumphrey, and S. Rudikoff, et al. 1988. The two-component, ATP-dependent Clp protease of *Escherichia coli.* Purification, cloning, and mutational analysis of the ATP-binding component. *J. Biol. Chem.* 263:15226–15236.

Kato, M., T. Tsunoda, and T. Takagi. 2000. Inferring genetic networks from DNA microarray data by multiple regression analysis. *Proc. Genome Inform. Ser. Workshop Genome Inform.* 11:118–128.

Kato-Maeda, M., J. T. Rhee, T. R. Gingeras, and H. Salamon, et al. 2001. Comparing genomes within the species *Mycobacterium tuberculosis. Genome Res.* 11:547–554.

Katz, B. A. 1997. Structural and mechanistic determinants of affinity and specificity of ligands discovered or engineered by phage display. *Annu. Rev. Biophys. Biomol. Struct.* 26:27–45.

Kaufmann, S. H. and P. Andersen. 1998. Immunity to mycobacteria with emphasis on tuberculosis: implications for rational design of an effective tuberculosis vaccine. *Chem. Immunol.* 70:21–59.

Kawarabayasi, Y., M. Sawada, H. Horikawa, and Y. Haikawa, et al. 1998. Complete sequence and gene organization of the genome of a hyper-thermophilic archaebacterium, *Pyrococcus horikoshii. DNA Res.* 5:55–76.

Kawarabayasi, Y., Y. Hino, H. Horikawa, and K. Jin-no, et al. 2001. Complete genome sequence of an aerobic thermoacidophilic crenarchaeon, *Sulfolobus tokodaii* strain7. *DNA Res.* 8:123–140.

Kawarabayasi, Y., Y. Hino, H. Horikawa, and S. Yamazaki, et al. 1999. Complete genome sequence of an aerobic hyper-thermophilic crenarchaeon, *Aeropyrum pernix* K1. *DNA Res.* 6:83–101.

Kawashima, T., N. Amano, H. Koike, and S. Makino, et al. 2000. Archaeal adaptation to higher temperatures revealed by genomic sequence of *Thermoplasma volcanium*. *Proc. Natl. Acad. Sci. USA* 97:14257–14262.

Keener, J. and M. Nomura. 1996. Regulation of ribosome synthesis. In F. C. Neidhardt, R. Curtiss III, J. L. Ingraham, and E. C. C. Lin, et al. (eds.) Escherichia coli *and* Salmonella*: Cellular and Molecular Biology*, 2nd ed., pp. 1417–1431. ASM Press, Washington, D.C.

Keggins, K. M., P. S. Lovett, and E. J. Duvall. 1978. Molecular cloning of genetically active fragments of *Bacillus* DNA in *Bacillus subtilis* and properties of the vector plasmid pUB110. *Proc. Natl. Acad. Sci. USA* 75:1423–1427.

Kelleher, N. L., S. V. Taylor, D. Grannis, and C. Kinsland, et al. 1998. Efficient sequence analysis of the six gene products (7–74 kDa) from the *Escherichia coli* thiamin biosynthetic operon by tandem high-resolution mass spectrometry. *Prot. Sci.* 7:1796–1801.

Kelley, J. M., C. E. Field, and M. B. Craven, et al. 1999. High throughput direct end sequencing of BAC clones. *Nucl. Acids Res.* 27:1539–1546.

Keogh, R. S., C. Seoighe, and K. H. Wolfe. 1998. Evolution of gene order and chromosome number in *Saccharomyces, Kluyveromyces* and related fungi. *Yeast* 14:443–457.

Kernodle, D. S., R. K. Voladri, B. E. Menzies, and C. C. Hager, et al. 1997. Expression of an antisense hla fragment in *Staphylococcus aureus* reduces alpha-toxin production in vitro and attenuates lethal activity in a murine model. *Infect. Immun.* 65:179–184.

Kerr, M. K. and G. A. Churchill. 2001a. Statistical design and the analysis of gene expression microarray data. *Genet. Res.* 77(2):123–128.

Kerr, M. K. and G. A. Churchill. 2001b. Experimental design for gene expression microarrays. *Biostatistics* 2:183–201.

Khodursky, A. B., B. J. Peter, N. R. Cozzarelli, and D. Botstein, et al. 2000. DNA microarray analysis of gene expression in response to physiological and genetic changes that affect tryptophan metabolism in *Escherichia coli. Proc. Natl. Acad. Sci. USA* 97:12170–12175.

Khrapko, K. R., Y. P. Lysov, A. A. Khorlyn, and I. B. Ivanov, et al. 1989. An oligonucleotide hybridization approach to DNA sequencing. *FEBS Lett.* 256:118–122.

Kiley, P. J. and H. Beinert. 1998. Oxygen sensing by the global regulator, FNR: the role of the iron-sulfur cluster. *FEMS Microbiol. Rev.* 22:341–352.

Kim, D., D. Xu, J. Guo, and K. Ellrott, et al. 2003. PROSPECT II: protein structure prediction program for genome-scale applications. *Protein Eng.* 16:641–650.

Kim, S. H. 1998. Shining a light on structural genomics. *Nat. Struct. Biol.* 5 Suppl:643–645.

Kim, S. K. 2001. http://*C. elegans*: mining the functional genomic landscape. *Nat. Rev. Genet.* 2:681–689.

Kimura, M. 1983. *The Neutral Theory of Molecular Evolution*. Cambridge Univ. Press, Cambridge, U.K.

King, H. C. and A. A. Sinha. 2001. Gene expression profile analysis by DNA microarrays: promise and pitfalls. *JAMA* 286:2280–2288.

Kingsley, M. T., T. Straub, D. R. Call, and D. S. Daly, et al. 2002. Fingerprinting closely related *Xanthomonas* pathovars with random nanamer oligonucleotide microarrays. *Appl. Environ. Microbiol.* 68:6361–6370.

Kingston, R. E. and G. J. Narlikar. 1999. ATP-dependent remodeling and acetylation as regulators of chromatin fluidity. *Genes Dev.* 13:2339–2352.

Kishimoto, N. and T. Tano. 1987. Acidophilic heterotrophic bacteria isolated from acidic mine drainage, sewage and soils. *J General Appl. Microbiol.* 33:11–25.

Kitano, H. 2002. Systems biology: a brief overview. *Science* 295:1662–1664.

Kleckner, N. 1989. Transposon Tn10. In D. E. Berg and M. M. Howe (eds.), *Mobile DNA*, pp. 227–268. American Society for Microbiology, Washington, D.C.

Klee, C. B., H. Ren, and X. Wang. 1998. Regulation of the calmodulin-stimulated protein phosphatase, calcineurin. *J. Biol. Chem.* 273:13367–13370.

Kleman-Leyer, K., D. W. Armbruster, and C. J. Daniels. 1997. Properties of *H. volcanii* tRNA intron endonuclease reveal a relationship between the archaeal and eucaryal tRNA intron processing systems. *Cell* 89:839–847.

Klenk, H. P., R. A. Clayton, J. F. Tomb, and O. White, et al. 1997. The complete genome sequence of the hyperthermophilic, sulphate-reducing archaeon *Archaeoglobus fulgidus*. *Nature* 394:364–370.

Klose, J. 1975. Protein mapping by combined isoelectric focusing and electrophoresis of mouse tissues. A novel approach to testing for induced point mutations in mammals. *Humangenetik* 26:231–243.

Knight, J. 2001. When the chips are down. *Nature* 410:860–861.

Knoblauch, N. T. M., S. Rudiger, H. J. Schonfeld, and A. J. M. Driessen, et al. 1999. Substrate specificity of the SecB chaperone. *J. Biol. Chem.* 274:34219–34225.

Knudtson, K. L. and F. C. Minion. 1993. Construction of Tn4001lac derivatives to be used as promoter probe vectors in mycoplasmas. *Gene* 137:217–222.

Kobayashi, K., M. Ogura, H. Yamaguchi, and K.-I. Yoshida, et al. 2001. Comprehensive DNA microarray analysis of *Bacillus subtilis* two-component regulatory systems. *J. Bacteriol.* 183:7365–7370.

Kobryn, K. and G. Chaconas. 2001. The circle is broken: telomere resolution in linear replicons. *Curr. Opin. Microbiol.* 4:558–564.

Kohonen, T. 1982a. Analysis of a simple self-organizing process. *Biolog. Cyber.* 44(2):135–140.

Kohonen, T. 1982b. Self-organizing formation of topologically correct feature maps. *Biolog. Cyber.* 43(1):59–69.

Kolber, Z. S., F. G. Plumley, A. S. Lang, and J. T. Beatty, et al. 2001. Contribution of aerobic photoheterotrophic bacteria to the carbon cycle in the ocean. *Science* 292:2492–2495.

Koller, A., M. P. Washburn, B. M. Lange, and N. L. Andon, et al. 2002. Proteomic survey of metabolic pathways in rice. *PNAS* 99:11969–11974.

Koonin, E. V. 2000. How many genes can make a cell: the minimal-gene-set concept. *Annu. Rev. Gen. Hum. Genet.* 1:99–116.

Koonin, E. V., A. R. Mushegian, and P. Bork. 1996. Non-orthologous gene displacement. *Trends Genet* 12:334–336.

Koonin, E. V., K. S. Makarova, and L. Aravind. 2001. Horizontal gene transfer in prokaryotes: quantification and classification. *Annu. Rev. Microbiol.* 55:709–742.

Koonin, E. V., L. Aravind, and A. S. Kondrashov. 2000. The impact of comparative genomics on our understanding of evolution. *Cell* 101:573–576.

Kornberg, T. B. and M. A. Krasnow. 2000. The *Drosophila* genome sequence: implications for biology and medicine. *Science* 287:2218–2220.

Korzheva, N., A. Mustaev, M. Kozlov, and A. Malhotra, et al. 2000. A structural model of transcription elongation. *Science* 289:619–625.

Kowalchuk, G. A. and J. R. Stephen. 2001. Ammonia-oxidizing bacteria: a model for molecular microbial ecology. *Annu. Rev. Microbiol.* 55:485–529.

Kozal, M. J., N. Shah, N. Shen, and R. Yang, et al. 1996. Extensive polymorphisms observed in HIV-1 clade B protease gene using high-density oligonucleotide arrays. *Nat. Med.* 2:753–759.

Kramer, A., T. Keitel, K. Winkler, and W. Stocklein, et al. 1997. Molecular basis for binding promiscuity of an anti-p24 (HIV-1) monoclonal antibody. *Cell* 91:799–809.

Kroos, L., B. Zhang, H. Ichikawa, and Y.-T. N. Yu. 1999. Control of σ factor activity during *Bacillus subtilis* sporulation. *Mol. Microbiol.* 31:1285–1294.

Kruger, E. and M. Hecker. 1998. The first gene of the *Bacillus subtilis* clpC operon, ctsR, encodes a negative regulator of its own operon and other class III heat shock genes. *J. Bacteriol.* 180:6681–6688.

Kumar, A., S. Agarwal, J. A. Heyman, and S. Matson, et al. 2002. Subcellular localization of the yeast proteome. *Genes Dev.* 16:707–719.

Kumar, S. and S. B. Hedges. 1998. A molecular timescale for vertebrate evolution. *Nature* 392:917–920.

Kuner, J. M. and D. Kaiser. 1981. Introduction of transposon Tn5 into *Myxococcus* for analysis of developmental and other nonselectable mutants. *Proc. Natl. Acad. Sci. USA* 78:425–429.

Kunst, F., N. Ogasawara, I. Moszer, and A. M. Albertini, et al. 1997. The complete genome sequence of the Gram-positive bacterium *Bacillus subtilis*. *Nature* 390:249–256.

Kuntz, I. D. 1992. Structure-based strategies for drug design and discovery. *Science* 257:1078–1082.

Kuntz, I. D., J. M. Blaney, S. J. Oatley, and R. Langridge, et al. 1982. A geometric approach to macromolecule-ligand interactions. *J. Mol. Biol.* 161:269–288.

Kuo, C. C., L. A. Jackson, L. A. Campbell, and J. T. Grayston. 1995. *Chlamydia pneumoniae* (TWAR). *Clin. Microbiol. Rev.* 8:451–461.

Kuo, Y.-P., D. K. Thompson, A. St. Jean, and R. L. Charlebois, et al. 1997. Characterization of two heat shock genes from *Haloferax volcanii*: A model system for transcription regulation in the Archaea. *J. Bacteriol.* 179:6318–6324.

Kurland, C. G. and S. G. Andersson. 2000. Origin and evolution of the mitochondrial proteome. *Microbiol. Mol. Biol. Rev.* 64:786–820.

Kuruvilla, F. G., A. F. Shamji, S. M. Sternson, and P. J. Hergenrother, et al. 2002. Dissecting glucose signalling with diversity-oriented synthesis and small-molecule microarrays. *Nature* 416:653–657.

Kuske, C. R., S. M. Barns, and J. D. Busch. 1997. Diverse uncultivated bacterial groups from soils of the arid southwestern United States that are present in many geographic regions. *Appl. Environ. Microbiol.* 63:3614–3621.

Kyoda, K. M., M. Morohashi, S. Onami, and H. Kitano. 2000. A gene network inference method from continuous-value gene expression data of wild-type and mutants. *Proc. Genome Inform. Ser. Workshop Genome Inform.*196–204.

Kyrpides, N. C. and G. J. Olsen. 1999. Archaeal and bacterial hyperthermophiles: horizontal gene exchange or common ancestry? *Trends Genet.* 15:298–299.

Le Chatelier, E., S. D. Ehrlich, and L. Janniere. 1996. Countertranscript-driven attenuation system of the pAM beta 1 repE gene. *Mol. Microbiol.* 20:1099–1112.

Lackman, M., T. Bucci, R. J. Mann, and L. A. Kravets, et al. 1996. Purification of a ligand for the EPH-like receptor HEK using a biosensor-based affinity detection approach. *Proc. Natl. Acad. Sci. USA* 93:2523–2527.

Lai, Y. C., H. L. Peng, and H. Y. Chang. 2001. Identification of genes induced in vivo during *Klebsiella pneumoniae* CG43 infection. *Infect. Immun.* 69:7140–7145.

Lake, J. A. and M. C. Rivera. 1994. Was the nucleus the first endosymbiont? *Proc. Natl. Acad. Sci. USA* 91:2880–2881.

Lakkis, M. M., M. F. DeCristofaro, H. J. Ahr, and T. A. Mansfield. 2002. Application of toxicogenomics to drug development. *Expert Rev. Mol. Diag.* 2:337–345.

Lampe, D. J., B. J. Akerley, E. J. Rubin, and J. J. Mekalanos, et al. 1999. Hyperactive transposase mutants of the Himar1 mariner transposon. *Proc. Natl. Acad. Sci. USA* 96:11428–11433.

Lamture, J. B., K. L. Beattie, B. E. Burke, and M. D. Eggers, et al. 1994. Direct detection of nucleic acid hybridization on the surface of a charge coupled device. *Nucl. Acids Res.* 22:2121–2125.

Landegren, U., M. Nilsson, and P. Y. Kwok. 1998. Reading bits of genetic information: methods for single-nucleotide polymorphism analysis. *Genome Res.* 8:769–776.

Lander, E. S. 1996. The new genomics: global views of biology. *Science* 274:536–539.

Lander, E. S., L. M. Linton, B. Birren, and C. Nusbaum, et al. 2001. Initial sequencing and analysis of the human genome. *Nature* 409:860–921.

Landy, A. 1989. Dynamic, structural, and regulatory aspects of site-specific recombination. *Annu. Rev. Biochem.* 58:913–949.

Lange, C. C., L. P. Wackett, K. W. Minton, and M. J. Daly. 1998. Engineering a recombinant *Deinococcus radiodurans* for organopollutant degradation in radioactive mixed waste environments. *Nat. Biotechnol.* 16:929–933.

Langen, H., B. Takacs, S. Evers, and P. Berndt, et al. 2000. Two-dimensional map of the proteome of *Haemophilus influenzae. Electrophoresis* 21:411–429.

Larsen, M.R. and P. Roepstorff. 2000. Mass spectrometric identification of proteins and characterization of their post-translational modifications in proteome analysis. *Fresenius J. Anal. Chem.* 366:677–690.

Lashkari, D. A., J. L. DeRisi, J. H. McCusker, and A. F. Namath, et al. 1997. Yeast microarrays for genome wide parallel genetic and gene expression analysis. *Proc. Natl. Acad. Sci. USA* 94:13057–13062.

Laskowski, R. A., N. M. Luscombe, M. B. Swindells, and J. M. Thornton. 1996. Protein clefts in molecular recognition and function. *Prot. Sci.* 5:2438–2452.

Lawrence, C. E., S. F. Altschul, M. S. Boguski, and J. S. Liu, 1993. Detecting subtle sequence signals: a Gibbs sampler strategy for multiple alignment. *Science* 262:208–214.

Lawrence, J. 1999. Selfish operons: the evolutionary impact of gene clustering in prokaryotes and eukaryotes. *Curr. Opin. Genet. Dev.* 9:642–648.

Lawrence, J. G. 1997. Selfish operons and speciation by gene transfer. *Trends Microbiol.* 5:355–359.

Lawrence, J. G. and H. Ochman. 1997. Amelioration of bacterial genomes: rates of change and exchange. *J. Mol. Evol.* 44:383–397.

Lawrence, J. G. and H. Ochman. 1998. Molecular archaeology of the *Escherichia coli* genome. *Proc. Natl. Acad. Sci. USA* 95:9413–9417.

Lawson, F. S., R. L. Charlebois, and J. A. Dillon. 1996. Phylogenetic analysis of carbamoylphosphate synthetase genes: complex evolutionary history includes an internal duplication within a gene which can root the tree of life. *Mol. Biol. Evol.* 13:970–977.

Lazazzera, B. A., T. Palmer, J. Quisel, and A. D. Grossman. 1999. Cell density control of gene expression and development in *Bacillus subtilis*. In G.M. Dunny, and S.C. Winans, (eds.), *Cell–Cell Signaling in Bacteria*, pp. 27–46. American Society for Microbiology, Washington, D.C.

Le Chatelier, E., S. D. Ehrlich, and L. Janniere L. 1996. Countertranscript-driven attenuation system of the pAM beta 1 repE gene. *Mol. Microbiol.* 20:1099–1112.

Lease, R. A. and M. Belfort M. 2000. A trans-acting RNA as a control switch in *Escherichia coli*: DsrA modulates function by forming alternative structures. *Proc. Natl. Acad. Sci. USA* 97:9919–9924.

Leblond, P., F. X. Francou, J. M. Simonet, and B. Decaris. 1990. Pulsed-field gel electrophoresis analysis of the genome of *Streptomyces ambofaciens* strains. *FEMS Microbiol. Lett.* 60:79–88.

Ledent, V. and M. Vervoort. 2001. The basic helix-loop-helix protein family: comparative genomics and phylogenetic analysis. *Genome Res.* 11:754–770.

Lederberg, J. and A. T. McCray. 2001. "Ome sweet omics"—a genealogical treasury of words. 15(7):8–8.

Lee, B. Y. and M. A. Horwitz. 1995. Identification of macrophage and stress-induced proteins of *Mycobacterium tuberculosis. J. Clin. Invest.* 96:245–249.

Lee, M. S. 1999. Molecular clock calibrations and metazoan divergence dates. *J. Mol. Evol.* 49:385–391.

Lee, S.-W., S. J. Berger, S. Martinović, and L. Paša-Tolić, et al. 2002. Direct mass spectrometric analysis of intact proteins of the yeast large ribosomal subunit using capillary LC/FTICR. *PNAS* 99:5942–5947.

Leenhouts, K. J., J. Kok, and G. Venema. 1991. Lactococcal plasmid pWV01 as an integration vector for lactococci. *Appl. Environ. Microbiol.* 57:2562–2567.

Leenhouts, K., G. Buist, A. Bolhuis, and A. ten Berge, et al. 1996. A general system for generating unlabelled gene replacements in bacterial chromosomes. *Mol. Gen. Genet.* 253:217–224.

LeGrain, P., J. Wojcik, and J.-M. Gauthier. 2001. Protein-protein interaction maps: a lead towards cellular functions. *Trends Genet.* 17:346–352.

Lehmann, J. M., L. B. Moore, T. A. Smith-Oliver, and W. O. Wilkison, et al. 1995. An antidiabetic thiazolidinedione is a high affinity ligand for peroxisome proliferator-activated receptor (PPAR). *J. Biol. Chem.* 270:12953–12956.

Lehoux, D. E. and R. C. Levesque. 2000. Detection of genes essential in specific niches by signature-tagged mutagenesis. *Curr. Opin. Biotechnol.* 11:434–439.

Lehoux, D. E., F. Sanschagrin, and R. C. Levesque. 2001. Discovering essential and infection-related genes. *Curr. Opin. Microbiol.* 4:515–519.

Lehoux, D. E., F. Sanschagrin, and R. C. Levesque. 2002. Identification of in vivo essential genes from *Pseudomonas aeruginosa* by PCR-based signature-tagged mutagenesis. *FEMS Microbiol. Lett.* 210:73–80.

Lehoux, D. E., F. Sanschagrin, and R. C. Levesque. 1999. Defined oligonucleotide tag pools and PCR screening in signature-tagged mutagenesis of essential genes from bacteria. *Biotechniques* 26:473–480.

Lei, B., S. Mackie, S. Lukomski, and J. M. Musser. 2000. Identification and immunogenicity of group A *Streptococcus* culture supernatant proteins. *Infect. Immun.* 68:6807–6818.

Leister, D. 2003. Chloroplast research in the genomic age. *Trends Genet.* 19:47–56.

Lesnik, E. A., R. Sampath, H. B. Levene, and T. J. Henderson, et al. 2001. Prediction of rho-independent transcriptional terminators in *Escherichia coli. Nucl. Acids Res.* 29(17):3583–3594.

Lessie, T. G., W. Hendrickson, B. D. Manning, and R. Devereux. 1996. Genomic complexity and plasticity of *Burkholderia cepacia. FEMS Microbiol. Lett.* 144:117–128.

Leung, M. Y., B. E. Blaisdell, C. Burge, and S. Karlin. 1991. An efficient algorithm for identifying matches with errors in multiple long molecular sequences. *J. Mol. Biol.* 221:1367–1378.

Levin, P. A. and A. D. Grossman. 1998. Cell cycle and sporulation in *Bacillus subtilis. Curr. Opin. Microbiol.* 1:630–635.

Levinson, G. and G. A. Gutman. 1987. Slipped-strand mispairing: a major mechanism for DNA sequence evolution. *Mol. Biol. Evol.* 4:203–221.

Li, W. 1999. Statistical properties of open reading frames in complete genome sequences. *Comp. Chem.* 23(3–4):283–301.

Li, W., C. L. Hendrickson, M. R. Emmett, and A. G. Marshall. 1999. Identification of intact proteins in mixtures by alternated capillary liquid chromatography electrospray ionization and LC ESI infrared multiphoton dissociation Fourier transform ion cyclotron resonance mass spectrometry. *Anal. Chem.* 71:4397–4402.

Li, W. H. 1993. Unbiased estimation of the rates of synonymous and nonsynonymous substitution. *J. Mol. Evol.* 36:96–99.

Li, W. H., D. L. Ellsworth, J. Krushkal, and B. H. Chang, et al. 1996. Rates of nucleotide substitution in primates and rodents and the generation-time effect hypothesis. *Mol. Phylogenet. Evol.* 5:182–187.

Li, W. H., Z. Gu, H. Wang, and A. Nekrutenko. 2001. Evolutionary analyses of the human genome. *Nature* 409:847–849.

Liang, S., S. Fuhrman, and R. Somogyi. 1998. REVEAL: a general reverse engineering algorithm for inference of genetic network architecture. *Proc. Pacific Symp. Biocompu.* 3:18–29.

Liao, D. 2000. Gene conversion drives within genic sequences: concerted evolution of ribosomal RNA genes in bacteria and archaea. *J. Mol. Evol.* 51:305–317.

Licklider, L. J., C. C. Thoreen, J. Peng, and S. P. Gygi. 2002. Automation of nanoscale microcapillary liquid chromatography-tandem mass spectrometry with a vented column. *Anal. Chem.* 74:3076–3083.

Liesack, W., P. H. Janssen, F. A. Rainey, and N. L. Ward-Rainey, et al. 1997. Microbial diversity in soil: the need for a combined approach using molecular and cultivation techniques. In van Elsas, J. D., Trevors, J. T., and Wellington E. M. H. (eds), *Modern Soil Microbiology*, pp. 375–439. Marcel Dekker, New York.

Liesak, W. and E. Stackebrandt. 1992. Occurrence of novel groups of the domain Bacteria as revealed by analysis of genetic material isolated from an Australian terrestrial environment. *J. Bacteriol.* 174:5072–5078.

Lim, A., E. T. Dimalanta, K. D. Potamousis, and G. Yen, et al. 2001. Shotgun optical maps of the whole *Escherichia coli* O157:H7 genome. *Genome Res.* 11:1584–1593.

Lin, J., R. Qi, C. Aston, and J. Jing, et al. 1999. Whole-genome shotgun optical mapping of *Deinococcus radiodurans*. *Science* 285:1558–1562.

Lin, L., P. Ayala, J. Larson, and M. Mulks, et al. 1997. The *Neisseria* type 2 IgA1 protease cleaves LAMP1 and promotes survival of bacteria within epithelial cells. *Mol. Microbiol.* 24:1083–1094.

Lin, L.-F., J. Posfai, R. J. Roberts, and H. Kong. 2001. Comparative genomics of the restriction-modification systems in *Helicobacter pylori*. *Proc. Natl. Acad. Sci. USA* 98:2740–2745.

Lin, Y. S., H. M. Kieser, D. A. Hopwood, and C. W. Chen. 1993. The chromosomal DNA of *Streptomyces lividans* 66 is linear. *Mol. Microbiol.* 10:923–933.

Lina, G., A. Quaglia, and M. E. Reverdy, et al. 1999. Distribution of genes encoding resistance to macrolides, lincosamides, and streptogramins among staphylococci. *Antimicrob. Agents Chemother.* 43:1062–1066.

Lindahl, E. and E. Elofsson. 2001. Identification of related proteins on family, superfamily and fold level. *J. Mol. Biol.* 295:613–625.

Lindsay, J. A., A. Ruzin, H. F. Ross, and N. Kurepina, et al. 1998. The gene for toxic shock toxin is carried by a family of mobile pathogenicity islands in *Staphylococcus aureus*. *Mol. Microbiol.* 29:527–543.

Link, A. J., L. G. Hays, E. B. Carmack, and J. R. Yates 3rd. 1997c. Identifying the major proteome components of *Haemophilus influenzae* type-strain NCTC 8143. *Electrophoresis* 18:1314–1334.

Link, A. J., D. Phillips, and G. M. Church. 1997a. Methods for generating precise deletions and insertions in the genome of wild-type *Escherichia coli*: application to open reading frame characterization. *J. Bacteriol.* 179:6228–6237.

Link, A. J., J. Eng, D. M. Schieltz, and E. Carmack, et al. 1999. Direct analysis of protein complexes using mass spectrometry. *Nat. Biotechnol.* 17:676–682.

Link, A. J., K. Robison, and G. M. Church. 1997b. Comparing the predicted and observed properties of proteins encoded in the genome of *Escherichia coli* K-12. *Electrophoresis* 18:1259–1313.

Lipshultz, R. J., S. P. A. Fodor, T. R. Gingeras, and D. J. Lockhart. 1999. High density synthetic oligonucleotide arrays. *Nat. Genet.* 21:20–24.

Lipton, M. S., L. Pasa-Tolic, G. A. Anderson, and D. J. Anderson, et al. 2002. Global analysis of the *Deinococcus radiodurans* proteome by using accurate mass tags. *Proc. Natl. Acad. Sci. USA* 99:11049–11054.

Lisitsyn, N. and M. Wigler. 1995. Representational difference analysis in detection of genetic lesions in cancer. *Methods Enzymol.* 254:291–304.

Lisitsyn, N., N. Lisitsyn, and M. Wigler. 1993. Cloning the differences between two complex genomes. *Science* 259:946–951.

Lisitsyn, N. A., N. M. Lisitsina, and G. Dalbagni, et al. 1995. Comparative genomic analysis of tumors: detection of DNA losses and amplification. *Proc. Natl. Acad. Sci. USA* 92:151–155.

Little, D. P., J. P. Speir, M. W. Senko, and P. B. O'Connor, et al. 1994. Infrared multiphoton dissociation of large multiply charged ions for biomolecule sequencing. *Anal. Chem.* 66:2809–2815.

Liu, H., D. Lin., and J. R. Yates. 2002. Multidimensional separations for protein/peptide analysis in the post-genomic era. *Biotechniques* 32:898–902.

Liu, Y., M. P. Patricelli, and B. F. Cravatt. 1999. Activity-based protein profiling: the serine hydrolases. *Proc. Natl. Acad. Sci. USA* 96:14694–14699.

Liu, Y., J. Zhou, M. V. Omelchenko, and A. Beliaev, et al. 2003. Transcriptome dynamics of *Deinococcus radiodurans* recovering from ionizing radiation. *Proc. Natl. Acad. Sci. USA* 100:4191–4196.

Lockhart, D. J. and E. A. Winzeler. 2000. Genomics, gene expression and DNA arrays. *Nature* 405:827–836.

Lockhart, D. J., H. Dong, M. C. Byrne, and M. T. Follettie, et al. 1996. Expression monitoring by hybridization to high-density oligonucleotide arrays. *Nat. Biotechnol.* 14:1675–1680.

Lofdahl, S., J. E. Sjostrom, and L. Philipson. 1978. Characterization of small plasmids from *Staphylococcus aureus. Gene* 3:145–159.

Loferer, H. 2000. Mining bacterial genomes for antimicrobial targets. *Mol. Med. Today* 6:470–474.

Loferer, H., A. Jacobi, and A. Posch, et al. 2000. Integrated bacterial genomics for the discovery of novel antimicrobials. *Drug Disc. Today* 5:107–114.

Logsdon, J. M. and D. M. Faguy. 1999. Thermotoga heats up lateral gene transfer. *Curr. Biol.* 9:R747–R751.

Long, A. D., H. J. Mangalam, B. Y. Chan, and L. Tolleri, et al. 2001. Improved statistical inference from DNA microarray data using analysis of variance and a Bayesian statistical framework. Analysis of global gene expression in Escherichia coli K12, *J Biol Chem.* 276(23):19937–19944.

Long, A. D., H. J. Mangalam, B. Y. P. Chan, L. Tolleri, G. W. Hatfield, and P. Baldi. 2001. Global gene expression profiling in *Escherichia coli* K12: improved statistical inference from DNA microarray data using analysis of variance and a Bayesian statistical framework. *J. Biol. Chem.* 276:19937–19944.

Loo, J. A., J. P. Quinn, S. I. Ryu, and K. D. Henry, et al. 1992. High-resolution tandem mass spectrometry of large biomolecules. *Proc. Natl. Acad. Sci. USA* 89:286–289.

Lopez-Garcia, P., F. Gaill, and D. Moreira. 2002. Wide bacterial diversity associated with tubes of the vent worm *Riftia pachyptila. Environ. Microbiol.* 4:204–215.

Loy, A., A. Lahner, N. Lee, and J. Adamczyk, et al. 2002. Oligonucleotide microarray for 16S rRNA gene-based detection of all recognized lineages of sulfate-reducing prokaryotes in the environment. *Appl. Environ. Microbiol.* 68:5064–5081.

Loreau, M., S. Naeem, P. Inchausti, and J. Bengtsson, et al. 2001. Biodiversity and ecosystem functioning: current knowledge and future challenges. *Science* 294:804–808.

Lorenz, M. G. and W. Wackernagel. 1994. Bacterial gene transfer by natural genetic transformation in the environment. *Microbiol. Rev.* 58:563–602.

Lovett, R. A. 2000. Toxicogenomics: toxicologists brace for genomics revolution. *Science* 289:536–537.

Lovley, D. 1991. Dissimilatory Fe(III) and Mn(IV) reduction. *Microbiol. Rev.* 55:259–287.

Lowe, A. M., D. T. Beattie, and R. L. Deresiewicz. 1998. Identification of novel staphylococcal virulence genes by in vivo expression technology. *Mol. Microbiol.* 27:967–976.

Lowe, M. and S. R Eddy. 1997. tRNAscan-SE: a program for improved detection of transfer RNA genes in genomic sequence. *Nucl. Acids Res.* 25:955–964.

Lowe, T. M. and S. R. Eddy. 1999. A computational screen for methylation guide snoRNAs in yeast. *Science* 283:1168–1171.

Lucito, R., M. Nakimura, and J. A. West, et al. 1998. Genetic analysis using genomic representations. *Proc. Natl. Acad. Sci. USA* 95:4487–4492.

Ludwig, R. A. 1987. Gene tandem-mediated selection of coliphage-receptive Agrobacterium, Pseudomonas, and Rhizobium strains. *Proc. Natl. Acad. Sci. USA* 84:3334–3348.

Luenberger, D. C. 1979. *Introduction to Dynamic Systems.* John Wiley & Sons, New York.

Luna, V. A., P. Coates, and E. A. Eady, et al. 1999. A variety of Gram-positive bacteria carry mobile mef genes. *J. Antimicrob. Chemother.* 44:19–25.

Luna, V. A., S. Cousin Jr., W. L. H. Whittington, and M. C. Roberts. 2000. Identification of the conjugative mef gene in clinical *Acinetobacter junii* and *Neisseria gonorrhoeae* isolates. *Antimicrob. Agents Chemother.* 44:2503–2506.

Lund, P. A. 2001. Microbial molecular chaperones. *Adv. Microb. Physiol.* 44:93–140.

Lutz, R. and H. Bujard. 1997. Independent and tight regulation of transcriptional units in *Escherichia coli* via the LacR/O, the TetR/O and AraC/I1-I2 regulatory elements. *Nucl. Acids Res.* 25:1203–1210.

Lynch, M. and A. Force. 2000. The probability of duplicate gene preservation by subfunctionalization. *Genetics* 154:459–473.

Lynch, M. and J. S. Conery. 2000. The evolutionary fate and consequences of duplicate genes. *Science* 290:1151–1155.

Lyons, T. J., A. P. Gasch, L. A. Gaither, and D. Botstein, et al. 2000. Genome-wide characterization of the Zap1p zinc-responsive regulon in yeast. *Proc. Natl. Acad. Sci. USA* 97:7957–7962.

Ma, B., J. Tromp, and M. Li. 2002. PatternHunter: faster and more sensitive homology search. *Bioinformatics* 18:440–445.

Ma, H., S. Kunes, P. J. Schatz, and D. Botstein. 1987. Plasmid construction by homologous recombination in yeast. *Gene* 58:201–216.

MacBeath, G. and S. L. Schreiber. 2000. Printing proteins as microarrays for high-throughput function determination. *Science* 289:1760–1763.

MacBeath, G., A. N. Koehler, and S. L. Schreiber. 1999. Printing small molecules as microarrays and detecting protein-ligand interactions en masse. *J. Am. Chem. Soc.* 121:7967–7968.

MacDiarmid, C. W., L. A. Gaither, and D. Eide. 2000. Zinc transporters that regulate vacuolar zinc storage in *Saccharomyces cerevisiae*. *EMBO J.* 19:2845–2855.

MacDougall, J. and I. Saint Girons. 1995. Physical map of the *Treponema denticola* circular chromosome. *J. Bacteriol.* 177:1805–1811.

Mace, M. L., J. Montagu, S. D. Rose, and G. McGuinnes. 2000. Novel microarray printing and detection technologies. In M. Schena (ed.), *Microarray Biochip Technology*, pp. 39–64. Eaton Publishing, Natick, Mass.

Maeda, S., M. Akanuma, Y. Mitsuno, and Y. Hirata, et al. 2001. Distinct mechanism of *Helicobacter pylori*-mediated NF-kappa B activation between gastric cancer cells and monocytic cells. *J. Biol. Chem.* 276:44856–44864.

Mager, W. H. and A. J. J. de Kruijff. 1995. Stress-induced transcriptional regulation. *Microbiol. Rev.* 89:506–531.

Maglich, J. M., A. Sluder, X. Guan, and Y. Shi, et al. 2001. Comparison of complete nuclear receptor sets from the human, *Caenorhabditis elegans* and *Drosophila* genomes. *Genome Biol.* 2:29.1–29.7.

Mahan, M. J., J. M. Slauch, and J. J. Mekalanos. 1993. Selection of bacterial virulence genes that are specifically induced in host tissues. *Science* 259:686–688.

Mahan, M. J., J. W. Tobias, and J. M. Slauch, et al. 1995. Antibiotic-based selection for bacterial genes that are specifically induced during infection of a host. *Proc. Natl. Acad. Sci. USA* 92:669–673.

Mahillon, J. and M. Chandler. 1998. Insertion sequences. *Microbiol. Mol. Biol. Rev.* 62:725–774.

Makarova, K. S., L. Aravind, Y. I. Wolf, and R. L. Tatusov, et al. 2001. Genome of the extremely radiation-resistant bacterium *Deinococcus radiodurans* viewed from the perspective of comparative genomics. *Microbiol. Mol. Biol. Rev.* 65:44–79.

Maloy, S. R. and W. D. Nunn. 1981. Selection for loss of tetracycline resistance by *Escherichia coli*. *J. Bacteriol.* 145:1110–1111.

Manca, C., L. Tsenova, C. E. Barry, 3rd, and A. Bergtold, et al. 1999. *Mycobacterium tuberculosis* CDC1551 induces a more vigorous host response in vivo and in vitro, but is not more virulent than other clinical isolates. *J. Immunol.* 162:6740–6746.

Mann, M., R. C. Hendrickson, and A. Pandey. 2001. Analysis of proteins and proteomes by mass spectrometry. *Annu. Rev. Biochem.* 70:437–473.

Mar Alba, M., M. F. Santibanez-Koref, and J. M. Hancock. 2001. The comparative genomics of polyglutamine repeats: extreme difference in the codon organization of repeat-encoding regions between mammals and *Drosophila*. *J. Mol. Evol.* 52:249–259.

Marcotte, E. M., I. Xenarios, A. M. van Der Bliek, and D. Eisenberg. 2000. Localizing proteins in the cell from their phylogenetic profiles. *Proc. Natl. Acad Sci. USA* 97(22):12115–12120.

Marcotte, E. M., M. Pellegrini, H. L. Ng, and D. W. Rice, et al. 1999a. Detecting protein function and protein-protein interactions from genome sequences. *Science* 285:751–753.

Marcotte, E. M., M. Pellegrini, M. J. Thompson, and T. O. Yeates, et al. 1999b. A combined algorithm for genome-wide prediction of protein function. *Nature* 402:83–86.

Margulis, L. 1970. *Origin of Eukaryotic Cells*. Yale University Press, New Haven, Conn.

Margulis, L. 1981. *Symbiosis in Cell Evolution*. Freeman, San Francisco, Calif.

Margulis, L. and R. Guerrero. 1991. Kingdoms in turmoil. *New Sci.* 1761:46–50.

Markham, P. N. and A. A. Neyfakh. 2001. Efflux-mediated drug resistance in Gram-positive bacteria. *Curr. Opin. Microbiol.* 4:509–514.

Marshall, A. 2001. Milestones in Fourier transform ion cyclotron resonance mass spectrometry technique development. *Int. J. Mass Spec.* 2001:331–356.

Marshall, A. G., C. L. Hendrickson, and G. S. Jackson. 1998. Fourier transform ion cyclotron resonance mass spectrometry: a primer. *Mass Spec. Rev.* 17:1–35.

Marshall, E. 2000. Genome sequencing. Claim and counterclaim on the human genome. *Science* 288:242–243.

Marshall, A. and J. Hodgson. 1998. DWA chips: an array of possibilities. *Nat. Biotechnol.* 16:27–31.

Martin, A. P. and S. R. Palumbi. 1993. Body size, metabolic rate, generation time, and the molecular clock. *Proc. Natl. Acad. Sci. USA* 90:4087–4091.

Martin, D. B. and P. S. Nelson. 2001. From genomics to proteomics: techniques and applications in cancer research. *Trends Cell Biol.* 11:S60–S65.

Martin, S. E., J. Shabanowitz, D. F. Hunt, and J. A. Marto. 2000. Subfemtomole MS and MS/MS peptide sequence analysis using nano-HPLC micro-ESI fourier transform ion cyclotron resonance mass spectrometry. *Anal. Chem.* 72:4266–4274.

Martin, W. 1999. Mosaic bacterial chromosomes: a challenge en route to a tree of genomes. *Bioessays* 21:99–104.

Martin, W. and M. Muller. 1998. The hydrogen hypothesis for the first eucaryote. *Nature* 392:37–41.

Martindale, J., D. Stroud, E. R. Moxon, and C. M. Tang. 2000. Genetic analysis of *Escherichia coli* K1 gastrointestinal colonization. *Mol. Microbiol.* 37:1293–1305.

Martinsky, T. and P. Haje. 2000. Microarray tools, kits, reagents, and services. In M. Schena (ed.), *Microarray Biochip Technology*, pp. 201–220. Eaton Publishing, Natick, Mass.

Marton, M. J., J. L. DeRisi, and H. A. Bennett, et al. 1998. Drug target validation and identification of secondary drug target effects using DNA microarrays. *Nat. Med.* 4:1293–1301.

Martzen, M. R., S. M. McCraith, S. L. Spinellis, and F. M. Torres, et al. 1999. A biochemical genomics approach for identifying genes by the activity of their products. *Science* 286:1153–1155.

Maruyama, I. N., H. I. Maruyama, and S. Brenner. 1994. 1 foo: A 1 phage vector for the expression of foreign proteins. *Proc. Natl. Acad. Sci. USA* 91:8273–8277.

Maslov, S. and K. Sneppen. 2002. Specificity and stability in topology of protein networks. *Science* 296:910–913.

Masse, E. and S. Gottesman. 2002. A small RNA regulates the expression of genes involved in iron metabolism in *Escherichia coli. Proc. Natl. Acad. Sci. USA* 99:4620–4625.

Matsushita, M. and K. D. Janda. 2002. Histidine kinases as targets for new antimicrobial agents. *Bioorg. Med. Chem.* 10:855–867.

Maule, J. 1998. Pulse field gel electrophoresis. *Molecular Biotechnology.* 9:107–126.

Maurelli, A. T., R. E. Fernandez, C. A. Bloch, and C. K. Rode, et al. 1998. "Black holes" and bacterial pathogenicity: a large genomic deletion that enhances the virulence of *Shigella* spp. and enteroinvasive *Escherichia coli. Proc. Natl. Acad. Sci. USA* 95:3943–3948.

Mawuenyega, K. G., H. Kaji, Y. Yamuchi, and T. Shinkawa, et al. 2003. Large-scale identification of *Caenorhabditis elegans* proteins by multidimensional liquid chromatography-tandem mass spectrometry. *J. Proteome Res.* 2:23–35.

Maxam, A. M. and W. Gilbert. 1977. A new method for sequencing DNA. *Proc. Natl. Acad. Sci. USA* 74:560–564.

Mayer, M. L. and P. Hieter. 2000. Protein networks-built by association. *Nat. Biotechnol.* 18:1242–1243.

Mayer, T. U., T. M. Kapoor, S. J. Haggarty, and R. W. King, et al. 1999. Small molecule inhibitor of mitotic spindle bipolarity identified in a phenotype-based screen. *Science* 286:971–974.

Mayr, E. 1990. A natural system of organisms. *Nature* 348:491.

McAdams, H. H. and L. Shapiro. 1995. Circuit simulation of genetic networks. *Science* 269:650–656.

McClintock, B. 1952. Chromosome organization and gene expression. *Cold Spring Harbor Symp. Quant. Biol.* 16:13.

McCormack, A. L., D. M. Schieltz, B. Goode, and S. Yang, et al. 1997. Direct analysis and identification of proteins in mixtures by LC/MS/MS and database searching at the low-femtomole level. *Anal. Chem.* 69:767–776.

McCraith, S., T. Holtzman, B. Moss, and S. Fields. 2000. Genome-wide analysis of *vaccinia* virus protein-protein interactions. *Proc. Natl. Acad. Sci. USA* 97:4879–4884.

McDaniel, T. K., K. G. Jarvis, M. S. Donnersberg, and J. B. Kaper. 1995. A genetic locus of enterocyte effacement conserved among diverse enterobacterial pathogens. *Proc. Natl. Acad. Sci. USA* 92:1664–1668.

McDevitt, D. and T. J. Foster. 1995. Variation in the size of the repeat region of the fibrinogen receptor (clumping factor) of *Staphylococcus aureus* strains. *Microbiology* 141:937–943.

McGall, G. H. and J. A. Fidanza. 2001. Photolithographic synthesis of high-density oligonucleotide arrays. In J. B. Rampal (ed.), *DNA Arrays—Methods and Protocols—Nuts & Bolts*, pp. 71–101. Humana Press, Totowa, N.J.

McKevitt, M., K. Patel, D. Smajs, and M. Marsh, et al. 2003. Systematic cloning of *Treponema pallidum* open reading frames for protein expression and antigen discovery. *Genome Res.* 13:1665–1674.

McKusick, V. A. 1997. Genomics: structural and functional studies of genomes. *Genomics* 45:244–249.

McLafferty, F. W., D. M. Horn, K. Breuker, and Y. Ge, et al. 2001. Electron capture dissociation of gaseous multiply charged ions by Fourier-transform ion cyclotron resonance. *J. Am. Soc. Mass Spectrom.* 12:245–249.

McLuckey, S. A. and J. L. Stephenson, Jr. 1998. Ion/ion chemistry of high-mass multiply charged ions. *Mass Spectrom. Rev.* 17:369–407.

Mdluli, K., R. A. Slayden, and Y. Zhu, et al. 1998. Inhibition of a *Mycobacterium tuberculosis* beta-ketoacyl ACP synthase by isoniazid. *Science* 280:1607–1610.

Mecsas, J. 2002. Use of signature-tagged mutagenesis in pathogenesis studies. *Curr. Opin. Microbiol.* 5:33–37.

Mecsas, J. and E. J. Strauss. 1996. Molecular mechanisms of bacterial virulence: type III secretion and pathogenicity islands. *Emerging Infect. Diseases* 2:271–288.

Medigue, C., T. Rouxel, P. Vigier, and A. Henaut, et al. 1991. Evidence for horizontal gene transfer in *Escherichia coli* speciation. *J. Mol. Biol.* 222:851–856.

Mei, J. M., F. Nourbakhsh, C. W. Ford, and D. W. Holden. 1997. Identification of *Staphylococcus aureus* virulence genes in a murine model of bacteraemia using signature-tagged mutagenesis. *Mol. Microbiol.* 26:399–407.

Mekalanos, J. J. 1992. Environmental signals controlling expression of virulence determinants in bacteria. *J. Bacteriol.* 174:1–7.

Mekalanos, J. J., D. J. Swartz, G. D. Pearson, and N. Harford, et al. 1983. Cholera toxin genes: nucleotide sequence, deletion analysis and vaccine development. *Nature* 306:551–557.

Meltzer, P. S. 2001. Spotting the target: microarrays for disease gene discovery. *Curr. Opin. Genet. Dev.* 11:258–263.

Meng, F., B. J. Cargile, L. M. Miller, and A. J. Forbes, et al. 2001. Informatics and multiplexing of intact protein identification in bacteria and the archaea. *Nat. Biotech.* 19:952–957.

Meng, F., B. J. Cargile, S. M. Patrie, and J. R. Johnson, et al. 2002. Processing complex mixtures of intact proteins for direct analysis by mass spectrometry. *Anal. Chem.* 74:2923–2929.

Merchant, M. and S. R.Weinberger. 2000. Recent advancements in surface-enhanced laser desorption/ionization-time of flight-mass spectrometry. *Electrophoresis* 21:1164–1167.

Mercier, G., Y. Denis, P. Marc, and L. Picard, et al. 2001. Transcriptional induction of repair genes during slowing of replication in irradiated *Saccharomyces cerevisiae*. *Mutation Res.* 487:157–172.

Merrifield, R. B. 1963. Solid phase peptide synthesis. I. The synthesis of a tetrapeptide. *J. Am. Chem. Soc.* 85:2149–2154.

Mewes, H. W., D. Frishman, C. Gruber, and B. Geier, et al. 2000. MIPS: a database for genomes and protein sequences. *Nucl. Acids Res.* 28:37–40.

Mewes, H. W., D. Frishman, U. Guldener, and G. Mannhaupt, et al. 2002. MIPS: a database for genomes and protein sequences. *Nucl. Acids Res.* 30:31–34.

Mewes, H. W., K. Albermann, K. Heumann, and S. Liebl, et al. 1997a. MIPS: a database for protein sequences, homology data and yeast genome information. *Nucl. Acids Res.* 25:28–30.

Mewes, H. W., K. Albermann, M. Bahr, and D. Frishman, et al. 1997b. Overview of the yeast genome. *Nature* 387:7–65.

Meyer, T. F., C. P. Gibbs, and R. Haas. 1990. Variation and control of protein expression in *Neisseria*. *Annu. Rev. Microbiol.* 44:451–477.

Michal, G. 1999. *Biochemical Pathways*. John Wiley & Sons, New York.

Miller, S. I., A. M. Kukral, and J. J. Mekalanos. 1989. A two-component regulatory system (phoP phoQ) controls *Salmonella typhimurium* virulence. *Proc. Natl. Acad. Sci. USA* 86:5054–5058.

Miller, V. L. and J. J. Mekalanos.1988. A novel suicide vector and its use in construction of insertion mutations: osmoregulation of outer membrane proteins and virulence determinants in *Vibrio cholerae* requires toxR. *J. Bacteriol.* 170:2575–2583.

Milohanic, E., P. Glaser, J. Y. Coppee, and L. Frangeul, et al. 2003. Transcriptome analysis of *Listeria monocytogenes* identifies three groups of genes differently regulated by PrfA. *Mol. Microbiol.* 47:1613–1625.

Milosavljevic, A., S. Savkovic, R. Crkvenjakov, and D. Salbego, et al. 1996. DNA sequence recognition to short oligomers: experimental verification of the method on the *E. coli* genome. *Genomics* 37:77–86.

Minton, K. W. 1996. Repair of ionizing-radiation damage in the radiation resistant bacterium *Deinococcus radiodurans*. *Mutat. Res.* 363:1–7.

Mir, K. U. 2000. The hypothesis is there is no hypothesis. *Trends Genet.* 16:63–64.

Mira, A., H. Ochman, and N. A. Moran. 2001. Deletional bias and the evolution of bacterial genomes. *Trends Genet.* 17:589–596.

Mira, A., L. Klasson, and S. G. Andersson. 2002. Microbial genome evolution: sources of variability. *Curr. Opin. Microbiol.* 5:506–512.

Missiakas, D., M. P. Mayer, M. Lemaire, and C. Georgopoulos, et al. 1997. Modulation of the *Escherichia coli* sigmaE (RpoE) heat-shock transcription-factor activity by the RseA, RseB and RseC proteins. *Mol. Microbiol.* 24:355–371.

Mitchison, T. J. 1994. Towards a pharmacological genetics. *Chem. Biol.* 1:3–6.

Mizunuma, M., D. Hirata, K. Miyahara, and E. Tsuchiya, et al. 1998. Role of calcineurin and Mpk1 in regulating the onset of mitosis in budding yeast. *Nature* 392:303–306.

Mohler, W. A. and H. M. Blau. 1996. Gene expression and cell fusion analyzed by lacZ complementation in mammalian cells. *Proc. Natl. Acad. Sci. USA* 93:12423–12427.

Moir, D. T., K. J. Shaw, R. S. Hare, and G. F. Vovis. 1999. Genomics and antimicrobial drug discovery. *Antimicrob. Agents Chemother.* 43:439–446.

Molina, F., D. Laune, C. Gougat, and B. Pau, et al. 1996. Improved performances of spot multiple peptide synthesis. *Pept. Res.* 9:151–155.

Molloy, M. P., B. R. Herbert, M. B. Slade, and T. Rabilloud, et al. 2000. Proteomic analysis of the *Escherichia coli* outer membrane. *Eur. J. Biochem.* 267:2871–2881.

Molloy, M. P., N. D. Phadke, J. R. Maddock, and P. C. Andrews. 2001. Two-dimensional electrophoresis and peptide mass fingerprinting of bacterial out membrane proteins. *Electrophoresis* 22:1686–1696.

Montelione, G. T. and S. Anderson. 1999. Structural genomics: keynote for a Human Proteome Project. *Nat. Struct. Biol.* 6:11–12.

Moore, P. B. 1999. Structural motifs in RNA. *Annu. Rev. Biochem.* 68:287–300.

Moran, N. A. 2002. Microbial minimalism: genome reduction in bacterial pathogens. *Cell* 108:583–586.

Moran, N. A. and J. J. Wernegreen. 2000. Lifestyle evolution in symbiotic bacteria: insights from genomics. *Trends Ecol. Evol.* 15:321–326.

Moran, N. A., M. A. Munson, P. Baumann, and H. Ishikawa. 1993. A molecular clock in endosymbiotic bacteria is calibrated using the insect hosts. *Proc. Roy. Soc. London* 253:167–171.

Moreira, D. and H. Philippe. 2000. Molecular phylogeny: pitfalls and progress. *Int. Microbiol.* 3:9–16.

Moreira, J. M. and S. Holmberg. 2000. Chromatin-mediated transcriptional regulation by the yeast architectural factors NHP6A and NHP6B. *EMBO J.* 19:6804–6813.

Morell, V. 1996. Life's last domain. *Science* 273:1043–1045.

Moreno, E. 1998. Genome evolution within the alpha Proteobacteria: why do some bacteria not possess plasmids and others exhibit more than one different chromosome? *FEMS Microbiol. Rev.* 22:255–275.

Morgan, D. O. 1997. Cyclin-dependent kinases: engines, clocks, and microprocessors. *Annu. Rev. Cell Dev. Biol.* 13:261–291.

Mori, H., R. E. Palmer, and P. W. Sternberg. 1994. The identification of a *Caenorhabditis elegans* homolog of p34cdc2 kinase. *Mol. Gen. Genet.* 245:781–786.

Morimoto, R. I., K. D. Sarge, and K. Abravaya. 1992. Transcriptional regulation of heat shock genes. *J. Biol. Chem.* 267:21987–21990.

Morino, Y. and E. E. Snell. 1967. A kinetic study of the reaction mechanism of tryptophanase-catalyzed reactions. *J. Biol. Chem.* 242:2793–2799.

Morrison, N., et al. Nature Genetics Microarray Meeting, Scottsdale, Ariz., 1999

Morton, T. A and D. G. Myszka. 1998. Kinetic analysis of macromolecular interactions using surface plasmon resonance biosensors. *Methods Enzymol.* 295:268–294.

Mortz, E., O. Vorm, M. Mann, and P. Roepstorff. 1994. Identification of proteins in polyacrylamide gels by mass spectrometric peptide mapping combined with database search. *Biol. Mass Spectrom.* 23:249–261.

Mortz, E., P. B. O'Connor, P. Roepstorff, and N. L. Kelleher, et al. 1996. Sequence tag identification of intact proteins by matching tandem mass spectral data against sequence data bases. *Proc. Natl. Acad. Sci. USA* 93:8264–8267.

Moser, D. and K. H. Nealson. 1996. Growth of the facultative anaerobe *Shewanella putrefaciens* by elemental sulfur reduction. *Appl. Environ. Microbiol.* 62:2100–2105.

Mosser, J., A. M. Douar, C. O. Sarde, and P. Kioschis, et al. 1993. Putative X-linked adrenoleukodystrophy gene shares unexpected homology with ABC transporters. *Nature* 361:726–730.

Moult, J. and E. Melamud. 2000. From fold to function. *Curr. Opin. Struct. Biol.* 10:384–389.

Mullins, T. D., T. B. Britschgi, R. L. Krest, and S. J. Giovannoni. 1995. Genetic comparisons reveal the same unknown bacterial lineages in Atlantic and Pacific bacterioplankton communities. *Limnol. Oceanogr.* 40:148–158.

Murray, A. E., C. M. Preston, R. Massana, and L. T. Taylor, et al. 1998. Seasonal and spatial variability of bacterial and archaeal assemblages in the coastal waters off Anvers Island, Antarctica. *Appl. Environ. Microbiol.* 64:2585–2595.

Movahedi, S. and W. Waites. 2000. A two-dimensional protein gel electrophoresis study of the heat stress response of *Bacillus subtilis* cells during sporulation. *J. Bacteriol.* 182:4758–4763.

Msadek, T., F. Kunst, and G. Rapoport. 1995. A signal transduction network in *Bacillus subtilis* includes the DegS/DegU and ComP/ComA two-component systems. In J. A. Hoch, and T. J.

Silhavy, (eds.), *Two-Component Signal Transduction*, pp. 447–471. American Society for Microbiology, Washington, D.C.

Muchardt, C. and M. Yaniv. 1999. ATP-dependent chromatin remodeling: SWI/SNF and Co. are on the job. *J. Mol. Biol.* 293:187–198.

Muller-Hill, B. 1996. *The lac Operon: A Short History of a Genetic Paradigm.* Walter de Gruyter & Co., Berlin.

Murphy, D. J., S. Hardy, and D. A. Engel. 1999. Human SWI-SNF component BRG1 represses transcription of the c-fos gene. *Mol. Cell. Biol.* 19:2724–2733.

Murphy, K. C. 1998. Use of bacteriophage lambda recombination functions to promote gene replacement in *Escherichia coli. J. Bacteriol.* 180:2063–2071.

Murphy, K. C., K. G. Campellone, and A. R. Poteete. 2000. PCR-mediated gene replacement in *Escherichia coli.* Gene 246:321–330.

Murray, A. E., D. Lies, G. Li, and K. Nealson, et al. 2001. DNA/DNA hybridization to microarrays reveals gene-specific differences between closely related microbial genomes. *Proc. Natl. Acad. Sci. USA* 98:9853–9858.

Murzin, A. G. 1993. OB(oligonucleotide/oligosaccharide binding)-fold: common structural and functional solution for non-homologous sequences. *EMBO J.* 12:861–867.

Murzin, A. G., S. E. Brenner, T. Hubbard, and C. Chothia. 1995. SCOP: a structural classification of proteins database for the investigation of sequences and structures. *J. Mol. Biol.* 247:536–540.

Mushegian, A. R. and E. V. Koonin. 1996. A minimal gene set for cellular life derived by comparison of complete bacterial genomes. *Proc. Natl. Acad. Sci. USA* 93:10268–10273.

Muto, A. and S. Osawa. 1987. The guanine and cytosine content of genomic DNA and bacterial evolution. *Proc. Natl. Acad. Sci. USA* 84:166–169.

Nadon, R. and J. Shoemaker. 2002. Statistical issues with microarrays: processing and analysis. *Trends Genet.* 18(5):265–271.

Nadon, R., P. Shi, A. Skandalis, and E. Woody, et al. 2001. Statistical inference methods for gene expression arrays. In M. L. Bittner et al. (eds.), *Microarrays: Optical Technologies and Informatics*, Vol 4266, pp. 46–55. SPIE, Bellingham, WA.

Nagahashi, S., H. Nakayama, K. Hamada, and H. Yang, et al. 1997. Regulation by tetracycline of gene expression in *Saccharomyces cerevisiae. Mol. Gen. Genet.* 255:372–375.

Nakanishi T., N. Okamoto, K. Tanaka, and A. Shimizu. 1994. Laser-desorption time-of-flight mass-spectrometric analysis of transferrin precipitated with antiserum—a unique simple method to identify molecular-weight variants. *Biol. Mass Spectrom.* 23:230–233.

Nakano, M. M. and P. Zuber. 1998. Anaerobic growth of a "strict aerobe" (*Bacillus subtilis*). *Annu. Rev. Microbiol.* 52:165–190.

Nakano, M. M., Y. P. Dailly, P. Zuber, and D. P. Clark. 1997. Characterization of anaerobic fermentative growth of *Bacillus subtilis*: identification of fermentation end products and genes required for growth. *J. Bacteriol.* 179:6749–6755.

Nakata, N., T. Tobe, I. Fukuda, and T. Suzuki, et al. 1993. The absence of a surface protease, OmpT, determines the intercellular spreading ability of *Shigella*: the relationship between the ompT and kcpA loci. *Mol. Microbiol.* 9:459–468.

Narahashi, T., J. W. Moore, and W. R. Scott. 1964. Tetrodotoxin blockage of sodium conductance increase in lobster giant axons. *J. Gen. Physiol.* 47:965–974.

Nealson, K. H. and D. A. Saffarini. 1994. Iron and manganese in anaerobic respiration: environmental significance, physiology, and regulation. *Ann. Rev. Microbiol.* 48:311–343.

Needleman, S. B. and C. D. Wunsch. 1970. A general method applicable to the search for similarities in the amino acid sequence of two proteins. *J. Mol. Biol.* 48:443–453.

Nei, M., P. Xu, and G. Glazko. 2001. Estimation of divergence times from multiprotein sequences for a few mammalian species and several distantly related organisms. *Proc. Natl. Acad. Sci. USA* 98:2497–2502.

Neidhardt, F. C. and H. E. Umbarger. 1996. Chemical composition of *Escherichia coli*. In F. C. Neidhardt, R. Curtiss III, J. L. Ingraham, and E. C. C. Lin, et al. (eds.), Escherichia coli *and* Salmonella*: Cellular and Molecular Biology*, 2nd ed., pp. 13–16. ASM Press, Washington, D.C.

Nelissen, B., R. De Wachter, and A. Goffeau. 1997. Classification of all putative permeases and other membrane plurispanners of the major facilitator superfamily encoded by the complete genome of *Saccharomyces cerevisiae*. *FEMS Microbiol. Rev.* 21:113–134.

Nelson, K. E., R. A. Clayton, S. R. Gill, and M. L. Gwinn, et al. 1999. Evidence for lateral gene transfer between Archaea and Bacteria from genome sequence of *Thermotoga maritima*. *Nature* 399:323–329.

Nelson, M., W. Humphrey, A. Gursoy, and A. Dalke, et al. 1996. NAMD—a parallel, object-oriented molecular dynamics program. *Internat. J. Supercomp. Appl. High Perf. Comput.* 10:251–268.

Nelson, M. L. and S. B. Levy. 1999. Reversal of tetracycline resistance mediated by different bacterial tetracycline resistance determinants by an inhibitor of the Tet(B) antiport protein. *Antimicrob. Agents Chemother.* 43:1719–1724.

Nelson, R. W., D. Nedelkov, and K. A. Tubbs. 2000. Biosensor chip mass spectrometry: a chip-based proteomics approach. *Electrophoresis* 21:1155–1163.

Nelson, R. W., J. R. Krone, and O. Jansson. 1997. Surface plasmon resonance biomolecular interaction analysis mass spectrometry. 1. Chip-based analysis. *Anal. Biochem.* 69:4363–4368.

Nemeth-Cawley, J. F. and J. C. Rouse. 2002. Identification and sequencing analysis of intact proteins via collision-induced dissociation and quadrupole time-of-flight mass spectrometry. *J. Mass Spectrom.* 37:270–282.

Neubauer, G., A. King, J. Rappsilber, and C. Calvio, et al. 1998. Mass spectrometry and EST-database searching allows characterization of the multi-protein spliceosome complex. *Nat. Genet.* 20:46–50.

Nevo, E. 2001. Evolution of genome-phenome diversity under environmental stress. *Proc. Natl. Acad. Sci. USA* 98:6233–6240.

Newland, J. W., B. A. Green, J. Foulds, and R. K. Holmes. 1985. Cloning of extracellular DNase and construction of a DNase-negative strain of *Vibrio cholerae*. *Infect. Immun.* 47:691–696.

Newman, J. R., E. Wolf, and P. S. Kim. 2000. A computationally directed screen identifying interacting coiled coils from *Saccharomyces cerevisiae*. *Proc. Natl. Acad. Sci. USA* 97:13203–13208.

Newman, J. R. S. and A. E. Keating. 2003. Comprehesive identification of human bZIP interactions with coiled-coil arrays. *Science* 300:2097–2101.

Ng, W. V., S. P. Kennedy, G. G. Mahairas, and B. Berquist, et al. 2000. Genome sequence of *Halobacterium* species NRC-1. *Proc. Natl. Acad. Sci. USA* 97:12176–12181.

Nicolas, P., L. Bize, F. Muri, and M. Hoebeke, et al. 2002. Mining *Bacillus subtilis* chromosome heterogeneities using hidden Markov models. *Nucl. Acids Res.* 30:1418–1426.

Nielsen, H., J. Engelbrecht, S. Brunak, and G. von Heijne. 1997. Identification of prokaryotic and eukaryotic signal peptides and prediction of their cleavage sites. *Prot. Eng.* 10:1–6.

Nilsson, M., L. Frykberg, J. I. Flock, and L. Pei, et al. 1998. A fibrinogen-binding protein of *Staphylococcus epidermidis*. *Infect. Immun.* 66:2666–2673.

Nordstrom, K. and E. G. Wagner. 1994. Kinetic aspects of control of plasmid replication by antisense RNA. *Trends Biochem. Sci.* 19:294–300.

Nogales, B., E. R. B. Moore, W. R. Abraham, and K. N. Timmis. 1999. Identification of the metabolically active members of a bacterial community in a polychlorinated biphenyl-polluted moorland soil. *Environ. Microbiol.* 1:199–212.

Normile, D. and E. Pennisi. 2002. The rice genome. Rice: boiled down to bare essentials. *Science* 296:32–36.

Nouwens, A. S., S. J. Cordwell, M. R. Larsen, and M. P. Molloy, et al. 2000. Complementing genomics with proteomics: the membrane subproteome of *Pseudomonas aeruginosa* PA01. *Electrophoresis* 21:3797–3809.

Novick, R. P. and F. C. Hoppensteadt. 1978. On plasmid incompatibility. *Plasmid* 1:421–434.

Nuwaysir, E. F., M. Bittner, and J. Trent, et al. 1999. Microarrays and toxicology: the advent of toxicogenomics. *Mol. Carcinog.* 24:153–159.

Nuwaysir, E. F., W. Huang, T. J. Albert, and J. Singh, et al. 2002. Gene expression analysis using oligonucleotide arrays produced by maskless photolithography. *Genome Res.* 12:1749–1755.

O'Donovan, C., R. Apweiler, and A. Bairoch. 2001. The human proteomics initiative (HPI). *Trends Biotechnol.* 19:178–181.

O'Neill, E. M., A. Kaffman, E. R. Jolly, and E. K. O'Shea. 1996. Regulation of PHO4 nuclear localization by the PHO80-PHO85 cyclin-CDK complex. *Science* 271:209–212.

Ochman, H. 2002. Distinguishing the ORFs from the ELFs: short bacterial genes and the annotation of genomes. *Trends Genet.* 18:335–337.

Ochman, H. and A. C. Wilson. 1987. Evolution in bacteria: evidence for a universal substitution rate in cellular genomes. *J. Mol. Evol.* 26:74–86.

Ochman, H. and I. B. Jones. 2000. Evolutionary dynamics of full genome content in *Escherichia coli*. *EMBO J.* 19:6637–6643.

Ochman, H. and N. A. Moran. 2001. Genes lost and genes found: evolution of bacterial pathogenesis and symbiosis. *Science* 292:1096–1099.

Ochman, H., J. G. Lawrence, and E. A. Groisman. 2000. Lateral gene transfer and the nature of bacterial innovation. *Nature* 405:299–304.

Ochman, H., S. Elwyn, and N. A. Moran. 1999. Calibrating bacterial evolution. *Proc. Natl. Acad. Sci. USA* 96:12638–12643.

Ochs, M. F. and G. Bidaut. 2002. Microarray data normalization. In S. Shah and G. Kamberova (eds.), *DNA Array Image Analysis—Nuts & Bolts*, pp. 131–154. DNA Press, Eagleville, Penn.

Oda, Y., K. Huang, F. R. Cross, and D. Cowburn, et al. 1999. Accurate quantitation of protein expression and site-specific phosphorylation. *Proc. Natl. Acad. Sci. USA* 96:6591–6596.

Odenbreit, S., M. Till, D. Hofreuter, and G. Faller, et al. 1999. Genetic and functional characterization of the alpAB gene locus essential for the adhesion of *Helicobacter pylori* to human gastric tissue. *Mol. Microbiol.* 31:1537–1548.

O'Donovan, C., M. Jesus Martin, E. Glemet, and J. J. Codani, et al. 1999. Removing redundancy in SWISS-PROT and TrEMBL. *Bioinformatics* 15:258–259.

Offringa, R. and F. van der Lee. 1995. Isolation and characterization of plant genomic DNA sequences via (inverse) PCR amplification. *Methods Mol. Biol.* 49:181–195.

Ogata, H., H. Bono, W. Fujibuchi, and S. Goto, et al. 1996. Analysis of binary relations and hierarchies of enzymes in the metabolic pathways. *Genome Inform.* 7:128–136.

Ogorzalek Loo, R. R., J. D. Cavalcoli, R. A. VanBogelen, and C. Mitchell, et al. 2001. Virtual 2-D gel electrophoresis: visualization and analysis of the *E. coli* proteome by mass spectrometry. *Anal. Chem.* 73:4063–4070.

Ogura, M., H. Yamaguchi, K.-I. Yoshida, and Y. Fujita, et al. 2001. DNA microarray analysis of *Bacillus subtilis* DegU, ComA and PhoP regulons: an approach to comprehensive analysis of *B. subtilis* two-component regulatory systems. *Nucl. Acids Res.* 29:3804–3813.

Ohlmeier, S., C. Scharf, and M. Hecker. 2000. Alkaline proteins of *Bacillus subtilis*: first steps towards a two-dimensional alkaline master gel. *Electrophoresis* 21:3701–3709.

Ohno, S. 1970. *Evolution by Gene Duplication*. Springer, New York.

Ohta, T. 1987. Very slightly deleterious mutations and the molecular clock. *J. Mol. Evol.* 26:1–6.

Ohta, T. 1995. Synonymous and nonsynonymous substitutions in mammalian genes and the nearly neutral theory. *J. Mol. Evol.* 40:56–63.

Ojaimi, C., C. Brooks, S. Casjens, and P. Rosa, et al. 2003. Profiling of temperature-induced changes in *Borrelia burgdorferi* gene expression by using whole genome arrays. *Infect. Immun.* 71:1689–1705.

Oka, A., K. Sugimoto, M. Takanami, and Y. Hirota. 1980. Replication origin of the *Escherichia coli* K-12 chromosome: the size and structure of the minimum DNA segment carrying the information for autonomous replication. *Mol. Gen. Genet.* 178:9–20.

Oliver, S. 1996. A network approach to the systematic analysis of yeast gene function. *Trends Genet.* 12:241–242.

Oliver, S. G. 2002. Functional genomics: lessons from yeast. *Phil. Trans. Roy. Soc. Lond. B Biol. Sci.* 357:17–23.

Olman, V., D. Xu, and Y. Xu. 2003. Identification of regulatory binding sites using minimum spanning trees. In *Proceedings of the 7th Pac. Symp. Biocomput.* 327–338.

Olsen, G. J. and C. R. Woese. 1997. Archaeal genomics: an overview. *Cell* 89:991–994.

Olsen, G. J., C. R. Woese, and R. Overbeek. 1994. The winds of (evolutionary) change: breathing new life into microbiology. *J. Bacteriol.* 176:1–6.

Olson, M. V. 1999. When less is more: gene loss as an engine of evolutionary change. *Am. J. Hum. Genet.* 64:18–23.

Olsson, L., M. E. Larsen, B. Ronnow, and J. D. Mikkelsen, et al. 1997. Silencing MIG1 in *Saccharomyces cerevisiae*: effects of antisense MIG1 expression and MIG1 gene disruption. *Appl. Environ. Microbiol.* 63:2366–2371.

Orengo, C. A., A. D. Michie, D. T. Jones, and M. B. Swindells, et al. 1998. CATH—a hierarchical classification of protein domain structures. *Structure* 5:1093–1108.

Ovreas, L. and V. V. Torsvik. 1998. Microbial Diversity and Community Structure in Two Different Agricultural Soil Communities. *Microb. Ecol.* 36:303–315.

Ovreas, L., F. L. Daae, M. Heldal, and F. Rodríguez-Valera, et al. 2001. Paper presented at the 9th International Symposium on Microbial Ecology: Interaction in the Microbial World, Amsterdam, 26 to 31 August.

Paabo, S. 2001. Genomics and society. The human genome and our view of ourselves. *Science* 291:1219–1220.

Pablos-Mendez, A., M. C. Raviglione, A. Laszlo, and N. Binkin, et al. 1998. Global surveillance for antituberculosis-drug resistance, 1994–1997. World Health Organization-International Union against tuberculosis and lung disease working group on anti-tuberculosis drug resistance surveillance. *N. Engl. J. Med.* 338:1641–1649.

Pace, N. R. 1997. A molecular view of microbial diversity and the biosphere. *Science* 276:734–740.

Pandey, A, A. V. Podtelejnikov, B. Blagoev, and X. R. Bustelo, et al. 2000. Analysis of receptor signal pathways by mass spectrometry: identification of vav-2 as a substrate of the epidermal and platelet-derived growth factor receptors. *Proc. Natl. Acad. Sci. USA* 97:179–184.

Pandey, A. and M. Mann. 2000. Proteomics to study genes and genomes. *Nature* 405:837–846.

Pao, S. S., I. T. Paulsen, and M. H. Saier, Jr. 1998. Major facilitator superfamily. *Microbiol. Mol. Biol. Rev.* 62:1–34.

Parish, T. and N. G. Stoker. 2000. Use of a flexible cassette method to generate a double unmarked *Mycobacterium tuberculosis* tlyA plcABC mutant by gene replacement. *Microbiology* 146:1969–1975.

Parke, D. 1990. Construction of mobilizable vectors derived from plasmids RP4, pUC18 and pUC19. *Gene* 93:135–137.

Paša-Tolić, L., P. K. Jensen, G. A. Anderson, and M. S. Lipton, et al. 1999. High-throughput proteome-wide precision measurements of protein expression using mass spectrometry. *J. Am. Chem. Soc.* 121:7949–7950.

Pasqualini, R. and E. Ruoslahti. 1996. Organ targeting in vivo using phage display peptide libraries. *Nature* 380:364–366.

Pastinen, T., A. Kurg, A. Metspalu, and L. Peltonen, et al. 1997. Minisequencing: a specific tool for DNA analysis and diagnostics on oligonucleotide arrays. *Genome Res.* 7:606–614.

Pastinen, T., M. Raitio, K. Lindroos, and P. Tainola, et al. 2000. A system for specific, high-throughput genotyping by allele-specific primer extension on microarrays. *Genome Res.* 10:1031–1042.

Pato, M. L. 1989. Bacteriophage Mu. In D. E. Berg and M. M. Howe (eds.), *Mobile DNA*, pp. 23–52. American Society for Microbiology, Washington, D.C.

Patrinos, A. 1996. *To Know Ourselves*. U. S. Dept. Energy, pp. 2–3. Washington, D.C.

Patterson, S. D. and R. Aebersold. 1995. Mass spectrometric approaches for the identification of gel-separated proteins. *Electrophoresis* 16:1791–1814.

Patti, J. M., B. L. Allen, M. J. McGavin, and M. Hook. 1994. MSCRAMM-mediated adherence of microorganisms to host tissues. *Annu. Rev. Microbiol.* 48:585–617.

Patti, J. M., K. Jonsson, B. Guss, and L. M. Switalski, et al. 1992. Molecular characterization and expression of a gene encoding a *S. aureus* collagen adhesin. *J. Biol. Chem.* 267:4766–4772.

Paustian, M. L., B. J. May, and V. Kapur. 2001. *Pasteurella multocida* gene expression in response to iron limitation. *Infect. Immun.* 69:4109–4115.

Pavesi, A., F. Conterio, A. Bolchi, and G. Dieci, et al. 1994. Identification of new eukaryotic tRNA genes in genomic DNA databases by a multistep weight matrix analysis of transcriptional control regions. *Nucl. Acids Res.* 22(7):1247–1256.

Pavletich, N. P. and C. O. Pabo. 1991. Zinc finger-DNA recognition: crystal structure of a Zif268-DNA complex at 2.1 A. *Science* 252:809–817.

Pawson, T. and J. D. Scott. 1997. Signaling through scaffold, anchoring, and adaptor proteins. *Science* 278:2075–2080.

Pawson, T. and P. Nash. 2000. Protein-protein interactions define specificity in signal transduction. *Genes Dev.* 14:1027–1047.

Pearson, W. R. 1990. Rapid and sensitive sequence comparison with FASTP and FASTA. *Methods Enzymol.* 183:63–98.

Pearson, W. R. and D. J. Lipman. 1988. Improved tools for biological sequence comparison. *Proc. Natl. Acad. Sci. USA* 85(8):2444–2448.

Pease, A. C., D. Solas, E. J. Sullivan, and M. T. Cronin, et al. 1994. Light-generated oligonucleotide arrays for rapid DNA sequence analysis. *Proc. Nat. Acad. Sci. USA* 91:5022–5026.

Peck, B., M. Ortkamp, K. D. Diehl, and E. Hundt, et al. 1999. Conservation, localization and expression of HopZ, a protein involved in adhesion of *Helicobacter pylori*. *Nucl. Acids Res.* 27:3325–3333.

Pedersen, A. G., P. Baldi, S. Brunak, and Y. Chauvin. 1996. Characterization of prokaryotic and eukaryotic promoters using hidden Markov models. *Proc. Int. Conf. Intell. Syst. Mol. Biol.* 4:182–191.

Pelicic, V., J.-M. Reyrat, and B. Gicquel. 1998. Genetic advances for studying *Mycobacterium tuberculosis* pathogenicity. *Mol. Microbiol.* 28:413–420.

Pellegrini, M., E. M. Marcotte, M. J. Thompson, and D. Eisenberg, et al. 1999. Assigning protein functions by comparative genome analysis: protein phylogenetic profiles. *Proc. Natl. Acad. Sci. USA* 96:4285–4288.

Pelletier, J. N., F.-X. Campbell-Valois, and S. W. Michnick. 1998. Oligomerization domain-directed reassembly of active dihydrofolate reductase from rationally designed fragments. *Proc. Natl. Acad. Sci. USA* 95:12141–12146.

Peng J. and S. P. Gygi. 2001. Proteomics: the move to mixtures. *J. Mass Spec.* 36:1083–1091.

Peng, J., J. E. Elias, C. C. Thoreen, and L. J. Licklider, et al. 2002. Evaluation of multidimensional chromatography coupled with tandem mass spectrometry (LC/LC-MS/MS) for large-scale protein analysis: the yeast proteome. *J. Proteome Res.* 2:43–50.

Penn, S. G., D. R. Rank, D. K. Hanzel, and D. L. Barker. 2000. Mining the human genome using microarrays of open reading frames. *Nat. Genet.* 26:315–318.

Pennie, W. D. 2000. Use of cDNA microarrays to probe and understand the toxicological consequences of altered gene expression. *Toxicol. Lett.* 112–113:473–477.

Pennington, S. R. and M. J. Dunn (eds.). 2001. *Proteomics: From Protein Sequence to Function.* Springer-Verlag, New York.

Pennisi, E. 1998. Genome data shake tree of life. *Science* 280:672–674.

Pennisi, E. 1999. Is it time to uproot the tree of life? *Science* 284:1305–1307.

Perkins, D. N., D. J. Pappin, D. M. Creasy, and J. S. Cottrell. 1999. Probability-based protein identification by searching sequence databases using mass spectrometry data. *Electrophoresis* 20:3551–3567.

Perna, N. T., G. Plunkett, 3rd, V. Burland, and B. Mau, et al. 2001. Genome sequence of enterohaemorrhagic *Escherichia coli* O157:H7. *Nature* 409:529–533.

Persson, C., E. G. Wagner, and K. Nordstrom. 1990. Control of replication of plasmid R1: formation of an initial transient complex is rate-limiting for antisense RNA—target RNA pairing. *EMBO J.* 9:3777–3785.

Pesole, G., C. Gissi, A. De Chirico, and C. Saccone. 1999. Nucleotide substitution rate of mammalian mitochondrial genomes. *J. Mol. Evol.* 48:427–434.

Pestova, E. V., L. S. Havarstein, and D. A. Morrison. 1996. Regulation of competence for genetic transformation in *Streptococcus pneumoniae* by an auto-induced peptide pheromone and a two-component regulatory system. *Mol. Microbiol.* 21:853–862.

Petersen, G., D. Song, B. Hugle-Dorr, and I. Oldenburg, et al. 1995. Mapping of linear epitopes recognized by monoclonal antibodies with gene-fragment phage display libraries. *Mol. Gen. Genet.* 249:425–431.

Petersohn, A., J. Bernhardt, U. Gerth, and D. Hoper, et al. 1999. Identification of σ^B-dependent genes in *Bacillus subtilis* using a promoter consensus-directed search and oligonucleotide hybridization. *J. Bacteriol.* 181:5718–5724.

Petersohn, A., M. Brigulla, S. Haas, and J. D. Hoheisel, et al. 2001. Global analysis of the general stress response of *Bacillus subtilis. J. Bacteriol.* 183:5617–5631.

Peterson, K. M. and J. J. Mekalanos. 1988. Characterization of the *Vibrio cholerae* ToxR regulon: identification of novel genes involved in intestinal colonization. *Infect. Immun.* 56:2822–2829.

Peterson, S., R. T. Cline, H. Tettelin, and V. Sharov, et al. 2000. Gene expression analysis of the *Streptococcus pneumoniae* competence regulons by use of DNA microarrays. *J. Bacteriol.* 182:6192–6202.

Peterson, S. N. and C. M. Fraser. 2001. The complexity of simplicity. Genome Biol 2: Comment2002.1-2002.8.

Petrov, A., S. Shah, S. Draghici, and S. Shams. 2002. Microarray image processing and quality control. In S. Shah and G. Kamberova (eds.), *DNA Array Image Analysis—Nuts & Bolts*, pp. 99–130. DNA Press, LLC, Eagleville, Penn.

Piddock, L. J. V. and R. Wise. 1987. Induction of the SOS response in *Escherichia coli* by 4-quinolone antimicrobial agents. *FEMS Microbiol. Lett.* 41:289–294.

Pierce, J. R. 1980. *An Introduction to Information Theory: Symbols, Signals and Noise.* Dover Publishing, New York.

Pinkel, D., R. Segraves, D. Sudar, and S. Clark, et al. 1998. High resolution analysis of DNA copy number variation using comparative genomic hybridization to microarrays. *Nat. Genet.* 20:207–211.

Pittard, J. 1996. The various strategies within the TyrR regulation of *Escherichia coli* to modulate gene expression. *Genes Cells* 1:717–725.

Pizza, M., V. Scarlato, V. Masignani, and M. M. Giuliani, et al. 2000. Identification of vaccine candidates against serogroup B *Meningococcus* by whole-genome sequencing. *Science* 287:1816–1820.

Polissi, A., A. Pontiggia, G. Feger, and M. Altieri, et al. 1998. Large-scale identification of virulence genes from *Streptococcus pneumoniae. Infect. Immun.* 66:5620–5629.

Pollack, J. R., C. M. Perou, A. A. Alizadeh, and M. B. Eisen, et al. 1999. Genome-wide analysis of DNA copy-number changes using cDNA microarrays. *Nat. Genet.* 23:41–46.

Pomposiello, P. J., B. K. Janes, and R. A. Bender. 1998. Two roles for the DNA recognition site of the *Klebsiella aerogenes* nitrogen assimilation control protein. *J. Bacteriol.* 180:578–585.

Ponting, C. P., L. Aravind, J. Schultz, and P. Bork, et al. 1999. Eukaryotic signaling domain homologues in Archaea and Bacteria. Ancient ancestry and horizontal gene transfer. *J. Mol. Biol.* 289:729–745.

Poole, K. 2001. Multidrug resistance in Gram-negative bacteria. *Curr. Opin. Microbiol.* 4:500–508.

Porcella, S. F. and T. G. Schwan. 2001. *Borrelia burgdorferi* and *Treponema pallidum*: a comparison of functional genomics, environmental adaptations, and pathogenic mechanisms. *J. Clin. Invest.* 107:651–656.

Portugaly, E. and M. Linial. 2000. Estimating the probability for a protein to have a new fold: a statistical computational model. *Proc. Natl. Acad. Sci. USA* 97(10):5161–5166.

Poltz, M. F. and C. M. Cavanaugh. 1998. Bias in template-to-product ratios in multitemplate PCR. *Appl. Environ. Microbiol.* 64:3724–3730.

Pradet-Balade, B., F. Boulme, H. Beug, and E. W. Mullner, et al. 2001. Translation control: bridging the gap between genomics and proteomics? *Trends Biochem. Sci.* 26:225–229.

Prestridge, D. S. 1991. SIGNAL SCAN: a computer program that scans DNA sequences for eukaryotic transcriptional elements. *CABIOS* 7:203–206.

Price, C. W. 2000. Protective function and regulation of the general stress response in *Bacillus subtilis* and related gram-positive bacteria. In G. Storz and R. Hengge-Aronis (eds.), *Bacterial Stress Responses*, pp. 179–197. ASM Press, Washington, D.C.

Price, C. W., P. Fawcett, H. Ceremonie, and N. Su, et al. 2001. Genome-wide analysis of the general stress response in *Bacillus subtilis. Mol. Microbiol.* 41:757–774.

Prim, R. C. 1957. Shortest connection networks and some generalizations. *Bell Syst. Tech. J.* 36:1389–1401.

Primig, M., R. M. Williams, E. A. Winzeler, and G. G. Tevzadze, et al. 2000. The core meiotic transcriptome in budding yeasts. *Nat. Genet.* 26:415–423.

Protchenko, O., T. Ferea, J. Rashford, and J. Tiedeman, et al. 2001. Three cell wall mannoproteins facilitate the uptake of iron in *Saccharomyce cerevisiae. J. Biol. Chem.* 276:49244–49250.

Ptashne, M. 1992. *A Genetic Switch*, 2nd ed. Blackwell Scientific Publications, Cambridge, Mass.

Puig, O., F. Caspary, G. Rigaut, and B. Rutz, et al. 2001. The tandem affinity purification (TAP) method: a general procedure of protein complex purification. *Methods* 24:218–229.

Pullen, S. S., T. T. A. Dang, J. J. Crute, and M. R. Kehry. 1999. CD40 signalling through tumor necrosis factor receptor-associated factors (TRAFs). *J. Biol. Chem.* 274:14246–14254.

Qiu, X. B., Y. L. Lin, K. C. Thome, and P. Pian, et al. 1998. An eukaryotic RuvB-like protein (RUVBL1) essential for growth. *J. Biol. Chem.* 273:27786–27793.

Quenzer, T. L., M. R. Emmett, C. L. Hendrickson, and P. H. Kelly, et al. 2001. High sensitivity Fourier transform ion cyclotron resonance mass spectrometry for biological analysis with nano-LC and microelectrospray ionization. *Anal. Chem.* 73:1721–1725.

Raamsdonk, L. M., B. Teusink, D. Broadhurst, and N. Zhang, et al. 2001. A functional genomics strategy that uses metabolome data to reveal the phenotype of silent mutations. *Nat. Biotechnol.* 19:45–50.

Rabiner, S. 1989. A tutorial on hidden Markov models and selected applications in speech recognition. *Proc. IEEE* 77(2):257–286.

Radelof, U., S. Hennig, P. Seranski, and M. Steinfath, et al. 1998. Preselection of shotgun clones by oligonucleotide fingerprinting: an efficient and high throughput strategy to reduce redundancy in large-scale sequencing projects. *Nucl. Acids Res.* 26:5358–5364.

Rader, C. and C. F. Barbas, III. 1997. Phage display of combinatorial antibody libraries. *Curr. Opin. Biotechnol.* 8:503–508.

Raghuraman, M. K., E. A. Winzeler, D. Collingwood, and S. Hunt, et al. 2001. Replication dynamics of the yeast genome. *Science* 294:115–121.

Ragno, S., M. Romano, S. Howell, and D. J. C. Pappin, et al. 2001. Changes in gene expression in macrophages infected with *Mycobacterium tuberculosis*: a combined transcriptomic and proteomic approach. *Immunology* 104:99–108.

Rahme, L. G., F. M. Ausubel, H. Cao, and E. Drenkard, et al. 2000. Plants and animals share functionally common bacterial virulence factors. *Proc. Natl. Acad. Sci. USA* 97:8815–8821.

Rain, J.-C., L. Selig, H. DeReuse, and V. Battaglia, et al. 2001. The protein-protein interaction map of *Helicobacter pylori*. *Nature* 409:211–215.

Raitio, M., K. Lindroos, M. Laukkanen, and T. Pastinen, et al. 2001. Y-chromosomal SNPs in Finno-Ugric-speaking populations analyzed by minisequencing on microarrays. *Genome Res.* 11:471–482.

Rajotte, D., W. Arap, M. Hagedorn, and E. Koivunen, et al. 1998. Molecular heterogeneity of the vascular endothelium revealed by in vivo phage display. *J. Clin. Invest.* 102:430–437.

Ramsay, G. 1998. DNA chips: state-of-the art. *Nat. Biotechnol.* 1:640–644.

Rappleye, C. A. and J. R. Roth. 1997. A Tn10 derivative (T-POP) for isolation of insertions with conditional (tetracycline-dependent) phenotypes. *J. Bacteriol.* 179:5827–5834.

Rappuoli, R. 2000. Pushing the limits of cellular microbiology: microarrays to study bacteria-host cell intimate contacts. *Proc. Natl. Acad. Sci. USA* 97:13467–13469.

Rasched, I. and E. Oberer. 1986. Ff coliphages: structural and functional relationships. *Microbiol. Rev.* 50:401–427.

Razin, S., D. Yogev, and Y. Naot. 1998. Molecular biology and pathogenicity of mycoplasmas. *Microbiol. Mol. Biol. Rev.* 62:1094–1156.

Read, T. D., S. R. Gill, H. Tettelin, and B. A. Dougherty. 2001. Finding drug targets in microbial genomes. *Drug Disc. Today* 6:887–892.

Read, T. D., S. N. Peterson, N. Tourasse, L. W. Baillie, et al. 2003. The genome sequence of *Bacillus anthracis* Ames and comparison to closely related bacteria. *Nature* 423:81–86.

Reanney, D. C. 1974. On the origin of prokaryotes. *J. Theor. Biol.* 48:243–251.

Reboul, J., P. Vaglio, J. F. Rual, and P. Lamesch, et al. 2003. *C. elegans* ORFeome version 1.1: experimental verification of the genome annotation and resource for proteome-scale protein expression. *Nat. Genet.* 34:35–41.

Redenbach, M., H. M. Kieser, and D. Denapaite, et al. 1996. A set of ordered cosmids and a detailed genetic and physical map for the 8 Mb *Streptomyces* coelicolor A3(2) chromosome. *Mol. Microbiol.* 21:77–96.

Reese, M., D. Kulp, H. L. Tammana, and D. Haussler. 2000. GENIE—gene finding in *Drosophila melanogaster. Genome Res.* 10(4):529–538.

Reese, M. G., N. L. Harris, and F. H. Eeckman. 1996. Large scale sequencing specific neural networks for promoter and splice site recognition. In L. Hunter and T. E. Klein (eds.), *Biocomputing: Proceedings of the 1996 Pacific Symposium.* World Scientific Publishing Co., Singapore, p. 737–738.

Reeve, J. N., K. Sandman, and C. J. Daniels. 1997. Archaeal histones, nulceosomes, and transcription initiation. *Cell* 89:999–1002.

Rehman, F. N., M. Audeh, E. S. Abrams, and P. W. Hammond, et al. 1999. Immobilization of acrylamide-modified oligonucleotides by co-polymerization. *Nucl. Acids Res.* 27:649–655.

Reich, K. A., L. Chovan, and P. Hessler. 1999. Genome scanning in *Haemophilus influenzae* for identification of essential genes. *J. Bacteriol.* 181:4961–4968.

Reid, G. E. and S. A. McLuckey. 2002. "Top down" protein characterization via tandem mass spectrometry. *J. Mass Spectrom.* 37:663–675.

Reid, S. D., C. J. Herbelin, A. C. Bumbaugh, and R. K. Selander, et al. 2000. Parallel evolution of virulence in pathogenic *Escherichia coli. Nature* 406:64–67.

Reineke, U., A. Kramer, and J. Schneider-Mergener. 1999a. Antigen sequence- and library-based mapping of linear and discontinuous protein-protein interaction sites by spot synthesis. *Curr. Top. Microbiol. Immunol.* 243:23–36.

Reineke, U., R. Sabat, R. Misselwitz, and H. Welfle, et al. 1999b. A synthetic mimic of a discontinuous binding site on interleukin-10. *Nat. Biotechnol.* 17:271–275.

Reineke, U., R. Volkmer-Engert, and J. Schneider-Mergener. 2001. Applications of peptide arrays prepared by the SPOT-technology. *Curr. Opin. Biotechnol.* 12:59–64.

Remy, I. and S. W. Michnick. 1999. Clonal selection and in vivo quantitation of protein interactions with protein-fragment complementation assays. *Proc. Natl. Acad. Sci. USA* 96:5394–5399.

Remy, I. and S. W. Michnick. 2001. Visualization of biochemical networks in living cells. *Proc. Natl. Acad. Sci. USA* 98:7678–7683.

Ren, Z. J., G. K. Lewis, P. T. Wingfield, and E. G. Locke, et al. 1996. Phage display of intact domains at high copy number: a system based on SOC, the smaller outer capsid protein of bacteriophage T4. *Prot. Sci.* 5:1833–1843.

Reynolds, A. E., J. Felton, and A. Wright. 1981. Insertion of DNA activates the cryptic bgl operon in *Escherichia coli* K-12. *Nature (Lond.)* 293:625–629.

Reynolds, K. J., X. Yao, and C. Fenselau. 2002. Proteolytic^{18}O labeling for comparative proteomics: evaluation of endoprotease Glu-C as the catalytic agent. *J. Proteome Res.* 1:27–33.

Reyrat, J. M., V. Pelicic, B. Gicquel, and R. Rappuoli. 1998. Counterselectable markers: untapped tools for bacterial genetics and pathogenesis. *Infect. Immun.* 66:4011–4017.

Rich, R. L. and D. G. Myszka. 2000. Advances in surface plasmon resonance biosensor analysis. *Curr. Opin. Biotechnol.* 11:54–61.

Richardson, D. J. 2000. Bacterial respiration: a flexible process for a changing environment. *Microbiology* 146:551–571.

Richardson, T. H., X. Tan, G. Frey, and W. Callen, et al. 2002. A novel, high performance enzyme for starch liquefaction: discovery and optimization of a low pH thermostable-amylase. *J. Biol. Chem.* 277:26501–26507.

Richmond, C. S., J. D. Glasner, R. Mau, and H. Jin, et al. 1999. Genome-wide expression profiling in *Escherichia coli* K-12. *Nucl. Acids Res.* 27:3821–3835.

Richmond, T. A. and C. R. Somerville. 2000. The cellulose synthase superfamily. *Plant Physiol.* 124:495–498.

Riechmann, L. and P. Holliger. 1997. The C-terminal domain of TolA is the coreceptor for filamentous phage infection of *E. coli. Cell* 90:351–360.

Rigaut, G., A. Shevchenko, B. Rutz, and M. Wilm, et al. 1999. A generic protein purification method for protein complex characterization and proteome exploration. *Nat. Biotechnol.* 17:1030–1032.

Ritter, A., G. Blum, L. Emody, and M. Kerenyi, et al. 1995. tRNA genes and pathogenicity islands: influence on virulence and metabolic properties of uropathogenic *Escherichia coli. Mol. Microbiol.* 17:109–121.

Rivera, M. C. and J. A. Lake. 1992. Evidence that eukaryotes and eocyte prokaryotes are immediate relatives. *Science* 257:74–76.

Rivera, M. C. R. Jain, J. E. Moore, and J. A. Lake. 1998. Genomic evidence for two functionally distinct gene classes. *Proc. Natl. Acad. Sci. USA* 95:6239–6244.

Robb, F. T., D. L. Maeder, J. R. Brown, and J. DiRuggiero, et al. 2001. Genomic sequence of hyperthermophile Pyrococcus furiosus: implications for physiology and enzymology. *Methods Enzymol.* 330:134–157.

Roberts, M., N. F. Fairweather, E. Leininger, and D. Pickard, et al. 1991. Construction and characterization of Bordetella pertussis mutants lacking the vir-regulated P69 outer membrane protein. *Mol. Microbiol.* 5:1393–1404.

Roberts, M. C. 1996. Tetracycline resistance determinants: mechanisms of action, regulation of expression, genetic mobility, and distribution. *FEMS Microbiol. Rev.* 19:1–24.

Roepstorff, P. and J. Fohlman. 1984. Proposal for a common nomenclature for sequence ions in mass-spectra of peptides. *Biomed. Mass Spectrum.* 11:601.

Rogers, Y., P. Jiang-Baucome, Z. Huange, and V. Bogdanov, et al. 1999. Immobilization of oligonucleotides onto a glass support via disulfide bonds: a method for preparation of DNA microarrays. *Ann. of Biochem.* 266:23–30.

Romero, P., Z. Obradovic, X. Li, and E. C. Garner, et al. 2001. Sequence complexity of disordered protein. *Proteins* 42(1):38–48.

Rondon, M. R., P. R. August, A. D. Bettermann, and S. F. Brady, et al. 2000. Cloning the soil metagenome: a strategy for accessing the genetic and functional diversity of uncultured microorganisms. *Appl. Environ. Microbiol.* 66:2541–2547.

Rondon, M. R., S. J. Raffel, R. M. Goodman, and J. Handelsman. 1999. Toward functional genomics in bacteria: analysis of gene expression in *Escherichia coli* from a bacterial artificial chromosome library of *Bacillus cereus. Proc. Natl. Acad. Sci. USA* 96:6451–6455.

Rondon, M. R., R. M. Goodman, and J. Handelsman. 1999. The Earth's bounty: assessing and accessing soil microbial diversity. *Trends Biotechnol.* 17:403–409.

Rosamond, J. and A. Allsop. 2000. Harnessing the power of the genome in the search for new antibiotics. *Science* 287:1973–1976.

Rose, D. 2000. Microfluidic technologies and instrumentation for printing DNA microarrays. In M. Schena (ed.), *Microarray Biochip Technology*, pp. 19–38. Eaton Publishing, Natick, Mass.

Rosenberg, A., G. Griffin, F. W. Studier, and M. McCormick, et al. 1996. T7Select phage display system: a powerful new protein display system based on bacteriophage T7. *Innovations* 6:1–6.

Rosenkrands, I., A. King, K. Weldingh, and M. Moniatte, et al. 2000. Towards the proteome of *Mycobacterium* tuberculosis. *Electrophoresis* 21:3740–3756.

Rosenkrands, I., K. Weldingh, S. Jacobsen, and C. V. Hansen, et al. 2000. Mapping and identification of *Mycobacterium tuberculosis* proteins by two-dimensional gel electrophoresis, microsequencing and immunodetection. *Electrophoresis* 21:935–948.

Rossello-Mora, R. and R. Amann. 2001. The species concept for prokaryotes. *FEMS Microbiol. Rev.* 25:39–67.

Rossi, F., C. A. Charlton, and H. M. Blau. 1997. Monitoring protein-protein interactions in intact eukaryotic cells by β-galactosidase complementation. *Proc. Natl. Acad. Sci. USA* 94:8405–8410.

Rossi, F. M. V., B. T. Blakely, and H. M. Blau. 2000. Interaction blues: protein interactions monitored in live mammalian cells by β-galactosidase complementation. *Trends Cell Biol.* 10:119–122.

Ross-Macdonald, P., P. S. Coelho, T. Roemer, and S. Agarwal, et al. 1999. Large-scale analysis of the yeast genome by transposon tagging and gene disruption. *Nature* 402:413–418.

Rout, M. P., J. D. Aitchinson, A. Suprapto, and K. Hjertaas, et al. 2000. The yeast nuclear pore complex: composition, architecture, and transport mechanism. *J. Cell Biol.* 148:635–651.

Rowe-Magnus, D. A., A. M. Guerout, and D. Mazel. 2002. Bacterial resistance evolution by recruitment of super-integron gene cassettes. *Mol. Microbiol.* 43:1657–1669.

Rowe-Magnus, D. A., A. M. Guerout, P. Ploncard, and B. Dychinco, et al. 2001. The evolutionary history of chromosomal super-integrons provides an ancestry for multiresistant integrons. *Proc. Natl. Acad. Sci. USA* 98:652–657.

Rowe-Magnus, D. A. and D. Mazel. 1999. Resistance gene capture. *Curr. Opin. Microbiol.* 2:483–488.

Rozwarski, D. A., G. A. Grant, D. H. Barton, and W. R. Jacobs Jr, et al. 1998. Modification of the NADH of the isoniazid target (InhA) from *Mycobacterium tuberculosis. Science* 279:98–102.

Rubin, G. M. and E. B. Lewis. 2000. A brief history of *Drosophila's* contributions to genome research. *Science* 287:2216–2218.

Rubin, G. M., M. D. Yandell, J. R. Wortman, and G. L. Gabor Miklos, et al. 2000. Comparative genomics of the eukaryotes. *Science* 287:2204–2215.

Rudgers, G. W. and T. Palzkill. 2001. Protein minimization by random fragmentation and selection. *Protein Eng.* 14:487–492.

Rudi, K., O. M. Skulberg, R. Skulberg, and K. S. Jakobsen. 2000. Application of sequence-specific labeled 16S rRNA gene oligonucleotide probes for genetic profiling of cyanbacterial abundance and diversity by array hybridization. *Appl. Environ. Microbiol.* 66:4004–4011.

Russell, C. B., D. S. Thaler, and F. W. Dahlquist. 1989. Chromosomal transformation of *Escherichia coli* recD strains with linearized plasmids. *J. Bacteriol.* 171:2609–2613.

Rutherford, J. C., S. Jaron, E. Ray, and P. O. Brown, et al. 2001. A second iron-regulatory system in yeast independent of Aft1p. *Proc. Natl. Acad. Sci. USA* 98:14322–14327.

Sacher, M., Y. Jiang, J. Barrowman, and A. Scarpa, et al. 1998. TRAPP, a highly conserved novel complex on the cis-Golgi that mediates vesicle docking and fusion. *EMBO J.* 17:2494–2503.

Sakakibara, Y., K. Yanagisawa, J. Katafuchi, and D. P. Ringer, et al. 1998. Molecular cloning, expression, and characterization of novel human RT SULT1C sulfotransferases that catalyze the sulfonation of RT N-hydroxy-2-acetylaminofluorene. *J. Biol. Chem.* 273:33929–33935.

Salama, N., K. Guillemin, T. K. McDaniel, and G. Sherlock, et al. 2000. A whole-genome microarray reveals genetic diversity among *Helicobacter pylori* strains. *Proc. Natl. Acad. Sci. USA* 97:14668–14673.

Sali, A. and T. L. Blundell. 1993. Comparative protein modelling by satisfaction of spatial restraints. *J. Mol. Biol.* 234:779–815.

Salunga, R. C., H. Guo, L. Luo, and A. Bittner, et al. 1999. Gene expression analysis via cDNA microarrays of laser capture microdissected cells from fixed tissue. In M. Schena (ed.), *DNA Microarrays: A Practical Approach*, pp. 121–136. Oxford University Press, New York.

Salzberg, S., A. Delcher, K. Fasman, and J. Henderson. 1998a. A decision-tree system for finding genes in DNA. *J. Comp. Biol.* 5:667–680.

Salzberg, S., A. Delcher, S. Kasif, and O. White. 1998b. Microbial gene identification using interpolated Markov models. *Nucl. Acids Res.* 26(2):544–548.

Salzberg, S. L., O. White, J. Peterson, and J. A. Eisen. 2001. Microbial genes in the human genome: lateral transfer or gene loss? *Science* 292:1903–1906.

Sanger, F., A. R. Coulson, G. F. Hong, and D. F. Hill, et al. 1982. Nucleotide sequence of bacteriophage lambda DNA. *J. Mol. Biol.* 162:729–773.

Sanger, F., S. Nicklen, and A. R. Coulson. 1977. DNA sequencing with chain-terminating inhibitors. *Proc. Natl. Acad. Sci. USA* 74:5463–5467.

Sanglard, D. 2001. Integrated antifungal drug discovery in *Candida albicans*. *Nat. Biotechnol.* 19:212–213.

Sankoff, D. 1999. Genome rearrangements with gene families. *Bioinformatics* 15:909–917.

Santagati, M., F. Iannelli, and M. R. Oggioni, et al. 2000. Characterization of a genetic element carrying the macrolide efflux gene mef(A) in *Streptococcus pneumoniae*. *Antimicrob. Agents Chemother.* 44:2585–2587.

Santini, C., D. Brennan, C. Mennuni, and R. H. Hoess, et al. 1998. Efficient display of an HCV cDNA expression library as C-terminal fusion to the capsid protein D of bacteriophage lambda. *J. Mol. Biol.* 282:125–135.

Sar, N., L. McCarter, M. Simon, and M. Silverman. 1990. Chemotactic control of the two flagellar systems of *Vibrio parahaemolyticus*. *J. Bacteriol.* 172:334–341.

Sassetti, C.M., D. H. Boyd, and E. J. Rubin. 2001. Comprehensive identification of conditionally essential genes in mycobacteria. *Proc. Natl. Acad. Sci. USA* 98:12712–12717.

Saunders, N. J. and E. R. Moxon. 1998. Implications of sequencing bacterial genomes for pathogenesis and vaccine development. *Curr. Opin. Biotechnol.* 9:618–623.

Sazuka, T., M. Yamaguchi, and O. Ohara. 1999. Cyano2Dbase updated: linkage of 234 protein spots to corresponding genes through N-terminal microsequencing. *Electrophoresis* 20:2160–2171.

Schachter, J. 1990. Chlamydial infections. *West. J. Med.* 153:523–534.

Schadt, E. E., C. Li, C. Su, and W. H. Wong. 2000. Analyzing high-density oligonucleotide gene expression array data. *J. Cell Biochem.* 80:192–202.

Schaffer, R., J. Landgraf, M. Perez-Amador, and E. Wisman. 2000. Monitoring genome-wide expression in plants. *Curr. Opin. Biotechnol.* 11:162–167.

Scharn, D., H. Wenschuh, U. Reineke, and J. Schneider-Mergener, et al. 2000. Spatially addressed synthesis of amino- and amino-oxy-substituted 1,3,5-triazine arrays on polymeric membranes. *J. Comb. Chem.* 2:361–369.

Sche, P. P., K. M. McKenzie, J. D. White, and D. J. Austin. 1999. Display cloning: functional identification of natural product receptors using cDNA-phage display. *Chem. Biol.* 6:707–716.

Scheele, G. A. 1975. Two-dimensional gel analysis of soluble proteins. Characterization of guinea pig exocrine pancreatic proteins. *J. Biol. Chem.* 250:5375–5385.

Schena, M. 1996. Genome analysis with gene expression microarrays. *Bioessays* 18:427–431.

Schena, M. 2002. *Microarray Analysis*. John Wiley & Sons, New York.

Schena, M. and R. W. Davis. 2000. Technology standards for microarray research. In M. Schena (ed.), *Microarray Biochip Technology*, pp. 1–18. Eaton Publishing, Natick, Mass.

Schena, M., D. Shalon, R. Heller, and A. Chai, et al. 1996. Parallel human genome analysis: microarray-based expression monitoring of 1000 genes. *Proc. Natl. Acad. Sci. USA* 93:10614–10619.

Schena, M., D. Shalon, R. W. Davis, and P. O. Brown. 1995. Quantitative monitoring of gene expression patterns with a complementary DNA microarray. *Science* 270:467–470.

Schena, M., R. Heller, T. Theriault, and K. Konrad, et al. 1998. Microarrays: biotechnology's disovery platform for functional genomics. *Trends Biotech.* 16:301–306.

Schermer, M. J. 1999. Confocal scanning microscopy in microarray detection. In M. Schena (ed.), *DNA Microarrays*, pp. 17–42. Oxford University Press, New York.

Schmid, M. B. 1998. Novel approaches to the discovery of antimicrobial agents. *Curr. Opin. Chem. Biol.* 2:529–534.

Schmid, M. B., N. Kapur, D. R. Isaacson, and P. Lindroos, et al. 1989. Genetic analysis of temperature-sensitive lethal mutants of *Salmonella typhimurium*. *Genetics* 123:625–633.

Schneider, T. D., G. D. Stormo, L. Gold, and A. Ehrenfeucht. 1986. Information content of binding sites on nucleotide sequences. *J. Mol. Biol.* 188:415–431.

Schopf, J. W. 1993. Microfossils of the Early Archean Apex chert: new evidence of the antiquity of life. *Science* 260:640–646.

Schreiber, S. L. 2000. Target-oriented and diversity-oriented organic synthesis in drug discovery. *Science* 287:1964–1969.

Schuchhardt, J., D. Beule, A. Malik, E. Wolski, H. Eickhoff, H. Lehrach, and H. Herzel. 2000. Normalization strategies for cDNA microarrays. *Nucl. Acids Res.* 28:e47, i–v.

Schuck, P. 1997. Use of surface plasmon resonance to probe the equilibrium and dynamic aspects of interactions between biological macromolecules. *Annu. Rev. Biophys. Biomol. Struct.* 26:541–566.

Schultz, G. E., and RH. Schmier. 1979. *Principles of Protein Function*. Springer-Verlag, New York.

Schultz, J., U. Hoffmuller, G. Krause, and J. Ashurst, et al. 1998. Specific interactions between the syntrophin PDZ domain and voltage-gated sodium channels. *Nat. Struct. Biol.* 5:19–24.

Schulz, A. and W. Schumann. 1996. hrcA, the first gene of the *Bacillus subtilis* dnaK operon, encodes a negative regulator of class I heat shock genes. *J. Bacteriol.* 178:1088–1093.

Schuster, M., C. P. Lostroh, T. Ogi, and E. P. Greenberg. 2003. Identification, timing, and signal specificity of *Pseudomonas aeruginosa* quorum-controlled genes: a transcriptome analysis. *J. Bacteriol.* 185:2066–2079.

Schut, G. J., J. Zhou, and M. W. W. Adams. 2001. DNA microarray analysis of the hyperthermophilic archaeon *Pyrococcus furiosus*: evidence for a new type of sulfur-reducing enzyme complex. *J. Bacteriol.* 183:7027–7036.

Schwan, W. R., S. N. Coulter, E. Y. Ng, and M. H. Langhorne, et al. 1998. Identification and characterization of the PutP proline permease that contributes to in vivo survival of *Staphylococcus aureus* in animal models. *Infect. Immun.* 66:567–572.

Schwartz, D. C., X. Li, L. Hernandez, and S. Ramnarain, et al. 1993. Ordered restriction maps of *Saccharomyces cerevisiae* chromosomes constructed by optical mapping. *Science* 262:110–114.

Schweizer, H. P. 1992. Allelic exchange in *Pseudomonas aeruginosa* using novel ColE1-type vectors and a family of cassettes containing a portable oriT and the counter-selectable *Bacillus subtilis* sacB marker. *Mol. Microbiol.* 6:1195–1204.

Schwikowski, B., P. Uetz, and S. Fields. 2000. A network of protein-protein interactions in yeast. *Nat. Biotechnol.* 18:1257–1261.

Sefton, A. M. 2002. Mechanisms of antimicrobial resistance: their clinical relevance in the new millennium. *Drugs* 62:557–566.

Segal, E. D., C. Lange, A. Covacci, and L. S. Tompkins, et al. 1997. Induction of host signal transduction pathways by *Helicobacter pylori*. *Proc. Natl. Acad. Sci. USA* 94:7595–7599.

Segal, E. D., S. Falkow, and L. S. Tompkins. 1996. *Helicobacter pylori* attachment to gastric cells induces cytoskeletal rearrangements and tyrosine phosphorylation of host cell proteins. *Proc. Natl. Acad. Sci. USA* 93:1259–1264.

Selkoe, D. J. 1998. The cell biology of beta-amyloid precursor protein and presenilin in Alzheimer's disease. *Trends Cell Biol.* 8:447–453.

Selkov, E., N. Maltsev, G. J. Olsen, and R. Overbeek, et al. 1997. A reconstruction of the metabolism of *Methanococcus jannaschii* from sequence data. *Gene* 197:GC11–GC26.

Senko, M. W., J. P. Spear, and F. W. McLafferty. 1994. Collisional activation of large multiply charged ions using Fourier transform mass spectrometry. *Anal. Chem.* 66:2801–2808.

Seoighe, C. and K. H. Wolfe. 1998. Extent of genomic rearrangement after genome duplication in yeast. *Proc. Natl. Acad. Sci. USA* 95:4447–4452.

Seoighe, C. and K. H. Wolfe. 1999. Updated map of duplicated regions in the yeast genome. *Gene* 238:253–261.

Shalom, G., J. G. Shaw, and M. S. Thomas. 2000. pGSTp: an IVET-compatible promoter probe vector conferring resistance to trimethoprim. *Biotechniques* 29:954–958.

Shalon, D., S. J. Smith, and P. O. Brown. 1996. A DNA microarray system for analyzing complex DNA samples using two-color fluorescent probe hybridization. *Genome Res.* 6:639–645.

Shani, N., P. A. Watkins, and D. Valle. 1995. PXA1, a possible *Saccharomyces cerevisiae* ortholog of the human adrenoleukodystrophy gene. *Proc. Natl. Acad. Sci. USA* 92:6012–6016.

Sharma, A. and A. Mondragon. 1995. DNA topoisomerases. *Curr. Opin. Struct. Biol.* 5:39–47.

Sharp, R. R. and J. C. Barrett. 2000. The environmental genome project: ethical, legal, and social implications. *Environ. Health Perspect.* 108:279–281.

Shchepinov, M. S., S. C. Case-Green, and E. M. Southern. 1997. Steric factors influencing hybridisation of nucleic acids to oligonucleotide arrays. *Nucl. Acids Res.* 25:1155–1161.

She, Q., R. K. Singh, F. Confalonieri, and Y. Zivanovic, et al. 2001. The complete genome of the crenarchaeon *Sulfolobus solfataricus* P2. *Proc. Natl. Acad. Sci. USA* 98:7835–7840.

Shea, J. E. and D. W. Holden. 2000. Signature-tagged mutagenesis helps identify virulence genes. *ASM News* 66:15–20.

Shea, J. E., J. D. Santangelo, and R. G. Feldman. 2000. Signature-tagged mutagenesis in the identification of virulence genes in pathogens. *Curr. Opin. Microbiol.* 3:451–458.

Shea, J. E., M. Hensel, C. Gleeson, and D. W. Holden. 1996. Identification of a virulence locus encoding a second type III secretion system in *Salmonella typhimurium. Proc. Natl. Acad. Sci. USA* 93:2593–2597.

Shen, Y., R. Zhao, M. E. Belov, and T. P. Conrads, et al. 2001. Packed capillary reversed-phase liquid chromatography with high-performance electrospray ionization Fourier transform ion cyclotron resonance mass spectrometry for proteomics. *Anal. Chem.* 73:1766–1775.

Sherman, D. R., M. Voskuil, D. Schnappinger, and R. Liao, et al. 2001. Regulation of the *Mycobacterium tuberculosis* hypoxic response gene encoding α-crystallin. *Proc. Natl. Acad. Sci. USA* 98:7534–7539.

Sherrat, D. 1989. Tn3 and related transposable elements: site-specific recombination and and transposition. In D. E. Berg and M. M. Howe (eds.), *Mobile DNA*, pp. 163–184. American Society for Microbiology, Washington, D.C.

Sherris, J. C. and J. J. Plorde. 1990. Staphylococci. In J. C. Sherris (ed.), *Medical Microbiology: An Introduction to Infectious Diseases*, pp. 275–289. Elsevier, New York.

Shevchenko, A., M. Wilm, O. Vorm, and M. Mann. 1996a. Mass spectrometric sequencing of proteins silver-stained polyacrylamide gels. *Anal. Chem.* 68:850–858.

Shevchenko, A., O. N. Jensen, A. V. Podtelejnikov, and F. Sagliocco, et al. 1996b. Linking genome and proteome by mass spectrometry: large-scale identification of yeast proteins from two dimensional gels. *Proc. Natl. Acad. Sci. USA* 93:14440–14445.

Shi, J., T. L. Blundell, and K. Mizuguchi. 2001. Fugue: sequence-structure homology recognition using environment-specific substitution tables and structure-dependent gap penalites. *J. Mol. Biol.* 310:243–257.

Shigenobu, S., H. Watanabe, M. Hattori, and K. Sakaki, et al. 2000. Genome sequence of the endocellular bacterial symbiont of aphids *Buchnera* sp. APS. *Nature* 7:81–86.

Shimkets, L. J. 1998. Structures and sizes of genomes of the Archaea and Bacteria. In F. J. Bruijn, J. R. Lupski, and G. M. Weinstock (eds), *Bacterial Genomes: Physical Structure and Analysis*, Chapman and Hall, New York, NY.

Shizuya, H., B. Birren, and U.-J. Kim, et al. 1992. Cloning and stable maintenance of 300-kilobase-pair fragments of human DNA in *Escherichia coli* using an F-factor-based vector. *Proc. Natl. Acad. Sci. USA* 89:8794–8797.

Shmulevich, I., E. R. Dougherty, S. Kim, and W. Zhang. 2002. Probabilistic Boolean networks: a rule-based uncertainty model for gene regulatory networks. *Bioinformatics* 18:261–274.

Shoemaker, D. D., D. A. Lashkari, D. Morris, and M. Mittmann, et al. 1996. Quantitative phenotypic analysis of yeast deletion mutants using a highly parallel molecular bar-coding strategy. *Nat. Genet.* 14:450–456.

Shriver-Lake, L. C. 1998. Silane-modified surfaces for biomaterial immobilization. In T. Cass and F. S. Ligler (eds.), *Immobilized Biomolecules in Analysis*, pp. 1–14. Oxford Press, Oxford, U.K.

Shuman, S. 1992a. DNA strand transfer reactions catalyzed by *vaccinia topoisomerase* I. *J. Biol. Chem.* 267:8620–8627.

Shuman, S. 1992b. Two classes of DNA end-joining reactions catalyzed by *vaccinia topoisomerase* I. *J. Biol. Chem.* 267:16755–16758.

Shuman, S. 1994. Novel approach to molecular cloning and polynucleotide synthesis using vaccinai DNA topoisomerase. *J. Biol. Chem.* 269:32678–32684.

Sibley, C. G. and J. E. Ahlquist. 1987. DNA hybridization evidence of hominoid phylogeny: results from an expanded data set. *J. Mol. Evol.* 26:99–121.

Sibley, C. G., J. A. Comstock, and J. E. Ahlquist. 1990. DNA hybridization evidence of hominoid phylogeny: a reanalysis of the data. *J. Mol. Evol.* 30:202–236.

Sidhu, S. S. 2000. Phage display in pharmaceutical biotechnology. *Curr. Opin. Biotechnol.* 11:610–616.

Silke, J. 1997. The majority of long non-stop reading frames on the antisense strand can be explained by biased codon usage. *Gene* 194(1):143–155.

Silver, L. L. and K. A. Bostian. 1993. Discovery and development of new antibiotics: the problem of antibiotic resistance. *Antimicrob. Agents Chemother.* 37:377–383.

Simonen, M. and I. Palva. 1993. Protein secretion in *Bacillus species*. *Microbiol. Rev.* 57:109–137.

Simons, R. W. 1988. Naturally occurring antisense RNA control—a brief review. *Gene* 72:35–44.

Simpson, J. C., R. Wellenreuther, A. Poustka, and R. Pepperkok, et al. 2000. Systematic subcellular localization of novel proteins identified by large-scale cDNA sequencing. *EMBO Rep.* 1:287–292.

Singh, I. R., R. A. Crowley, and P. O. Brown. 1997. High-resolution functional mapping of a cloned gene by genetic footprinting. *Proc. Natl. Acad. Sci. USA* 94:1304–1309.

Singh-Gasson, S., R. D. Green, Y. J. Yue, and C. Nelson, et al. 1999. Maskless fabrication of light-directed oligonucleotide microarrays using a digital micromirror array. *Nat. Biotechnol.* 17(10):974–978.

Skorupski, K. and R. K. Taylor. 1996. Positive selection vectors for allelic exchange. *Gene* 169:47–52.

Skovgaard, M., L. J. Jensen, S. Brunak, and D. Ussery, et al. 2001. On the total number of genes and their length distribution in complete microbial genomes. *Trends Genet.* 17(8):425–428.

Slesarev, A. I., K. V. Mezhevaya, K. S. Makarova, and N. N. Polushin, et al. 2002. The complete genome of hyperthermophile *Methanopyrus kandleri* AV19 and monophyly of archaeal methanogens. *Proc. Natl. Acad. Sci. USA* 99:4644–4649.

Small, J., D. R. Call, F. J. Brockman, and T. M. Straub, et al. 2001. Direct detection of 16S rRNA in soil extracts by using oligonucleotide microarrays. *Appl. Environ. Microbiol.* 67:4708–4716.

Smit, A. F. A. 1999. Interspersed repeats and other mementos of transposable elements in the mammalian genomes. *Curr. Opin. Genet. Devel.* 9:657–663.

Smith, D. L., L. A. Doucette-Stamm, C. Deloughery, and Lee, et al. 1997a. Complete genome sequence of *Methanobacterium thermoautotrophicum* ΔH: Functional analysis and comparative genomics. *J. Bacteriol.* 179:7135–7155.

Smith, D. L., Y. Z. Deng, and Z. Q. Zhang. 1997b. Probing the non-covalent structure of proteins by amide hydrogen exchange and mass spectrometry. *J. Mass Spectrum.* 32:135–146.

Smith, G. P. 1985. Filamentous fusion phage: novel expression vectors that display cloned antigens on the virion surface. *Science* 228:1315–1317.

Smith, G. P. and V. A. Petrenko. 1997. Phage display. *Chem. Rev.* 97:391–410.

Smith, R. D. and T. F. Smith. 1992. Pattern-induced multi-sequence alignment (PIMA) algorithm employing secondary structure-dependent gap penalties for use in comparative modeling. *Prot. Eng.* 5(1):35–41.

Smith, R. D., G. A. Anderson, M. S. Lipton, and C. Masselon, et al. 2002a. The use of accurate mass tags for high-throughput microbial proteomics. *OMICS* 6:61–90.

Smith, R. D., G. A. Anderson, M. S. Lipton, and L. Paša-Tolić, et al. 2002b. An accurate mass tag strategy for quantitative and high-throughput proteome measurements. *Proteomics* 2:513–523.

Smith, T. F. and M. Waterman. 1982. Comparison of biosequences. *Adv. Appl. Math.* 2:482–489.

Smoot, L. M., J. C. Smoot, M. R. Graham, and G. A. Somerville, et al. 2001. Global differential gene expression in response to growth temperature alteration in group A *Streptococcus*. *Proc. Natl. Acad. Sci. USA* 98:10416–10421.

Snel, B., P. Bork, and M. A. Huynen. 1999. Genome phylogeny based on gene content. *Nat. Genet.* 21:108–110.

Snyder, E. and G. Stormo. 1993. Identification of coding regions in genomic DNA sequences: an application of dynamic programming and neural networks. *Nucl. Acid. Res.* 21:607–613.

Sogin, M. 1997. History assignment: when was the mitochondrion founded? *Curr. Opin. Genet. Dev.* 7:792–799.

Sogin, M. L. 1991. Early evolution and the origin of eukaryotes. *Curr. Opin. Genet. Dev.* 1:457–463.

Solomen, P., D. A. Baxter, and J. H. Byrne. 2000. Mathematical modeling of gene regulatory network review. *Neuron* 26:567–580.

Someya, Y., A. Yamaguchi, and T. Sawai. 1995. A novel glycycline, 9-(N,N-dimethylglycylamido)-6-deoxytetracycline, is neither transported nor recognized by the transposon Tn10-encoded metal-tetracycline/H + antiporter. *Antimicrob. Agents Chemother.* 39:247–249.

Southern, E., Mir, K., Shchepinov, M. 1999. Molecular interactions on microarrays. *Nat Genet.* 21(Suppl. S):5–9.

Southern, E. M. 2001. DNA microarrays. History and overview. *Methods Mol. Biol.* 170:1–15.

Southern, E. M., S. C. Case-Green, J. K. Elder, and M. Johnsone, et al. 1994. Arrays of complementary oligonucleotides for analyzing the hybridizaiton behavior of nucleic acids. *Nucl. Acids Res.* 22:1368–1373.

Southern, E. M., U. Maskos, and J. K. Elder. 1992. Analyzing and comparing nucleic acid sequences by hybridization to arrays of oligonucleotides: evaluation using experimental models. *Genomics* 13:1108–1107.

Spahr, C. S., M. T. Davis, M. D. McGinley, and J. H. Robinson, et al. 2001. Towards defining the urinary proteome using liquid chromatography-tandem mass spectrometry. I. Profiling an unfractionated tryptic digest. *Proteomics* 1:93–107.

Spellman, P. T., G. Sherlock, M. Q. Zhang, and V. R. Iyer, et al. 1998. Comprehensive identification of cell cycle-regulated genes of the yeast *Saccharomyces cerevisiae* by microarray hybridization. *Mol. Biol. Cell* 9:3273–3297.

Sreevatsan, S., X. Pan, K. E. Stockbauer, and N. D. Connell, et al. 1997. Restricted structural gene polymorphism in the *Mycobacteium tuberculosis* complex indicates evolutionarily recent global dissemination. *Proc. Natl. Acad. Sci. USA* 94:9869–9874.

Stackebrandt, E., W. Frederiksen, G. M. Garrity, and P. A. Grimont, et al. 2002. Report of the ad hoc committee for the re-evaluation of the species definition in bacteriology. *Int. J. Syst. Evol. Microbiol.* 52:1043–1047.

Stackebrandt, E. and B. M. Goebel. 1994. Taxonomic note: A place for DNA-DNA reassociation and 16S rRNA sequence analysis in the present species definition in Bacteriology. *Int. J. Syst. Bacteriol.* 44:846–849.

Staley, J. T. and A. Konopka. 1985. Measurement of in-situ activities of nonphotosynthetic microorganisms in aquatic and terrestrial habitats. *Annu. Rev. Microbiol.* 39:321–346.

Stanhope, M. J., A. Lupas, M. J. Italia, and K. K. Koretke et al. 2001. Phylogenetic analyses do not support horizontal gene transfers from bacteria to vertebrates. *Nature* 411:940–944.

Stears, R. L., R. C. Getts, and S. R. Gullans. 2000. A novel, sensitive detection system for high-density microarrays using dendrimer technology. *Physiol. Genomics* 3:93–99.

Steel, A. B., R. L. Levicky, T. M. Herne, and M. J. Tarlo. 2000. Immobilization of nucleic acids at solid surfaces: effect of oligonucleotide length on layer assembly. *Biophysics* J79:975–981.

Steinberg, T. H., L. J. Jones, R. P. Haugland, and V. L. Singer. 1996a. SYPRO orange and SYPRO red protein gel stains: one-step fluorescent staining of denaturing gels for detection of nanogram levels of protein. *Anal. Biochem.* 239:223–237.

Steinberg, T. H., R. P. Haugland, and V. L. Singer. 1996b. Applications of SYPRO orange and SYPRO red protein gel stains. *Anal. Biochem.* 239:238–245.

Stephanopoulos, G. N., A. A. Aristidou, and J. Nielsen. 1998. *Metabolic Engineering: Principles and Methodologies*. Academic Press, San Diego, Calif.

Stephens, R. S., S. Kalman, C. Lammel, and J. Fan, et al. 1998. Genome sequence of an obligate intracellular pathogen of humans: *Chlamydial trachomatis*. *Science* 282:754–759.

Sterky, F. and J. Lundeberg. 2000. Sequence analysis of genes and genomes. *J. Biotechnol.* 76:1–31.

Stetter, K. O. 1999. Extremophiles and their adaptation to hot environments. *FEBS Lett.* 452:22–25.

Stibitz, S. 1994. Use of conditionally counterselectable suicide vectors for allelic exchange. *Methods Enzymol.* 235:458–465.

Stieger, M., B. Wohlgensinger, M. Kamber, and R. Lutz, et al. 1999. Integrational plasmids for the tetracycline-regulated expression of genes in *Streptococcus pneumoniae*. *Gene* 226:243–251.

Stillman, B. A. and J. L. Tonkinson. 2000. FAST slides: a novel surface for microarrays. *Biotechniques* 29:630–635.

Stock, A. M., V. L. Robinson, and P. N. Goudreau. 2000. Two-component signal transduction. *Annu. Rev. Biochem.* 69:183–215.

Stokes, H. W., A. J. Holmes, B. S. Nield, and M. P. Holley, et al., 2001. Gene cassette PCR: sequence-independent recovery of entire genes from environmental DNA. *Appl. Environ. Microbiol.* 67:5240–5246.

Storck, T., M. C. von Brevern, C. K. Behrens, and J. Scheel, et al. 2002. Transcriptomics in predictive toxicology. *Curr. Opin. Drug Disc. Devel.* 5:90–97.

Stormo, G. D. and G. W. Hartzell. 1989. Identifying protein-binding sites from unaligned DNA fragments. *Proc. Natl. Acad. Sci. USA* 86:1091–1100.

Stougaard, P., S. Molin, and K. Nordstrom. 1981. RNAs involved in copy-number control and incompatibility of plasmid R1. *Proc. Natl. Acad. Sci. USA* 78:6008–6012.

Stover, C. K., X. Q. Pham, A. L. Erwin, and S. D. Mizoguchi, et al. 2000. Complete genome sequence of *Pseudomonas aeruginosa* PAO1, an opportunistic pathogen. *Nature* 406:959–964.

Stragier, P. and R. Losick. 1996. Molecular genetics of sporulation in *Bacillus subtilis. Annu. Rev. Genet.* 30:297–341.

Strauss, E. J. and S. Falkow. 1997. Microbial pathogenesis: genomics and beyond. *Science* 276:707–712.

Sturino, J. M. and T. R. Klaenhammer. 2002. Expression of antisense RNA targeted against *Streptococcus thermophilus* bacteriophages. *Appl. Environ. Microbiol.* 68:588–596.

Subramanian, G., E. V. Koonin, and L. Aravind. 2000. Comparative genome analysis of the pathogenic spirochetes *Borrelia burgdorferi* and *Treponema pallidum. Infect. Immun.* 68:1633–1648.

Sudarsanam, P., V. R. Iyer, P. O. Brown, and F. Winston. 2000. Whole-genome expression analysis of snf/swi mutants of *Saccharomyces cerevisiae. Proc. Natl. Acad. Sci. USA* 97:3364–3369.

Suerbaum, S., J. M. Smith, K. Bapumia, and G. Morelli, et al. 1998. Free recombination within *Helicobacter pylori. Proc. Natl. Acad. Sci. USA* 95:12619–12624.

Sun, Y. H., S. Bakshi, R. Chalmers, and C. M. Tang. 2000. Functional genomics of *Neisseria meningitidis* pathogenesis. *Nat. Med.* 6:1269–1273.

Surdo, P. L., M. J. Bottomley, A. Arcaro, and G. Siegal, et al. 1999. Structural and biochemical evaluation of the interaction of the phosphatidylinositol 3-kinase p85 Src homology 2 domains with phosphoinositides and inositol polyphosphates. *J. Biol. Chem.* 274:15678–15685.

Sutcliffe, J., A. Tait-Kamradt, and L. Wondrack. 1996. *Streptococcus pneumoniae* and *Streptococcus pyogenes* resistant to macrolides but sensitive to clindamycin: a common resistance mechanism mediated by an efflux system. *Antimicrob. Agents Chemother.* 40:1817–1824.

Suzuki, M. T. and S. J. Giovannoni. 1996. Bias caused by template annealing in the amplification of mixtures of 16S rRNA genes by PCR. *Appl. Environ. Microbiol.* 62:625–630.

Suzuki, M. T., M. S. Rappe, Z. W. Haimberger, and H. Winfield, et al., 1997. Bacterial diversity among small-subunit rRNA gene clones and cellular isolates from the same seawater sample. *Appl. Environ. Microbiol.* 63:983–989.

Swenson, D. L., N. O. Bukanov, D. E. Berg, and R. A. Welch. 1996. Two pathogenicity islands in uropathogenic *Escherichia coli* J96: cosmid cloning and sample sequencing. *Infect. Immun.* 64:3736–3743.

Syvanen, M. 1994. Horizontal gene transfer: evidence and possible consequences. *Annu. Rev. Genet.* 28:237–261.

Szyperski, T., C. Fernandez, C. Mumenthaler, and K. Wuthrich. 1998. Structure comparison of human glioma pathogenesis-related protein GliPR and the plant pathogenesis-related protein P14a indicates a functional link between the human immune system and a plant defense system. *Proc. Natl. Acad. Sci. USA* 95:2262–2266.

Tabb, D. L., W. Hayes-McDonald, and J. R. Yates. 2002. DTASelect and contrast: tools for assembling and comparing protein identifications from shotgun proteomics. *J. Proteome Res.* 1:21–26.

Tait-Kamradt, A., J. Clancy, and M. Cronan, et al. 1997. MefE is necessary for the erythromycin-resistant M phenotype in *Streptococcus pneumoniae. Antimicrob. Agents Chemother.* 41:2251–2255.

Takai, K., D. P. Moser, M. DeFlaun, and T. C. Onstott, et al. 2001. Archaeal diversity in waters from deep south African gold mines. *Appl. Envir. Microbiol.* 67:5750–5760.

Tamayo, P., D. Slonim, J. Mesirov, and Q. Zhu, et al. 1999. Interpreting patterns of gene expression with self-organizing maps: methods and application to hematopoietic differentiation. *Proc. Natl. Acad. Sci. USA* 96:2907–2912.

Tan, H. M., H. Y. Tang, C. L. Joannou, and N. H. Abdel-Wahab, et al. 1993. The *Pseudomonas putida* ML2 plasmid-encoded genes for benzene dioxygenase are unusual in codon usage and low in G + C content. *Gene* 130:33–39.

Tanida, I., A. Hasegawa, H. Iida, and Y. Ohya, et al. 1995. Cooperation of calcineurin and vacuolar $H^{(+)}$-ATPase in intracellular Ca^{2+} homeostasis of yeast cells. *J. Biol. Chem.* 270:10113–10119.

Taniguchi, M., K. Miura, H. Iwao, and S. Yamanaka. 2001. Quantitative assessment of DNA microarrays—comparison with Northern blot analyses. *Genomics* 71:34–39.

Tao, H., C. Bausch, C. Richmond, and F. R. Blattner, et al. 1999. Functional genomics: expression analysis of *Escherichia coli* growing on minimal and rich media. *J. Bacteriol.* 181:6425–6440.

Taroncher-Oldenburg, G., Griner, E. M., Francis, C. A., and Ward, B. B. (2003). Oligonucleotide microarray for the study of functional gene diversity in the nitrogen cycle in the environment. *Appl. Environ. Microbiol.* 69:1159–1171.

Tatusov, R. L., A. R. Mushegian, P. Bork, and N. P. Brown, et al. 1996. Metabolism and evolution of *Haemophilus influenzae* deduced from a whole- genome comparison with *Escherichia coli. Curr. Biol.* 6:279–291.

Tatusov, R. L., E. V. Koonin, and D. J. Lipman. 1997. A genomic perspective on protein families. *Science* 278:631–637.

Tatusov, R. L., D. A. Natale, I. V. Garkavtsev, and T. A. Tatusova, et al. 2001. The COG database: new developments in phylogenetic classification of proteins from complete genomes. *Nucl. Acids Res.* 29:22–28.

Taylor, J. K. and N. M. Dean. 1999. Regulation of pre-mRNA spicing by antisense oligonucleotides. *Curr. Opin. Drug Disc. Dev.* 2:147–151.

Taylor, J. S., Y. Van de Peer, and A. Meyer. 2001. Genome duplication, divergent resolution and speciation. *Trends Genet*, 17:299–301.

Taylor, R. K., V. L. Miller, D. B. Furlong, and J. J. Mekalanos. 1987. Use of phoA gene fusions to identify a pilus colonization factor coordinately regulated with cholera toxin. *Proc. Natl. Acad. Sci. USA* 84:2833–2837.

Tekaia, F., A. Lazcano, and B. Dujon. 1999. The genomic tree as revealed from whole proteome comparisons. *Genome Res.* 9:550–557.

Tennant, R. W. 2002. The national center for toxicogenomics: using new technologies to inform mechanistic toxicology. *Environ. Health Persp.* 110:A8–A10.

ter Linde, J. J., H. Liang, R. W. Davis, and H. Y. Steensma, et al. 1999. Genome-wide transcriptional analysis of aerobic and anaerobic chemostat cultures of *Saccharomyces cerevisiae. J. Bacteriol.* 181:7409–7413.

Terai, G., T. Takagi, and K. Nakai. 2001. Prediction of co-regulated genes in *Bacillus subtilis* on the basis of upstream elements conserved across three closely related species. *Genome Biol.* 2: RESEARCH0048.

The *Arabidopsis* Genome Initiative. 2000. Analysis of the genome sequence of the flowering plant *Arabidopsis thaliana. Nature* 408:796–815.

The *C. elegans* Sequencing Consortium. 1998. Genome sequence of the nematode *C. elegans*: a platform for investigating biology. *Science* 282:2012–2018.

The News and Editorial Staffs. 1998. Breakthrough of the year: the runners-up. *Science* 282:2157–2161.

Thieffry, D. and R. Thomas. 1995. Dynamical behavior of biological regulatory networks. II, immunity control in bacteriophage lamda. *Bull. Math. Biol.* 57:277.

Thieffry, D. and R. Thomas. 1998. Qualitative analysis of gene networks. *Pac. Symp. Biocomput.* 77–88.

Thiele, D. J. 1988. ACE1 regulates expression of the *Saccharomyces cerevisiae* metallothionein gene. *Mol. Cell. Biol.* 8:2745–2752.

Thompson, D. K., A. S. Beliaev, C. S., Giometti, and S. L., Tollaksen, et al. 2002a. Transcriptional and proteomic analysis of a ferric uptake regulator (Fur) mutant of *Shewanella oneidensis*: possible involvement of Fur in energy metabolism, transcriptional regulation, and oxidative stress. *Appl. Environ. Microbiol.* 68:881–892.

Thompson, J. F. and A. Landy, 1989. Regulation of lambda site-specific recombination. In D. E. Berg and M. M. Howe (eds.), *Mobile DNA*, pp. 1–22. American Society for Microbiology, Washington, D.C.

Thompson, J. R., L. A. Marcelino, and M. F. Polz. 2002b. Heteroduplexes in mixed-template amplifications: formation, consequence and elimination by 'reconditioning PCR'. *Nucleic Acids Res.* 30:2083–2088.

Thompson, L. J., D. S. Merrell, B. A. Neilan, and H. Mitchell, et al. 2003. Gene expression profiling of *Helicobacter pylori* reveals a growth-phase-dependent switch in virulence gene expression. *Infect. Immun.* 71:2643–2655.

Tiedje, J. M. 1995. Approaches to the comprehensive evaluation of prokaryotic diversity of a habitat. *In* Microbial diversity and ecosystem function. Allsopp D., Colwell R. R., and Hawksworth D. L. (eds). CAB International.

Tiedje, J. M. 2000. 20 years since Dunedin: The past and the future of microbial ecology. *In* Microbial Biosystems: New Frontiers. Proceedings of the 8th International symposium on microbial ecology. Bell C., Brylinsky M, and Johnson-Green P. (eds). Atlantic Canada society for microbial ecology, Halifax, Canada, 2000.

Tiedje, J. M., S. Asuming-Brempong, K. Nusslein, and T. L. Marsh et al. 1999. Opening the black box of soil microbial diversity. *Applied Soil Ecology.* 13:109–122.

Timberlake, W. E. 1998. Agricultural genomics comes of age. Nat. *Biotechnol.* 16:116–117.

Tinsley, C. R. and X. Nassif. 1996. Analysis of the genetic differences between *Neisseria meningitidis* and *Neisseria gonorrhoeae*: two closely related bacteria expressing two different pathogenicities. *Proc. Natl. Acad. Sci. USA* 93:11109–11114.

Tjalsma, H., A. Bolhuis, M. L. van Roosmalen, and T. Wiegert, et al. 1998. Functional analysis of the secretory precursor processing machinery of *Bacillus subtilis*: identification of a eubacterial homolog of archaeal and eukaryotic signal peptidases. *Genes Dev.* 12:2318–2331.

Tjalsma, H., M. A. Noback, S. Bron, and G. Venema, et al. 1997. *Bacillus subtilis* contains four closely related type I signal peptidases with overlapping substrate specificities. Constitutive and temporarily controlled expression of different sip genes. *J. Biol. Chem.* 272:25983–25992.

Tjalsma, H., V. P. Kontinen, Z. Pragai, and H. Wu, et al. 1999. The role of lipoprotein processing by signal peptidase II in the Gram-positive eubacterium *Bacillus subtilis*. Signal peptidase II is required for the efficient secretion of alpha-amylase, a non-lipoprotein. *J. Biol. Chem.* 274:1698–1707.

Toepert, F., J. R. Pires, C. Landgraf, and H. Oschkinat, et al. 2001. Synthesis of an array comprising 837 variants of the hYAP WW protein domain. *Angew. Chem. Int. Ed.* 40:897–900.

Tolley, L., J. W. Jorgenson, and M. A. Moseley. 2001. Very high pressure gradient LC/MS/MS. *Anal. Chem.* 73:2985–2991.

Tomb, J.-F., O. White, A. R. Kerlavage, and R. A. Clayton, et al. 1997. The complete genome sequence of the gastric pathogen *Helicobacter pylori*. *Nature* 388:539–547.

Tomizawa, J., T. Itoh, G. Selzer, and T. Som. 1981. Inhibition of ColE1 RNA primer formation by a plasmid-specified small RNA. *Proc. Natl. Acad. Sci. USA* 78:1421–1425.

Tong, A. H. Y., B. Drees, G. Nardelli, and G. D. Bader, et al. 2002. A combined experimental and computational strategy to define protein interaction networks for peptide recognition modules. *Science* 295:321–324.

Torsvik, V. and L. Øvreås. 2002. Microbial diversity and function in soil: from genes to ecosystems. *Curr. Opin. Microbiol.* 5:240–245.

Torsvik, V., F. L. Daae, R. A. Sandaa, and L. Ovreas. 1998. Novel techniques for analysing microbial diversity in natural and perturbed environments. *J. Biotechnol.* 64:53–62.

Torsvik, V., J. Goksoyr, and F. L. Daae. 1990. High diversity of DNA in soil bacteria. *Appl. Environ. Microbiol.* 56:782–787.

Treves, D. S., B. Xia, J. Zhou, and J. M. Tiedje. 2003. A two-species test of the hypothesis that spatial isolation influences microbial diversity in soil. *Microb. Ecol.* 45:20–28.

Trouche, D., C. Le Chalony, C. Muchardt, and M. Yaniv, et al. 1997. RB and hbrm cooperate to repress the activation functions of E2F1. *Proc. Natl. Acad. Sci. USA* 94:11268–11273.

Tseng, G. C., M. K. Oh, L. Rohlin, and J. C. Liao, et al. 2001. Issues in cDNA microarray analysis: quality filtering, channel normalization, models of variations and assessment of gene effects. *Nucl. Acids Res.* 29:2549–2557.

Tsoka, S. and C. A. Ouzounis. 2000. Recent developments and future directions in computational genomics. *FEBS Lett.* 480:42–48.

Tunlid, A. 1999. Molecular biology: a linkage between microbial ecology, general ecology and organismal biology. *OIKOS* 85:177–189.

Uberbacher, E. C. and R. J. Mural. 1991. Locating protein-coding regions in human DNA sequences by a multiple sensors-neural network approach. *Proc. Natl. Acad. Sci. USA* 88:11261–11265.

Uberbacher, E. C., Y. Xu, and R. J. Mural. 1996. Discovering and understanding genes in human DNA sequence using GRAIL. *Methods Enzymol.* 266:259–281.

Uchiki, T., R. Hettich, V. Gupta, and C. Dealwis. 2002. Characterization of monomeric and dimeric forms of recombinant sml1p-histag protein by electrospray-MS. *Anal. Biochem.* 301:35–48.

Ueda, K., T. Seki, T. Kudo, and T. Yoshida, et al. 1999. Two distinct mechanisms cause heterogeneity of 16S rRNA. *J. Bacteriol.* 181:78–82.

Uetz, P., L. Giot, G. Cagney, and T. A. Mansfield, et al. 2000. A comprehensive analysis of protein-protein interactions in *Saccharomyces cerevisiae*. *Nature* 403:623–627.

Urakawa, H., P. A. Noble, S. El Fantroussi, and J. J. Kelly, et al. 2002. Single-base-pair discrimination of terminal mismatches by using oligonucleotide microarrays and neural network analyses. *Appl. Environ. Microbiol.* 68:235–244.

Ushinsky, S. C., H. Bussey, A. A. Ahmed, and Y. Wang, et al. 1997. Histone H1 in *Saccharomyces cerevisiae*. *Yeast* 13:151–161.

Valdes, I., A. Pitarch, C. Gil, and A. Bermudez, et al. 2000. Novel procedure for the identification of proteins by mass fingerprinting combining two-dimensional electrophoresis with fluorescent SYPRO red staining. *J. Mass Spectrom.* 35:672–682.

Valdivia, R. H. and S. Falkow. 1997. Fluorescence-based isolation of bacterial genes expressed within host cells. *Science* 277:2007–2011.

Valls, M. and V. de Lorenzo. 2002. Exploiting the genetic and biochemical capacities of bacteria for the remediation of heavy metal pollution. *FEMS Microbiol. Rev.* 26:327–338.

van Alphen, L., L. Geelen-van den Broek, L. Blaas, and M. van Ham, et al. 1991. Blocking of fimbria-mediated adherence of *Haemophilus influenzae* by sialyl gangliosides. *Infect. Immun.* 59:4473–4477.

van Belkum, A., S. Scherer, L. van Alphen, and H. Verbrugh. 1998. Short-sequence DNA repeats in prokaryotic genomes. *Microbiol. Mol. Biol. Rev.* 62:275–293.

Vandamme, P., B. Pot, M. Gillis, and P. de Vos, et al. 1996. Polyphasic taxonomy, a consensus approach to bacterial systematics. *Microbiol. Rev.* 60:407–438.

van den Berg, W. A., W. M. van Dongen, and C. Veeger. 1991. Reduction of the amount of periplasmic hydrogenase in *Desulfovibrio vulgaris* (Hildenborough) with antisense RNA: direct evidence for an important role of this hydrogenase in lactate metabolism. *J. Bacteriol.* 173:3688–3694.

Van der Ende, A., C. T. Hopman, S. Zaat, and B. B. Essink, et al. 1995. Variable expression of class I outer membrane protein in *Neisseria meningitidis* is caused by variation in the spacing of the -10 and -35 regions of the promoter. *J. Bacteriol.* 177:2475–2480.

van der Heeft, E., G. J. ten Hove, C. A. Herberts, and H. D. Meiring, et al. 1998. A microcapillary column switching HPLC-electrospray ionization MS system for the direct identification of peptides presented by major histocompatibility complex class I molecules. *Anal Chem.* 70:3742–3751.

Van Dijl, J. M., A. deJong, G. Venema, and S. Bron. 1992. Signal peptidase I of *Bacillus subtilis*: patterns of conserved amino acids in prokaryotic and eukaryotic type I signal peptidases. *EMBO J.* 11:2819–2828.

Van Ham, S. M., L. van Alphen, F. R. Mooi, and J. P. M. van Putten. 1993. Phase variation of *H. influenzae* fimbriae: transcriptional control of two divergent genes through a variable combined promoter region. *Cell* 73:1187–1196.

Van Ham, S. M., L. van Alphen, F. R. Mooi, and J. P. M. van Putten. 1994. The fimbrial gene cluster of *H. influenzae* type b. *Mol. Microbiol.* 13:673–684.

van Montfort, B. A., B. Canas, R. Duurkens, and J. Godovac-Zimmermann, et al. 2002. Improved in-gel approaches to generate peptide maps of integral membrane proteins with matrix-assisted laser desorption/ionization time-of-flight mass spectrometry. *J. Mass Spectrom.* 37:322–330.

van Ommen, G. J. 2001. Medical genomics. *Eur. J. Hum. Genet.* 9:729.

Varga, L. V., S. Toth, I. Novak, and A. Falus. 1999. Antisense strategies: functions and applications in immunology. *Immunol. Lett.* 69:217–224.

Varma, A. and B. O. Palsson. 1994. Stoichiometric flux balance models quantitatively predict growth and metabolic by-product secretion in wild-type *Escherichia coli* W3110. *Appl. Environ. Microbiol.* 60:3724–3731.

Vasiliskov, A., E. Timofeev, S. Surzhikov, and A. Drobyshev, et al. 1999. Fabricaiton of microarray of gel-immobilized compounds on a chip by copolymerization. *Biotechniques* 27:592–606.

Vazquez-Boland, J. A., M. Kuhn, P. Berche, and T. Chakraborty, et al. 2001. *Listeria* pathogenesis and molecular virulence determinants. *Clin. Microbiol. Rev.* 14:584–640.

Venables, W. and B. Ripley, Modern Applied Statistics with S-Plus. New York: Springer-Verlag, 1994.

Venkateswaran, A., S. C. McFarlan, D. Ghostal, and K. W. Minton, et al. 2000. Physiologic determinants of radiation resistance in *Deinococcus radiodurans*. *Appl. Environ. Microbiol.* 66:2620–2626.

Venkateswaran, K., D. P. Moser, M. E. Dollhopf, and D. P. Lies, et al. 1999. Polyphasic taxonomy of the genus *Shewanella* and description of *Shewanella oneidensis* sp. nov. *Int. J. Syst. Bacteriol.* 49:705–724.

Venter, J. C., H. O. Smith, and C. M. Fraser. 1999. Microbial genomics: in the beginning. *ASM News* 65:322–327.

Venter, J. C., M. D. Adams, E. W. Myers, and P. W. Li, et al. 2001. The sequence of the human genome. *Science* 291:1304–1351.

VerBerkmoes, N. C., J. L. Bundy, L. Hauser, and K. G. Asano, et al. 2002. Integrating "top-down" and "bottom-up" mass spectrometric approaches for proteomic analysis of *Shewanella oneidensis*. *J. Proteome Res.* 1:239–252.

Verdnik, D., S. Handran, and S. Pickett. 2002. Key considerations for accurate microarray scanning and image analysis. In S. Shah and G. Kamberova (eds.), *DNA Array Image Analysis—Nuts & Bolts*, pp. 83–98. DNA Press, LLC, Eagleville, Penn.

Vertibi, A. J. and J. K. Omura. 1979. *Princples of Digital Communication and Coding*. McGraw-Hill, New York.

Vestal, M. L. 2001. Methods of ion generation. *Chem. Rev.* 101:361–375.

Vidal, M. and H. Endoh. 1999. Prospects for drug screening using the reverse two-hybrid system. *TIBTECH* 17:374–381.

Vidan, S. and M. Snyder. 2001. Large-scale mutagenesis: yeast genetics in the genome era. *Curr. Opin. Biotechnol.* 12:28–34.

Vingron, M. and P. Argos. 1989. A fast and sensitive multiple sequence alignment algorithm. *Comput. Appl. Biosci.* 5:115–121.

Volff, J. N. and J. Altenbuchner. 2000. A new beginning with new ends: linearisation of circular chromosomes during bacterial evolution. *FEMS Microbiol. Lett.* 186:143–150.

Volfovsky, N., B. J. Haas, and S. L. Salzberg. 2001. A clustering method for repeat analysis in DNA sequences. *Genome Bio.* 2:Research 0027.1–0027.11.

von Hippel, P. H. 1992. An integrated model of the transcription complex in elongation, termination and editing. *Science* 281:660–665.

von Mering, C., R. Krause, B. Snel, and M. Cornell, et al. 2002. Comparative assessment of large-scale data sets of protein-protein interactions. *Nature* 417:399–403.

Voordouw, G. 1998. Reverse sample genome probing of microbial community dynamics. *ASM News* 64:627–633.

Voordouw, G., J. K. Voordouw, R. R. Karkhoff-Schweizer, P. M. Fedorak, and D. W. S. Westlake. 1991. Reverse sample genome probing, a new technique for identification of bacteria in environmental samples by DNA hybridization, and its application to the identification of sulfate-reducing bacteria in oil field samples. *Appl. Environ. Microbiol.* 57:3070–3078.

Wagner, A. 2001. Birth and death of duplicated genes in completely sequenced eukaryotes. *Trends Genet.* 17:237–239.

Wagner, E. G. and R. W. Simons. 1994. Antisense RNA control in bacteria, phages, and plasmids. *Annu. Rev. Microbiol.* 48:713–742.

Wagner, E. G., S. Altuvia, and P. Romby P. 2002a. Antisense RNAs in bacteria and their genetic elements. *Adv. Genet.* 46:361–398.

Wagner, M. A., M. Eschenbrenner, T. A. Horn, and J. A. Kraycer, et al. 2002b. Global analysis of the *Brucella melitensis* proteome: identification of proteins expressed in laboratory-grown culture. *Proteomics* 2:1047–1060.

Wagner, V. E., D. Bushnell, L. Passador, and A. I. Brooks, et al. 2003. Microarray analysis of *Pseudomonas aeruginosa* quorum-sensing regulons: effects of growth phase and environment. *J. Bacteriol.* 185:2080–2095.

Walhout, A. J. M., R. Sordella, X. Lu, and J. L. Hartley, et al. 2000. Protein interaction mapping in *C. elegans* using proteins involved in vulval development. *Science* 287:116–122.

Walker, E. M., J. K. Howell, Y. You, and A. R. Hoffmaster, et al. 1995. Physical map of the genome of *Treponema pallidum subsp. pallidum* (Nichols). *J. Bacteriol.* 177:1797–1804.

Walker, J. M. 1994. The Dansyl–Edman method for peptide sequencing. *Methods Mol. Biol.* 32:329–334.

Wallace, A. C., R. A. Laskowski, and J. M Thornton. 1996. Derivation of 3D coordinate templates for searching structural databases: application to Ser-His-Asp catalytic triads in the serine proteases and lipases. *Prot. Sci.* 5:1001–1013.

Wang, G. C. Y. and D Y. Wang. 1996. The frequency of chimeric molecules as a consequence of PCR coamplification of 16S rRNA genes from different bacterial species. *Microbiology* 142:1107–1114.

Wang, D. G., J. B. Fan, C. J. Siao, and A. Berno, et al. 1998. Large-scale identification, mapping, and genotyping of single-nucleotide polymorphisms in the human genome. *Science* 280:1077–1082.

Wang, J. C. and A. S. Lynch. 1993. Transcription and DNA supercoiling. *Curr. Opin. Genet. Dev.* 3:764–768.

Wang, Q. P. and J. M. Kaguni. 1989. A novel sigma factor is involved in expression of the rpoH gene of *Escherichia coli*. *J. Bacteriol.* 171:4248–4253.

Wang, Z. X. 1998. A re-estimation for the total numbers of protein folds and superfamilies. *Prot. Eng.* 11:621–626.

Ward, S. J. 2001. Impact of genomics in drug discovery. *Biotechniques* 31:626–634.

Warrington, J. A., S. Dee, and M. Trulson. 2000. Large-scale genomic analysis using Affymetrix GeneChip® probe arrays. In M. Schena (ed.), *Microarray Biochip Technology*, pp. 119–148. Eaton Publishing, Natick, Mass.

Washburn, M. P., D. Wolters, and J. R. Yates, 3rd. 2001. Large-scale analysis of the yeast proteome by multidimensional protein identification technology. *Nat. Biotechnol.* 19:242–247.

Washburn, M. P., R. Ulaszek, C. Deciu, and D. M. Schieltz, et al. 2002. Analysis of quantitative proteomic data generated via multidimensional protein identification technology. *Anal. Chem.* 74:1650–1657.

Wasinger, V. C., J. D. Polack, and I. Humphery-Smith. 2000. The proteome of *Mycoplasma genitalium*. Chaps-soluble component. *Eur. J. Biochem.* 267:1571–1582.

Wasinger, V. C., S. J. Cordwell, A. Cerpa-Poljak, and J. X. Yan, et al. 1995. Progress with gene-product mapping of the mollicutes: *Mycoplasma genitalium*. *Electrophoresis* 16:1090–1094.

Watanabe, K., Y. Kodama, N. Hamamura, and N. Kaku. 2002. Diversity, abundance, and activity of archaeal populations in oil-contaminated groundwater accumulated at the bottom of an underground crude oil storage cavity. *Appl. Envir. Microbiol.* 68:3899–3907.

Watson, N., D. S. Dunyak, E. L. Rosey, and J. L. Slonczewski, et al. 1992. Identification of elements involved in transcriptional regulation of the *Escherichia coli* cad operon by external pH. *J. Bacteriol.* 174:530–540.

Wei, Y., J.-M. Lee, C. Richmond, and F. R. Blattner, et al. 2001. High-density microarray-mediated gene expression profiling of *Escherichia coli*. *J. Bacteriol.* 183:545–556.

Weiner, P. 1973. Linear pattern matching algorithms. In *Proceedings of 14th Annual Symposium on Switching and Automata Theory*, pp. 1–11, IEEE Press.

Weinstock, G. M. 2000. Genomics and bacterial pathogenesis. *Emerging Infect. Dis.* 6:496–504.

Weiser, J. N., J. M. Love, and E. R. Moxon. 1989. The molecular mechanism of phase variation of *H. influenzae* lipopolysaccharide. *Cell* 59:657–665.

Welch, R. A., V. Burland, G. Plunkett 3rd, and P. Redford, et al. 2002. Extensive mosaic structure revealed by the complete genome sequence of uropathogenic *Escherichia coli*. *Proc. Natl. Acad. Sci.* USA 99:17020–17024.

Wen, S., S. Fuhrman, G. S. Carr, and S. Smith, et al. 1998. Large-scale temporal gene expression mapping of central nervous system development. *Proc. Natl. Acad. Sci. USA* 95:334–339.

Westbrook, J., Z. Feng, S. Jain, and T. N. Bhat, et al. 2000. The Protein Data Bank: unifying the archive. *Nucl. Acids Res.* 30(1):245–248.

White, A. P., S. K. Collinson, J. Burian, and S. C. Clouthier, et al. 1999a. High efficiency gene replacement in *Salmonella enteritidis*: chimeric fimbrins containing a T-cell epitope from *Leishmania major*. *Vaccine* 17:2150–2161.

White, O., J. A. Eisen, J. F. Heidelberg, and E. K. Hickey, et al. 1999b. Genome sequence of the radioresistant bacterium *Deinococcus radiodurans* R1. *Science* 286:1571–1577.

Whiteley, M., M. G. Bangera, R. E. Bumgarner, and M. R. Parsek, et al. 2001. Gene expression in *Pseudomonas aeruginosa* biofilms. *Nature* 413:860–864.

Whitman, W. B., D. C. Coleman, and W. J. Wiebe. 1998. Prokaryotes: the unseen majority. *Proc. Natl. Acad. Sci. USA* 95:6578–6583.

Whittam, T. S. and A. C. Bumbaugh. 2002. Inferences from whole-genome sequences of bacterial pathogens. *Curr. Opin. Genet. Dev.* 12:719–725.

Wigge, P. A., O. N. Jensen, S. Holmes, and S. Soues, et al. 1998. Analysis of the *Saccharomyces* spindle pole by matrix-assisted laser desorption/ionization (MALDI) mass spectrometry. *J. Cell Biol.* 141:967–977.

Wildgruber, R., A. Harder, C. Obermaier, and G. Boguth, et al. 2000. Towards higher resolution: two-dimensional electrophoresis of *Saccharomyces cerevisiae* proteins using overlapping narrow immobilized pH gradients. *Electrophoresis* 21:2610–2616.

Williams, C. and T. A. Addona. 2000. The integration of SPR biosensors with mass spectrometry: possible applications for proteome analysis. *Trends Biotechnol.* 18:45–48.

Williamson, R. A., D. Peretz, C. Pinilla, and H. Ball, et al. 1998. Mapping the prion protein using recombinant antibodies. *J. Virol.* 72:9413–9418.

Wilm, M., A. Shevchenko, T. Houthaeve, and S. Breit, et al. 1996. Femtomole sequencing of proteins from polyacrylamide gels by nano-electrospray mass spectrometry. *Nature* 379:466–469.

Wilson, A. C., H. Ochman, and E. M. Prager. 1987. Molecular time scale for evolution. *TIG* 3:241–247.

Wilson, G. G. and N. E. Murray. 1991. Restriction and modification systems. *Annu. Rev. Genet.* 25:585–627.

Wilson, K. H., W. J. Wilson, J. L. Radosevich, and T. Z. DeSantis, et al. 2002. High-density microarray of small-subunit ribosomal DNA probes. *Appl. Environ. Microbiol.* 68:2535–2541.

Wilson, M., J. DeRisi, and H.-H. Kristensen, et al. 1999. Exploring drug-induced alterations in gene expression in *Mycobacterium tuberculosis* by microarray hybridization. *Proc. Natl. Acad. Sci. USA* 96:12833–12838.

Winter, D., A. V. Podtelejnikiov, M. Mann, and R. Li. 1997. The complex containing actin-related proteins Arp2 and Arp3 is required for the motility and integrity of yeast actin patches. *Curr. Biol.* 7:519–529.

Wintzingerode, F. V., U. B. Gobel, and E. Stachebrandt. 1997. Determination of microbial diversity in environmental samples: Pitfalls of PCR-based rRNA analysis. *FEMS Microbiol. Rev.* 21:213–229.

Winzeler, E. A., D. D. Shoemaker, A. Astromoff, and H. Liang, et al. 1999. Functional characterization of the *S. cerevisiae* genome by gene deletion and parallel analysis. *Science* 285:901–906.

Wodicka, L., H. Dong, M. Mittmann, and M. H. Ho, et al. 1997. Genome-wide expression monitoring in *Saccharomyces cerevisiae*. *Nat. Biotechnol.* 15:1359–1367.

Woese, C. 1998. The universal ancestor. *Proc. Natl. Acad. Sci. USA* 95:6854–6859.

Woese, C. R. 1987. Bacterial evolution. *Microbiol. Rev.* 51:221–271.

Woese, C. R. 2000. Interpreting the universal phylogenetic tree. *Proc. Natl. Acad. Sci. USA* 97:8392–8396.

Woese, C. R. and G. E. Fox. 1977. Phylogenetic structure of the prokaryotic domain: the primary kingdoms. *Proc. Natl. Acad. Sci. USA* 74:5088–5090.

Woese, C. R., O. Kandler, and M. L. Wheelis. 1990a. Towards a natural system of organisms: proposal for the domains Archaea, Bacteria, and Eucarya. *Proc. Natl. Acad. Sci. USA* 87:4576–4579.

Woese, C. R., S. Winker, and R. R. Gutell. 1990b. Architecture of ribosomal RNA: constraints on the sequence of "tetra-loops." *Proc. Natl Acad. Sci. USA* 87:8467–8471.

Wolf, E., P. S. Kim, and B. Berger. 1997. MultiCoil: a program for predicting two- and three-stranded coiled coils. *Prot. Sci.* 6:1179–1189.

Wolf, Y. I., I. B. Rogozin, A. S. Kondrashov, and E. V. Koonin. 2001. Genome alignment, evolution of prokaryotic genome organization, and prediction of gene function using genomic context. *Genome Res.* 11:356–372.

Wolf, Y. I., L. Aravind, N. V. Grishin, and E. V. Koonin. 1999. Evolution of aminoacyl-tRNA synthetases—analysis of unique domain architectures and phylogenetic trees reveals a complex history of horizontal gene transfer events. *Genome Res.* 9:689–710.

Wolf, Y. I., N. V. Grishin, and E. V Koonin. 2000. Estimating the number of protein folds and families from complete genome data. *J. Mol. Biol.* 299(4):897–905.

Wolf, Y. I., I. B. Rogozin, N. V. Grishin, and E. V. Koonin. 2002. Genome trees and the tree of life. *Trends Genet.* 18:472–479.

Wolfe, K. H. 2001. Yesterday's polyploids and the mystery of diploidization. *Nat. Rev. Genet.* 2:333–341.

Wolfe, K. H. and D. C. Shields. 1997. Molecular evidence for an ancient duplication of the entire yeast genome. *Nature* 387:708–713.

Wolska, K. I., J. Paciorek, and K. Kardys. 1999. Physiological consequences of mutations in *Escherichia coli* heat shock dnaK and dnaJ genes. *Microbios* 97:55–67.

Wolter, D. A., M. P. Washburn, and J. R. Yates, 3rd. 2001. An automated multidimensional protein identification technology for shotgun proteomics. *Anal. Chem.* 73:5683–5690.

Wong, D. K., B.-Y. Lee, M. A. Horwitz, and B. W. Gibson. 1999. Identification of Fur, aconitase, and other proteins expressed by *Mycobacterium tuberculosis* under conditions of low and high concentrations of iron by combined two-dimensional gel electrophoresis and mass spectrometry. *Infect. Immun.* 67:327–336.

Wong, S. M. and J. J. Mekalanos. 2000. Genetic footprinting with mariner-based transposition in *Pseudomonas aeruginosa. Proc. Natl. Acad. Sci. USA* 97:10191–10196.

Workman, J. L. and R. E. Kingston. 1998. Alteration of nucleosome structure as a mechanism of transcriptional regulation. *Annu. Rev. Biochem.* 67:545–579.

Worley, J., K. Bechtol, S. Penn, and D. Roach, et al. 2000. A systems approach to fabricating and analyzing DNA microarrays. In M. Schena (ed.), *Microarray Biochip Technology*, pp. 65–85. Eaton Publishing, Natick, Mass.

Wray, G., J. Levinton, and L. Shapiro. 1996. Molecular evidence for deep Precambrian divergences among Metazoan phyla. *Science* 274:568–573.

Wright, G. L., L. H. Cazares, S.-M. Leung, and S. Nasim, et al. 2000. ProteinChip surface enhanced laser desorption/ionization (SELDI) mass spectrometry: a novel protein biochip technology for detection of prostate cancer biomarkers in complex protein mixtures. *Prostate Canc. Prostatic Dis.* 2:264–276.

Wu, C. I., and W. H. Li. 1985. Evidence for higher rates of nucleotide substitution in rodents than in man. *Proc. Natl. Acad. Sci. USA* 82:1741–1745.

Wu, L.Y., D. Thompson, G.-S. Li, and R. Hurt, et al. 2001. Development and evaluation of functional gene arrays for detection of selected genes in the environment. *Appl. Environ. Microbiol.* 67:5780–5790.

Wu, T. D. 2001. Analyzing gene expression data from DNA microarrays to identify candidate genes. *J. Pathol.* 195:53–65.

Wu, T. Y., L. Liono, S. L. Chen, and C. Y. Chen, et al. 2000. Expression of highly controllable genes in insect cells using a modified tetracycline-regulated gene expression system. *J. Biotechnol.* 80:75–83.

Xenarios, I., D. W. Rice, L. Salwinski, and M. K. Baron, et al. 2000. DIP: The Database of Interacting Proteins. *Nucl. Acids Res.* 28:289–291.

Xiang, Z., S. Censini, P. F. Bayeli, and J. L. Telford, et al. 1995. Analysis of expression of CagA and VacA virulence factors in 43 strains of *Helicobacter pylori* reveals that clinical isolates can be divided into two major types and that CagA is not necessary for expression of the vacuolating cytotoxin. *Infect. Immun.* 63:94–98.

Xu, D., G. Li, L. Wu, J.-Z. Zhou, and Y. Xu. 2002. PRIMEGENS: a computer program for robust and efficient design of gene-specific targets on microarrays. *Bioinformatics* 18(11):1432–1437.

Xu, Q., R. D. Morgan, R. J. Roberts, and M. J. Blaser. 2000. Identification of type II restriction and modification systems in *Helicobacter pylori* reveals their substantial diversity among strains. *Proc. Natl. Acad. Sci. USA* 97:9671–9676.

Xu, Y. and D. Xu. 2000. Protein threading using PROSPECT: design and evaluation. *Prot. Struct. Funct. Genet.* 40:343–354.

Xu, Y. and E. C. Uberbacher. 1997. Automated gene identification in large-scale genomic sequences. *J. Comp. Biol.* 4(3):325–338.

Xu, Y., D. Xu, and H. N. Gabow. 2000. Protein domain decomposition using a graph-theoretic approach. *Bioinformatics* 16:1091–1104.

Xu, Y., J. R. Einstein, R. J. Mural, and M. Shah, et al. 1994. An improved system for exon recognition and gene modeling in human DNA sequences. In R. Altman et al. (eds.), *Proceedings of the Second International Conference on Intelligent Systems for Molecular Biology*, pp. 376–383. AAAI Press, Cambirdge, U.K.

Xu, Y., R. J. Mural, and E. C. Uberbacher. 1996a. An iterative algorithm for correcting DNA sequencing errors in coding regions. *J. Comp. Biol.* 3(3):333–344.

Xu, Y., R. J. Mural, J. R. Einstein, and M. Shah, *et al.* 1996b. GRAIL: a multi-agent neural network system for gene identification. *Proc. IEEE* 84(10):1544–1552.

Xu, Y., V. Olman, and D. Xu. 2001. Minimum spanning trees for gene-expression data clustering. In *Proceedings of the 12th International Conference on Genomics Informatics, Tokyo, Japan*, pp. 24–33.

Xu, Y., V. Olman, and D. Xu. 2002. Clustering gene expression data using a graph-theoretic approach: an application of minimum spanning trees. *Bioinformatics* 18(4):536–545.

Yakobson, E., and D. G. Guiney. 1984. Conjugal transfer of bacterial chromosomes mediated by the RK2 plasmid transfer origin cloned into transposon Tn5. *J. Bacteriol.* 160:451–453.

Yamaguchi-Iwai, Y., A. Dancis, and R. D. Klausner. 1995. AFT1: a mediator of iron regulated transcriptional control in *Saccharomyces cerevisiae. EMBO J.* 14:1231–1239.

Yamaguchi-Iwai, Y., R. Stearman, A. Dancis, and R. D. Klausner. 1996. Iron-regulated DNA binding by the AFT1 protein controls the iron regulon in yeast. *EMBO J.* 15:3377–3384.

Yamanaka, K. and M. Inouye. 1997. Growth-phase-dependent expression of cspD, encoding a member of the CspA family in *Escherichia coli. J. Bacteriol.* 179:5126–5130.

Yang, Y. H. and T. Speed. 2002. Design issues for cDNA microarray experiments. *Nat. Rev. Genet.* 3:579–588.

Yang, Z. 1996. Phylogenetic analysis using parsimony and likelihood methods. *J. Mol. Evol.* 42:294–307.

Yang, Z. N., T. C. Mueser, F. D. Bushman, and C. C. Hyde. 2000. Crystal structure of an active two-domain derivative of Rous sarcoma virus integrase. *J. Mol Biol.* 296:535.

Yao, X., A. Freas, J. Ramirez, and P. A. Demirev, et al. 2001. Proteolytic 18O labeling for comparative proteomics: model studies with two serotypes of adenovirus. *Anal. Chem.* 73:2836–2842.

Yao, Z.-J., M. C. C. Kao, and M. C. M. Chung. 1995. Epitope identification by polyclonal antibody from phage-displayed random peptide library. *J. Prot. Chem.* 14:161–166.

Yates, J. R. 2000. Mass spectrometry—from genomics to proteomics. *Trends Gen.* 16:5–8.

Ye, R., W. Tao, L. Bedzyk, and T. Young, et al. 2000. Global gene expression profiles of *Bacillus subtilis* grown under anaerobic conditions. *J. Bacteriol.* 182:4458–4465.

Yershov, G., V. Barsky, A. Belgovskiy, and E. Kirillov, et al. 1996. DNA analysis and diagnostics on oligonucleotide microchips. *Proc. Natl. Acad. Sci. USA* 93:4913–4918.

Yin, D. and Y. Ji. 2002. Genomic analysis using conditional phenotypes generated by antisense RNA. *Curr. Opin. Microbiol.* 5:330–333.

Yoshida, K., K. Kobayashi, Y. Miwa, and C. M. Kang, et al. 2001. Combined transcriptome and proteome analysis as a powerful approach to study genes under glucose repression in *Bacillus subtilis*. *Nucl. Acids Res.* 29:683–692.

Young, M. M., N. Tang, J. C. Hempel, and C. M. Oshiro, et al. 2000. High throughput protein fold identification by using experimental constraints derived from intramolecular cross-links and mass spectrometry. *Proc. Natl. Acad. Sci. USA* 97:5802–5806.

Young, R. A. and R. W. Davis. 1983. Efficient isolation of genes using by using antibody probes. *Proc. Natl. Acad. Sci USA* 80:1194–1198.

Yu, B. J., B. H. Sung, M. D. Koob, and C. H. Lee, et al. 2002a. Minimization of the *Escherichia coli* genome using Tn5-targeted Cre/loxP excision system. *Nat. Biotech.* 20:1018–1023.

Yu, D., H. M. Ellis, E.-C. Lee, and N. A. Jenkins, et al. 2000. An efficient recombination system for chromosome engineering of *Escherichia coli*. *Proc. Natl. Acad. Sci. USA* 97:5978–5983.

Yu, J., S. Hu, J. Wang, and G. K. Wong, et al. 2002b. A draft sequence of the rice genome (*Oryza sativa L.* ssp. *indica*). *Science* 296:79–92.

Yun, C. W., T. Ferea, J. Rashford, and O. Ardon, et al. 2000. Desferrioxamine-mediated iron uptake in *Saccharomyces cerevisiae*. *J. Biol. Chem.* 275:10709–10715.

Zagursky, R. J. and D. Russell. 2001. Bioinformatics: use in bacterial vaccine discovery. *Biotechniques* 31:636–659.

Zammatteo, N., L. Jeanmart, S. Hamels, and S. Courtois, et al. 2000. Comparison between different strategies of covalent attachment of DNA to glass surfaces to build DNA microarrays. *Anal. Biochem.* 280:143–150.

Zgurskaya, H. I. and Nikaido, H. 1999. Bypassing the periplasm: reconstitution of the AcrAB multidrug efflux pump of *Escherichia coli*. *Proc. Natl. Acad. Sci. USA* 96:7190–7195.

Zgurskaya, H. I. and Nikaido, H. 2000. Multidrug resistance mechanisms: drug efflux across two membranes. *Mol. Microbiol.* 37:219–225.

Zhang, B., L. Rychlewski, K. Pawlowski, and J. S. Fetrow, et al. 1999. From fold predictions to function predictions: automation of functional site conservation analysis for functional genome predictions. *Prot. Sci.* 8(5):1104–1115.

Zhang, J. P. and S. Normark. 1996. Induction of gene expression in *Escherichia coli* after pilus-mediated adherence. *Science* 273:1234–1236.

Zhang, M. Q. 1997. Identification of protein coding regions in the human genome by quadratic discriminant analysis. *Proc. Natl. Acad. Sci. USA* 94:565–568.

Zhang, Z., S. Schwartz, L. Wagner, and W. Miller. 2000. A greedy algorithm for aligning DNA sequences. *J. Comp. Biol.* 7:203–214.

Zhao, H. and D. J. Eide. 1997. Zap1p, a metalloregulatory protein involved in zinc-responsive transcriptional regulation in *Saccharomyces cerevisiae*. *Mol. Cell. Biol.* 17:5044–5052.

Zhao, H., E. Butler, J. Rodgers, and T. Spizzo, et al. 1998. Regulation of zinc homeostasis in yeast by binding of the ZAP1 transcriptional activator to zinc-responsive promoter elements. *J. Biol. Chem.* 273:28713–28720.

Zhao, X., S. Nampalli, A. J. Serino, and S. Kumar. 2001. Immobilization of oligodeoxyribonucleotides with multiple anchors to microchips. *Nucl. Acids Res.* 29:955–959.

Zheng, Y., J. D. Szustakowski, L. Fortnow, and R. J. Roberts, et al. 2002. Computational identification of operons in microbial genomes. *Genome Res.* 12(8):1221–1230.

Zhou, J. and D. K. Thompson. 2002. Challenges in applying microarrays to environmental studies. *Curr. Opin. Biotechnol.* 13:204–207.

Zhou, J., B. Xia, D. S. Treves, and L. Y. Wu, et al. 2002. Spatial and resource factors influencing high microbial diversity in soil. *Appl. Environ. Microbiol.* 68:326–334.

Zhou, J.-Z., B. Xia, H. Huang, and D. S. Treves, et al. 2003. Phylogenetic diversity and a novel candidate division of two humid region sandy surface soils. *Soil Biol. Biochem.* 35:915–924.

Zhou, J. Z., A. V. Palumbo, and J. M. Tiedje. 1997. Sensitive detection of a novel class of toluene-degrading denitrifiers, *Azoarcus tolulyticus*, using SSU rRNA primers and probes. *Appl. Environ. Microbiol.* 63:2384–2390.

Zhou, J.-Z. 2003. Microarrays for bacterial detection and microbial community analysis. *Curr Opin. Microbiol.* 6:288–294.

Zhou, Y.-N., N. Kusukawa, J. W. Erickson, and C. A. Gross, et al. 1988. Isolation and characterization of *Escherichia coli* mutants that lack the heat shock sigma factor σ^{32}. *J. Bacteriol.* 170:3640–3649.

Zhou, Y.-X., P. Kalocsai, J.-Y. Chen, and S. Shams. 2000. Information processing issues and solutions associated with microarray technology. In M. Schena (ed.), *Microarray Biochip Technology*, pp. 167–200. Eaton Publishing, Natick, Mass.

Zhu, G., P. T. Spellman, T. Volpe, and P. O. Brown, et al. 2000a. Two yeast forkhead genes regulate the cell cycle and pseudohyphal growth. *Nature* 406:90–94.

Zhu, H., J. F. Klemic, S. Chang, and P. Bertone, et al. 2000b. Analysis of yeast protein kinases using protein chips. *N. Genet.* 26:283–289.

Zhu, H., M. Bilgin, R. Bangham, and D. Hall, et al. 2001. Global analysis of protein activities using proteome chips. *Science* 293:2101–2105.

Zillig, W. 1991. Comparative biochemistry of Archaea and Bacteria. *Curr. Opin. Genet. Dev.* 1:544–551.

Zillig, W., H.-P. Klenk, P. Palm, and H. Leffers, et al. 1989. Did eukaryotes originate by a fusion event? *Endocytobiosis Cell Res.* 6:1–25.

Zimmer, D. P., E. Soupene, H. L. Lee, and V. F. Wendisch, et al. 2000. Nitrogen regulatory protein C-controlled genes of *Escherichia coli*: scavenging as a defense against nitrogen limitation. *Proc. Natl. Acad. Sci. USA* 97:14674–14679.

Zlatanova, J. and A. Mirzabekov. 2001. Gel-immobilized microarrays of nucleic acids and proteins: production and application for macromolecular research. *Methods Mol. Biol.* 170:17–38.

Zozulya, S., M. Lioubin, R. J. Hill, and C. Abram, et al. 1999. Mapping signal transduction pathways by phage display. *Nat. Biotechnol.* 17:1193–1198.

Zuckerkandl, E. and L. Pauling. 1962. Horizons in biochemistry. In M. P. Kasha (ed.), *Horizons in Biochemistry*, pp. 97–166. Academic, New York.

Zuckerkandl, E. and L. Pauling. 1965. *Evolving Genes and Proteins*. Academic Press, New York.

Zuker, M. 1989. Computer prediction of RNA structure. *Methods Enzymol.* 180:262–288.

Zumft, W. G. 1997. Cell biology and molecular basis of denitrification. Microbiol. *Mol. Biol. Rev.* 61:533–616.

Index

Microbial Functional Genomics, Edited by Jizhong Zhou, Dorothea K. Thompson, Ying Xu, and James M. Tiedje.
ISBN 0-471-07190-0   © 2004 John Wiley & Sons, Inc.